186 318

The Art of Digital Video

For Chrissie

The Art of Digital Video

Third Edition

John Watkinson

NORWICH CITY COLLEGE LIBRARY			
Stock No.	186318		
Class	621.38833 WAT		
Cat.		Proc.	

Focal Press

OXFORD AUCKLAND BOSTON JOHANNESBURG MELBOURNE NEW DELHI

Focal Press
An imprint of Butterworth-Heinemann
Linacre House, Jordan Hill, Oxford OX2 8DP
225 Wildwood Avenue, Woburn, MA 01801-2041
A division of Reed Educational and Professional Publishing Ltd

A member of the Reed Elsevier plc group

First published 1990
Second edition 1994
Third edition 2000

© John Watkinson 2000

All rights reserved. No part of this publication may be reproduced in any material form (including photocopying or storing in any medium by electronic means and whether or not transiently or incidentally to some other use of this publication) without the written permission of the copyright holder except in accordance with the provisions of the Copyright, Designs and Patents Act 1988 or under the terms of a licence issued by the Copyright Licensing Agency Ltd, 90 Tottenham Court Road, London, England W1P 0LP. Applications for the copyright holder's written permission to reproduce any part of this publication should be addressed to the publishers

British Library Cataloguing in Publication Data
A catalogue record for this book is available from the British Library

Library of Congress Cataloguing in Publication Data
A catalogue record for this book is available from the Library of Congress

ISBN 0 240 51586 2

Composition by Genesis Typesetting, Rochester, Kent
Printed and bound in Great Britain

Contents

Preface xiii

Acknowledgements xv

Chapter 1 Introducing digital video **1**

1.1 What is a video signal? 1
1.2 Comparing analog and digital 2
1.3 Why binary? 6
1.4 Colour 9
1.5 Why digital? 10
1.6 Video as data 11
1.7 Some digital video processes outlined 14
1.8 The frame store 16
1.9 The synchronizer 16
1.10 Time compression 17
1.11 Error correction and concealment 20
1.12 Product codes 22
1.13 Shuffling 25
1.14 Channel coding 26
1.15 Video compression and MPEG 26
1.16 Disk-based recording 28
1.17 Rotary head digital recorders 30
1.18 DVD and DVHS 30
1.19 Digital television broadcasting 30
1.20 Internet video 32
1.21 Digital audio 32
References 33

Chapter 2 Video principles **34**

2.1 The eye 34
2.2 Gamma 38
2.3 Scanning 40
2.4 Synchronizing 42
2.5 Black-level clamping 46
2.6 Sync separation 47
2.7 The monochrome cathode ray tube 49
2.8 The monochrome TV camera 53
2.9 Bandwidth and definition 55
2.10 MTF, contrast and sharpness 56
2.11 Aperture effect 58
2.12 Colour vision 60
2.13 Colorimetry 61
2.14 Colour displays 69
2.15 Colour difference signals 71
2.16 Motion portrayal and dynamic resolution 76
2.17 Progressive or interlaced scan? 77

Chapter 3 Digital principles **80**

3.1 Pure binary code 80
3.2 Two's complement 84
3.3 Introduction to digital processing 87
3.4 Logic elements 88
3.5 Storage elements 91
3.6 The phase-locked loop 93
3.7 Binary adding 95
3.8 The computer 98
3.9 The processor 100
3.10 Interrupts 101
3.11 Programmable timers 104
3.12 Timebase correction 104
3.13 Multiplexing principles 111
3.14 Packets 111
3.15 Statistical multiplexing 112
3.16 Gain control by multiplication 113
3.17 Digital faders and controls 114
3.18 Filters 117
3.19 Transforms 122
3.20 FIR and IIR filters 126
3.21 FIR filters 126

3.22 Sampling-rate conversion 132
3.23 The Fourier transform 143
3.24 The discrete cosine transform (DCT) 153
3.25 The wavelet transform 154
3.26 Modulo-*n* arithmetic 157
3.27 The Galois field 158
3.28 Noise and probability 161
References 163

Chapter 4 Conversion 165

4.1 Introduction to conversion 165
4.2 Sampling and aliasing 165
4.3 Reconstruction 169
4.4 Filter design 171
4.5 Two-dimensional sampling spectra 172
4.6 Choice of sampling rate 176
4.7 Sampling clock jitter 179
4.8 Quantizing 182
4.9 Quantizing error 184
4.10 Introduction to dither 188
4.11 Requantizing and digital dither 191
4.12 Dither techniques 193
4.12.1 Rectangular pdf dither 193
4.12.2 Triangular pdf dither 195
4.12.3 Gaussian pdf dither 196
4.13 Basic digital-to-analog conversion 197
4.14 Basic analog-to-digital conversion 199
4.15 Factors affecting convertor quality 201
4.16 Oversampling 203
4.17 Gamma in the digital domain 210
4.18 Colour in the digital domain 210
References 216

Chapter 5 Digital video processing 218

5.1 A simple digital vision mixer 218
5.2 Concentrators/combiners 221
5.3 Blanking 224
5.4 Keying 225
5.5 Chroma keying 227
5.6 Simple effects 228

5.7 Planar digital video effects 232
5.8 Mapping 234
5.9 Separability and transposition 235
5.10 Address generation and interpolation 237
5.11 Skew and rotation 244
5.12 Perspective rotation 246
5.13 DVE backgrounds 256
5.14 Non-planar effects 259
5.15 Controlling effects 260
5.16 Graphics 262
5.17 Graphic art/paint systems 269
5.18 Applications of motion compensation 270
5.19 Motion-estimation techniques 272
5.19.1 Block matching 272
5.19.2 Gradient matching 274
5.19.3 Phase correlation 275
5.20 A phase-correlation motion-estimation system 280
5.21 Motion-compensated standards conversion 286
5.22 Motion-compensated telecine system 292
5.23 Camera shake compensation 294
5.24 De-interlacing 297
5.25 Noise reduction 300
References 302

Chapter 6 Video compression and MPEG 303

6.1 Introduction to compression 303
6.2 What is MPEG? 311
6.3 Spatial and temporal redundancy in MPEG 314
6.4 *I* and *P* coding 323
6.5 Bidirectional coding 324
6.6 Coding applications 327
6.7 Spatial compression 327
6.8 Scanning and run-length/variable-length coding 332
6.9 A bidirectional coder 336
6.10 Slices 340
6.11 Handling interlaced pictures 341
6.12 An MPEG-2 coder 346
6.13 The Elementary Stream 348
6.14 An MPEG-2 decoder 349
6.15 Coding artifacts 352
6.16 Processing MPEG-2 and concatenation 354
References 361

Chapter 7 Digital audio in video 362

7.1 What is sound? 362
7.2 The ear 363
7.3 Level and loudness 366
7.4 Critical bands 367
7.5 Choice of sampling rate for audio 371
7.6 Basic digital-to-analog conversion 373
7.7 Basic analog-to-digital conversion 380
7.8 Alternative convertors 385
7.9 Oversampling and noise shaping 388
7.10 One-bit convertors 396
7.11 Operating levels in digital audio 398
7.12 Digital audio in VTRs 400
7.13 MPEG audio compression 401
7.14 Dolby AC-3 413
References 414

Chapter 8 Digital recording principles 417

8.1 Introduction to the channel 417
8.2 Types of recording medium 418
8.3 Magnetic recording 419
8.4 Azimuth recording and rotary heads 424
8.5 Optical disks 426
8.6 Magneto-optical disks 428
8.7 Equalization 429
8.8 Data separation 430
8.9 Slicing 431
8.10 Jitter rejection 435
8.11 Channel coding 437
8.12 Group codes 446
8.13 EFM code in D-3/D-5 449
8.14 EFM code of CD and DVD 459
8.15 Error detection in group codes 461
8.16 Tracking signals 462
8.17 Randomizing 462
8.18 Partial response 465
8.19 Synchronizing 470
References 473

Chapter 9 Error correction 474

9.1 Sensitivity of message to error 474
9.2 Error mechanisms 476
9.3 Basic error correction 477
9.4 Error handling 478
9.5 Concealment by interpolation 479
9.6 Parity 480
9.7 Block and convolutional codes 482
9.8 Hamming code 485
9.9 Hamming distance 487
9.10 Cyclic codes 491
9.11 Punctured codes 496
9.12 Applications of cyclic codes 497
9.13 Burst correction 497
9.14 Introduction to the Reed–Solomon codes 502
9.15 R–S calculations 504
9.16 Correction by erasure 510
9.17 Interleaving 512
9.18 Product codes 514
9.19 Editing interleaved recordings 517
References 520
Appendix 9.1 Calculation of Reed–Solomon generator polynomials 521

Chapter 10 Digital interfaces 522

10.1 Introduction 522
10.2 Interface principles 523
10.3 Serial digital interface (SDI) 527
10.4 SDTI 536
10.5 Serial digital routing 538
10.6 Timing in digital installations 539
10.7 HD component parallel interface 540
10.8 HD fibre-optic interface 540
10.9 Testing digital video interfaces 542
10.10 Introduction to the AES/EBU interface 546
10.11 The electrical interface 547
10.12 Frame structure 549
10.13 Timing tolerance of serial interfaces 555
10.14 Embedded audio in SDI 557
10.15 FireWire 564
References 566

Chapter 11 Digital video tape 568

11.1 Introduction 568
11.2 Compression in DVTRs 569
11.3 Why rotary heads? 570
11.4 Helical geometry 570
11.5 Track and head geometry 578
11.6 Track-following systems 583
11.7 Time compression and segmentation 584
11.8 The basic rotary head transport 588
11.9 Servos 589
11.10 The cassette 594
11.11 Loading the cassette 596
11.12 Operating modes of a digital recorder 598
11.13 Editing 599
11.14 Variable-speed replay 600
11.15 DVTR signal systems 609
11.16 Product codes and segmentation 612
11.17 Distribution 613
11.18 The shuffle strategy 613
11.19 The track structure 617
11.20 Digital Betacam 622
11.21 DVC and DVCPRO 629
11.22 The D-9 format 640
References 641

Chapter 12 Disks 642

12.1 Types of disk 642
12.2 Structure of disk 645
12.3 Magnetic disk terminology 645
12.4 Principle of flying head 645
12.5 Magnetic reading and writing 647
12.6 Moving the heads 649
12.7 Rotation 654
12.8 Servo-surface disks 655
12.9 Soft sectoring 658
12.10 Winchester technology 660
12.11 Rotary positioners 664
12.12 Floppy disks 666
12.13 The disk controller 669
12.14 Defect handling 673
12.15 RAID arrays 678

12.16 Disk servers 679
12.17 Disks and compression 680
12.18 Optical disk principles 681
12.19 Optical pickups 689
12.20 Focus systems 695
12.21 Tracking systems 699
12.22 Structure of a DVD player 702
References 706

Chapter 13 Digital video editing 707

13.1 Introduction 707
13.2 Linear and non-linear editing 708
13.3 Online and offline editing 708
13.4 Timecode 709
13.5 Editing on recording media 709
13.6 A digital tape editor 711
13.7 The non-linear workstation 714
13.8 Locating the edit point 715
13.9 Editing with disk drives 717
13.10 Digitally assisted film making 720
13.11 Automation 721
13.12 The future 722
References 725

Chapter 14 Digital television broadcasting 726

14.1 Background 726
14.2 Overall system block 727
14.3 Packets and time stamps 729
14.4 Program Clock Reference 731
14.5 Program Specific Information (PSI) 733
14.6 Multiplexing 734
14.7 Remultiplexing 736
14.8 Modulation techniques 738
14.9 OFDM 743
14.10 Error correction 747
14.11 DVB 748
14.12 The DVB receiver 750
14.13 ATSC 752

Index 759

Preface

In the ten years since the first edition of this book appeared the applications of digital video have changed out of all recognition, and that must be reflected in some changes to this volume. The rapid adoption of digital video has been a direct consequence of the continuously falling cost of digital processing and storage equipment alongside continuously increasing performance.

Ten years ago, in most homes a moving picture would most likely have come from an analog broadcast television station, possibly delayed by an analog VCR. The developments in compression, recording and communications mean that today's moving picture might be from a digital television broadcast station, delayed by a digital VCR which might contain a hard drive. However, it could equally be on a computer screen, sourced from a digital video disk or the Internet.

It follows immediately that the readership of this book has also enlarged, not least because of the convergence between film, digital imaging and computing.

This new edition reflects these changes not just by treating new subjects, but also by adopting an approach which will be of benefit to readers from a wide range of backgrounds. Essentially fewer assumptions have been made about the reader so that the book begins at a more straightforward level and has more to say about the human visual system which defines the quality needed.

Whilst a great deal has changed, the original approach to the subject has not changed at all. Every subject is explained in plain English with the minimum of equations and numerous illustrations. Even transform coding has been explained in this way.

With the rapid introduction of a wide range of products and services based on digital imaging technology, a proper understanding of the principles has never been more important.

Acknowledgements

I must first thank David Kirk, who invited me to write a series on digital video for *Broadcast Systems Engineering*, which he then edited, and Ron Godwyn, who succeeded him. The response to that series provided the impetus to turn it into a book. *BSE* is no longer published, a victim of recession, but it deserves a place in our memories.

The publications of the Society of Motion Picture and Television Engineers, the European Broadcasting Union and the Audio Engineering Society have been extremely useful.

I am indebted to the many people who have found time to discuss complex subjects and to suggest reference works. Particular thanks go to David Lyon, Roderick Snell, Tim Borer, Bruce Devlin, Takeo Eguchi, Jim Wilkinson, David Huckfield and John Ive, Steve Owen, Yoshinobu Oba, Mark Schubin, Luigi Gallo, Mikael Reichel, Graham Roe, David Stone, Robin Caine, Dave Trytko, Joe Attard, John Watney, Fraser Morrison, Roger Wood, Tom Cavanagh, Steve Lyman, Kees Schouhamer Immink, Peter de With, Gerard de Haan, Ab Weber, Tom MacMahon, Dave Marsh, John Mallinson and Peter Kraniauskas.

Thanks are also due to Margaret Riley of Focal Press for her endless support.

John Watkinson
Burghfield Common, England

1

Introducing digital video

1.1 What is a video signal?

Video signals are electrical waveforms which allow moving pictures to be conveyed from one place to another. Observing the real world with the human eye results in a two-dimensional image on the retina. This image changes with time and so the basic information is three-dimensional. With two eyes a stereoscopic view can be obtained and stereoscopic television is possible with suitable equipment. However, this is restricted to specialist applications and has yet to be exploited in broadcasting.

An electrical waveform is two-dimensional in that it carries a voltage changing with respect to time. In order to convey three-dimensional picture information down a two-dimensional cable it is necessary to resort to scanning. Instead of attempting to convey the brightness of all parts of a picture at once, scanning conveys the brightness of a single point which moves with time.

The scanning process converts spatial resolution on the image into the temporal frequency domain. The higher the resolution of the image, the more lines are necessary to resolve the vertical detail. The line rate is increased along with the number of cycles of modulation which need to be carried in each line. If the frame rate remains constant, the bandwidth goes up as the square of the resolution.

The first methods used in television transmission and recording were understandably analog, and the signal formats were essentially determined by requirements of the cathode ray tube as a display, so that the receiver might be as simple as possible, and be constructed with a minimum number of vacuum tubes. Following the development of magnetic audio recording during World War II, a need for television recording was perceived. This was initially due to the various time zones across the United States. Without recording, popular television programmes had to be performed live several times so they could be seen at peak viewing time in each time zone. Ampex eventually succeeded in recording monochrome video in the 1950s, and the fundamentals of the Quadruplex machine were so soundly based that modern analog video recorders only refine the process.[1] Study of every shortcoming in the analog production process has led to the development of some measure to reduce it, and current equipment is capable of excellent performance. The point to appreciate, however, is that analog production equipment became a mature technology, and almost reached limits determined by the laws of physics. The process of refinement then produced increasingly small returns.

1.2 Comparing analog and digital

As there is now another video technology, known as digital, the previous technology has to be referred to as analog. It is appropriate to begin by contrasting the two technologies. Comparison is only possible in areas where the two technologies overlap. The real strength of digital video lies in areas where no comparison is possible because such processes are simply impossible in the analog domain.

In an analog system, information is conveyed by some infinite variation of a continuous parameter such as the voltage on a wire or the strength or frequency of flux on a tape. In a recorder, distance along the medium is a further, continuous, analog of time. It does not matter at what point a recording is examined along its length, a value will be found for the recorded signal. That value can itself change with infinite resolution within the physical limits of the system.

Those characteristics are the main weakness of analog signals. Within the allowable bandwidth, *any* waveform is valid. If the speed of the medium is not constant, one valid waveform is changed into another valid waveform; a timebase error cannot be detected in an analog system. In addition, a voltage error simply changes one valid voltage into another; noise cannot be detected in an analog system. We might suspect noise, but how is one to know what proportion of the received signal is noise and what is the original? If the transfer function of a system is not linear, distortion results, but the distorted waveforms are still valid; an analog system cannot detect distortion. Again we might suspect distortion, but how are we to know how much of the energy at a given frequency is due to the distortion and how much was actually present in the original signal?

It is a characteristic of analog systems that degradations cannot be separated from the original signal, so nothing can be done about them. At the end of a system a signal carries the sum of all degradations introduced at each stage through which it passed. This sets a limit to the number of stages through which a signal can be passed before it is useless. Alternatively, if many stages are envisaged, each piece of equipment must be far better than necessary so that the signal is still acceptable at the end. The equipment will naturally be more expensive.

One of the vital concepts to grasp is that digital video is simply an alternative means of carrying a video waveform. Although there are a number of ways in which this can be done, there is one system, known as pulse code modulation (PCM) which is in virtually universal use.[2] Figure 1.1 shows how PCM works. Instead of being continuous, the time axis is represented in a discrete, or stepwise manner. The waveform is not carried by continuous representation, but by measurement at regular intervals. This process is called sampling and the frequency with which samples are taken is called the sampling rate or sampling frequency F_s. It should be stressed that sampling is an analog process. Each sample still varies infinitely as the original waveform did. To complete the conversion to PCM, each sample is then represented to finite accuracy by a discrete number in a process known as quantizing.

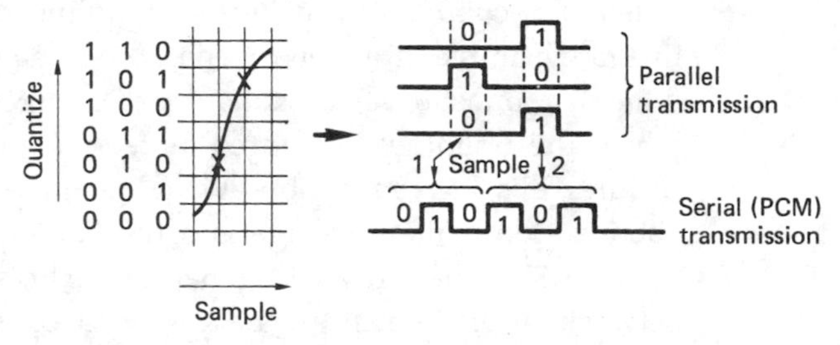

Figure 1.1 When a signal is carried in numerical form, either parallel or serial, the mechanisms of Figure 1.5 ensure that the only degradation is in the conversion process.

In television systems the input image which falls on the camera sensor will be continuous in time, and continous in two spatial dimensions corresponding to the height and width of the sensor. In analog video systems, the time axis is sampled into frames, and the vertical axis is sampled into lines. Digital video simply adds a third sampling process along the lines.

There is a direct connection between the concept of temporal sampling, where the input changes with respect to time at some frequency and is sampled at some other frequency, and spatial sampling, where an image changes a given number of times per unit distance and is sampled at some other number of times per unit distance. The connection between

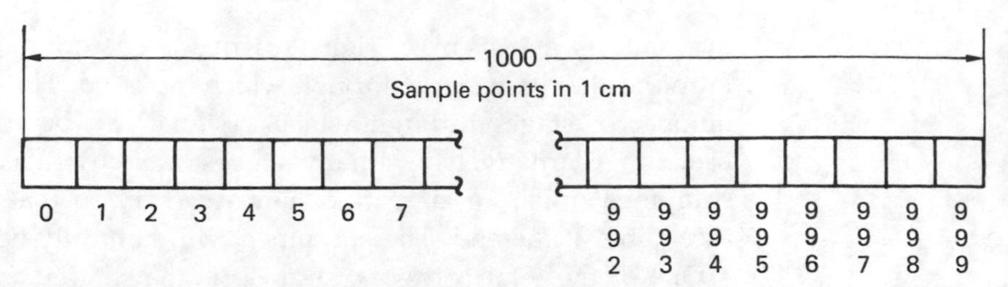

Figure 1.2 If the above spatial sampling arrangement of 1000 points per centimetre is scanned in 1 millisecond, the sampling rate will become 1 MHz.

the two is the process of scanning. Temporal frequency can be obtained by multiplying spatial frequency by the speed of the scan. Figure 1.2 shows a hypothetical image sensor which has 1000 discrete sensors across a width of one centimetre. The spatial sampling rate of this sensor is thus 1000 per centimetre. If the sensors are measured sequentially during a scan which takes one millisecond to go across the one centimetre width, the result will be a temporal sampling rate of 1 MHz. In the spatial domain we refer to resolution, whereas in the time domain we refer to frequency response. The two are linked by scanning.

Whilst any sampling rate which is high enough could be used for video, it is common to make the sampling rate a whole multiple of the line rate. Samples are then taken in the same place on every line. If this is done, a monochrome digital image is a rectangular array of points at which the brightness is stored as a number. The points are known as picture cells, generally abbreviated to pixels, although sometimes the abbreviation is more savage and they are known as pels. As shown in Figure 1.3(a), the array will generally be arranged with an even spacing between pixels, which are in rows and columns. By placing the pixels close together, it is hoped that the observer will perceive a continuous image. Obviously the finer the pixel spacing, the greater the resolution of the picture will be, but the amount of data needed to store one picture will increase as the square of the resolution, and with it the costs.

At the ADC (analog-to-digital convertor), every effort is made to rid the sampling clock of jitter, or time instability, so every sample is taken at an exactly even time step. Clearly if there is any subsequent timebase error, the instants at which samples arrive will be changed and the effect can be detected. If samples arrive at some destination with an irregular timebase, the effect can be eliminated by storing the samples temporarily in a memory and reading them out using a stable, locally generated clock. This process is called timebase correction and all properly engineered digital video systems must use it.

Those who are not familiar with digital principles often worry that sampling takes away something from a signal because it is not taking

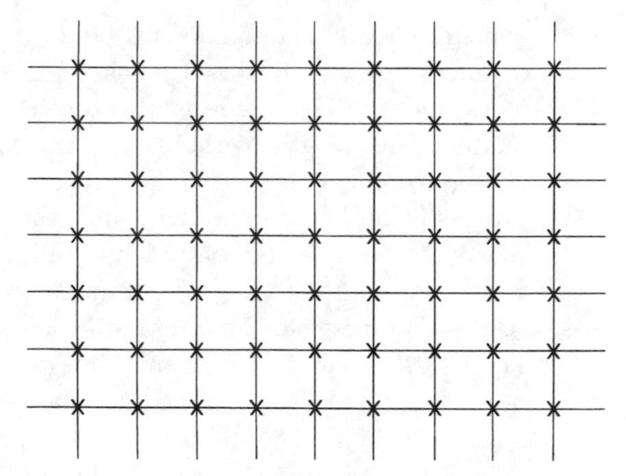

Figure 1.3(a) A picture can be stored digitally by representing the brightness at each of the above points by a binary number. For a colour picture each point becomes a vector and has to describe the brightness, hue and saturation of that part of the picture. Samples are usually but not always formed into regular arrays of rows and columns, and it is most efficient if the horizontal and vertical spacing are the same.

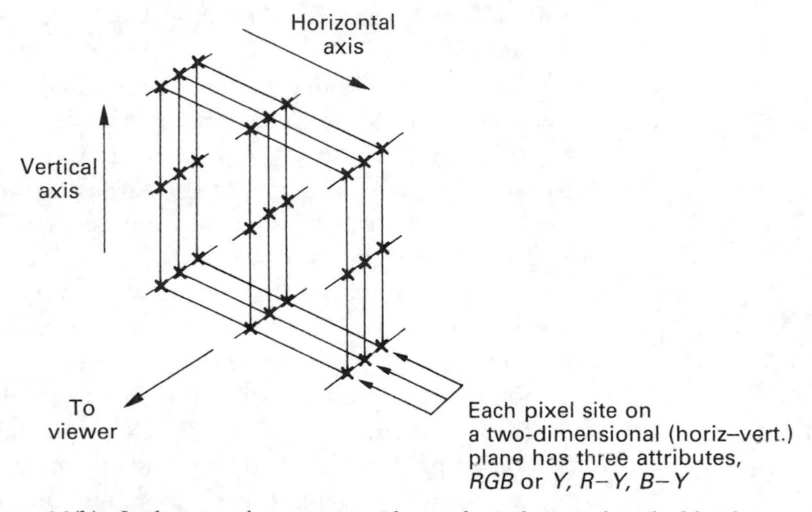

Figure 1.3(b) In the case of component video, each pixel site is described by three values and so the pixel becomes a vector quantity.

notice of what happened between the samples. This would be true in a system having infinite bandwidth, but no analog signal can have infinite bandwidth. All analog signal sources from cameras, VTRs and so on have a resolution or frequency response limit, as indeed do devices such as CRTs and human vision. When a signal has finite bandwidth, the rate at which it can change is limited, and the way in which it changes becomes predictable. When a waveform can only change between samples in one way, it is then only necessary to convey the samples and the original waveform can unambiguously be reconstructed from them. A more detailed treatment of the principle will be given in Chapter 4.

As stated, each sample is also discrete, or represented in a stepwise manner. The length of the sample, which will be proportional to the voltage of the video signal, is represented by a whole number. This process is known as quantizing and results in an approximation, but the size of the error can be controlled until it is negligible. If, for example, we were to measure the height of humans to the nearest metre, virtually all adults would register 2 metres high and obvious difficulties would result. These are generally overcome by measuring height to the nearest centimetre. Clearly there is no advantage in going further and expressing our height in a whole number of millimetres or even micrometres. The point is that an appropriate resolution can also be found for video signals, and a higher figure is not beneficial. The link between video quality and sample resolution is explored in Chapter 4. The advantage of using whole numbers is that they are not prone to drift. If a whole number can be carried from one place to another without numerical error, it has not changed at all. By describing video waveforms numerically, the original information has been expressed in a way which is better able to resist unwanted changes.

Essentially, digital video carries the original waveform numerically. The number of the sample is an analog of time, which itself is an analog of position across the screen, and the magnitude of the sample is (in the case of luminance) an analog of the brightness at the appropriate point in the image. In fact the succession of samples in a digital system is actually *an analog* of the original waveform. This sounds like a contradiction and as a result some authorities prefer the term 'numerical video' to 'digital video' and in fact the French word is numérique. The term 'digital' is so well established that it is unlikely to change.

As both axes of the digitally represented waveform are discrete, the waveform can accurately be restored from numbers as if it were being drawn on graph paper. If we require greater accuracy, we simply choose paper with smaller squares. Clearly more numbers are then required and each one could change over a larger range.

1.3 Why binary?

Humans insist on using numbers expressed to the base of ten, having evolved with that number of digits. Other number bases exist; most people are familiar with the duodecimal system which uses the dozen and the gross. The most minimal system is binary, which has only two digits, 0 and 1. BInary digiTS are universally contracted to bits. These are readily conveyed in switching circuits by an 'on' state and an 'off' state.

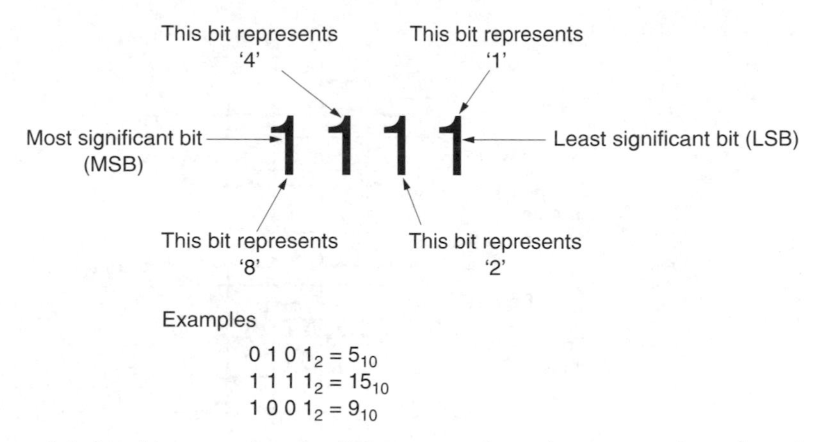

Figure 1.4 In a binary number, the digits represent increasing powers of two from the LSB. Also defined here are MSB and wordlength. When the wordlength is eight bits, the word is a byte. Binary numbers are used as memory addresses, and the range is defined by the address wordlength. Some examples are shown here.

With only two states, there is little chance of error. Had humans evolved with only one finger, we might have adopted binary more commonly.

In decimal systems, the digits in a number (counting from the right, or least significant end) represent ones, tens, hundreds and thousands, etc. Figure 1.4 shows that in binary, the bits represent one, two, four, eight, sixteen, etc. A multi-digit binary number is commonly called a word, and the number of bits in the word is called the wordlength. The right-hand bit is called the Least Significant Bit (LSB) whereas the bit on the left-hand end of the word is called the Most Significant Bit (MSB). Clearly more digits are required in binary than in decimal, but they are more easily handled. A word of eight bits is called a byte, which is a contraction of 'by eight'. The capacity of memories and storage media is measured in bytes, but to avoid large numbers, kilobytes, megabytes and gigabytes are often used. As memory addresses are themselves binary numbers, the wordlength limits the address range. The range is found by raising two to the power of the wordlength. Thus a four-bit word has sixteen combinations, and could address a memory having sixteen locations. A ten-bit word has 1024 combinations, which is close to 1000. In digital terminology, 1K = 1024, so a kilobyte of memory contains 1024 bytes. A megabyte (1MB) contains 1024 kilobytes and a gigabyte contains 1024 megabytes.

In a digital video system, the whole number representing the length of the sample is expressed in binary. The signals sent have two states, and change at predetermined times according to some stable clock. Figure 1.5 shows the consequences of this form of transmission. If the binary signal is degraded by noise, this will be rejected by the receiver, which judges the signal solely by whether it is above or below the half-way threshold,

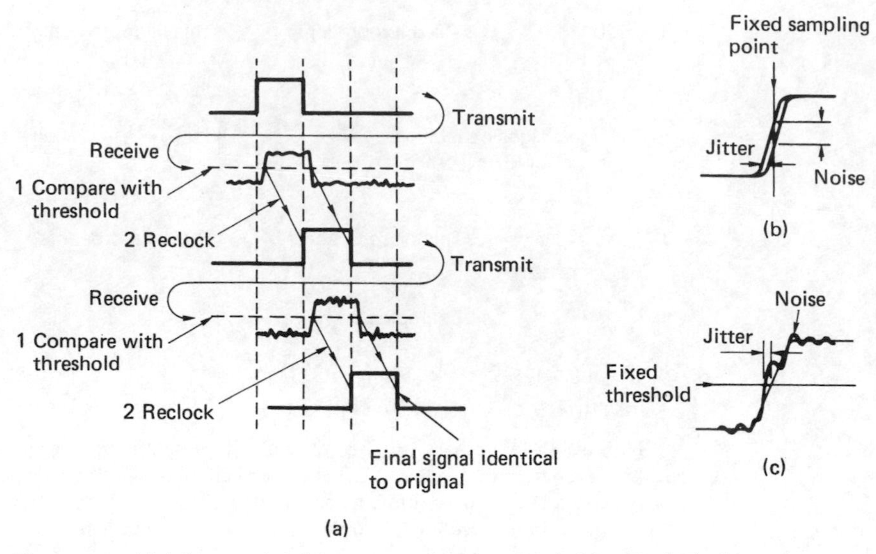

Figure 1.5 (a) A binary signal is compared with a threshold and relocked on receipt; thus the meaning will be unchanged. (b) Jitter on a signal can appear as noise with respect to fixed timing. (c) Noise on a signal can appear as jitter when compared with a fixed threshold.

a process known as slicing. The signal will be carried in a channel with finite bandwidth, and this limits the slew rate of the signal; an ideally upright edge is made to slope. Noise added to a sloping signal can change the time at which the slicer judges that the level passed through the threshold. This effect is also eliminated when the output of the slicer is reclocked. However many stages the binary signal passes through, it still comes out the same, only later.

Video samples which are represented by whole numbers can reliably be carried from one place to another by such a scheme, and if the number is correctly received, there has been no loss of information en route.

There are two ways in which binary signals can be used to carry samples and these are also shown in Figure 1.1. When each digit of the binary number is carried on a separate wire this is called parallel transmission. The state of the wires changes at the sampling rate. Using multiple wires is cumbersome and it is preferable to use a single wire in which successive digits from each sample are sent serially. This is the definition of Pulse Code Modulation. Clearly the clock frequency must now be higher than the sampling rate. Whilst the transmission of video by such a scheme is advantageous in that noise and timebase error have been eliminated, there is a penalty that a high-quality standard definition moving colour picture requires around 200 million bits per second. Clearly digital video production could only become commonplace when such a data rate could be handled economically. Consumer applications could only become possible when compression technology became available to reduce the data rate.

1.4 Colour

Colorimetry will be treated in depth in Chapter 2 and it is intended to introduce only the basics here. Colour is created by the additive mixing in the display of three primary colours, red, green and blue. Effectively the display needs to be supplied with three video signals, each representing a primary colour. Since practical colour cameras generally also have three separate sensors, one for each primary colour, a camera and a display can be directly connected. *RGB* consists of three parallel signals each having the same spectrum, and is used where the highest accuracy is needed. *RGB* is seldom used for broadcast applications because of the high cost.

If *RGB* is used in the digital domain, it will be seen from Figure 1.3(b) that each image consists of three superimposed layers of samples, one for each primary colour. The pixel is no longer a single number representing a scalar brightness value, but a vector which describes in some way the brightness, hue and saturation of that point in the picture. In *RGB*, the pixels contain three unipolar numbers representing the proportion of each of the three primary colours at that point in the picture.

Some saving of bandwidth can be obtained by using colour difference working. The human eye relies on brightness to convey detail, and much less resolution is needed in the colour information. Accordingly R, G and B are matrixed together to form a luminance (and monochrome-compatible) signal Y which needs full bandwidth. The eye is not equally sensitive to the three primary colours, as can be seen in Figure 1.6, and so the luminance signal is a weighted sum.

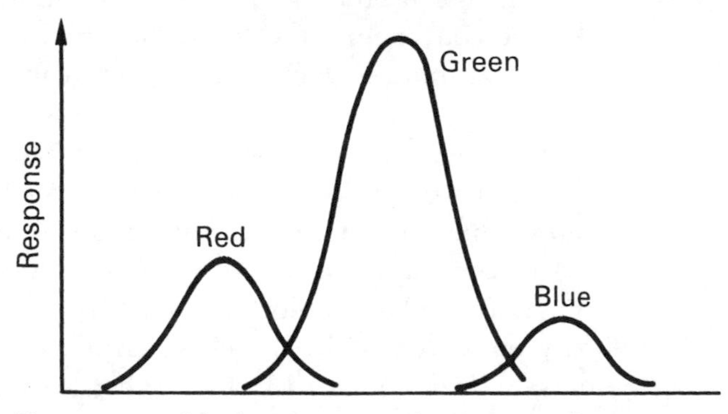

Figure 1.6 The response of the human eye to colour is not uniform.

The matrix also produces two colour difference signals, *R-Y* and *B-Y*. Colour difference signals do not need the same bandwidth as *Y*, because the eye's acuity does not extend to colour vision. One half or one quarter of the bandwidth will do depending on the application.

In the digital domain, each pixel again contains three numbers, but one of these is a unipolar number representing the luminance and the other two are bipolar numbers representing the colour difference values. As the colour difference signals need less bandwidth, in the digital domain this translates to the use of a lower data rate, typically between one half and one sixteenth the bit rate of the luminance.

For monochrome-compatible analog colour television broadcasting, the NTSC, PAL and SECAM systems interleave into the spectrum of a monochrome signal a subcarrier which carries two colour difference signals of restricted bandwidth. The subcarrier is intended to be invisible on the screen of a monochrome television set. A subcarrier-based colour system is generally referred to as composite video, and the modulated subcarrier is called chroma. Composite video is effectively a compression technique because it allows colour pictures in the same bandwidth as monochrome.

In the digital domain the use of composite video is not appropriate for broadcasting and the most common equivalent process is the use of an MPEG compression scheme. From a broadcasting standpoint MPEG is simply a more efficient digital replacement for composite video.

1.5 Why digital?

There are two main answers to this question, and it is not possible to say which is the most important, as it will depend on one's standpoint.

(a) The quality of reproduction of a well-engineered digital video system is independent of the medium and depends only on the quality of the conversion processes and of any compression scheme.
(b) The conversion of video to the digital domain allows tremendous opportunities which were denied to analog signals.

Someone who is only interested in picture quality will judge the former the most relevant. If good-quality convertors can be obtained, all the shortcomings of analog recording and transmission can be eliminated to great advantage. One's greatest effort is expended in the design of convertors, whereas those parts of the system which handle data need only be workmanlike. Timebase error, tape noise, print-through, drop-outs, and moiré are all history. When a digital recording is copied, the same numbers appear on the copy: it is not a dub, it is a clone. If the copy is indistinguishable from the original, there has been no generation loss. Digital recordings can be copied indefinitely without loss of quality. This is, of course wonderful for the production process, but when the technology becomes available to the consumer the issue of copyright becomes of great importance.

In the real world everything has a cost, and one of the greatest strengths of digital technology is low cost. If copying causes no quality loss, recorders do not need to be far better than necessary in order to withstand generation loss. They need only be adequate on the first generation whose quality is then maintained. There is no need for the great size and extravagant tape consumption of professional analog recorders. When the information to be recorded is discrete numbers, they can be packed densely on the medium without quality loss. Should some bits be in error because of noise or dropout, error correction can restore the original value. Digital recordings take up less space than analog recordings for the same or better quality. Tape costs are far less and storage costs are reduced.

Digital circuitry costs less to manufacture. Switching circuitry which handles binary can be integrated more densely than analog circuitry. More functionality can be put in the same chip. Analog circuits are built from a host of different component types which have a variety of shapes and sizes and are costly to assemble and adjust. Digital circuitry uses standardized component outlines and is easier to assemble on automated equipment. Little if any adjustment is needed.

Digital equipment can have self-diagnosis programs built-in. The machine points out its own failures. The days of chasing a signal with an oscilloscope are over. Even if a faulty component in a digital circuit could be located with such a primitive tool, it is nearly impossible to replace a chip having 60 pins soldered through a six-layer circuit board. The cost of finding the fault may be more than the board is worth.

Routine, mind-numbing adjustment of analog circuits to counteract drift is no longer needed. The cost of maintenance falls. A small operation may not need maintenance staff at all; a service contract is sufficient. A larger organization will still need maintenance staff, but they will be fewer in number and their skills will be oriented more to systems than devices.

1.6 Video as data

Digital video does far more than merely compare favourably with analog. Its most exciting aspects are the tremendous possibilities which are denied to analog technology. Error correction, compression, motion estimation and interpolation are difficult or impossible in the analog domain, but are straightforward in the digital domain. Once video is in the digital domain, it becomes data, and as such is indistinguishable from any other type of data.

The worlds of digital video, digital audio, communication and computation are closely related, and that is where the real potential lies. The time when television was a specialist subject which could evolve in isolation from other disciplines has gone, digital technology has made sure of that. Video has now become a branch of data processing.

Systems and techniques developed in other industries for other purposes can be used to store, process and transmit video. Computer equipment is available at low cost because the volume of production is far greater than that of professional video equipment. Disk drives and memories developed for computers can be put to use in video products.

As the power of processors increases, it becomes possible to perform under software control processes which previously required dedicated hardware. This causes a dramatic reduction in hardware cost. Inevitably the very nature of broadcast equipment and the ways in which it is used is changing along with the manufacturers who supply it. The computer industry is competing with traditional broadcast manufacturers, but they have the economics of mass production on their side.

Tape is a linear medium and it is necessary to wait for the tape to wind to a desired part of the recording. In contrast, the head of a hard disk drive can access any stored data in milliseconds. This is known in computers as direct access and in broadcasting as non-linear access. As a result the non-linear editing workstation based on hard drives has eclipsed the use of videotape for editing.

Communications networks developed to handle data can happily carry digital video and accompanying audio over indefinite distances without quality loss. Techniques such as ADSL allow compressed digital video to travel over a conventional telephone line to the consumer.

Digital TV broadcasting uses coding techniques to eliminate the interference, fading and multipath reception problems of analog broadcasting. At the same time, more efficient use is made of available bandwidth.

One of the fundamental requirements of computer communication is that it is bidirectional. When this technology becomes available to the consumer, services such as video-on-demand and interactive video become possible. Television programs may contain metadata which allows the viewer rapidly to access web sites relating to items mentioned in the program. When the TV set is a computer there is no difficulty in displaying both on the same screen.

Increasingly the viewer will be deciding what to watch instead of passively accepting the broadcaster's output. With a tape-based VCR, the

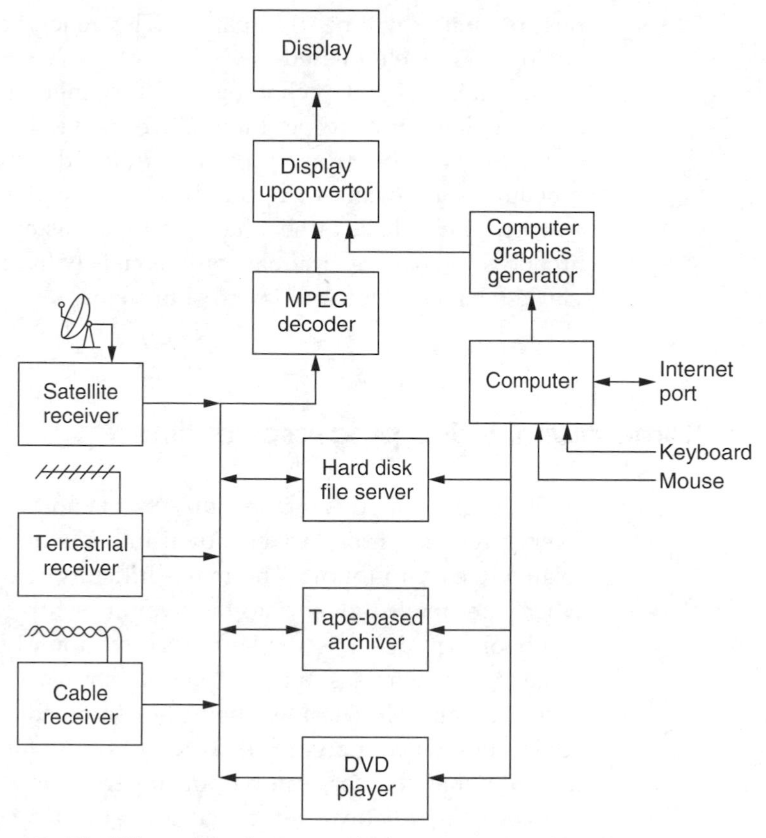

Figure 1.7 The TV set of the future may look into something like this.

consumer was limited to time-shifting broadcast programs. When the hard drive-based consumer VCR is available, the consumer has more power. For example, he or she may never watch another TV commercial again. The consequences of this technology are far-reaching.

Figure 1.7 shows what the television set of the future may look like. MPEG compressed signals may arrive in real time by terrestrial or satellite broadcast, via a cable, or on media such as DVD. The TV set is simply a display, and the heart of the system is a hard drive-based server. This can be used to time-shift broadcast programs, to skip commercial breaks or to assemble requested movies transmitted in non-real time at low bit rates. If equipped with a web browser, the server may explore the web looking for material which is of the same kind the viewer normally watches. As the cost of storage falls, the server may download this material speculatively.

Note that when the hard drive is used to time shift or record, it simply stores the MPEG bitstream. On playback the bitstream is decoded and the picture quality will be unimpaired. The generation loss due to using an analog VCR is eliminated.

Ultimately digital technology will change the nature of television broadcasting out of recognition. Once the viewer has non-linear storage technology and electronic program guides, the broadcaster's transmitted schedule is irrelevant. Increasingly viewers will be able to choose what is watched and when, rather than the broadcaster deciding for them. The broadcasting of conventional commercials will cease to be effective when viewers have the technology to skip them.

1.7 Some digital video processes outlined

Whilst digital video is a large subject, it is not necessarily a difficult one. Every process can be broken down into smaller steps, each of which is relatively easy to follow. The main difficulty with study is to appreciate where the small steps fit into the overall picture. Subsequent chapters of this book will describe the key processes found in digital technology in some detail, whereas this chapter illustrates why these processes are necessary and shows how they are combined in various ways in real equipment. Once the general structure of digital devices is appreciated, the following chapters can be put in perspective.

Figure 1.8(a) shows a minimal digital video system. This is no more than a point-to-point link which conveys analog video from one place to another. It consists of a pair of convertors and hardware to serialize and de-serialize the samples. There is a need for standardization in serial transmission so that various devices can be connected together. These standards for digital interfaces are described in Chapter 10.

Analog video entering the system is converted in the analog-to-digital convertor (ADC) to samples which are expressed as binary numbers. A typical sample would have a wordlength of eight bits. The sample is connected in parallel into an output register which controls the cable drivers. The cable also carries the sampling rate clock. The data are sent to the other end of the line where a slicer relects noise picked up on each signal. Sliced data are then loaded into a receiving register by the clock, and sent to the digital-to-analog convertor (DAC), which converts the sample back to an analog voltage.

Following a casual study one might conclude that if the convertors were of transparent quality, the system must be ideal. Unfortunately this is incorrect. As Figure 1.5 showed, noise can change the timing of a sliced signal. Whilst this system rejects noise which threatens to change the numerical value of the samples, it is powerless to prevent noise from causing jitter in the receipt of the word clock. Noise on the word clock

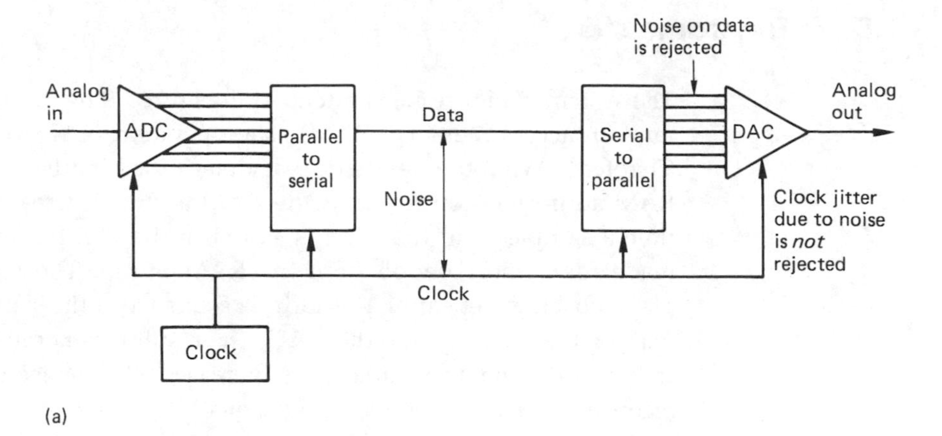

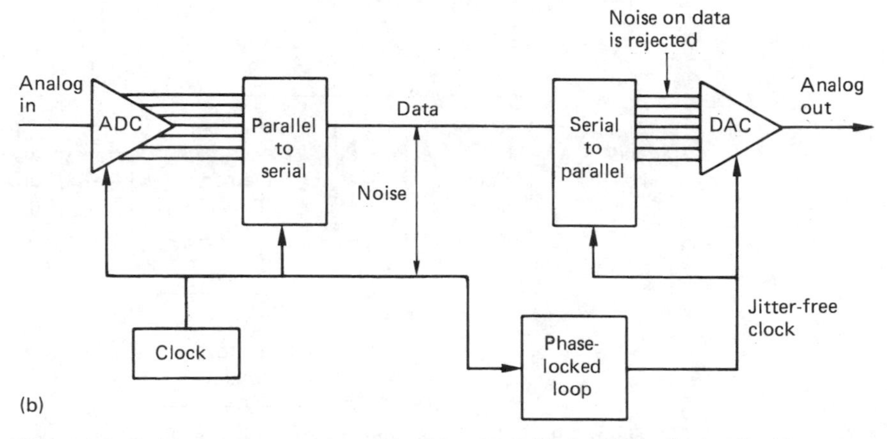

Figure 1.8 In (a) two convertors are joined by a serial link. Although simple, this system is deficient because it has no means to prevent noise on the clock lines causing jitter at the receiver. In (b) a phase-locked loop is incorporated, which filters jitter from the clock.

means that samples are not converted with a regular timebase and the impairment caused can be noticeable. Stated another way, analog characteristics of the interconnect are not prevented from affecting the reproduced waveform and so the system is not truly digital.

The jitter problem is overcome in Figure 1.8(b) by the inclusion of a phase-locked loop which is an oscillator that synchronizes itself to the *average* frequency of the clock but which filters out the instantaneous jitter. The operation of a phase-locked loop is analogous to the function of the flywheel on a piston engine. The samples are then fed to the convertor with a regular spacing and the impairment is no longer audible. Chapter 4 shows why the effect occurs and deduces the clock accuracy needed for accurate conversion.

1.8 The frame store

The system of Figure 1.8 is extended in Figure 1.9 by the addition of some random access memory (RAM). The operation of RAM is described in Chapter 3. What the device does is determined by the way in which the RAM address is controlled. If the RAM address increases by one every time a sample from the ADC is stored in the RAM, a recording can be made for a short period until the RAM is full. The recording can be played back by repeating the address sequence at the same clock rate but reading the memory into the DAC. The result is generally called a frame store.[3] If the memory capacity is increased, the device can be used for recording. At a rate of 200 million bits per second, each frame needs a megabyte of memory and so the RAM recorder will be restricted to a fairly short playing time.

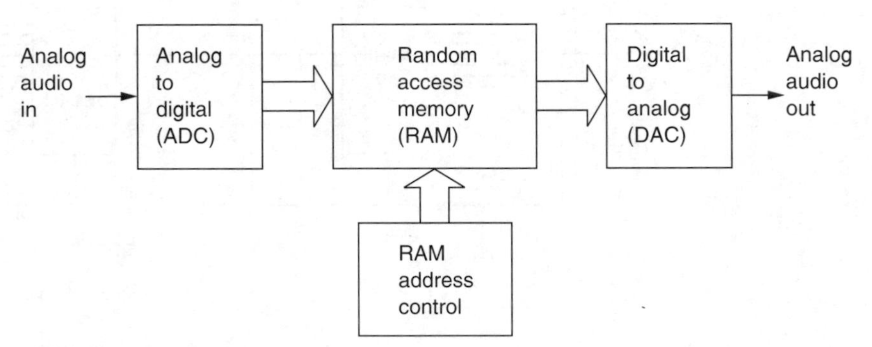

Figure 1.9 In the frame store, the recording medium is a random access memory (RAM). Recording time available is short compared with other media, but access to the recording is immediate and flexible as it is controlled by addressing the RAM.

Using compression, the playing time of a RAM-based recorder can be extended. For pre-determined images such as test patterns and station IDs, read only memory (ROM) can be used instead as it is non-volatile.

1.9 The synchronizer

If the RAM is used in a different way, it can be written and read at the same time. The device then becomes a synchronizer which allows video interchange between two systems that are not genlocked. Controlling the relationship between the addresses makes the RAM into a variable delay. The addresses are generated by counters which overflow to zero after they have reached a maximum count at the end of a frame. As a result the

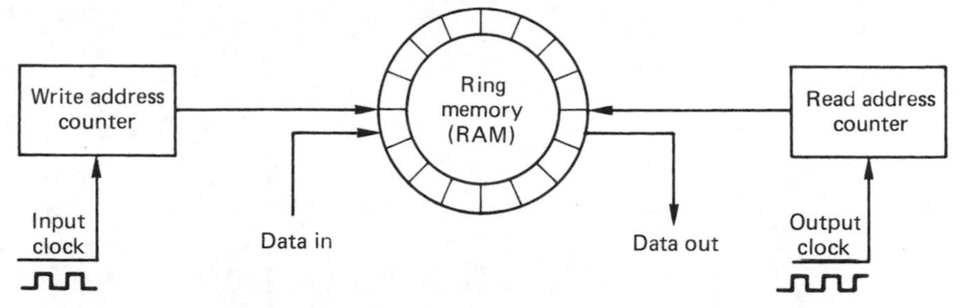

Figure 1.10 If the memory address is arranged to come from a counter which overflows, the memory can be made to appear circular. The write address then rotates endlessly, overwriting previous data once per revolution. The read address can follow the write address by a variable distance (not exceeding one revolution) and so a variable delay takes place between reading and writing.

memory space appears to be circular as shown in Figure 1.10. The read and write addresses chase one another around the circle. If the read address follows close behind the write address, the delay is short. If it stays just ahead of the write address, the maximum delay is reached. If the input and output have an identical frame rate, the address relationship will be constant, but if there is a drift then the address relationship will change slowly. Eventually the addresses will coincide and then cross. Properly handled, this results in a frame being omitted or repeated.

The issue of signal timing has always been critical in analog video, but the adoption of digital routing relaxes the requirements considerably. Analog vision mixers need to be fed by equal-length cables from the router to prevent propagation delay variation. In the digital domain this is no longer an issue as delay is easily obtained and each input of a digital vision mixer can have its own local synchronizer. A synchronizer with less than a frame of RAM can be used to remove static timing errors due, for example, to propagation delays in large systems. The finite RAM capacity gives a finite range of timing error which can be accommodated. This is known as the window. Provided signals are received having timing within the window of the inputs, all inputs are retimed to the same phase within the mixer. Chapter 10 deals with synchronizing large systems.

1.10 Time compression

When samples are converted, the ADC must run at a constant clock rate and it outputs an unbroken stream of samples during the active line.

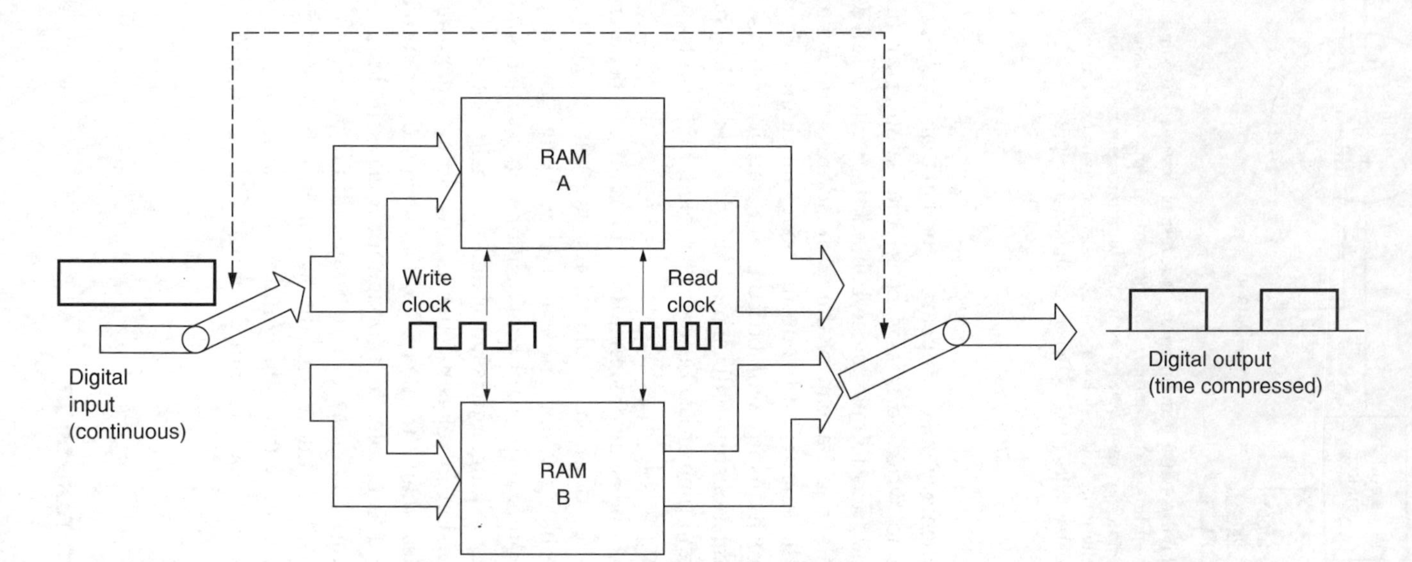

Figure 1.11 In time compression, the unbroken real-time stream of samples from an ADC is broken up into discrete blocks. This is accomplished by the configuration shown here. Samples are written into one RAM at the sampling rate by the write clock. When the first RAM is full, the switches change over, and writing continues into the second RAM whilst the first is read using a higher-frequency clock. The RAM is read faster than it was written and so all of the data will be output before the other RAM is full. This opens spaces in the data flow which are used as described in the text.

Following a break during blanking, the sample stream resumes. Time compression allows the sample stream to be broken into blocks for convenient handling.

Figure 1.11 shows an ADC feeding a pair of RAMs. When one is being written by the ADC, the other can be read, and vice versa. As soon as the first RAM is full, the ADC output switched to the input of the other RAM so that there is no loss of samples. The first RAM can then be read at a higher clock rate than the sampling rate. As a result the RAM is read in less time than it took to write it, and the output from the system then pauses until the second RAM is full. The samples are now time compressed. Instead of being an unbroken stream which is difficult to handle, the samples are now arranged in blocks with convenient pauses in between them. In these pauses numerous processes can take place. A rotary head recorder might spread the data from a frame over several tape tracks; a hard disk might move its heads to another track. In all types of recording and transmission, the time compression of the samples allows time for synchronizing patterns, subcode and error-correction words to be inserted.

In digital VTRs, the video data are time compressed so that part of the track is left for audio data. Figure 1.12 shows that heavy time compression of the audio data raises the data rate up to that of the video data so that the same tracks, heads and much common circuitry can be used to record both.

Subsequently, any time compression can be reversed by time expansion. Samples are written into a RAM at the incoming clock rate, but read out at the standard sampling rate. Unless there is a design fault, time compression is totally undetectable. In a recorder, the time expansion stage can be combined with the timebase correction stage so that speed variations in the medium can be eliminated at the same time. The use of time compression is universal in digital recording and widely used in

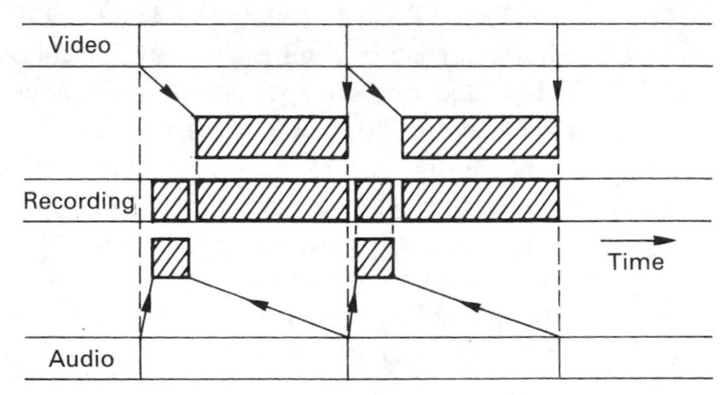

Figure 1.12 Time compression is used to shorten the length of track needed by the video. Heavily time-compressed audio samples can then be recorded on the same track using common circuitry.

transmission. In general the *instantaneous* data rate at the medium is not the same as the rate at the convertors, although clearly the *average* rate must be the same.

Another application of time compression is to allow several channels of audio to be carried along with video on a single cable. This technique is used in the serial digital interface (SDI) which is explained in Chapter 10.

1.11 Error correction and concealment

All practical recording and transmission media are imperfect. Magnetic media, for example, suffer from noise and dropouts. In a digital recording of binary data, a bit is either correct or wrong, with no intermediate stage. Small amounts of noise are rejected, but inevitably, infrequent noise impulses cause some individual bits to be in error. Dropouts cause a larger number of bits in one place to be in error. An error of this kind is called a burst error. Whatever the medium and whatever the nature of the mechanism responsible, data are either recovered correctly, or suffer some combination of bit errors and burst errors. In optical disks, random errors can be caused by imperfections in the moulding process, whereas burst errors are due to contamination or scratching of the disk surface.

The visibility of a bit error depends upon which bit of the sample is involved. If the LSB of one sample was in error in a detailed, contrasty picture, the effect would be totally masked and no-one could detect it. Conversely, if the MSB of one sample was in error in a flat field, no-one could fail to notice the resulting spot. Clearly a means is needed to render errors from the medium inaudible. This is the purpose of error correction.

In binary, a bit has only two states. If it is wrong, it is necessary only to reverse the state and it must be right. Thus the correction process is trivial and perfect. The main difficulty is in identifying the bits which are in error. This is done by coding the data by adding redundant bits. Adding redundancy is not confined to digital technology, airliners have several engines and cars have twin-braking systems. Clearly the more failures which have to be handled, the more redundancy is needed. If a four-engined airliner is designed to fly normally with one engine failed, three of the engines have enough power to reach cruise speed, and the fourth one is redundant. The amount of redundancy is equal to the amount of failure which can be handled. In the case of the failure of two engines, the plane can still fly, but it must slow down; this is graceful degradation. Clearly the chances of a two-engine failure on the same flight are remote.

In digital recording, the amount of error which can be corrected is proportional to the amount of redundancy, and it will be shown in

Chapter 9 that within this limit, the samples are returned to exactly their original value. Consequently *corrected* samples are indetectable. If the amount of error exceeds the amount of redundancy, correction is not possible, and, in order to allow graceful degradation, concealment will be used. Concealment is a process where the value of a missing sample is estimated from those nearby. The estimated sample value is not necessarily exactly the same as the original, and so under some circumstances concealment can be audible, especially if it is frequent. However, in a well-designed system, concealments occur with negligible frequency unless there is an actual fault or problem.

Concealment is made possible by rearranging the sample sequence prior to recording. This is shown in Figure 1.13 where odd-numbered samples are separated from even-numbered samples prior to recording. The odd and even sets of samples may be recorded in different places on the medium, so that an uncorrectable burst error affects only one set. On replay, the samples are recombined into their natural sequence, and the error is now split up so that it results in every other sample being lost in a two-dimensional structure. The picture is now described half as often, but can still be reproduced with some loss of accuracy. This is better than not being reproduced at all even if it is not perfect. Many digital video

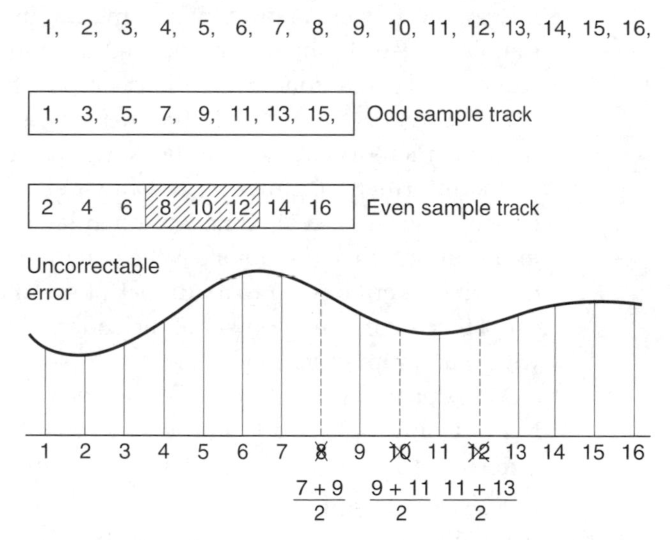

Figure 1.13 In cases where the error correction is inadequate, concealment can be used provided that the samples have been ordered appropriately in the recording. Odd and even samples are recorded in different places as shown here. As a result an uncorrectable error causes incorrect samples to occur singly, between correct samples. In the example shown, sample 8 is incorrect, but samples 7 and 9 are unaffected and an approximation to the value of sample 8 can be had by taking the average value of the two. This interpolated value is substituted for the incorrect value.

recorders use such an odd/even distribution for concealment. Clearly if any errors are fully correctable, the distribution is a waste of time; it is needed only if correction is not possible.

The presence of an error-correction system means that the video (and audio) quality is independent of the medium/head quality within limits. There is no point in trying to assess the health of a machine by watching a monitor or listening to the audio, as this will not reveal whether the error rate is normal or within a whisker of failure. The only useful procedure is to monitor the frequency with which errors are being corrected, and to compare it with normal figures. Professional DVTRs have an error rate display for this purpose and in addition most allow the error correction system to be disabled for testing.

1.12 Product codes

Digital systems such as broadcasting, optical disks and magnetic recorders are prone to burst errors. Adding redundancy equal to the size of expected bursts to every code is inefficient. Figure 1.14(a) shows that the efficiency of the system can be raised using interleaving. Sequential samples from the ADC are assembled into codes, but these are not recorded/transmitted in their natural sequence. A number of sequential codes are assembled along rows in a memory. When the memory is full, it is copied to the medium by reading down columns. Subsequently, the samples need to be de-interleaved to return them to their natural sequence. This is done by writing samples from tape into a memory in columns, and when it is full, the memory is read in rows. Samples read from the memory are now in their original sequence so there is no effect on the information. However, if a burst error occurs as is shown shaded on the diagram, it will damage sequential samples in a vertical direction in the de-interleave memory. When the memory is read, a single large error is broken down into a number of small errors whose size is exactly equal to the correcting power of the codes and the correction is performed with maximum efficiency.

An extension of the process of interleave is where the memory array has not only rows made into codewords, but also columns made into codewords by the addition of vertical redundancy. This is known as a product code. Figure 1.14(b) shows that in a product code the redundancy calculated first and checked last is called the outer code, and the redundancy calculated second and checked first is called the inner code. The inner code is formed along tracks on the medium. Random errors due to noise are corrected by the inner code and do not impair the burst correcting power of the outer code. Burst errors are declared uncorrectable by the inner code which flags the bad samples on the way into the

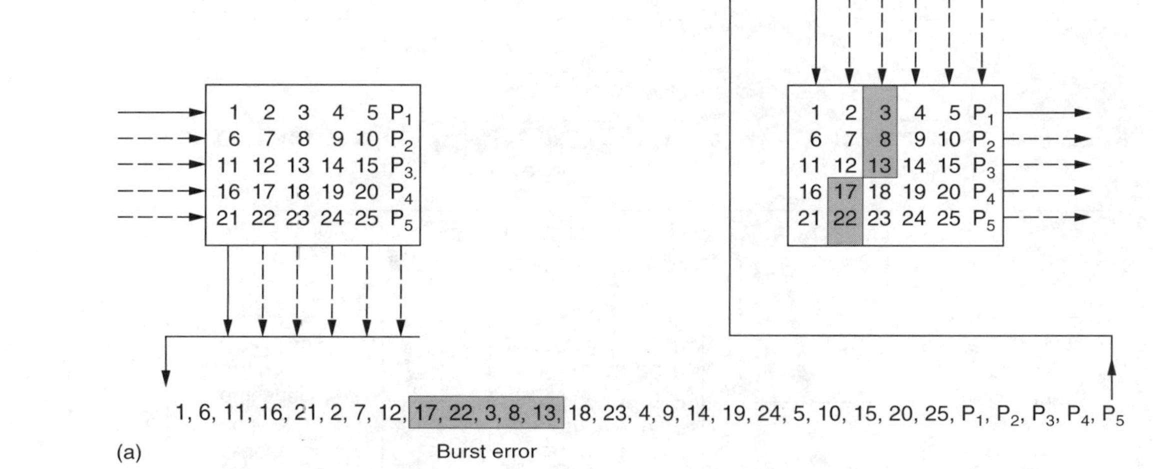

Figure 1.14(a) Interleaving is essential to make error correction schemes more efficient. Samples written sequentially in rows into a memory have redundancy P added to each row. The memory is then read in columns and the data sent to the recording medium. On replay the non-sequential samples from the medium are de-interleaved to return them to their normal sequence. This breaks up the burst error (shaded) into one error symbol per row in the memory, which can be corrected by the redundancy P.

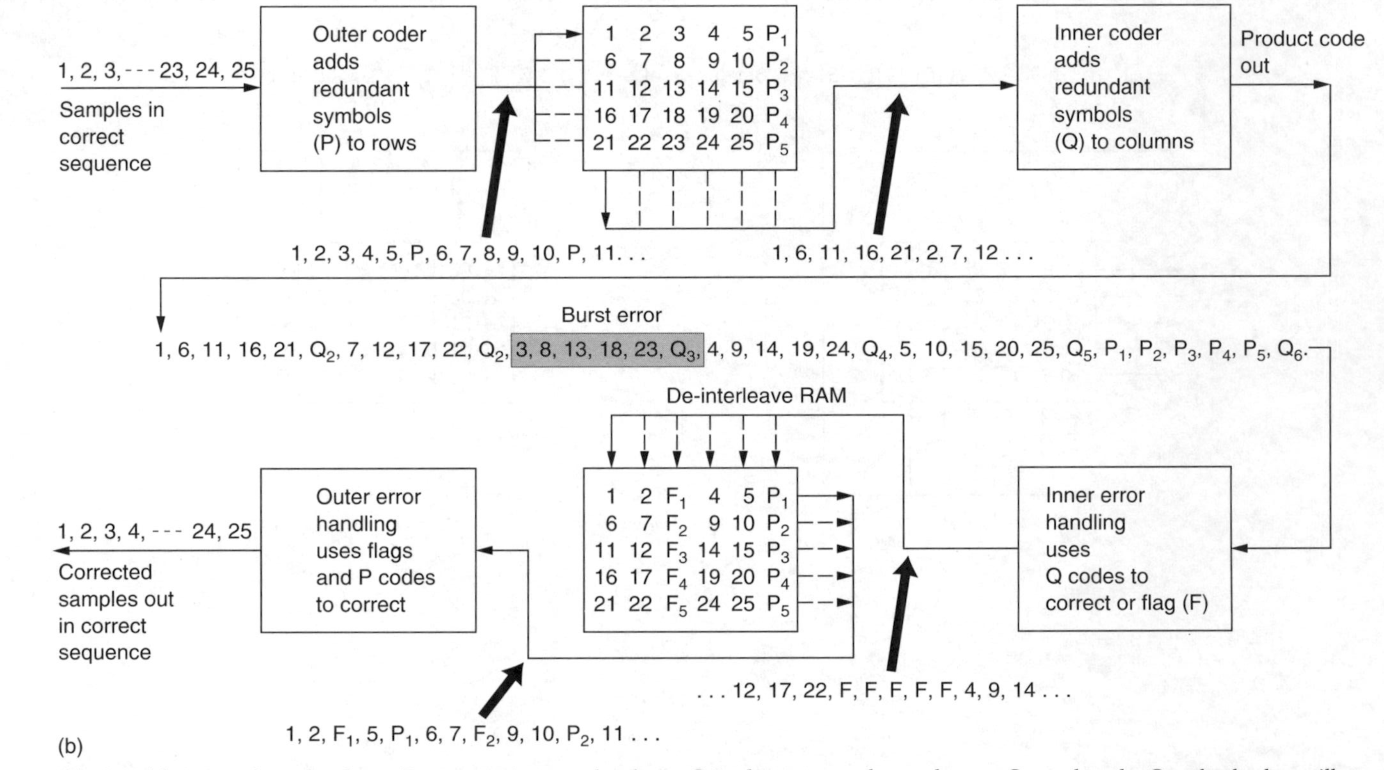

Figure 1.14(b) In addition to the redundancy P on rows, inner redundancy Q is also generated on columns. On replay, the Q code checker will pass on flag F if it finds an error too large to handle itself. The flags pass through the de-interleave process and are used by the outer error correction to identify which symbol in the row needs correcting with P redundancy. The concept of crossing two codes in this way is called a product code.

de-interleave memory. The outer code reads the error flags in order to locate the erroneous data. As it does not have to compute the error locations, the outer code can correct more errors.

The interleave, de-interleave, time-compression and timebase-correction processes inevitably cause delay.

1.13 Shuffling

When a product code based recording suffers an uncorrectable error the result is a rectangular block of failed sample values which require concealment. Such a regular structure would be visible even after concealment, and an additional process is necessary to reduce the visibility. Figure 1.15 shows that a shuffle process is performed prior to

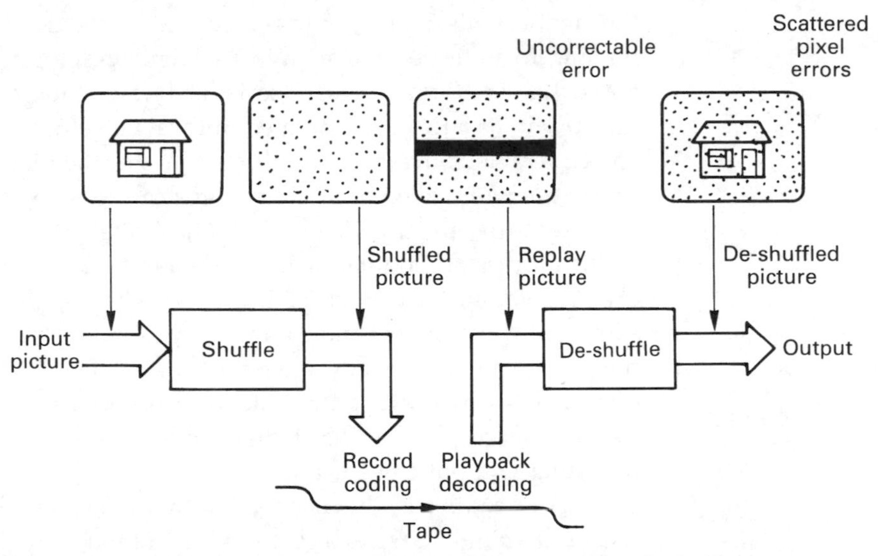

Figure 1.15 The shuffle before recording and the corresponding de-shuffle after playback cancel out as far as the picture is concerned. However, a block of errors due to dropout only experiences the de-shuffle, which spreads the error randomly over the screen. The pixel errors are then easier to correct.

product coding in which the pixels are moved around the picture in a pseudo-random fashion. The reverse process is used on replay, and the overall effect is nullified. However, if an uncorrectable error occurs, this will only pass through the de-shuffle and so the regular structure of the failed data blocks will be randomized. The errors are spread across the picture as individual failed pixels in an irregular structure. Chapter 11 treats shuffling techniques in more detail.

1.14 Channel coding

In most recorders used for storing digital information, the medium carries a track which reproduces a single waveform. Clearly data words representing video contain many bits and so they have to be recorded serially, a bit at a time. Some media, such as optical or magnetic disks, have only one active track, so it must be totally self-contained. DVTRs may have one, two or four tracks read or written simultaneously. At high recording densities, physical tolerances cause phase shifts, or timing errors, between tracks and so it is not possible to read them in parallel. Each track must still be self-contained until the replayed signal has been timebase corrected.

Recording data serially is not as simple as connecting the serial output of a shift register to the head. In digital video, samples may contain strings of identical bits. If a shift register is loaded with such a sample and shifted out serially, the output stays at a constant level for the period of the identical bits, and nothing is recorded on the track. On replay there is nothing to indicate how many bits were present, or even how fast to move the medium. Clearly, serialized raw data cannot be recorded directly, it has to be modulated into a waveform which contains an embedded clock irrespective of the values of the bits in the samples. On replay a circuit called a data separator can lock to the embedded clock and use it to separate strings of identical bits.

The process of modulating serial data to make it self-clocking is called channel coding. Channel coding also shapes the spectrum of the serialized waveform to make it more efficient. With a good channel code, more data can be stored on a given medium. Spectrum shaping is used in optical disks to prevent the data from interfering with the focus and tracking servos, and in hard disks and in certain DVTRs to allow re-recording without erase heads.

Channel coding is also needed to broadcast digital television signals where shaping of the spectrum is an obvious requirement to avoid interference with other services.

The techniques of channel coding for recording are covered in detail in Chapter 8, whereas the modulation schemes for digital television are described in Chapter 14.

1.15 Video compression and MPEG

In its native form, digital video suffers from an extremely high data rate, particularly in high definition. One approach to the problem is to use compression which reduces that rate significantly with a moderate loss of subjective quality of the picture. The human eye is not equally sensitive to all spatial frequencies, so some coding gain can be obtained by using fewer bits to describe the frequencies which are less visible. Video images typically contain a great deal of redundancy where flat areas contain the same pixel value repeated many times. Furthermore, in many cases there is little difference between one picture and the next, and compression can be achieved by sending only the differences.

Whilst these techniques may achieve considerable reduction in bit rate, it must be appreciated that compression systems reintroduce the generation loss of the analog domain to digital systems. As a result high compression factors are only suitable for final delivery of fully produced material to the viewer.

For production purposes, compression may be restricted to exploiting the redundancy within each picture individually and then with a mild compression factor. This allows simple algorithms to be used and also permits multiple generation work without artifacts being visible. A similar approach is used in disk-based workstations. Where offline editing is used (see Chapter 13) higher compression factors can be used as the impaired pictures are not seen by the viewer.

Clearly a consumer DVTR needs only single-generation operation and has simple editing requirements. A much greater degree of compression can then be used, which may takes advantage of redundancy between fields. The same is true for broadcasting, where bandwidth is at a premium. A similar approach may be used in disk-based camcorders which are intended for ENG purposes.

The future of television broadcasting (and of any high-definition television) lies completely in compression technology. Compression requires an encoder prior to the medium and a compatible decoder after it. Extensive consumer use of compression could not occur without suitable standards. The ISO-MPEG coding standards were specifically designed to allow wide interchange of compressed video data. Digital television broadcasting and the digital video disk both use MPEG standard bitstreams and these are detailed in Chapter 6.

Figure 1.16 shows that the output of a single compressor is called an *Elementary Stream*. In practice audio and video streams of this type can be combined using multiplexing. The *program stream* is optimized for recording and the multiplexing is based on blocks of arbitrary size. The *transport stream* is optimized for transmission and is based on blocks of constant size. In production equipment such as workstations and VTRs which are designed for editing, the MPEG standard is less useful and many successful products use non-MPEG compression.

Compression and the corresponding decoding are complex processes and take time, adding to existing delays in signal paths. Concealment of uncorrectable errors is also more difficult on compressed data.

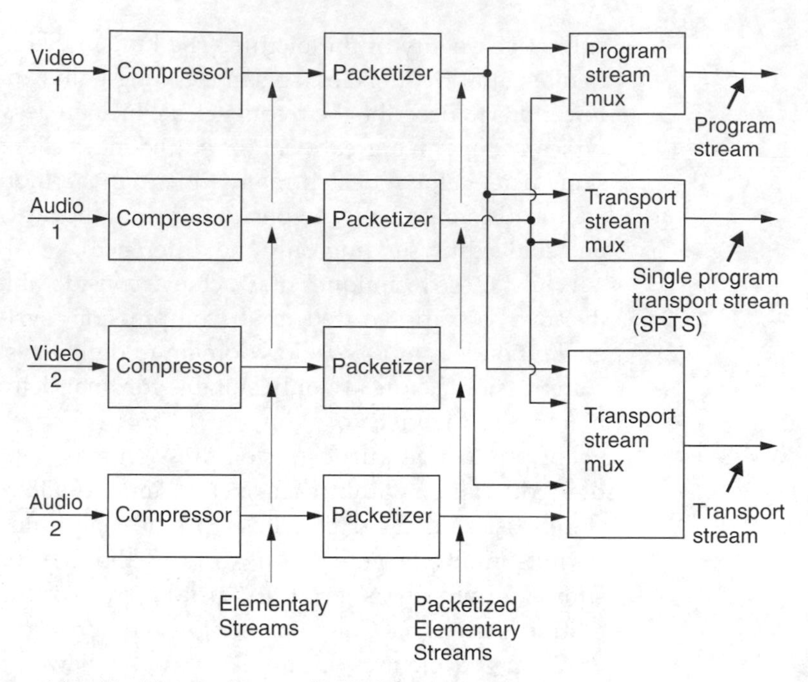

Figure 1.16 The bitstream types of MPEG-2. See text for details.

1.16 Disk-based recording

The magnetic disk drive was perfected by the computer industry to allow rapid random access to data, and so it makes an ideal medium for editing. As will be seen in Chapter 12, the heads do not touch the disk, but are supported on a thin air film which gives them a long life but which restricts the recording density. Thus disks cannot compete with tape for lengthy recordings, but for short-duration work such as commercials or animation they have no equal. The data rate of digital video is too high for a single disk head, and so a number of solutions have been explored. One obvious solution is to use compression, which cuts the data rate and extends the playing time. Another approach is to operate a large array of conventional drives in parallel. The highest-capacity magnetic disks are not removable from the drive.

The disk drive suffers from intermittent data transfer owing to the need to reposition the heads. Figure 1.17 shows that disk-based devices rely on a quantity of RAM acting as a buffer between the real-time video environment and the intermittent data environment.

Figure 1.18 shows the block diagram of a camcorder based on hard disks and compression. The recording time and picture quality will not compete with full-bandwidth tape-based devices, but following acquisi-

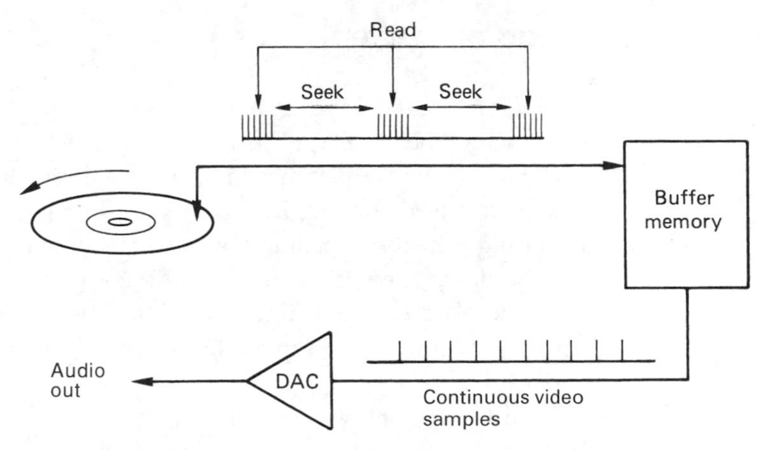

Figure 1.17 In a hard disk recorder, a large-capacity memory is used as a buffer or timebase corrector between the convertors and the disk. The memory allows the convertors to run constantly despite the interruptions in disk transfer caused by the head moving between tracks.

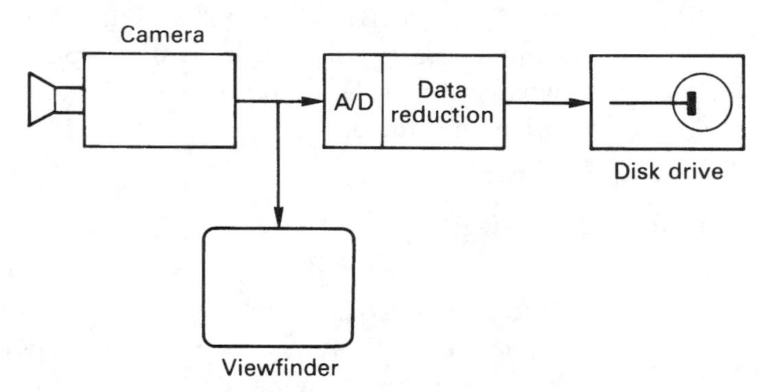

Figure 1.18 In a disk-based camcorder, the PCM data rate from the camera is too high for direct recording on disk. Data reduction is used to cut the bit rate and extend playing time. If a standard file structure is used, disks may physically be transferred to an edit system after recording.

tion the disks can directly be used in an edit system, allowing a useful time saving in ENG applications.

Development of the optical disk was stimulated by the availability of low-cost lasers. Optical disks are available in many different types, some which can only be recorded once, some which are erasable. These will be contrasted in Chapter 12. Optical disks have in common the fact that access is generally slower than with magnetic drives and it is difficult to obtain high data rates, but most of them are removeable and can act as interchange media.

1.17 Rotary head digital recorders

The rotary head recorder has the advantage that the spinning heads create a high head to tape speed offering a high bit rate recording without high tape speed. Whilst mechanically complex, the rotary head transport has been raised to a high degree of refinement and offers the highest recording density and thus lowest cost per bit of all digital recorders.[4] Rotary head transports are considered in Chapter 11. Digital VTRs segment incoming fields into several tape tracks and invisibly reassemble them in memory on replay in order to keep the tracks reasonably short.

Figure 1.19 shows a representative block diagram of a DVTR. Following the convertors, a compression process may be found. In an uncompressed recorder, there will be distribution of odd and even samples and a shuffle process for concealment purposes. An interleaved product code will be formed prior to the channel coding stage which produces the recorded waveform. On replay the data separator decodes the channel code and the inner and outer codes perform correction as in section 1.11. Following the de-shuffle the data channels are recombined and any necessary concealment will take place. Any compression will be decoded prior to the output convertors. Chapter 11 treats rotary head recorders in some detail.

1.18 DVD and DVHS

DVD (digital video disk) and DVHS (digital VHS) are formats intended for home use. DVD is a pre-recorded optical disk which carries an MPEG program stream containing a moving picture and one or more audio channels. DVD in many respects is a higher density development of compact disc technology and it will be detailed in Chapter 12.

DVHS is development of the VHS system which records MPEG data. In a digital television broadcast receiver, an MPEG transport stream is demodulated from the off-air signal. Transport streams contain several TV programs and the one required is selected by demultiplexing. As DVHS can record a transport stream, it can record more than one program simultaneously, with the choice being made on playback.

1.19 Digital television broadcasting

Although it has given good service for many years, analog television broadcasting is extremely inefficient because the transmitted waveform is directly compatible with the CRT display, and nothing is transmitted

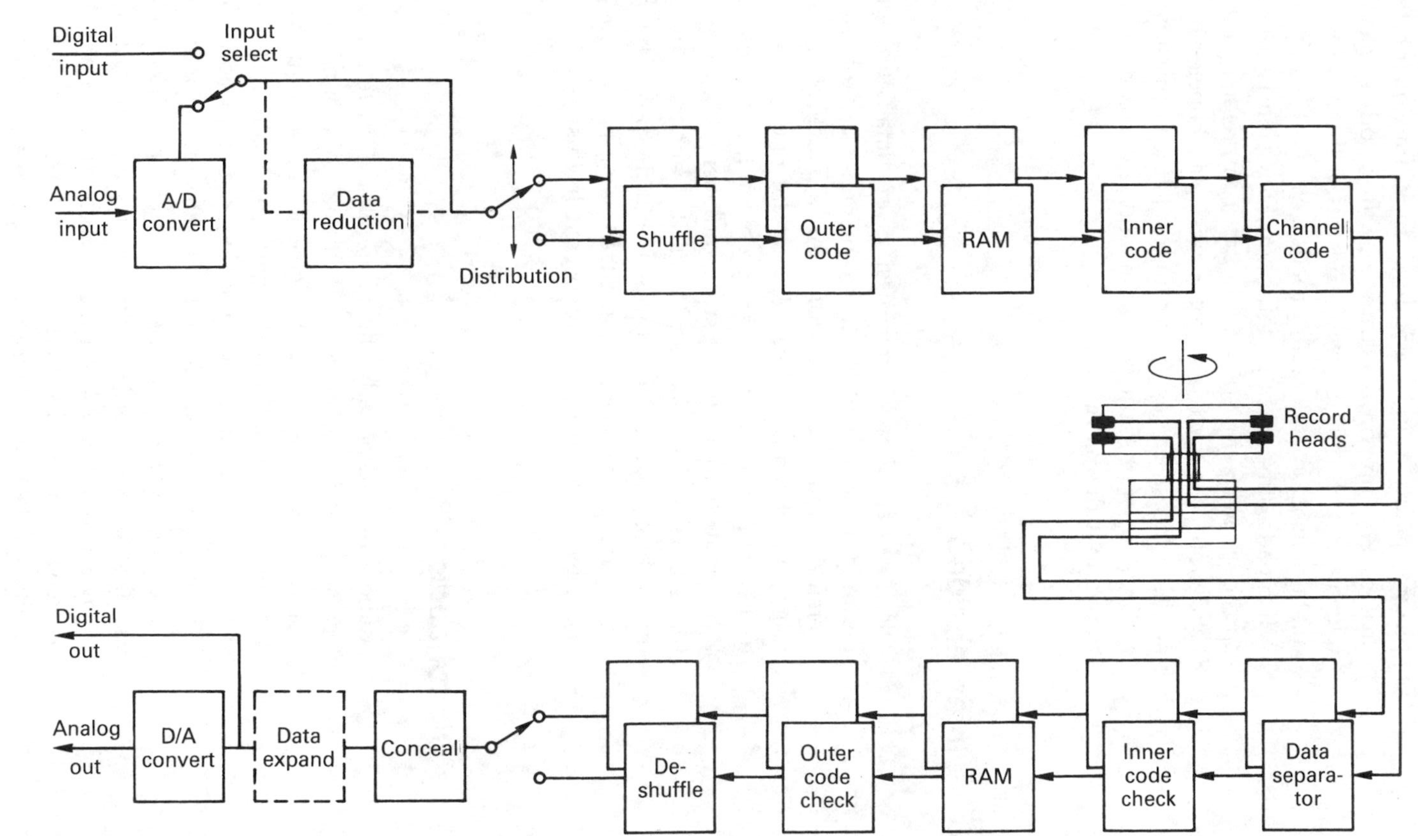

Figure 1.19 Block diagram of a DVTR. Note optional data reduction unit which may be used to allow a common transport to record a variety of formats.

during the blanking periods whilst the beam retraces. Using compression, digital modulation and error-correction techniques, the same picture quality can be obtained in a fraction of the bandwidth of analog. Pressure on spectrum use from other users such as mobile telephones will only increase and this will ensure rapid changeover to digital television and radio broadcasts.

If broadcasting of high-definition television is ever to become widespread it will do so via digital compressed transmission as the bandwidth required will otherwise be hopelessly uneconomic. In addition to conserving spectrum, digital transmission is (or should be) resistant to multipath reception and gives consistent picture quality throughout the service area.

1.20 Internet video

The Internet allows transmission of data files whose content or meaning is irrelevant to the transmission medium. These files can therefore contain digital video. As a practical matter, most Internet users suffer from a relatively limited bit rate and if moving pictures are to be sent, a very high compression factor will have to be used. Pictures with a relatively small number of lines and much reduced frame rates will be needed. Whilst the quality does not compare with that of broadcast television, this is not the point. Internet video allows a wide range of services which traditional broadcasting cannot provide and phenomenal growth is expected in this area.

1.21 Digital audio

Audio was traditionally the poor relation in television, with the poor technical quality of analog TV sound compounded by the miserable loudspeaker fitted in many TV sets. The introduction of the compact disc served to raise the consumer's expectations in audio and this was shortly followed in many countries by the introduction of the NICAM 728 system which added stereo digital audio to analog broadcast TV.

The development of hi-fi sound in the VHS format was another useful step. Pre-recorded VHS movies became available which carried Dolby Surround Sound on their audio tracks, and a significant number of viewers bought decoders to have surround sound in the home.

Loudspeakers in television sets have improved, but in some cases these are disabled because external surround speakers are employed.

With the advent of digital television broadcasting, the audio is naturally digital, but also compressed, with the choice of format divided between the ISO-MPEG audio coding standards and the Dolby AC-3 system. These are compared in Chapter 7.

References

1. Ginsburg, C.P., Comprehensive description of the Ampex video tape recorder. *SMPTE Journal*, **66** 177–182 (1957)
2. Devereux, V.G., Pulse code modulation of video signals: 8 bit coder and decoder. *BBC Res. Dept. Rept.*, **EL-42** No.25 (1970)
3. Pursell, S. and Newby, H., Digital frame store for television video. *SMPTE Journal*, **82** 402–403 (1973)
4. Baldwin, J.L.E., Digital television recording – history and background. *SMPTE Journal*, **95** 1206–1214 (1986)

2

Video principles

2.1 The eye

All television signals ultimately excite some response in the eye and the viewer can only describe the result subjectively. Familiarity with the operation and limitations of the eye is essential to an understanding of television principles.

The simple representation of Figure 2.1 shows that the eyeball is nearly spherical and is swivelled by muscles so that it can track movement. This has a large bearing on the way moving pictures are reproduced. The space between the cornea and the lens is filled with transparent fluid known as *aqueous humour.* The remainder of the eyeball is filled with a transparent jelly known as *vitreous humour.* Light enters the cornea, and the amount of light admitted is controlled by the pupil in the iris. Light entering is involuntarily focused on the retina by the lens in a process called *visual accommodation*. The lens is the only part of the eye which is not nourished by the bloodstream and its centre is technically dead. In a young person the lens is flexible and muscles distort it to perform the focusing action. In old age the lens loses some flexibility and causes *presbyopia* or limited accommodation. In some people the length of the eyeball is incorrect resulting in *myopia* (shortsightedness) or *hypermetropia* (longsightedness). The cornea should have the same curvature in all meridia, and if this is not the case, *astigmatism* results.

The retina is responsible for light sensing and contains a number of layers. The surface of the retina is covered with arteries, veins and nerve fibres and light has to penetrate these in order to reach the sensitive layer. This contains two types of discrete receptors known as *rods* and *cones* from their shape. The distribution and characteristics of these two receptors are quite different. Rods dominate the periphery of the retina

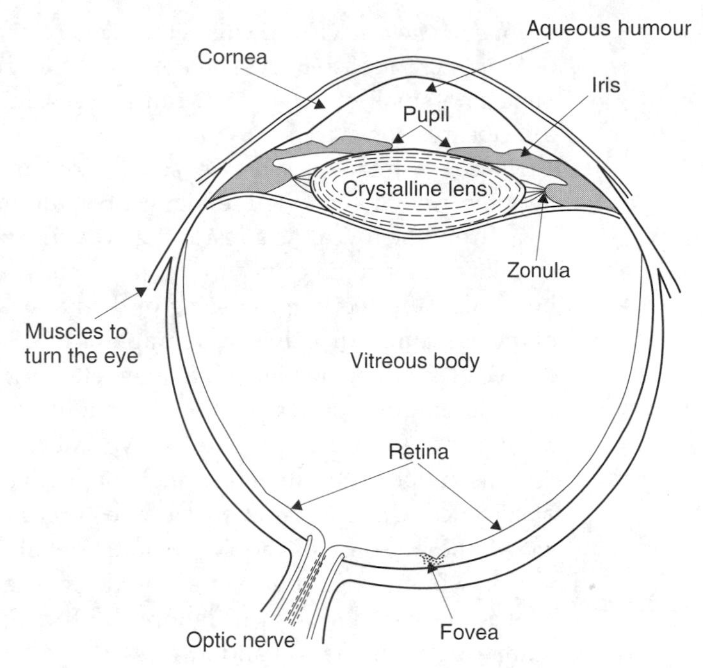

Figure 2.1 A simple representation of an eyeball; see text for details.

whereas cones dominate a central area known as the *fovea* outside which their density drops off. Vision using the rods is monochromatic and has poor resolution but remains effective at very low light levels, whereas the cones provide high resolution and colour vision but require more light. Figure 2.2 shows how the sensitivity of the retina slowly increases in response to entering darkness. The first part of the curve is the adaptation

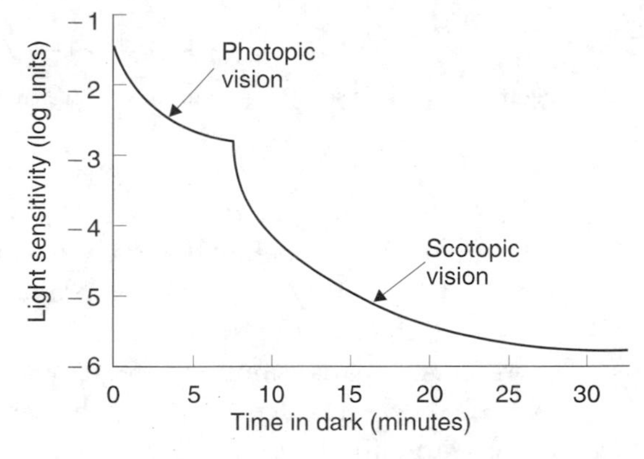

Figure 2.2 Retinal sensitivity changes after sudden darkness. The initial curve is due to adaptation of cones. At very low light levels cones are blind and monochrome rod vision takes over.

of cone or *photopic* vision. This is followed by the greater adaptation of the rods in *scotopic* vision. At such low light levels the fovea is essentially blind and small objects which can be seen in the peripheral rod vision disappear when stared at.

The cones in the fovea are densely packed and directly connected to the nervous system allowing the highest resolution. Resolution then falls off away from the fovea. As a result the eye must move to scan large areas of detail. The image perceived is not just a function of the retinal response, but is also affected by processing of the nerve signals. The overall acuity of the eye can be displayed as a graph of the response plotted against the degree of detail being viewed. Image detail is generally measured in lines per millimetre or cycles per picture height, but this takes no account of the distance from the image to the eye. A better unit for eye resolution is one based upon the subtended angle of detail as this will be independent of distance. Units of cycles per degree are then appropriate. Figure 2.3 shows the response of the eye to static detail. Note that the response to very low frequencies is also attenuated. An extension of this characteristic allows the vision system to ignore the fixed pattern of shadow on the retina due to the nerves and arteries.

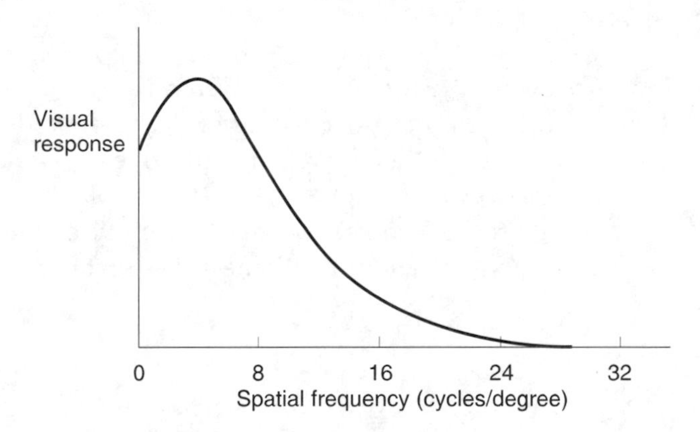

Figure 2.3 Response of the eye to different degrees of detail.

The retina does not respond instantly to light, but requires between 0.15 and 0.3 second before the brain perceives an image. The resolution of the eye is primarily a spatio-temporal compromise. The eye is a spatial sampling device; the spacing of the rods and cones on the retina represents a spatial sampling frequency. The measured acuity of the eye exceeds the value calculated from the sample site spacing because a form of oversampling is used.

The eye is in a continuous state of unconscious vibration called saccadic motion. This causes the sampling sites to exist in more than one location, effectively increasing the spatial sampling rate provided there is a temporal filter which is able to integrate the information from the various different positions of the retina.

This temporal filtering is responsible for 'persistence of vision'. Flashing lights are perceived to flicker until the critical flicker frequency (CFF) is reached; the light appears continuous for higher frequencies. The CFF is not constant but varies with brightness. Note that the field rate of European television at 50 fields per second is marginal with bright images. Film projected at 48 Hz works because cinemas are darkened and the screen brightnes is actually quite low. Figure 2.4 shows the two-dimensional or spatio-temporal response of the eye.

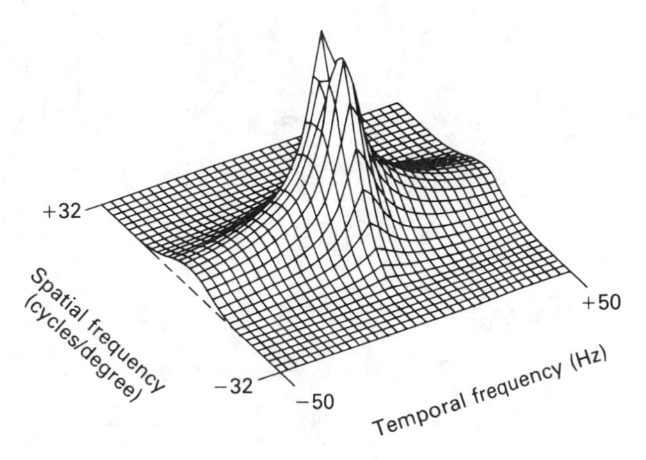

Figure 2.4 The response of the eye shown with respect to temporal and spatial frequencies. Note that even slow relative movement causes a serious loss of resolution. The eye tracks moving objects to prevent this loss.

If the eye were static, a detailed object moving past it would give rise to temporal frequencies, as Figure 2.5(a) shows. The temporal frequency is given by the detail in the object, in lines per millimetre, multiplied by the speed. Clearly a highly detailed object can reach high temporal frequencies even at slow speeds, yet Figure 2.4 shows that the eye cannot respond to high temporal frequencies.

However, the human viewer has an interactive visual system which causes the eyes to track the movement of any object of interest. Figure 2.5(b) shows that when eye tracking is considered, a moving object is rendered stationary with respect to the retina so that temporal frequencies fall to zero and much the same acuity to detail is available despite motion. This is known as dynamic resolution and it's how humans judge the detail in real moving pictures. Dynamic resolution will be considered in section 2.16.

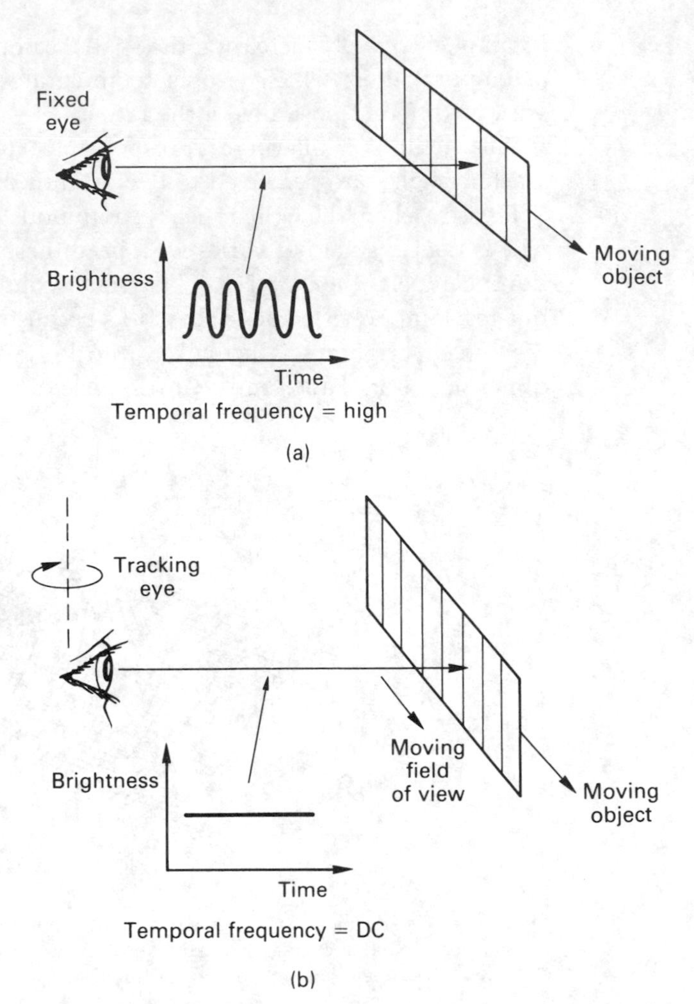

Figure 2.5 In (a) a detailed object moves past a fixed eye, causing temporal frequencies beyond the response of the eye. This is the cause of motion blur. In (b) the eye tracks the motion and the temporal frequency becomes zero. Motion blur cannot then occur.

The contrast sensitivity of the eye is defined as the smallest brightness difference which is visible. In fact the contrast sensitivity is not constant, but increases proportionally to brightness. Thus whatever the brightness of an object, if that brightness changes by about 1 per cent it will be equally detectable.

2.2 Gamma

The true brightness of a television picture can be affected by electrical noise on the video signal. As contrast sensitivity is proportional to brightness, noise is more visible in dark picture areas than in bright areas.

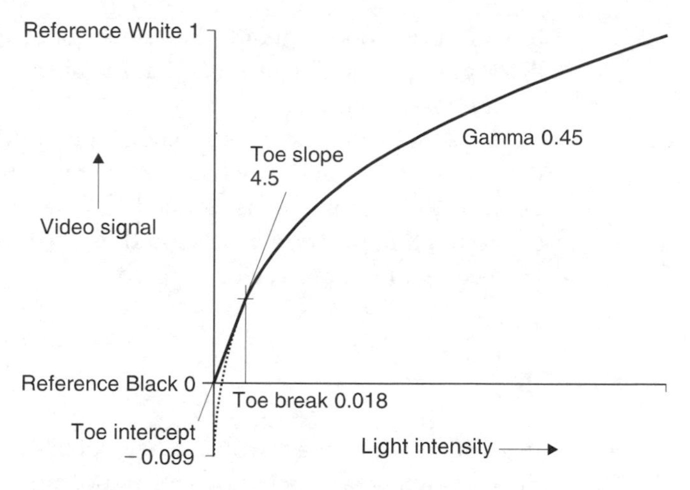

Figure 2.6 CCIR Rec.709 reverse gamma function used at camera has a straight line approximation at the lower part of the curve to avoid boosting camera noise. Note that the output amplitude is greater for modulation near black.

For economic reasons, video signals have to be made non-linear to render noise less visible. An inverse gamma function takes place at the camera so that the video signal is non-linear for most of its journey. Figure 2.6 shows a reverse gamma function. As a true power function requires infinite gain near black, a linear segment is substituted. It will be seen that contrast variations near black result in larger signal amplitude than variations near white. The result is that noise picked up by the video signal has less effect on dark areas than bright areas. After a gamma function at the display, noise at near-black levels is compressed with respect to noise near-white levels. Thus a video transmission system using gamma has a lower perceived noise level than one without. Without gamma, vision signals would need around 30 dB better signal-to-noise ratio for the same perceived quality and digital video samples would need five or six extra bits.

In practice the system is not rendered perfectly linear by gamma correction and a slight overall exponential effect is usually retained in order further to reduce the effect of noise in darker parts of the picture. A gamma correction factor of 0.45 may be used to achieve this effect.

Clearly image data which is intended to be displayed on a video system must have the correct gamma characteristic or the grey scale will not correctly be reproduced. Image data from computer systems often have gamma characteristics which are incompatible with the standards adopted in video and a gamma conversion process will be required to obtain a correct display. This may take the form of a lookup table.

Electrical noise has no DC component and so cannot shift the average video voltage. However, on extremely noisy signals, the non-linear effect of gamma is to exaggerate the white-going noise spikes more than the black-going ones. The result is that the black level appears to rise and the picture loses contrast.

As all television signals, analog and digital, are subject to gamma correction, it is technically incorrect to refer to the *Y* signal as luminance, because this parameter is defined as linear in colorimetry. It has been suggested that the term *luma* should be used to describe luminance which has been gamma corrected.

2.3 Scanning

It is difficult to convey two-dimensional images from one place to another directly, whereas electrical and radio signals are easily carried. The problem is to convert a two-dimensional image into a single voltage

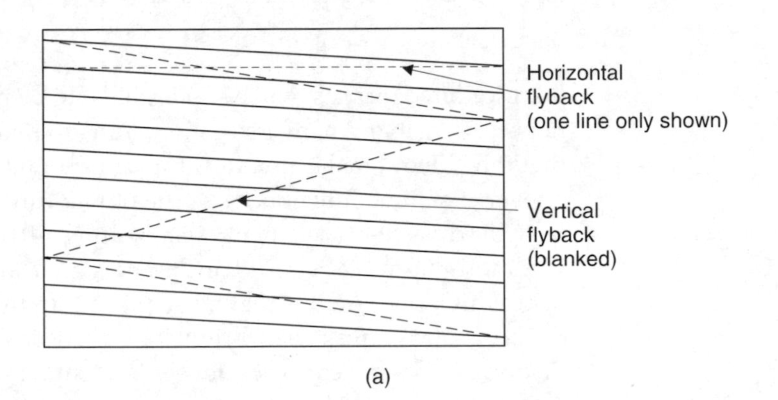

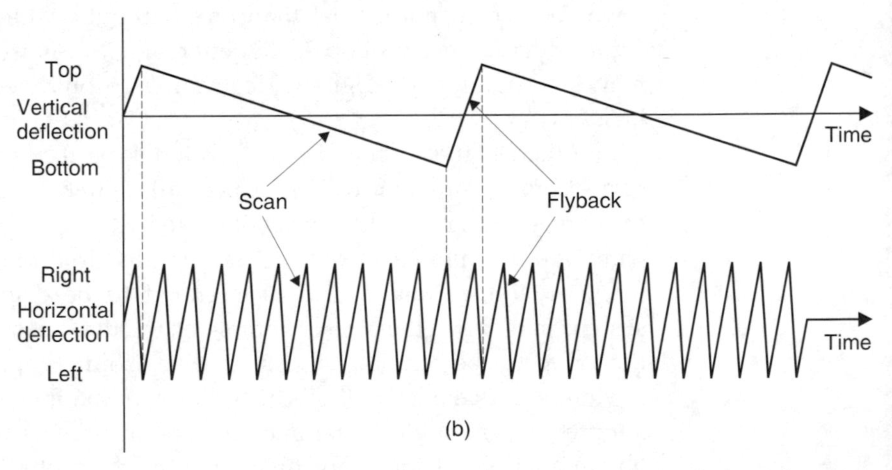

Figure 2.7 Scanning converts two-dimensional images into a signal which can be sent electrically. In (a) the scanning of camera and display must be identical. The scanning is controlled by horizontal and vertical sawtooth waveforms (b).

changing with time. The solution is to use the principle of scanning. Figure 2.7(a) shows that the monochrome camera produces a *video signal* whose voltage is a function of the image brightness at a single point on the sensor. This voltage is converted back to the brightness of the same point on the display. The points on the sensor and display must be scanned synchronously if the picture is to be recreated properly. If this is done rapidly enough it is largely invisible to the eye. Figure 2.7(b) shows that the scanning is controlled by a triangular or *sawtooth* waveform in each dimension which causes a constant speed forward scan followed by a rapid return or *flyback*. As the horizontal scan is much more rapid than the vertical scan the image is broken up into lines which are not quite horizontal.

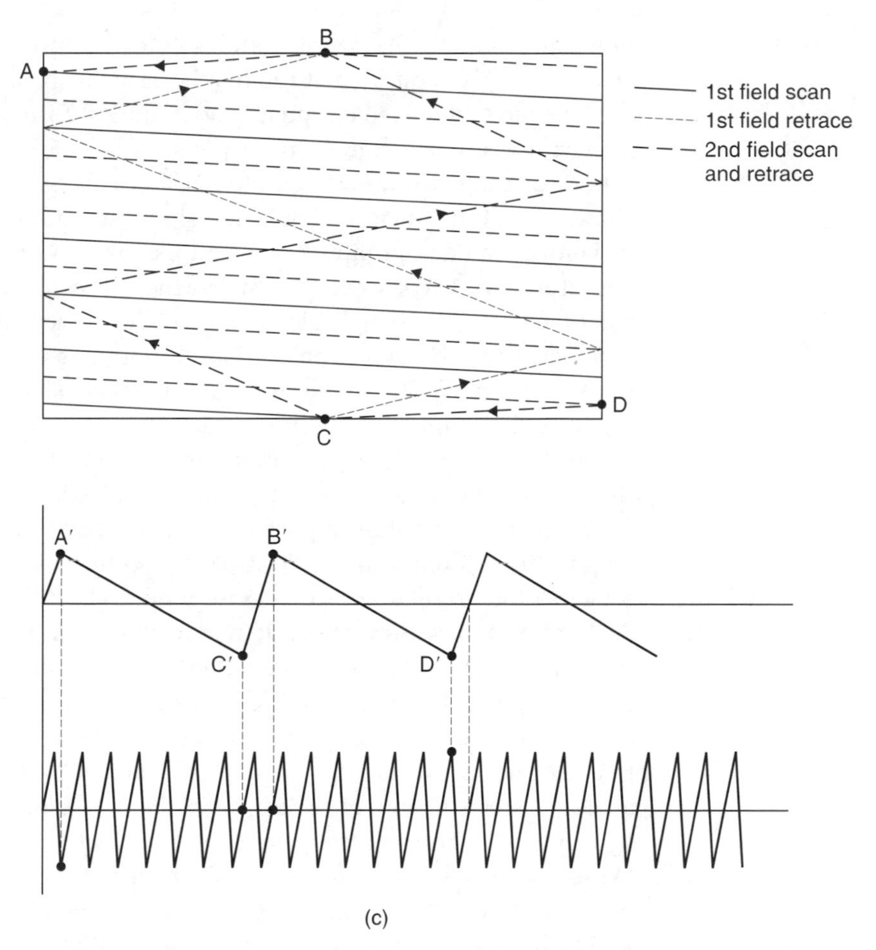

Figure 2.7 (*Continued*) Where two vertical scans are needed to complete a whole number of lines, the scan is interlaced as shown in (c). The frame is now split into two fields.

In the example of Figure 2.7(b), the horizontal scanning frequency or *line rate*, F_h, is an integer multiple of the vertical scanning frequency or *frame rate* and a *progressive scan* system results in which every frame is identical. Figure 2.7(c) shows an *interlaced scan* system in which there is an integer number of lines in *two* vertical scans or *fields*. The first field begins with a full line and ends on a half line and the second field begins with a half line and ends with a full line. The lines from the two fields interlace or mesh on the screen. Current analog broadcast systems such as PAL and NTSC use interlace, although in MPEG systems it is not necessary. The additional complication of interlace has both merits and drawbacks which will be discussed in section 2.17.

2.4 Synchronizing

It is vital that the horizontal and vertical scanning at the camera is simultaneously replicated at the display. This is the job of the *synchronizing* or sync system which must send timing information to the display alongside the video signal. In very early television equipment this was achieved using two quite separate or *non-composite* signals. Figure 2.8(a) shows one of the first (US) television signal standards in which the video waveform had an amplitude of 1 Volt pk–pk and the sync signal had an amplitude of 4 Volts pk–pk. In practice, it was more convenient to combine both into a single electrical waveform then called *composite video* which carries the synchronizing information as well as the scanned brightness signal. The single signal is effectively shared by using some of the flyback period for synchronizing.

The 4 Volt sync signal was attenuated by a factor of ten and added to the video to produce a 1.4 Volt pk–pk signal. This was the origin of the 10:4 video:sync relationship of US analog television practice. Later the amplitude was reduced to 1 Volt pk–pk so that the signal had the same range as the original non-composite video. The 10:4 ratio was retained. As Figure 2.8(b) shows, this ratio results in some rather odd voltages and to simplify matters, a new unit called the *IRE* unit (after the Institute of Radio Engineers) was devised. Originally this was defined as 1 per cent of the video voltage swing, independent of the actual amplitude in use, but it came in practice to mean 1 per cent of 0.714 Volt. In European analog systems shown in Figure 2.8(c) the messy numbers were avoided by using a 7:3 ratio and the waveforms are always measured in milliVolts. Whilst such a signal was originally called composite video, today it would be referred to as monochrome video or *Ys*, meaning luma carrying syncs although in practice the *s* is often omitted.

Figure 2.8(a) shows how the two signals are separated. The voltage swing needed to go from black to peak white is less than the total swing

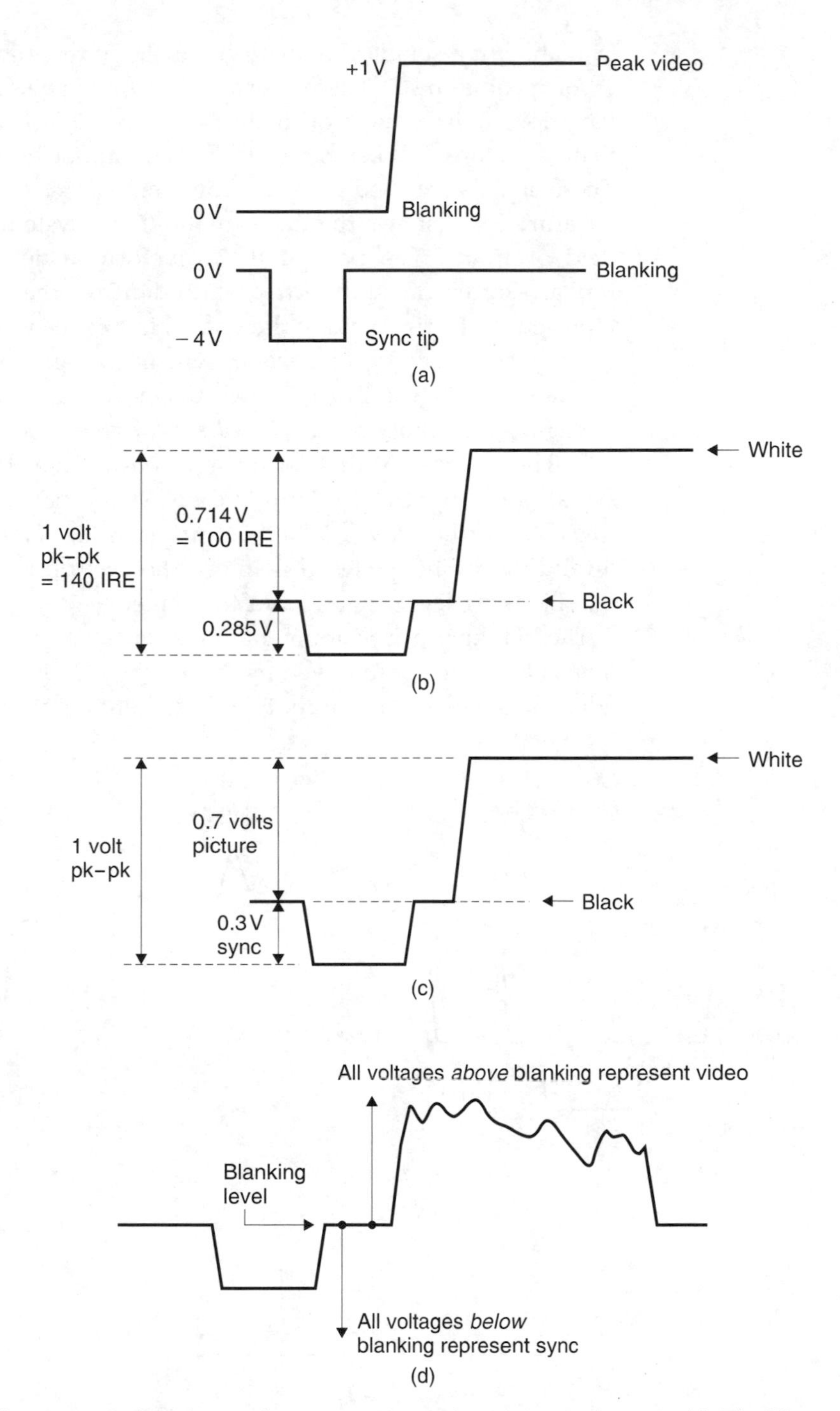

Figure 2.8 Early video used separate vision and sync signals shown in (a). The US one volt video waveform in (b) has 10:4 video/sync ratio. (c) European systems use 7:3 ratio to avoid odd voltages. (d) Sync separation relies on two voltage ranges in the signal.

available. In a standard analog video signal the maximum amplitude is 1 Volt peak-to-peak. The upper part of the voltage range represents the variations in brightness of the image from black to white. Signals below that range are 'blacker than black' and cannot be seen on the display. These signals are used for synchronizing.

Figure 2.9(a) shows the line synchronizing system part-way through a field or frame. The part of the waveform which corresponds to the forward scan is called the *active line* and during the active line the voltage represents the brightness of the image. In between the active line periods are *horizontal blanking intervals* in which the signal voltage will be at or below black. Figure 2.9(b) shows that in some systems the active line voltage is superimposed on a *pedestal* or *black level set-up* voltage of 7.5 *IRE*. The purpose of this set-up is to ensure that the blanking interval signal is below black on simple displays so that it is guaranteed to be invisible on the screen. When set-up is used, black level and blanking level differ by the pedestal height. When set-up is not used, black level and blanking level are one and the same.

The blanking period immediately after the active line is known as the *front porch*, which is followed by the *leading edge of sync*. When the leading edge of sync passes through 50 per cent of its own amplitude, the

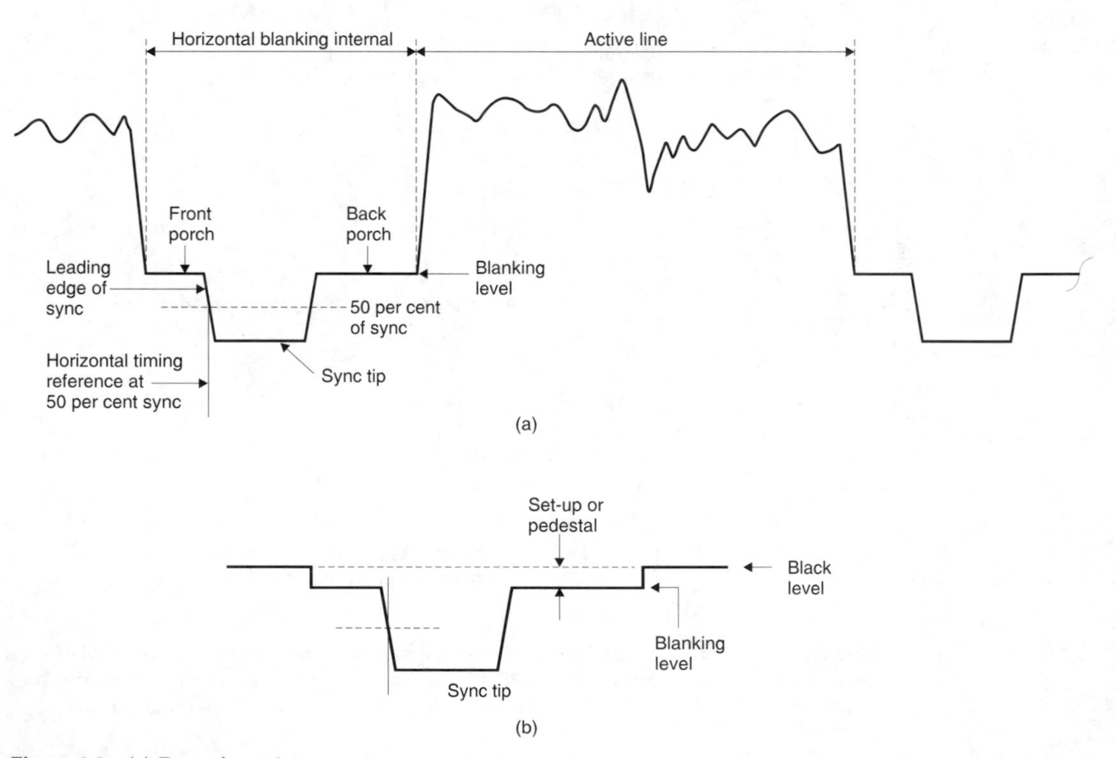

Figure 2.9 (a) Part of a video waveform with important features named. (b) Use of pedestal or set-up.

horizontal retrace pulse is considered to have occurred. The flat part at the bottom of the horizontal sync pulse is known as *sync tip* and this is followed by the trailing edge of sync which returns the waveform to blanking level. The signal remains at blanking level during the *back porch* during which the display completes the horizontal flyback. The sync pulses have sloping edges because if they were square they would contain high frequencies which would go outside the allowable channel bandwidth on being broadcast.

The vertical synchronizing system is more complex because the vertical flyback period is much longer than the horizontal line period and horizontal synchronization must be maintained throughout it. The vertical synchronizing pulses are much longer than horizontal pulses so that they are readily distinguishable. Figure 2.10(a) shows a simple approach to vertical synchronizing. The signal remains predominantly at sync tip for several lines to indicate the vertical retrace, but returns to

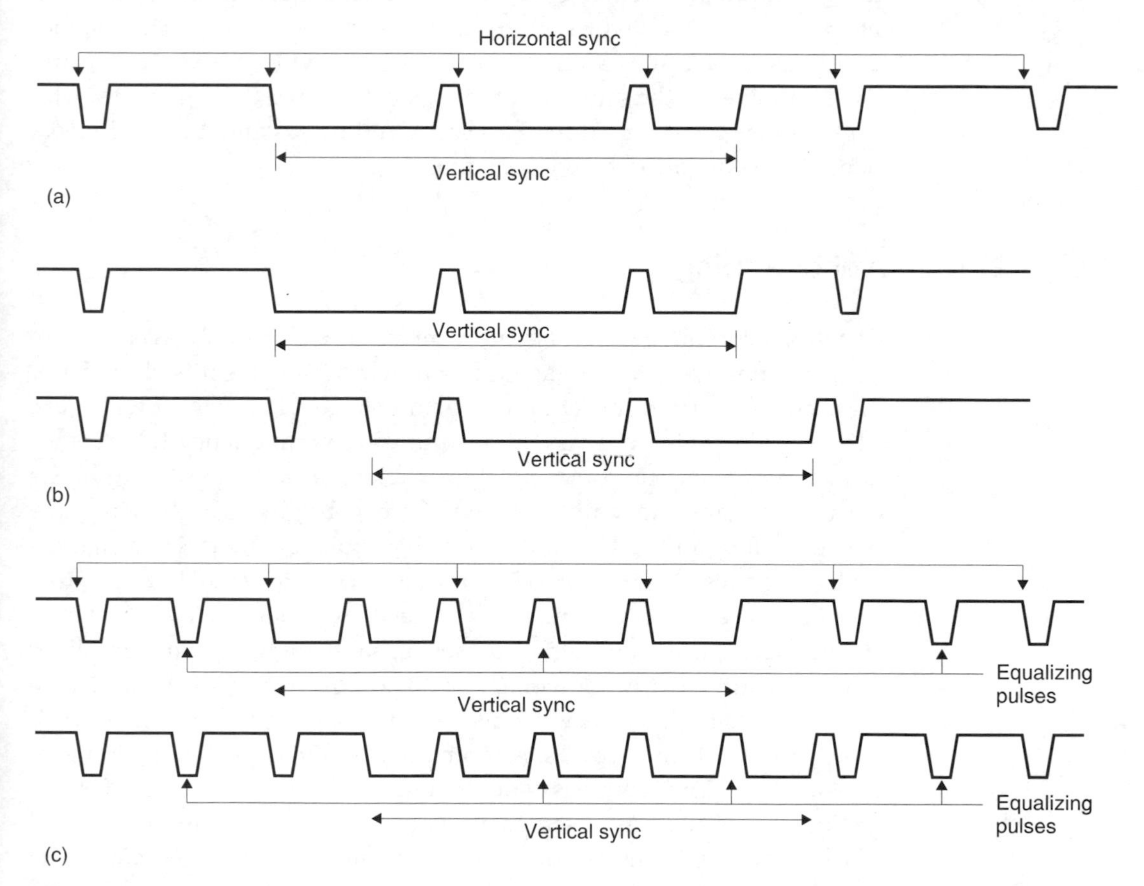

Figure 2.10 (a) A simple vertical pulse is longer than a horizontal pulse. (b) In an interlaced system there are two relationships between H and V. (c) The use of equalizing pulses to balance the DC component of the signal.

blanking level briefly immediately prior to the leading edges of the horizontal sync, which continues throughout. Figure 2.10(b) shows that the presence of interlace complicates matters, as in one vertical interval the vertical sync pulse coincides with a horizontal sync pulse whereas in the next the vertical sync pulse occurs half-way down a line.

In practice the long vertical sync pulses were found to disturb the average signal voltage too much and to reduce the effect extra *equalizing pulses* were put in, half-way between the horizontal sync pulses. The horizontal timebase system can ignore the equalizing pulses because it contains a flywheel circuit which expects pulses only roughly one line period apart. Figure 2.10(c) shows the final result of an interlaced system with equalizing pulses. The vertical blanking interval can be seen, with the vertical pulse itself towards the beginning.

In digital video signals it is possible to synchronize simply by digitizing the analog sync pulses. However, this is inefficient because many samples are needed to describe them. In practice the analog sync pulses are used to generate timing reference signals (TRS) which are special codes inserted into the video data which indicate the picture timing. In a manner analogous to the analog approach of dividing the video voltage range into two, one for syncs, the solution in the digital domain is the same: certain bit combinations are reserved for TRS codes and these cannot occur in legal video. TRS codes are detailed in Chapter 10.

2.5 Black-level clamping

As the synchronizing and picture content of the video waveform are separated purely by the voltage range in which they lie, it is clear that if any accidental drift or offset of the signal voltage takes place it will cause difficulty. Unwanted offsets may result from low-frequency interference such as power line hum picked up by cabling. The video content of the signal also varies in amplitude with scene brightness, changing the average voltage of the signal. When such a signal passes down a channel not having a response down to DC, the baseline of the signal can wander. Such offsets can be overcome using a black-level clamp which is shown in Figure 2.11. The video signal passes through an operational amplifier which can add a correction voltage or DC offset to the waveform. At the output of the amplifier the video waveform is sampled by a switch which closes briefly during the back porch when the signal should be at blanking level. The sample is compared with a locally generated reference blanking level and any discrepancy is used to generate an error signal which drives the integrator producing the correction voltage. The correction voltage integrator will adjust itself until the error becomes zero.

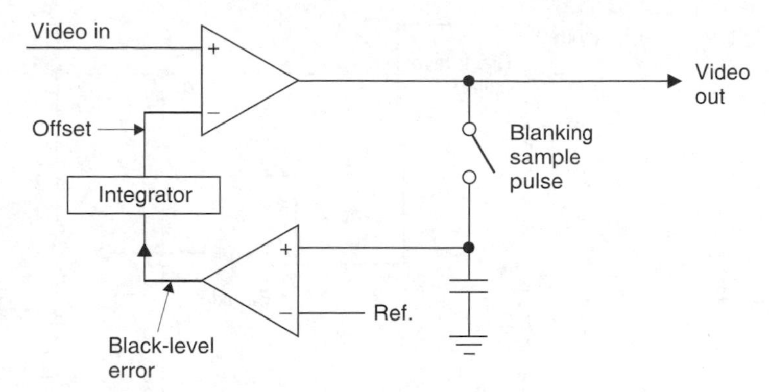

Figure 2.11 Black-level clamp samples video during blanking and adds offset until the sample is at black level.

2.6 Sync separation

It is essential accurately to extract the timing or synchronizing information from a sync or *Ys* signal in order to control some process such as the generation of a digital sampling clock. Figure 2.12(a) shows a block diagram of a simple sync separator. The first stage will generally consist of a black level clamp which stabilizes the DC conditions in the separator. Figure 2.12(b) shows that if this is not done the presence of a DC shift on a sync edge can cause a timing error.

The sync time is defined as the instant when the leading edge passes through the 50 per cent level. The incoming signal should ideally have a sync amplitude of either 0.3 Volt pk–pk or 40 *IRE*, in which case it can be *sliced* or converted to a binary waveform by using a comparator with a reference of either 0.15 Volt or 20 *IRE*. However, if the sync amplitude is for any reason incorrect, the slicing level will be wrong. Figure 2.12(a) shows that the solution is to measure both blanking and sync tip voltages and to derive the slicing level from them with a potential divider. In this way the slicing level will always be 50 per cent of the input amplitude. In order to measure the sync tip and blanking levels, a coarse sync separator is required, which is accurate enough to generate sampling pulses for the voltage measurement system. Figure 2.12(c) shows the timing of the sampling process.

Once a binary signal has been extracted from the analog input, the horizontal and vertical synchronizing information can be separated. All falling edges are potential horizontal sync leading edges, but some are due to equalizing pulses and these must be rejected. This is easily done because equalizing pulses occur part-way down the line. A flywheel oscillator or phase-locked loop will lock to genuine horizontal sync pulses because they always occur exactly one line period apart. Edges at

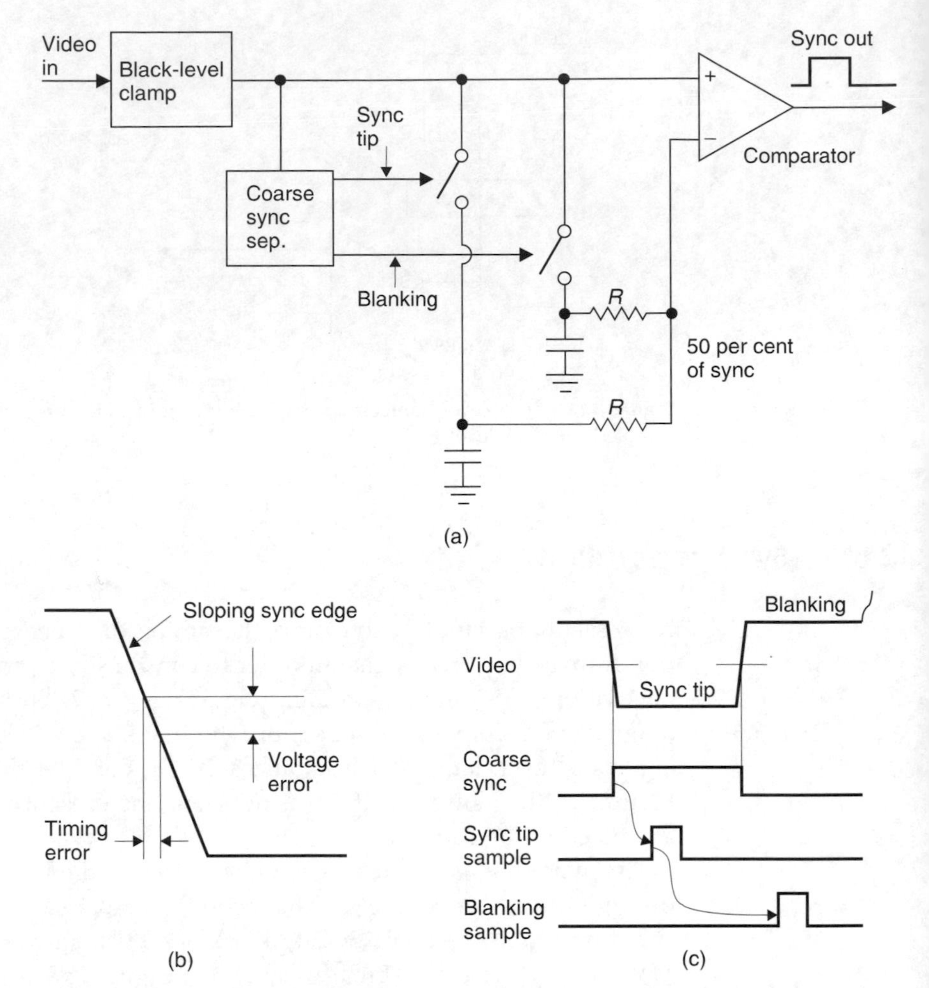

Figure 2.12 (a) Sync separator block diagram; see text for details. (b) Slicing at the wrong level introduces a timing error. (c) The timing of the sync separation process.

other spacings are eliminated. Vertical sync is detected with a timer whose period exceeds that of a normal horizontal sync pulse. If the sync waveform is still low when the timer expires, there must be a vertical pulse present. Once again a phase-locked loop may be used which will continue to run if the input is noisy or disturbed. This may take the form of a counter which counts the number of lines in a frame before resetting.

The sync separator can determine which type of field is beginning because in one the vertical and horizontal pulses coincide whereas in the other the vertical pulse begins in the middle of a line.

2.7 The monochrome cathode ray tube

The CRT is a relative of the vacuum tube and is shown in Figure 2.13. Inside a glass envelope a vacuum is formed initially by pumping and completed by igniting a material called a *getter* which burns any remaining oxygen to form a harmless solid. The pressure differential across the tube face results in considerable force. For example, the atmosphere exerts a force of about a ton on the face of a 20-inch diagonal tube. As glass is weak in tension the tube is strengthened with a steel band which is stretched around the perimeter of the screen. As the tube face is slightly domed outwards, the pressure load is converted to a radial out-thrust to which the steel band provides a reaction.

The cathode is coated with a barium compound and contains an insulated heating element which raises its temperature. This heating causes the coating to emit electrons. The electrons have negative charge and so are attracted towards an anode which is supplied with a positive voltage. Between the cathode and the anode is a wire mesh grid. If this grid is held at a suitable negative voltage with respect to the cathode it will repel electrons from the cathode and they will be prevented from

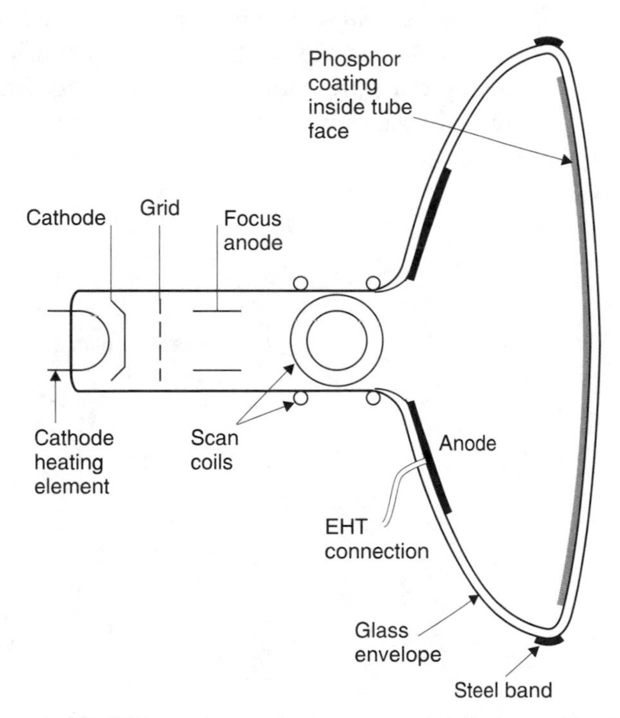

Figure 2.13 A simple CRT contains an electron gun and deflection mechanism for scanning. The cathode-to-grid voltage determines the beam current and hence brightness.

reaching the anode. If the grid voltage is reduced the effect diminishes and some electrons can pass. The voltage on the grid controls the current.

The anode contains a hole through which the electrons emerge in a beam. They are further accelerated by more electrodes at successively higher voltages, the last of these being the EHT (extra high tension) electrode which runs at 15–25 KiloVolts. The electron beam strikes the inside of the tube face which is coated with material known as a phosphor. The impact of energetic electrons causes electrons in the phosphor to be driven to higher, unstable valence levels, and when they drop back, photons of a specific wavelength are released. By mixing a number of phosphor materials white light can be obtained.

The intensity of the light is effectively controlled by the intensity of the electron beam which is in turn controlled by the grid voltage. As it is the relative voltage between the cathode and the grid which determines the beam current, some tubes are driven by holding the grid at a constant voltage and varying the voltage on the cathode. The electron impact may also generate low-level X-rays and the face of the tube will be made from lead glass to absorb most of them.

The relationship between the tube drive voltage and the phosphor brightness is not linear, but an exponential function where the power is known as *gamma*. The power is the same for all CRTs as it is a function of the physics of the electron gun and it has a value of around 2.8. It is a happy but pure coincidence that the gamma function of a CRT follows roughly the same curve as human contrast sensitivity.

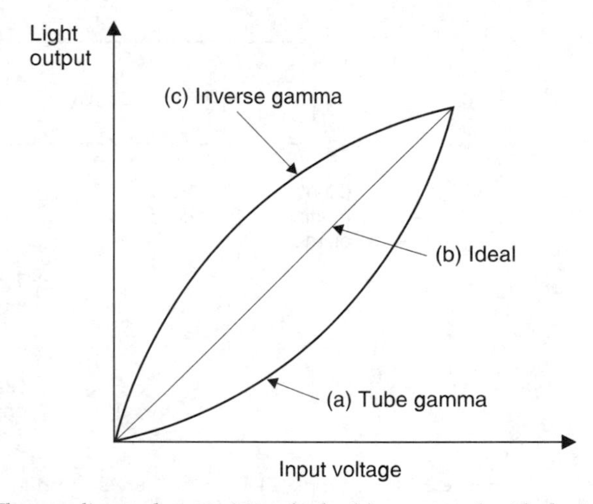

Figure 2.14 The non-linear characteristic of tube (a) contrasted with the ideal response (b). Non-linearity may be opposed by gamma correction with a response (c).

Consequently if video signals are pre-distorted at source by an inverse gamma, the gamma characteristic of the CRT will linearize the signal. Figure 2.14 shows the principle. CRT gamma is not a nuisance, but is actually used to enhance the noise performance of a system. If the CRT had no gamma characteristic, a gamma circuit would have been necessary ahead of it. As all standard video signals are inverse gamma processed, it follows that if a non-CRT display such as a plasma or LCD device is to be used, some gamma conversion will be required at the display.

Virtually all CRT-based displays are fitted with two controls traditionally and misleadingly marked *brightness* and *contrast*. Figure 2.15(a) shows what the 'brightness' control actually does. When correctly set, the lowest drive voltage, i.e. blanking level, results in the electron beam being just cut off so that the CRT displays black. If the 'brightness' is set too low, as in (b), the CRT cuts off prematurely and all inputs below a certain level are reproduced as black. The symptom is decribed as *black crushing*. If the

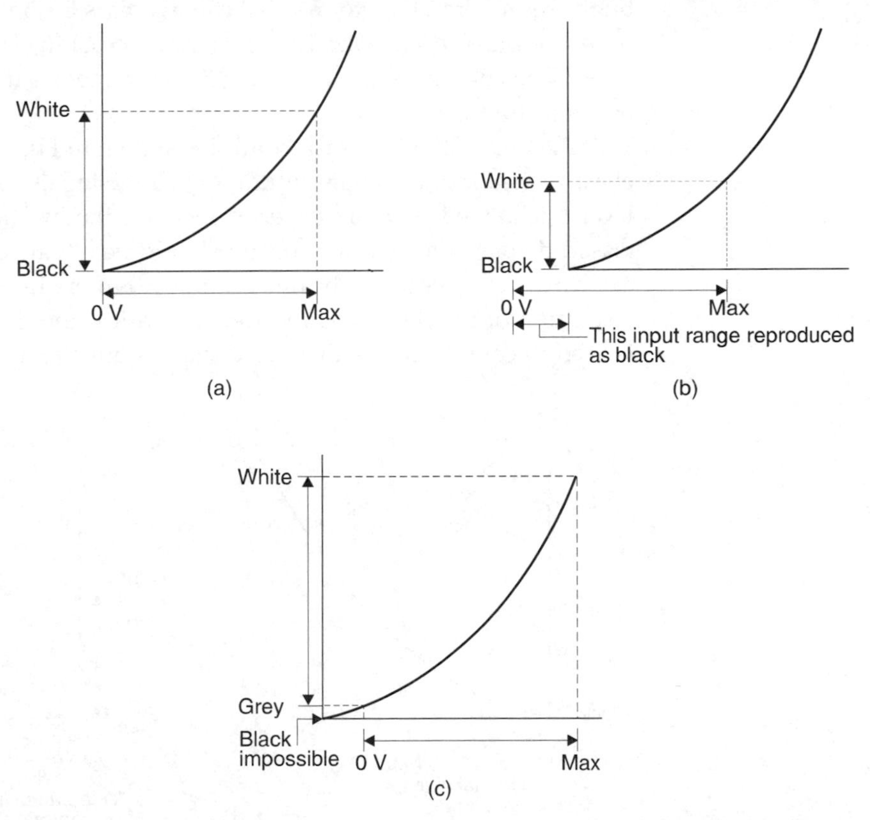

Figure 2.15 (a) Correctly set brightness control cuts off the electron beam at blanking level. (b) Control set too low causes black crushing. (c) Control set too high causes grey pedestal on picture, spoiling contrast. Brightness control is a misnomer as it has only one correct setting.

control is set too high, as in (c), video blanking results in a substantial light output such that all displayed images are superimposed on a grey level.

It should be clear that there is only one correct setting for a brightness control because it is in fact a tube bias control. In order correctly to set a brightness control the grey stepped scale of a test card is used. The brightness control is advanced until the black part of the scale appears obviously grey, and then it is turned down until the the black part of the scale is just displayed as truly black, but not so far that the darkest grey step next to it becomes black as well. Once set in this way the CRT is correctly biased and further adjustment will only be needed if component values drift.

Special test signals exist to assist with monitor alignment. One of these is known as PLUGE (Picture Line-Up GEnerator) pronounced 'plooj'. The PLUGE signal contains a black area in which there are two 'black' bars which sit typically ±20 mV above and below black level. If the 'brightness' control is adjusted downwards from an initial excessively bright seting, it will be found that the two bars become indistinguishable from the black background, due to black crushing, at slightly different times. The correct setting is achieved when one bar has vanished but the other is still visible.

The action of the 'contrast' control is shown in Figure 2.16. This has the effect of increasing the amplitude of white signals whilst leaving black level unchanged. Thus in order to increase the brightness of a correctly biased display, the contrast control should be advanced. If the contrast is excessive the electron beam becomes larger in diameter and the resolution of the display is reduced. In critical monitoring, a light meter is used in conjunction with a peak-white input signal to allow a standard

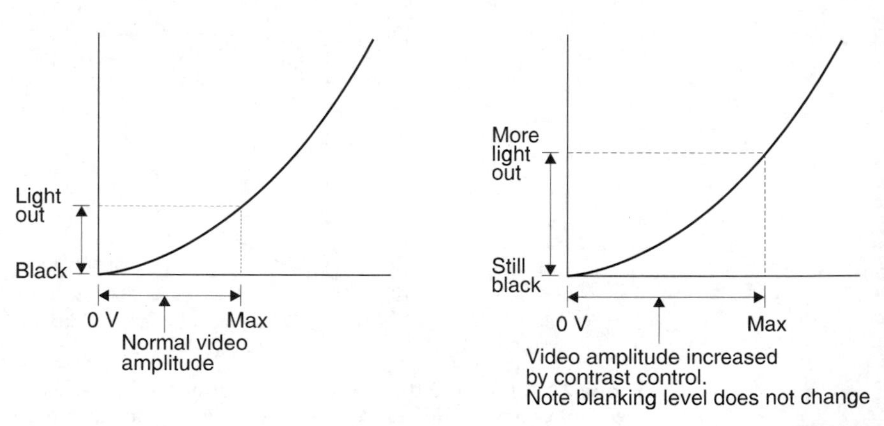

Figure 2.16 Action of contrast control. Paradoxically, the picture is made brighter by increasing contrast.

brightness to be achieved. The PLUGE signal contains a peak white area to assist with this.

In practice the contrast of a CRT is also affected quite badly by ambient lighting. With black input voltage cutting off the beam, the brightness of a CRT cannot fall below the brightness of reflected ambient light. Thus ambient light reduces contrast. For best results all display technologies should be viewed in subdued lighting where the best combination of contrast and resolution will be obtained. As all CRT tubes are reflective to some extent, it is important to ensure no bright objects are positioned where they could be seen reflected in the screen. Some CRTs are provided with non-reflective coatings which have a beneficial effect.

The CRT phosphor continues to emit light in an exponentially decaying curve for some time after the electron beam has passed. This is known as *persistence*. The persistence time has to be short compared to that of oscilloscope CRTs because excessive values cause smearing on moving images.

The electron beam is an electric current and this can be deflected electrostatically by voltages applied to plates at the neck of the tube, or magnetically by coils outside the tube. Electrostatic deflection is fine for the small tubes used in oscilloscopes, but the large deflection angles needed in TV tubes can only be obtained with magnetic deflection.

The horizontal deflection coils are generally driven by a transformer. During the flyback period the flux in the transformer changes rapidly and this can be used to generate the EHT supply by providing an additional winding on the transformer which feeds a high-voltage rectifier.

There is some evidence to suggest that the X-rays or magnetic fields from CRTs may present a health hazard after prolonged exposure.

The rate at which the electron beam can retrace is limited by the inductance of the scan coils which limit the rate at which the deflection current can change. The active line has to be made considerably shorter than the line period so that sufficient retrace time is available. Normally the horizontal and vertical retrace is invisible, but certain monitors have the facility to shift the scan phase with respect to incoming syncs by part of a line and part of a field. The result is that the blanking periods become visible for inspection on the screen in the form of a cross, hence the term *pulse-cross monitor.*

2.8 The monochrome TV camera

There are two types of television camera, the older type based on an electron tube and the newer type based on charge-coupled devices (CCDs). The tube camera is basically a small CRT as described above, except that the tube face is not coated with a light-producing phosphor. Instead it is coated with a material which emits electrons when excited by light. The lens system focuses an image onto the tube face. As the tube contains a vacuum and the light-sensitive material is non-conductive the result is that a replica of the image is created in the distribution of electronic charge.

The face of the tube is scanned by an electron beam which forms the only conductive path by which electronic charge can leave. The variations in charge result in variations in tube current as the image is scanned and these are amplified to produce a video signal.

The shape and size of the image is determined entirely by the dimensions of the raster scan. Image distortions caused by the lens can be compensated by distorting the scan in the same way. Thus if the lens results in an image having pincushion distortion, the scan will also be given a degree of pincushion distortion so that the video signal represents a rectangular image.

The tubes used in cameras suffer from a number of shortcomings. They need a warm-up period, have a finite life and require periodic replacement. Their signal-to-noise ratio is marginal, and powerful lighting is required. Possibly the worst characteristic of tubes is that they suffer from smear on moving objects because the charge built up in an image cannot be discharged by a single pass of the electron beam.

The CCD camera breaks up the image area into a large number of discrete picture elements or pixels which are arranged in rows or columns. Each pixel is a small photosensor which builds up charge when exposed to light. When a frame pulse is generated, the charge in every pixel in a line is transferred to the various stages of an analog shift register based on analog switches and capacitors. The scanning process is replicated by shifting and the video signal appears at the end of the register.

The CCD element is small, light and has an indefinite life. It requires litle power and does not need to warm up. The motion smear of tube cameras is eliminated as each pixel is completely reset every scan. As the CCD camera samples the image in one instant before scanning, it is possible to fit a shutter so that the exposure time is less than the field period. This can be used in the same way as the exposure control on a still camera. By shortening the exposure, sharper pictures of moving objects can be obtained, or the depth of field can be reduced in bright light. The CCD elements are low-noise devices and can be made very sensitive so that economies in studio lighting can be made. Outdoor working in poor lighting conditions is possible. For most broadcast purposes the CCD camera has made the tube camera obsolete.

One drawback of the CCD camera is that the scanning pattern is determined by the geometry of the CCD chip and it is not possible to correct for lens distortions by scan correction. This puts extra demands on the lens performance.

Both tube and CCD cameras are nearly linear and this produces signals of excessive amplitude if there is a great deal of brightness range in the scene. Attenuating the signal to accommodate the range causes loss of contrast in the wanted range. Instead the camera signal is made non-linear so that a certain range of brightness is responsible for most of the signal range, whereas at the ends of the range the signal is compressed. Film is made with this characteristic and television cameras attempt to simulate the process electronically with varying degrees of success.

2.9 Bandwidth and definition

As the conventional analog television picture is made up of lines, the line structure determines the *definition* or the fineness of detail which can be portrayed in the vertical axis. The limit is reached in theory when alternate lines show black and white. In a 625-line picture there are roughly 600 unblanked lines. If 300 of these are white and 300 are black then there will be 300 complete cycles of detail in one picture height. One unit of resolution, which is a unit of spatial frequency, is c/ph or cycles per picture height. In practical displays the contrast will have fallen to virtually nothing at this ideal limit and the resolution actually achieved is around 70 per cent of the ideal, or about 210 c/ph. The degree to which the ideal is met is known as the *Kell factor* of the display.

Definition in one axis is wasted unless it is matched in the other and so the horizontal axis should be able to offer the same performance. As the aspect ratio of conventional television is 4:3 then it should be possible to display 400 cycles in one picture width, reduced to about 300 cycles by the Kell factor. As part of the line period is lost due to flyback, 300 cycles per picture width becomes about 360 cycles per line period.

In 625-line television, the frame rate is 25 Hz and so the line rate F_h will be:

$$F_h = 625 \times 25 = 15\,625\,\text{Hz}$$

If 360 cycles of video waveform must be carried in each line period, then the bandwidth required will be given by:

$$15\,625 \times 360 = 5.625\ \text{MegaHertz}$$

In the 525-line system, there are roughly 500 unblanked lines allowing 250 c/ph theoretical definition, or 175 lines allowing for the Kell factor. Allowing for the aspect ratio, equal horizontal definition requires about 230 cycles per picture width. Allowing for horizontal blanking this requires about 280 cycles per line period.

In 525-line video, $F_h = 525 \times 30 = 15\,750\,\text{Hz}$ Thus the bandwidth required is:

$$15\,750 \times 280 = 4.4\ \text{MegaHertz}$$

If it is proposed to build a high-definition television system, one might start by doubling the number of lines and hence double the definition. Thus in a 1250-line format about 420 c/ph might be obtained. To achieve equal horizontal definition, bearing in mind the aspect ratio is now 16:9, then nearly 750 cycles per picture width will be needed. Allowing for horizontal blanking, then around 890 cycles per line period will be needed. The line frequency is now given by:

$$F_h = 1250 \times 25 = 31\,250\,\text{Hz}$$

and the bandwidth required is given by:

$$31\,250 \times 890 = 28\,\text{MHz}$$

Note the dramatic increase in bandwidth. In general the bandwidth rises as the square of the resolution because there are more lines and more cycles needed in each line. It should be clear that, except for research purposes, high-definition television will never be broadcast as a conventional analog signal because the bandwidth required is simply uneconomic. If and when high-definition broadcasting becomes common, it will be compelled to use digital compression techniques to make it economic.

2.10 MTF, contrast and sharpness

The modulation transfer function (MTF) is a way of describing the ability of an imaging system to carry detail. It is the spatial equivalent of frequency response in electronics. Prior to describing the MTF it is necessary to define some terms used in assessing image quality.

Spatial frequency is measured in cycles per millimetre (mm^{-1}). Contrast index (CI) is shown in Figure 2.17(a). The luminance variation across an image has peaks and troughs and the relative size of these is used to calculate the contrast index as shown. A test image can be made having the same contrast index over a range of spatial frequencies as shown in (b). If a non-ideal optical system is used to examine the test image, the output will have a contrast index which falls with rising spatial frequency.

The ratio of the output CI to the input CI is the MTF as shown in (c). In the special case where the input CI is unity the output CI is identical

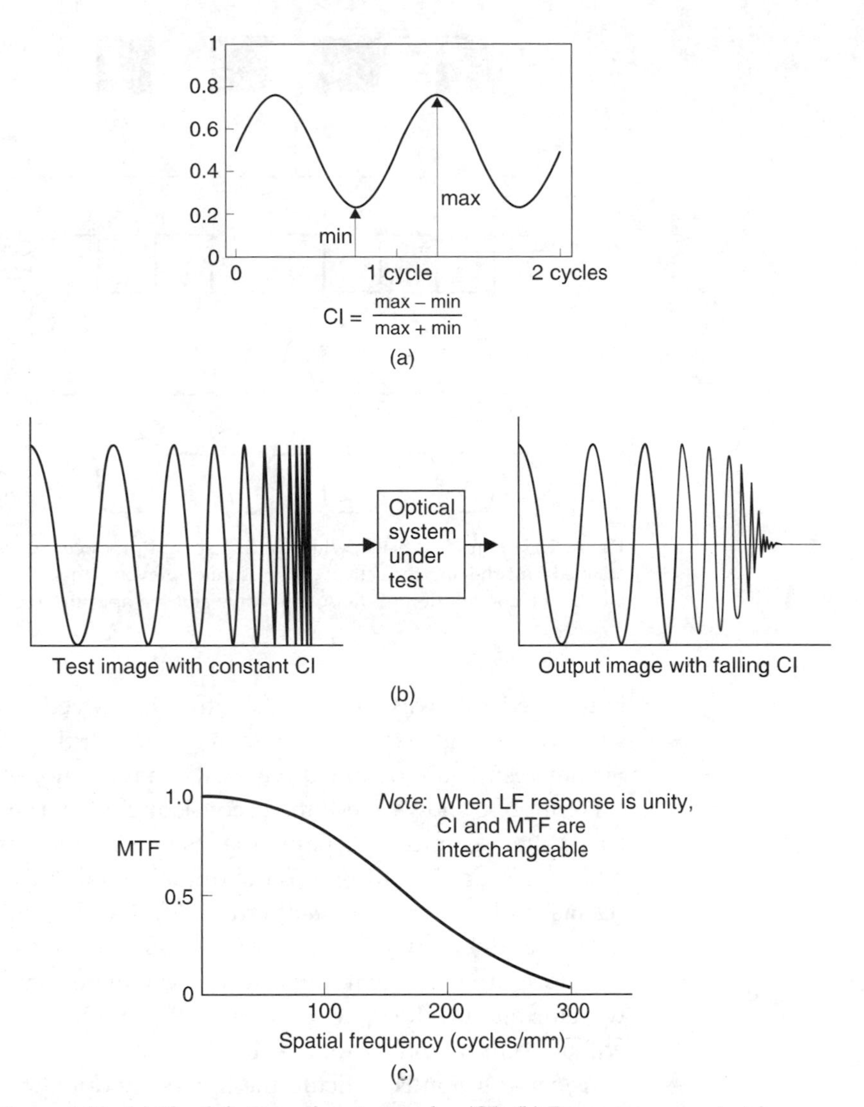

Figure 2.17 (a) The definition of contrast index (CI). (b) Frequency sweep test image having constant CI. (c) MTF is the ratio of output and input CIs.

to the output MTF. It is common to measure resolution by quoting the frequency at which the MTF has fallen to one half. This is known as the 50 per cent MTF frequency. The limiting resolution is defined as the point where the MTF has fallen to 10 per cent.

Whilst MTF resolution testing is objective, human vision is subjective and gives an impression we call sharpness. However, the assessment of sharpness is affected by contrast. Increasing the contrast of an image will result in an increased sensation of sharpness even though the MTF is unchanged. When CRTs having black areas between the phosphors were

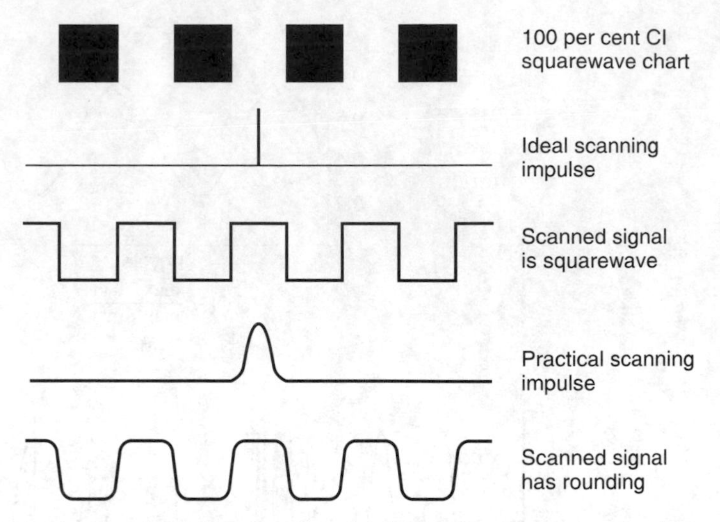

Figure 2.18 An ideal black/white grating should produce a square wave when scanned, but the aperture effect of the scanner prevents this. The waveform produced will be the convolution of the square wave and the aperture function.

introduced, it was found that the improved contrast resulted in subjectively improved sharpness even though the MTF was unchanged. Similar results are obtained with CRTs having non-reflective coatings.

Figure 2.18 shows a test image consisting of alternating black and white bars. When scanned by a point, the result is a square wave which contains odd harmonics in addition to the fundamental. However, in practice the scanner will suffer from an aperture effect and its aperture function will not be a delta function but will be Gaussian or rectangular. In the waveform domain the impulse response will be convolved with the input waveform to produce the output waveform which will not be square, but will be rounded off to some extent.

As convolution is difficult, the frequency domain can be used instead. The spectrum of the output can be obtained by multiplying the input spectrum by the frequency response. The MTF is essentially an optical frequency response and is a function of depth of contrast with respect to spatial frequency. The MTF is given by the Fourier transform of the aperture function (see Chapter 3). As the Fourier transform of a Gaussian impulse is also Gaussian, a cathode ray tube with spot having a Gaussian intensity distribution will also have a Gaussian frequency response.

2.11 Aperture effect

The aperture effect will show up in many aspects of television in both the sampled and continuous domains. The image sensor has a finite aperture function. In tube cameras and in CRTs, the beam will have a finite radius with a Gaussian distribution of energy across its diameter. This results in a Gaussian spatial frequency response. Tube cameras often contain an *aperture corrector* which is a filter designed to boost the higher spatial frequencies which are attenuated by the Gaussian response. The horizontal filter is simple enough, but the vertical filter will require line delays in order to produce points above and below the line to be corrected. Aperture correctors also amplify aliasing products and an over-corrected signal may contain more vertical aliasing than resolution.

Some digital-to-analog convertors keep the signal constant for a substantial part of or even the whole sample period. In CCD cameras, the sensor is split into elements which may almost touch in some cases. The element integrates light falling on its surface. In both cases the aperture will be rectangular. The case where the pulses have been extended in width to become equal to the sample period is known as a zero-order hold system and has a 100 per cent aperture ratio.

Rectangular apertures have a $\sin x/x$ spectrum which is shown in Figure 2.19. With a 100 per cent aperture ratio, the frequency response falls to a null at the sampling rate, and as a result is about 4 dB down at the edge of the baseband.

The temporal aperture effect varies according to the equipment used. Tube cameras have a long integration time and thus a wide temporal aperture. Whilst this reduces temporal aliasing, it causes smear on moving objects. CCD cameras do not suffer from lag and as a result their temporal response is better. Some CCD cameras deliberately have a short temporal aperture as the time axis is resampled by a shutter. The intention is to reduce smear, hence the popularity of such devices

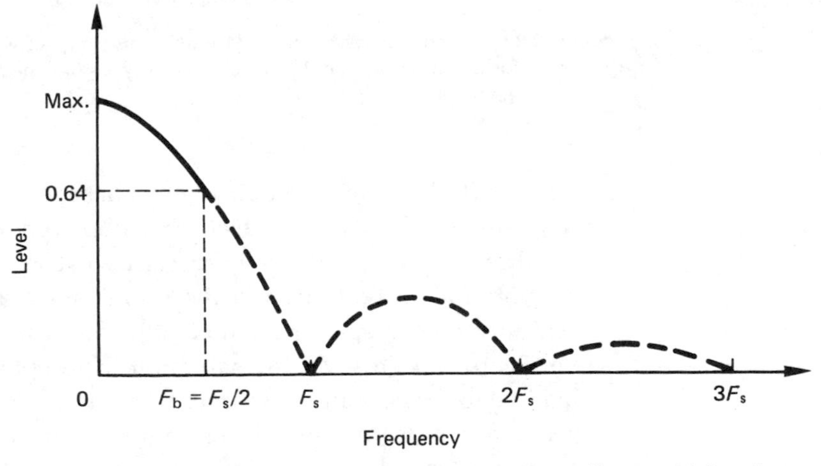

Figure 2.19 Frequency response with 100 per cent aperture nulls at multiples of sampling rate. Area of interest is up to half sampling rate.

for sporting events, but there will be more aliasing on certain subjects.

The eye has a temporal aperture effect which is known as persistence of vision, and the phosphors of CRTs continue to emit light after the electron beam has passed. These produce further temporal aperture effects in series with those in the camera.

2.12 Colour vision

Colour vision is made possible by the cones on the retina which occur in three different types, responding to different colours. Figure 2.20 shows that human vision is restricted to range of light wavelengths from 400 nm to 700 nm. Shorter wavelengths are called ultra-violet and longer wavelengths are called infra-red. Note that the response is not uniform, but peaks in the area of green. The response to blue is very poor and makes a nonsense of the traditional use of blue lights on emergency vehicles.

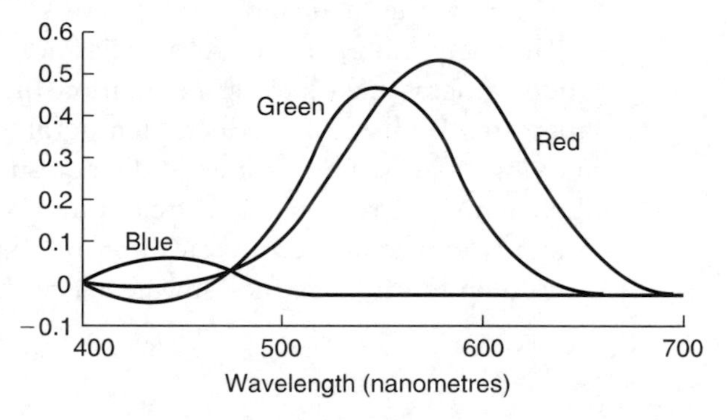

Figure 2.20 All human vision takes place over this range of wavelengths. The response is not uniform, but has a central peak. The three types of cone approximate to the three responses shown to give colour vision.

Figure 2.20 shows an approximate response for each of the three types of cone. If light of a single wavelength is observed, the relative responses of the three sensors allows us to discern what we call the colour of the light. Note that at both ends of the visible spectrum there are areas in which only one receptor responds; all colours in those areas look the same. There is a great deal of variation in receptor response from one individual to the next and the curves used in television are the average of a great many tests. In a surprising number of people the single receptor zones are extended and discrimination between, for example, red and orange is difficult.

The full resolution of human vision is restricted to brightness variations. Our ability to resolve colour details is only about a quarter of that.

2.13 Colorimetry

The triple receptor characteristic of the eye is extremely fortunate as it means that we can generate a range of colours by adding together light sources having just three different wavelengths in various proportions. This process is known as *additive colour matching* which should be clearly distinguished from the subtractive colour matching that occurs with paints and inks. Subtractive matching begins with white light and selectively removes parts of the spectrum by filtering. Additive matching uses coloured light sources which are combined.

An effective colour television system can be made in which only three pure or single wavelength colours or *primaries* can be generated. The primaries need to be similar in wavelength to the peaks of the three receptor responses, but need not be identical. Figure 2.21 shows a rudimentary colour television system. Note that the colour camera is in fact three cameras in one, where each is fitted with a different coloured filter. Three signals, *R*, *G* and *B* must be transmitted to the display which produces three images which must be superimposed to obtain a colour picture.

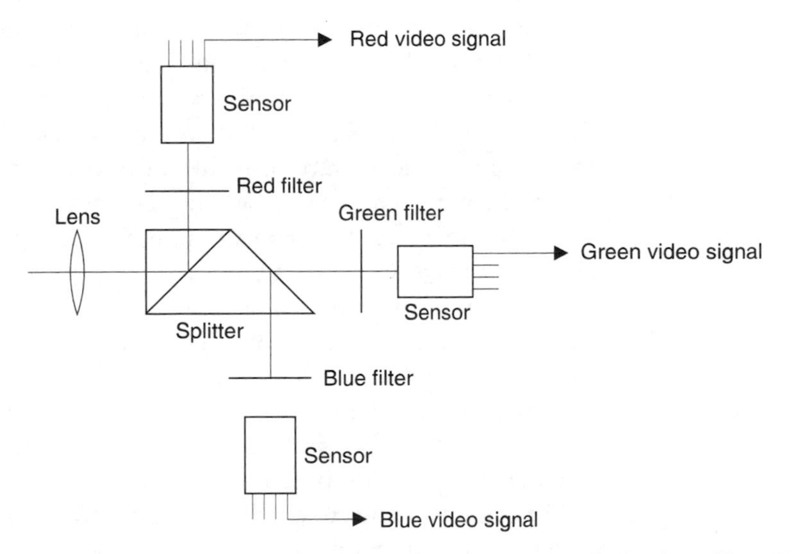

Figure 2.21 Simple colour television system. Camera image is split by three filters. Red, green and blue video signals are sent to three primary coloured displays whose images are combined.

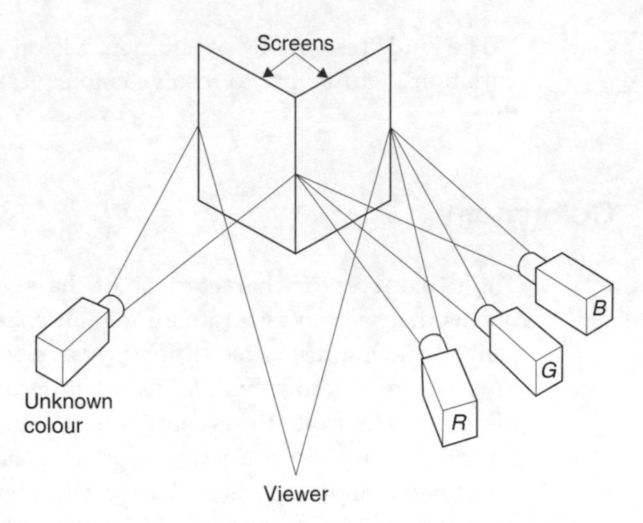

Figure 2.22 Simple colorimeter. Intensities of primaries on the right screen are adjusted to match the test colour on the left screen.

In practice the primaries must be selected from available phosphor compounds. Once the primaries have been selected, the proportions needed to reproduce a given colour can be found using a colorimeter. Figure 2.22 shows a colorimeter which consists of two adjacent white screens. One screen is illuminated by three light sources, one of each of the selected primary colours. Initially, the second screen is illuminated with white light and the three sources are adjusted until the first screen displays the same white. The sources are then calibrated. Light of a single wavelength is then projected onto the second screen. The primaries are once more adjusted until both screens appear to have the same colour. The proportions of the primaries are noted. This process is repeated for the whole visible spectrum, resulting in *colour mixture curves* shown in Figure 2.23. In some cases it will not be possible to find a match because an impossible negative contribution is needed. In this case we can simulate a negative contribution by shining some primary colour onto the test screen until a match is obtained. If the primaries were ideal, monochromatic or single wavelength sources, it would be possible to find three wavelengths at which two of the primaries were completely absent. However, practical phosphors are not monochromatic, but produce a distribution of wavelengths around the nominal value, and in order to make them spectrally pure other wavelengths have to be subtracted.

The colour-mixing curves dictate what the response of the three sensors in the colour camera must be. The primaries are determined in this way because it is easier to make camera filters to suit available CRT phosphors rather than the other way round.

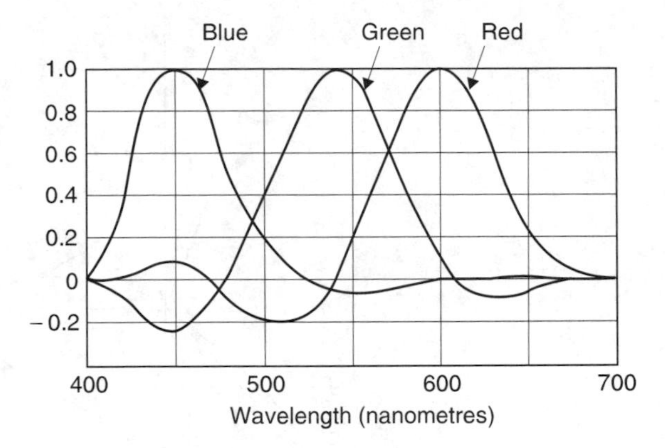

Figure 2.23 Colour mixture curves show how to mix primaries to obtain any spectral colour.

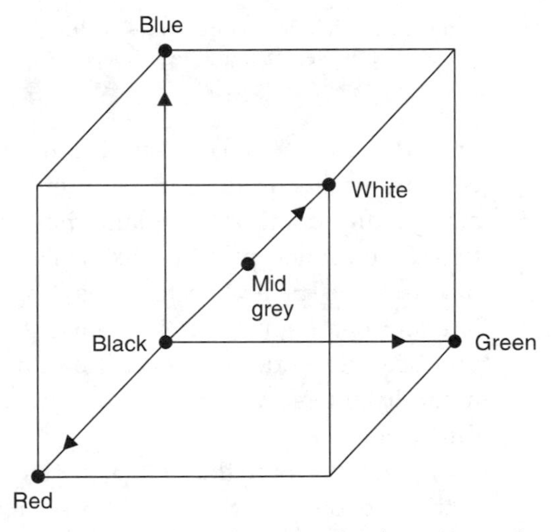

Figure 2.24 *RGB* colour space is three-dimensional and not easy to draw.

As there are three signals in a colour television system, they can only be simultaneously depicted in three dimensions. Figure 2.24 shows the *RGB* colour space which is basically a cube with black at the origin and white at the diagonally opposite corner. Figure 2.25 shows the colour mixture curves plotted in *RGB* space. For each visible wavelength a vector exists whose direction is determined by the proportions of the three primaries. If the brightness is allowed to vary this will affect all three primaries and thus the length of the vector in the same proportion.

Depicting and visualizing the *RGB* colour space is not easy and it is also difficult to take objective measurements from it. The solution is to modify the diagram to allow it to be rendered in two dimensions on flat paper. This is done by eliminating luminance (brightness) changes and depicting

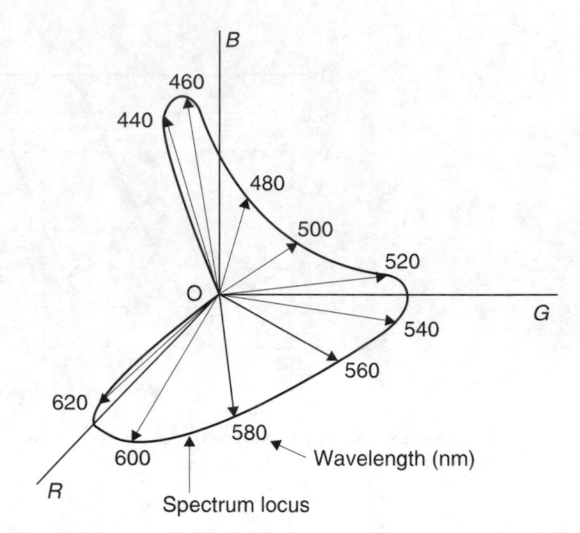

Figure 2.25 Colour mixture curves plotted in *RGB* space result in a vector whose locus moves with wavelength in three dimensions.

only the colour at constant brightness. Figure 2.26(a) shows how a constant luminance unit plane intersects the *RGB* space at unity on each axis. At any point on the plane the three components add up to one. A two-dimensional plot results when vectors representing all colours intersect the plane. Vectors may be extended if necessary to allow intersection. Figure 2.26(b) shows that the 500 nm vector has to be produced (extended) to meet the unit plane, whereas the 580 nm vector naturally intersects. Any colour can now uniquely be specified in two dimensions.

The points where the unit plane intersects the axes of *RGB* space form a triangle on the plot. The horseshoe-shaped locus of pure spectral colours goes outside this triangle because, as was seen above, the colour mixture curves require negative contributions for certain colours.

Having the spectral locus outside the triangle is a nuisance, and a larger triangle can be created by postulating new coordinates called *X*, *Y* and *Z* representing hypothetical primaries that cannot exist. This representation is shown in Figure 2.26(c).

The Commission Internationale d'Eclairage (CIE) standard *chromaticity diagram* shown in Figure 2.26(d) is obtained in this way by projecting the unity luminance plane onto the *X*, *Y* plane. This projection has the effect of bringing the red and blue primaries closer together. Note that the curved part of the locus is due to spectral or single wavelength colours. The straight base is due to non-spectral colours obtained by additively mixing red and blue.

As negative light is impossible, only colours within the triangle joining the primaries can be reproduced and so practical television systems

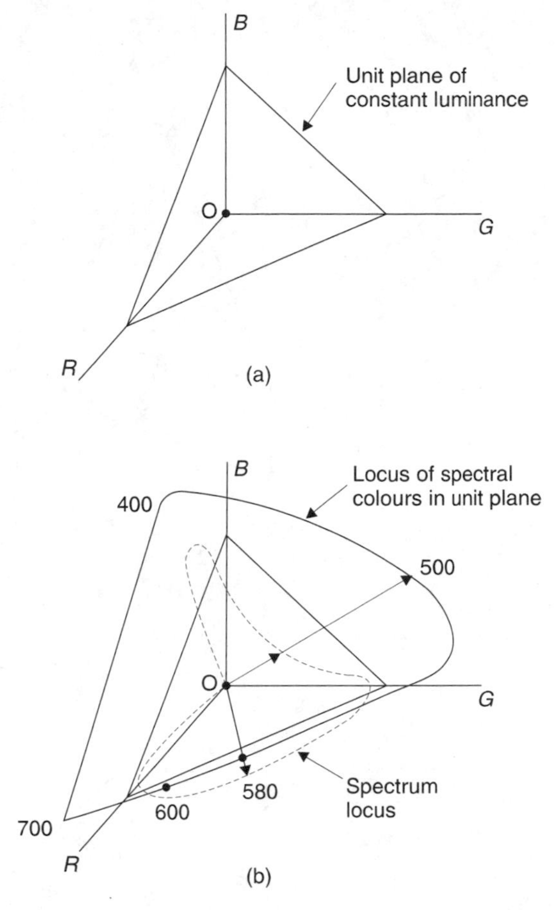

Figure 2.26 (a) A constant luminance plane intersects *RGB* space, allowing colours to be studied in two dimensions only. (b) The intersection of the unit plane by vectors joining the origin and the spectrum locus produces the locus of spectral colours which requires negative values of *R*, *G* and *B* to describe it.

cannot reproduce all possible colours. Clearly efforts should be made to obtain primaries which embrace as large an area as possible. Figure 2.27 shows how the colour range or gamut of television compares with paint and printing inks and illustrates that the comparison is favourable. Most everyday scenes fall within the colour gamut of television. Exceptions include saturated turquoise, spectrally pure iridescent colours formed by interference in duck's feathers or reflections in Compact Discs. For special purposes displays have been made having four primaries to give a wider colour range, but these are uncommon.

Figure 2.28 shows the primaries initially selected for NTSC. However, manufacturers looking for brighter displays substituted more efficient phosphors having a smaller colour range. This was later standardized as the SMPTE C phosphors which were also adopted for PAL.

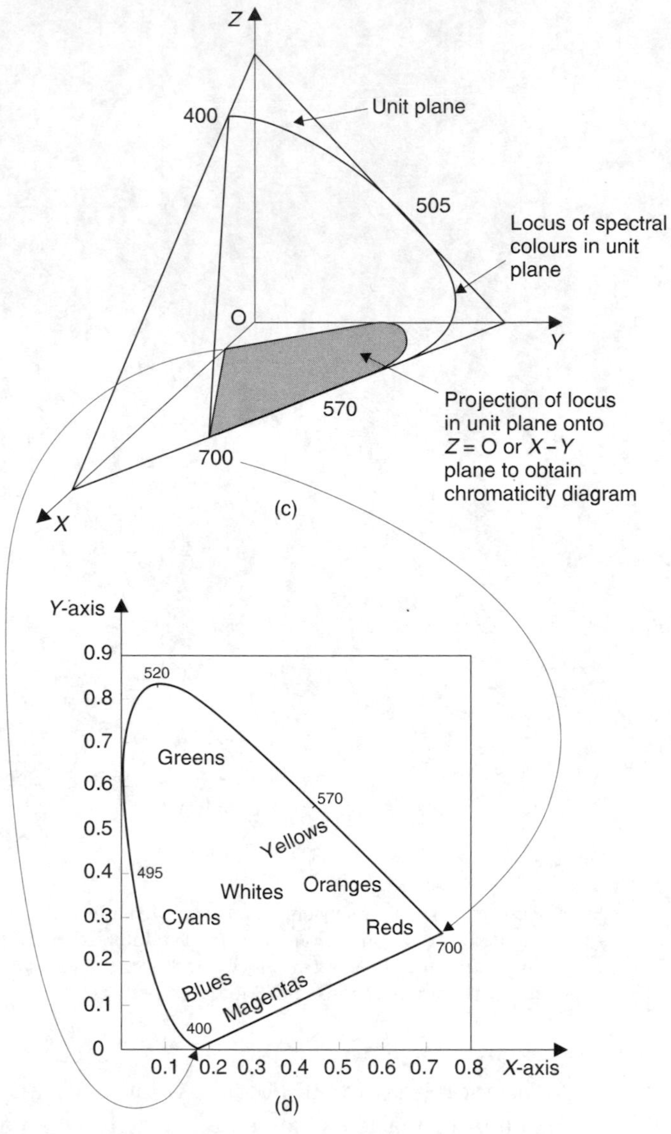

Figure 2.26 (*Continued*) In (c) a new coordinate system, *X*, *Y*, *Z*, is used so that only positive values are required. The spectrum locus now fits entirely in the triangular space where the unit plane intersects these axes. To obtain the CIE chromaticity diagram (d), the locus is projected onto the *X*–*Y* plane.

Whites appear in the centre of the chromaticity diagram corresponding to roughly equal amounts of primary colour. Two terms are used to describe colours: *hue* and *saturation*. Colours having the same hue lie on a straight line between the white point and the perimeter of the primary triangle. The saturation of the colour increases with distance from the white point. As an example, pink is a desaturated red.

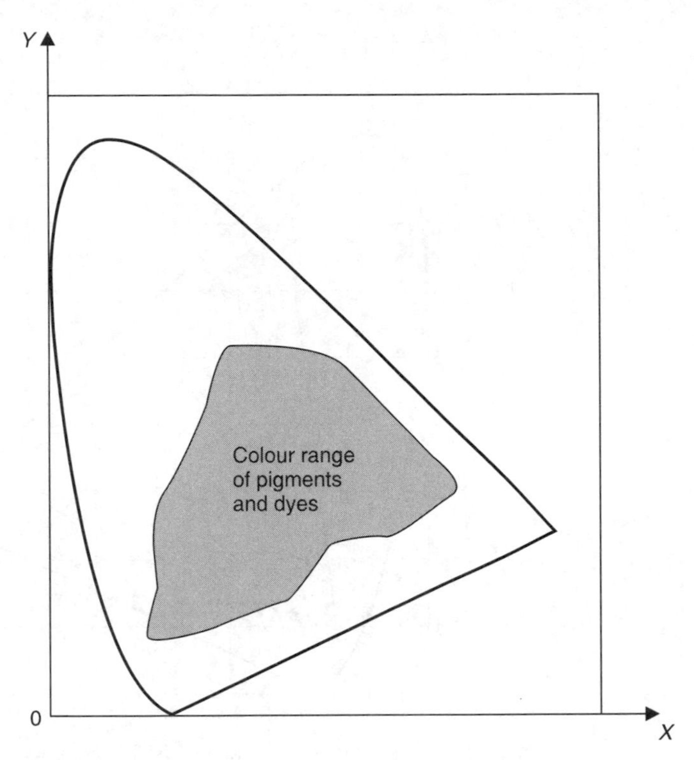

Figure 2.27 The colour range of television compares well with printing and photography.

The apparent colour of an object is also a function of the illumination. The 'true colour' will only be revealed under ideal white light which in practice is uncommon. An ideal white object reflects all wavelengths equally and simply takes on the colour of the ambient illumination. Figure 2.29 shows the location of three 'white' sources or *illuminants* on the chromaticity diagram. Illuminant A corresponds to a tungsten filament lamp, illuminant B corresponds to midday sunlight and illuminant C corresponds to typical daylight which is bluer because it consists of a mixture of sunlight and light scattered by the atmosphere. In everyday life we accommodate automatically to the change in apparent colour of objects as the sun's position or the amount of cloud changes and as we enter artificially lit buildings, but colour cameras accurately reproduce these colour changes. Attempting to edit a television program from recordings made at different times of day or indoors and outdoors would result in obvious and irritating colour changes unless some steps are taken to keep the white balance reasonably constant.

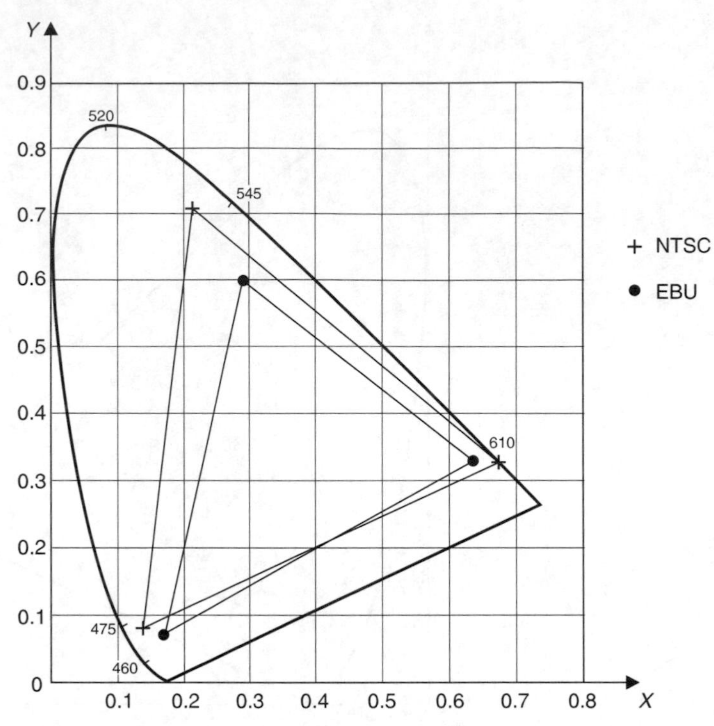

Figure 2.28 The primary colours for NTSC were initially as shown. These were later changed to more efficient phosphors which were also adopted for PAL. See text.

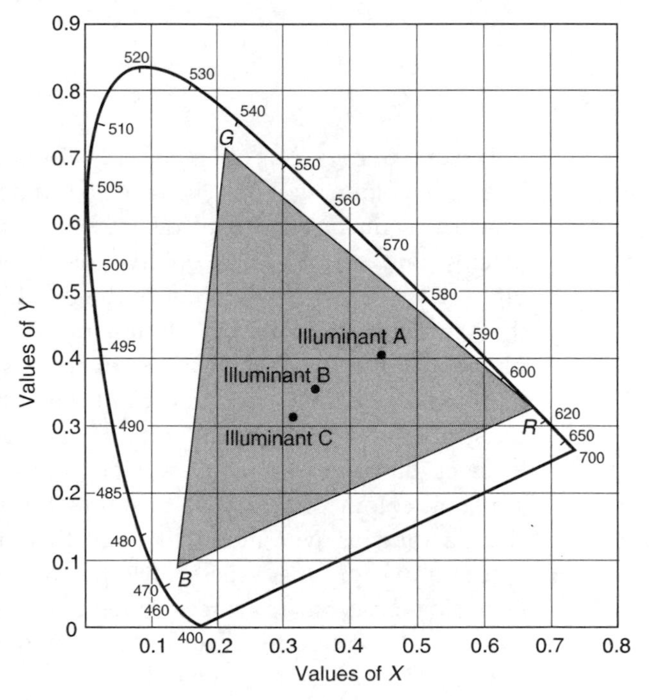

Figure 2.29 Position of three common illuminants on chromaticity diagram.

2.14 Colour displays

In order to display colour pictures, three simultaneous images must be generated, one for each primary colour. The colour CRT does this geometrically. Figure 2.30(a) shows that three electron beams pass down the tube from guns mounted in a triangular or delta array. Immediately before the tube face is mounted a perforated metal plate known as a shadow mask. The three beams approach holes in the shadow mask at a slightly different angle and so fall upon three different areas of phosphor which each produce a different primary colour. The sets of three phosphors are known as triads. Figure 2.30(b) shows an alternative arrangement in which the three electron guns are mounted in a straight line and the shadow mask is slotted and the triads are rectangular. This is known as a PIL (precision-in-line) tube. The triads can easily be seen upon close inspection of an operating CRT.

During the manufacturing process the shadow mask is fitted and the tube is assembled except for the electron guns. The inside of the tube is coated with photoresist and a light source is positioned at the point where

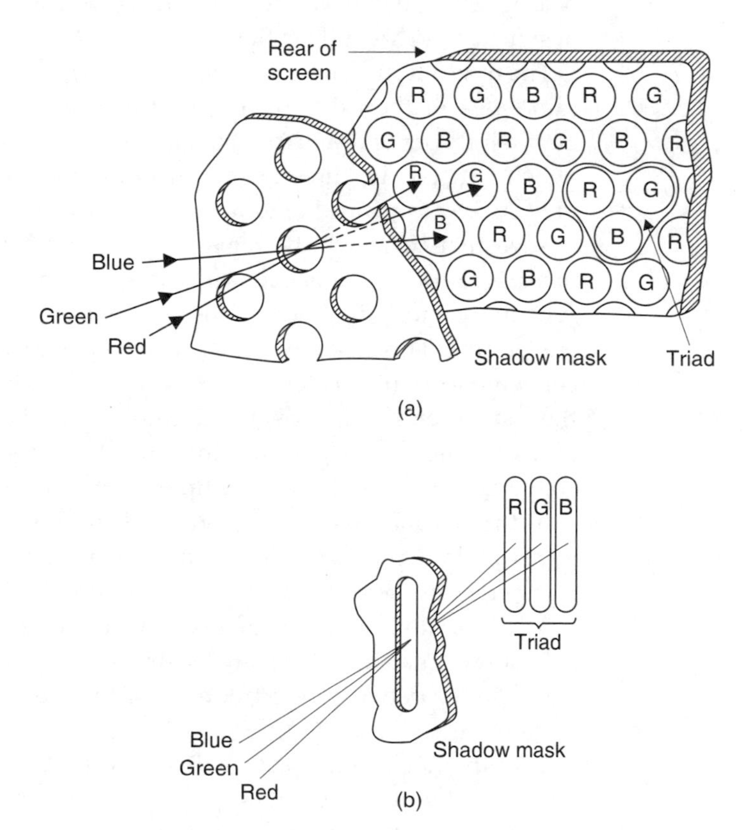

Figure 2.30 (a) Triads of phosphor dots are triangular and electron guns are arranged in a triangle. (b) Inline tube has strips of phosphor side by side.

the scan coils would deflect an electron beam such that the resist is exposed in all locations where one colour of phosphor should be deposited. The process is repeated for each phosphor.

In early tubes the space between the phosphor dots was grey. Later tubes replaced this with black in order to reduce reflection of ambient light and thereby increase contrast and apparent sharpness.

When the tube is completed, the scan coils have to be installed in exactly the right place otherwise the correct beam geometry will not result and beams may fall on part of the wrong phosphor in the triads. Adjusting the scan coils to ensure correct triad registration is called the *purity* adjustment. The shadow mask is heated by electron impact in service, and is not readily cooled because it is in a vacuum. Should the shadow mask overheat, it may distort due to thermal expansion and damage the purity. This effect limits the brightness of shadow mask CRTs. Purity can also be damaged by stray magnetic fields which may magnetize the shadow mask. Most monitors incorporate coils which degauss the shadow mask when the unit is first switched on. Many loudspeakers produce stray magnetic fields sufficiently strong to affect nearby CRTs and it is advisable to use loudspeakers which have been designed to contain their fields.

The three electron beams must also be arranged to scan over exactly the same area of the tube so that the three images correctly superimpose. Static shifts of the beams can be obtained by the *static convergence* controls which register the three beams in the tube centre. All three beams are deflected by the same horizontal and vertical scan coils, and the geometry is such that there will be registration errors between the beams which increase with the deflection. These errors are cancelled by providing subsidiary individual scan coils for two of the beams with correction waveforms. This is known as *dynamic convergence*. In order to adjust the convergence, a test pattern which produces a white cross-hatch is used. If the convergence is incorrect, separate coloured lines can be seen instead of one white line. The inline tube has the advantage that the dynamic convergence waveforms are simpler to generate.

In the plasma display the source of light is an electrical discharge in a gas at low pressure. This generates ultra-violet light which excites phosphors in the same way that a fluorescent light operates. Each pixel consists of three such elements, one for each primary colour. Figure 2.31 shows that the pixels are controlled by arranging the discharge to take place between electrodes which are arranged in rows and columns.

The advantage of the plasma display is that it can be made perfectly flat and it is very thin, even in large screen sizes. There is a size limit in CRTs beyond which they become very heavy. Plasma displays allow this limit to be exceeded.

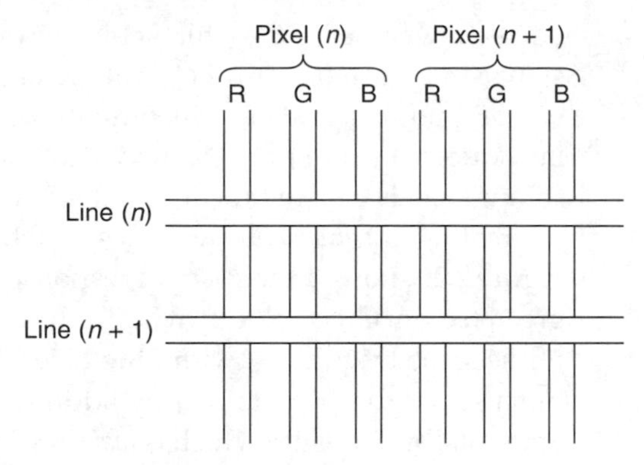

Figure 2.31 When a voltage is applied between a line or row electrode and a pixel electrode, a plasma discharge occurs. This excites a phosphor to produce visible light.

The great difficulty with the plasma display is that the relationship between light output and drive voltage is highly non-linear. Below a certain voltage there is no discharge at all. Consequently the only way that the brightness can be varied is to modulate the time for which the discharge takes place. The electrode signals are pulse width modulated.

Eight-bit digital video has 256 different brightnesses and it is difficult to obtain such a scale by pulse width modulation as the increments of pulse length would need to be generated by a clock of fantastic frequency. It is common practice to break up the picture period into many pulses, each of which is modulated in width. Despite this plasma displays often show contouring or posterizing, indicating a lack of sufficient brightness levels. Multiple pulse drive also has some temporal effects which may be visible on moving material unless motion compensation is used.

2.15 Colour difference signals

There are many different ways in which television signals can be carried and these will be considered here. A monochrome camera produces a single luma signal *Y* or *Ys* whereas a colour camera produces three signals, or *components*, *R*, *G* and *B* which are essentially monochrome video signals representing an image in each primary colour. In some systems sync is present on a separate signal (*RGBS*), rarely it is present on all three components, whereas most commonly it is only present on the green component leading to the term *RGsB*. The use of the green component for sync has led to suggestions that the components should be called *GBR*. As the original and long-standing term *RGB* or *RGsB* correctly reflects the sequence of the colours in the spectrum it remains to be seen whether *GBR* will achieve common usage. Like luma, *RGsB* signals may use 0.7 or 0.714 Volt signals, with or without set-up.

RGB and *Y* signals are incompatible, yet when colour television was introduced it was a practical necessity that it should be possible to display colour signals on a monochrome display and vice versa.

Creating or *transcoding* a luma signal from *R*, *Gs* and *B* is relatively easy. Figure 2.20 showed the spectral response of the eye which has a peak in the green region. Green objects will produce a larger stimulus than red objects of the same brightness, with blue objects producing the least stimulus. A luma signal can be obtained by adding *R*, *G* and *B* together, not in equal amounts, but in a sum which is *weighted* by the relative response of the eye. Thus:

$$Y = 0.299R + 0.587G + 0.114B$$

Syncs may be regenerated, but will be identical to those on the *Gs* input and when added to *Y* result in *Ys* as required.

If *Ys* is derived in this way, a monochrome display will show nearly the same result as if a monochrome camera had been used in the first place. The results are not identical because of the non-linearities introduced by gamma correction.

As colour pictures require three signals, it should be possible to send *Ys* and two other signals which a colour display could arithmetically convert back to *R*, *G* and *B*. There are two important factors which restrict the form which the other two signals may take. One is to achieve reverse compatibility. If the source is a monochrome camera, it can only produce *Ys* and the other two signals will be completely absent. A colour display should be able to operate on the *Ys* signal only and show a monochrome picture. The other is the requirement to conserve bandwidth for economic reasons.

These requirements are met by sending two *colour difference signals* along with *Ys*. There are three possible colour difference signals, $R - Y$, $B - Y$ and $G - Y$. As the green signal makes the greatest contribution to *Y*, then the amplitude of $G - Y$ would be the smallest and would be most susceptible to noise. Thus $R - Y$ and $B - Y$ are used in practice as Figure 2.32 shows.

R and *B* are readily obtained by adding *Y* to the two colour difference signals. *G* is obtained by rearranging the expression for *Y* above such that:

$$G = \frac{Y - 0.3R - 0.11B}{0.59}$$

If a colour CRT is being driven, it is possible to apply inverted luma to the cathodes and the $R - Y$ and $B - Y$ signals directly to two of the grids

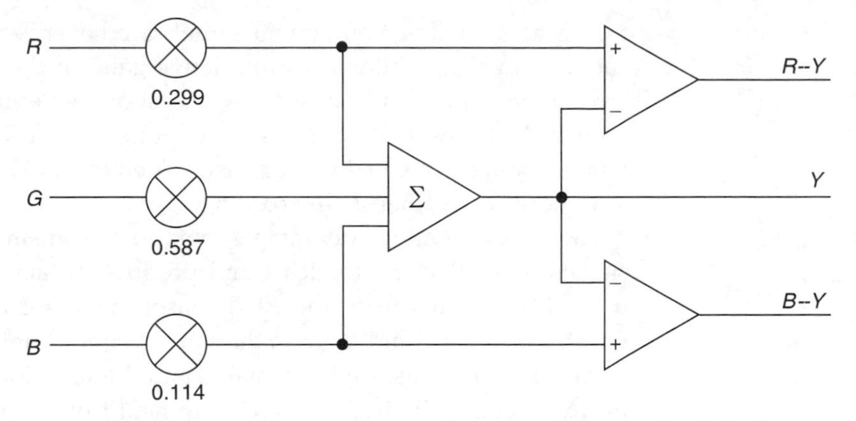

Figure 2.32 Colour components are converted to colour difference signals by the transcoding shown here.

so that the tube performs some of the matrixing. It is then only necessary to obtain $G - Y$ for the third grid, using the expression:

$$G - Y = -0.51(R - Y) - 0.186(B - Y)$$

If a monochrome source having only a *Ys* output is supplied to a colour display, $R - Y$ and $B - Y$ will be zero. It is reasonably obvious that if there are no colour difference signals the colour signals cannot be different from one another and $R = G = B$. As a result the colour display can only produce a neutral picture.

The use of colour difference signals is essential for compatibility in both directions between colour and monochrome, but it has a further advantage which follows from the way in which the eye works. In order to produce the highest resolution in the fovea, the eye will use signals from all types of cone, regardless of colour. In order to determine colour the stimuli from three cones must be compared. There is evidence that the nervous system uses some form of colour difference processing to make this possible. As a result the acuity of the human eye is only available in monochrome. Differences in colour cannot be resolved so well. A further factor is that the lens in the human eye is not achromatic and this means that the ends of the spectrum are not well focused. This is particularly noticeable on blue.

If the eye cannot resolve colour very well there is no point is expending valuable bandwidth sending high-resolution colour signals. Colour difference working allows the luma to be sent separately at full bandwidth. This determines the subjective sharpness of the picture. The colour difference signals can be sent with considerably reduced bandwidth, as little as one quarter that of luma, and the human eye is unable to tell.

In practice analog component signals are never received perfectly, but suffer from slight differences in relative gain. In the case of *RGB* a gain error in one signal will cause a colour cast on the received picture. A gain error in *Y* causes no colour cast and gain errors in $R - Y$ or $B - Y$ cause much smaller perceived colour casts. Thus colour difference working is also more robust than *RGB* working.

The overwhelming advantages obtained by using colour difference signals mean that in broadcast and production facilities *RGB* is seldom used. The outputs from the *RGB* sensors in the camera are converted directly to Y, $R - Y$ and $B - Y$ in the camera control unit and output in that form. Standards exist for both analog and digital colour difference signals to ensure compatibility between equipment from various manufacturers. The M-II and Betacam formats record analog colour difference signals, and there are a number of colour difference digital formats.

Whilst signals such as *Y, R, G* and *B* are *unipolar* or positive only, it should be stressed that colour difference signals are *bipolar* and may meaningfully take on levels below 0 Volts.

The wide use of colour difference signals has led to the development of test signals and equipment to display them. The most important of the test signals is the ubiquitous *colour bars*. Colour bars are used to set the

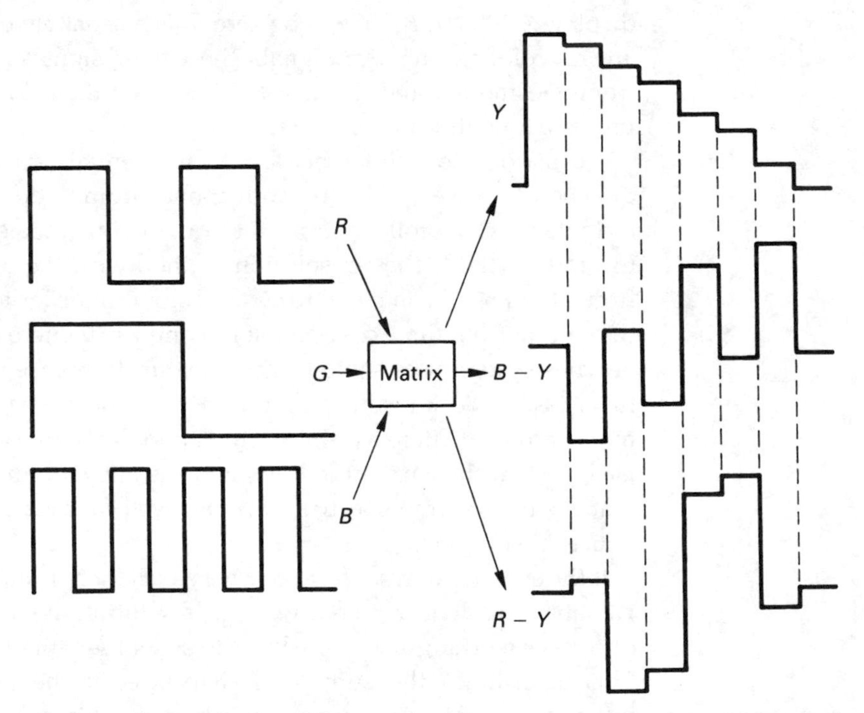

Figure 2.33 Origin of colour difference signals representing colours bars. Adding *R, G* and *B* according to the weighting factors produces an irregular luminance staircase.

gains and timing of signal components and to check that matrix operations are performed using the correct weighting factors. Further details will be found in Chapter 4. The origin of the colour bar test signal is shown in Figure 2.33. In *100 per cent amplitude bars*, peak amplitude binary *RGB* signals are produced, having one, two and four cycles per screen width. When these are added together in a weighted sum, an eight-level luma staircase results because of the unequal weighting. The matrix also produces two colour difference signals, $R - Y$ and $B - Y$ as shown. Sometimes *75 per cent amplitude bars* are generated by suitably reducing the *RGB* signal amplitude. Note that in both cases the colours are fully saturated; it is only the brightness which is reduced to 75 per cent. Sometimes the white bar of a 75 per cent bar signal is raised to 100 per cent to make calibration easier. Such a signal is sometimes erroneously called a 100 per cent bar signal.

Figure 2.34(a) shows an SMPTE/EBU standard colour difference signal set in which the signals are called *Ys*, P_b and P_r. 0.3 Volt syncs are on luma only and all three video signals have a 0.7 Volt pk–pk swing with 100 per cent bars. In order to obtain these voltage swings, the following gain corrections are made to the components:

$$P_r = 0.71327(R - Y) \text{ and } P_b = 0.56433(B - Y)$$

Within waveform monitors, the colour difference signals may be offset by 350 mV as in Figure 2.34(b) to match the luma range for display purposes.

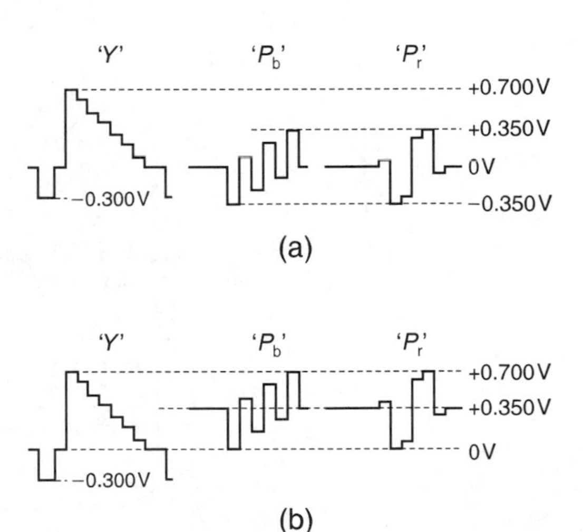

Figure 2.34 (a) 100 per cent colour bars represented by SMPTE/EBU standard colour difference signals. (b) Level comparison is easier in waveform monitors if the $B-Y$ and $R-Y$ signals are offset upwards.

2.16 Motion portrayal and dynamic resolution

As the eye uses involuntary tracking at all times, the criterion for measuring the definition of moving image portrayal systems has to be dynamic resolution, defined as the apparent resolution perceived by the viewer in an object moving within the limits of accurate eye tracking. The traditional metric of static resolution in film and television has to be abandoned as unrepresentative.

Figure 2.35(a) shows that when the moving eye tracks an object on the screen, the viewer is watching with respect to the optic flow axis, not the time axis, and these are not parallel when there is motion. The optic flow axis is defined as an imaginary axis in the spatio-temporal volume which joins the same points on objects in successive frames. Clearly when many objects move independently there will be one optic flow axis for each.

The optic flow axis is identified by motion-compensated standards convertors to eliminate judder and also by MPEG compressors because the greatest similarity from one picture to the next is along that axis.

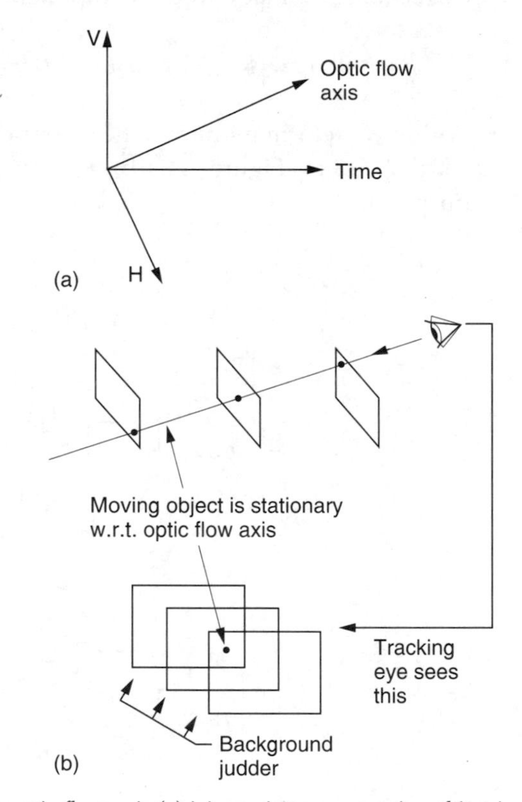

Figure 2.35 The optic flow axis (a) joins points on a moving object in successive pictures. (b) When a tracking eye follows a moving object on a screen, that screen will be seen in a different place at each picture. This is the origin of background strobing.

The success of these devices is testimony to the importance of the theory.

Figure 2.35(b) shows that when the eye is tracking, successive pictures appear in different places with respect to the retina. In other words if an object is moving down the screen and followed by the eye, the raster is actually moving up with respect to the retina. Although the tracked object is stationary with respect to the retina and temporal frequencies are zero, the object is moving with respect to the sensor and the display and in those units high temporal frequencies will exist. If the motion of the object on the sensor is not correctly portrayed, dynamic resolution will suffer.

In real-life eye tracking, the motion of the background will be smooth, but in an image-portrayal system based on periodic presentation of frames, the background will be presented to the retina in a different position in each frame. The retina seperately perceives each impression of the background leading to an effect called background strobing.

The criterion for the selection of a display frame rate in an imaging system is sufficient reduction of background strobing. It is a complete myth that the display rate simply needs to exceed the critical flicker frequency. Manufacturers of graphics displays which use frame rates well in excess of those used in film and television are doing so for a valid reason: it gives better results! Note that the display rate and the transmission rate need not be the same in an advanced system.

Dynamic resolution analysis confirms that both interlaced television and conventionally projected cinema film are both seriously sub-optimal. In contrast, progressively scanned television systems have no such defects.

2.17 Progressive or interlaced scan?

Interlaced scanning is a crude compression technique which was developed empirically in the 1930s as a way of increasing the picture rate to reduce flicker without increasing the video bandwidth. Instead of transmitting entire frames, the lines of the frame are sorted into odd lines and even lines. Odd lines are transmitted in one field, even lines in the next. A pair of fields will interlace to produce a frame. Vertical detail such as an edge may only be present in one field of the pair and this results in frame rate flicker called 'interlace twitter'.

Figure 2.36(a) shows a dynamic resolution analysis of interlaced scanning. When there is no motion, the optic flow axis and the time axis are parallel and the apparent vertical sampling rate is the number of lines in a frame. However, when there is vertical motion, (b), the optic flow axis turns. In the case shown, the sampling structure due to

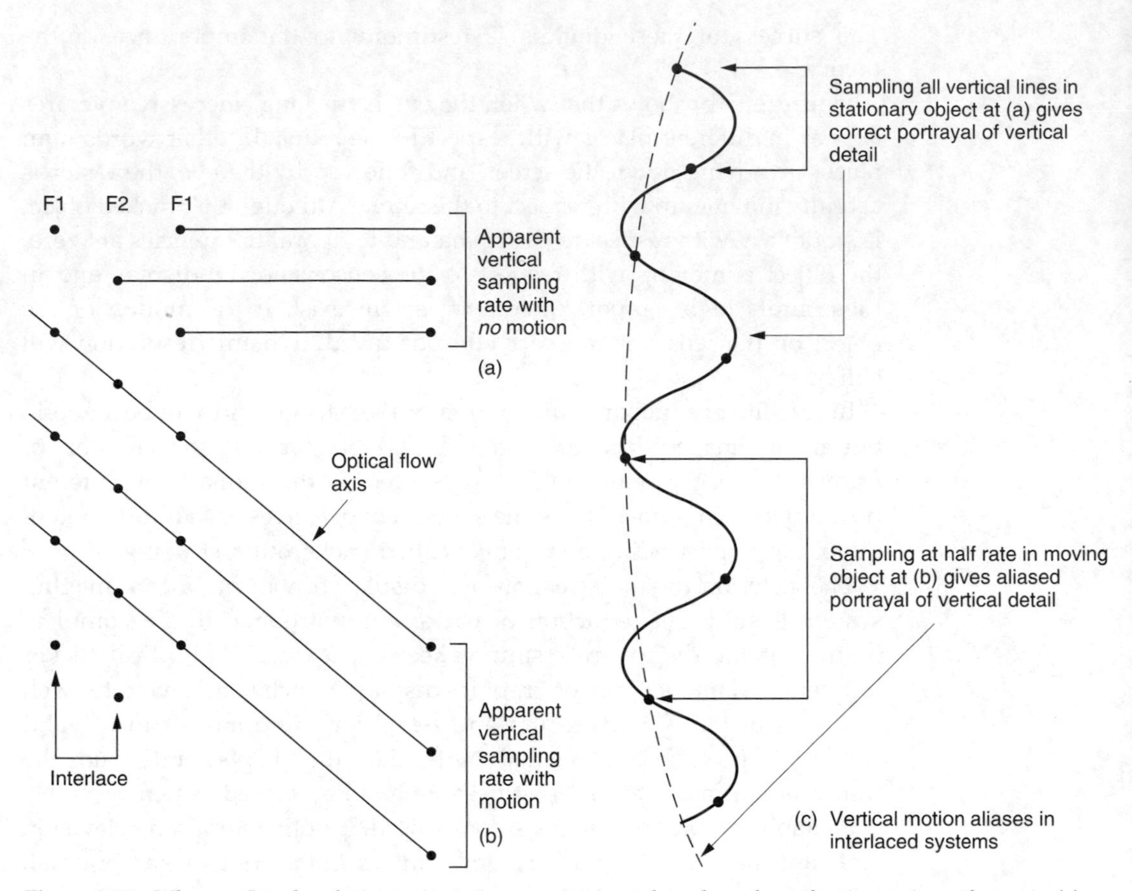

Figure 2.36 When an Interlaced picture is stationary, viewing takes place along the time axis as shown in (a). When a vertical component of motion exists, viewing takes place along the optic flow axis. (b) The vertical sampling rate falls to one half its stationary value. (c) The halving in sampling rate causes high spatial frequencies to alias.

interlace results in the vertical sampling rate falling to one half of its stationary value.

Consequently interlace does exactly what would be expected from a half-bandwidth filter. It halves the vertical resolution when any motion with a vertical component occurs. In a practical television system, there is no anti-aliasing filter in the vertical axis and so when the vertical sampling rate of an interlaced system is halved by motion, high spatial frequencies will alias or heterodyne causing annoying artifacts in the picture. This is easily demonstrated.

Figure 2.36(c) shows how a vertical spatial frequency well within the static resolution of the system aliases when motion occurs. In a progressive scan system this effect is absent and the dynamic resolution due to scanning can be the same as the static case.

This analysis also illustrates why interlaced television systems have to have horizontal raster lines. This is because in real life, horizontal motion is more common than vertical. It is easy to calculate the vertical image motion velocity needed to obtain the half-bandwidth speed of interlace, because it amounts to one raster line per field. In 525/60 (NTSC) there are about 500 active lines, so motion as slow as one picture height in 8 seconds will halve the dynamic resolution. In 625/50 (PAL) there are about 600 lines, so the half-bandwidth speed falls to one picture height in 12 seconds. This is why NTSC, with fewer lines and lower bandwidth, doesn't look as soft as it should compared to PAL, because it has better dynamic resolution.

The situation deteriorates rapidly if an attempt is made to use interlaced scanning in systems with a lot of lines. In 1250/50, the resolution is halved at a vertical speed of just one picture height in 24 seconds. In other words on real moving video a 1250/50 interlaced system has the same dynamic resolution as a 625/50 progressive system. By the same argument a 1080 *I* system has the same performance as a 480 *P* system.

3 Digital principles

3.1 Pure binary code

For digital video use, the prime purpose of binary numbers is to express the values of the samples which represent the original analog video waveform. Figure 3.1 shows some binary numbers and their equivalent in decimal. The radix point has the same significance in binary: symbols to the right of it represent one half, one quarter and so on. Binary is convenient for electronic circuits, which do not get tired, but numbers expressed in binary become very long, and writing them is tedious and error-prone. The octal and hexadecimal notations are both used for writing binary since conversion is so simple. Figure 3.1 also shows that a binary number is split into groups of three or four digits starting at the least significant end, and the groups are individually converted to octal or hexadecimal digits. Since sixteen different symbols are required in hex the letters A–F are used for the numbers above nine.

There will be a fixed number of bits in a PCM video sample, and this number determines the size of the quantizing range. In the eight-bit samples used in much digital video equipment, there are 256 different numbers. Each number represents a different analog signal voltage, and care must be taken during conversion to ensure that the signal does not go outside the convertor range, or it will be clipped. In Figure 3.2(a) it will be seen that in an eight-bit pure binary system, the number range goes from 00 hex, which represents the smallest voltage, through to FF hex, which represents the largest positive voltage. The video waveform must be accommodated within this voltage range, and (b) shows how this can be done for a PAL composite signal. A luminance signal is shown in (c). As component digital systems handle only the active line, the quantizing range is optimized to suit the gamut of the unblanked luminance. There is a small offset in order to handle slightly misadjusted inputs.

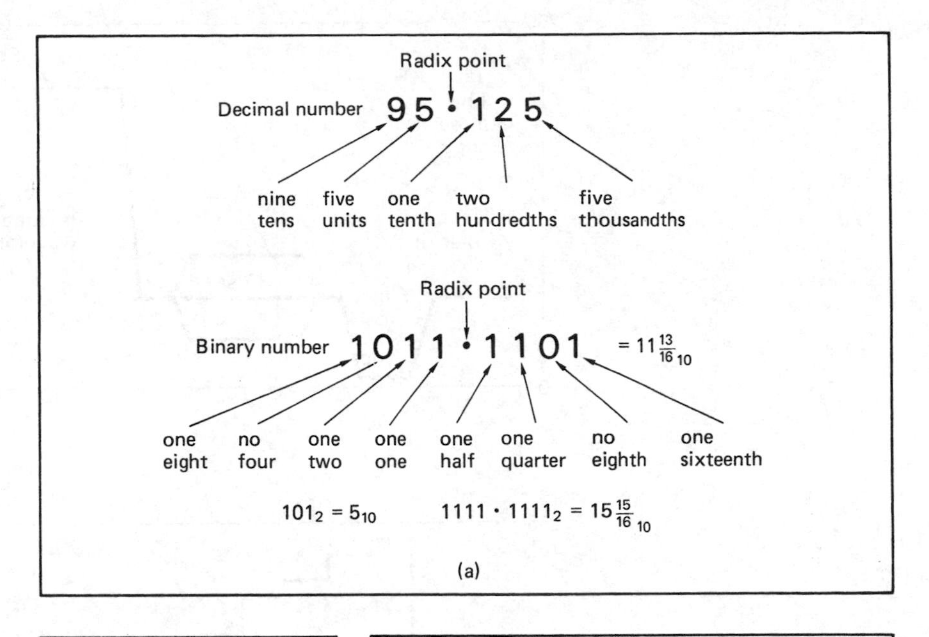

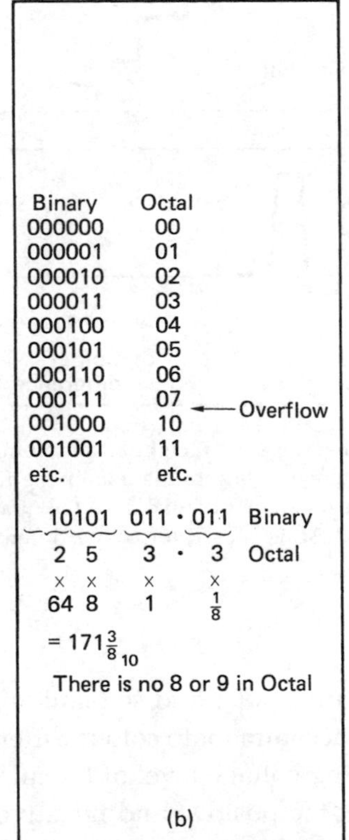

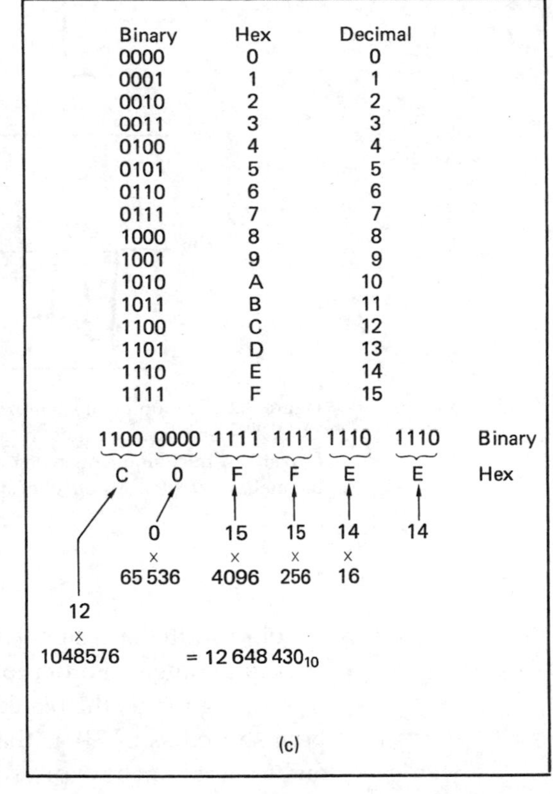

Figure 3.1 (a) Binary and decimal, (b) In octal, groups of three bits make one symbol 0–7. (c) In hex, groups of four bits make one symbol O–F. Note how much shorter the number is in hex.

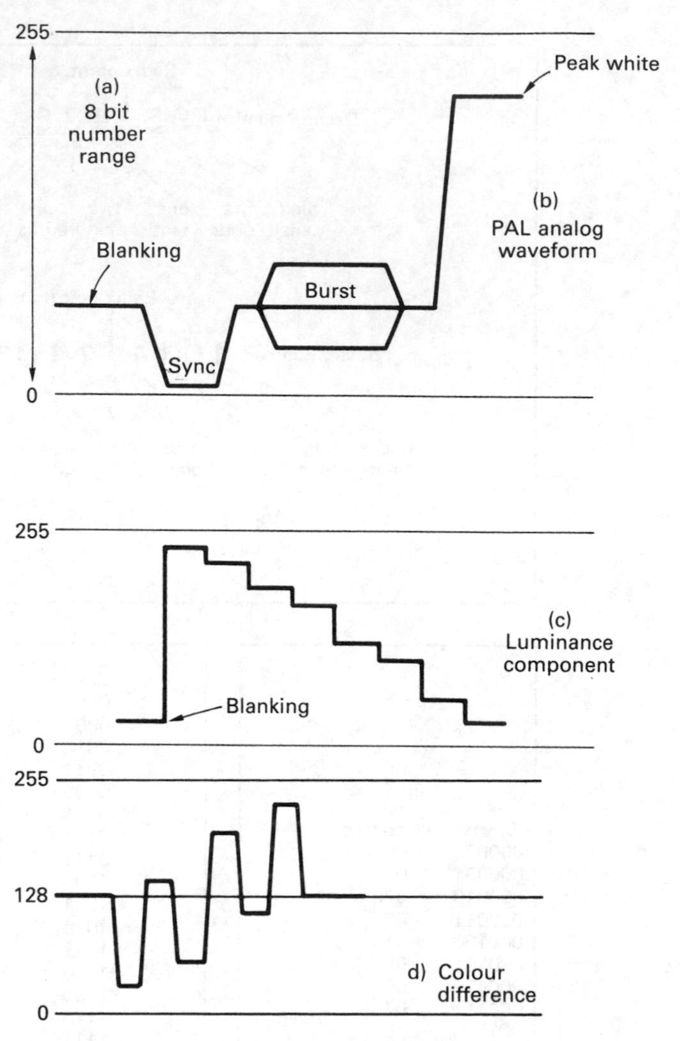

Figure 3.2 The unipolar quantizing range of an eight-bit pure binary system is shown at (a). The analog input must be shifted to fit into the quantizing range, as shown for PAL at (b). In component, sync pulses are not digitized, so the quantizing intervals can be smaller as at (c). An offset of half scale is used for colour difference signals (d).

Colour difference signals are bipolar and so blanking is in the centre of the signal range. In order to accommodate colour difference signals in the quantizing range, the blanking voltage level of the analog waveform has been shifted as in (d) so that the positive and negative voltages in a real audio signal can be expressed by binary numbers which are only positive. This approach is called offset binary. Strictly speaking both the composite and luminance signals are also offset binary because the blanking level is part-way up the quantizing scale.

Offset binary is perfectly acceptable where the signal has been digitized only for recording or transmission from one place to another, after which it will be converted directly back to analog. Under these conditions it is not necessary for the quantizing steps to be uniform, provided both ADC and DAC are constructed to the same standard. In practice, it is the requirements of signal processing in the digital domain which make both non-uniform quantizing and offset binary unsuitable.

Figure 3.3 shows that analog video signal voltages are referred to blanking. The level of the signal is measured by how far the waveform deviates from blanking, and attenuation, gain and mixing all take place around blanking level. Digital vision mixing is achieved by adding sample values from two or more different sources, but unless all the quantizing intervals are of the same size and there is no offset, the sum of two sample values will not represent the sum of the two original analog voltages. Thus sample values which have been obtained by non-uniform or offset quantizing cannot readily be processed because the binary numbers are not proportional to the signal voltage.

If two offset binary sample streams are added together in an attempt to perform digital mixing, the result will be that the offsets are also added and this may lead to an overflow. Similarly, if an attempt is made to attenuate by, say, 6.02 dB by dividing all the sample values by two, Figure 3.4 shows that the offset is also divided and the waveform suffers a shifted baseline. This problem can be overcome with digital luminance signals simply by subtracting the offset from each sample before processing as this results in numbers truly proportional to the luminance voltage. This approach is not suitable for colour difference or composite signals because negative numbers would result when the analog voltage goes below blanking and pure binary coding cannot handle them. The problem with offset binary is that it works with reference to one end of the range. What is needed is a numbering system which operates symmetrically with reference to the centre of the range.

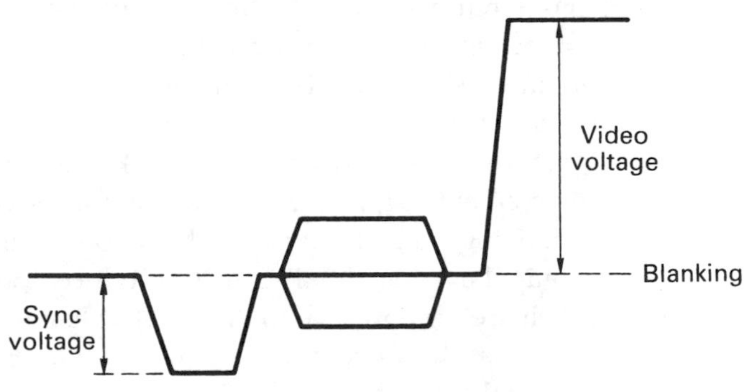

Figure 3.3 All video signal voltages are referred to blanking and must be added with respect to that level.

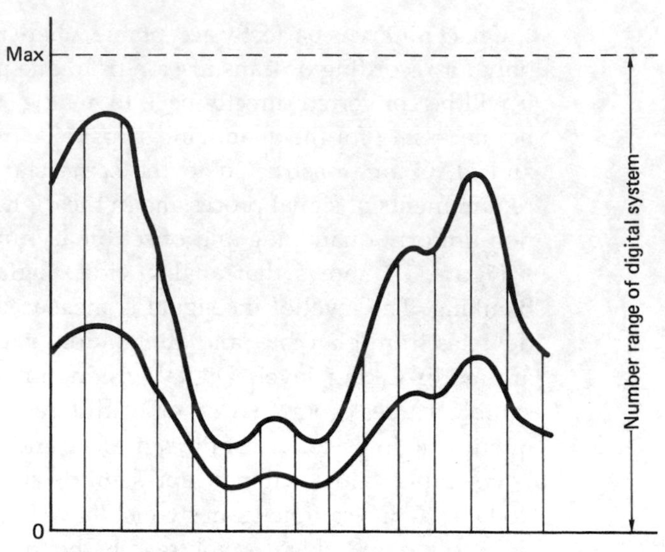

Figure 3.4 The result of an attempted attenuation in pure binary code is an offset. Pure binary cannot be used for digital video processing.

3.2 Two's complement

In the two's complement system, the upper half of the pure binary number range has been redefined to represent negative quantities. If a pure binary counter is constantly incremented and allowed to overflow, it will produce all the numbers in the range permitted by the number of available bits, and these are shown for a four-bit example drawn around the circle in Figure 3.5. As a circle has no real beginning, it is possible to consider it to start wherever it is convenient. In two's complement, the quantizing range represented by the circle of numbers does not start at zero, but starts on the diametrically opposite side of the circle. Zero is midrange, and all numbers with the MSB (most significant bit) set are considered negative. The MSB is thus the equivalent of a sign bit where 1 = minus. Two's complement notation differs from pure binary in that the most significant bit is inverted in order to achieve the half-circle rotation.

Figure 3.6 shows how a real ADC is configured to produce two's complement output. At (a) an analog offset voltage equal to one half the quantizing range is added to the bipolar analog signal in order to make it unipolar as at (b). The ADC produces positive only numbers at (c) which are proportional to the input voltage. The MSB is then inverted at (d) so that the all-zeros code moves to the centre of the quantizing range. The analog offset is often incorporated into the ADC as is the MSB inversion. Some convertors are designed to be used in either pure binary

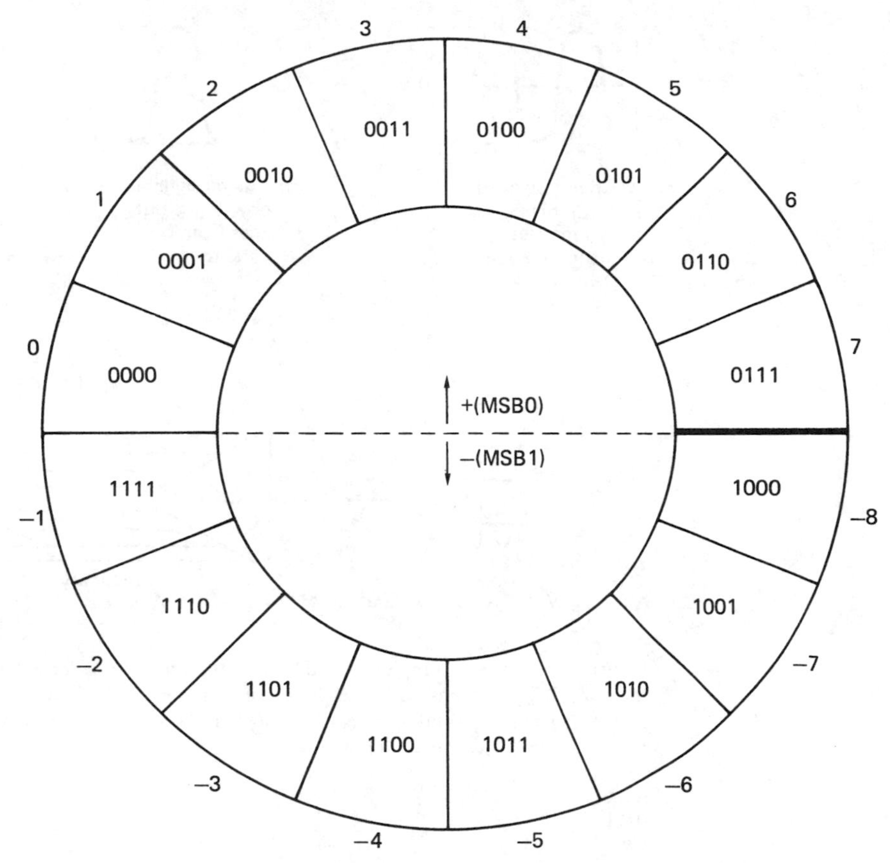

Figure 3.5 In this example of a four-bit two's complement code, the number range is from –8 to +7. Note that the MSB determines polarity.

or two's complement mode. In this case the designer must arrange the appropriate DC conditions at the input. The MSB inversion may be selectable by an external logic level. In the digital video interface standards the colour difference signals use offset binary because the codes of all zeros and all ones are at the end of the range and can be reserved for synchronizing. A digital vision mixer simply inverts the MSB of each colour difference sample to convert it to two's complement.

The two's complement system allows two sample values to be added, or 'mixed' in video parlance, and the result will be referred to the system midrange; this is analogous to adding analog signals in an operational amplifier.

Figure 3.7 illustrates how adding two's complement samples simulates a bipolar mixing process. The waveform of input A is depicted by solid black samples, and that of B by samples with a solid outline. The result of mixing is the linear sum of the two waveforms obtained by adding pairs of sample values. The dashed lines depict the output values. Beneath each

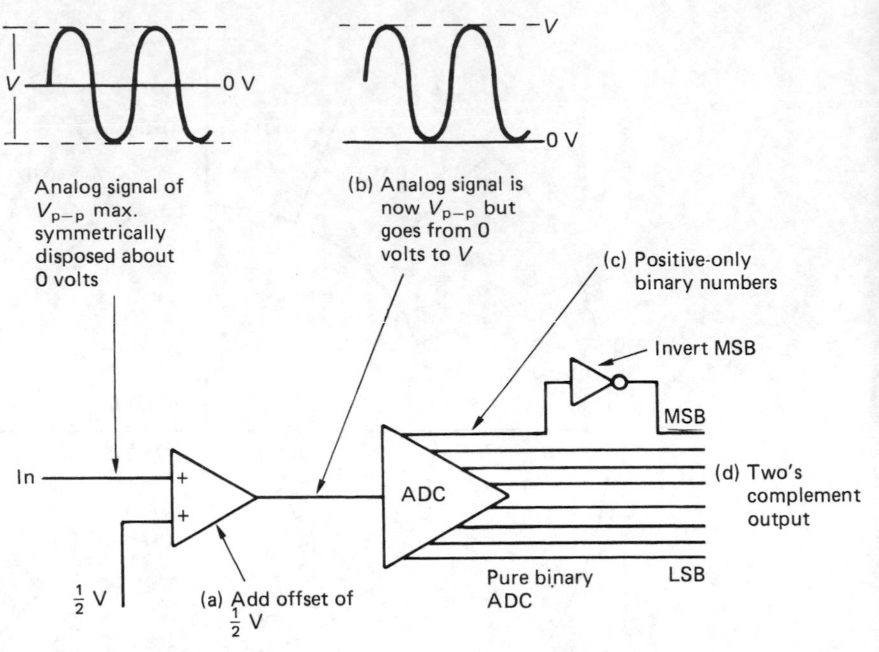

Figure 3.6 A two's complement ADC. At (a) an analog offset voltage equal to one-half the quantizing range is added to the bipolar analog signal in order to make it unipolar as at (b). The ADC produces positive only numbers at (c), but the MSB is then inverted at (d) to give a two's complement output.

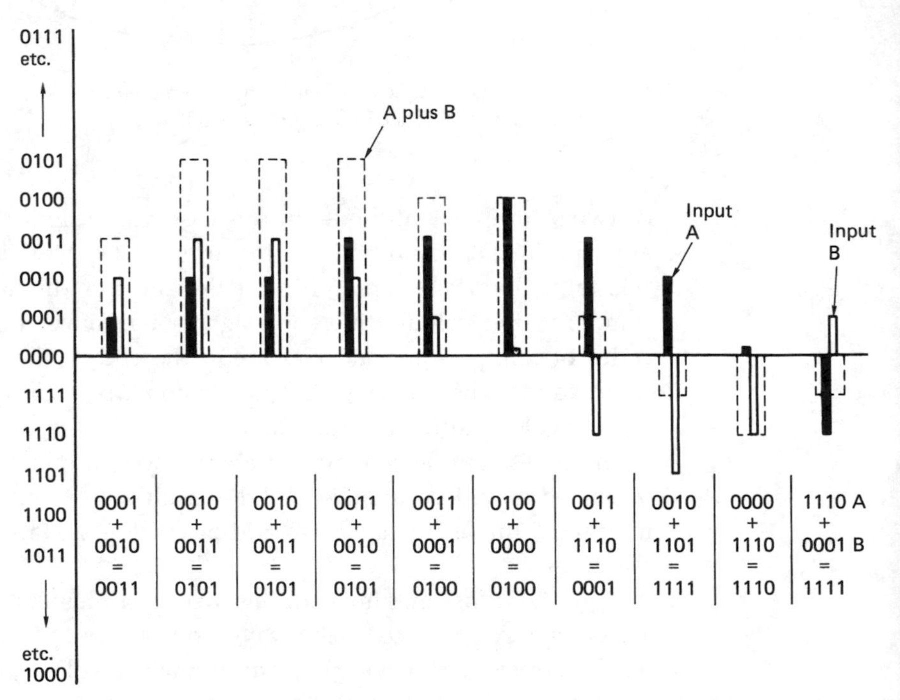

Figure 3.7 Using two's complement arithmetic, single values from two waveforms are added together with respect to midrange to give a correct mixing function.

set of samples is the calculation which will be seen to give the correct result. Note that the calculations are pure binary. No special arithmetic is needed to handle two's complement numbers.

It is sometimes necessary to phase reverse or invert a digital signal. The process of inversion in two's complement is simple. All bits of the sample value are inverted to form the one's complement, and one is added. This can be checked by mentally inverting some of the values in Figure 3.5. The inversion is transparent and performing a second inversion gives the original sample values.

Using inversion, signal subtraction can be performed using only adding logic. The inverted input is added to perform a subtraction, just as in the analog domain. This permits a significant saving in hardware complexity, since only carry logic is necessary and no borrow mechanism need be supported.

In summary, two's complement notation is the most appropriate scheme for bipolar signals, and allows simple mixing in conventional binary adders. It is in virtually universal use in digital video and audio processing.

Two's complement numbers can have a radix point and bits below it just as pure binary numbers can. It should, however, be noted that in two's complement, if a radix point exists, numbers to the right of it are added. For example, 1100.1 is not –4.5, it is –4 + 0.5 = –3.5.

3.3 Introduction to digital processing

However complex a digital process, it can be broken down into smaller stages until finally one finds that there are really only two basic types of element in use, and these can be combined in some way and supplied with a clock to implement virtually any process. Figure 3.8 shows that the first type is a *logic* element. This produces an output which is a logical function of the input with minimal delay. The second type is a *storage* element which samples the state of the input(s) when clocked and holds or delays that state. The strength of binary logic is that the signal has only two states, and considerable noise and distortion of the binary waveform can be tolerated before the state becomes uncertain. At every logic element, the signal is compared with a threshold, and can thus can pass through any number of stages without being degraded. In addition, the use of a storage element at regular locations throughout logic circuits eliminates time variations or jitter. Figure 3.8 shows that if the inputs to a logic element change, the output will not change until the *propagation delay* of the element has elapsed. However, if the output of the logic element forms the input to a storage element, the output of that element will not change until the input is sampled *at the next clock edge*. In this way

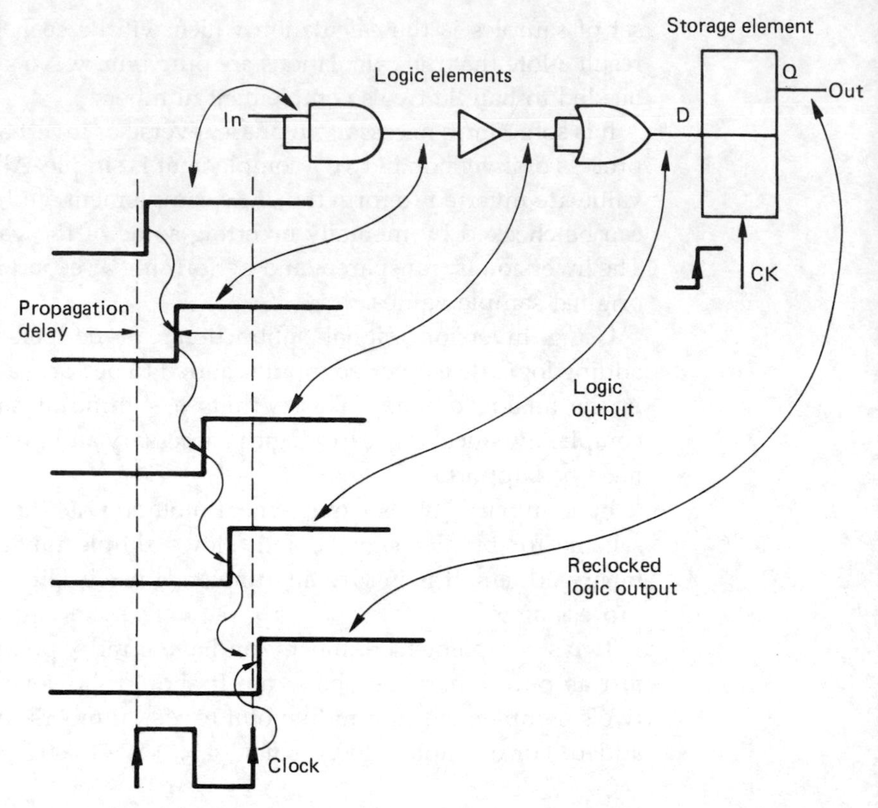

Figure 3.8 Logic elements have a finite propagation delay between input and output and cascading them delays the signal an arbitrary amount. Storage elements sample the input on a clock edge and can return a signal to near coincidence with the system clock. This is known as reclocking. Reclocking eliminates variations in propagation delay in logic elements.

the signal edge is aligned to the system clock and the propagation delay of the logic becomes irrelevant. The process is known as reclocking.

3.4 Logic elements

The two states of the signal when measured with an oscilloscope are simply two voltages, usually referred to as high and low. The actual voltage levels will depend on the type of logic family in use, and on the supply voltage used. Within logic, these levels are not of much consequence, and it is only necessary to know them when interfacing between different logic families or when driving external devices. The pure logic designer is not interested at all in these voltages, only in their meaning. Just as the electrical waveform from a microphone represents sound velocity, so the waveform in a logic circuit represents the truth of some statement. As there are only two states, there can only be *true* or *false* meanings. The true state of the signal can be assigned by the designer to either voltage state. When a high voltage represents a true logic condition and a low voltage represents a false condition, the system is known as *positive*, or *high true* logic. This is the usual system, but sometimes the low voltage represents the true condition and the high voltage represents the false condition. This is known as *negative* or *low true* logic. Provided that everyone is aware of the logic convention in use, both work equally well.

Negative logic is often found in the TTL (Transistor Transistor Logic) family, because in this technology it is easier to sink current to ground than to source it from the power supply. Figure 3.9 shows that if it is

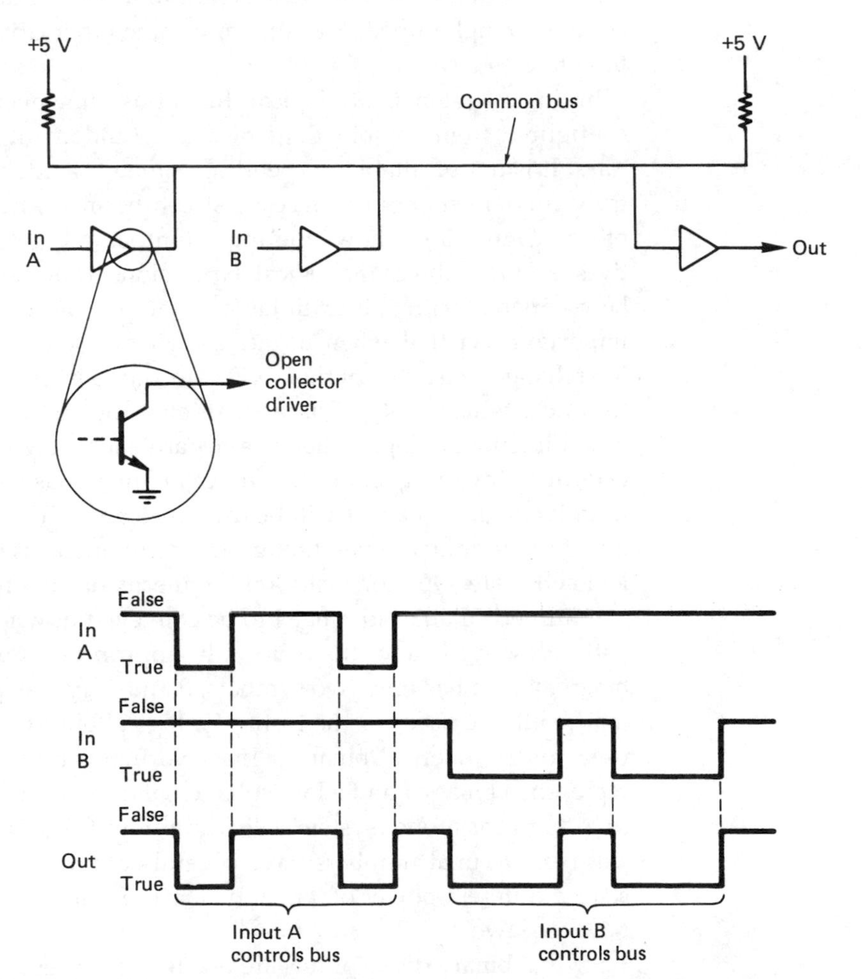

Figure 3.9 Using open-collector drive, several signal sources can share one common bus. If negative logic is used, the bus drivers turn off their output transistors with a false input, allowing another driver to control the bus. This will not happen with positive logic.

necessary to connect several logic elements to a common bus so that any one can communicate with any other, an open collector system is used, where high levels are provided by pull-up resistors and the logic elements only pull the common line down. If positive logic were used, when no device was operating the pull-up resistors would cause the common line to take on an absurd true state; whereas if negative logic is used, the common line pulls up to a sensible false condition when there is no device using the bus. Whilst the open collector is a simple way of obtaining a shared bus system, it is limited in frequency of operation due to the time constant of the pull-up resistors charging the bus capacitance. In the so-called tri-state bus systems, there are both active pull-up and pull-down devices connected in the so-called totem-pole output configuration. Both devices can be disabled to a third state, where the output assumes a high impedance, allowing some other driver to determine the bus state.

In logic systems, all logical functions, however complex, can be configured from combinations of a few fundamental logic elements or *gates*. It is not profitable to spend too much time debating which are the truly fundamental ones, since most can be made from combinations of others. Figure 3.10 shows the important simple gates and their derivatives, and introduces the logical expressions to describe them, which can be compared with the truth-table notation. The figure also shows the important fact that when negative logic is used, the OR gate function interchanges with that of the AND gate. Sometimes schematics are drawn to reflect which voltage state represents the true condition. In the so-called intentional logic scheme, a negative logic signal always starts and ends at an inverting 'bubble'. If an AND function is required between two negative logic signals, it will be drawn as an AND symbol with bubbles on all the terminals, even though the component used will be a positive logic OR gate. Opinions vary on the merits of intentional logic.

If numerical quantities need to be conveyed down the two-state signal paths described here, then the only appropriate numbering system is binary, which has only two symbols, 0 and 1. Just as positive or negative logic could be used for the truth of a logical binary signal, it can also be used for a numerical binary signal. Normally, a high voltage level will represent a binary 1 and a low voltage will represent a binary 0, described as a 'high for a one' system. Clearly a 'low for a one' system is just as feasible. Decimal numbers have several columns, each of which represents a different power of ten; in binary the column position specifies the power of two.

Several binary digits or bits are needed to express the value of a binary video sample. These bits can be conveyed at the same time by several signals to form a parallel system, which is most convenient inside equipment or for short distances because it is inexpensive, or one at a

Positive logic name	*Boolean expression*	*Positive logic symbol*	*Positive logic truth table*	*Plain English*
Inverter or NOT gate	$Q = \overline{A}$	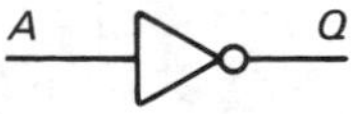	A | Q: 0 | 1; 1 | 0	Output is opposite of input
AND gate	$Q = A \cdot B$	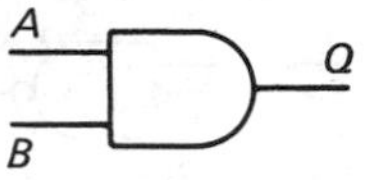	A B | Q: 0 0 | 0; 0 1 | 0; 1 0 | 0; 1 1 | 1	Output true when both inputs are true only
NAND (Not AND) gate	$Q = \overline{A \cdot B} = \overline{A} + \overline{B}$	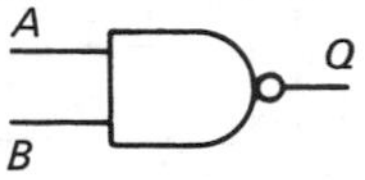	A B | Q: 0 0 | 1; 0 1 | 1; 1 0 | 1; 1 1 | 0	Output false when both inputs are true only
OR gate	$Q = A + B$	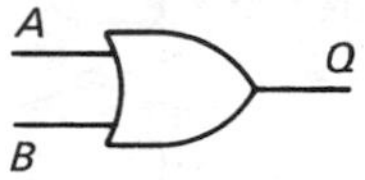	A B | Q: 0 0 | 0; 0 1 | 1; 1 0 | 1; 1 1 | 1	Output true if either or both inputs true
NOR (Not OR) gate	$Q = \overline{A + B} = \overline{A} \cdot \overline{B}$	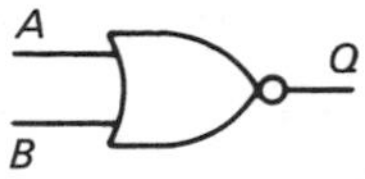	A B | Q: 0 0 | 1; 0 1 | 0; 1 0 | 0; 1 1 | 0	Output false if either or both inputs true
Exclusive OR (XOR) gate	$Q = A \oplus B$	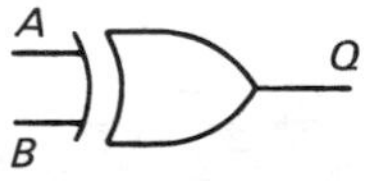	A B | Q: 0 0 | 0; 0 1 | 1; 1 0 | 1; 1 1 | 0	Output true if inputs are different

Figure 3.10 The basic logic gates compared.

time down a single signal path, which is more complex, but convenient for cables between pieces of equipment because the connectors require fewer pins. When a binary system is used to convey numbers in this way, it can be called a digital system.

3.5 Storage elements

The basic memory element in logic circuits is the latch, which is constructed from two gates as shown in Figure 3.11(a), and which can be set or reset. A more useful variant is the D-type latch shown in (b) which

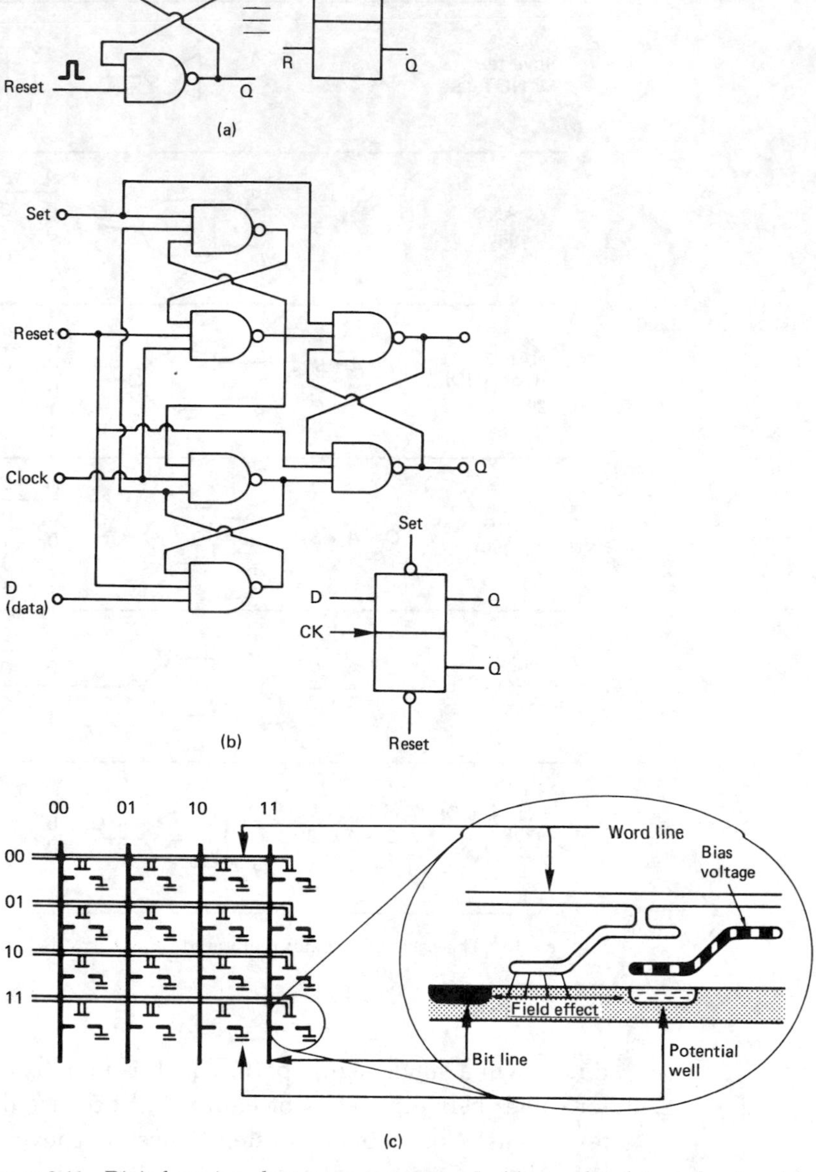

Figure 3.11 Digital semiconductor memory types. In (a), one data bit can be stored in a simple set–reset latch, which has little application because the D-type latch in (b) can store the state of the single data input when the clock occurs. These devices can be implemented with bipolar transistors or FETs, and are called static memories because they can store indefinitely. They consume a lot of power.

In (c), a bit is stored as the charge in a potential well in the substrate of a chip. It is accessed by connecting the bit line with the field effect from the word line. The single well where the two lines cross can then be written or read. These devices are called dynamic RAMs because the charge decays, and they must be read and rewritten (refreshed) periodically.

remembers the state of the input at the time a separate clock either changes state for an edge-triggered device, or after it goes false for a level-triggered device. D-type latches are commonly available with four or eight latches to the chip. A shift register can be made from a series of latches by connecting the Q output of one latch to the D input of the next and connecting all the clock inputs in parallel. Data are delayed by the number of stages in the register. Shift registers are also useful for converting between serial and parallel data transmissions.

Where large numbers of bits are to be stored, cross-coupled latches are less suitable because they are more complicated to fabricate inside integrated circuits than dynamic memory, and consume more current.

In large random access memories (RAMs), the data bits are stored as the presence or absence of charge in a tiny capacitor as shown in Figure 3.11(c). The capacitor is formed by a metal electrode, insulated by a layer of silicon dioxide from a semiconductor substrate, hence the term MOS (metal oxide semiconductor). The charge will suffer leakage, and the value would become indeterminate after a few milliseconds. Where the delay needed is less than this, decay is of no consequence, as data will be read out before they have had a chance to decay. Where longer delays are necessary, such memories must be refreshed periodically by reading the bit value and writing it back to the same place. Most modern MOS RAM chips have suitable circuitry built in. Large RAMs store thousands of bits, and it is clearly impractical to have a connection to each one. Instead, the desired bit has to be addressed before it can be read or written. The size of the chip package restricts the number of pins available, so that large memories use the same address pins more than once. The bits are arranged internally as rows and columns, and the row address and the column address are specified sequentially on the same pins.

3.6 The phase-locked loop

All digital video systems need to be clocked at the appropriate rate in order to function properly. Whilst a clock may be obtained from a fixed frequency oscillator such as a crystal, many operations in video require *genlocking* or synchronizing the clock to an external source. The phase-locked loop excels at this job, and many others, particularly in connection with recording and transmission.

In phase-locked loops, the oscillator can run at a range of frequencies according to the voltage applied to a control terminal. This is called a voltage-controlled oscillator or VCO. Figure 3.12 shows that the VCO is driven by a phase error measured between the output and some reference. The error changes the control voltage in such a way that the error is reduced, such that the output eventually has the same frequency

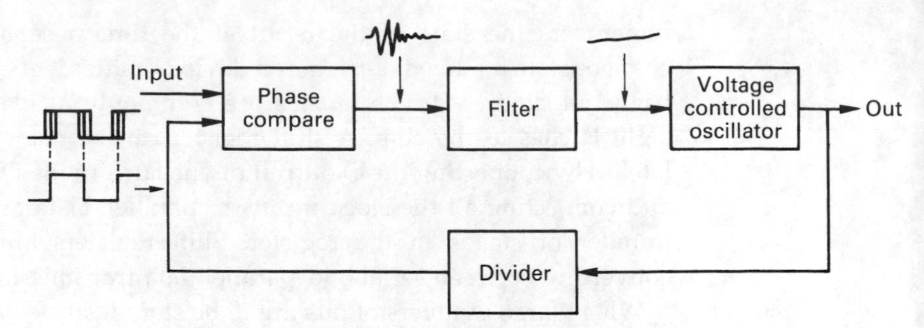

Figure 3.12 A phase-locked loop requires these components as a minimum. The filter in the control voltage serves to reduce clock jitter.

as the reference. A low-pass filter is fitted in the control voltage path to prevent the loop becoming unstable. If a divider is placed between the VCO and the phase comparator, as in the figure, the VCO frequency can be made to be a multiple of the reference. This also has the effect of making the loop more heavily damped, so that it is less likely to change frequency if the input is irregular.

In digital video, the frequency multiplication of a phase-locked loop is extremely useful. Figure 3.13 shows how the 13.5 MHz clock of component digital video is obtained from the sync pulses of an analog reference by such a multiplication process.

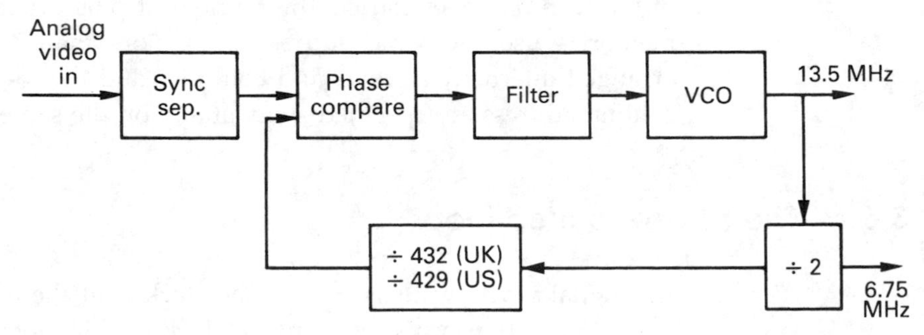

Figure 3.13 In order to obtain 13.5 MHz from input syncs, a PLL with an appropriate division ratio is required.

Figure 3.14 shows the NLL or numerically locked loop. This is similar to a phase-locked loop, except that the two phases concerned are represented by the state of a binary number. The NLL is useful to generate a remote clock from a master. The state of a clock count in the master is periodically transmitted to the NLL which will recreate the same clock frequency. The technique is used in MPEG transport streams.

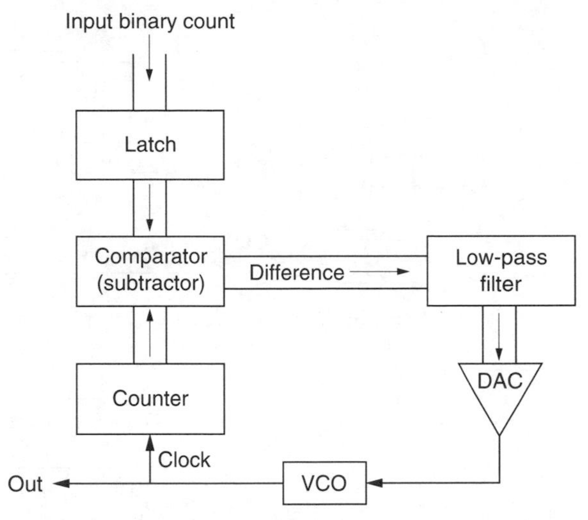

Figure 3.14 The numerically locked loop (NLL) is a digital version of the phase-locked loop.

3.7 Binary adding

The circuitry necessary for adding pure binary or two's complement numbers is shown in Figure 3.15. Addition in binary requires two bits to be taken at a time from the same position in each word, starting at the least significant bit. Should both be ones, the output is zero, and there is a *carry-out* generated. Such a circuit is called a half adder, shown in Figure 3.15(a) and is suitable for the least-significant bit of the calculation. All higher stages will require a circuit which can accept a carry input as well as two data inputs. This is known as a full adder (Figure 3.15(b)). Multibit full adders are available in chip form, and have carry-in and carry-out terminals to allow them to be cascaded to operate on long wordlengths. Such a device is also convenient for inverting a two's complement number, in conjunction with a set of inverters. The adder chip has one set of inputs grounded, and the carry-in permanently held true, such that it adds one to the one's complement number from the inverter.

When mixing by adding sample values, care has to be taken to ensure that if the sum of the two sample values exceeds the number range the result will be clipping rather than wraparound. In two's complement, the action necessary depends on the polarities of the two signals. Clearly if one positive and one negative number are added, the result cannot exceed the number range. If two positive numbers are added, the symptom of positive overflow is that the most significant bit sets, causing an erroneous negative result, whereas a negative overflow results in the most significant bit clearing. The overflow control circuit will be designed

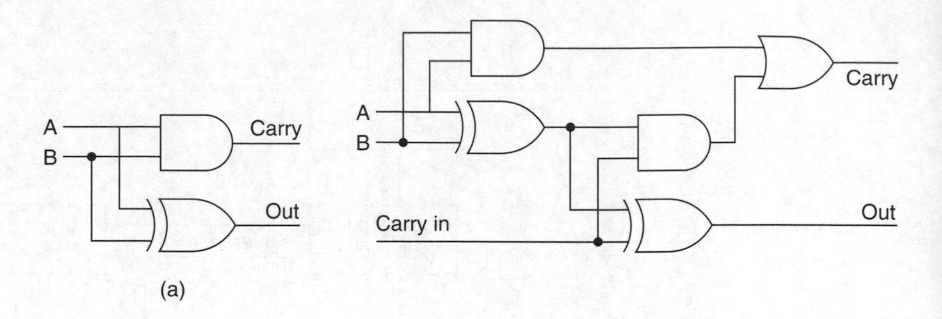

Data A	Bits B	Carry in	Out	Carry out
0	0	0	0	0
0	0	1	1	0
0	1	0	1	0
0	1	1	0	1
1	0	0	1	0
1	0	1	0	1
1	1	0	0	1
1	1	1	1	1

(b)

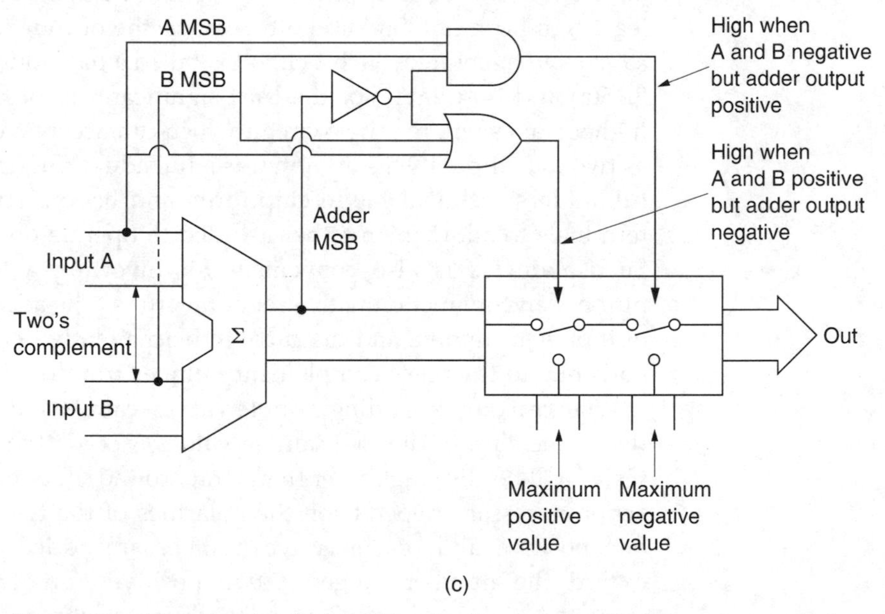

Figure 3.15 (a) Half adder; (b) full-adder circuit and truth table; (c) comparison of sign bits prevents wraparound on adder overflow by substituting clipping level.

to detect these two conditions, and override the adder output. If the MSB of both inputs is zero, the numbers are both positive, thus if the sum has the MSB set, the output is replaced with the maximum positive code (0111 . . .). If the MSB of both inputs is set, the numbers are both negative, and if the sum has no MSB set, the output is replaced with the maximum

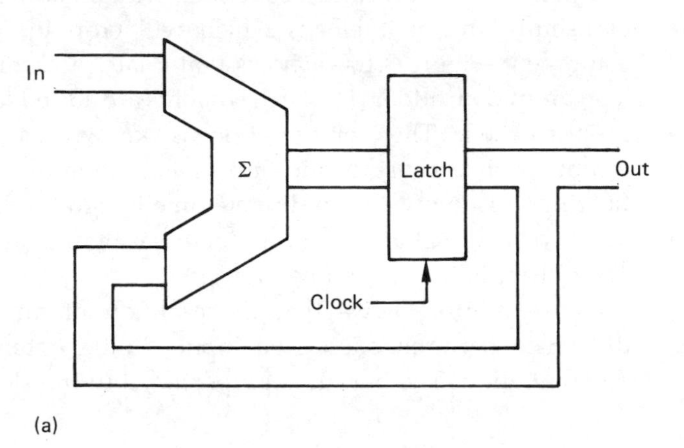

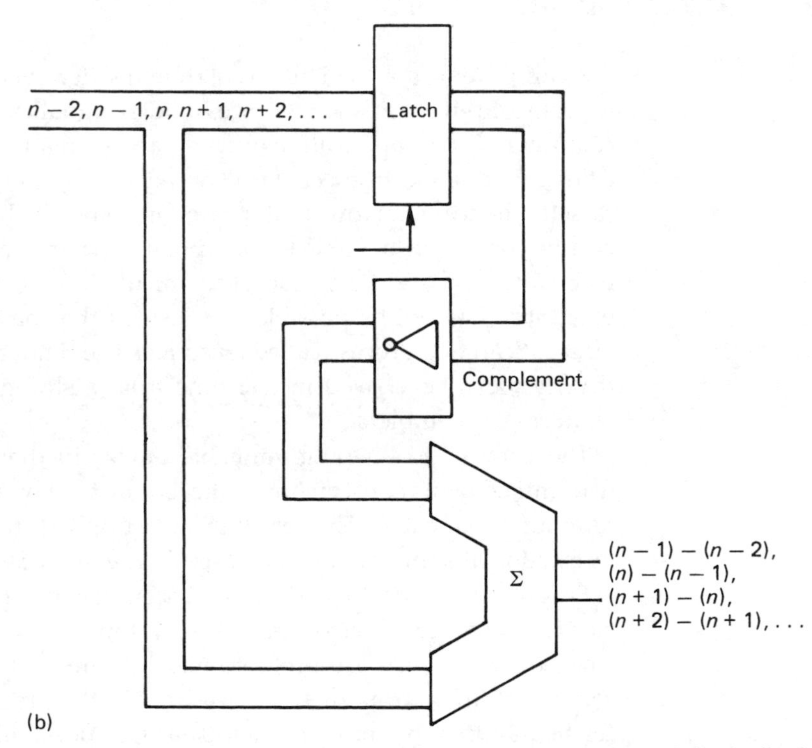

Figure 3.16 Two configurations which are common in processing. In (a) the feedback around the adder adds the previous sum to each input to perform accumulation or digital integration. In (b) an inverter allows the difference between successive inputs to be computed. This is differentiation.

negative code (1000 . . .). These conditions can also be connected to warning indicators. Figure 3.15(c) shows this system in hardware. The resultant clipping on overload is sudden, and sometimes a PROM is included which translates values around and beyond maximum to soft-clipped values below or equal to maximum.

A storage element can be combined with an adder to obtain a number of useful functional blocks which will crop up frequently in audio equipment. Figure 3.16(a) shows that a latch is connected in a feedback loop around an adder. The latch contents are added to the input each time it is clocked. The configuration is known as an accumulator in computation because it adds up or accumulates values fed into it. In filtering, it is known as an discrete time integrator. If the input is held at some constant value, the output increases by that amount on each clock. The output is thus a sampled ramp.

Figure 3.16(b) shows that the addition of an invertor allows the difference between successive inputs to be obtained. This is digital differentiation. The output is proportional to the slope of the input.

3.8 The computer

The computer is now a vital part of digital video systems, being used both for control purposes and to process video signals as data. In control, the computer finds applications in database management, automation, editing, and in electromechanical systems such as tape drives and robotic cassette handling. Now that processing speeds have advanced sufficiently, computers are able to manipulate certain types of digital video in real time. Where very complex calculations are needed, real-time operation may not be possible and instead the computation proceeds as fast as it can in a process called *rendering*. The rendered data are stored so that they can be viewed in real time from a storage medium when the rendering is complete.

The computer is a programmable device in that its operation is not determined by its construction alone, but instead by a series of *instructions* forming a *program*. The program is supplied to the computer one instruction at a time so that the desired sequence of events takes place.

Programming of this kind has been used for over a century in electromechanical devices, including automated knitting machines and street organs which are programmed by punched cards. However, the computer differs from these devices in that the program is not fixed, but can be modified by the computer itself. This possibility led to the creation of the term *software* to suggest a contrast to the constancy of hardware.

Computer instructions are binary numbers each of which is interpreted in a specific way. As these instructions don't differ from any other kind of data, they can be stored in RAM. The computer can change its own instructions by accessing the RAM. Most types of RAM are volatile, in that they lose data when power is removed. Clearly if a program is stored entirely in this way, the computer will not be able to recover fom a power failure. The solution is that a very simple starting or *bootstrap* program is stored in non-volatile ROM which will contain instructions to bring in the main program from a storage system such as a disk drive after power is applied. As programs in ROM cannot be altered, they are sometimes referred to as *firmware* to indicate that they are classified between hardware and software.

Making a computer do useful work requires more than simply a program which performs the required computation. There is also a lot of mundane activity which does not differ significantly from one program to the next. This includes deciding which part of the RAM will be occupied by the program and which by the data, producing commands to the storage disk drive to read the input data from a file and write back the results. It would be very inefficient if all programs had to handle these processes themselves. Consequently the concept of an *operating system* was developed. This manages all the mundane decisions and creates an environment in which useful programs or *applications* can execute.

The ability of the computer to change its own instructions makes it very powerful, but it also makes it vulnerable to abuse. Programs exist which are deliberately written to do damage. These *viruses* are generally attached to plausible messages or data files and enter computers through storage media or communications paths.

There is also the possibility that programs contain logical errors such that in certain combinations of circumstances the wrong result is obtained. If this results in the unwitting modification of an instruction, the next time that instruction is accessed the computer will crash. In consumer grade software, written for the vast personal computer market, this kind of thing is unfortunately accepted.

For critical applications, software must be *verified*. This is a process which can prove that a program can recover from absolutely every combination of circumstances and keep running properly. This is a non-trivial process, because the number of combinations of states a computer can get into is staggering. As a result most software is unverified.

It is of the utmost importance that networked computers which can suffer virus infection or computers running unverified software are never used in a life-support or critical application.

Figure 3.17 shows a simple computer system. The various parts are linked by a bus which allows binary numbers to be transferred from one place to another. This will generally use tri-state logic (see section 3.4) so that when one device is sending to another, all other devices present a high impedance to the bus.

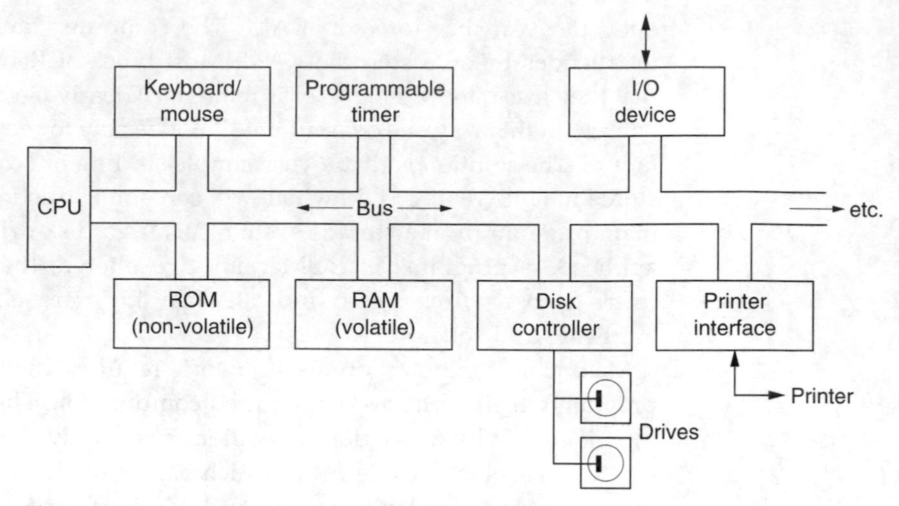

Figure 3.17 A simple computer system. All components are linked by a single data/address/control bus. Although cheap and flexible, such a bus can only make one connection at a time, so it is slow.

The ROM stores the start-up program, the RAM stores the operating system, applications programs and the data to be processed. The disk drive stores large quantities of data in a non-volatile form. The RAM only needs to be able to hold part of one program as other parts can be brought from the disk as required. A program executes by *fetching* one instruction at a time from the RAM to the processor along the bus.

The bus also allows keyboard/mouse inputs and outputs to the display and printer. Inputs and outputs are generally abbreviated to I/O. Finally a programmable timer will be present which acts as a kind of alarm clock for the processor.

3.9 The processor

The processor or CPU (central processing unit) is the heart of the system. Figure 3.18 shows a simple example of a CPU. The CPU has a bus interface which allows it to generate bus addresses and input or output data. Sequential instructions are stored in RAM at contiguously increasing locations so that a program can be executed by fetching instructions from a RAM address specified by the program counter (PC) to the instruction register in the CPU. As each instruction is completed, the PC is incremented so that it points to the next instruction. In this way the time taken to execute the instruction can vary.

The processor is notionally divided into data paths and control paths. Figure 3.18 shows the data path. The CPU contains a number of general-

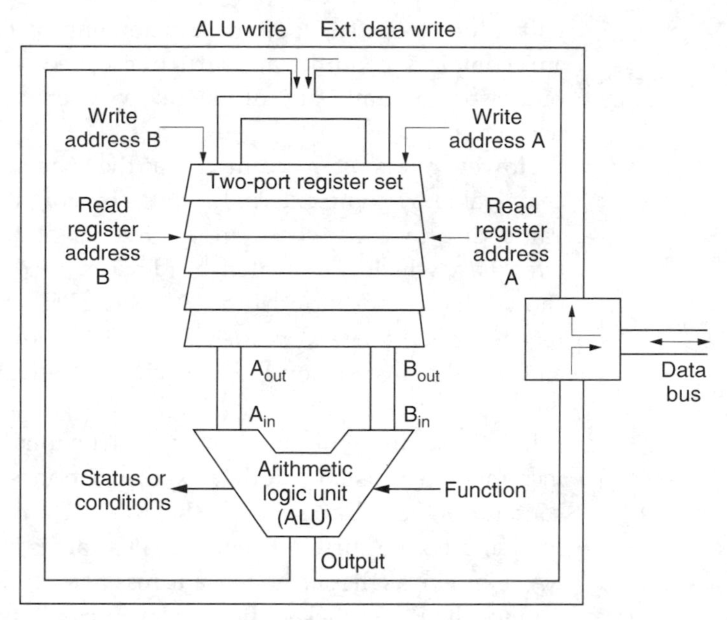

Figure 3.18 The data path of a simple CPU. Under control of an instruction, the ALU will perform some function on a pair of input values from the registers and store or output the result.

purpose registers or scratchpads which can be used to store partial results in complex calculations. Pairs of these registers can be addressed so that their contents go to the ALU (arithmetic logic unit). This performs various arithmetic (add, subtract, etc.) or logical (and, or, etc.) functions on the input data. The output of the ALU may be routed back to a register or output. By reversing this process it is possible to get data into the registers from the RAM. The ALU also outputs the conditions resulting from the calculation, which can control conditional instructions.

Which function the ALU performs and which registers are involved are determined by the instruction currently in the instruction register that is decoded in the control path. One pass through the ALU can be completed in one cycle of the processor's clock. Instructions vary in complexity as do the number of clock cycles needed to complete them. Incoming instructions are decoded and used to access a look-up table which converts them into *microinstructions*, one of which controls the CPU at each clock cycle.

3.10 Interrupts

Ordinarily instructions are executed in the order that they are stored in RAM. However, some instructions direct the processor to jump to a new memory location. If this is a jump to an earlier instruction, the program will enter a loop. The loop must increment a count in a register each time, and contain a conditional instruction called a branch, which allows the processor to jump out of the loop when a predetermined count is reached.

However, it is often required that the sequence of execution should be changeable by some external event. This might be the changing of some value due to a keyboard input. Events of this kind are handled by *interrupts*, which are created by devices needing attention. Figure 3.19 shows that in addition to the PC, the CPU contains another dedicated register called the *stack pointer*. Figure 3.20 shows how this is used. At the end of every instruction the CPU checks to see if an interrupt is asserted on the bus.

If it is, a different set of microinstructions are executed. The PC is incremented as usual, but the next instruction is not executed. Instead, the contents of the PC are stored so that the CPU can resume execution when it has handled the current event. The PC state is stored in a reserved area of RAM known as the *stack*, at an address determined by the stack pointer.

Once the PC is stacked, the processor can handle the interrupt. It issues a bus interrupt acknowledge, and the interrupting device replies with an unique code identifying itself. This is known as a *vector* which steers the processor to a RAM address containing a new program counter. This is the RAM address of the first instruction of the *subroutine* which is the program that will handle the interrupt. The CPU loads this address into the PC and begins execution of the subroutine.

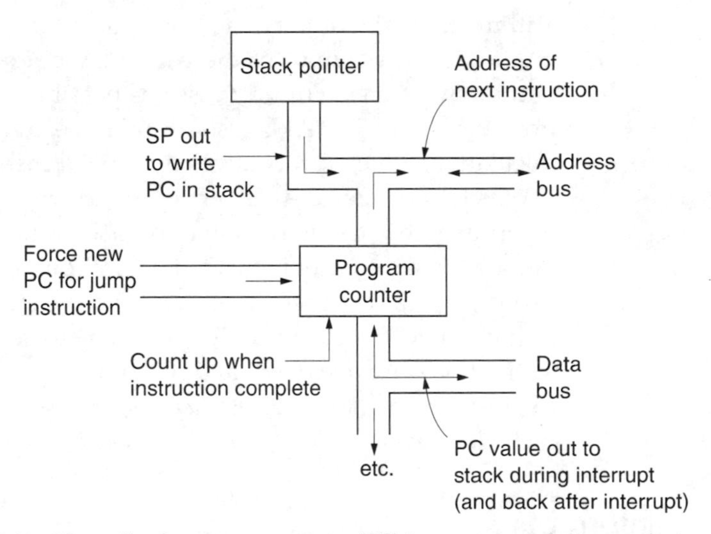

Figure 3.19 Normally the program counter (PC) increments each time an instruction is completed in order to select the next instruction. However, an interrupt may cause the PC state to be stored in the stack area of RAM prior to the PC being forced to the start address of the interrupt subroutine. Afterwards the PC can get its original value back by reading the stack.

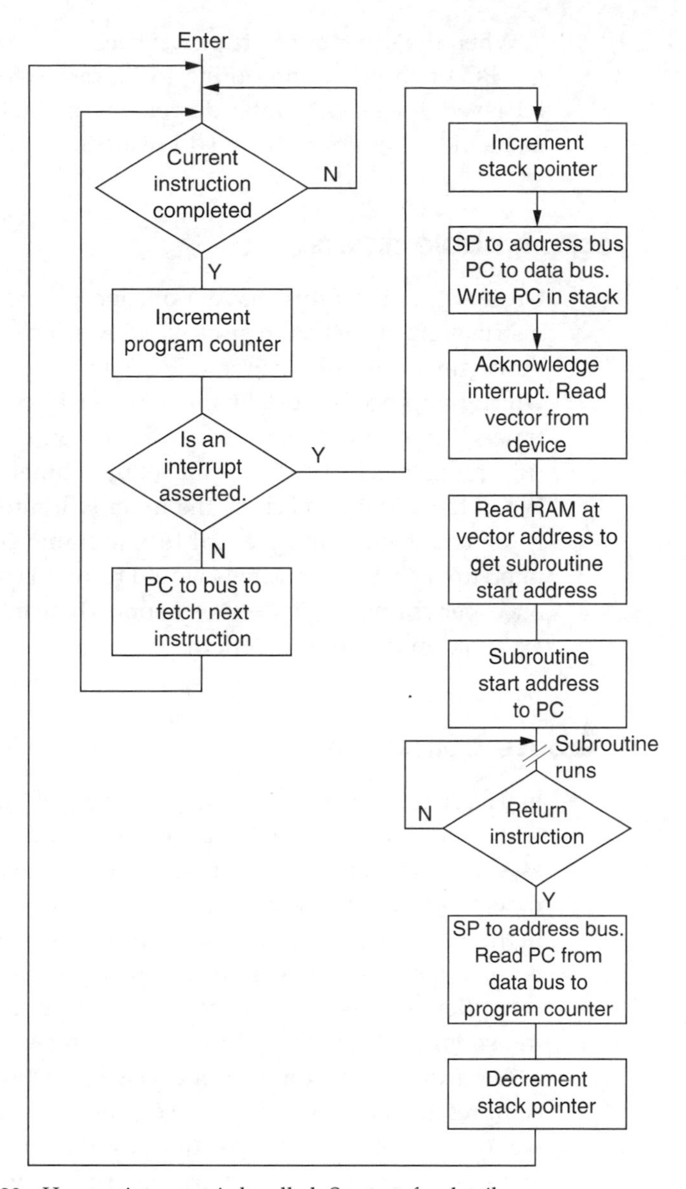

Figure 3.20 How an interrupt is handled. See text for details.

At the end of the subroutine there will be a return instruction. This causes the CPU to use the stack pointer as a memory address in order to read the return PC state from the stack. With this value loaded into the PC, the CPU resumes execution where it left off.

The stack exists so that subroutines can themselves be interrupted. If a subroutine is executing when a higher-priority interrupt occurs, the subroutine can be suspended by incrementing the stack pointer and storing the current PC in the next location in the stack.

When the second interrupt has been serviced, the stack pointer allows the PC of the first subroutine to be retrieved. Whenever a stack PC is retrieved, the stack pointer decrements so that it always points to the PC of the next item of unfinished business.

3.11 Programmable timers

Ordinarily processors have no concept of time and simply execute instructions as fast as their clock allows. This is fine for general-purpose processing, but not for time-critical processes such as video. One way in which the processor can be made time conscious is to use programmable timers. These are devices which reside on the computer bus and which run from a clock. The CPU can set up a timer by loading it with a count. When the count is reached, the timer will interrupt. To give an example, if the count were to be equal to one frame period, there would be one interrupt per frame, and this would result in the execution of a subroutine once per frame, provided, of course, that all the instructions could be executed in one frame period.

3.12 Timebase correction

In Chapter 1 it was stated that a strength of digital technology is the ease with which delay can be provided. Accurate control of delay is the essence of timebase correction, necessary whenever the instantaneous time of arrival or rate from a data source does not match the destination. In digital video, the destination will almost always have perfectly regular timing, namely the sampling rate clock of the final DAC. Timebase correction consists of aligning jittery signals from storage media or transmission channels with that stable reference.

A further function of timebase correction is to reverse the time compression applied prior to recording or transmission. As was shown in section 1.8, digital recorders compress data into blocks to facilitate editing and error correction as well as to permit head switching between blocks in rotary-head machines. Owing to the spaces between blocks, data arrive in bursts on replay, but must be fed to the output convertors in an unbroken stream at the sampling rate.

In computer hard-disk drives, which are used in digital video workstations, time compression is also used, but a converse problem also arises. Data from the disk blocks arrive at a reasonably constant rate, but cannot necessarily be accepted at a steady rate by the logic because of contention for the use of buses and memory by the different parts of the system. In this case the data must be buffered by a relative of the timebase corrector which is usually referred to as a silo.

Although delay is easily implemented, it is not possible to advance a data stream. Most real machines cause instabilities balanced about the correct timing: the output jitters between too early and too late. Since the information cannot be advanced in the corrector, only delayed, the solution is to run the machine in advance of real time. In this case, correctly timed output signals will need a nominal delay to align them with reference timing. Early output signals will receive more delay, and late output signals will receive less delay.

Section 3.5 showed the principles of digital storage elements which can be used for delay purposes. The shift-register approach and the RAM approach to delay are very similar, as a shift register can be thought of as a memory whose address increases automatically when clocked. The data rate and the maximum delay determine the capacity of the RAM required. Figure 3.21 shows that the addressing of the RAM is by a

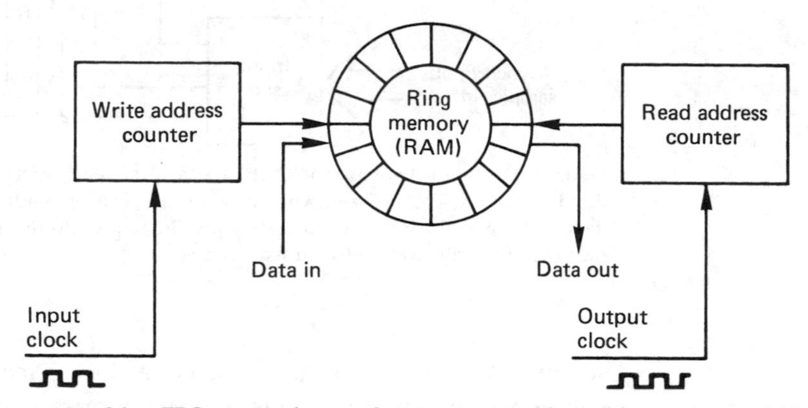

Figure 3.21 Most TBCs are implemented as a memory addressed by a counter which periodically overflows to give a ring structure. The memory allows the read and write sides to be asynchronous.

counter that overflows endlessly from the end of the memory back to the beginning, giving the memory a ring-like structure. The write address is determined by the incoming data, and the read address is determined by the outgoing data. This means that the RAM has to be able to read and write at the same time. The switching between read and write involves not only a data multiplexer but also an address multiplexer. In general, the arbitration between read and write will be done by signals from the stable side of the TBC as Figure 3.22 shows. In the replay case the stable clock will be on the read side. The stable side of the RAM will read a sample when it demands, and the writing will be locked out for that period. However, the input data cannot be interrupted in many applications so a small buffer silo is installed before the memory, which fills up as the writing is locked out, and empties again as writing is permitted. Alternatively, the memory will be split into blocks as was

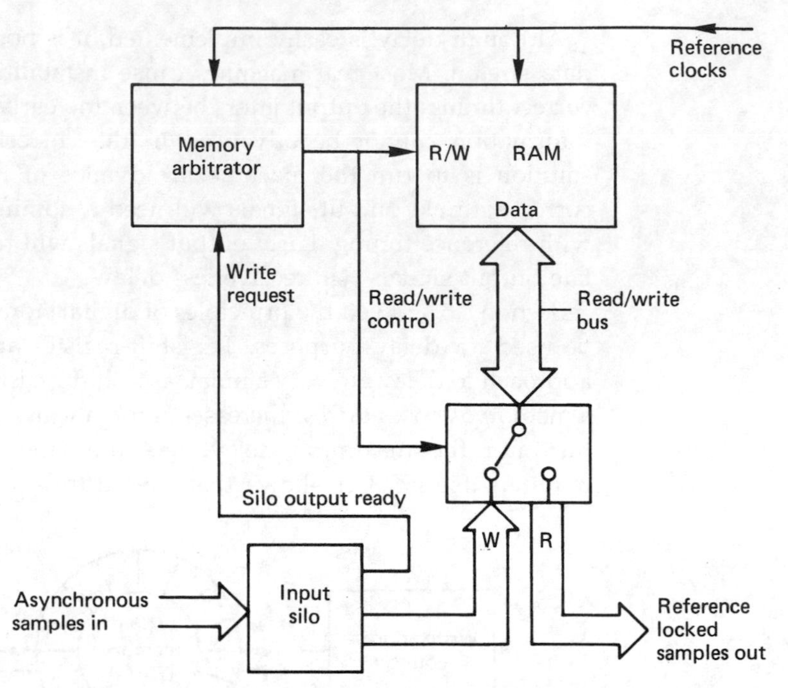

Figure 3.22 In a RAM-based TBC, the RAM is reference synchronous, and an arbitrator decides when it will read and when it will write. During reading, asynchronous input data back up in the input silo, asserting a write request to the arbitrator. Arbitrator will then cause a write cycle between read cycles.

shown in Chapter 1, such that when one block is reading a different block will be writing and the problem does not arise.

In most digital video applications, the sampling rate exceeds the rate at which economically available RAM chips can operate. The solution is to arrange several video samples into one longer word, known as a superword, and to construct the memory so that it stores superwords in parallel.

Figure 3.23 shows the operation of a FIFO chip, colloquially known as a silo because the data are tipped in at the top on delivery and drawn off at the bottom when needed. Each stage of the chip has a data register and a small amount of logic, including a data-valid or V bit. If the input register does not contain data, the first V bit will be reset, and this will cause the chip to assert 'input ready'. If data are presented at the input, and clocked into the first stage, the V bit will set, and the 'input ready' signal will become false. However, the logic associated with the next stage sees the V bit set in the top stage, and if its own V bit is clear, it will clock the data into its own register, set its own V bit, and clear the input V bit, causing 'input ready' to reassert, when another word can be fed in. This process then continues as the word moves down the silo, until it

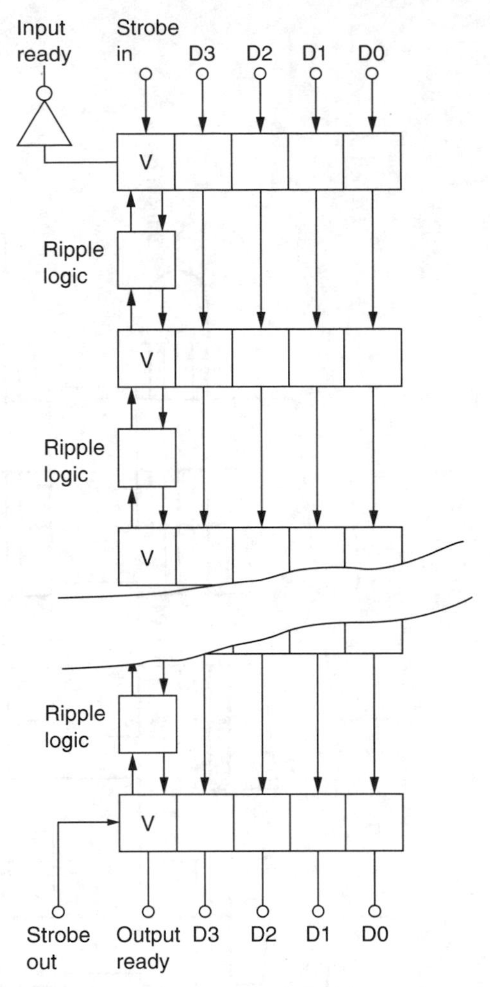

Figure 3.23 Structure of FIFO or silo chip. Ripple logic controls propagation of data down silo.

arrives at the last register in the chip. The V bit of the last stage becomes the 'output ready' signal, telling subsequent circuitry that there are data to be read. If this word is not read, the next word entered will ripple down to the stage above. Words thus stack up at the bottom of the silo. When a word is read out, an external signal must be provided which resets the bottom V bit. The 'output ready' signal now goes false, and the logic associated with the last stage now sees valid data above, and loads down the word when it will become ready again. The last register but one will now have no V bit set, and will see data above itself and bring that down. In this way a reset V bit propagates up the chip while the data ripple down, rather like a hole in a semiconductor going the opposite way to the electrons. Silo chips are usually available in four-bit

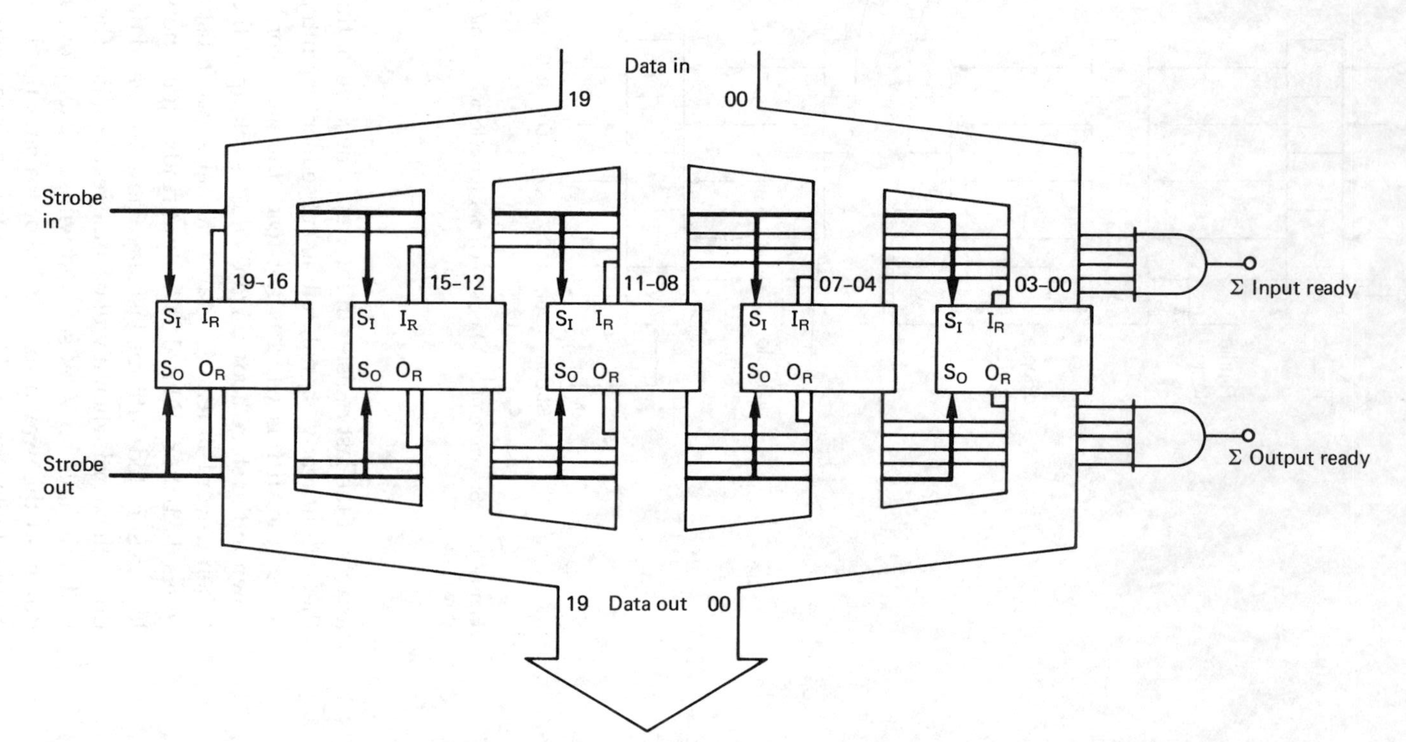

Figure 3.24 In this example, a twenty-bit wordlength silo is made from five parallel FIFO chips. The asynchronous ripple action of FIFOs means that it is necessary to 'AND' together the ready signals.

wordlengths, but can easily be connected in parallel to form superwords. Silo chips are asynchronous, and paralleled chips will not necessarily all work at the same speed. This problem is easily overcome by 'anding' together all the input-ready and output-ready signals and parallel-connecting the strobes. Figure 3.24 shows this mode of operation.

When used in a hard-disk system, a silo will allow data to and from the disk, which is turning at constant speed. When reading the disk, Figure 3.25(a) shows that the silo starts empty, and if there is bus contention, the silo will start to fill. Where the bus is free, the disk controller will attempt to empty the silo into the memory. The system can take advantage of the interblock gaps on the disk, containing headers, preambles and redundancy, for in these areas there are no data to transfer, and there is some breathing space to empty the silo before the next block. In practice the silo need not be empty at the start of every block, provided it never becomes full before the end of the transfer. If this happens some data are lost and the function must be aborted. The block containing the silo overflow will generally be re-read on the next revolution. In sophisticated systems, the

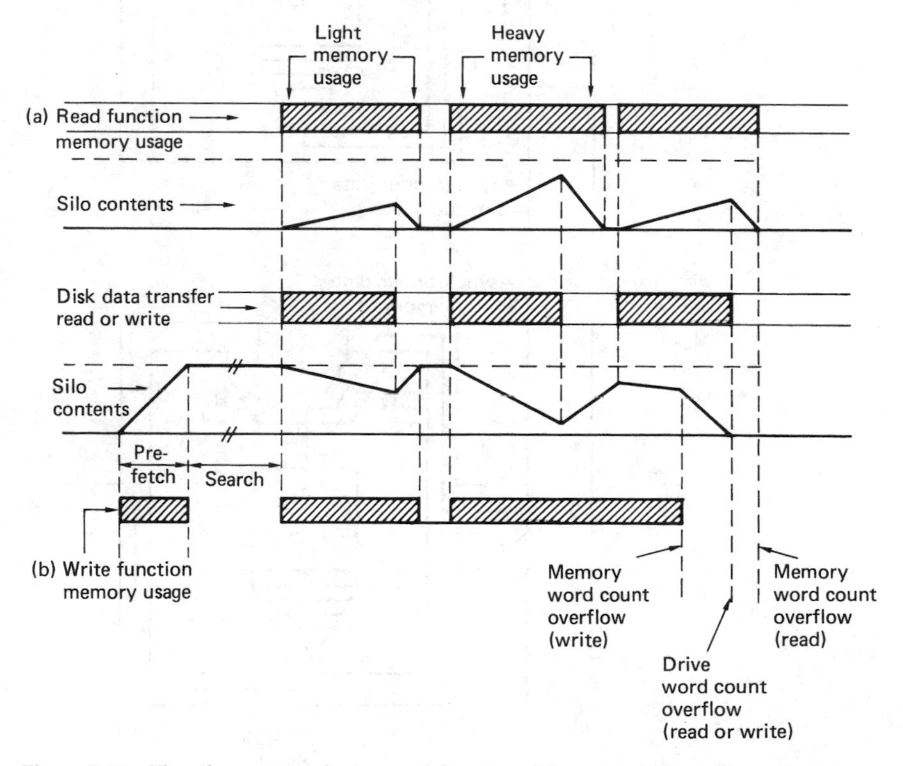

Figure 3.25 The silo contents during read functions (a) appear different from those during write functions (b). In (a), the control logic attempts to keep the silo as empty as possible; in (b) the logic prefills the silo and attempts to keep it full until the memory word count overflows.

silo has a kind of dipstick, and can interrupt the CPU if the data get too deep. The CPU can then suspend some bus activity to allow the disk controller more time to empty the silo.

When the disk is to be written, as in Figure 3.25(b), a continuous data stream must be provided during each block, as the disk cannot stop. The silo will be pre-filled before the disk attempts to write, and the disk controller attempts to keep it full. In this case all will be well if the silo does not become empty before the end of the transfer. Figure 3.26 shows the silo of a typical disk controller with the multiplexers necessary to put it in the read data stream or the write data stream.

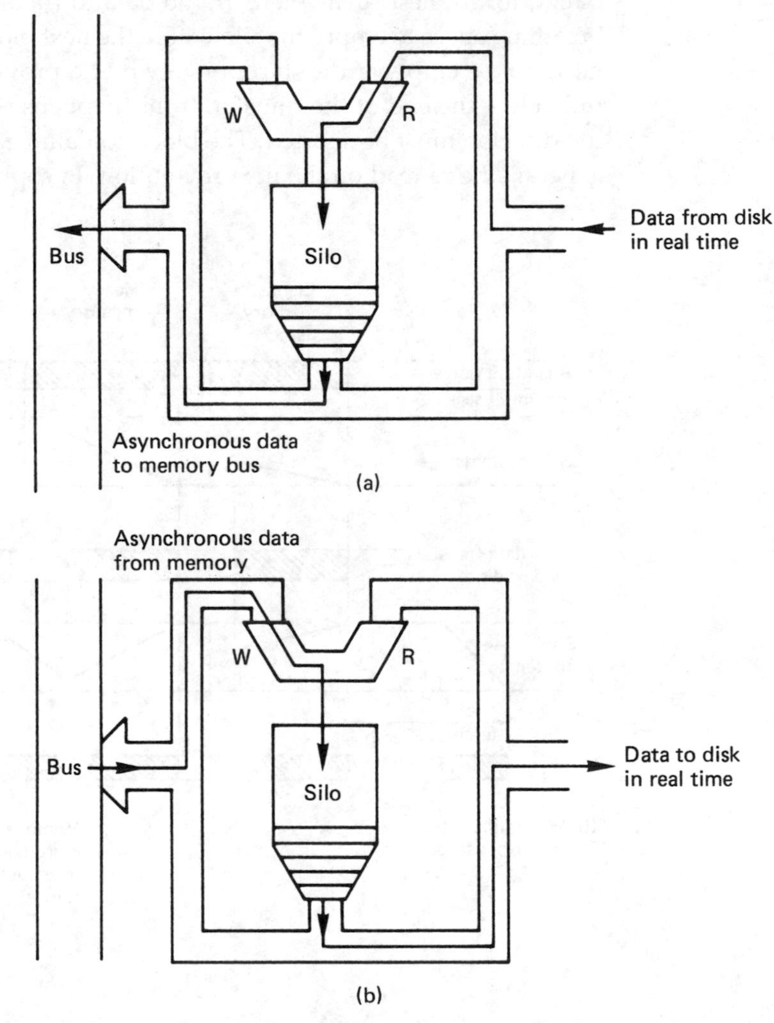

Figure 3.26 In order to guarantee that the drive can transfer data in real time at regular intervals (determined by disk speed and density) the silo provides buffering to the asynchronous operation of the memory access process. At (a) the silo is configured for a disk read. The same silo is used at (b) for a disk write.

3.13 Multiplexing principles

Multiplexing is used where several signals are to be transmitted down the same channel. The channel bit rate must be the same as or greater than the sum of the source bit rates. Figure 3.27 shows that when multiplexing is used, the data from each source has to be time compressed. This is done by buffering source data in a memory at the multiplexer. It is written into the memory in real time as it arrives, but will be read from the memory with a clock which has a much higher rate. This means that the readout occurs in a smaller timespan. If, for example, the clock frequency is raised by a factor of ten, the data for a given signal will be transmitted in a tenth of the normal time, leaving time in the multiplex for nine more such signals.

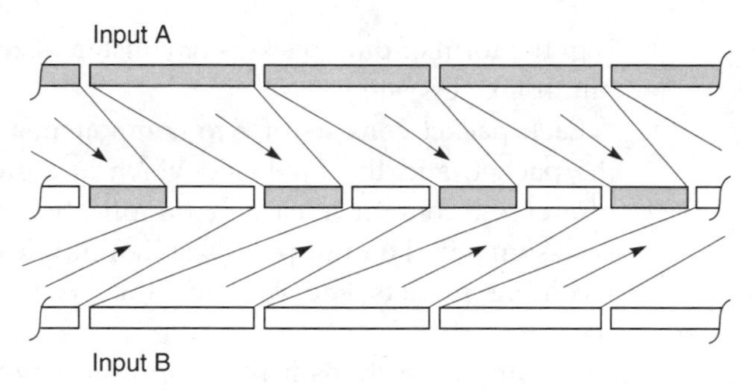

Figure 3.27 Multiplexing requires time compression on each input.

In the demultiplexer another buffer memory will be required. Only the data for the selected signal will be written into this memory at the bit rate of the multiplex. When the memory is read at the correct speed, the data will emerge with its original timebase.

In practice it is essential to have mechanisms to identify the separate signals to prevent them being mixed up and to convey the original signal clock frequency to the demultiplexer. In time-division multiplexing the timebase of the transmission is broken into equal slots, one for each signal. This makes it easy for the demultiplexer, but forces a rigid structure on all the signals such that they must all be locked to one another and have an unchanging bit rate. Packet multiplexing overcomes these limitations.

3.14 Packets

The multiplexer must switch between different time-compressed signals to create the bitstream and this is much easier to organize if each signal

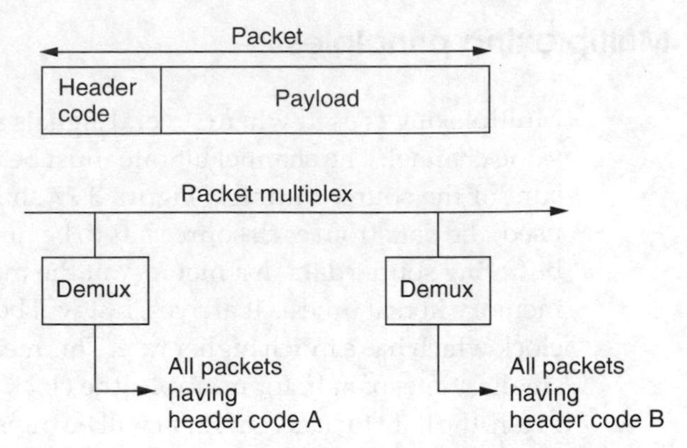

Figure 3.28 Packet multiplexing relies on headers to identify the packets.

is in the form of data packets of constant size. Figure 3.28 shows a packet multiplexing system.

Each packet consists of two components: the header, which identifies the packet, and the payload, which is the data to be transmitted. The header will contain at least an identification code (ID) which is unique for each signal in the multiplex. The demultiplexer checks the ID codes of all incoming packets and discards those which do not have the wanted ID.

In complex systems it is common to have a mechanism to check that packets are not lost or repeated. This is the purpose of the packet continuity count which is carried in the header. For packets carrying the same ID, the count should increase by one from one packet to the next. Upon reaching the maximum binary value, the count overflows and recommences.

3.15 Statistical multiplexing

Packet multiplexing has advantages over time-division multiplexing because it does not set the bit rate of each signal. A demultiplexer simply checks packet IDs and selects all packets with the wanted code. It will do this however frequently such packets arrive. Consequently it is practicable to have variable bit rate signals in a packet multiplex. The multiplexer has to ensure that the total bit rate does not exceed the rate of the channel, but that rate can be allocated arbitrarily between the various signals.

As a practical matter is is usually necessary to keep the bit rate of the multiplex constant. With variable rate inputs this is done by creating null packets which are generally called *stuffing* or *packing*. The headers of these packets contain an unique ID which the demultiplexer does not recognize and so these packets are discarded on arrival.

In an MPEG environment, statistical multiplexing can be extremely useful because it allows for the varying difficulty of real program material. In a multiplex of several television programs, it is unlikely that all the programs will encounter difficult material simultaneously. When one program encounters a detailed scene or frequent cuts which are hard to compress, more data rate can be allocated at the allowable expense of the remaining programs which are handling easy material.

3.16 Gain control by multiplication

When making a digital recording, the gain of the analog input will usually be adjusted so that the quantizing range is fully exercised in order to make a recording of maximum signal-to-noise ratio. During post-production, the recording may be played back and mixed with other signals, and the desired effect can only be achieved if the level of each can be controlled independently. Gain is controlled in the digital domain by multiplying each sample value by a coefficient. If that coefficient is less than one, attenuation will result; if it is greater than one, amplification can be obtained.

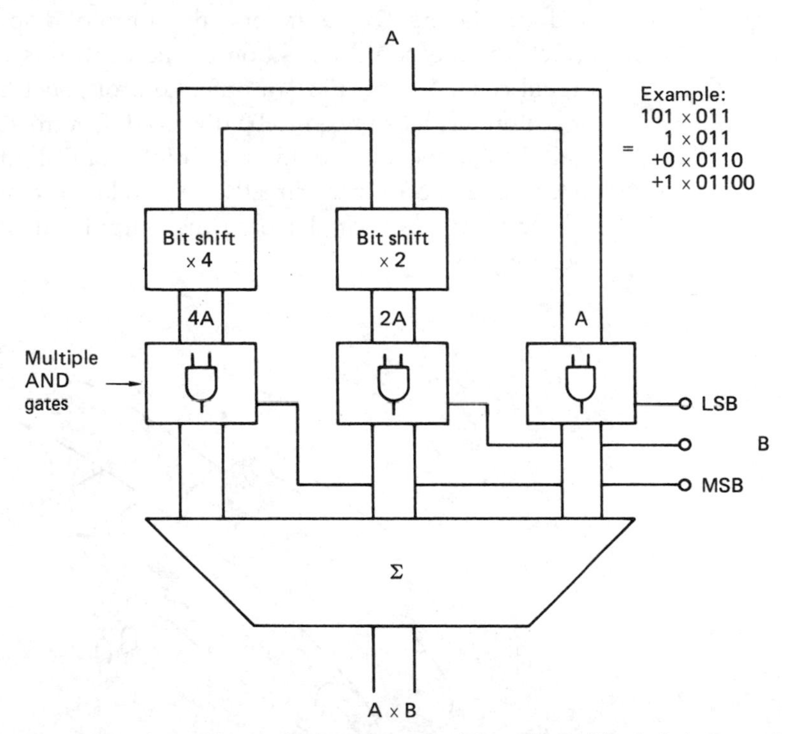

Figure 3.29 Structure of fast multiplier: the input A is multiplied by 1, 2, 4, 8, etc., by bit shifting. The digits of the B input then determine which multiples of A should be added together by enabling AND gates between the shifters and the adder. For long wordlengths, the number of gates required becomes enormous, and the device is best implemented in a chip.

Multiplication in binary circuits is difficult. It can be performed by repeated adding, but this is too slow to be of any use. In fast multiplication, one of the inputs will simultaneously be multiplied by one, two, four, etc., by hard-wired bit shifting. Figure 3.29 shows that the other input bits will determine which of these powers will be added to produce the final sum, and which will be neglected. If multiplying by five, the process is the same as multiplying by four, multiplying by one, and adding the two products. This is achieved by adding the input to itself shifted two places. As the wordlength of such a device increases, the complexity increases exponentially, so this is a natural application for an integrated circuit. It is probably true that digital video would not have been viable without such chips.

3.17 Digital faders and controls

In a digital mixer, the gain coefficients will originate in hand-operated faders, just as in analog. Analog faders may be retained and used to produce a varying voltage which is converted to a digital code or gain coefficient in an ADC, but it is also possible to obtain coefficients directly in digital faders. Digital faders are a form of displacement transducer in which the mechanical position of the control is converted directly to a digital code. The position of other controls, such as jog wheels on VTRs or editors, will also need to be digitized. Controls can be linear or rotary, and absolute or relative. In an absolute control, the position of the knob determines the output directly. In a relative control, the knob can be moved to increase or decrease the output, but its absolute position is meaningless.

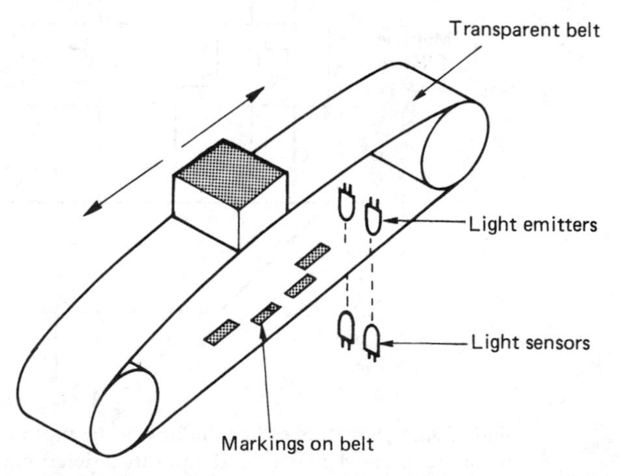

Figure 3.30 An absolute linear fader uses a number of light beams which are interrupted in various combinations according to the position of a grating. A Gray code shown in Figure 3.31 must be used to prevent false codes.

Figure 3.30 shows an absolute linear fader. A grating is moved with respect to several light beams, one for each bit of the coefficient required. The interruption of the beams by the grating determines which photocells are illuminated. It is not possible to use a pure binary pattern on the grating because this results in transient false codes due to mechanical tolerances. Figure 3.31 shows some examples of these false codes. For example, on moving the fader from 3 to 4, the MSB goes true slightly

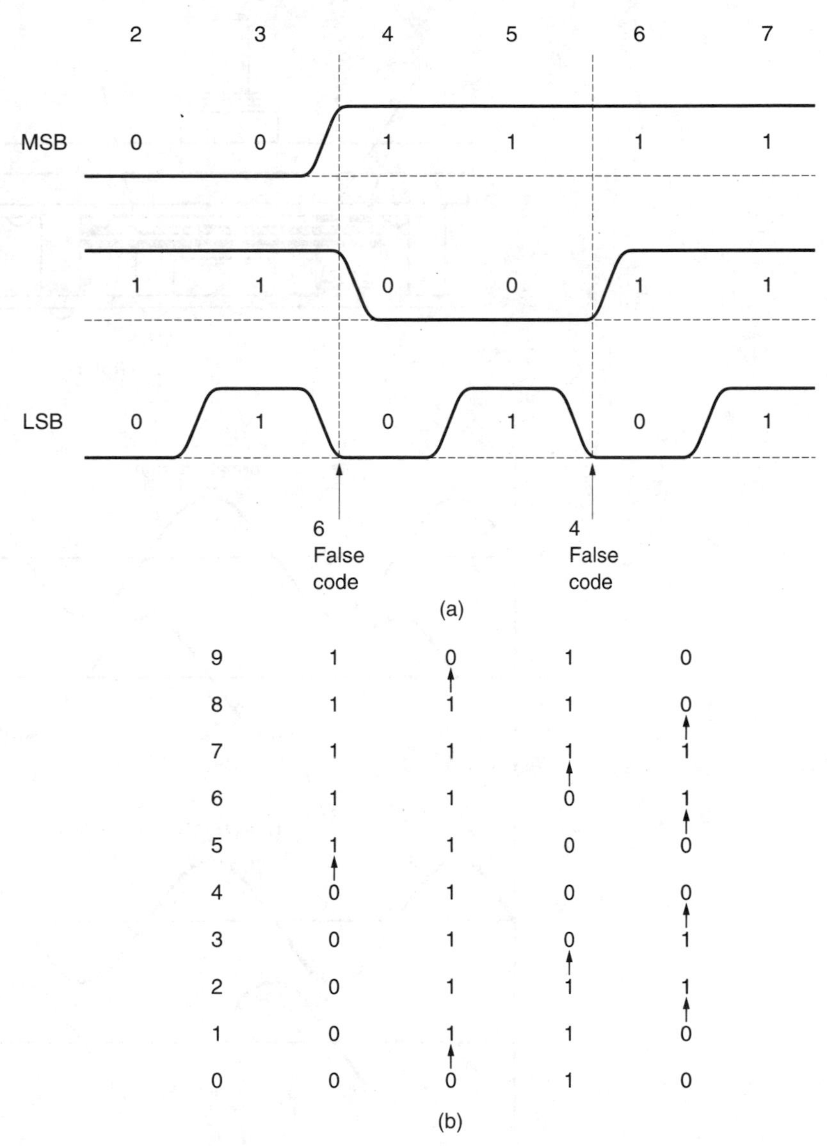

Figure 3.31 (a) Binary cannot be used for position encoders because mechanical tolerances cause false codes to be produced. (b) In Gray code, only one bit (arrowed) changes in between positions, so no false codes can be generated.

before the middle bit goes false. This results in a momentary value of 4 + 2 = 6 between 3 and 4. The solution is to use a code in which only one bit ever changes in going from one value to the next. One such code is the Gray code which was devised to overcome timing hazards in relay logic but is now used extensively in position encoders.

Gray code can be converted to binary in a suitable PROM or gate array. These are available as industry-standard components.

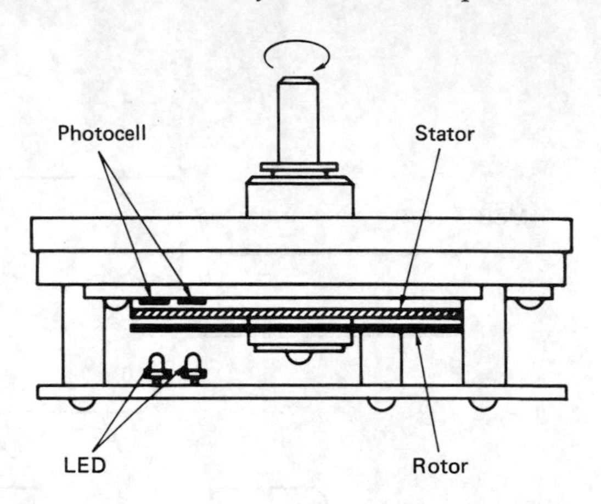

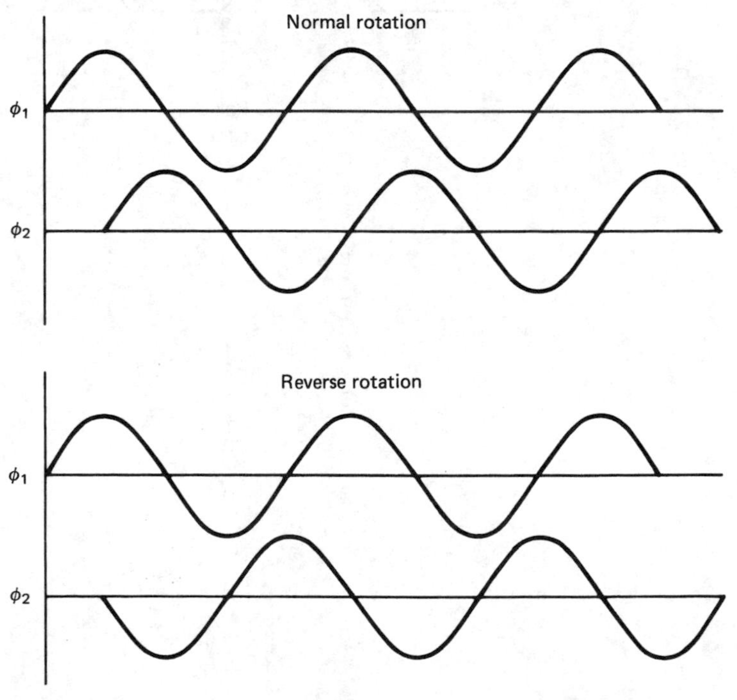

Figure 3.32 The fixed and rotating gratings produce moiré fringes which are detected by two light paths as quadrature sinusoids. The relative phase determines the direction, and the frequency is proportional to speed of rotation.

Figure 3.32 shows a rotary incremental encoder. This produces a sequence of pulses whose number is proportional to the angle through which it has been turned. The rotor carries a radial grating over its entire perimeter. This turns over a second fixed radial grating whose bars are not parallel to those of the first grating. The resultant moiré fringes travel inward or outward depending on the direction of rotation. Two suitably positioned light beams falling on photocells will produce outputs in quadrature. The relative phase determines the direction and the frequency is proportional to speed. The encoder outputs can be connected to a counter whose contents will increase or decrease according to the direction the rotor is turned. The counter provides the coefficient output.

The wordlength of the gain coefficients requires some thought as they determine the number of discrete gains available. If the coefficient wordlength is inadequate, the gain control becomes 'steppy' particularly towards the end of a fadeout. A compromise between performance and the expense of high-resolution faders is to insert a digital interpolator having a low-pass characteristic between the fader and the gain control stage. This will compute intermediate gains to higher resolution than the coarse fader scale so that the steps cannot be discerned.

3.18 Filters

Filtering is inseparable from digital video and audio. Analog or digital filters, and sometimes both, are required in ADCs, DACs, in the data channels of digital recorders and transmission systems and in sampling rate convertors and equalizers. Optical systems used in disk recorders also act as filters.[1] There are many parallels between analog, digital and optical filters, which this section treats as a common subject. The main difference between analog and digital filters is that in the digital domain very complex architectures can be constructed at low cost in LSI and that arithmetic calculations are not subject to component tolerance or drift.

Filtering may modify the frequency response of a system, and/or the phase response. Every combination of frequency and phase response determines the impulse response in the time domain. Figure 3.33 shows that impulse response testing tells a great deal about a filter. In a perfect filter, all frequencies should experience the same delay. If some groups of frequencies experience a different delay to others, there is a group-delay error. As an impulse contains an infinite spectrum, a filter suffering from group-delay error will separate the different frequencies of an impulse along the time axis. A pure delay will cause a phase shift proportional to frequency, and a filter with this characteristic is said to be phase-linear. The impulse response of a phase-linear filter is symmetrical. If a filter

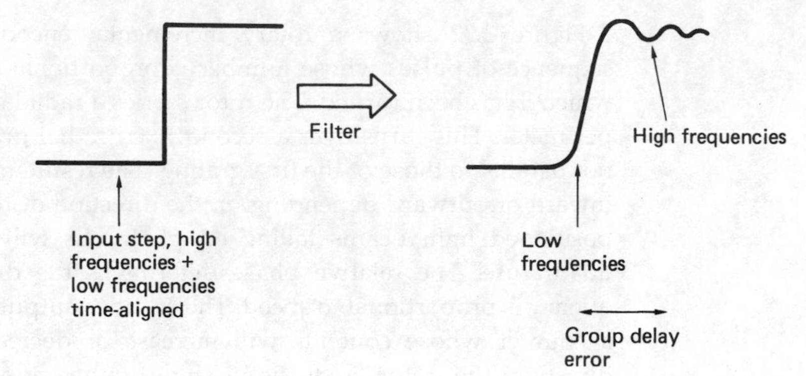

Figure 3.33 Group delay time-displaces signals as a function of frequency.

suffers from group-delay error it cannot be phase-linear. It is almost impossible to make a perfectly phase-linear analog filter, and many filters have a group-delay equalization stage following them which is often as complex as the filter itself. In the digital domain it is straightforward to make a phase-linear filter, and phase equalization becomes unnecessary.

Because of the sampled nature of the signal, whatever the response at low frequencies may be, all digital channels (and sampled analog channels) act as low-pass filters cutting off at the Nyquist limit, or half the sampling frequency.

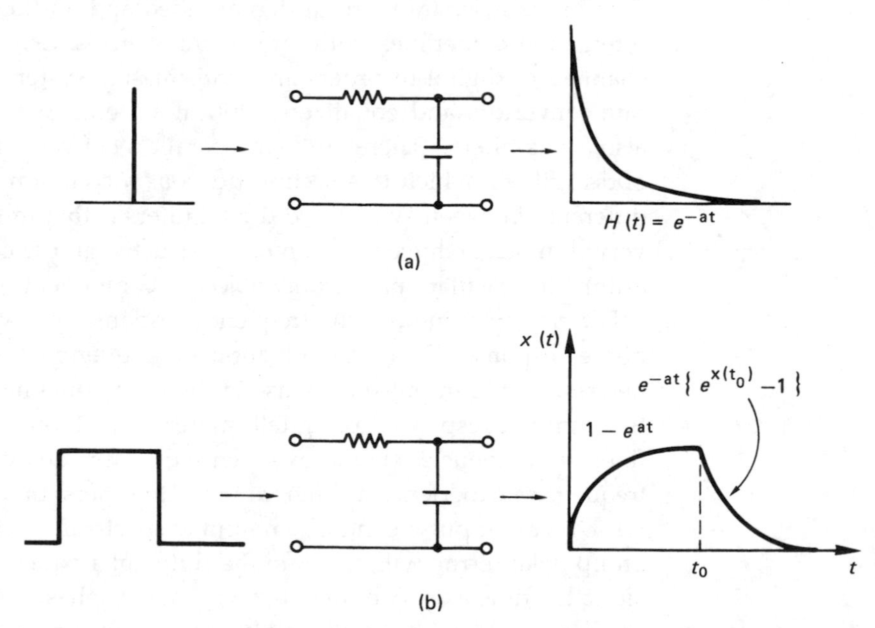

Figure 3.34 (a) The impulse response of a simple RC network is an exponential decay. This can used to calculate the response to a square wave, as in (b).

Figure 3.34(a) shows a simple RC network and its impulse response. This is the familiar exponential decay due to the capacitor discharging through the resistor (in series with the source impedance which is assumed here to be negligible). The figure also shows the response to a squarewave at (b). These responses can be calculated because the inputs involved are relatively simple. When the input waveform and the impulse response are complex functions, this approach becomes almost impossible.

In any filter, the time domain output waveform represents the convolution of the impulse response with the input waveform. Convolution can be followed by reference to a graphic example in Figure 3.35. Where the impulse response is asymmetrical, the decaying tail occurs *after* the input. As a result it is necessary to reverse the impulse response in time so that it is mirrored prior to sweeping it through the input waveform. The output voltage is proportional to the shaded area shown where the two impulses overlap.

The same process can be performed in the sampled, or discrete time domain as shown in Figure 3.36. The impulse and the input are now a set of discrete samples which clearly must have the same sample spacing. The impulse response only has value where impulses coincide. Elsewhere it is zero. The impulse response is therefore stepped through the input one sample period at a time. At each step, the area is still proportional to the output, but as the time steps are of uniform width, the area is proportional to the impulse height and so the output is obtained by adding up the lengths of overlap. In mathematical terms, the output samples represent the convolution of the input and the impulse response by summing the coincident cross-products.

As a digital filter works in this way, perhaps it is not a filter at all, but just a mathematical simulation of an analog filter. This approach is quite useful in visualizing what a digital filter does.

Somewhere between the analog filter and the digital filter is the switched capacitor filter. This uses analog quantities, namely the charges on capacitors, but the time axis is discrete because the various charges are routed using electronic switches which close during various phases of the sampling rate clock. Switched capacitor filters have the same characteristics as digital filters with infinite precision. They are often used in preference to continuous time analog filters in integrated circuit convertors because they can be implemented with the same integration techniques. Figure 3.37(a) shows a switched capacitor delay. There are two clock phases and during the first the input voltage is transferred to the capacitor. During the second phase the capacitor voltage is transferred to the output. Combining delay with operational amplifier summation allows frequency dependent circuitry to be realized. Figure 3.37(b) shows a simple switched capacitor filter. The delay causes a phase shift which is dependent on frequency. The frequency response is sinusoidal.

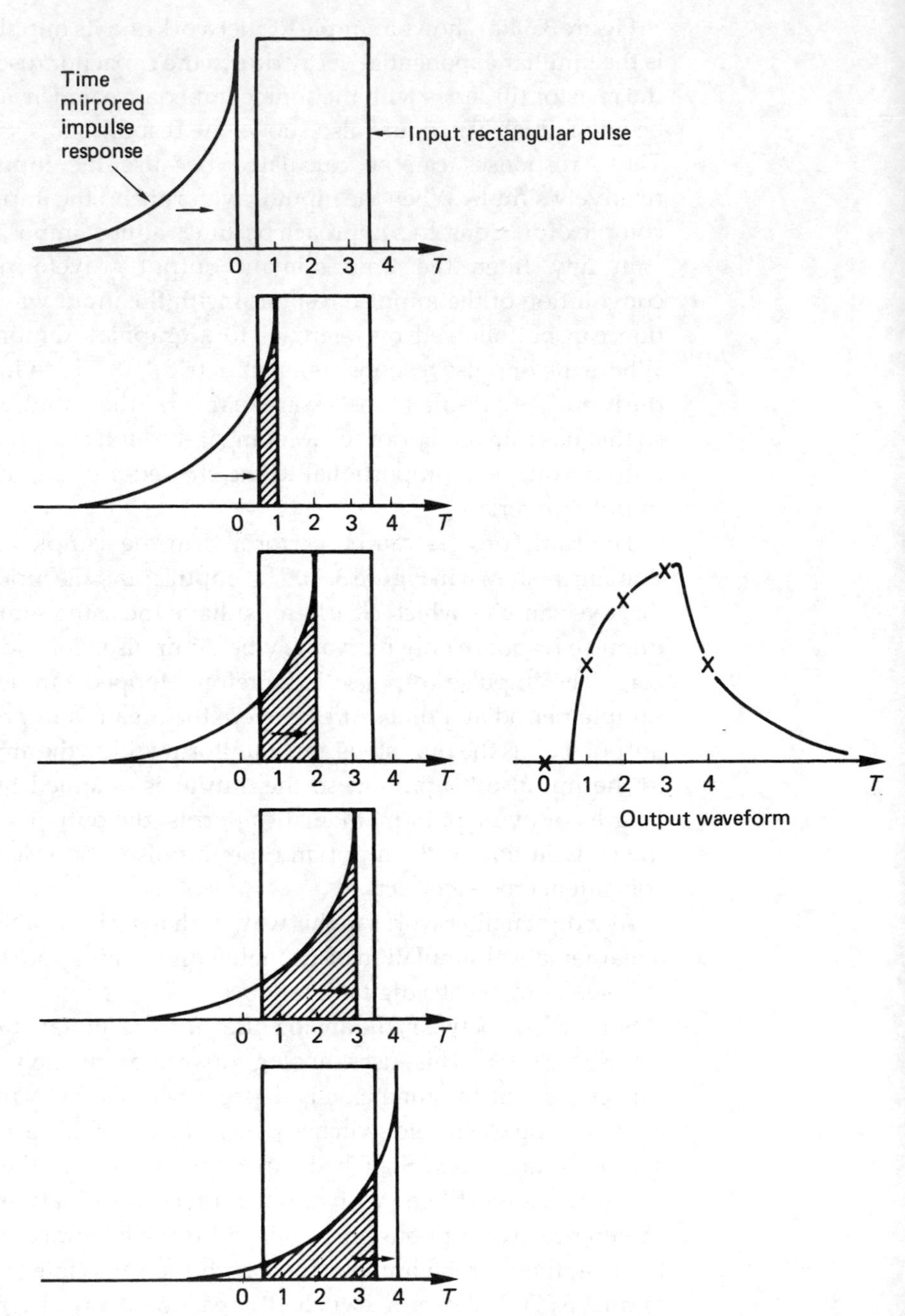

Figure 3.35 In the convolution of two continuous signals (the impulse response with the input), the impulse must be time reversed or mirrored. This is necessary because the impulse will be moved from left to right, and mirroring gives the impulse the correct time-domain response when it is moved past a fixed point. As the impulse response slides continuously through the input waveform, the area where the two overlap determines the instantaneous output amplitude. This is shown for five different times by the crosses on the output waveform.

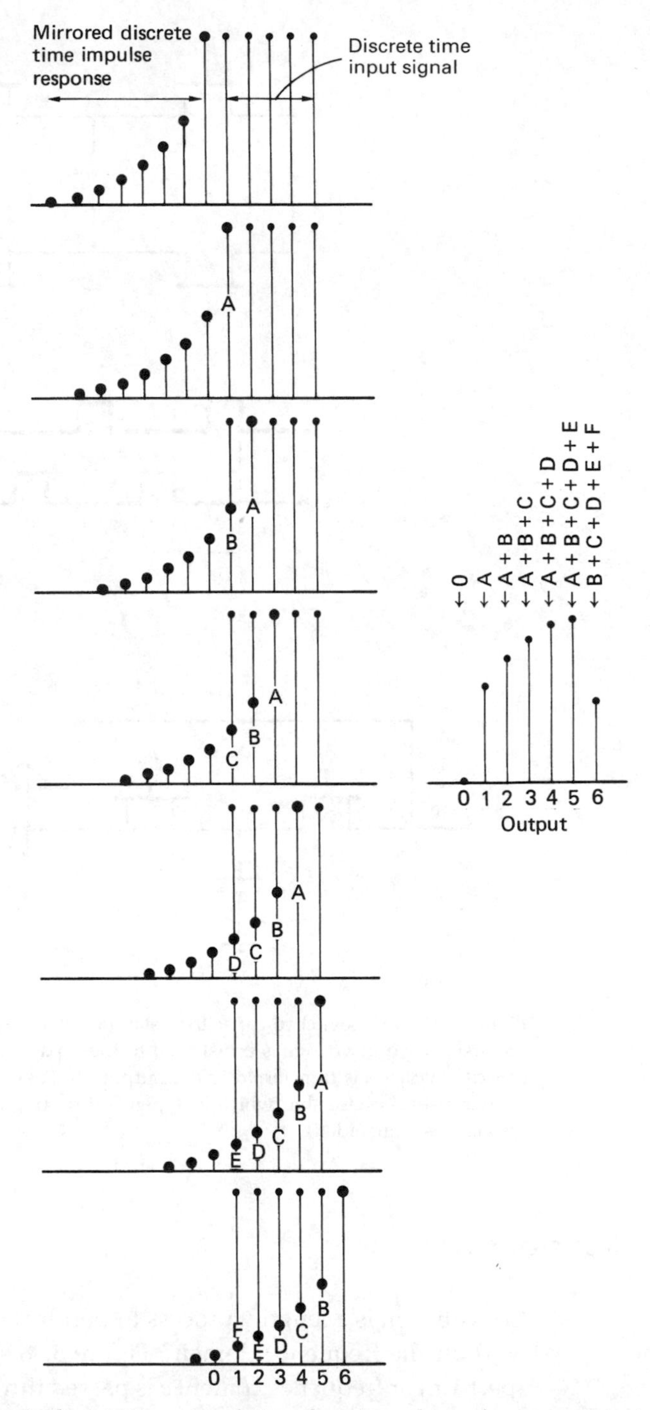

Figure 3.36 In discrete time convolution, the mirrored impulse response is stepped through the input one sample period at a time. At each step, the sum of the cross-products is used to form an output value. As the input in this example is a constant height pulse, the output is simply proportional to the sum of the coincident impulse response samples. This figure should be compared with Figure 3.35.

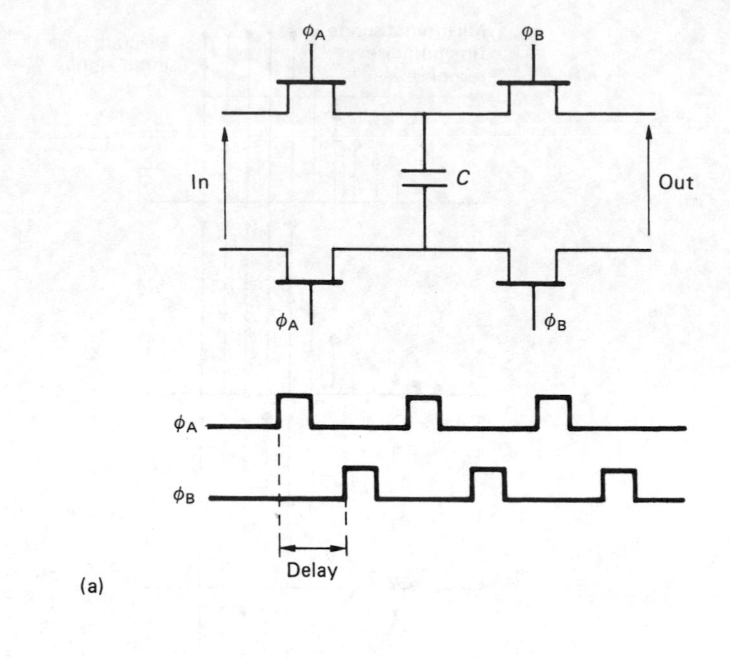

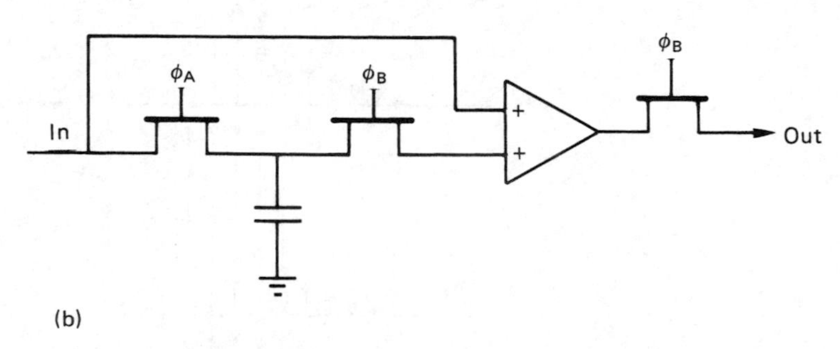

Figure 3.37 In a switched capacitor delay (a), there are two clock phases, and during the first the input voltage is transferred to the capacitor. During the second phase the capacitor voltage is transferred to the output. (b) A simple switched capacitor filter. The delay causes a phase shift which is dependent on frequency and the resultant frequency response is sinusoidal.

3.19 Transforms

Convolution is a lengthy process to perform on paper. It is much easier to work in the frequency domain. Figure 3.38 shows that if a signal with a spectrum or frequency content *a* is passed through a filter with a frequency response *b* the result will be an output spectrum which is simply the product of the two. If the frequency responses are drawn on logarithmic scales (i.e. calibrated in dB) the two can be simply added because the addition of logs is the same as multiplication. Whilst frequency in audio

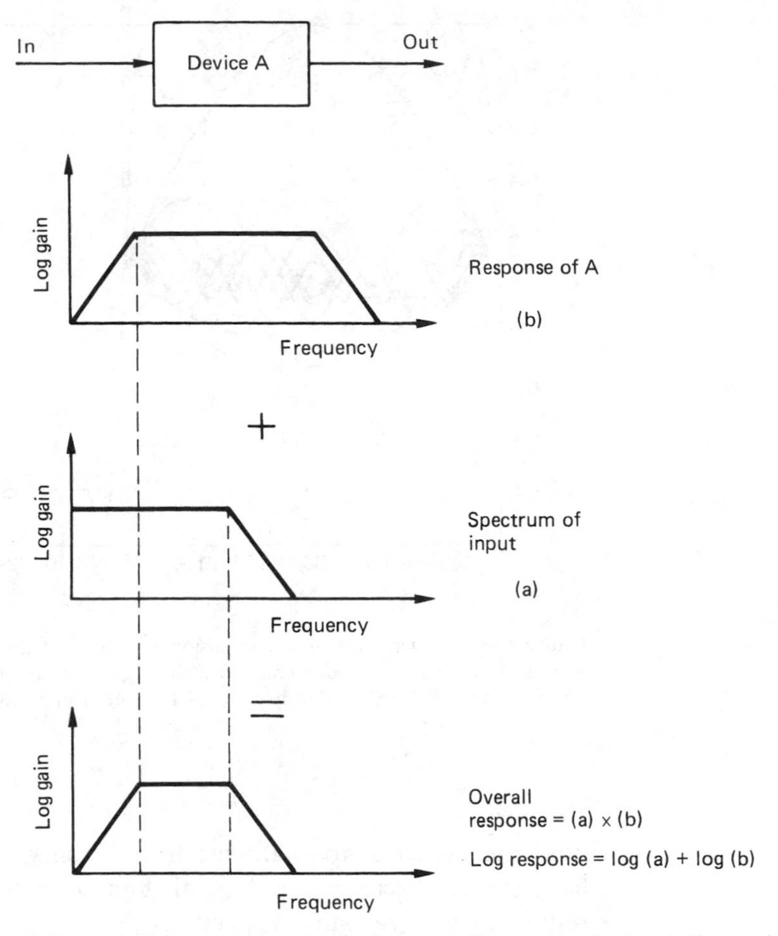

Figure 3.38 In the frequency domain, the response of two series devices is the product of their individual responses at each frequency. On a logarithmic scale the responses are simply added.

has traditionally meant temporal frequency measured in Hertz, frequency in optics can also be spatial and measured in lines per millimetre (mm^{-1}). Multiplying the spectra of the responses is a much simpler process than convolution.

In order to move to the frequency domain or spectrum from the time domain or waveform, it is necessary to use the Fourier transform, or in sampled systems, the discrete Fourier transform (DFT). Fourier analysis holds that any waveform can be reproduced by adding together an arbitrary number of harmonically related sinusoids of various amplitudes and phases. Figure 3.39 shows how a square wave can be built up of harmonics. The spectrum can be drawn by plotting the amplitude of the harmonics against frequency. It will be seen that this gives a spectrum which is a decaying wave. It passes through zero at all even multiples of the

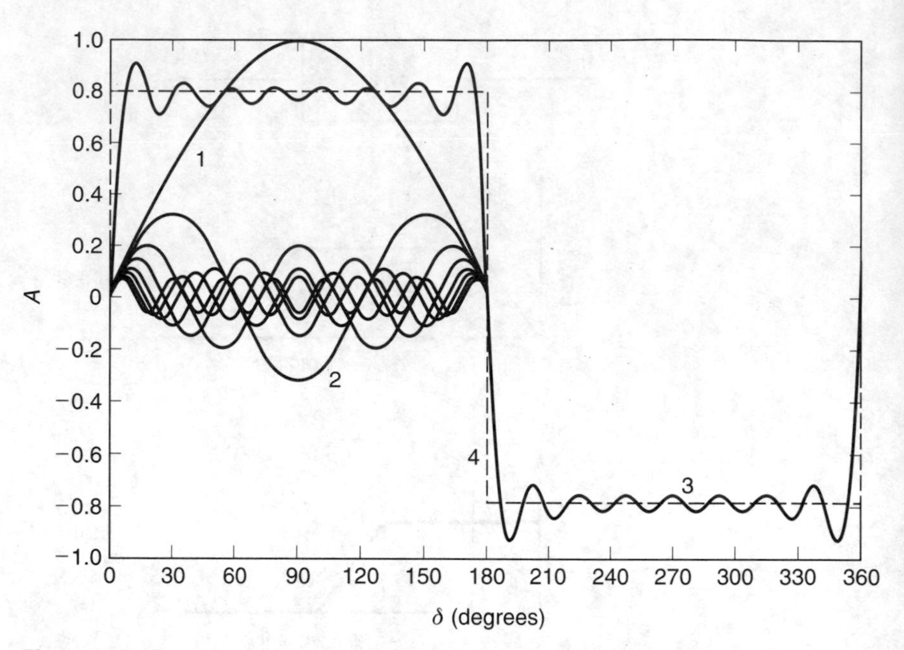

Figure 3.39 Fourier analysis of a square wave into fundamental and harmonics. A, amplitude; δ, phase of fundamental wave in degrees; 1, first harmonic (fundamental); 2, odd harmonics 3–15; 3, sum of harmonics 1–15; 4, ideal square wave.

fundamental. The shape of the spectrum is a $\sin x/x$ curve. If a square wave has a $\sin x/x$ spectrum, it follows that a filter with a rectangular impulse response will have a $\sin x/x$ spectrum.

A low-pass filter has a rectangular spectrum, and this has a $\sin x/x$ impulse response. These characteristics are known as a transform pair. In transform pairs, if one domain has one shape of the pair, the other domain will have the other shape. Thus a square wave has a $\sin x/x$ spectrum and a $\sin x/x$ impulse has a square spectrum. Figure 3.40 shows a number of transform pairs. Note the pulse pair. A time domain pulse of infinitely short duration has a flat spectrum. Thus a flat waveform, i.e. DC, has only zero in its spectrum. Interestingly the transform of a Gaussian response in still Gaussian. The impulse response of the optics of a laser disk has a $\sin^2 x/x^2$ function, and this is responsible for the triangular falling frequency response of the pickup.

The spectrum of a pseudo-random sequence is not flat because it has a finite sequence length. The rate at which the sequence repeats is visible in the spectrum. Where pseudo-random sequences are to be used in sample manipulation, i.e. where their effects can be visible, it is essential that the sequence length should be long enough to prevent the periodicity being discerned.

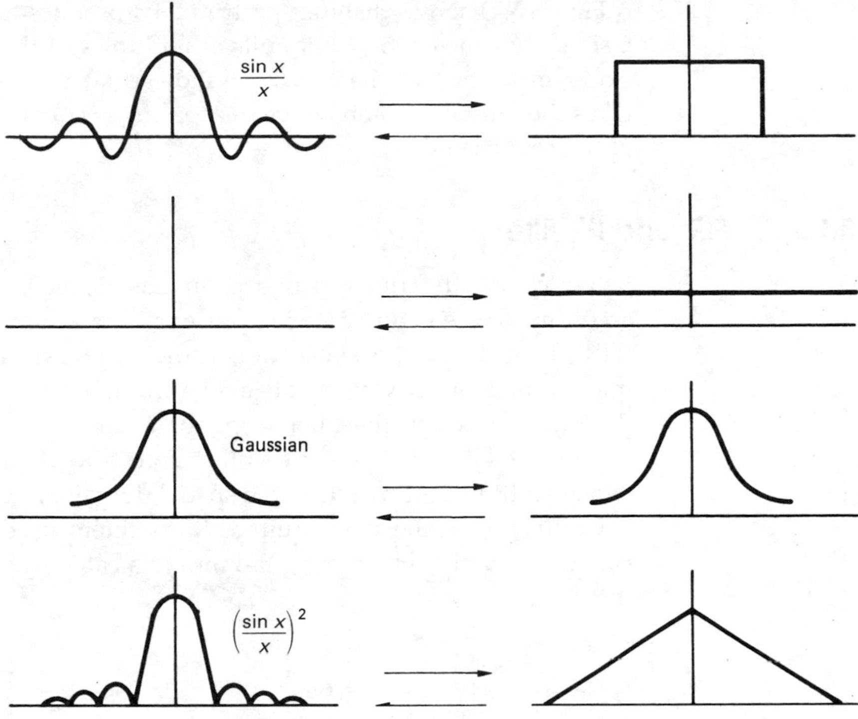

Figure 3.40 The concept of transform pairs illustrates the duality of the frequency (including spatial frequency) and time domains.

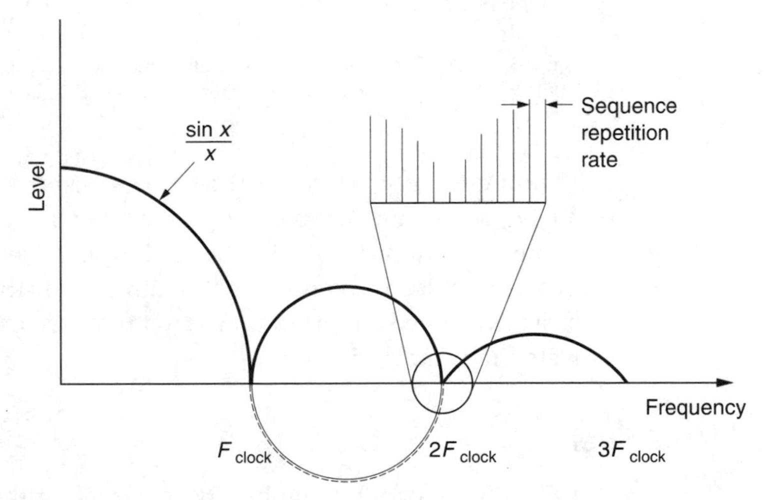

Figure 3.41 The spectrum of a pseudo-random sequence has a $\sin x/x$ characteristic, with nulls at multiples of the clock frequency. The spectrum is not continuous, but resembles a comb where the spacing is equal to the repetition rate of the sequence.

Figure 3.41 shows that the spectrum of a pseudo-random sequence has a $\sin x/x$ characteristic, with nulls at multiples of the clock frequency. A closer inspection of the spectrum shows that it is not continuous, but takes the form of a comb where the spacing is equal to the repetition rate of the sequence.

3.20 FIR and IIR filters

Filters can be described in two main classes, as shown in Figure 3.42, according to the nature of the impulse response. Finite-impulse response (FIR) filters are always stable and, as their name suggests, respond to an impulse once, as they have only a forward path. In the temporal domain, the time for which the filter responds to an input is finite, fixed and readily established. The same is therefore true about the distance over which a FIR filter responds in the spatial domain. FIR filters can be made perfectly phase linear if required. Most filters used for sampling rate conversion and oversampling fall into this category.

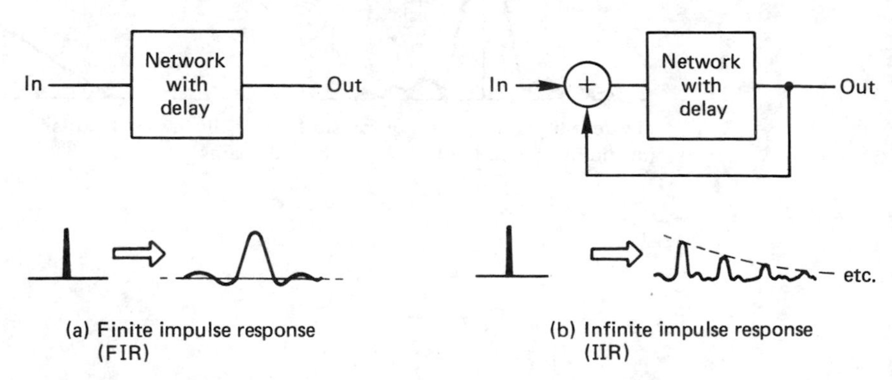

Figure 3.42 An FIR filter (a) responds only once to an input, whereas the output of an IIR filter (b) continues indefinitely rather like a decaying echo.

Infinite-impulse response (IIR) filters respond to an impulse indefinitely and are not necessarily stable, as they have a return path from the output to the input. For this reason they are also called recursive filters. As the impulse response in not symmetrical, IIR filters are not phase linear. Noise reduction units may employ recursive filters and will be treated in Chapter 5.

3.21 FIR filters

A FIR filter works by graphically constructing the impulse response for every input sample. It is first necessary to establish the correct impulse response. Figure 3.43(a) shows an example of a low-pass filter which cuts

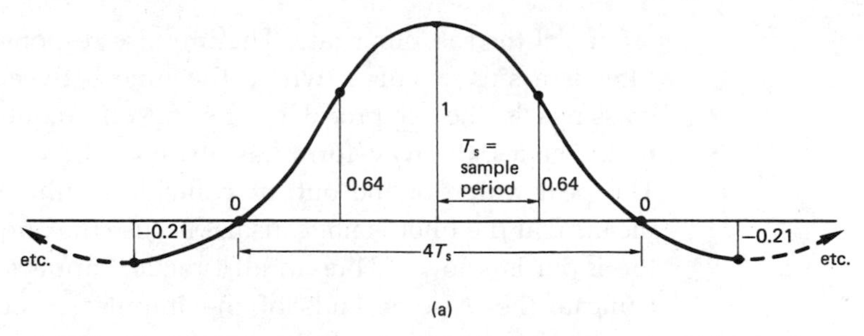

Figure 3.43(a) The impulse response of an LPF is a sinx/x curve which stretches from $-\infty$ to $+\infty$ in time. The ends of the response must be neglected, and a delay introduced to make the filter causal.

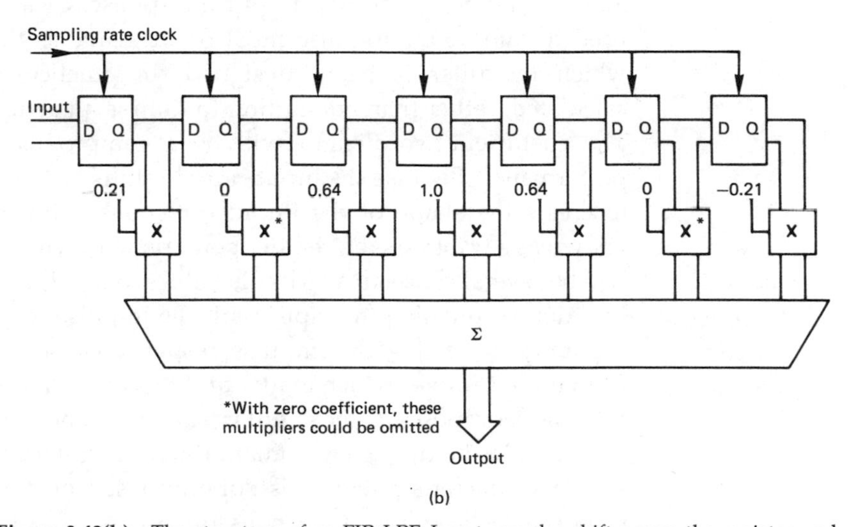

Figure 3.43(b) The structure of an FIR LPF. Input samples shift across the register and at each point are multiplied by different coefficients.

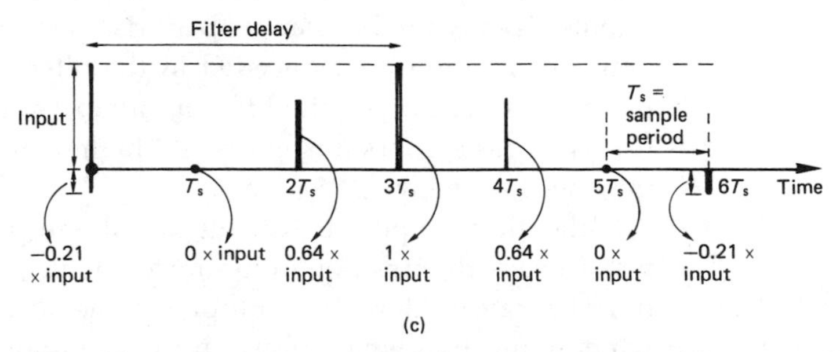

Figure 3.43(c) When a single unit sample shifts across the circuit of Figure 3.43(b), the impulse response is created at the output as the impulse is multiplied by each coefficient in turn.

off at $\frac{1}{4}$ of the sampling rate. The impulse response of a perfect low-pass filter is a $\sin x/x$ curve, where the time between the two central zero crossings is the reciprocal of the cut-off frequency. According to the mathematics, the waveform has always existed, and carries on for ever. The peak value of the output coincides with the input impulse. This means that the filter is not causal, because the output has changed before the input is known. Thus in all practical applications it is necessary to truncate the extreme ends of the impulse response, which causes an aperture effect, and to introduce a time delay in the filter equal to half the duration of the truncated impulse in order to make the filter causal. As an input impulse is shifted through the series of registers in Figure 3.43(b), the impulse response is created, because at each point it is multiplied by a coefficient as in (c). These coefficients are simply the result of sampling and quantizing the desired impulse response. Clearly the sampling rate used to sample the impulse must be the same as the sampling rate for which the filter is being designed. In practice the coefficients are calculated, rather than attempting to sample an actual impulse response. The coefficient wordlength will be a compromise between cost and performance. Because the input sample shifts across the system registers to create the shape of the impulse response, the configuration is also known as a transversal filter. In operation with real sample streams, there will be several consecutive sample values in the filter registers at any time in order to convolve the input with the impulse response.

Simply truncating the impulse response causes an abrupt transition from input samples which matter and those which do not. Truncating the filter superimposes a rectangular shape on the time domain impulse response. In the frequency domain the rectangular shape transforms to a $\sin x/x$ characteristic which is superimposed on the desired frequency response as a ripple. One consequence of this is known as Gibb's phenomenon; a tendency for the response to peak just before the cut-off frequency.[2,3] As a result, the length of the impulse which must be considered will depend not only on the frequency response, but also on the amount of ripple which can be tolerated. If the relevant period of the impulse is measured in sample periods, the result will be the number of points or multiplications needed in the filter. Figure 3.44 compares the performance of filters with different numbers of points. Video filters may use as few as eight points whereas a high-quality digital audio FIR filter may need as many as 96 points.

Rather than simply truncate the impulse response in time, it is better to make a smooth transition from samples which do not count to those that do. This can be done by multiplying the coefficients in the filter by a window function which peaks in the centre of the impulse. Figure 3.45 shows some different window functions and their responses. The rectangular window is the case of truncation, and the response is shown

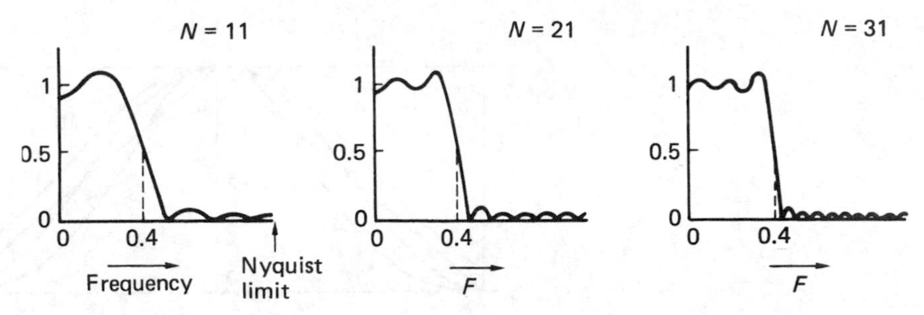

Figure 3.44 The truncation of the impulse in an FIR filter caused by the use of a finite number of points (N) results in ripple in the response. Shown here are three different numbers of points for the same impulse response. The filter is an LPF which rolls off at 0.4 of the fundamental interval. (Courtesy *Philips Technical Review.*)

at I. A linear reduction in weight from the centre of the window to the edges characterizes the Bartlett window II, which trades ripple for an increase in transition-region width. At III is shown the Hamming window, which is essentially a raised cosine shape. Not shown is the similar Hamming window, which offers a slightly different trade-off between ripple and the width of the main lobe. The Blackman window introduces an extra cosine term into the Hamming window at half the period of the main cosine period, reducing Gibb's phenomenon and ripple level, but increasing the width of the transition region. The Kaiser window is a family of windows based on the Bessel function, allowing various trade-offs between ripple ratio and main lobe width. Two of these are shown in IV and V. The drawback of the Kaiser windows is that they are complex to implement.

Filter coefficients can be optimized by computer simulation. One of the best-known techniques used is the Remez exchange algorithm, which converges on the optimum coefficients after a number of iterations.

In the example of Figure 3.46, the low-pass filter of Figure 3.43 is shown with a Bartlett window. Acceptable ripple determines the number of significant sample periods embraced by the impulse. This determines in turn both the number of points in the filter and the filter delay. As the impulse is symmetrical, the delay will be half the impulse period. The impulse response is a $\sin x/x$ function, and this has been calculated in the figure. The $\sin x/x$ response is next multiplied by the window function to give the windowed impulse response.

If the coefficients are not quantized finely enough, it will be as if they had been calculated inaccurately, and the performance of the filter will be less than expected. Figure 3.47 shows an example of quantizing coefficients. Conversely, raising the wordlength of the coefficients increases cost.

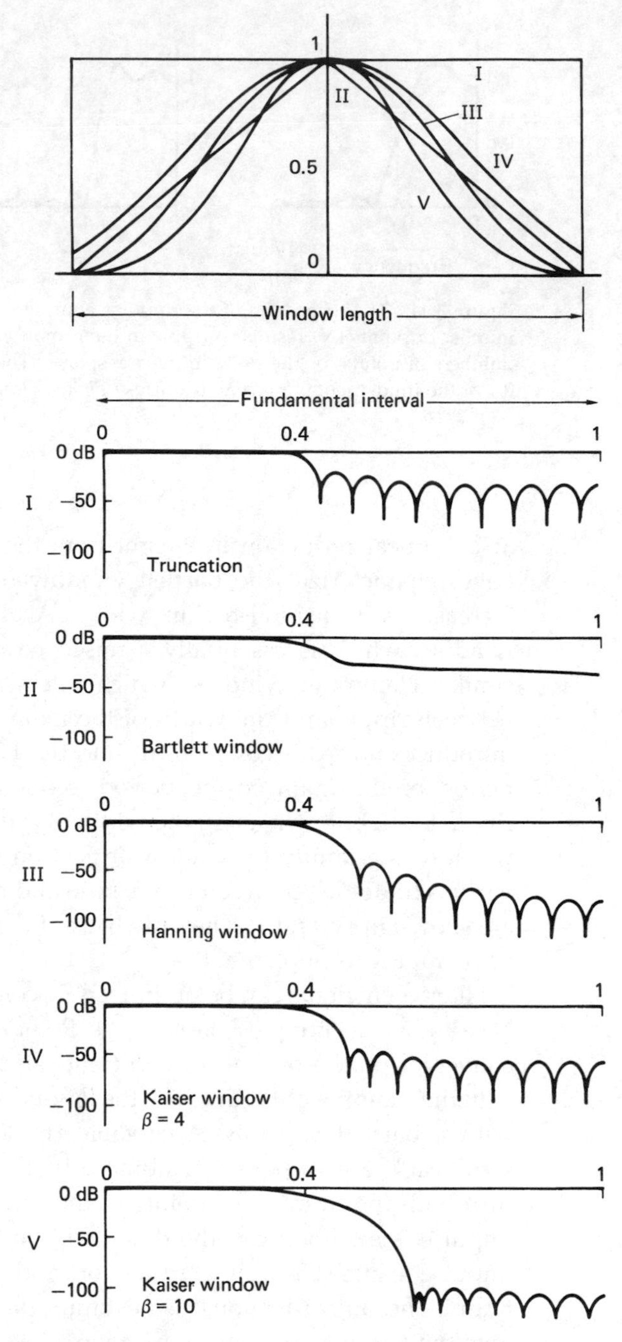

Figure 3.45 The effect of window functions. At top, various window functions are shown in continuous form. Once the number of samples in the window is established, the continuous functions shown here are sampled at the appropriate spacing to obtain window coefficients. These are multiplied by the truncated impulse response coefficients to obtain the actual coefficients used by the filter. The amplitude responses I–V correspond to the window functions illustrated. (Responses courtesy *Philips Technical Review.*)

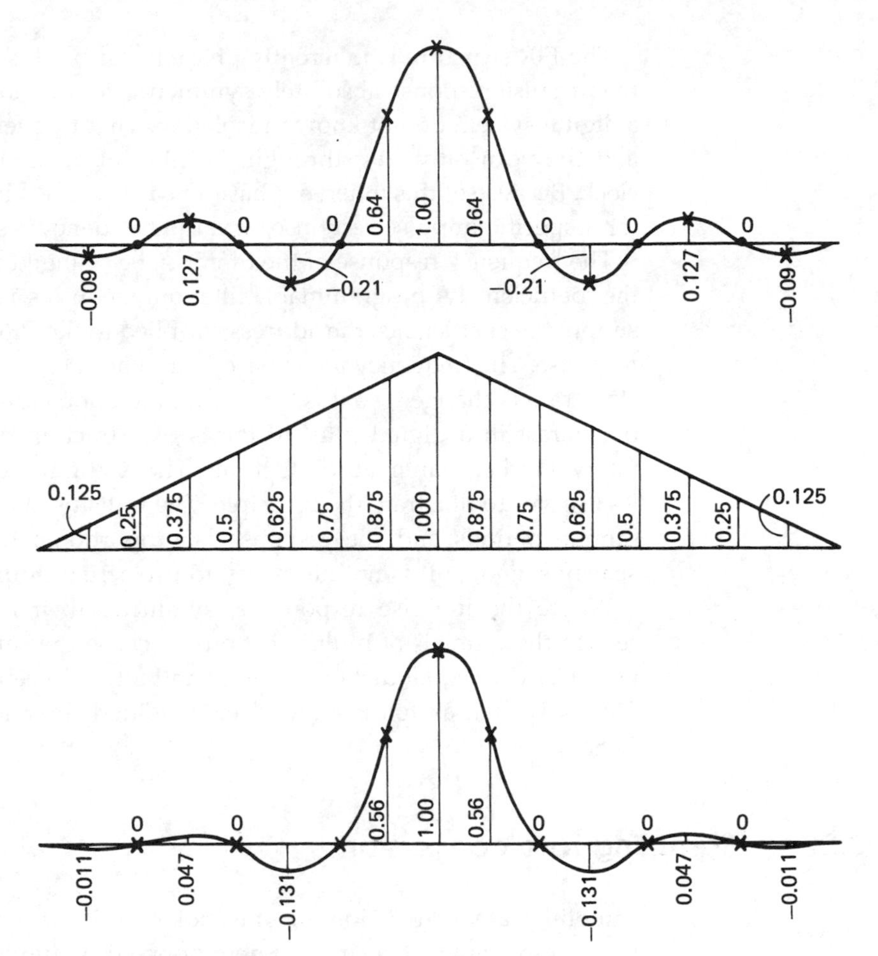

Figure 3.46 A truncated sinx/x impulse (top) is multiplied by a Bartlett window function (centre) to produce the actual coefficients used (bottom).

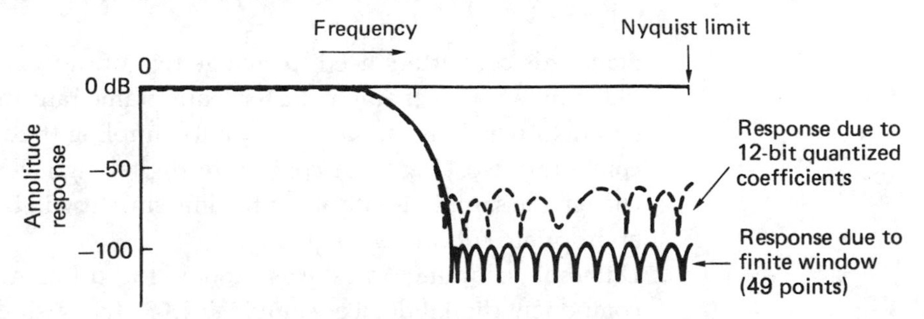

Figure 3.47 Frequency response of a 49-point transversal filter with infinite precision (solid line) shows ripple due to finite window size. Quantizing coefficients to twelve bits reduces attenuation in the stopband. (Responses courtesy *Philips Technical Review*.)

The FIR structure is inherently phase linear because it is easy to make the impulse response absolutely symmetrical. The individual samples in a digital system do not know in isolation what frequency they represent, and they can only pass through the filter at a rate determined by the clock. Because of this inherent phase-linearity, a FIR filter can be designed for a specific impulse response, and the frequency response will follow.

The frequency response of the filter can be changed at will by changing the coefficients. A programmable filter only requires a series of PROMs to supply the coefficients; the address supplied to the PROMs will select the response. The frequency response of a digital filter will also change if the clock rate is changed, so it is often less ambiguous to specify a frequency of interest in a digital filter in terms of a fraction of the fundamental interval rather than in absolute terms. The configuration shown in Figure 3.43 serves to illustrate the principle. The units used on the diagrams are sample periods and the response is proportional to these periods or spacings, and so it is not necessary to use actual figures.

Where the impulse response is symmetrical, it is often possible to reduce the number of multiplications, because the same product can be used twice, at equal distances before and after the centre of the window. This is known as folding the filter. A folded filter is shown in Figure 3.48.

3.22 Sampling-rate conversion

Sampling-rate conversion or interpolation is an important enabling technology on which a large number of practical digital video devices are based. In digital video, the sampling rate takes on many guises. When analog video is sampled in real time, the sampling rate is temporal, but where pixels form a static array, the sampling rate is a spatial frequency.

Some of the applications of interpolation are set out here:

1 Standards convertors need to change two of the sampling rates of the video they handle, namely the temporal frame rate and the vertical line spacing, which is in fact a spatial sampling frequency. Standards convertors working with composite digital signals will also need to change the sampling rate along the line since it will be a multiple of the appropriate subcarrier frequency.
2 Different sampling rates exist today for different purposes. Most component digital devices sample at 13.5 MHz, using the 4:2:2 format, but other variations are possible, such as 3:1:1. Composite machines sample at a multiple of the subcarrier frequency of their line standard. Rate conversion allows material to be exchanged freely between such

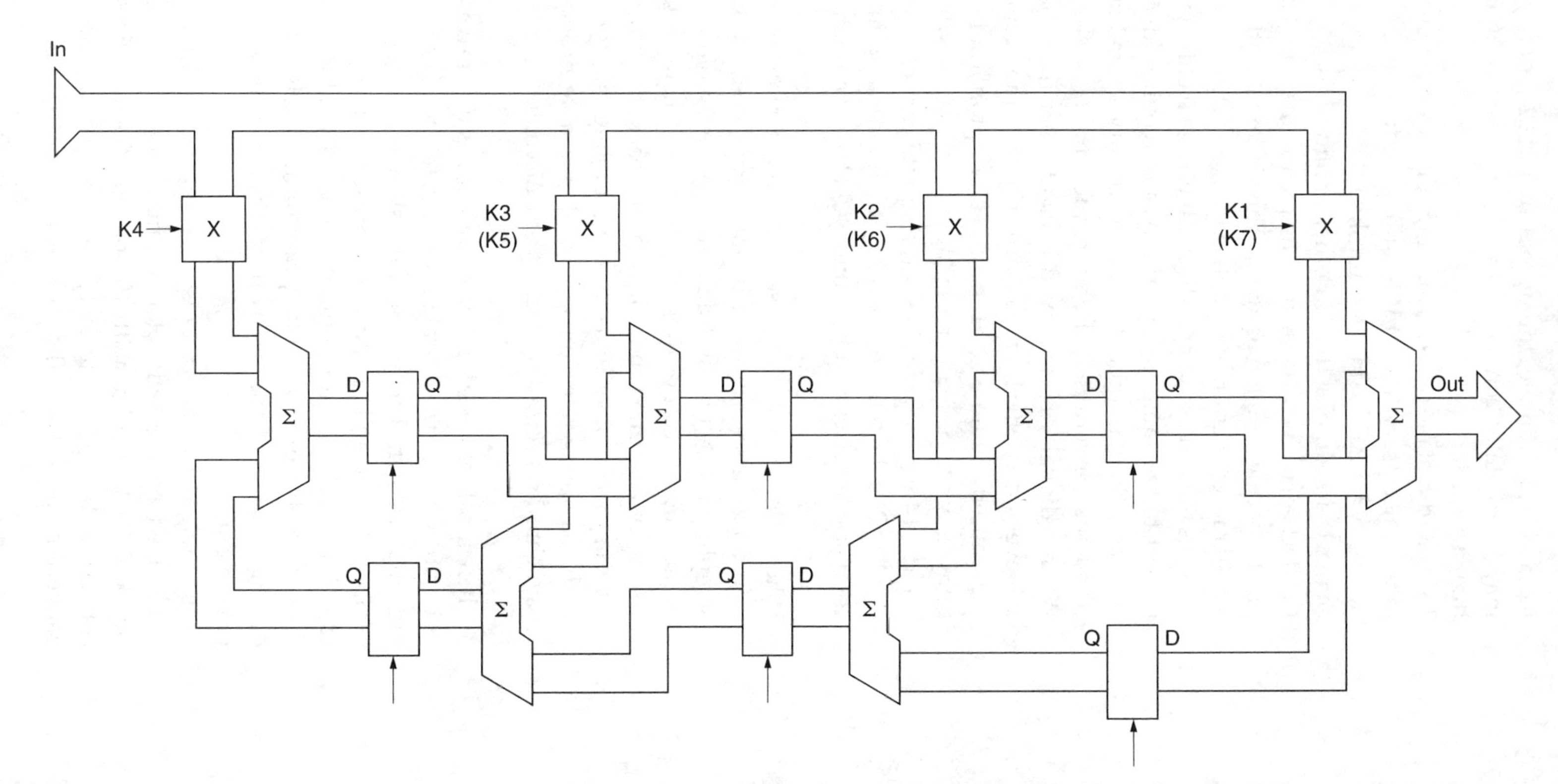

Figure 3.48 A seven-point folded filter for a symmetrical impulse response. In this case K1 and K7 will be identical, and so the input sample can be multiplied once, and the product fed into the output shift system in two different places. The centre coefficient K4 appears once. In an even-numbered filter the centre coefficient would also be used twice.

formats. For example, the output of a 4:2:2 paint system at 13.5 MHz may be digitally converted to $4\,F_{sc}$ for use as input to a composite digital recorder.

3 To take advantage of oversampling convertors, an increase in sampling rate is necessary for DACs and a reduction in sampling rate is necessary for ADCs. In oversampling the factors by which the rates are changed are simpler than in other applications.

4 In effects machines, the size of the picture may be changed without the pixel spacing being changed. This is exactly the converse of the standards convertor, which leaves the picture size unchanged and changes the pixel spacing. Alternatively the picture may be shifted with respect to the sampling matrix by any required distance to sub-pixel accuracy. Similar processes are necessary in motion estimation for standards convertors and data reduction.

5 When a digital VTR is played back at other than the correct speed to achieve some effect or to correct pitch, the sampling rate of the reproduced audio signal changes in proportion. If the playback samples are to be fed to a digital mixing console which works at some standard frequency, audio sampling rate conversion will be necessary. Whilst DVTRs universally use an audio sampling rate of 48 kHz, compact disc uses 44.1 kHz and 32 kHz is common for broadcast use (e.g. DVB).

6 When digital audio is used in conjunction with film or video, difficulties arise because it is not always possible to synchronize the sampling rate with the frame rate. An example of this is where the digital audio recorder uses its internally generated sampling rate, but also records studio timecode. On playback, the timecode can be made the same as on other units, or the sampling rate can be locked, but not both. Sampling-rate conversion allows a recorder to play back an asynchronous recording locked to timecode.

There are three basic but related categories of rate conversion, as shown in Figure 3.49. The most straightforward (a) changes the rate by an integer ratio, up or down. The timing of the system is thus simplified because all samples (input and output) are present on edges of the higher-rate sampling clock. Such a system is generally adopted for oversampling convertors; the exact sampling rate immediately adjacent to the analog domain is not critical, and will be chosen to make the filters easier to implement.

Next in order of difficulty is the category shown at (b) where the rate is changed by the ratio of two small integers. Samples in the input periodically time-align with the output. Such devices can be used for converting from $4 \times F_{sc}$ to $3 \times F_{sc}$, in the vertical processing of standards convertors, or between the various rates of CCIR-601.

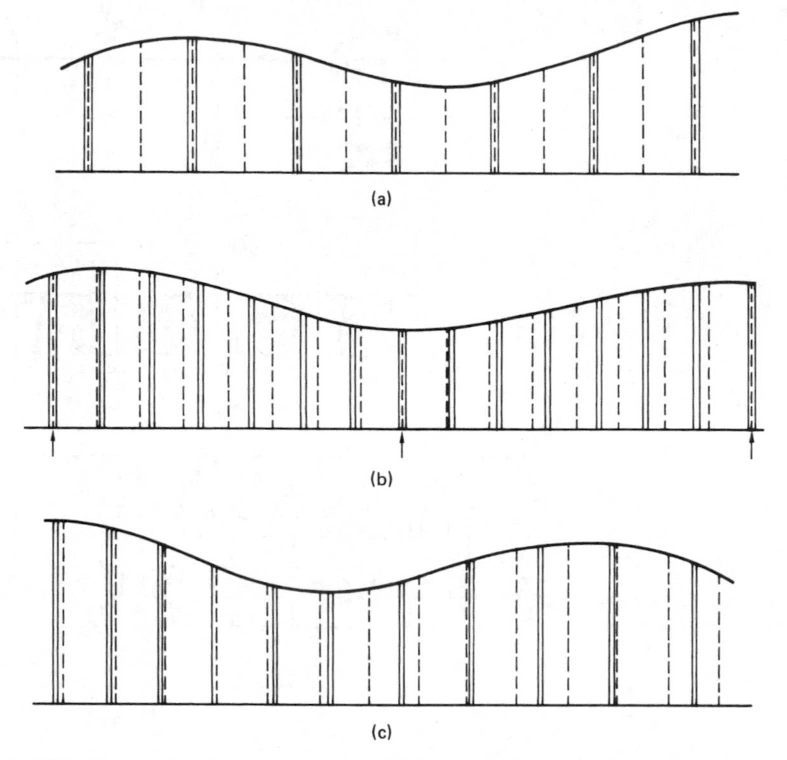

Figure 3.49 Categories of rate conversion. (a) Integer-ratio conversion, where the lower-rate samples are always coincident with those of the higher rate. There are a small number of phases needed. (b) Fractional-ratio conversion, where sample coincidence is periodic. A larger number of phases is required. Example here is conversion from 50.4 kHz to 44.1 kHz (8/7). (c) Variable-ratio conversion, where there is no fixed relationship, and a large number of phases are required.

The most complex rate-conversion category is where there is no simple relationship between input and output sampling rates, and in fact they may vary. This situation, shown in (c), is known as variable-ratio conversion. The temporal or spatial relationship of input and output samples is arbitrary. This problem will be met in effects machines which zoom or rotate images.

The technique of integer-ratio conversion is used in conjunction with oversampling convertors in digital video and audio and in motion estimation and compression systems where sub-sampled or reduced resolution versions of an input image are required. These applications will be detailed in Chapter 5. Sampling-rate reduction by an integer factor is dealt with first.

Figure 3.50(a) shows the spectrum of a typical sampled system where the sampling rate is a little more than twice the analog bandwidth. Attempts to reduce the sampling rate simply by omitting samples, a

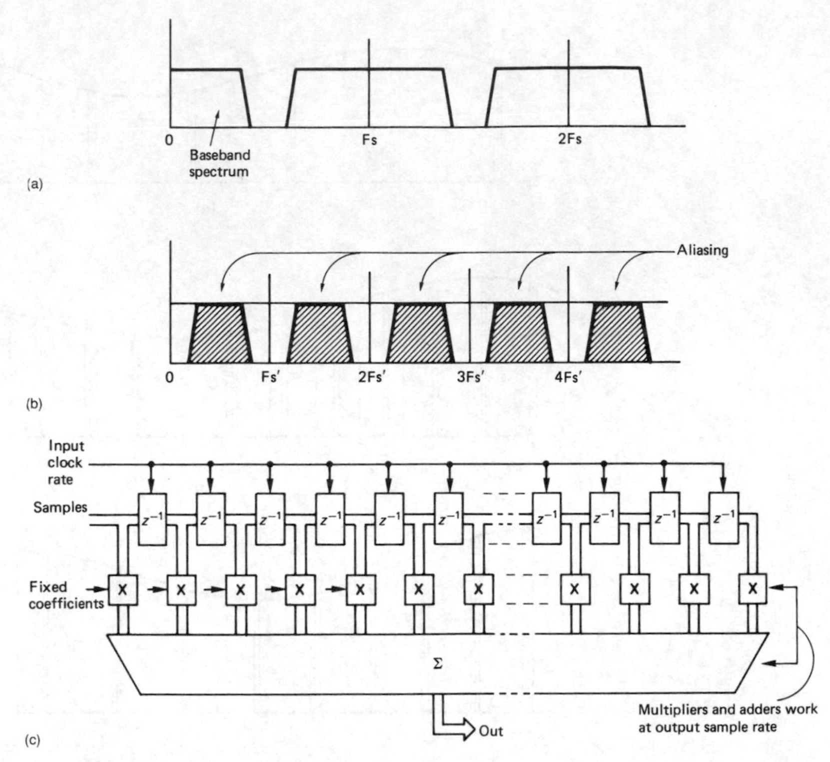

Figure 3.50 The spectrum of a typical digital sample stream at (a) will be subject to aliasing as in (b) if the baseband width is not reduced by an LPF. At (c) an FIR low-pass filter prevents aliasing. Samples are clocked transversely across the filter at the input rate, but the filter only computes at the output sample rate. Clearly this will only work if the two are related by an integer factor.

process known as decimation, will result in aliasing, as shown in (b). Intuitively it is obvious that omitting samples is the same as if the original sampling rate was lower. In order to prevent aliasing, it is necessary to incorporate low-pass filtering into the system where the cut-off frequency reflects the new, lower, sampling rate. An FIR type low-pass filter could be installed, as described earlier in this chapter, immediately prior to the stage where samples are omitted, but this would be wasteful, because for much of its time the FIR filter would be calculating sample values which are to be discarded. The more effective method is to combine the low-pass filter with the decimator so that the filter only calculates values to be retained in the output sample stream. Figure 3.50(c) shows how this is done. The filter makes one accumulation for every output sample, but that accumulation is the result of multiplying all relevant input samples in the filter window by an appropriate coefficient. The number of points in the filter is determined by the number of *input* samples in the period of the filter window, but the number of multiplications per second is obtained by multiplying that figure by the *output* rate. If the filter is not integrated with the decimator, the number of points has to be multiplied by the input rate. The larger the rate-reduction factor, the more advantageous the decimating filter ought to be, but this is not quite the case, as the greater the reduction in rate, the longer the filter window will need to be to accommodate the broader impulse response.

When the sampling rate is to be increased by an integer factor, additional samples must be created at even spacing between the existing ones. There is no need for the bandwidth of the input samples to be reduced since, if the original sampling rate was adequate, a higher one must also be adequate.

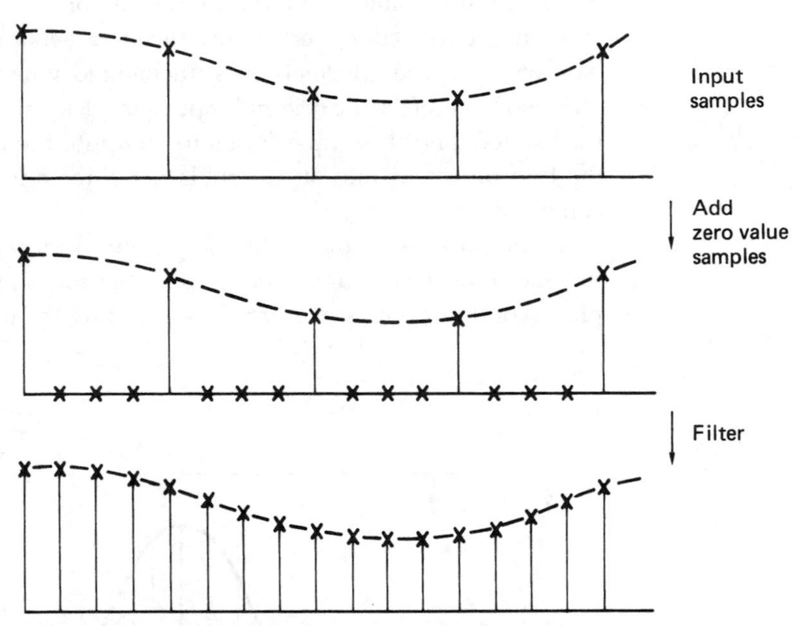

Figure 3.51 In integer-ratio sampling, rate increase can be obtained in two stages. First, zero-value samples are inserted to increase the rate, and then filtering is used to give the extra samples real values. The filter necessary will be an LPF with a response which cuts off at the Nyquist frequency of the input samples.

Figure 3.51 shows that the process of sampling-rate increase can be thought of in two stages. First, the correct rate is achieved by inserting samples of zero value at the correct instant, and then the additional samples are given meaningful values by passing the sample stream through a low-pass filter which cuts off at the Nyquist frequency of the original sampling rate. This filter is known as an interpolator, and one of its tasks is to prevent images of the input spectrum from appearing in the extended baseband of the output spectrum.

How do interpolators work? Remember that, according to sampling theory, all sampled systems have finite bandwidth. An individual digital sample value is obtained by sampling the instantaneous voltage of the original analog waveform, and because it has zero duration, it must contain an infinite spectrum. However, such a sample can never be heard in that form because of the reconstruction process, which limits the spectrum of the impulse to the Nyquist limit. After reconstruction, one infinitely short digital sample ideally represents a $\sin x/x$ pulse whose central peak width is determined by the response of the reconstruction filter, and whose amplitude is proportional to the sample value. This implies that, in reality, one sample value has meaning over a considerable timespan, rather than just at the sample instant. If this were not true, it would be impossible to build an interpolator.

As in rate reduction, performing the steps separately is inefficient. The bandwidth of the information is unchanged when the sampling rate is increased; therefore the original input samples will pass through the filter unchanged, and it is superfluous to compute them. The combination of the two processes into an interpolating filter minimizes the amount of computation.

As the purpose of the system is purely to increase the sampling rate, the filter must be as transparent as possible, and this implies that a linear-phase configuration is mandatory, suggesting the use of an FIR structure.

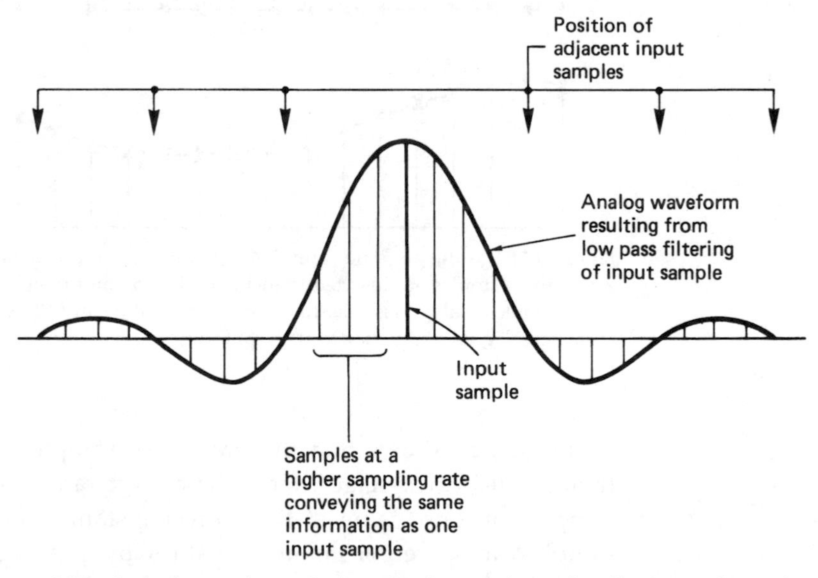

Figure 3.52 A single sample results in a $\sin x/x$ waveform after filtering in the analog domain. At a new, higher, sampling rate, the same waveform after filtering will be obtained if the numerous samples of differing size shown here are used. It follows that the values of these new samples can be calculated from the input samples in the digital domain in an FIR filter.

Figure 3.52 shows that the theoretical impulse response of such a filter is a $\sin x/x$ curve which has zero value at the position of adjacent input samples. In practice this impulse cannot be implemented because it is infinite. The impulse response used will be truncated and windowed as described earlier. To simplify this discussion, assume that a $\sin x/x$

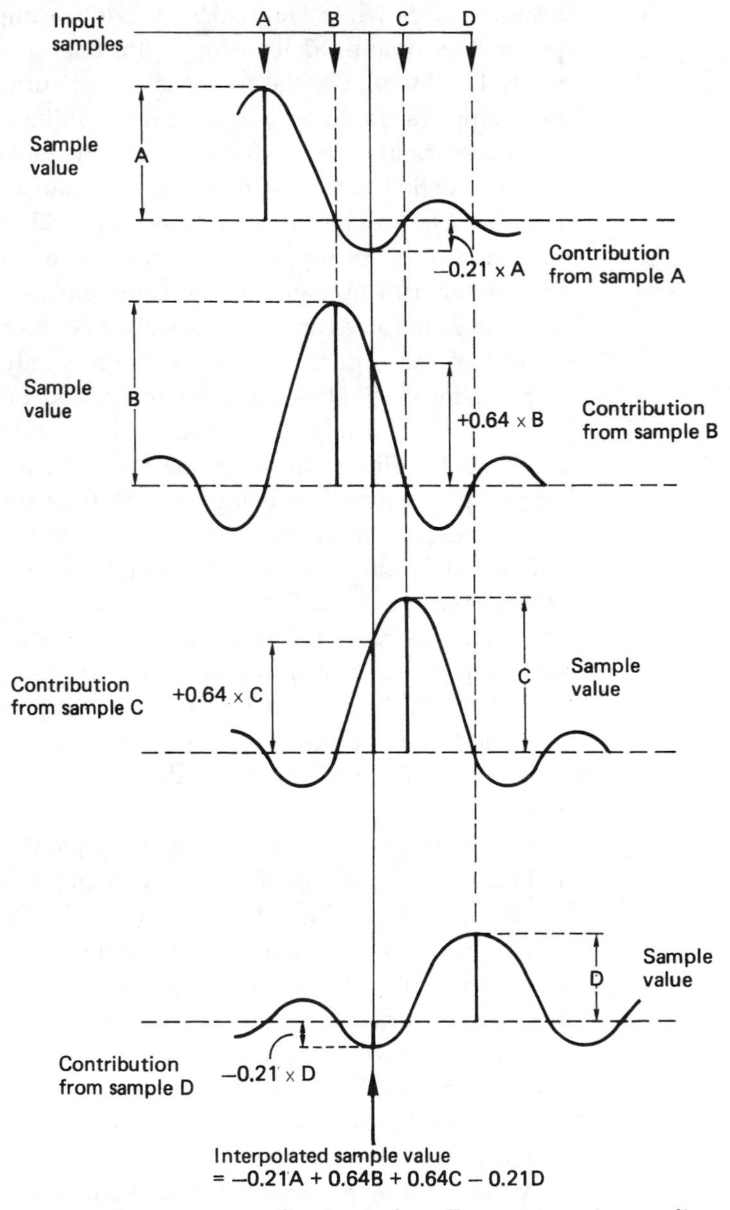

Figure 3.53 A two times oversampling interpolator. To compute an intermediate sample, the input samples are imagined to be $\sin x/x$ impulses, and the contributions from each at the point of interest can be calculated. In practice, rather more samples on either side need to be taken into account.

impulse is to be used. There is a strong parallel with the operation of a DAC where the analog voltage is returned to the time-continuous state by summing the analog impulses due to each sample. In a digital interpolating filter, this process is duplicated.[4]

If the sampling rate is to be doubled, new samples must be interpolated exactly halfway between existing samples. The necessary impulse response is shown in Figure 3.53; it can be sampled at the *output* sample period and quantized to form coefficients. If a single input sample is multiplied by each of these coefficients in turn, the impulse response of that sample at the new sampling rate will be obtained. Note that every other coefficient is zero, which confirms that no computation is necessary on the existing samples; they are just transferred to the output. The intermediate sample is computed by adding together the impulse responses of every input sample in the window. The figure shows how this mechanism operates. If the sampling rate is to be increased by a factor of four, three sample values must be interpolated between existing input samples. Figure 3.54 shows that it is only necessary to sample the impulse response at one-quarter the period of input samples to obtain three sets of coefficients which will be used in turn. In hardware-implemented filters, the input sample which is passed straight to the output is transferred by using a fourth filter phase where all coefficients are zero except the central one, which is unity.

Fractional ratio conversion allows interchange between different CCIR-601 rates such as 4:2:2 and 4:2:0. Fractional ratios also occur in the vertical axis of standards convertors. Figure 3.49 showed that when the two sampling rates have a simple fractional relationship m/n, there is a periodicity in the relationship between samples in the two streams. It is possible to have a system clock running at the least-common multiple frequency which will divide by different integers to give each sampling rate.[5]

The existence of a common clock frequency means that a fractional-ratio convertor could be made by arranging two integer-ratio convertors in series. This configuration is shown in Figure 3.55(a). The input-sampling rate is multiplied by m in an interpolator, and the result is divided by n in a decimator. Although this system would work, it would be grossly inefficient, because only one in n of the interpolator's outputs would be used. A decimator followed by an interpolator would also offer the correct sampling rate at the output, but the intermediate sampling rate would be so low that the system bandwidth would be quite unacceptable.

As has been seen, a more efficient structure results from combining the processes. The result is exactly the same structure as an integer-ratio interpolator, and requires an FIR filter. The impulse response of the filter is determined by the lower of the two sampling rates, and as before it

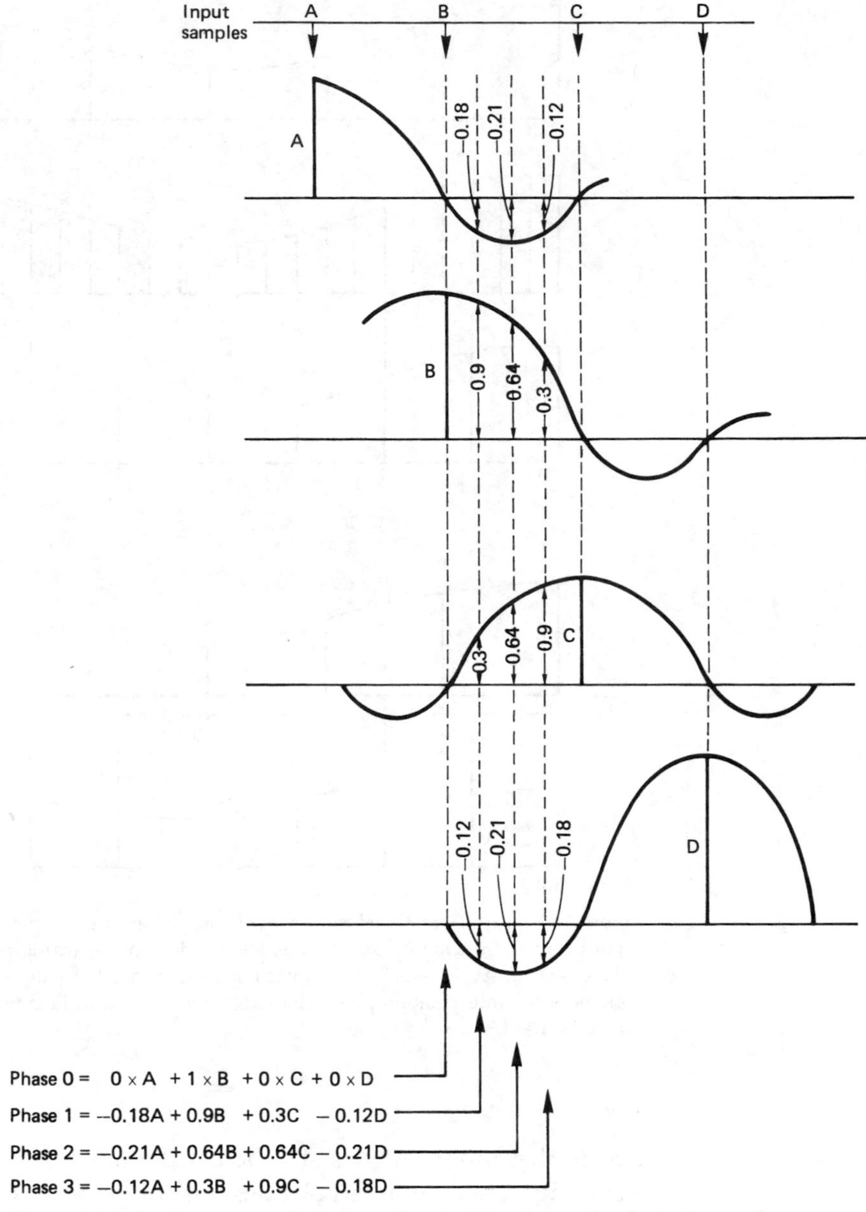

Figure 3.54 In 4× oversampling, for each set of input samples, four phases of coefficients are necessary, each of which produces one of the oversampled values.

prevents aliasing when the rate is being reduced, and prevents images when the rate is being increased. The interpolator has sufficient coefficient phases to interpolate *m* output samples for every input sample, but not all these values are computed; only interpolations which coincide with an output sample are performed. It will be seen in Figure

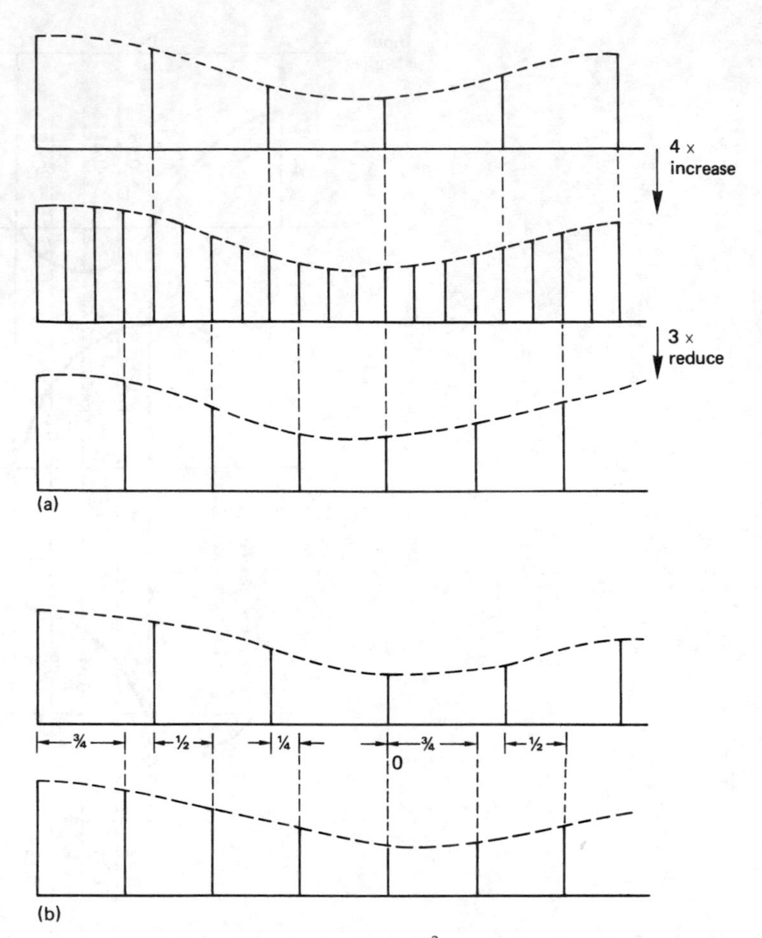

Figure 3.55 At (a), fractional-ratio conversion of $\frac{3}{4}$ in this sample is by increasing to 4× input prior to reducing by 3×. The inefficiency due to discarding previously computed values is clear. At (b), efficiency is raised since only needed values will be computed. Note how the interpolation phase changes for each output. Fixed coefficients can no longer be used.

3.55(b) that input samples shift across the transversal filter at the input sampling rate, but interpolations are only performed at the output sample rate. This is possible because a different filter phase will be used at each interpolation.

In the previous examples, the sample rate or spacing of the filter output had a constant relationship to the input, which meant that the two rates had to be phase-locked. This is an undesirable constraint in some applications, including image manipulators. In a variable-ratio interpolator, values will exist for the points at which input samples were made, but it is necessary to compute what the sample values would have been at absolutely any point between available samples. The general concept of the interpolator is the same as for the fractional-ratio convertor, except that an infinite number of filter phases is ideally necessary. Since a realizable filter will have a finite number of phases, it is necessary to study the degradation this causes. The desired continuous temporal or spatial axis of the interpolator is quantized by the phase spacing, and a sample value needed at a particular point will be replaced by a value for the nearest available filter phase. The number of phases in the filter therefore determines the accuracy of the interpolation. The effects of calculating a value for the wrong point are identical to those of sampling with clock jitter, in that an error occurs proportional to the slope of the signal. The result is program-modulated noise. The higher the noise specification, the greater the desired time accuracy and the greater the number of phases required. The number of phases is equal to the number of sets of coefficients available, and should not be confused with the number of points in the filter, which is equal to the number of coefficients in a set (and the number of multiplications needed to calculate one output value).

The sampling jitter accuracy necessary for eight-bit working is measured in picoseconds. This implies that something like 32 filter phases will be required for adequate performance in an eight-bit sampling-rate convertor.

3.23 The Fourier transform

Figure 3.39 showed that if the amplitude and phase of each frequency component is known, linearly adding the resultant components in an inverse transform results in the original waveform. In digital systems the waveform is expressed as a number of discrete samples. As a result the Fourier transform analyses the signal into an equal number of discrete frequencies. This is known as a Discrete Fourier Transform or DFT in which the number of frequency coefficients is equal to the number of input samples. The fast Fourier transform is no more than an efficient way of computing the DFT.[6] As was seen in the previous section, practical systems must use windowing to create short-term transforms.

It will be evident from Figure 3.56 that the knowledge of the phase of the frequency component is vital, as changing the phase of any component will seriously alter the reconstructed waveform. Thus the DFT must accurately analyse the phase of the signal components.

There are a number of ways of expressing phase. Figure 3.57 shows a point which is rotating about a fixed axis at constant speed. Looked at from the side, the point oscillates up and down at constant frequency. The waveform of that motion is a sinewave, and that is what we would see if the rotating point were to translate along its axis whilst we continued to look from the side.

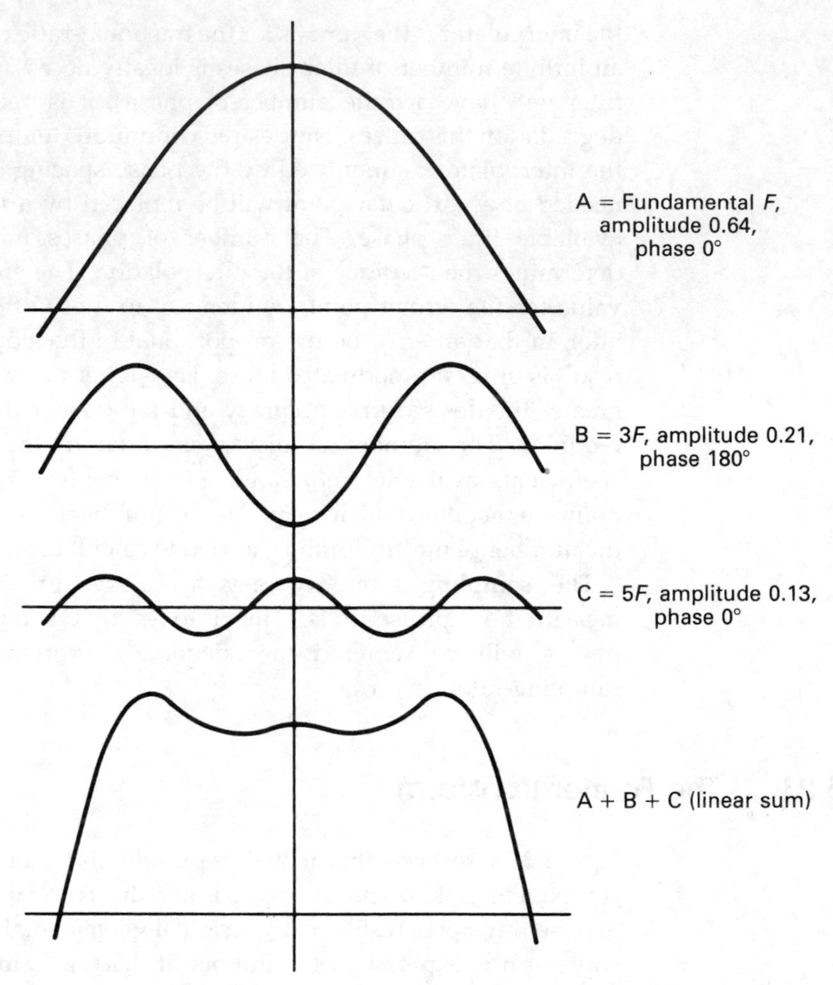

Figure 3.56 Fourier analysis allows the synthesis of any waveform by the addition of discrete frequencies of appropriate amplitude and phase.

One way of defining the phase of a waveform is to specify the angle through which the point has rotated at time zero ($T = 0$). If a second point is made to revolve at 90° to the first, it would produce a cosine wave when translated. It is possible to produce a waveform having arbitrary phase by adding together the sine and cosine wave in various proportions and polarities. For example, adding the sine and cosine waves in equal proportion results in a waveform lagging the sine wave by 45°.

Figure 3.57 shows that the proportions necessary are respectively the sine and the cosine of the phase angle. Thus the two methods of describing phase can be readily interchanged.

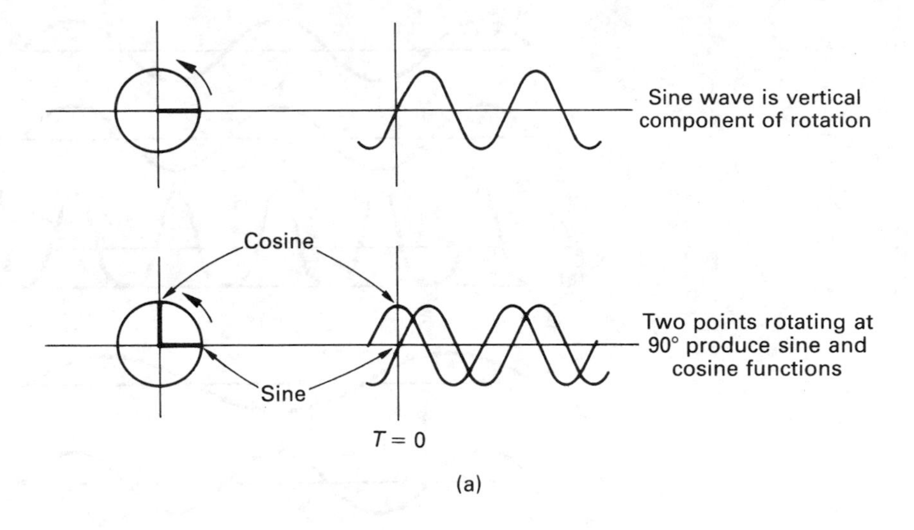

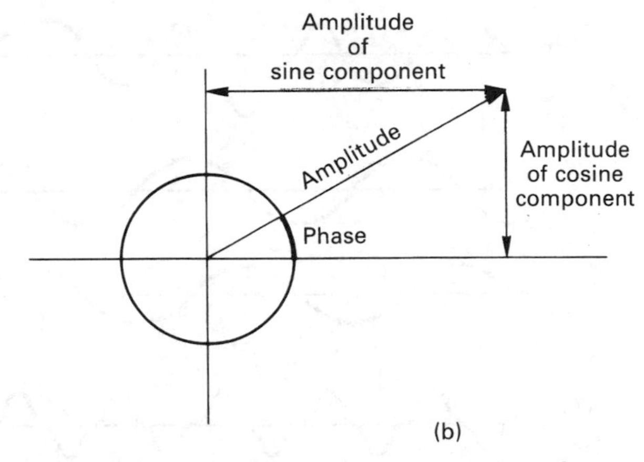

Figure 3.57 The origin of sine and cosine waves is to take a particular viewpoint of a rotation. Any phase can be synthesized by adding proportions of sine and cosine waves.

The discrete Fourier transform spectrum-analyses a string of samples by searching separately for each discrete target frequency. It does this by multiplying the input waveform by a sine wave, known as the basis function, having the target frequency and adding up or integrating the products. Figure 3.58(a) shows that multiplying by basis functions gives a non-zero integral when the input frequency is the same, whereas (b) shows that with a different input frequency (in fact all other different frequencies) the integral is zero showing that no component of the target frequency exists. Thus from a real waveform containing many frequencies all frequencies except the target frequency are excluded. The magnitude of the integral is proportional to the amplitude of the target component.

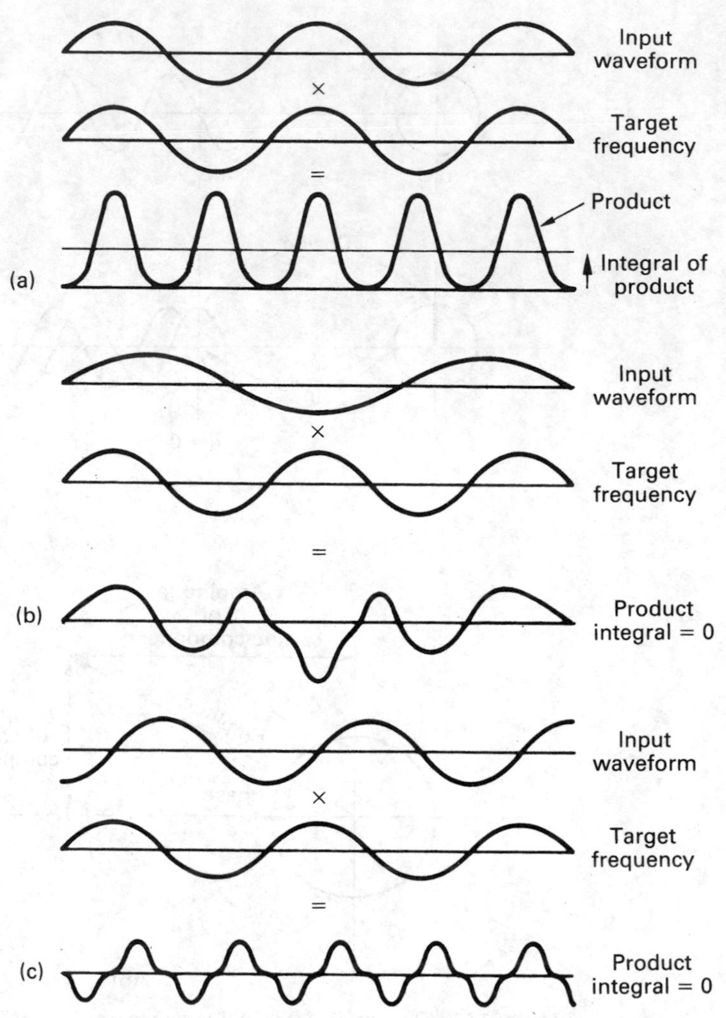

Figure 3.58 The input waveform is multiplied by the target frequency and the result is averaged or integrated. At (a) the target frequency is present and a large integral results. With another input frequency the integral is zero as at (b). The correct frequency will also result in a zero integral shown at (c) if it is at 90° to the phase of the search frequency. This is overcome by making two searches in quadrature.

Figure 3.58(c) shows that the target frequency will not be detected if it is phase shifted 90° as the product of quadrature waveforms is always zero. Thus the discrete Fourier transform must make a further search for the target frequency using a cosine basis function. It follows from the arguments above that the relative proportions of the sine and cosine integrals reveal the phase of the input component. Thus each discrete frequency in the spectrum must be the result of a pair of quadrature searches.

Searching for one frequency at a time as above will result in a DFT, but only after considerable computation. However, a lot of the calculations are repeated many times over in different searches. The fast Fourier transform gives the same result with less computation by logically gathering together all the places where the same calculation is needed and making the calculation once.

The amount of computation can be reduced by performing the sine and cosine component searches together. Another saving is obtained by

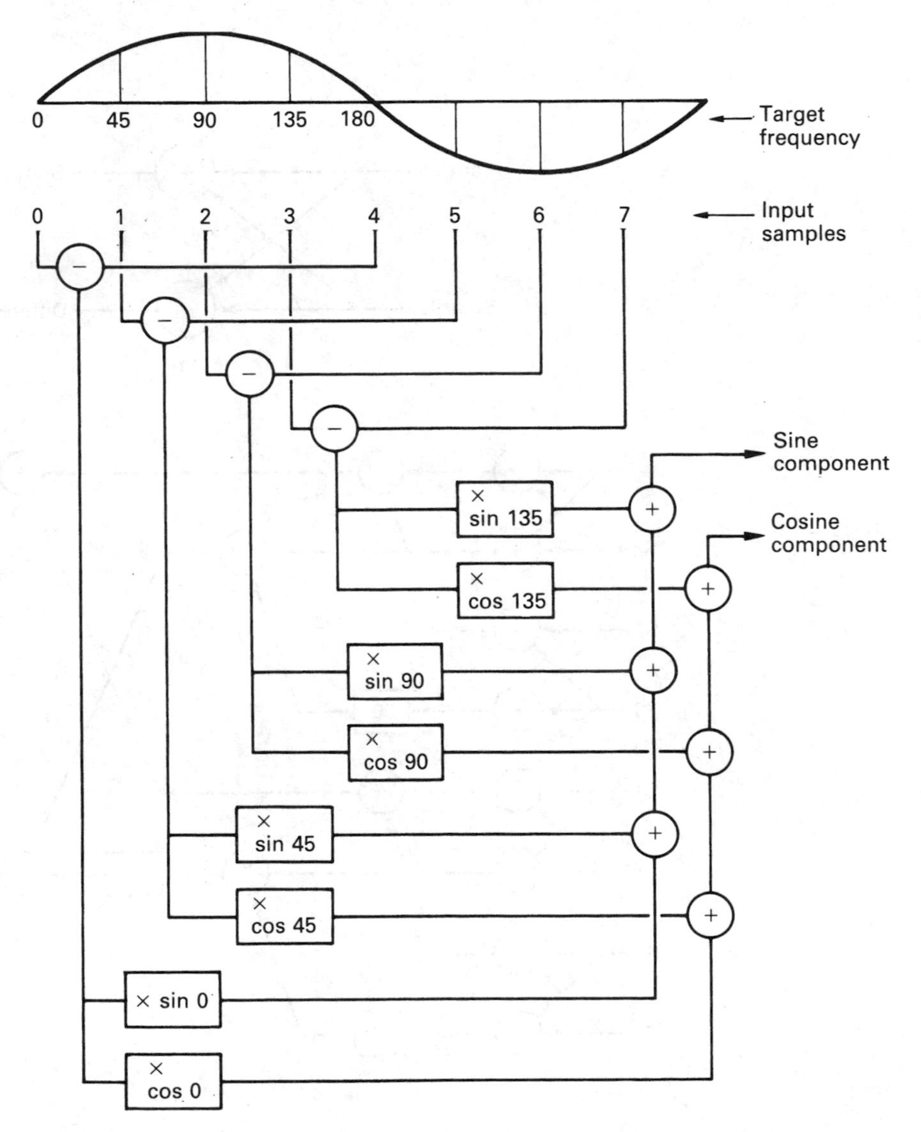

Figure 3.59 An example of a filtering search. Pairs of samples are subtracted and multiplied by sampled sine and cosine waves. The products are added to give the sine and cosine components of the search frequency.

noting that every 180° the sine and cosine have the same magnitude but are simply inverted in sign. Instead of performing four multiplications on two samples 180° apart and adding the pairs of products it is more economical to subtract the sample values and multiply twice, once by a sine value and once by a cosine value.

The first coefficient is the arithmetic mean which is the sum of all the sample values in the block divided by the number of samples. Figure 3.59 shows how the search for the lowest frequency in a block is performed. Pairs of samples are subtracted as shown, and each

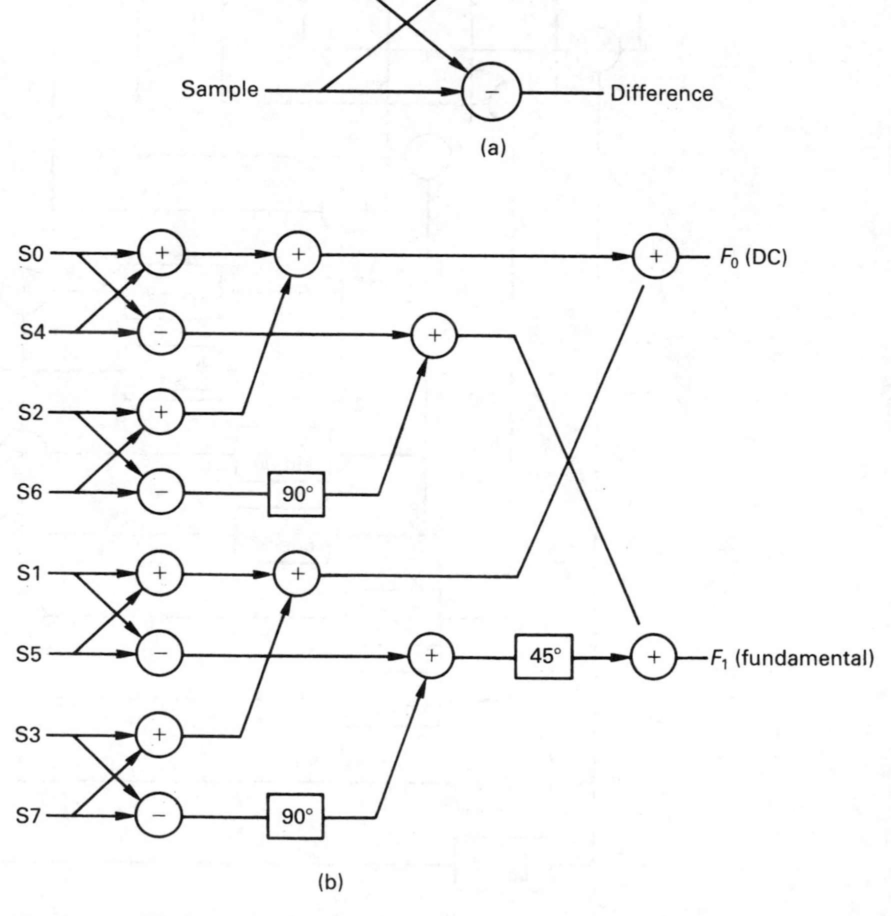

Figure 3.60 The basic element of an FFT is known as a butterfly as at (a) because of the shape of the signal paths in a sum and difference system. The use of butterflies to compute the first two coefficients is shown in (b).

difference is then multiplied by the sine and the cosine of the search frequency. The process shifts one sample period, and a new sample pair are subtracted and multiplied by new sine and cosine factors. This is repeated until all the sample pairs have been multiplied. The sine and cosine products are then added to give the value of the sine and cosine coefficients respectively.

It is possible to combine the calculation of the DC component which requires the sum of samples and the calculation of the fundamental which requires sample differences by combining stages shown in Figure 3.60(a) which take a pair of samples and add and subtract them. Such a stage is called a butterfly because of the shape of the schematic. Figure 3.60(b) shows how the first two components are calculated. The phase rotation

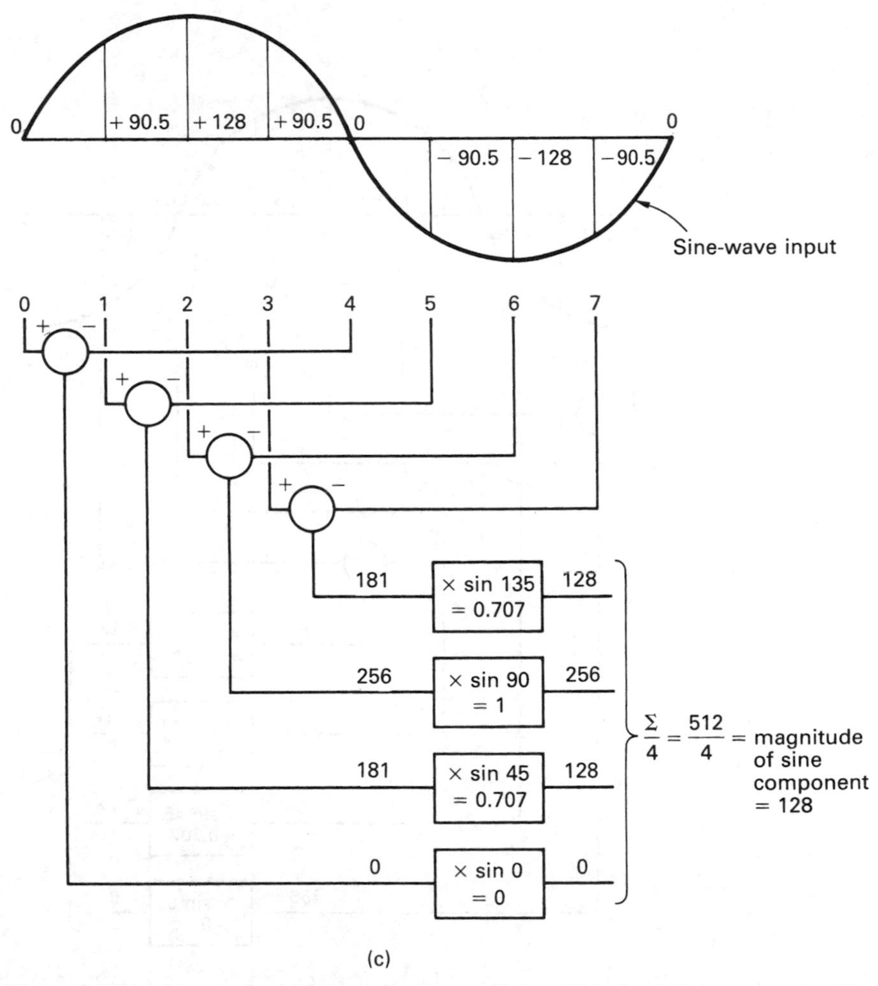

Figure 3.60(c) An actual calculation of a sine coefficient. This should be compared with the result shown in (d).

boxes attribute the input to the sine or cosine component outputs according to the phase angle. As shown, the box labelled 90° attributes nothing to the sine output, but unity gain to the cosine output. The 45° box attributes the input equally to both components.

Figure 3.60(c) shows a numerical example. If a sinewave input is considered where 0° coincides with the first sample, this will produce a zero sine coefficient and non-zero cosine coefficient while (d) shows the same input waveform shifted by 90°. Note how the coefficients change over.

Figure 3.60(e) shows how the next frequency coefficient is computed. Note that exactly the same first-stage butterfly outputs are used, reducing the computation needed.

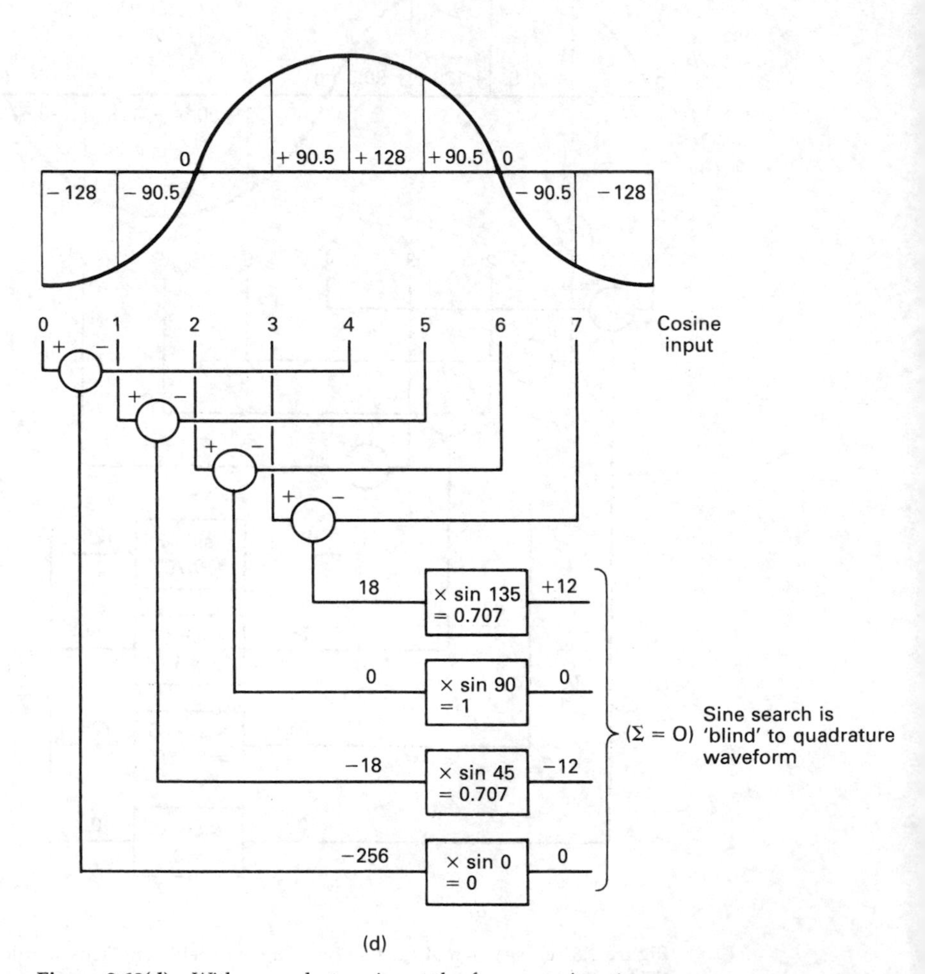

Figure 3.60(d) With a quadrature input the frequency is not seen.

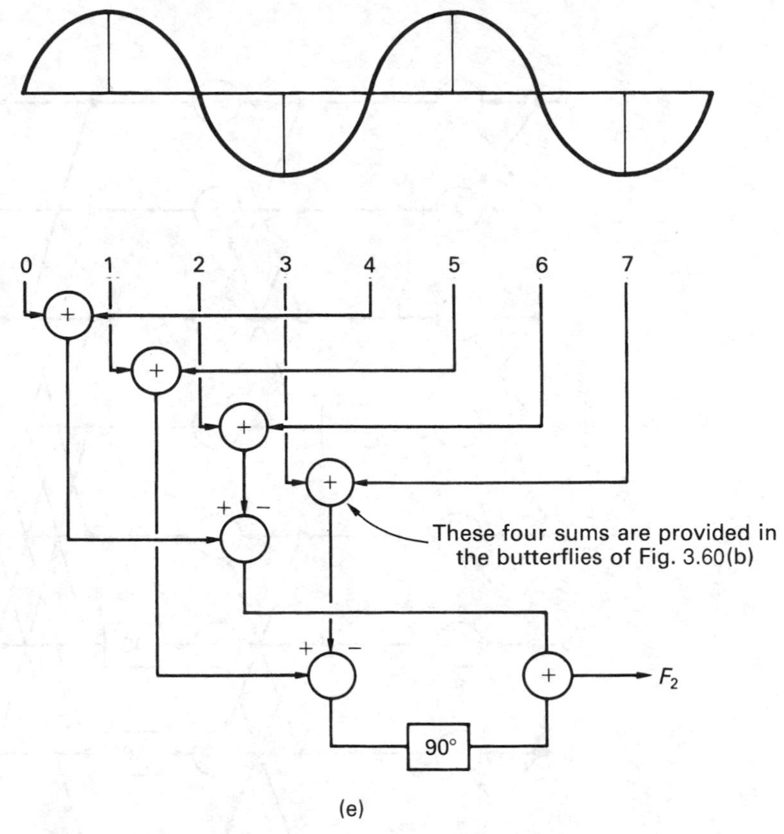

Figure 3.60(e) The butterflies used for the first coefficients form the basis of the computation of the next coefficient.

A similar process may be followed to obtain the sine and cosine coefficients of the remaining frequencies. The full FFT diagram for eight samples is shown in Figure 3.61(a). The spectrum this calculates is shown in (b). Note that only half of the coefficients are useful in a real band-limited system because the remaining coefficients represent frequencies above one half of the sampling rate.

In STFTs the overlapping input sample blocks must be multiplied by window functions. The principle is the same as for the application in FIR filters shown in section 3.21. Figure 3.62 shows that multiplying the search frequency by the window has exactly the same result except that this need be done only once and much computation is saved. Thus in the STFT the basis function is a windowed sine or cosine wave.

The FFT is used extensively in such applications as phase correlation, where the accuracy with which the phase of signal components can be analysed is essential. It also forms the foundation of the discrete cosine transform.

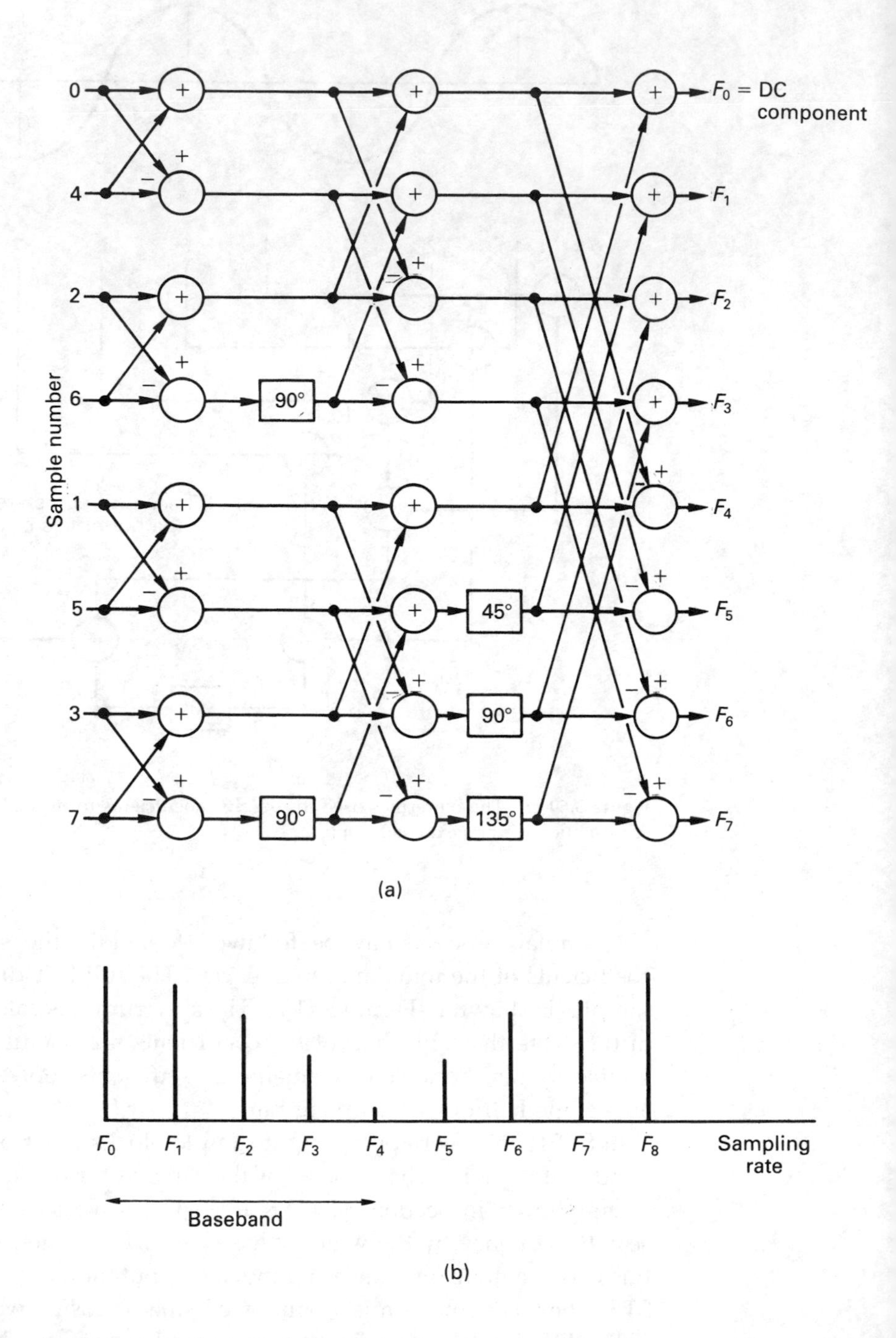

Figure 3.61 At (a) is the full butterfly diagram for an FFT. The spectrum this computes is shown at (b).

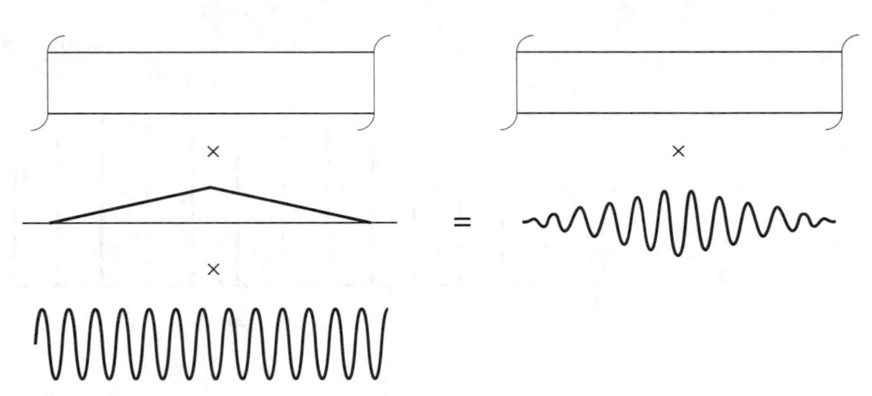

Figure 3.62 Multiplication of a windowed block by a sine wave basis function is the same as multiplying the raw data by a windowed basis function but requires less multiplication as the basis function is constant and can be pre-computed.

3.24 The discrete cosine transform (DCT)

The DCT is a special case of a discrete Fourier transform in which the sine components of the coefficients have been eliminated leaving a single number. This is actually quite easy. Figure 3.63(a) shows a block of input samples to a transform process. By repeating the samples in a time-reversed order and performing a discrete Fourier transform on the double-length sample set a DCT is obtained. The effect of mirroring the input waveform is to turn it into an even function whose sine coefficients are all zero. The result can be understood by considering the effect of individually transforming the input block and the reversed block.

Figure 3.63(b) shows that the phase of all the components of one block are in the opposite sense to those in the other. This means that when the components are added to give the transform of the double-length block all the sine components cancel out, leaving only the cosine coefficients, hence the name of the transform.[7] In practice the sine component calculation is eliminated. Another advantage is that doubling the block length by mirroring doubles the frequency resolution, so that twice as many useful coefficients are produced. In fact a DCT produces as many useful coefficients as input samples.

For image processing two-dimensional transforms are needed. In this case for every horizontal frequency, a search is made for all possible vertical frequencies. A two-dimensional DCT is shown in Figure 3.64. The DCT is separable in that the two-dimensional DCT can be obtained by computing in each dimension separately. Fast DCT algorithms are available.[8]

Figure 3.65 shows how a two-dimensional DCT is calculated by multiplying each pixel in the input block by terms which represent sampled cosine waves of various spatial frequencies. A given DCT

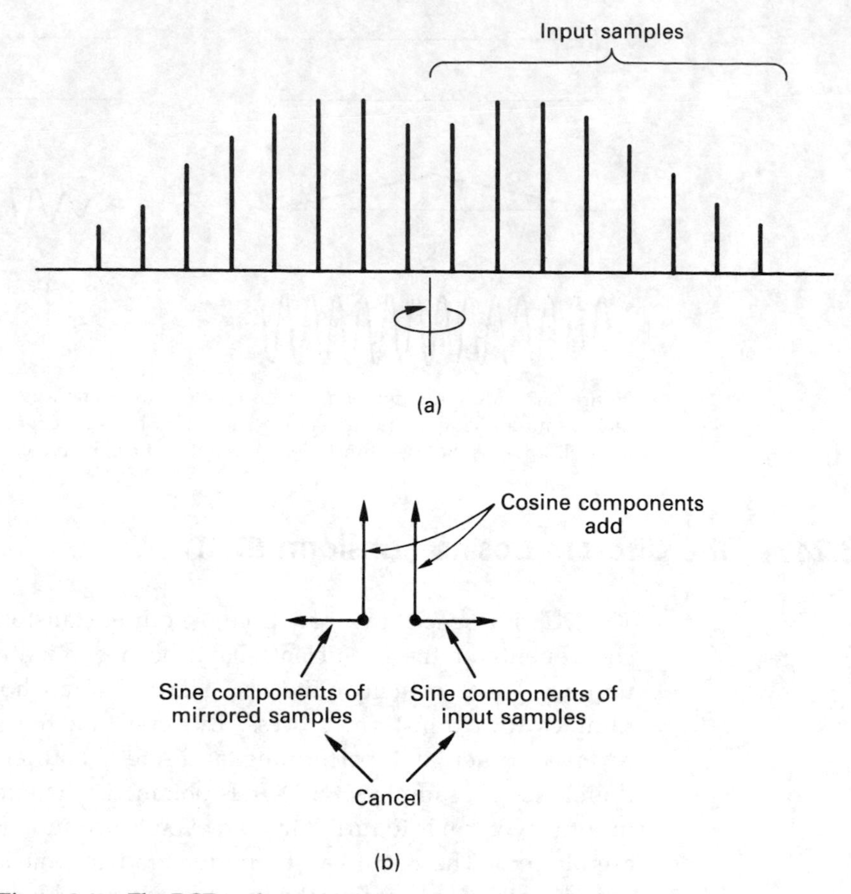

Figure 3.63 The DCT is obtained by mirroring the input block as shown at (a) prior to an FFT. The mirroring cancels out the sine components as at (b), leaving only cosine coefficients.

coefficient is obtained when the result of multiplying every input pixel in the block is summed. Although most compression systems, including JPEG and MPEG, use square DCT blocks, this is not a necessity and rectangular DCT blocks are possible and are used in, for example, Digital Betacam, SX and DVC.

The DCT is primarily used in MPEG-2 because it converts the input waveform into a form where redundancy can be easily detected and removed. More details of the DCT can be found in Chapter 5.

3.25 The wavelet transform

The wavelet transform was not discovered by any one individual, but has evolved via a number of similar ideas and was only given a strong mathematical foundation relatively recently.[9–12] The wavelet transform is

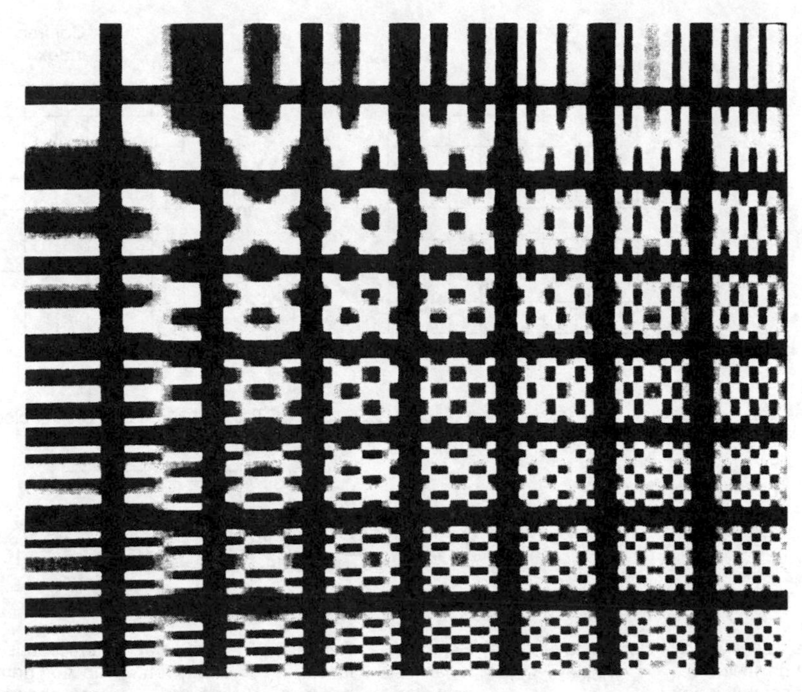

Figure 3.64 The discrete cosine transform breaks up an image area into discrete frequencies in two dimensions. The lowest frequency can be seen here at the top left corner. Horizontal frequency increases to the right and vertical frequency increases downwards.

similar to the Fourier transform in that it has basis functions of various frequencies which are multiplied by the input waveform to identify the frequencies it contains. However, the Fourier transform is based on periodic signals and endless basis functions and requires windowing. The wavelet transform is fundamentally windowed, as the basis functions employed are not endless sinewaves, but are finite on the time axis; hence the name. Wavelet transforms do not use a fixed window, but instead the window period is inversely proportional to the frequency being analysed. As a result a useful combination of time and frequency resolutions is obtained. High frequencies corresponding to transients in audio or edges in video are transformed with short basis functions and therefore are accurately located. Low frequencies are transformed with long basis functions which have good frequency resolution.

Figure 3.66 shows that that a set of wavelets or basis functions can be obtained simply by scaling (stretching or shrinking) a single wavelet on the time axis. Each wavelet contains the same number of cycles such that as the frequency reduces, the wavelet gets longer. Thus the frequency discrimination of the wavelet transform is a constant fraction of the signal frequency. In a filter bank such a characteristic would be described as

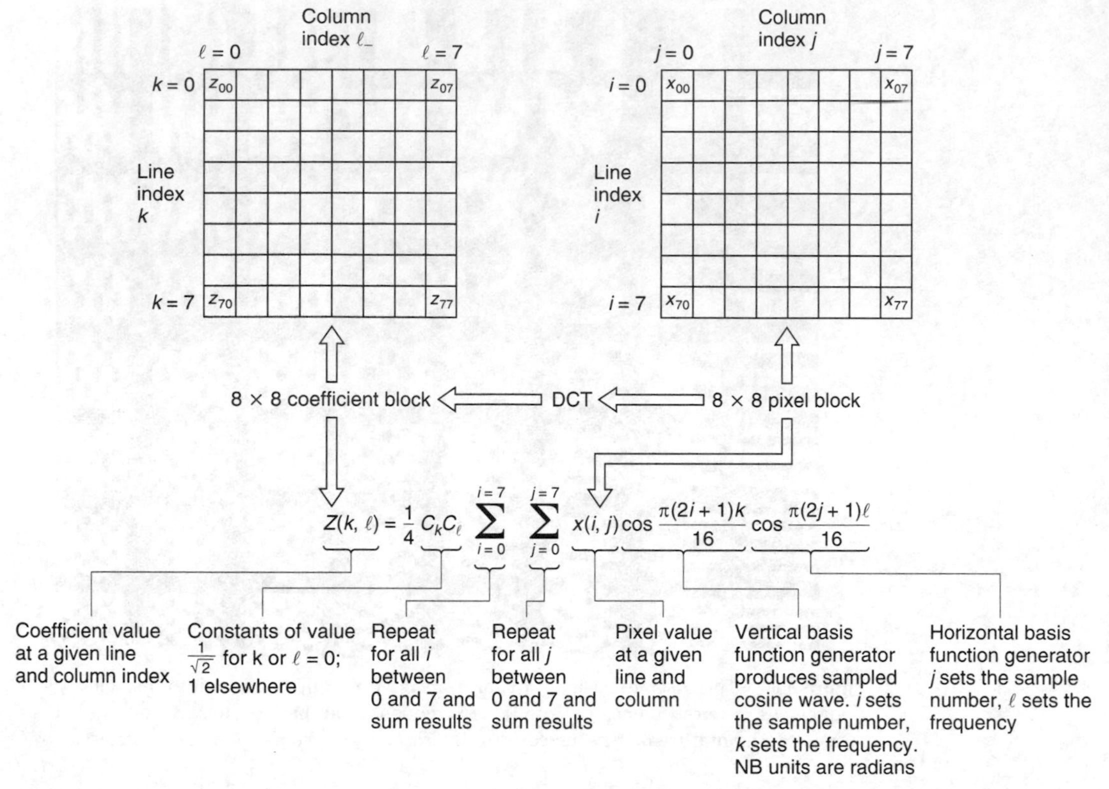

Figure 3.65 A two-dimensional DCT is calculated as shown here. Starting with an input pixel block one calculation is necessary to find a value for each coefficient. After 64 calculations using different basis functions the coefficient block is complete.

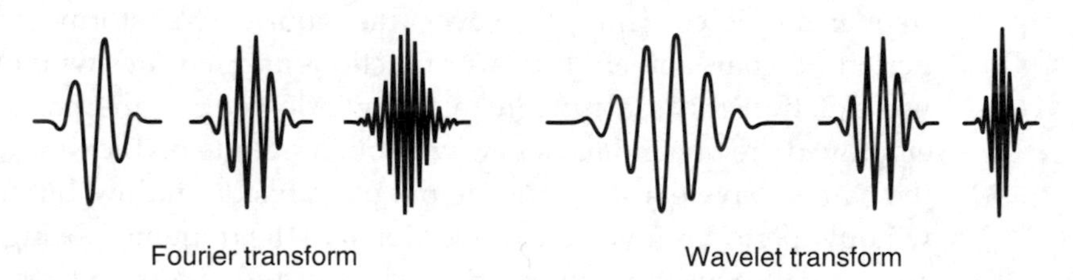

Figure 3.66 Unlike discrete Fourier transforms, wavelet basis functions are scaled so that they contain the same number of cycles irrespective of frequency. As a result their frequency discrimination ability is a constant proportion of the centre frequency.

'constant *Q*'. Figure 3.67 shows that the division of the frequency domain by a wavelet transform is logarithmic whereas in the Fourier transform the division is uniform. The logarithmic coverage is effectively dividing the frequency domain into octaves and as such parallels the frequency discrimination of human hearing.

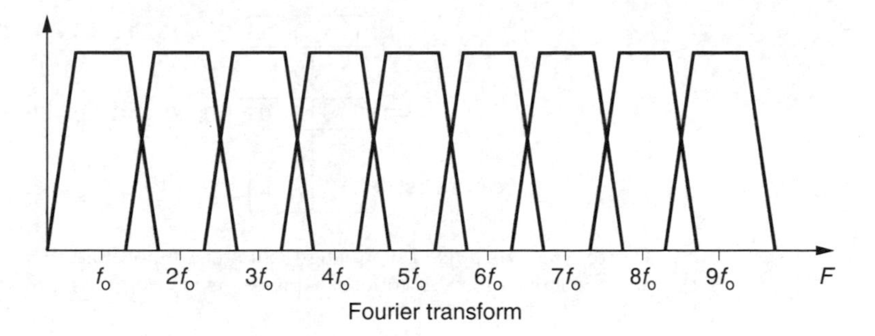

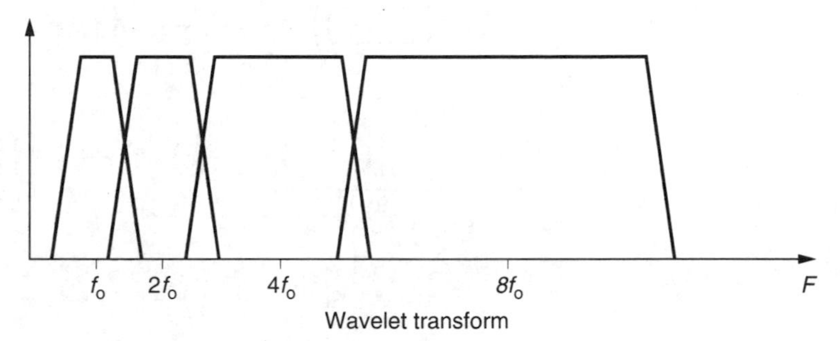

Figure 3.67 Wavelet transforms divide the frequency domain into octaves instead of the equal bands of the Fourier transform.

As it is relatively recent, the wavelet transform has yet to be widely used although it shows great promise. It has successfully been used in audio and in commercially available non-linear video editors and in other fields such as radiology and geology.

In video, wavelet compression does not display the 'blocking' of DCT-based coders at high compression factors; instead compression error is spread over the spectrum and appears as white noise.[13] It is naturally a multi-resolution transform allowing scalable decoding

3.26 Modulo-*n* arithmetic

Conventional arithmetic which is in everyday use relates to the real world of counting actual objects, and to obtain correct answers the concepts of borrow and carry are necessary in the calculations. There is an alternative type of arithmetic which has no borrow or carry which is known as modulo arithmetic. In modulo-n no number can exceed n. If it does, n or whole multiples of n are subtracted until it does not. Thus 25 modulo-16 is 9 and 12 modulo-5 is 2. The count shown in Figure 3.68 is from a four-bit device which overflows when it reaches 1111 because the carry out is ignored. If a number of clock pulses m are applied from the zero state, the state of the counter will be given by m Mod.16. Thus modulo arithmetic is appropriate

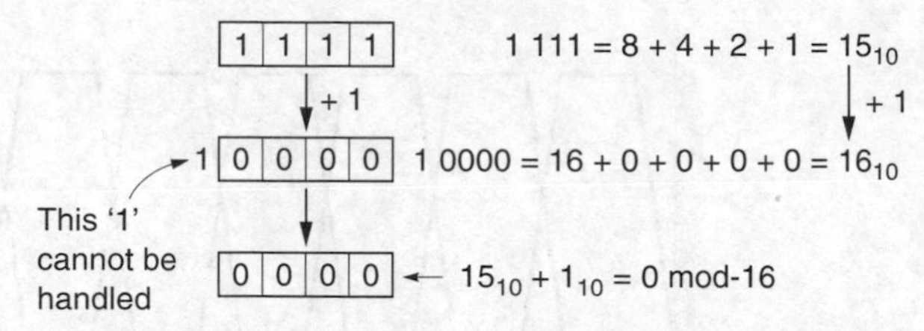

Figure 3.68 As a fixed wordlength counter cannot hold the carry-out bit, it will resume at zero. Thus a four-bit counter expresses every count as a modulo-16 number.

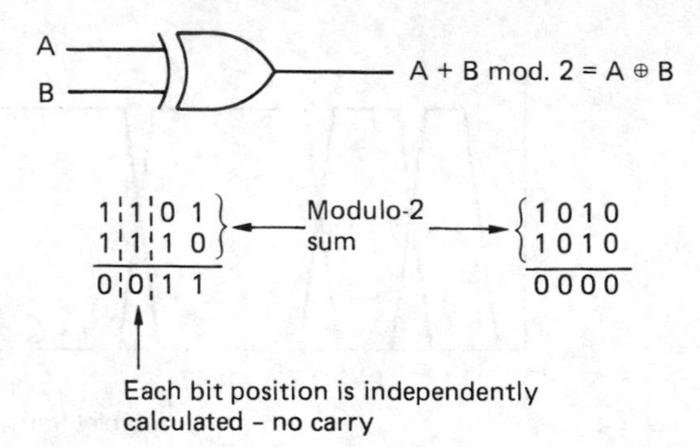

Figure 3.69 In modulo-2 calculations, there can be no carry or borrow operations and conventional addition and subtraction become identical. The XOR gate is a modulo-2 adder.

to systems in which there is a fixed wordlength and this means that the range of values the system can have is restricted by that wordlength. A number range which is restricted in this way is called a finite field.

Modulo-2 is a numbering scheme which is used frequently in digital processes. Figure 3.69 shows that in modulo-2 the conventional addition and subtraction are replaced by the XOR function such that:

A + B Mod.2 = A XOR B.

When multi-bit values are added Mod.2, each column is computed quite independently of any other. This makes Mod.2 circuitry very fast in operation as it is not necessary to wait for the carries from lower-order bits to ripple up to the high-order bits.

Modulo-2 arithmetic is not the same as conventional arithmetic and takes some getting used to. For example, adding something to itself in Mod.2 always gives the answer zero.

3.27 The Galois field

Figure 3.70 shows a simple circuit consisting of three D-type latches which are clocked simultaneously. They are connected in series to form a shift register. At (a) a feedback connection has been taken from the output

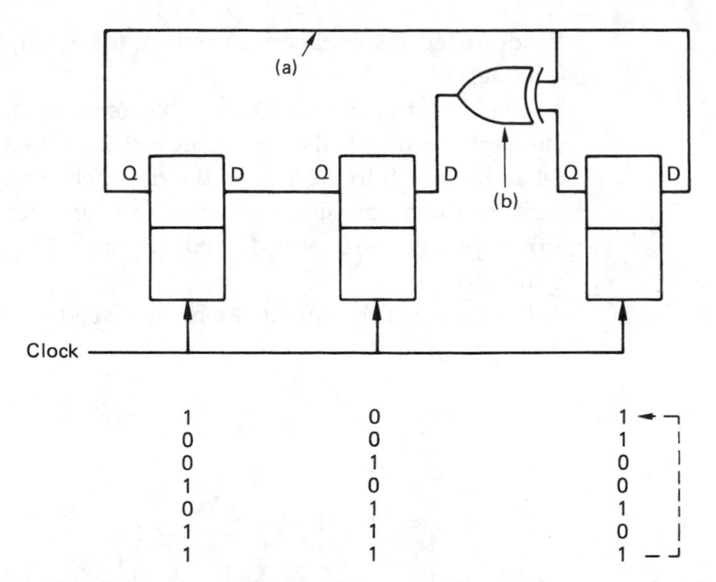

Figure 3.70 The circuit shown is a twisted-ring counter which has an unusual feedback arrangement. Clocking the counter causes it to pass through a series of non-sequential values. See text for details.

to the input and the result is a ring counter where the bits contained will recirculate endlessly. At (b) one XOR gate is added so that the output is fed back to more than one stage. The result is known as a twisted-ring counter and it has some interesting properties. Whenever the circuit is clocked, the left-hand bit moves to the right-hand latch, the centre bit moves to the left-hand latch and the centre latch becomes the XOR of the two outer latches. The figure shows that whatever the starting condition of the three bits in the latches, the same state will always be reached again after seven clocks, except if zero is used. The states of the latches form an endless ring of non-sequential numbers called a Galois field after the French mathematical prodigy Evariste Galois who discovered them. The states of the circuit form a maximum length sequence because there are as many states as are permitted by the wordlength. As the states of the sequence have many of the characteristics of random numbers, yet are repeatable, the result can also be called a pseudo-random sequence (prs). As the all-zeros case is disallowed, the length of a maximum length sequence generated by a register of m bits cannot exceed $(2^m - 1)$ states. The Galois field, however, includes the zero term. It is useful to explore the bizarre mathematics of Galois fields which use modulo-2 arithmetic. Familiarity with such manipulations is helpful when studying error correction, particularly the Reed–Solomon codes used in recorders and treated in Chapter 9. They will also be found in processes which require pseudo-random numbers such as digital dither, treated in Chapter 5, and randomized channel codes used in, for example, DVB and discussed in Chapter 14.

The circuit of Figure 3.70 can be considered as a counter and the four points shown will then be representing different powers of 2 from the MSB on the left to the LSB on the right. The feedback connection from the MSB to the other stages means that whenever the MSB becomes 1, two other powers are also forced to one so that the code of 1011 is generated.

Each state of the circuit can be described by combinations of powers of x, such as

$$x^2 = 100$$

$$x = 010$$

$$x^2 + x = 110, \text{ etc.}$$

The fact that three bits have the same state because they are connected together is represented by the Mod.2 equation:

$$x^3 + x + 1 = 0$$

Let $x = a$, which is a primitive element.

Now

$$a^3 + a + 1 = 0 \tag{3.1}$$

In modulo-2

$$a + a = a^2 + a^2 = 0$$

$$a = x = 010$$

$$a^2 = x^2 = 100$$

$$a^3 = a + 1 = 011 \text{ from (3.1)}$$

$$a^4 = a \times a^3 = a(a + 1) = a^2 + a = 110$$

$$a^5 = a \times a^4 = a(a^2 + a) = a^3 + a^2 = 111$$

$$a^6 = a^3 \times a^3 = (a + 1)^2 = a^2 + a + a + 1$$

$$= a^2 + 1 = 101$$

$$a^7 = a(a^2 + 1) = a^3 + a$$

$$= a + 1 + a = 1 = 001$$

In this way it can be seen that the complete set of elements of the Galois field can be expressed by successive powers of the primitive element. Note that the twisted-ring circuit of Figure 3.70 simply raises a to higher and higher powers as it is clocked. Thus the seemingly complex multibit changes caused by a single clock of the register become simple to calculate using the correct primitive and the appropriate power.

The numbers produced by the twisted-ring counter are not random; they are completely predictable if the equation is known. However, the sequences produced are sufficiently similar to random numbers that in many cases they will be useful. They are thus referred to as pseudo-random sequences. The feedback connection is chosen such that the expression it implements will not factorize. Otherwise a maximum-length sequence could not be generated because the circuit might sequence around one or other of the factors depending on the initial condition. A useful analogy is to compare the operation of a pair of meshed gears. If the gears have a number of teeth which is relatively prime, many revolutions are necessary to make the same pair of teeth touch again. If the number of teeth have a common multiple, far fewer turns are needed.

Figure 3.71 shows the pseudo-random sequence generator used in DVB. Its purpose is to modify the transmitted spectrum so that the amount of energy transmitted is as uniform as possible across the channel.

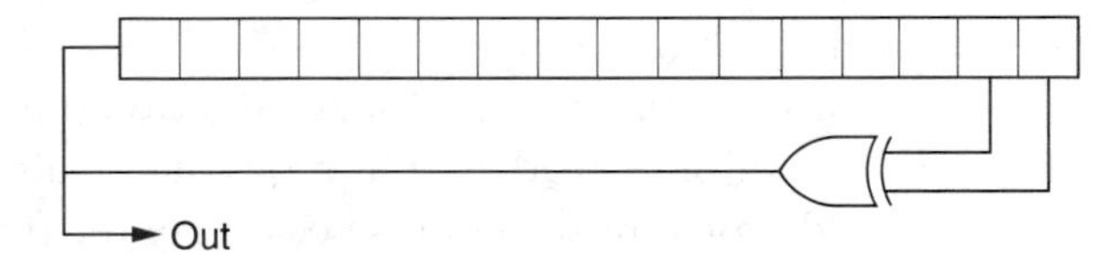

Figure 3.71 The PRS generator of DVB.

3.28 Noise and probability

Probability is a useful concept when dealing with processes which are not completely predictable. Thermal noise in electronic components is random, and although under given conditions the noise power in a system may be constant, this value only determines the heat that would be developed in a resistive load. In digital systems, it is the instantaneous voltage of noise which is of interest, since it is a form of interference which could alter the state of a binary signal if it were large enough. Unfortunately the instantaneous voltage cannot be predicted; indeed if it could the interference could not be called noise. Noise can only be

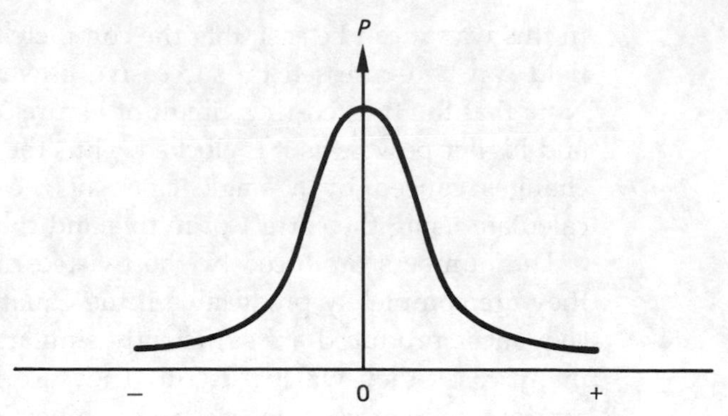

Figure 3.72 White noise in analog circuits generally has the Gaussian amplitude distribution shown.

quantified statistically, by measuring or predicting the likelihood of a given noise amplitude.

Figure 3.72 shows a graph relating the probability of occurrence to the amplitude of noise. The noise amplitude increases away from the origin along the horizontal axis, and for any amplitude of interest, the probability of that noise amplitude occurring can be read from the curve. The shape of the curve is known as a Gaussian distribution, which crops up whenever the overall effect of a large number of independent phenomena is considered. Thermal noise is due to the contributions from countless molecules in the component concerned. Magnetic recording depends on superimposing some average magnetism on vast numbers of magnetic particles.

If it were possible to isolate an individual noise-generating microcosm of a tape or a head on the molecular scale, the noise it could generate would have physical limits because of the finite energy present. The noise distribution might then be rectangular as shown in Figure 3.73(a), where all amplitudes below the physical limit are equally likely. The output of a twisted-ring counter such as that in Figure 3.70 can have a uniform probability. Each value occurs once per sequence. The outputs are positive only but do not include zero, but every value from 1 up to 2^{n-1} is then equally likely.

The output of a prs generator can be made into the two's complement form by inverting the MSB. This has the effect of exchanging the missing all zeros value for a missing fully negative value as can be seen by considering the number ring in Figure 3.5. In this example, inverting the MSB causes the code of 1000 representing –8 to become 0000. The result is a four-bit prs generating uniform probability from –7 to +7 as shown in Figure 3.73(a).

If the combined effect of two of these uniform probability processes is considered, clearly the maximum amplitude is now doubled, because the

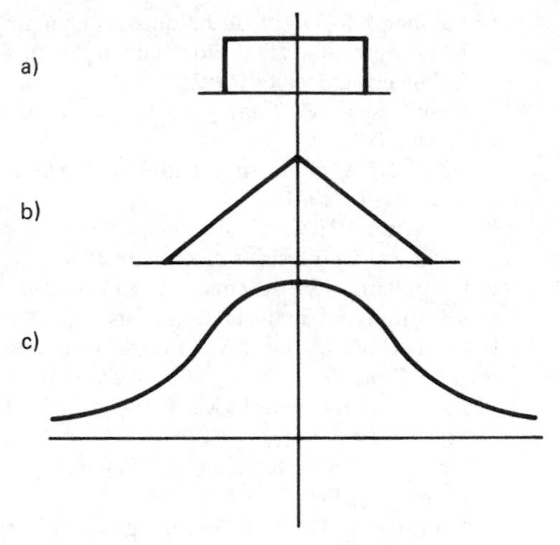

Figure 3.73 At (a) is a rectangular probability; all values are equally likely but between physical limits. At (b) is the sum of two rectangular probabilities, which is triangular, and at (c) is the Gaussian curve which is the sum of an infinite number of rectangular probabilities.

two effects can add, but provided the two effects are uncorrelated, they can also subtract, so the probability is no longer rectangular, but becomes triangular as in (b). The probability falls to zero at peak amplitude because the chances of two independent mechanisms reaching their peak value with the same polarity at the same time are understandably small.

If the number of mechanisms summed together is now allowed to increase without limit, the result is the Gaussian curve shown in (c), where it will be seen that the curve has no amplitude limit, because it is just possible that all mechanisms will simultaneously reach their peak value together, although the chances of this happening are incredibly remote. Thus the Gaussian curve is the overall probability of a large number of uncorrelated uniform processes.

References

1. Ray, S.F., *Applied Photographic Optics*, Oxford: Focal Press (1988) (Ch. 17)
2. van den Enden, A.W.M. and Verhoeckx, N.A.M., Digital signal processing: theoretical background. *Philips Tech. Rev.*, **42**, 110–144, (1985)
3. McClellan, J.H., Parks, T.W. and Rabiner, L.R., A computer program for designing optimum FIR linear-phase digital filters. *IEEE Trans. Audio and Electroacoustics*, **AU-21**, 506–526 (1973)
4. Crochiere, R.E. and Rabiner, L.R., Interpolation and decimation of digital signals – a tutorial review. *Proc. IEEE*, **69**, 300–331 (1981)

5. Rabiner, L.R., Digital techniques for changing the sampling rate of a signal. In B. Blesser, B. Locanthi and T.G. Stockham Jr (eds), *Digital Audio*, pp. 79–89, New York: Audio Engineering Society (1982)
6. Kraniauskas, P., *Transforms in signals and systems*, Chapter 6. Wokingham: Addison Wesley (1992)
7. Ahmed, N., Natarajan, T. and Rao, K., Discrete Cosine Transform. *IEEE Trans. Computers*, **C-23** 90–93 (1974)
8. De With, P.H.N., *Data compression techniques for digital video recording*, PhD thesis, Technical University of Delft (1992)
9. Goupillaud, P., Grossman, A. and Morlet, J., Cycle-Octave and related transforms in seismic signal analysis. *Geoexploration*, **23**, 85–102, Elsevier Science (1984/5)
10. Daubechies, I., The wavelet transform, time–frequency localisation and signal analysis. *IEEE Trans. Info. Theory*, **36**, No.5, 961–1005 (1990)
11. Rioul, O. and Vetterli, M., Wavelets and signal processing. *IEEE Signal Process. Mag.*, 14–38 (Oct. 1991)
12. Strang, G. and Nguyen, T., Wavelets and Filter Banks, Wellesly, MA: Wellesley-Cambridge Press (1996)
13. Huffman, J., Wavelets and image compression. Presented at 135th SMPTE Tech. Conf. (Los Angeles 1993), Preprint No. 135–198

4

Conversion

4.1 Introduction to conversion

There are a number of ways in which a video waveform can digitally be represented, but the most useful and therefore common is Pulse Code Modulation or PCM which was introduced in Chapter 1. The input is a continuous-time, continuous-voltage video waveform, and this is converted into a discrete-time, discrete-voltage format by a combination of sampling and quantizing. As these two processes are orthogonal (a $64 000 word for at right angles to one another) they are totally independent and can be performed in either order. Figure 4.1(a) shows an analog sampler preceding a quantizer, whereas (b) shows an asynchronous quantizer preceding a digital sampler. Ideally, both will give the same results; in practice each has different advantages and suffers from different deficiencies. Both approaches will be found in real equipment.

The independence of sampling and quantizing allows each to be discussed quite separately in some detail, prior to combining the processes for a full understanding of conversion.

Whilst sampling an analog video waveform takes place in the time domain in an electrical ADC, this is because the analog waveform is the result of scanning an image. In reality the image has been spatially sampled in two dimensions (lines and pixels) and temporally sampled into pictures along a third dimension. Sampling in a single dimension will be considered before moving on to more dimensions.

4.2 Sampling and aliasing

Sampling is no more than periodic measurement, and it will be shown here that there is no theoretical need for sampling to be detectable.

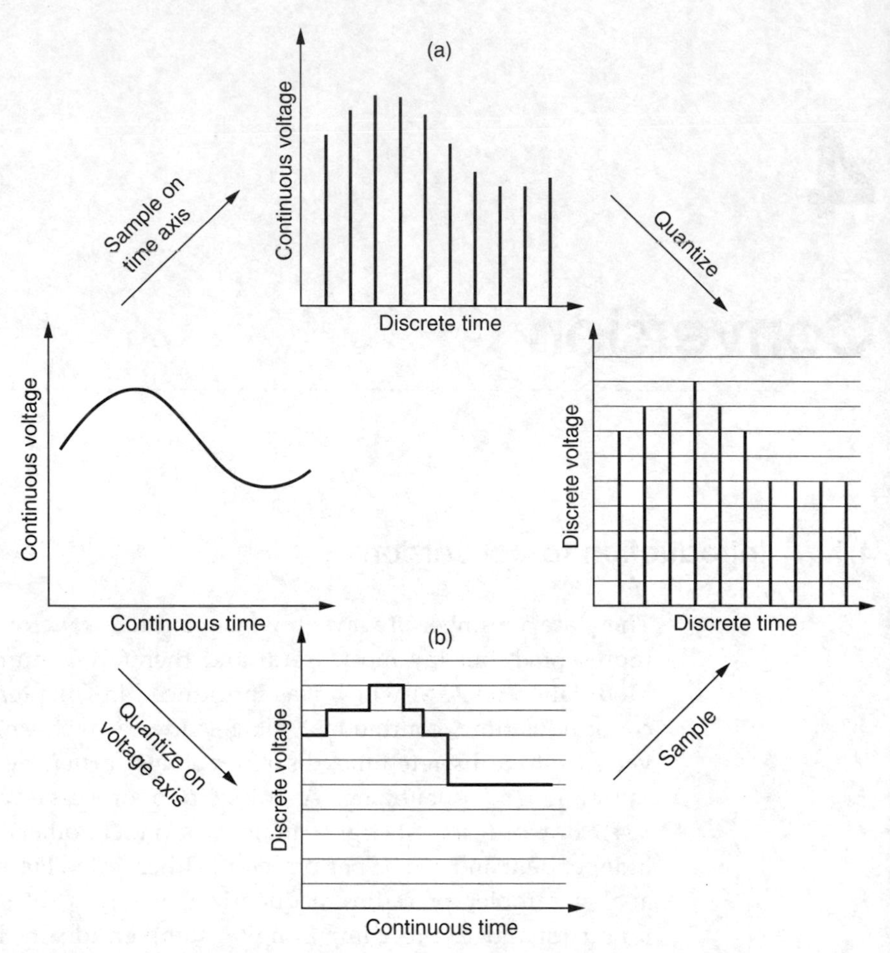

Figure 4.1 Since sampling and quantizing are orthogonal, the order in which they are performed is not important. In (a) sampling is performed first and the samples are quantized. This is common in audio convertors. In (b) the analog input is quantized into an asynchronous binary code. Sampling takes place when this code is latched on sampling clock edges. This approach is universal in video convertors.

Practical television equipment is, of course, less than ideal, particularly in the case of temporal sampling.

Video sampling must be regular, because the process of timebase correction prior to conversion back to a conventional analog waveform assumes a regular original process as was shown in Chapter 1. The sampling process originates with a pulse train which is shown in Figure 4.2(a) to be of constant amplitude and period. The video waveform amplitude-modulates the pulse train in much the same way as the carrier is modulated in an AM radio transmitter. One must be careful to avoid over-modulating the pulse train as shown in (b) and this is achieved by

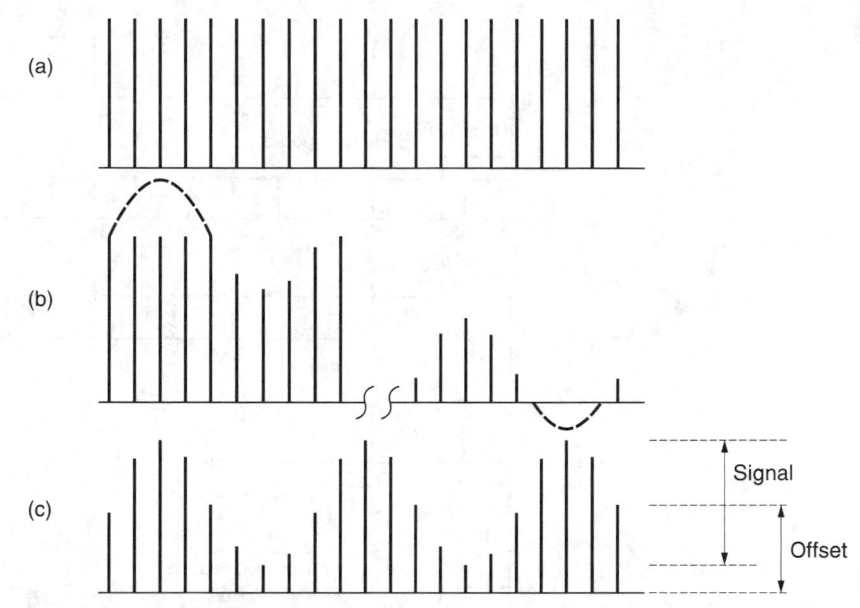

Figure 4.2 The sampling process requires a constant-amplitude pulse train as shown in (a). This is amplitude modulated by the waveform to be sampled. If the input waveform has excessive amplitude or incorrect level, the pulse train clips as shown in (b). For a bipolar waveform, the greatest signal level is possible when an offset of half the pulse amplitude is used to centre the waveform as shown in (c).

applying a DC offset to the analog waveform so that blanking corresponds to a level part-way up the pulses as in (c).

In the same way that AM radio produces sidebands or images above and below the carrier, sampling also produces sidebands although the carrier is now a pulse train and has an infinite series of harmonics as shown in Figure 4.3(a). The sidebands repeat above and below each harmonic of the sampling rate as shown in (b).

The sampled signal can be returned to the continuous-time domain simply by passing it into a low-pass filter. This filter has a frequency response which prevents the images from passing, and only the baseband signal emerges, completely unchanged. If considered in the frequency domain, this filter can be called an anti-image filter; if considered in the time domain it can be called a reconstruction filter. It can also be considered as a spatial filter if a sampled still image is being returned to a continuous image. Such a filter will be two-dimensional.

If an input is supplied having an excessive bandwidth for the sampling rate in use, the sidebands will overlap (Figure 4.3(c)) and the result is aliasing, where certain output frequencies are not the same as their input frequencies but instead become difference frequencies (d). It will be seen from Figure 4.3 that aliasing does not occur when the input frequency is equal to or less than half the sampling rate, and this derives the most

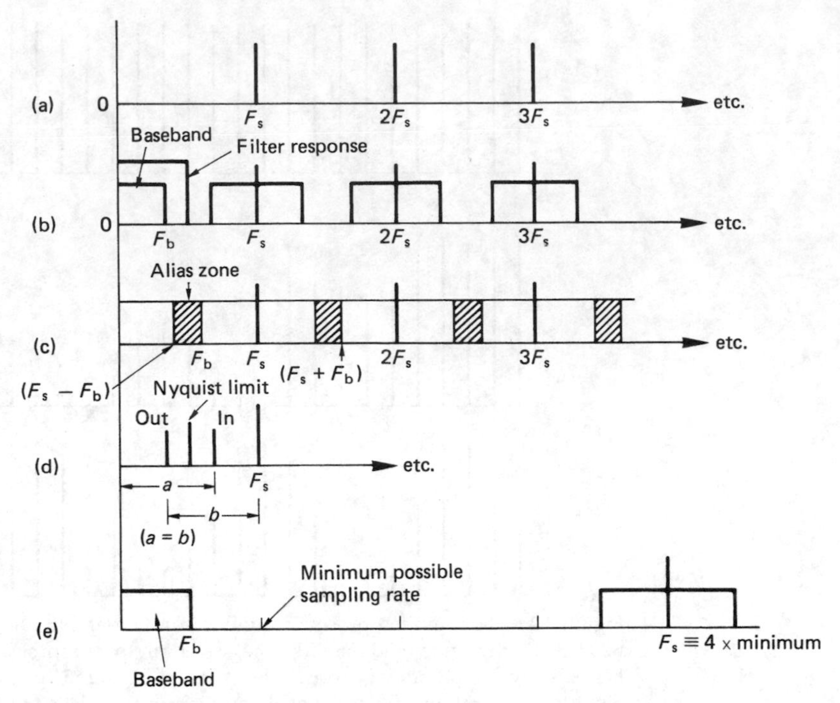

Figure 4.3 (a) Spectrum of sampling pulses. (b) Spectrum of samples. (c) Aliasing due to sideband overlap. (d) Beat-frequency production. (e) 4 × oversampling.

fundamental rule of sampling, which is that the sampling rate must be at least twice the input bandwidth. Sampling theory is usually attributed to Shannon[1,2] who applied it to information theory at around the same time as Kotelnikov in Russia. These applications were pre-dated by Whittaker. Despite that it is often referred to as Nyquist's theorem.

Whilst aliasing has been described above in the frequency domain, it can be described equally well in the time domain. In Figure 4.4(a) the sampling rate is obviously adequate to describe the waveform, but in (b) it is inadequate and aliasing has occurred.

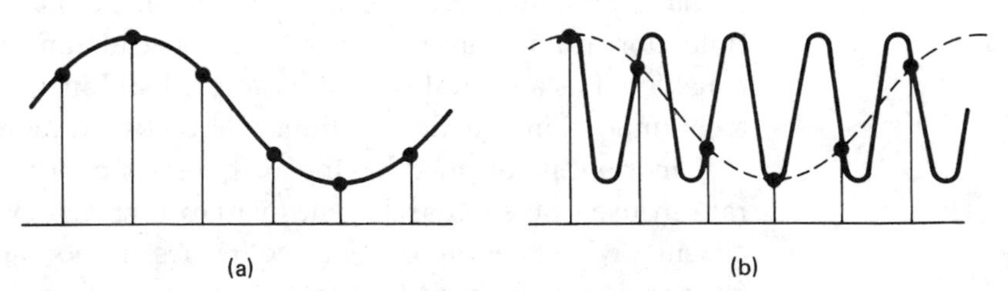

Figure 4.4 In (a), the sampling is adequate to reconstruct the original signal. In (b) the sampling rate is inadequate, and reconstruction produces the wrong waveform (detailed). Aliasing has taken place.

One often has no control over the spectrum of input signals and in practice it is necessary also to have a low-pass filter at the input to prevent aliasing. This anti-aliasing filter prevents frequencies of more than half the sampling rate from reaching the sampling stage. The requirement for an anti-aliasing filter extends to two-dimensional sampling devices such as CCD sensors, as will be seen in section 3.6.

Whilst electrical or optical anti-aliasing filters are quite feasible, there is no corresponding device which can precede the image sampling at frame or field rate in film or TV cameras and as a result aliasing is commonly seen on television and in the cinema, owing to the relatively low frame rates used.

With a frame rate of 24 Hz, a film camera will alias on any object changing at more than 12 Hz. Such objects include the spokes of stagecoach wheels. When the spoke-passing frequency reaches 24 Hz the wheels appear to stop. Temporal aliasing in television is less visible than might be thought because of the way in which the eye perceives motion. This was discussed in Chapter 2.

4.3 Reconstruction

If ideal low-pass anti-aliasing and anti-image filters are assumed, having a vertical cut-off slope at half the sampling rate, an ideal spectrum shown in Figure 4.5(a) is obtained. It was shown in Chapter 2 that the impulse response of a phase linear ideal low-pass filter is a $\sin x/x$ waveform in the time domain, and this is repeated in (b). Such a waveform passes through 0 Volts periodically. If the cut-off frequency of the filter is one-half of the sampling rate, the impulse passes through zero *at the sites of all other samples*. It can be seen from Figure 4.5(c) that at the output of such a filter, the voltage at the centre of a sample is due to that sample alone, since the value of *all* other samples is zero at that instant. In other words the continuous time output waveform must join up the tops of the input samples. In between the sample instants, the output of the filter is the sum of the contributions from many impulses, and the waveform smoothly joins the tops of the samples. If the time domain is being considered, the anti-image filter of the frequency domain can equally well be called the reconstruction filter. It is a consequence of the band-limiting of the original anti-aliasing filter that the filtered analog waveform could only travel between the sample points in one way. As the reconstruction filter has the same frequency response, the reconstructed output waveform must be identical to the original band-limited waveform prior to sampling. A rigorous mathematical proof of reconstruction can be found in Betts.[3]

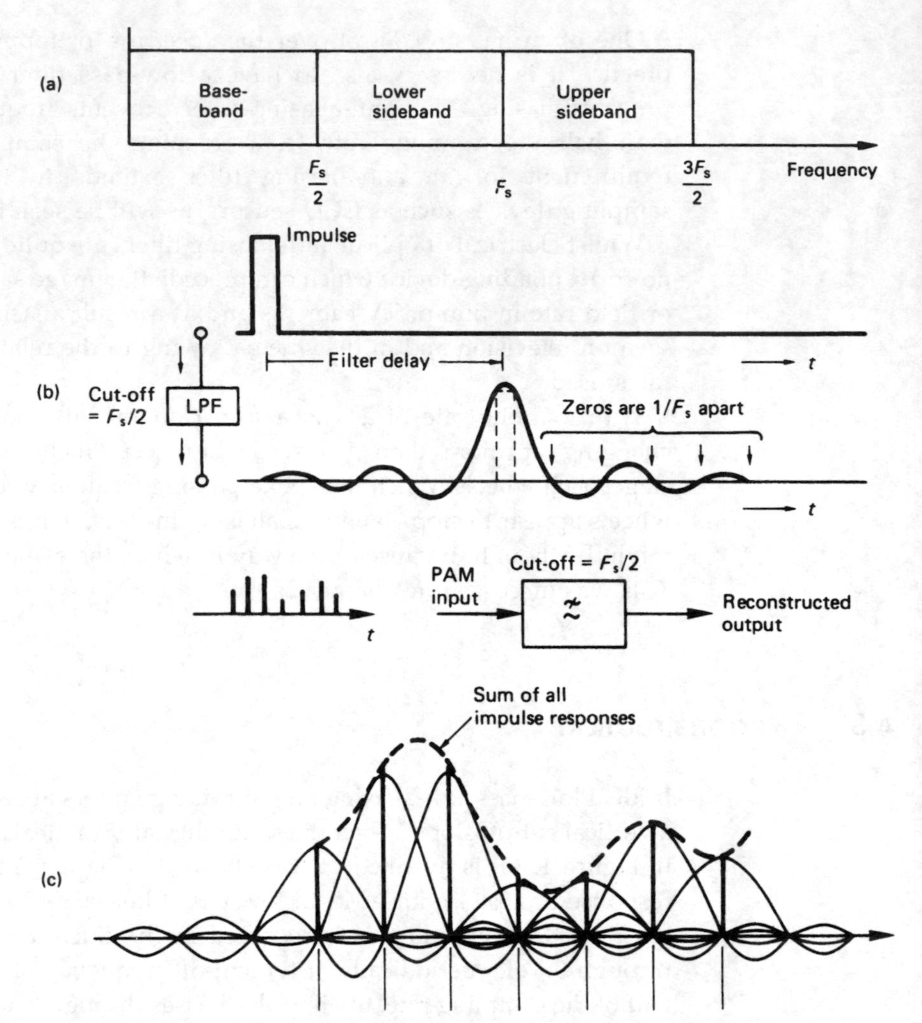

Figure 4.5 If ideal 'brick wall' filters are assumed, the efficient spectrum of (a) results. An ideal low-pass filter has an impulse response shown in (b). The impulse passes through zero at intervals equal to the sampling period. When convolved with a pulse train at the sampling rate, as shown in (c), the voltage at each sample instant is due to that sample alone as the impulses from all other samples pass through zero there.

The ideal filter with a vertical 'brick-wall' cut-off slope is difficult to implement. As the slope tends to vertical, the delay caused by the filter goes to infinity. In practice, a filter with a finite slope has to be accepted as shown in Figure 4.6. The cut-off slope begins at the edge of the required band, and consequently the sampling rate has to be raised a little to drive aliasing products to an acceptably low level. There is no absolute factor by which the sampling rate must be raised; it depends upon the

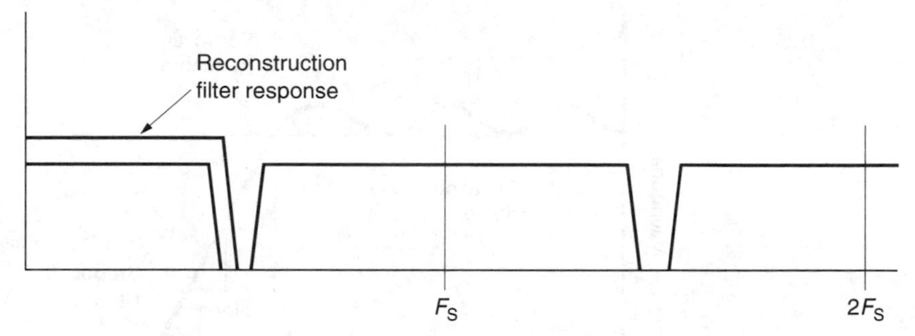

Figure 4.6 As filters with finite slope are needed in practical systems, the sampling rate is raised slightly beyond twice the highest frequency in the baseband.

filters which are available and the level of aliasing products which are acceptable. The latter will depend upon the wordlength to which the signal will be quantized.

4.4 Filter design

It is not easy to specify anti-aliasing and reconstruction filters, particularly the amount of stopband rejection needed. The resulting aliasing would depend on, among other things, the amount of out-of-band energy in the input signal. Very little is known about the energy in typical source material outside the usual frequency range. As a further complication, an out-of-band signal will be attenuated by the response of the anti-aliasing filter to that frequency, but the residual signal will then alias, and the reconstruction filter will reject it according to its attenuation at the new frequency to which it has aliased. To take the opposite extreme, if a camera were used which had no response at all above the video band, no anti-aliasing filter would be needed.

It would also be acceptable to bypass one of the filters involved in a copy from one digital machine to another via the analog domain, although a digital transfer is, of course, to be preferred.

The nature of the filters used has a great bearing on the subjective quality of the system. Entire books have been written about analog filters, and they will be treated only briefly here.

Figure 4.7 shows the terminology used to describe the common elliptic low-pass filter. These are popular because they can be realized with fewer components than other filters of similar response. It is a characteristic of these elliptic filters that there are ripples in the passband and stopband. In much equipment the anti-aliasing filter and the reconstruction filter will have the same specification, so that the passband ripple is doubled. Sometimes slightly different filters are used to reduce the effect.

Active filters can simulate inductors using op-amp techniques, but they tend to suffer non-linearity at high frequencies where the falling

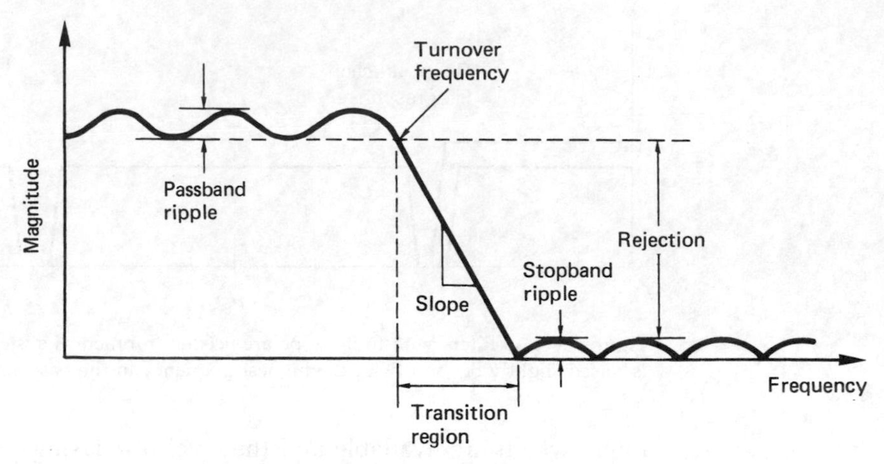

Figure 4.7 The important features and terminology of low-pass filters used for anti-aliasing and reconstruction.

open-loop gain reduces the effect of feedback. Active filters also can contribute noise, but this is not necessarily a bad thing in controlled amounts, since it can act as a dither source.

For video applications, the phase response of such filters must be linear (see Chapter 2). Since a sharp cut-off is generally achieved by cascading many filter sections which cut at a similar frequency, the phase responses of these sections will accumulate. The phase may start to leave linearity at only a half of the passband frequency, and near the cut-off frequency the phase error may be severe. Effective group delay equalization is necessary.

It is possible to construct a ripple-free phase-linear filter with the required stopband rejection, but it may be expensive due to the amount of design effort needed and the component complexity, and it might drift out of specification as components age. The money may be better spent in avoiding the need for such a filter. Much effort can be saved in analog filter design by using oversampling. Chapter 2 showed that digital filters are inherently phase-linear and, using LSIs, can be inexpensive to construct. The technical superiority of oversampling convertors along with economics means that they are increasingly used, which is why the subject is more prominent in this book than the treatment of filter design.

4.5 Two-dimensional sampling spectra

Analog video samples in the time domain and vertically, whereas a two-dimensional still image such as a photograph must be sampled horizontally and vertically. In both cases a two-dimensional spectrum will result, one vertical/temporal and one vertical/horizontal.

Figure 4.8(a) shows a square matrix of sampling sites which has an identical spatial sampling frequency both vertically and horizontally. The corresponding spectrum is shown in (b). The baseband spectrum is in the centre of the diagram, and the repeating sampling sideband spectrum extends vertically and horizontally. The star-shaped spectrum results from viewing an image of a man-made object such as a building containing primarily horizontal and vertical elements. A more natural scene such as foliage would result in a more circular or elliptical spectrum.

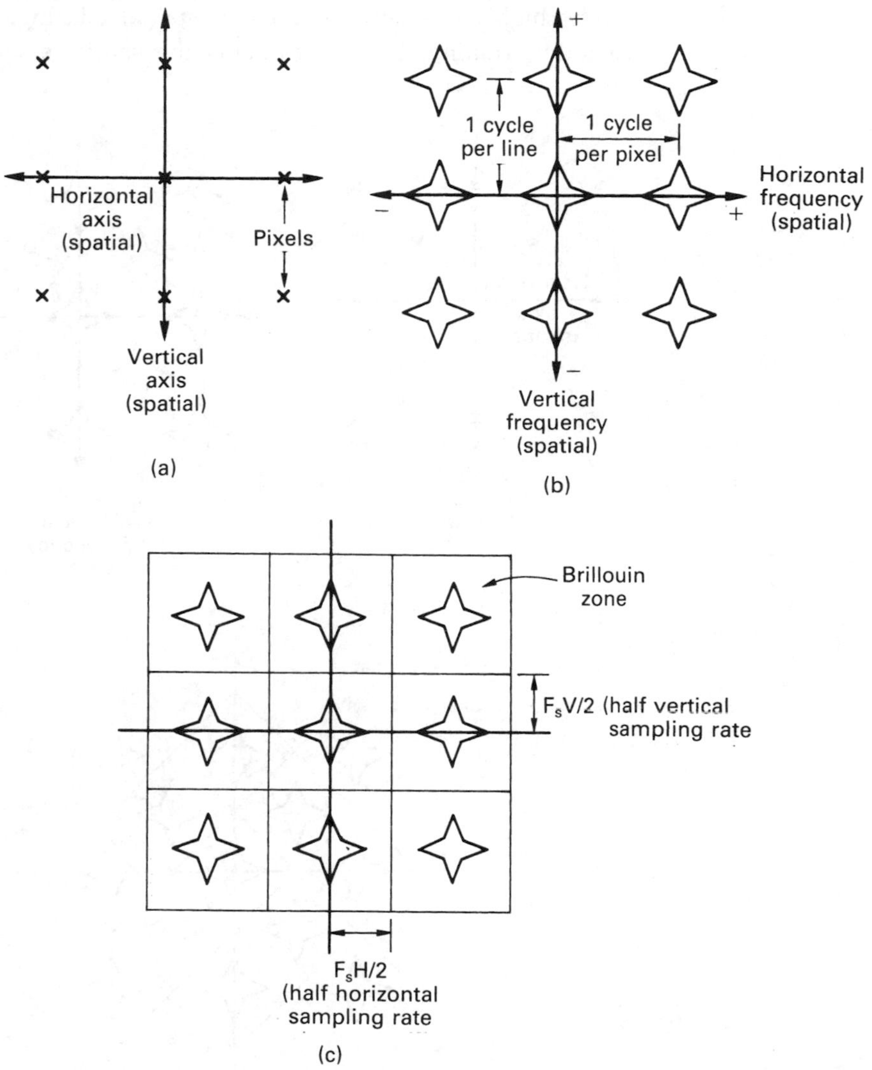

Figure 4.8 Image sampling spectra. The rectangular array of (a) has a spectrum shown at (b) having a rectangular repeating structure. Filtering to return to the baseband requires a two-dimensional filter whose response lies within the Brillouin zone shown at (c).

In order to return to the baseband image, the sidebands must be filtered out with a two-dimensional spatial filter. The shape of the two-dimensional frequency response shown in (c) is known as a Brillouin zone.

Figure 4.8(d) shows an alternative sampling site matrix known as quincunx sampling because of the similarity to the pattern of five dots on a die. The resultant spectrum has the same characteristic pattern as shown in (e). The corresponding Brillouin zones are shown in (f). Quincunx sampling offers a better compromise between diagonal and horizontal/vertical resolution but is complex to implement.

It is highly desirable to prevent spatial aliasing, since the result is visually irritating. In tube cameras the spatial aliasing will be in the

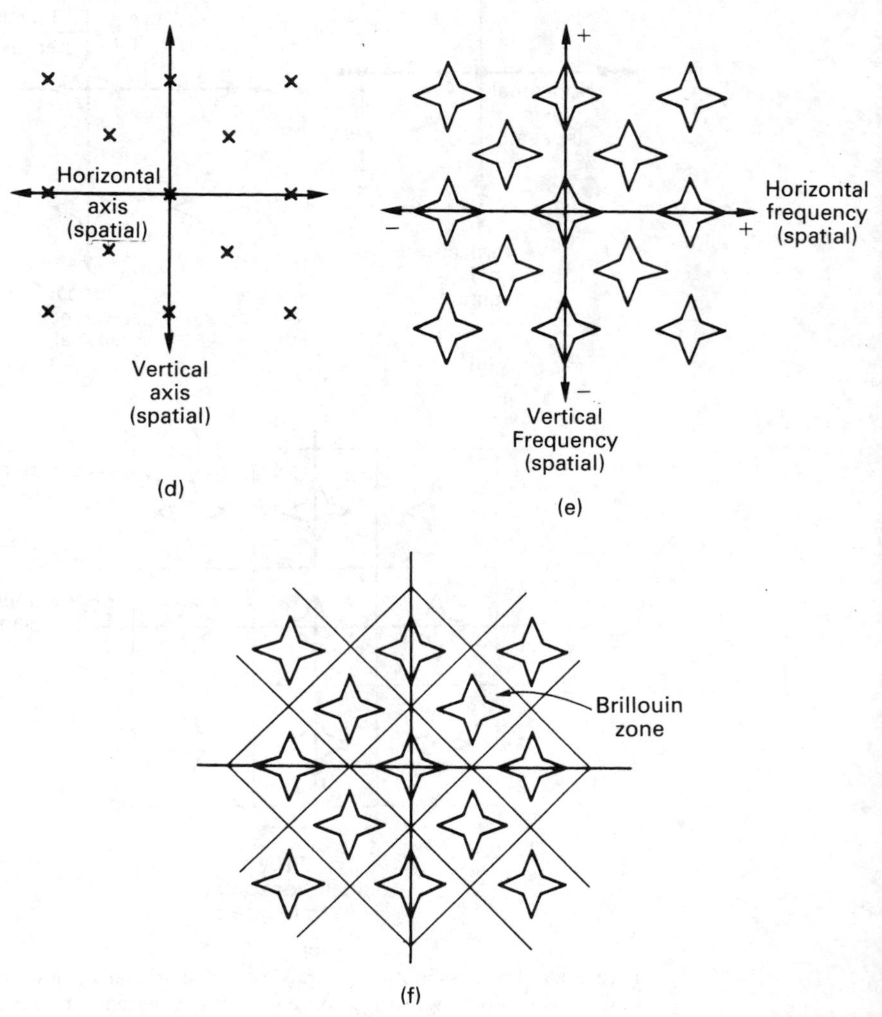

Figure 4.8 continued Quincunx sampling is shown at (d) to have a similar spectral structure (e). An appropriate Brillouin zone is required as at (f).

vertical dimension only, since the horizontal dimension is continuously scanned. Such cameras seldom attempt to prevent vertical aliasing. CCD sensors can, however, alias in both horizontal and vertical dimensions, and so an anti-aliasing optical filter is generally fitted between the lens and the sensor. This takes the form of a plate which diffuses the image formed by the lens. Such a device can never have a sharp cut-off nor will the aperture be rectangular. The aperture of the anti-aliasing plate is in series with the aperture effect of the CCD elements, and the combination of the two effectively prevents spatial aliasing, and generally gives a good balance between horizontal and vertical resolution, allowing the picture a natural appearance.

With a conventional approach, there are effectively two choices. If aliasing is permitted, the theoretical information rate of the system can be approached. If aliasing is prevented, realizable anti-aliasing filters cannot sharp cut, and the information conveyed is below system capacity.

These considerations also apply at the television display. The display must filter out spatial frequencies above one half the sampling rate. In a conventional CRT this means that vertical optical filter should be fitted in front of the screen to render the raster invisible. Again the aperture of a simply realizable filter would attenuate too much of the wanted spectrum, and so the technique is not used.

Figure 4.9 shows the spectrum of analog monochrome video (or of an analog component). The use of interlace has a similar effect on the

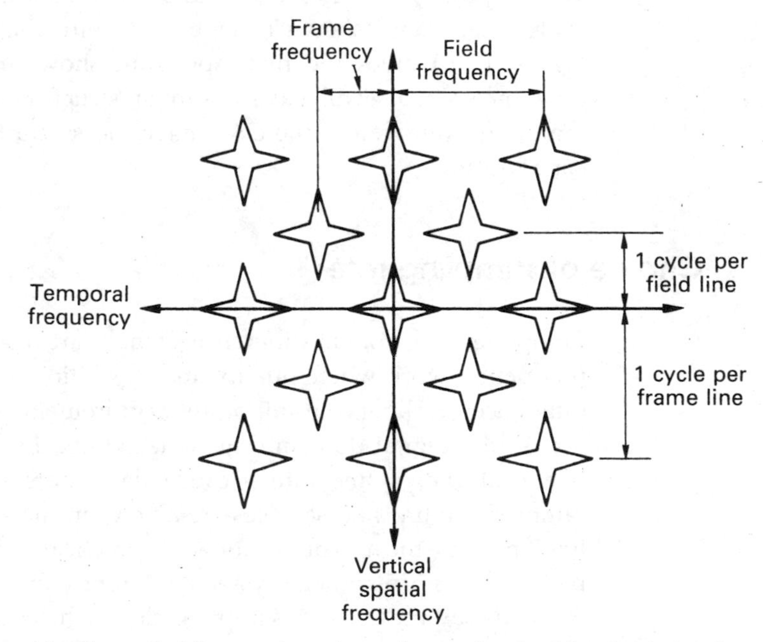

Figure 4.9 The vertical/temporal spectrum of monochrome video due to interlace.

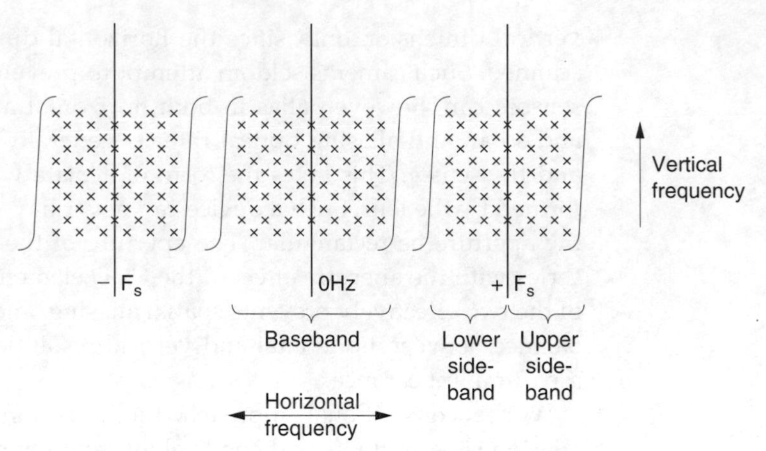

Figure 4.10 Spectrum of digital luminance is the baseband spectrum repeating around multiples of the sampling rate.

vertical/temporal spectrum as the use of quincunx sampling on the vertical/horizontal spectrum. The concept of the Brillouin zone cannot really be applied to reconstruction in the spatial/temporal domains. This is partly due to there being two different units in which the sampling rates are measured and partly because the temporal sampling process cannot prevent aliasing in real systems.

Sampling conventional video along the line to create pixels makes the horizontal axis of the three-dimensional spectrum repeat at multiples of the sampling rate. Thus combining the three-dimensional spectrum of analog luminance shown in Figure 4.9 with the sampling spectrum of Figure 4.3(b) gives the final spectrum shown in Figure 4.10. Colour difference signals will have a similar structure but often use a lower sampling rate and thereby have less horizontal resolution or bandwidth.

4.6 Choice of sampling rate

Component or colour difference signals are used primarily for post-production work where quality and flexibility are paramount. In colour difference working, the important requirement is for image manipulation in the digital domain. This is facilitated by a sampling rate which is a multiple of line rate because then there is a whole number of samples in a line and samples are always in the same position along the line and can form neat columns. A practical difficulty is that the line period of the 525 and 625 systems is slightly different. The problem was overcome by the use of a sampling clock which is an integer multiple of both line rates.

ITU-601 (formerly CCIR-601) recommends the use of certain sampling rates which are based on integer multiples of the carefully chosen fundamental frequency of 3.375 MHz. This frequency is normalized to 1 in the document.

In order to sample 625/50 luminance signals without quality loss, the lowest multiple possible is 4 which represents a sampling rate of 13.5 MHz. This frequency line-locks to give 858 samples per line period in 525/59.94 and 864 samples per line period in 625/50.

In the component analog domain, the colour difference signals used for production purposes typically have one half the bandwidth of the luminance signal. Thus a sampling rate multiple of 2 is used and results in 6.75 MHz. This sampling rate allows respectively 429 and 432 samples per line.

Component video sampled in this way has a 4:2:2 format. Whilst other combinations are possible, 4:2:2 is the format for which the majority of digital component production equipment is constructed and is the only component format for which parallel and serial interface standards exist. The D-1, D-5, D-9, and Digital Betacam DVTRs operate with 4:2:2 format data. Figure 4.11 shows the spatial arrangement given by 4:2:2 sampling. Luminance samples appear at half the spacing of colour difference samples, and every other luminance sample is co-sited with a pair of colour difference samples. Co-siting is important because it allows all attributes of one picture point to be conveyed with a three-sample vector

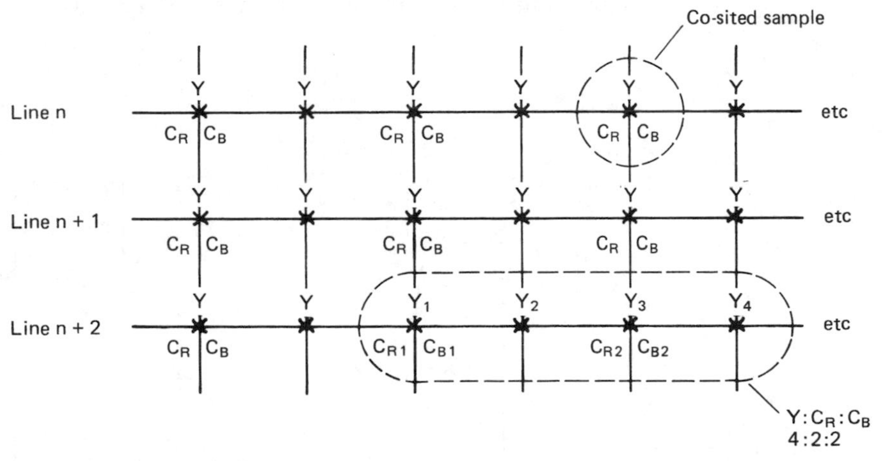

Figure 4.11 In CCIR-601 sampling mode 4:2:2, the line synchronous sampling rate of 13.5 MHz results in samples having the same position in successive lines, so that vertical columns are generated. The sampling rates of the colour difference signals C_R, C_B are one-half of that of luminance, i.e. 6.75 MHz, so that there are alternate Y only samples and co-sited samples which describe Y, C_R and C_B. In a run of four samples, there will be four Y samples, two C_R samples and two C_B samples, hence 4:2:2.

quantity. Modification of the three samples allows such techniques as colour correction to be performed. This would be difficult without co-sited information. Co-siting is achieved by clocking the three ADCs simultaneously.

For lower bandwidths, particularly in prefiltering operations prior to compression, the sampling rate of the colour difference signal can be halved. 4:1:1 delivers colour bandwidth in excess of that required by analog composite video.

In 4:2:2 the colour difference signals are sampled horizontally at half the luminance sampling rate, yet the vertical colour difference sampling rates are the same as for luminance. Whilst this is not a problem in a production application, this disparity of sampling rates represents a data rate overhead which is undesirable in a compression environment. In this case it is possible to halve the vertical sampling rate of the colour difference signals as well.

Figure 4.12 shows that in MPEG-2 4:2:0 sampling, the colour difference signals are downsampled so that the same vertical and horizontal resolution is obtained.

The chroma samples in 4:2:0 are positioned half-way between luminance samples in the vertical axis so that they are evenly spaced when an interlaced source is used. To obtain a 4:2:2 output from 4:2:0 data a vertical interpolation process will be needed in addition to low-pass filtering.

The sampling rates of ITU-601 are based on commonality between 525- and 625-line systems. However, the consequence is that the pixel spacing is different in the horizontal and vertical axes. This is incompatible with

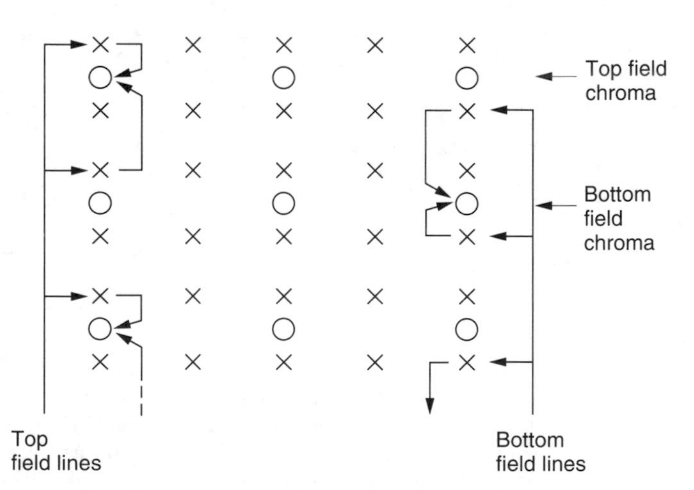

Figure 4.12 In 4:2:0 coding the colour difference pixels are downsampled vertically as well as horizontally. Note that the sample sites need to be vertically interpolated so that when two interlaced fields are combined the spacing is even.

computer graphics in which so-called 'square' pixels are used. This means that the horizontal and vertical spacing is the same, giving the same resolution in both axes. However, high-definition TV and computer graphics formats universally use 'square' pixels. Converting between square and non-square pixel data will require a rate conversion process as described in section 3.22.

The traditional TV screen has an aspect ratio of 4:3, whereas an aspect ratio of 16:9 has now been adopted. Expressing 4:3 as 12:9 makes it clear that the 16:9 picture is 16/12 or 4/3 times as wide. There are two ways of handling 16:9 pictures in the digital domain. One is to retain the standard sampling rate of 13.5 Mhz, which results in the horizontal resolution falling to 3/4 of its previous value, the other is to increase the sampling rate in proportion to the screen width. This results in a luminance sampling rate of $13.5 \times 4/3$ MHz or 18.0 MHz.

4.7 Sampling clock jitter

The instants at which samples are taken in an ADC and the instants at which DACs make conversions must be evenly spaced, otherwise unwanted signals can be added to the video. Figure 4.13 shows the effect of sampling clock jitter on a sloping waveform. Samples are taken at the wrong times. When these samples have passed through a system, the timebase correction stage prior to the DAC will remove the jitter, and the result is shown in (b). The magnitude of the unwanted signal is proportional to the slope of the audio waveform and so the amount of jitter which can be tolerated falls at 6 dB per octave. As the resolution of the system is increased by the use of longer sample wordlength, tolerance to jitter is further reduced. The nature of the unwanted signal depends on the spectrum of the jitter. If the jitter is random, the effect is noise-like and relatively benign unless the amplitude is excessive. Figure 4.14 shows the effect of differing amounts of random jitter with respect to the noise floor of various wordlengths. Note that even small amounts of jitter can degrade a ten-bit convertor to the performance of a good eight-bit unit. There is thus no point in upgrading to higher-resolution convertors if the clock stability of the system is insufficient to allow their performance to be realized.

Clock jitter is not necessarily random. Figure 4.15 shows that one source of clock jitter is crosstalk or interference on the clock signal, although a balanced clock line will be more immune to such crosstalk. The unwanted additional signal changes the time at which the sloping clock signal appears to cross the threshold voltage of the clock receiver. This is simply the same phenomenon as that of Figure 4.13 but in reverse. The threshold itself may be changed by ripple on the clock receiver power

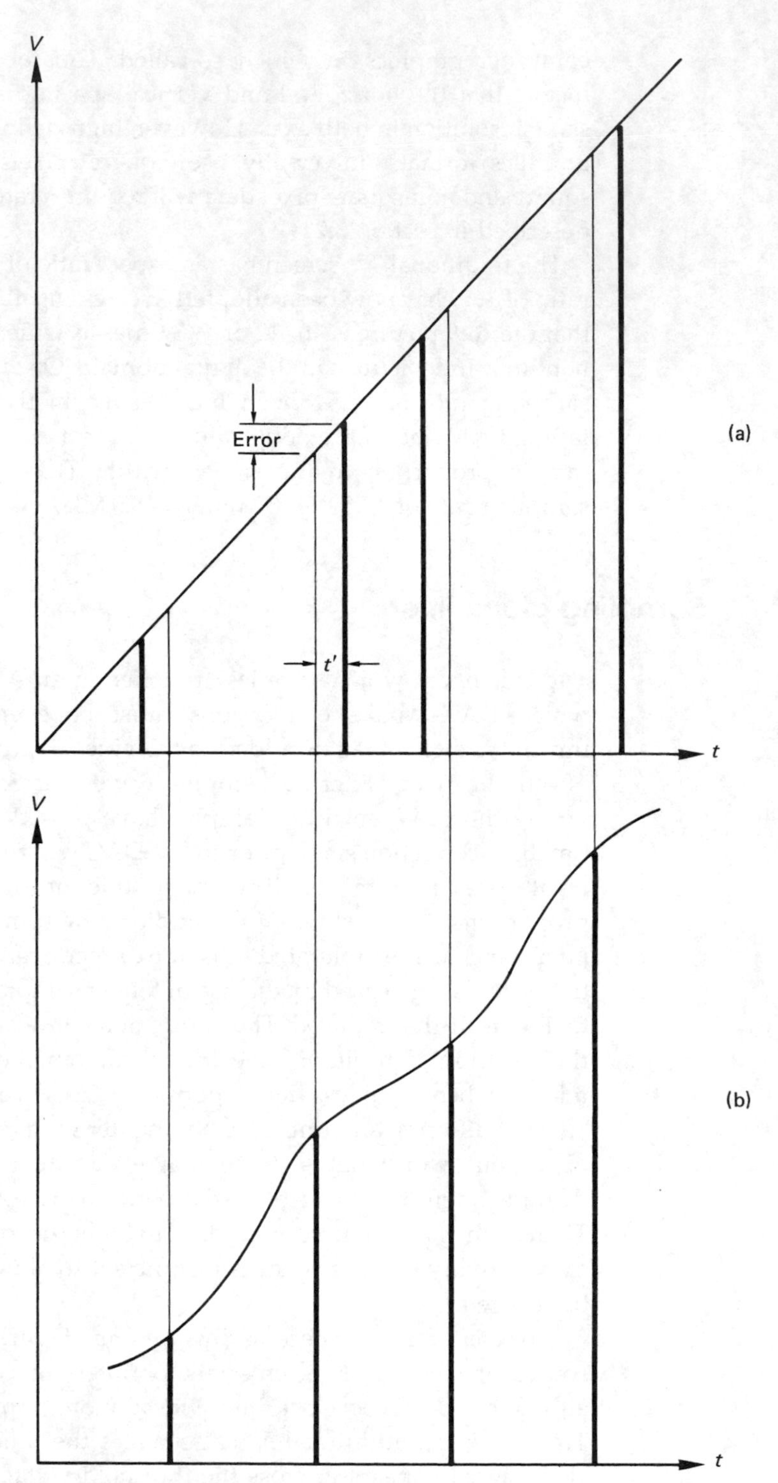

Figure 4.13 The effect of sampling timing jitter on noise. At (a) a sloping signal sampled with jitter has error proportional to the slope. When jitter is removed by reclocking, the result at (b) is noise.

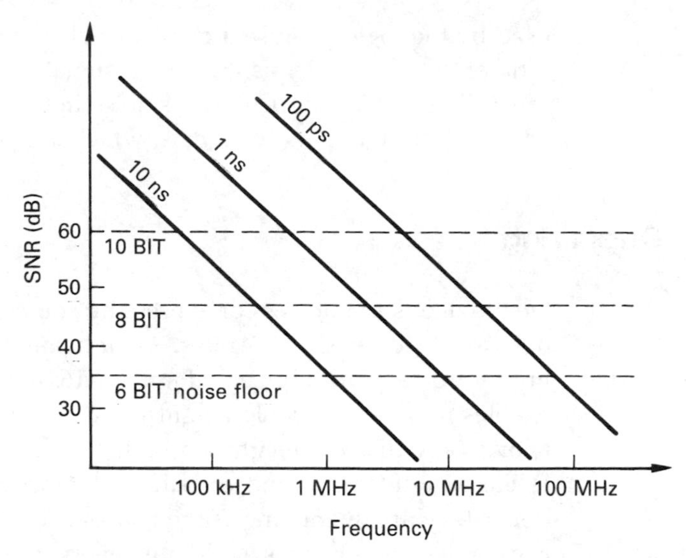

Figure 4.14 The effect of sampling clock jitter on signal-to-noise ratio at various frequencies, compared with the theoretical noise floors with different wordlengths.

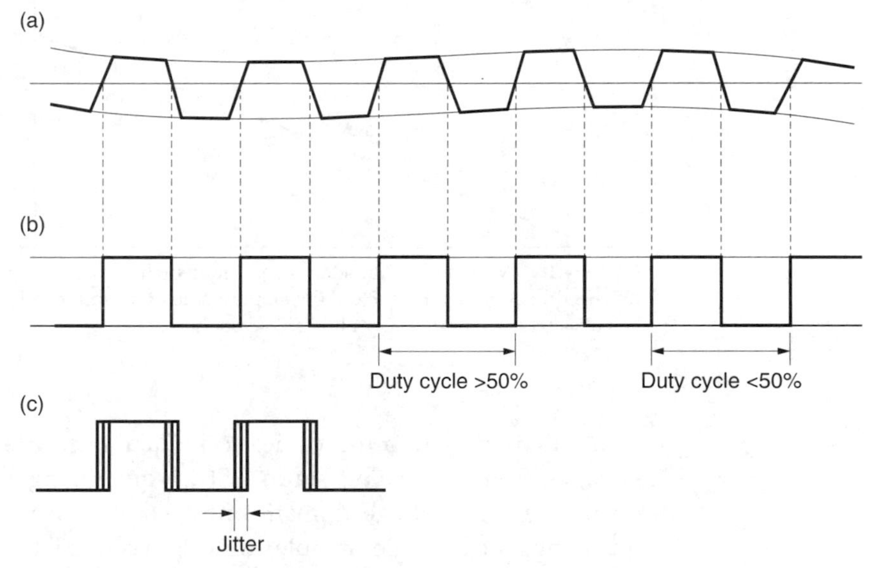

Figure 4.15 Crosstalk in transmission can result in unwanted signals being added to the clock waveform. It can be seen here that a low-frequency interference signal affects the slicing of the clock and causes a periodic jitter.

supply. There is no reason why these effects should be random; they may be periodic and potentially visible.[4]

The allowable jitter is measured in picoseconds and clearly steps must be taken to eliminate it by design. Convertor clocks must be generated from clean power supplies which are well decoupled from the power used by the logic because a convertor clock must have a signal-to-noise ratio of the same order as that of the signal. Otherwise noise on the clock causes jitter which in turn causes noise in the video. The same effect will be found in digital audio signals, which are perhaps more critical.

4.8 Quantizing

Quantizing is the process of expressing some infinitely variable quantity by discrete or stepped values. Quantizing turns up in a remarkable number of everyday guises. Figure 4.16 shows that an inclined ramp enables infinitely variable height to be achieved, whereas a step-ladder allows only discrete heights to be had. A step-ladder quantizes height. When accountants round off sums of money to the nearest pound or dollar they are quantizing. Time passes continuously, but the display on a digital clock changes suddenly every minute because the clock is quantizing time.

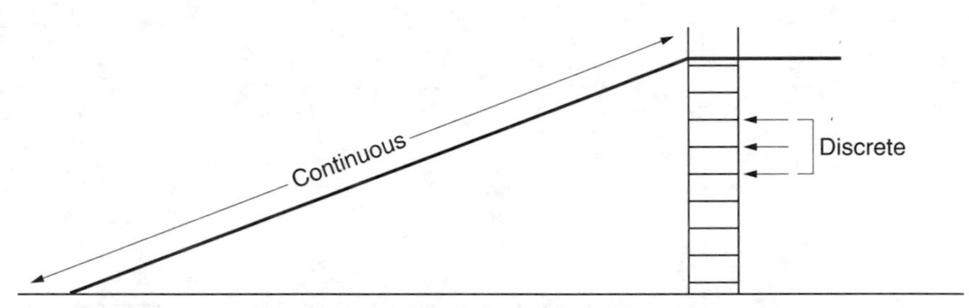

Figure 4.16 An analog parameter is continuous whereas a quantized parameter is restricted to certain values. Here the sloping side of a ramp can be used to obtain any height whereas a ladder only allows discrete heights.

In video and audio the values to be quantized are infinitely variable voltages from an analog source. Strict quantizing is a process which operates in the voltage domain only. For the purpose of studying the quantizing of a single sample, time is assumed to stand still. This is achieved in practice either by the use of a track-hold circuit or the adoption of a quantizer technology such as a flash convertor which operates before the sampling stage.

Figure 4.17(a) shows that the process of quantizing divides the voltage range up into quantizing intervals Q, also referred to as steps S. In applications such as telephony these may advantageously be of differing size, but for digital video the quantizing intervals are made as identical as possible. If this is done, the binary numbers which result are truly proportional to the original analog voltage, and the digital equivalents of

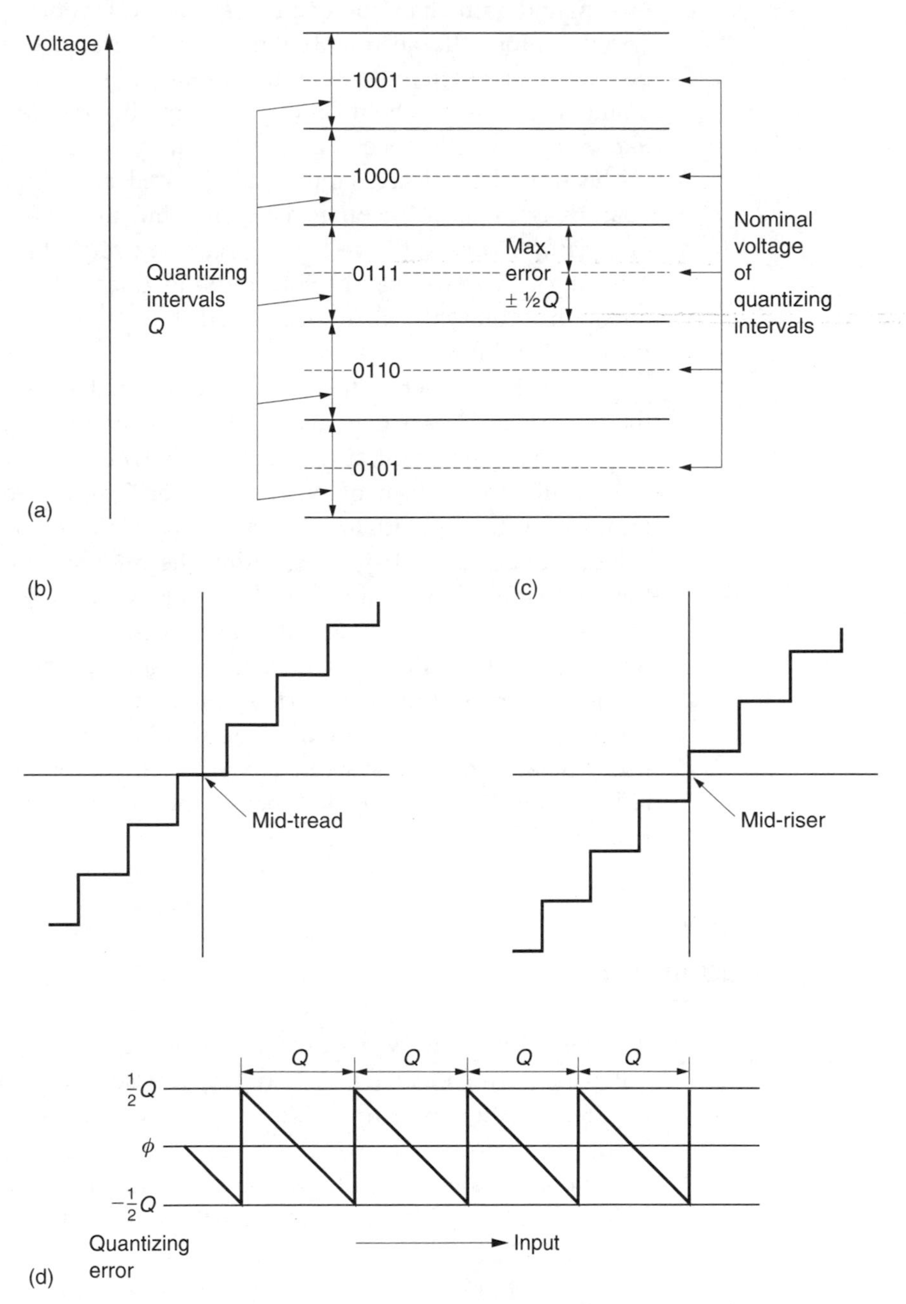

Figure 4.17 Quantizing assigns discrete numbers to variable voltages. All voltages within the same quantizing interval are assigned the same number which causes a DAC to produce the voltage at the centre of the intervals shown by the dashed lines in (a). This is the characteristic of the mid-tread quantizer shown in (b). An alternative system is the mid-riser system shown in (c). Here 0 Volts analog falls between two codes and there is no code for zero. Such quantizing cannot be used prior to signal processing because the number is no longer proportional to the voltage. Quantizing error cannot exceed $\pm\frac{1}{2}Q$ as shown in (d).

mixing and gain changing can be performed by adding and multiplying sample values. If the quantizing intervals are unequal this cannot be done. When all quantizing intervals are the same, the term uniform quantizing is used. The term linear quantizing will be found, but this is, like military intelligence, a contradiction in terms.

The term LSB (least significant bit) will also be found in place of quantizing interval in some treatments, but this is a poor term because quantizing works in the voltage domain. A bit is not a unit of voltage and can only have two values. In studying quantizing voltages within a quantizing interval will be discussed, but there is no such thing as a fraction of a bit.

Whatever the exact voltage of the input signal, the quantizer will locate the quantizing interval in which it lies. In what may be considered a separate step, the quantizing interval is then allocated a code value which is typically some form of binary number. The information sent is the number of the quantizing interval in which the input voltage lies. Whereabouts that voltage lies within the interval is not conveyed, and this mechanism puts a limit on the accuracy of the quantizer. When the number of the quantizing interval is converted back to the analog domain, it will result in a voltage at the centre of the quantizing interval as this minimizes the magnitude of the error between input and output. The number range is limited by the wordlength of the binary numbers used. In an eight-bit system, 256 different quantizing intervals exist, although in digital video the ones at the extreme ends of the range are reserved for synchronizing.

4.9 Quantizing error

It is possible to draw a transfer function for such an ideal quantizer followed by an ideal DAC, and this is also shown in Figure 4.17. A transfer function is simply a graph of the output with respect to the input. In audio, when the term linearity is used, this generally means the overall straightness of the transfer function. Linearity is a goal in video and audio, yet it will be seen that an ideal quantizer is anything but linear.

Figure 4.17(b) shows that the transfer function is somewhat like a staircase, and the blanking level is half-way up a quantizing interval, or on the centre of a tread. This is the so-called mid-tread quantizer which is universally used in video and audio. Figure 4.17(c) shows the alternative mid-riser transfer function which causes difficulty because it does not have a code value at blanking level and as a result the numerical code value is not proportional to the analog signal voltage.

Quantizing causes a voltage error in the sample which is given by the difference between the actual staircase transfer function and the ideal straight line. This is shown in (d) to be a sawtooth-like function which is periodic in Q. The amplitude cannot exceed $\pm\frac{1}{2}Q$ peak-to-peak unless the input is so large that clipping occurs.

Quantizing error can also be studied in the time domain where it is better to avoid complicating matters with the aperture effect of the DAC. For this reason it is assumed here that output samples are of negligible duration. Then impulses from the DAC can be compared with the original analog waveform and the difference will be impulses representing the quantizing error waveform. This has been done in Figure 4.18. The horizontal lines in the drawing are the boundaries between the quantizing intervals, and the curve is the input waveform. The vertical bars are the quantized samples which reach to the centre of the quantizing interval. The quantizing error waveform shown in (b) can be thought of as an unwanted signal which the quantizing process adds to the perfect original. If a very small input signal remains within one quantizing interval, the quantizing error *is* the signal.

As the transfer function is non-linear, ideal quantizing can cause distortion. As a result practical digital video equipment deliberately uses non-ideal quantizers to achieve linearity. The quantizing error of an ideal quantizer is a complex function, and it has been researched in great depth.[5–8] It is not intended to go into such depth here. The characteristics of an ideal quantizer will only be pursued far enough to convince the reader that such a device cannot be used in quality video or audio applications.

As the magnitude of the quantizing error is limited, its effect can be minimized by making the signal larger. This will require more quantizing intervals and more bits to express them. The number of quantizing intervals multiplied by their size gives the quantizing range of the convertor. A signal outside the range will be clipped. Provided that clipping is avoided, the larger the signal, the less will be the effect of the quantizing error.

Where the input signal exercises the whole quantizing range and has a complex waveform (such as from a contrasty, detailed scene), successive samples will have widely varying numerical values and the quantizing error on a given sample will be independent of that on others. In this case the size of the quantizing error will be distributed with equal probability between the limits. Figure 4.18(c) shows the resultant uniform probability density. In this case the unwanted signal added by quantizing is an additive broadband noise uncorrelated with the signal, and it is appropriate in this case to call it quantizing noise. This is not quite the same as thermal noise which has a Gaussian probability shown in (d) (see section 3.28 for a treatment of statistics). The difference is of no

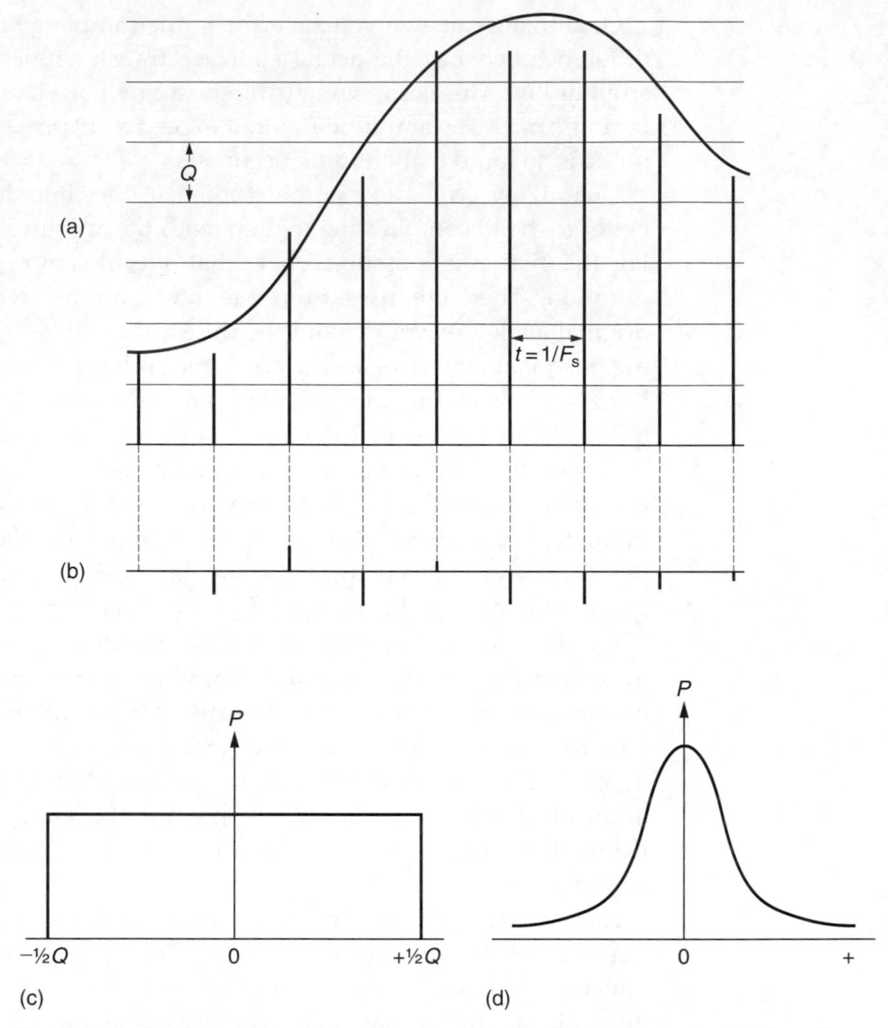

Figure 4.18 In (a) an arbitrary signal is represented to finite accuracy by PAM needles whose peaks are at the centre of the quantizing intervals. The errors caused can be thought of as an unwanted signal (b) added to the original. In (c) the amplitude of a quantizing error needle will be from $-\frac{1}{2}Q$ to $+\frac{1}{2}Q$ with equal probability. Note, however, that white noise in analog circuits generally has Gaussian amplitude distribution, shown in (d).

consequence as in the large signal case the noise is masked by the signal. Under these conditions, a meaningful signal-to-noise ratio can be calculated as follows.

In a system using n-bit words. there will be 2^n quantizing intervals. The largest sinusoid which can fit without clipping will have this peak-to-peak amplitude. The peak amplitude will be half as great, i.e. $2^{n-1}Q$ and the rms amplitude will be this value divided by $\sqrt{2}$. The quantizing error has an amplitude of $\frac{1}{2}Q$ peak which is the equivalent of $\frac{Q}{\sqrt{12}}$ rms. The signal-to-noise ratio for the large signal case is then given by:

$$
\begin{aligned}
20 \log_{10} \frac{\sqrt{12} \times 2^{n-1}}{\sqrt{2}} \text{ dB} &= 20 \log_{10} (\sqrt{6} \times 2^{n-1}) \text{ dB} \\
&= 20 \log (2^n \times \frac{\sqrt{6}}{2} \text{ dB} \\
&= 20n \log 2 + 20 \log \frac{\sqrt{6}}{2} \text{ dB} \\
&= 6.02n + 1.76 \text{ dB} \qquad (4.1)
\end{aligned}
$$

By way of example, an eight-bit system will offer very nearly 50 dB SNR.

Whilst the above result is true for a large complex input waveform, treatments which then assume that quantizing error is *always* noise give results which are at variance with reality. The expression above is only valid if the probability density of the quantizing error is uniform. Unfortunately at low depths of modulations, and particularly with flat fields or simple pictures, this is not the case.

At low modulation depth, quantizing error ceases to be random, and becomes a function of the input waveform and the quantizing structure as Figure 4.18 shows. Once an unwanted signal becomes a deterministic function of the wanted signal, it has to be classed as a distortion rather than a noise. Distortion can also be predicted from the non-linearity, or staircase nature, of the transfer function. With a large signal, there are so many steps involved that we must stand well back, and a staircase with 256 steps appears to be a slope. With a small signal there are few steps and they can no longer be ignored.

The effect can be visualized readily by considering a television camera viewing a uniformly painted wall. The geometry of the lighting and the coverage of the lens means that the brightness is not absolutely uniform, but falls slightly at the ends of the TV lines. After quantizing, the gently sloping waveform is replaced by one which stays at a constant quantizing level for many sampling periods and then suddenly jumps to the next quantizing level. The picture then consists of areas of constant brightness with steps between, resembling nothing more than a contour map, hence the use of the term *contouring* to describe the effect.

Needless to say, the occurrence of contouring precludes the use of an ideal quantizer for high-quality work. There is little point in studying the adverse effects further as they should be and can be eliminated completely in practical equipment by the use of dither. The importance of correctly dithering a quantizer cannot be emphasized enough, since failure to dither irrevocably distorts the converted signal: there can be no process which will subsequently remove that distortion. The signal-to-noise ratio derived above has no relevance to practical applications as it will be modified by the dither.

4.10 Introduction to dither

At high signal levels, quantizing error is effectively noise. As the depth of modulation falls, the quantizing error of an ideal quantizer becomes more strongly correlated with the signal and the result is distortion, visible as contouring. If the quantizing error can be decorrelated from the input in some way, the system can remain linear but noisy. Dither performs the job of decorrelation by making the action of the quantizer unpredictable and gives the system a noise floor like an analog system.[9,10]

In one approach, pseudo-random noise (see Chapter 2) with rectangular probability and a peak-to-peak amplitude of Q was added to the input signal prior to quantizing, but was subtracted after reconversion to analog. This is known as subtractive dither and was investigated by Schuchman[11] and much later by Sherwood.[12] Subtractive dither has the advantages that the dither amplitude is non-critical, the noise has full statistical independence from the signal[13] and has the same level as the quantizing error in the large signal undithered case.[14] Unfortunately, it suffers from practical drawbacks, since the original noise waveform must accompany the samples or must be synchronously re-created at the DAC. This is virtually impossible in a system where the signal may have been edited or where its level has been changed by processing, as the noise needs to remain synchronous and be processed in the same way. All practical digital video systems use non-subtractive dither where the dither signal is added prior to quantization and no attempt is made to remove it at the DAC.[15] The introduction of dither prior to a conventional quantizer inevitably causes a slight reduction in the signal-to-noise ratio attainable, but this reduction is a small price to pay for the elimination of non-linearities.

The ideal (noiseless) quantizer of Figure 4.17 has fixed quantizing intervals and must always produce the same quantizing error from the same signal. In Figure 4.19 it can be seen that an ideal quantizer can be dithered by linearly adding a controlled level of noise either to the input signal or to the reference voltage which is used to derive the quantizing intervals. There are several ways of considering how dither works, all of which are equally valid.

The addition of dither means that successive samples effectively find the quantizing intervals in different places on the voltage scale. The

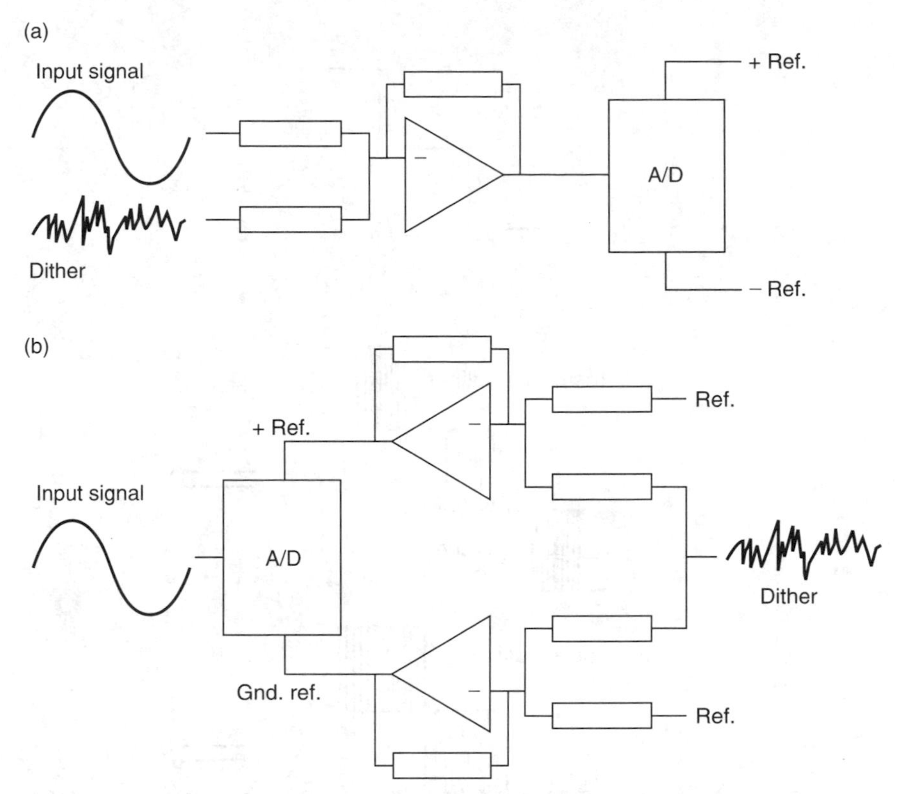

Figure 4.19 Dither can be applied to a quantizer in one of two ways. In (a) the dither is linearly added to the analog input signal, whereas in (b) it is added to the reference voltages of the quantizer.

quantizing error becomes a function of the dither, rather than a predictable function of the input signal. The quantizing error is not eliminated, but the subjectively unacceptable distortion is converted into a broadband noise which is more benign.

Some alternative ways of looking at dither are shown in Figure 4.20. Consider the situation where a low-level input signal is changing slowly within a quantizing interval. Without dither, the same numerical code is output for a number of sample periods, and the variations within the interval are lost. Dither has the effect of forcing the quantizer to switch between two or more states. The higher the voltage of the input signal within a given interval, the more probable it becomes that the output code will take on the next higher value. The lower the input voltage within the interval, the more probable it is that the output code will take the next lower value. The dither has resulted in a form of duty cycle modulation, and the resolution of the system has been extended indefinitely instead of being limited by the size of the steps.

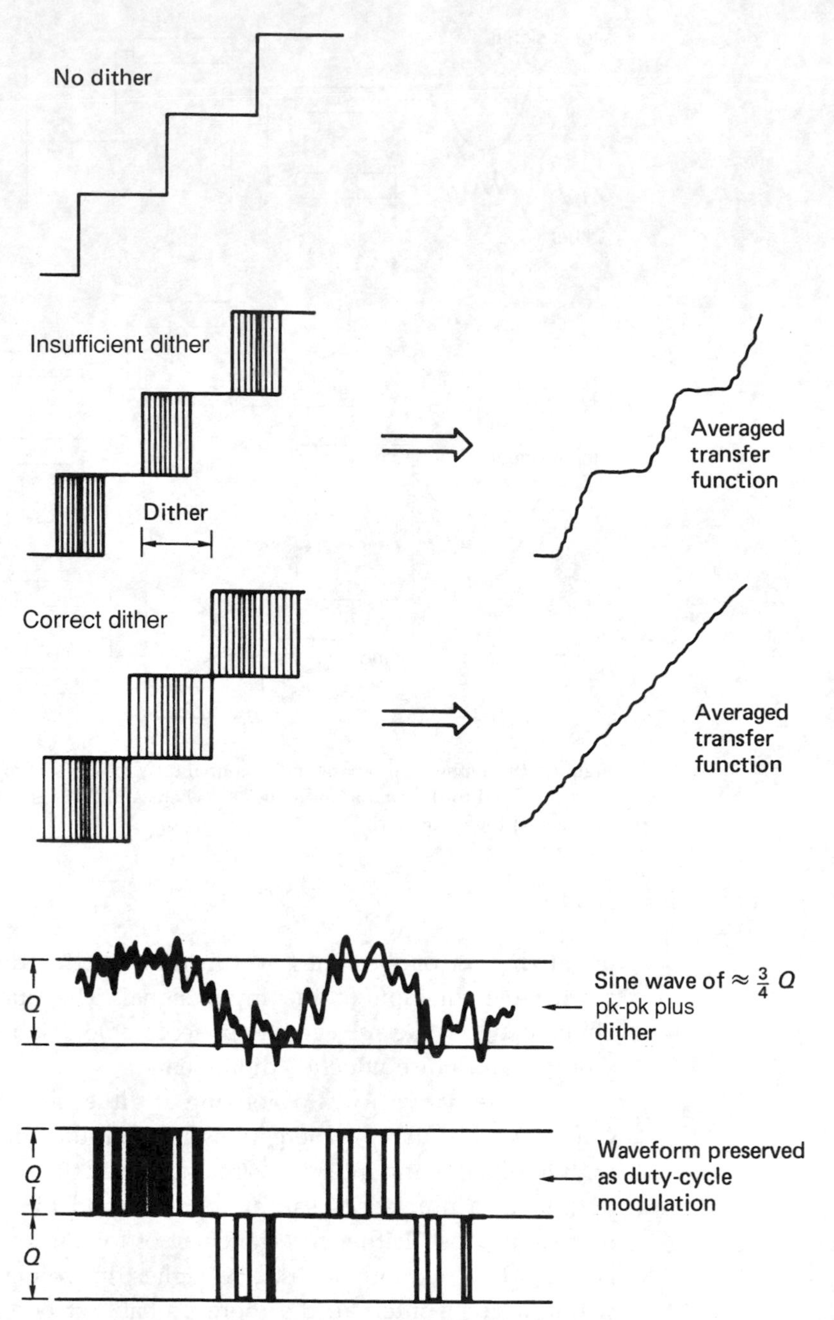

Figure 4.20 Wideband dither of the appropriate level linearizes the transfer function to produce noise instead of distortion. This can be confirmed by spectral analysis. In the voltage domain, dither causes frequent switching between codes and preserves resolution in the duty cycle of the switching.

Dither can also be understood by considering what it does to the transfer function of the quantizer. This is normally a perfect staircase, but in the presence of dither it is smeared horizontally until with a certain amplitude the average transfer function becomes straight.

4.11 Requantizing and digital dither

Recent ADC technology allows the resolution of video samples to be raised from eight bits to ten or even twelve bits. The situation then arises that an existing eight-bit device such as a digital VTR needs to be connected to the output of an ADC with greater wordlength. The words need to be shortened in some way.

It will be seen in Chapter 5 that when a sample value is attenuated, the extra low-order bits which come into existence below the radix point preserve the resolution of the signal and the dither in the least significant bit(s) which linearizes the system. The same word extension will occur in any process involving multiplication, such as digital filtering. It will subsequently be necessary to shorten the wordlength. Low-order bits must be removed in order to reduce the resolution whilst keeping the signal magnitude the same. Even if the original conversion was correctly dithered, the random element in the low-order bits will now be some way below the end of the intended word. If the word is simply truncated by discarding the unwanted low-order bits or rounded to the nearest integer the linearizing effect of the original dither will be lost.

Shortening the wordlength of a sample reduces the number of quantizing intervals available without changing the signal amplitude. As Figure 4.21 shows, the quantizing intervals become larger and the original signal is *requantized* with the new interval structure. This will introduce requantizing distortion having the same characteristics as quantizing distortion in an ADC. It then is obvious that when shortening the wordlength of a ten-bit convertor to eight bits, the two low-order bits must be removed in a way that displays the same overall quantizing structure as if the original convertor had been only of eight-bit wordlength. It will be seen from Figure 4.21 that truncation cannot be used because it does not meet the above requirement but results in signal-dependent offsets because it always rounds in the same direction. Proper numerical rounding is essential in video applications because it accurately simulates analog quantizing to the new interval size. Unfortunately the ten-bit convertor will have a dither amplitude appropriate to quantizing intervals one quarter the size of an eight-bit unit and the result will be highly non-linear.

In practice, the wordlength of samples must be shortened in such a way that the requantizing error is converted to noise rather than distortion.

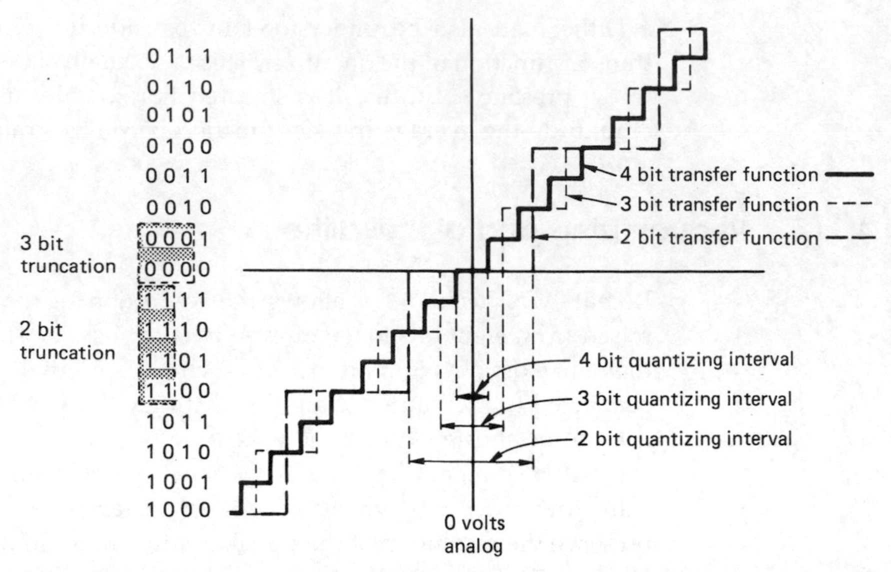

Figure 4.21 Shortening the wordlength of a sample reduces the number of codes which can describe the voltage of the waveform. This makes the quantizing steps bigger hence the term requantizing. It can be seen that simple truncation or omission of the bits does not give analogous behaviour. Rounding is necessary to give the same result as if the larger steps had been used in the original conversion.

One technique which meets this requirement is to use digital dithering[16] prior to rounding. This is directly equivalent to the analog dithering in an ADC.

Digital dither is a pseudo-random sequence of numbers. If it is required to simulate the analog dither signal of Figures 4.19 and 20, then it is obvious that the noise must be bipolar so that it can have an average voltage of zero. Two's complement coding must be used for the dither values.

Figure 4.22 shows a simple digital dithering system (i.e. one without noise shaping) for shortening sample wordlength. The output of a two's complement pseudo-random sequence generator (see Chapter 3) of appropriate wordlength is added to input samples prior to rounding. The most significant of the bits to be discarded is examined in order to determine whether the bits to be removed sum to more or less than half a quantizing interval. The dithered sample is either rounded down, i.e. the unwanted bits are simply discarded, or rounded up, i.e. the unwanted bits are discarded but one is added to the value of the new short word. The rounding process is no longer deterministic because of the added dither which provides a linearizing random component.

If this process is compared with that of Figure 4.19 it will be seen that the principles of analog and digital dither are identical; the processes

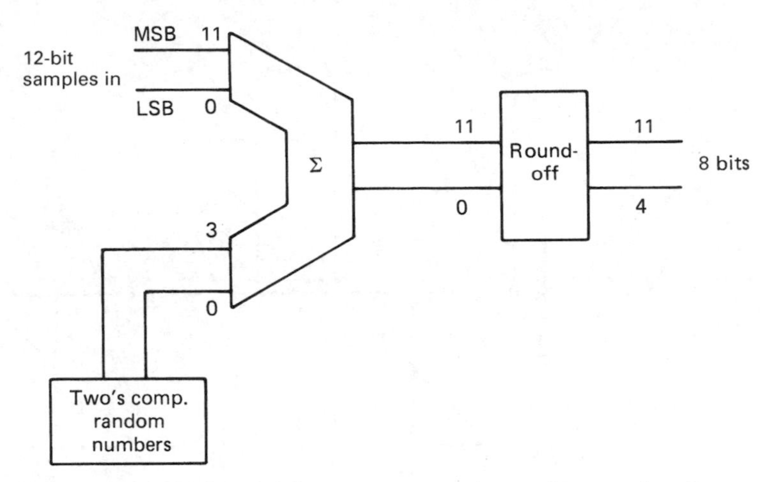

Figure 4.22 In a simple digital dithering system, two's complement values from a random number generator are added to low-order bits of the input. The dithered values are then rounded up or down according to the value of the bits to be removed. The dither linearizes the requantizing.

simply take place in different domains using two's complement numbers which are rounded or voltages which are quantized as appropriate. In fact quantization of an analog-dithered waveform is identical to the hypothetical case of rounding after bipolar digital dither where the number of bits to be removed is infinite, and remains identical for practical purposes when as few as eight bits are to be removed. Analog dither may actually be generated from bipolar digital dither (which is no more than random numbers with certain properties) using a DAC.

4.12 Dither techniques

The intention here is to treat the processes of analog and digital dither as identical except where differences need to be noted. The characteristics of the noise used are rather important for optimal performance, although many sub-optimal but nevertheless effective systems are in use. The main parameters of interest are the peak-to-peak amplitude, the amplitude probability distribution function (pdf) and the spectral content.

The most comprehensive ongoing study of non-subtractive dither has been that of Vanderkooy and Lipshitz[15–17] and the treatment here is based largely upon their work.

4.12.1 Rectangular pdf dither

Chapter 2 showed that the simplest form of dither (and therefore the easiest to generate) is a single sequence of random numbers which have uniform or rectangular probability. The amplitude of the dither is critical.

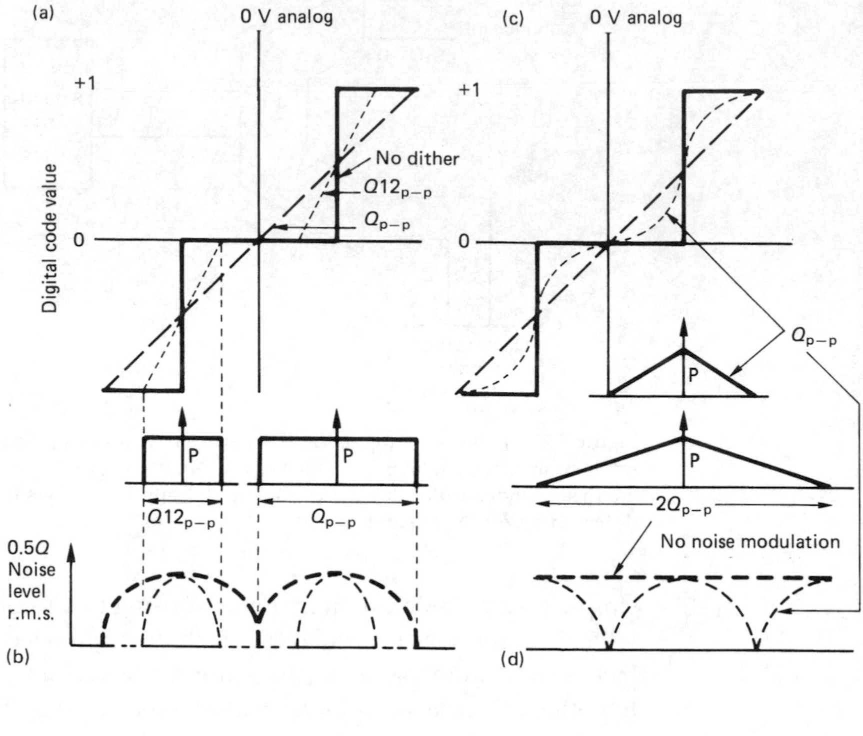

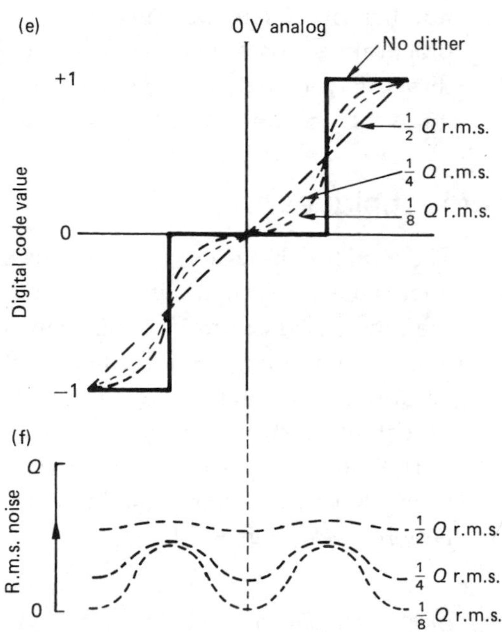

Figure 4.23 (a) Use of rectangular probability dither can linearize, but noise modulation (b) results. Triangular pdf dither (c) linearizes, and noise modulation is eliminated as at (d). Gaussian dither (e) can also be used, almost eliminating noise modulation at (f).

Figure 4.23(a) shows the time-averaged transfer function of one quantizing interval in the presence of various amplitudes of rectangular dither. The linearity is perfect at an amplitude of $1\,Q$ peak-to-peak and then deteriorates for larger or smaller amplitudes. The same will be true of all levels which are an integer multiple of Q. Thus there is no freedom in the choice of amplitude.

With the use of such dither, the quantizing noise is not constant. Figure 4.23(b) shows that when the analog input is exactly centred in a quantizing interval, (such that there is no quantizing error) the dither has no effect and the output code is steady. There is no switching between codes and thus no noise. On the other hand, when the analog input is exactly at a riser or boundary between intervals, there is the greatest switching between codes and the greatest noise is produced. Mathematically speaking, the first moment or mean error is zero but the second moment, which in this case is equal to the variance, is not constant. From an engineering standpoint, the system is linear but suffers noise modulation: the noise floor rises and falls with the signal content and this is audible in the presence of low-frequency signals. The dither adds an average noise amplitude of $\frac{Q}{\sqrt{12}}$ rms to the quantizing noise of the same level. In order to find the resultant noise level it is necessary to add the powers as the signals are uncorrelated. The total power is given by:

$$2 \times \frac{Q^2}{12} = \frac{Q^2}{6}$$

and the rms voltage is $Q/\sqrt{6}$. Another way of looking at the situation is to consider that the noise power doubles and so the rms noise voltage has increased by 3 dB in comparison with the undithered case. Thus for an n-bit wordlength, using the same derivation as expression (4.1) above, the signal-to-noise ratio for Q pk-pk rectangular dither will be given by:

$$6.02\,n - 1.24\,\text{dB} \tag{4.2}$$

Unlike the undithered case, this is a true signal-to-noise ratio and linearity is maintained at all signal levels. By way of example, for a ten-bit system nearly 59 dB SNR is achieved. The 3 dB loss compared to the undithered case is a small price to pay for linearity.

4.12.2 *Triangular pdf dither*

The noise modulation due to the use of rectangular-probability dither is undesirable. It comes about because the process is too simple. The undithered quantizing error is signal dependent and the dither represents a single uniform-probability random process. This is only capable of decorrelating the quantizing error to the extent that its mean value is zero, rendering the system linear. The signal dependence is not eliminated, but is displaced to the next statistical moment. This is the variance and the result is noise modulation. If a further uniform-probability random process is introduced into the system, the signal dependence is displaced to the next moment and the second moment or variance becomes constant.

Adding together two statistically independent rectangular probability functions produces a triangular probability function. A signal having this characteristic can be used as the dither source.

Figure 4.23(c) shows the averaged transfer function for a number of dither amplitudes. Linearity is reached with a pk-pk amplitude of $2Q$ and at this level there is no noise modulation. The lack of noise modulation is another way of stating that the noise is constant. The triangular pdf of the dither matches the triangular shape of the quantizing error function.

The dither adds two noise signals with an amplitude of $\frac{Q}{\sqrt{12}}$ rms to the quantizing noise of the same level. In order to find the resultant noise level it is necessary to add the powers as the signals are uncorrelated. The total power is given by:

$$3 \times \frac{Q^2}{12} = \frac{Q^2}{4}$$

and the rms voltage is $Q/\sqrt{4}$. Another way of looking at the situation is to consider that the noise power is increased by 50 per cent in comparison to the rectangular dithered case and so the rms noise voltage has increased by 1.76 dB. Thus for an n-bit wordlength, using the same derivation as expressions (4.1) and (4.2) above, the signal-to-noise ratio for Q pk–pk rectangular dither will be given by:

$$6.02n - 3\,\text{dB} \tag{4.3}$$

Continuing the use of a ten-bit example, a SNR of 57.2 dB is available which is 4.8 dB worse than the SNR of an undithered quantizer in the large signal case. It is a small price to pay for perfect linearity and an unchanging noise floor.

4.12.3 Gaussian pdf dither

Adding more uniform probability sources to the dither makes the overall probability function progressively more like the Gaussian distribution of analog noise. Figure 4.23(d) shows the averaged transfer function of a quantizer with various levels of Gaussian dither applied. Linearity is reached with $\frac{1}{2}Q$ rms and at this level noise modulation is negligible. The total noise power is given by:

$$\frac{Q^2}{4} + \frac{Q^2}{12} = \frac{3 \times Q^2}{12} + \frac{Q^2}{12} = \frac{Q^2}{3}$$

and so the noise level will be $Q/\sqrt{3}$ rms. The noise level of an undithered quantizer in the large signal case is $Q/\sqrt{12}$ and so the noise is higher by a factor of:

$$\frac{Q}{\sqrt{3}} \times \frac{\sqrt{12}}{Q} = \frac{Q}{\sqrt{3}} \times \frac{2\sqrt{3}}{Q} = 2 = 6.02\,\text{dB} \qquad (4.4)$$

Thus the SNR is given by $6.02(n - 1) + 1.76$ dB. A ten-bit system with correct Gaussian dither has a SNR of 56 dB.

This is inferior to the figure in expression (4.3) by 1.1 dB. In digital dither applications, triangular probability dither of 2Q pk–pk is optimum because it gives the best possible combination of nil distortion, freedom from noise modulation and SNR. Using dither with more than two rectangular processes added is detrimental. Whilst this result is also true for analog dither, it is not practicable to apply it to a real ADC as all real analog signals contain thermal noise which is Gaussian. If triangular dither is used on a signal containing Gaussian noise, the results derived above are not obtained. ADCs should therefore use Gaussian dither of $Q/2$ rms and performance will be given by expression (4.4).

4.13 Basic digital-to-analog conversion

This direction of conversion will be discussed first, since ADCs often use embedded DACs in feedback loops. The purpose of a digital-to-analog convertor is to take numerical values and reproduce the continuous waveform that they represent. Figure 4.24 shows the major elements of a conventional conversion subsystem, i.e. one in which oversampling is not employed. The jitter in the clock needs to be removed with a VCO or VCXO. Sample values are buffered in a latch and fed to the convertor element which operates on each cycle of the clean clock. The output is then a voltage proportional to the number for at least a part of the sample period. A resampling stage may be found next, in order to remove switching transients, reduce the aperture ratio or allow the use of a convertor which takes a substantial part of the sample period to operate.

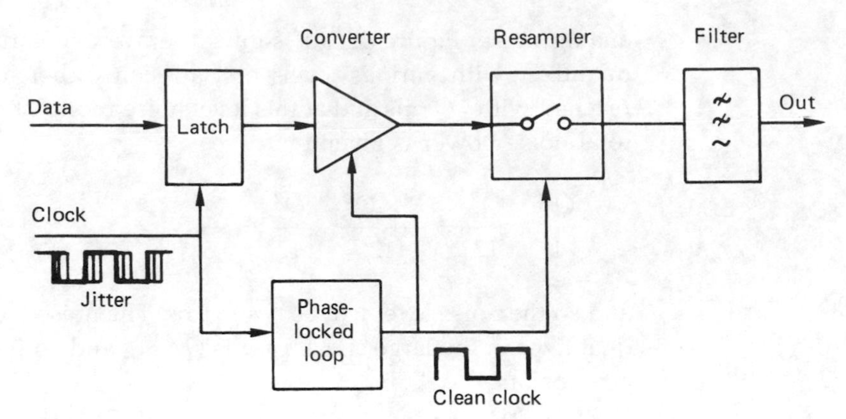

Figure 4.24 The components of a conventional convertor. A jitter-free clock drives the voltage conversion, whose output may be resampled prior to reconstruction.

The resampled waveform is then presented to a reconstruction filter which rejects frequencies above the audio band.

This section is primarily concerned with the implementation of the convertor element. The most common way of achieving this conversion is to control binary-weighted currents and sum them in a virtual earth. Figure 4.25 shows the classical *R*-2*R* DAC structure. This is relatively simple to construct, but the resistors have to be extremely accurate. To see why this is so, consider the example of Figure 4.26. At (a) the binary code is about to have a major overflow, and all the low-order currents are

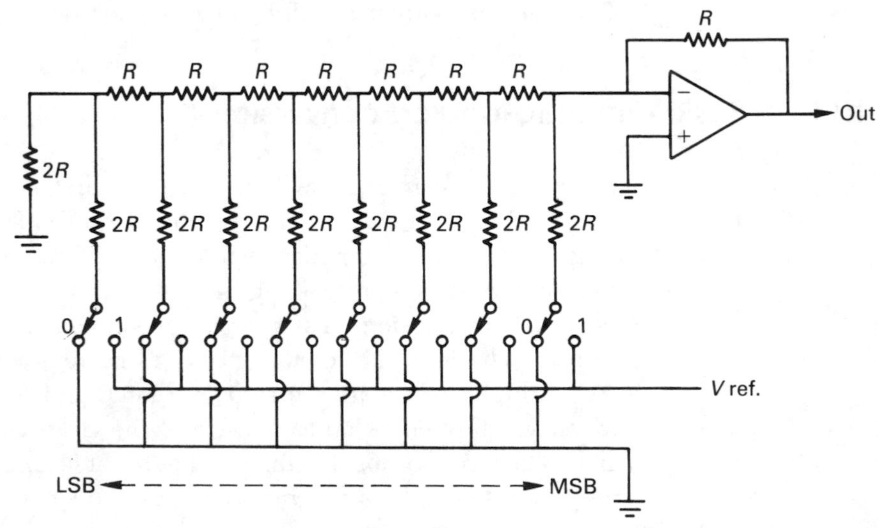

Figure 4.25 The classical *R*–2*R* DAC requires precise resistance values and 'perfect' switches.

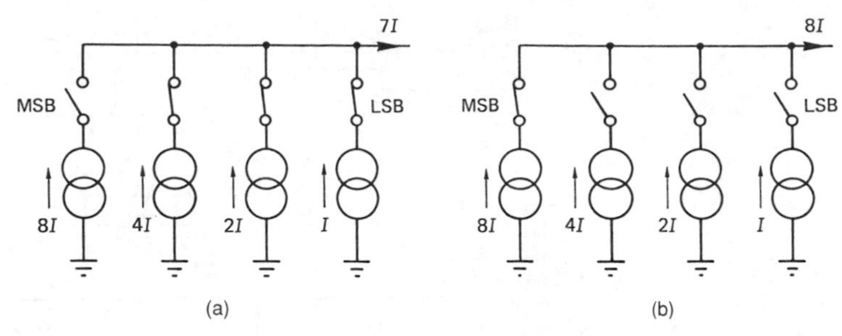

Figure 4.26 At (a) current flow with an input of 0111 is shown. At (b) current flow with input code one greater.

flowing. At (b), the binary input has increased by one, and only the most significant current flows. This current must equal the sum of all the others plus one. The accuracy must be such that the step size is within the required limits. In this eight-bit example, if the step size needs to be a rather casual 10 per cent accurate, the necessary accuracy is only one part in 2560, but for a ten-bit system it would become one part in 10 240. This degree of accuracy is difficult to achieve and maintain in the presence of ageing and temperature change.

4.14 Basic analog-to-digital conversion

The general principle of a quantizer is that different quantized voltages are compared with the unknown analog input until the closest quantized voltage is found. The code corresponding to this becomes the output. The comparisons can be made in turn with the minimal amount of hardware, or simultaneously with more hardware.

The flash convertor is probably the simplest technique available for PCM video conversion. The principle is shown in Figure 4.27. The threshold voltage of every quantizing interval is provided by a resistor chain which is fed by a reference voltage. This reference voltage can be varied to determine the sensitivity of the input. There is one voltage comparator connected to every reference voltage, and the other input of all the comparators is connected to the analog input. A comparator can be considered to be a one-bit ADC. The input voltage determines how many of the comparators will have a true output. As one comparator is necessary for each quantizing interval, then, for example, in an eight-bit system there will be 255 binary comparator outputs, and it is necessary to use a priority encoder to convert these to a binary code. Note that the quantizing stage is asynchronous; comparators change state as and when the variations in the input waveform result in a reference voltage being crossed. Sampling takes place when the comparator outputs are clocked

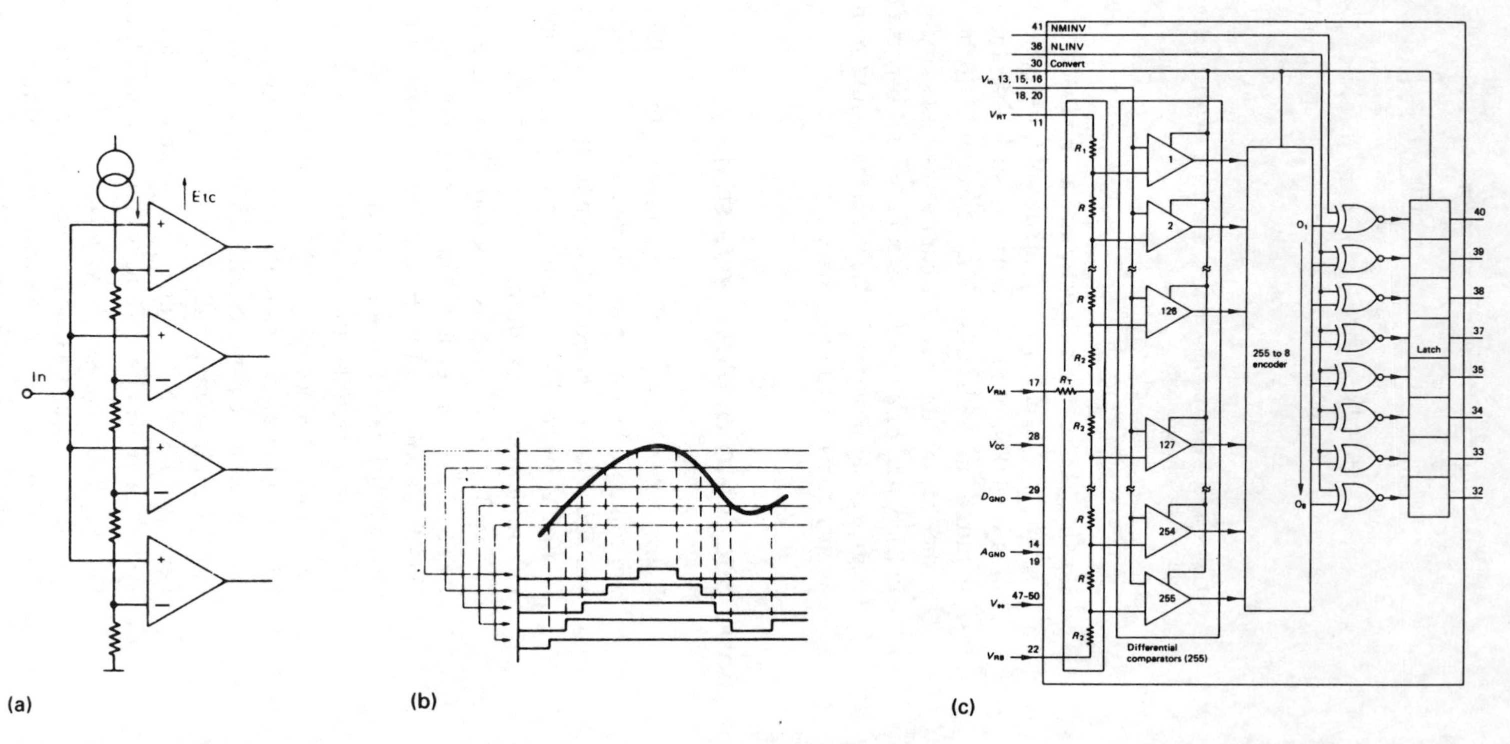

Figure 4.27 The flash convertor. In (a) each quantizing interval has its own comparator, resulting in waveforms of (b). A priority encoder is necessary to convert the comparator outputs to a binary code. Shown in (c) is a typical eight-bit flash convertor primarily intended for video applications. (Courtesy TRW).

into a subsequent latch. This is an example of quantizing before sampling as was illustrated in Figure 4.1. Although the device is simple in principle, it contains a lot of circuitry and can only be practicably implemented on a chip. The analog signal has to drive many inputs which results in a significant parallel capacitance, and a low-impedance driver is essential to avoid restricting the slewing rate of the input. The extreme speed of a flash convertor is a distinct advantage in oversampling. Because computation of all bits is performed simultaneously, no track/hold circuit is required, and droop is eliminated. Figure 4.27(c) shows a flash convertor chip. Note the resistor ladder and the comparators followed by the priority encoder. The MSB can be selectively inverted so that the device can be used either in offset binary or two's complement mode.

The flash convertor is ubiquitous in digital video because of the high speed necessary. For audio purposes, many more conversion techniques are available and these are considered in Chapter 7.

4.15 Factors affecting convertor quality

In theory the quality of a digital audio system comprising an ideal ADC followed by an ideal DAC is determined at the ADC. The ADC parameters such as the sampling rate, the wordlength and any noise shaping used put limits on the quality which can be achieved. Conversely the DAC itself may be transparent, because it only converts data whose quality is already determined back to the analog domain. In other words, the ADC determines the system quality and the DAC does not make things any worse.

In practice both ADCs and DACs can fall short of the ideal, but with modern convertor components and attention to detail the theoretical limits can be approached very closely and at reasonable cost. Shortcomings may be the result of an inadequacy in an individual component such as a convertor chip, or due to incorporating a high-quality component in a poorly though-out system. Poor system design can destroy the performance of a convertor. Whilst oversampling is a powerful technique for realizing high-quality convertors, its use depends on digital interpolators and decimators whose quality affects the overall conversion quality.

ADCs and DACs have the same transfer function, since they are only distinguished by the direction of operation, and therefore the same terminology can be used to classify the possible shortcomings of both. Figure 4.28 shows the transfer functions resulting from the main types of convertor error:

(a) *Offset error.* A constant appears to have been added to the digital signal. This has a serious effect in video systems because it alters the

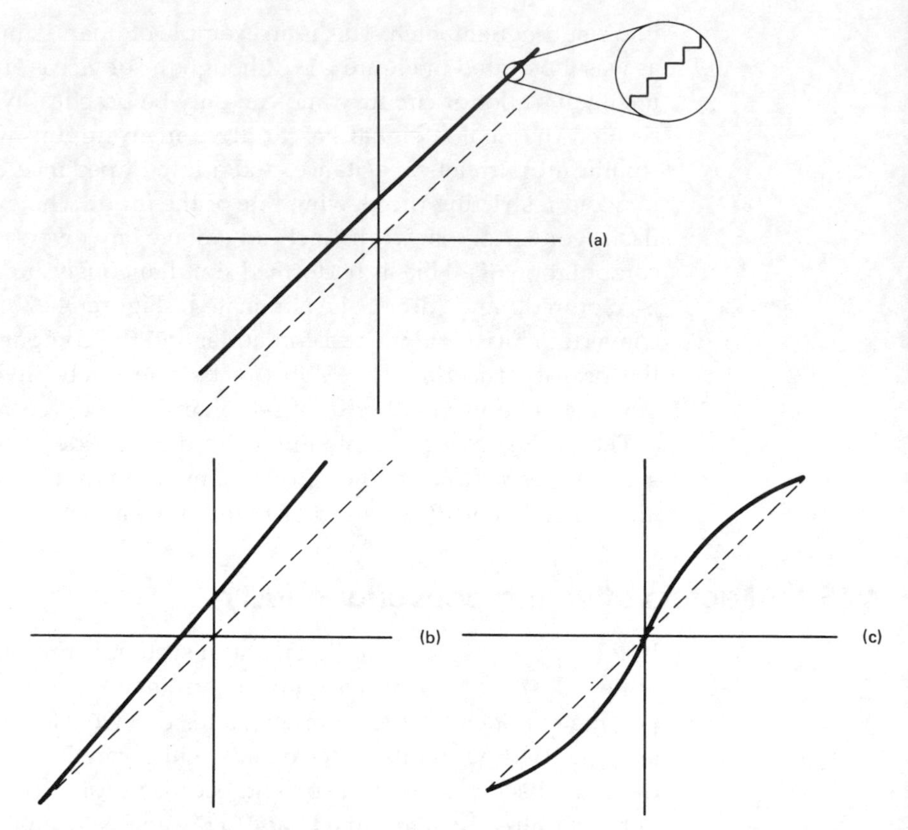

Figure 4.28 Main convertor errors (solid line) compared with perfect transfer function (dotted line). These graphs hold for ADCs and DACs, and the axes are interchangeable; if one is chosen to be analog, the other will be digital.

black level. Offset error is sometimes cancelled by digitally sampling the convertor output during blanking and feeding it back to the analog input as a small control voltage.

(b) *Gain error.* The slope of the transfer function is incorrect. Since convertors are often referred to one end of the range, gain error causes an offset error. Severe gain error causes clipping.

(c) *Integral linearity.* This is the deviation of the dithered transfer function from a straight line. It has exactly the same significance and consequences as linearity in analog circuits, since if it is inadequate, harmonic distortion will be caused.

(d) *Differential non-linearity* is the amount by which adjacent quantizing intervals differ in size. This is usually expressed as a fraction of a quantizing interval.

(e) *Monotonicity* is a special case of differential non-linearity. Non-monotonicity means that the output does not increase for an increase in input. Figure 4.26 showed how this can happen in a DAC. With a convertor input code of 01111111 (127 decimal), the seven low-order current sources of the convertor will be on. The next code is 10000000 (128 decimal), where only the eighth current source is operating. If the current it supplies is in error on the low side, the analog output for 128 may be less than that for 127. In an ADC non-monotonicity can result in missing codes. This means that certain binary combinations within the range cannot be generated by any analog voltage. If a device has better than $\frac{1}{2}Q$ linearity it must be monotonic. It is not possible for a one-bit convertor to be non-monotonic.

(f) *Absolute accuracy.* This is the difference between actual and ideal output for a given input. For video and audio it is rather less important than linearity. For example, if all the current sources in a convertor have good thermal tracking, linearity will be maintained, even though the absolute accuracy drifts.

4.16 Oversampling

Oversampling means using a sampling rate which is greater (generally substantially greater) than the Nyquist rate. Neither sampling theory nor quantizing theory *require* oversampling to be used to obtain a given signal quality, but Nyquist rate conversion places extremely high demands on component accuracy when a convertor is implemented. Oversampling allows a given signal quality to be reached without requiring very close tolerance, and therefore expensive, components.

Figure 4.29 shows the main advantages of oversampling. At (a) it will be seen that the use of a sampling rate considerably above the Nyquist rate allows the anti-aliasing and reconstruction filters to be realized with a much more gentle cut-off slope. There is then less likelihood of phase linearity and ripple problems in the passband. Figure 4.29(b) shows that information in an analog signal is two-dimensional and can be depicted as an area which is the product of bandwidth and the linearly expressed signal-to-noise ratio. The figure also shows that the same amount of information can be conveyed down a channel with a SNR of half as much (6 dB less) if the bandwidth used is doubled, with 12 dB less SNR if bandwidth is quadrupled, and so on, provided that the modulation scheme used is perfect.

The information in an analog signal can be conveyed using some analog modulation scheme in any combination of bandwidth and SNR which yields the appropriate channel capacity. If bandwidth is replaced by sampling rate and SNR is replaced by a function of wordlength, the same must be true for a digital signal as it is no more than a numerical analog. Thus raising the sampling rate potentially allows the wordlength of each sample to be reduced without information loss.

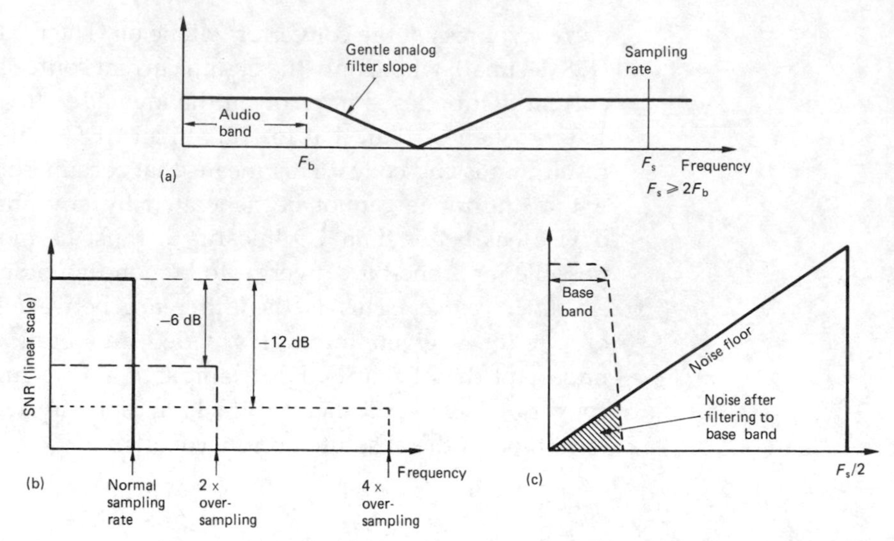

Figure 4.29 Oversampling has a number of advantages. In (a) it allows the slope of analog filters to be relaxed. In (b) it allows the resolution of convertors to be extended. In (c) *a noise shaped* convertor allows a disproportionate improvement in resolution.

Information theory predicts that if a signal is spread over a much wider bandwidth by some modulation technique, the SNR of the demodulated signal can be higher than that of the channel it passes through, and this is also the case in digital systems. The concept is illustrated in Figure 4.30. At (a) four-bit samples are delivered at sampling rate F. As four bits have sixteen combinations, the information rate is $16F$. At (b) the same information rate is obtained with three-bit samples by raising the sampling rate to $2F$ and at (c) two-bit samples having four combinations require to be delivered at a rate of $4F$. Whilst the information rate has been maintained, it will be noticed that the bit-rate of (c) is twice that of (a). The reason for this is shown in Figure 4.31. A single binary digit can have only two states; thus it can only convey two pieces of information, perhaps 'yes' or 'no'. Two binary digits together can have four states, and can thus convey four pieces of information, perhaps 'spring summer autumn or winter', which is two pieces of information per bit. Three binary digits grouped together can have eight combinations, and convey eight pieces of information, perhaps 'doh re mi fah so lah te or doh', which is nearly three pieces of information per digit. Clearly the further this principle is taken, the greater the benefit. In a sixteen-bit system, each bit is worth 4K pieces of information. It is always more efficient, in information-capacity terms, to use the combinations of long binary words than to send single bits for every piece of information. The greatest

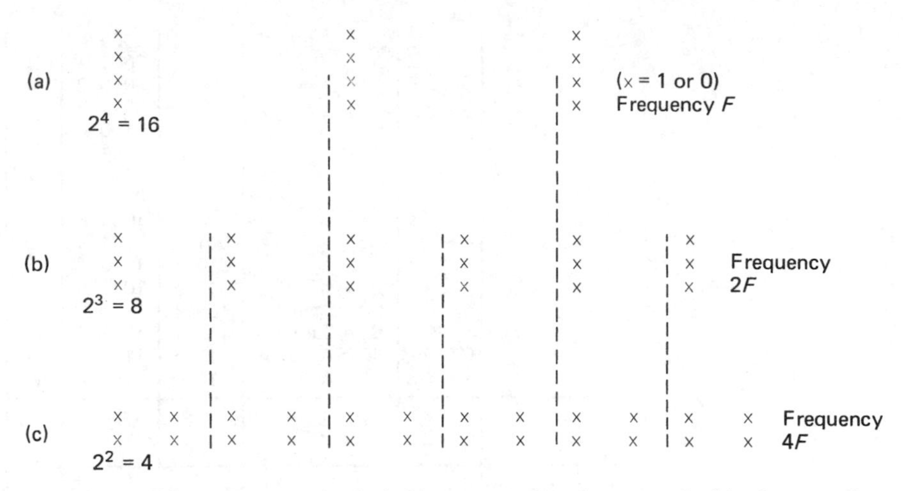

Figure 4.30 Information rate can be held constant when frequency doubles by removing one bit from each word. In all cases here it is 16*F*. Note bit rate of (c) is double that of (a). Data storage in oversampled form is inefficient.

efficiency is reached when the longest words are sent at the slowest rate which must be the Nyquist rate. This is one reason why PCM recording is more common than delta modulation, despite the simplicity of implementation of the latter type of convertor. PCM simply makes more efficient use of the capacity of the binary channel.

As a result, oversampling is confined to convertor technology where it gives specific advantages in implementation. The storage or transmission system will usually employ PCM, where the sampling rate is a little more than twice the input bandwidth. Figure 4.32 shows a digital VTR using oversampling convertors. The ADC runs at *n* times the Nyquist rate, but once in the digital domain the rate needs to be reduced in a type of digital filter called a *decimator*. The output of this is conventional Nyquist rate PCM, according to the tape format, which is then recorded. On replay the sampling rate is raised once more in a further type of digital filter called an *interpolator*. The system now has the best of both worlds: using oversampling in the convertors overcomes the shortcomings of analog anti-aliasing and reconstruction filters and the wordlength of the convertor elements is reduced making them easier to construct; the recording is made with Nyquist rate PCM which minimizes tape consumption.

Oversampling is a method of overcoming practical implementation problems by replacing a single critical element or bottleneck by a number of elements whose overall performance is what counts. As Hauser[18] properly observed, oversampling tends to overlap the operations which are quite distinct in a conventional convertor. In earlier sections of this

	0 = No 1 = Yes	00 = Spring 01 = Summer 10 = Autumn 11 = Winter	000 do 001 re 010 mi 011 fa 100 so 101 la 110 te 111 do	0000 0 0001 1 0010 2 0011 3 0100 4 0101 5 0110 6 0111 7 1000 8 1001 9 1010 A 1011 B 1100 C 1101 D 1110 E 1111 F	≈	0000 ⋮ FFFF Digital audio sample values
No of bits	1	2	3	4	≈	16
Information per word	2	4	8	16	≈	65 536
Information per bit	2	2	≈3	4	≈	4096

Figure 4.31 The amount of information per bit increases disproportionately as wordlength increases. It is always more efficient to use the longest words possible at the lowest word rate. It will be evident that sixteen bit PCM is 2048 times as efficient as delta modulation. Oversampled data are also inefficient for storage.

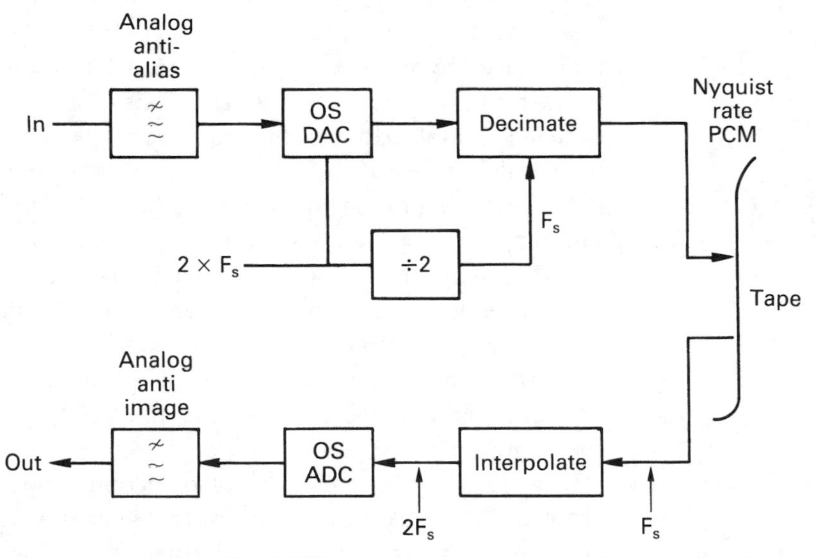

Figure 4.32 An oversampling DVTR. The convertors run faster than sampling theory suggests to ease analog filter design. Sampling-rate reduction allows efficient PCM recording on tape.

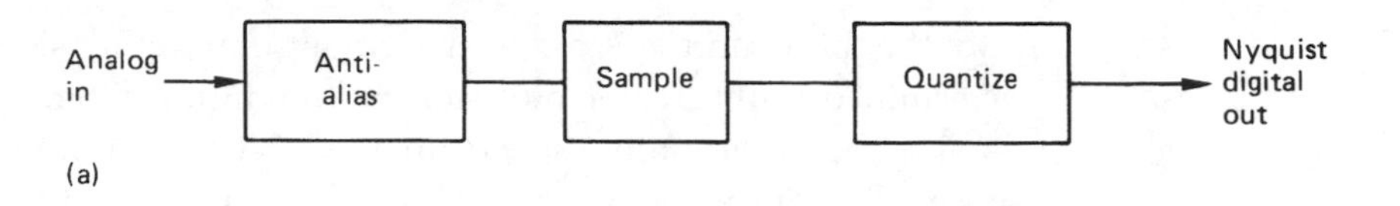

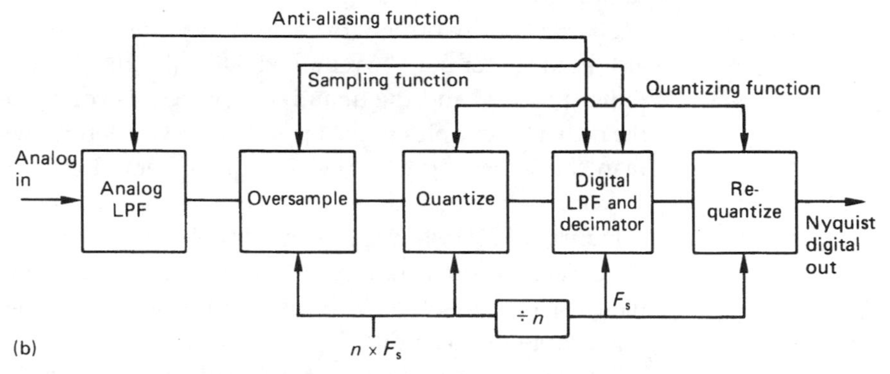

Figure 4.33 A conventional ADC performs each step in an identifiable location as in (a). With oversampling, many of the steps are distributed as shown in (b).

chapter, the vital subjects of filtering, sampling, quantizing and dither have been treated almost independently. Figure 4.33(a) shows that it is possible to construct an ADC of predictable performance by a taking a suitable anti-aliasing filter, a sampler, a dither source and a quantizer and assembling them like building bricks. The bricks are effectively in series and so the performance of each stage can only limit the overall performance. In contrast (b) shows that with oversampling the overlap of operations allows different processes to augment one another allowing a synergy which is absent in the conventional approach.

If the oversampling factor is n, the analog input must be bandwidth limited to $n.F_s/2$ by the analog anti-aliasing filter. This unit need only have flat frequency response and phase linearity within the audio band. Analog dither of an amplitude compatible with the quantizing interval size is added prior to sampling at $n.F_s$ and quantizing.

Next, the anti-aliasing function is completed in the digital domain by a low-pass filter which cuts off at $F_s/2$. Using an appropriate architecture this filter can be absolutely phase linear and implemented to arbitrary accuracy. Such filters were discussed in Chapter 2. The filter can be considered to be the demodulator of Figure 4.29 where the SNR improves as the bandwidth is reduced. The wordlength can be expected to increase. As Chapter 2 illustrated, the multiplications taking place within the filter extend the wordlength considerably more than the bandwidth reduction alone would indicate. The analog filter serves only to prevent aliasing into the baseband at the oversampling rate; the signal spectrum is determined with greater precision by the digital filter.

With the information spectrum now Nyquist limited, the sampling process is completed when the rate is reduced in the decimator. One sample in n is retained.

The excess wordlength extension due to the anti-aliasing filter arithmetic must then be removed. Digital dither is added, completing the dither process, and the quantizing process is completed by requantizing the dithered samples to the appropriate wordlength which will be greater than the wordlength of the first quantizer. Alternatively noise shaping may be employed.

Figure 4.34(a) shows the building-brick approach of a conventional DAC. The Nyquist rate samples are converted to analog voltages and then a steep-cut analog low-pass filter is needed to reject the sidebands of the sampled spectrum.

Figure 4.34(b) shows the oversampling approach. The sampling rate is raised in an interpolator which contains a low-pass filter which restricts the baseband spectrum to the audio bandwidth shown. A large frequency gap now exists between the baseband and the lower sideband. The multiplications in the interpolator extend the wordlength considerably and this must be reduced within the capacity of the DAC element by the addition of digital dither prior to requantizing.

Oversampling may also be used to considerable benefit in other dimensions. Figure 4.35 shows how vertical oversampling can be used to increase the resolution of a TV system. A 1250-line camera, for example,

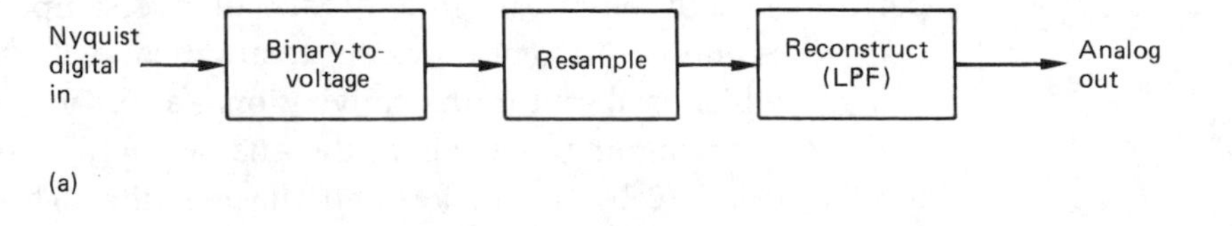

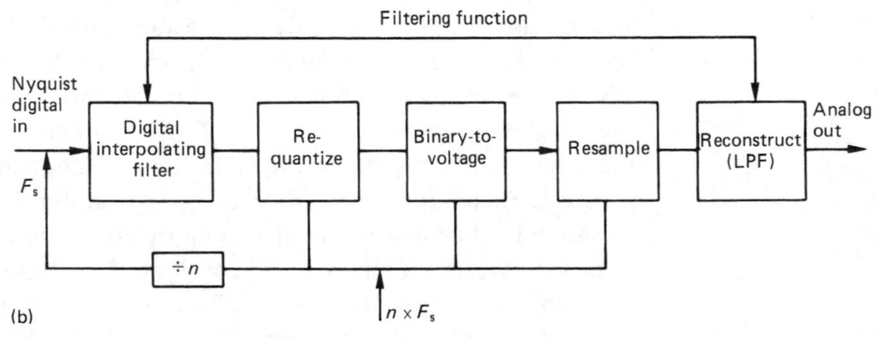

Figure 4.34 A conventional DAC in (a) is compared with the oversampling implementation in (b).

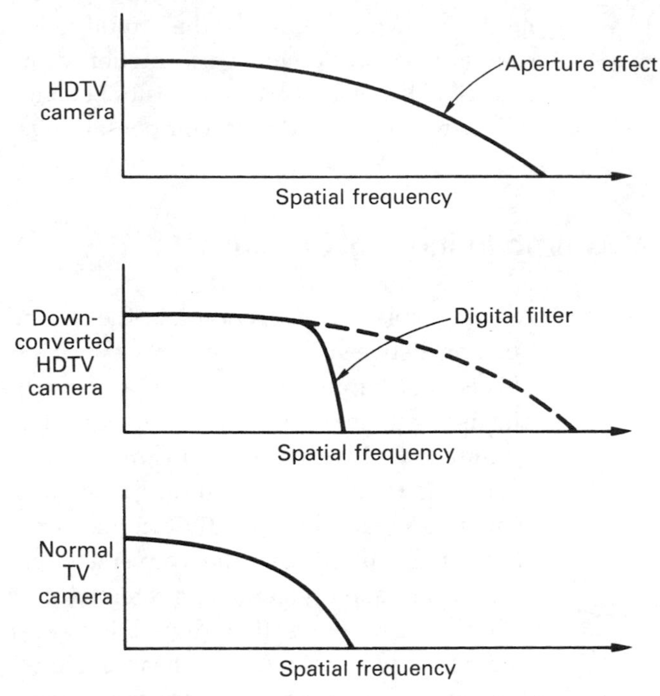

Figure 4.35 Using an HDTV camera with down-conversion is a form of oversampling and gives better results than a normal camera because the aperture effect is overcome.

is used as the input device, but the 1250-line signal is fed to a standards convertor which reduces the number of lines to 625. The standards convertor must incorporate a vertical low-pass spatial filter to prevent aliasing when the vertical sampling rate is effectively halved. Such a filter was described in Chapter 3. As it is a digital filter, it can have arbitrarily accurate peformance, including a flat passband and steep cut-off slope. The combination of the vertical aperture effect of the 1250-line camera and the vertical LPF in the standards convertor gives a better spatial frequency response than could be achieved with a 625-line camera. The improvement in subjective quality is quite noticeable in practice.

In the case of display technology, oversampling can also be used, this time to render the raster invisible and to improve the vertical aperture of the display. Once more a standards convertor is required, but this now doubles the number of input lines using interpolation. Again the filter can have arbitrary accuracy. The vertical aperture of the 1250-line display does not affect the passband of the input signal because of the use of oversampling.

Oversampling can also be used in th time domain in order to reduce or eliminate display flicker. A different type of standards convertor is necessary which doubles the input field rate by interpolation. The standards convertor must use motion compensation otherwise moving objects will not be correctly positioned in intermediate fields and will suffer from judder. Motion compensation is considered in Chapter 5.

4.17 Gamma in the digital domain

As was explained in section 2.2, the use of gamma makes the transfer function between luminance and the analog video voltage non-linear. This is done because the eye is more sensitive to noise at low brightness. The use of gamma allows the receiver to compress the video signal at low brightness and with it any transmission noise.

There is a strong argument to retain gamma in the digital domain for analog compatibility. In the digital domain transmission noise is eliminated, but instead the conversion process introduces quantizing noise. Consequently gamma is retained in the digital domain.

Figure 4.36 shows that digital luma can be considered in several equivalent ways. At (a) a linear analog luminance signal is passed through a gamma corrector to create luma and this is then quantized uniformly. At (b) the linear analog luminance signal is fed directly to a non-uniform quantizer. At (c) the linear analog luminance signal is uniformly quantized to produce digital luminance. This is converted to digital luma by a digital process having a non-linear transfer function.

Whilst the three techniques shown give the same result, (a) is the simplest, (b) requires a special ADC with gamma-spaced quantizing steps, and (c) requires a high-resolution ADC of perhaps fourteen–sixteen bits because it works in the linear luminance domain where noise is highly visible. Technique (c) is used in digital processing cameras where long wordlength is common practice.

As digital luma with eight-bit resolution gives the same subjective performace as digital luminance with fourteen-bit resolution it will be seen that gamma can also be considered to be an effective perceptive compression technique.

4.18 Colour in the digital domain

Colour cameras and most graphics computers produce three signals, or components, *R*, *G* and *B* which are essentially monochrome video signals representing an image in each primary colour. Figure 4.37 shows that the three primaries are spaced apart in the chromaticity diagram and the only colours which can be generated fall within the resultant triangle.

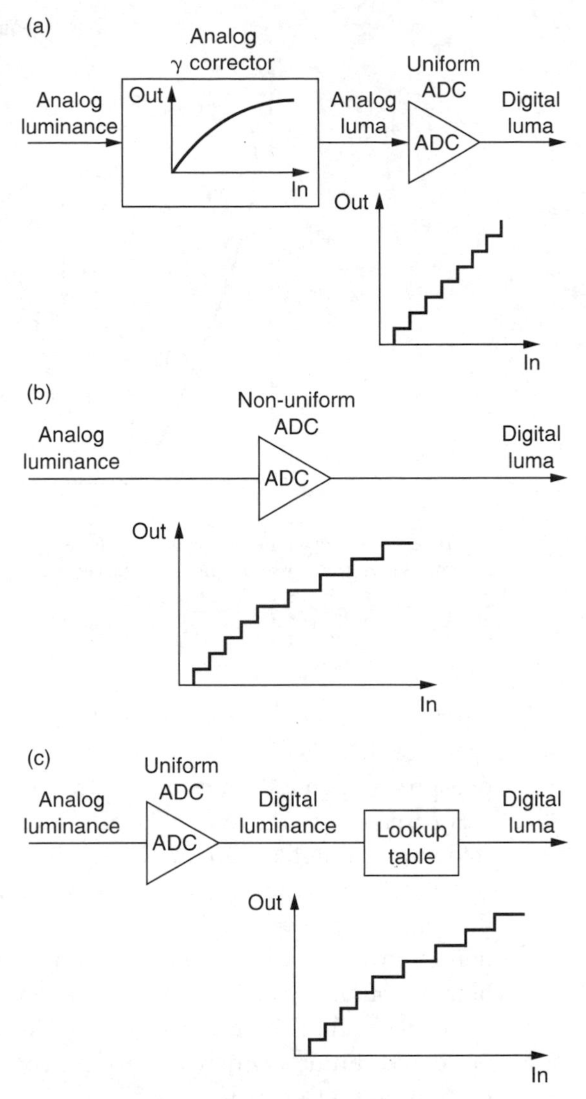

Figure 4.36 (a) Analog γ correction prior to ADC. (b) Non-uniform quantizer gives direct γ conversion. (c) Digital γ correction using look-up table

RGB signals are only strictly compatible if the colour primaries assumed in the source are present in the display. If there is a primary difference the reproduced colours are different. Clearly broadcast television must have a standard set of primaries. The EBU television systems have only ever had one set of primaries. NTSC started off with one set and then adopted another because the phosphors were brighter. Computer displays have any number of standards because initially all computer colour was false, i.e. synthetic, and the concept of accurate

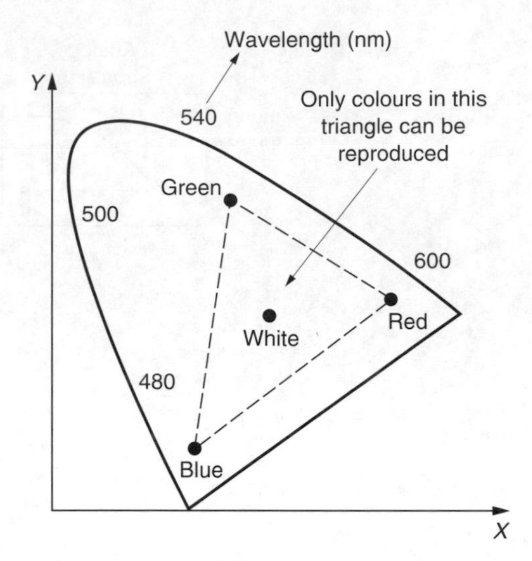

Figure 4.37 Additive mixing colour systems can only reproduce colours within a triangle where the primaries lie on each vertex.

reproduction didn't arise. Now that computer displays are going to be used for television it will be necessary for them to adopt standard phosphors or to use colorimetric transcoding on the signals.

Fortunately the human visual system is quite accommodating. The colour of daylight changes throughout the day, so everything changes colour with it. Humans, however, accommodate to that. We tend to see the colour we expect rather than the actual colour. The colour reproduction of printing and photographic media is pretty appalling, but so is human colour memory, and so it's acceptable.

On the other hand, our ability to discriminate between colours presented simultaneously is razor sharp, hence the difficulty car repairers have in getting paint to match.

RGB and *Y* signals are incompatible, yet when colour television was introduced it was a practical necessity that it should be possible to display colour signals on a monochrome display and vice versa.

Creating or transcoding a luminance signal from *RGB* is relatively easy. The spectral response of the eye has a peak in the green region. Green objects will produce a larger stimulus than red objects of the same brightness, with blue objects producing the least stimulus. A luminance signal can be obtained by adding *R*, *G* and *B* together, not in equal amounts, but in a sum which is weighted by the relative response of the human visual system. Thus:

$$Y = 0.299\mathrm{R} + 0.587G + 0.114B$$

Note that the factors add up to one. If Y is derived in this way, a monochrome display will show nearly the same result as if a monochrome camera had been used in the first place. The results are not identical because of the non-linearities introduced by gamma correction.

As colour pictures require three signals, it should be possible to send Y and two other signals which a colour display could arithmetically convert back to *RGB*. There are two important factors which restrict the form that the other two signals may take. One is to achieve reverse compatibility. If the source is a monochrome camera, it can only produce Y and the other two signals will be completely absent. A colour display should be able to operate on the Y signal only and show a monochrome picture. The other is the requirement to conserve bandwidth for economic reasons.

These requirements are met by creating two colour difference signals along with Y. There are three possible colour difference signals, $R - Y$, $B - Y$ and $G - Y$. As the green signal makes the greatest contribution to Y, then the amplitude of $G - Y$ would be the smallest and would be most susceptible to noise. Thus $R - Y$ and $B - Y$ are used in practice as Figure 4.38 shows. In the digital domain $R - Y$ is known as C_r and $B - Y$ is known as C_b.

Whilst signals such as Y, R, G and B are unipolar or positive only, colour difference signals are bipolar and may meaningfully take on negative values. Figure 4.38(a) shows the colour space available in eight-bit *RGB*. In computers, eight-bit *RGB* is common and we often see claims that 16 million different colours are possible. This is utter nonsense.

A colour is a given combination of hue and saturation and is independent of brightness. Consequently all sets of *RGB* values having the same ratios produce the same colour. For example, $R = G = B$ always gives the same colour whether the pixel value is 0 or 255. Thus there are 256 brightnesses which have the same colour allowing a more believable 65 000 different colours.

Figure 4.38(c) shows the *RGB* cube mapped into eight-bit colour difference space so that it is no longer a cube. Now the grey axis goes straight up the middle because greys correspond to both C_r and C_b being zero. To visualize colour difference space, imagine looking down along the grey axis. This makes the black and white corners coincide in the centre. The remaining six corners of the legal colour difference space now correspond to the six boxes on a component vectorscope. Although there are still 16 million combinations, many of these are now illegal. For example, as black or white are approached, the colour differences must fall to zero.

From an information theory standpoint, colour difference space is redundant. With some tedious geometry, it can be shown that fewer than a quarter of the codes are legal. The luminance resolution remains the same, but there is about half as much information in each colour axis. This due to

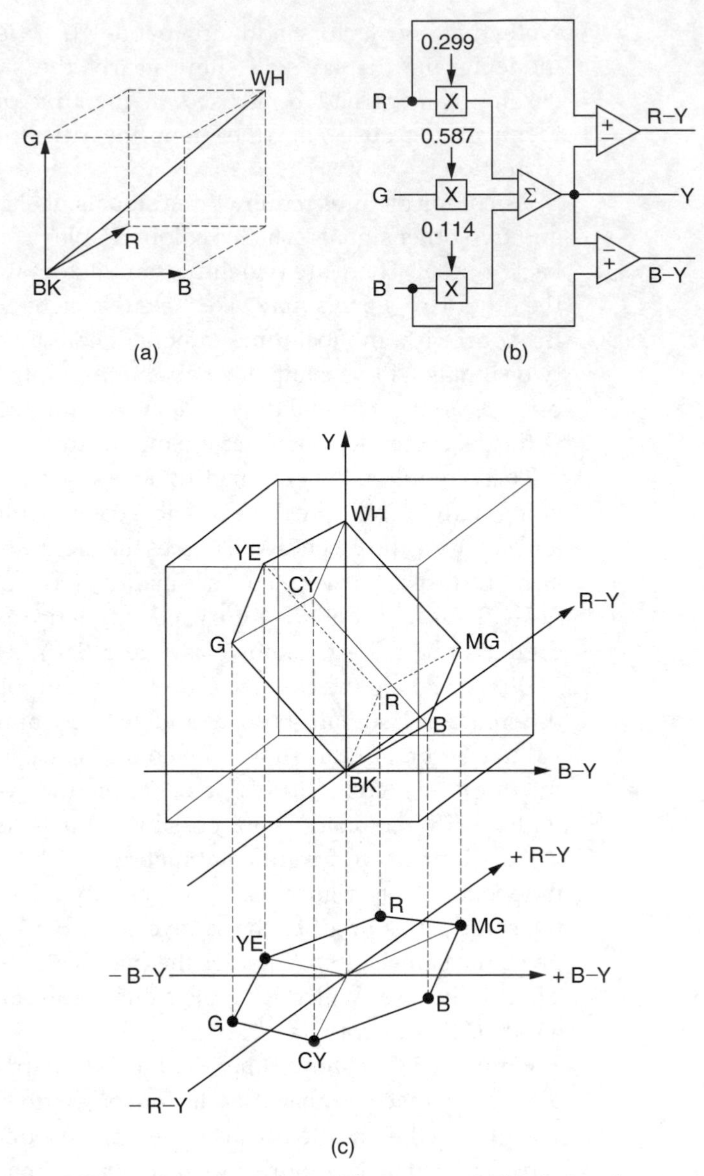

Figure 4.38 *RGB* transformed to colour difference space. This is done because *R–Y* and *B–Y* can be sent with reduced bandwidth. (a) *RGB* cube. WH–BK axis is diagonal. All locations within cube are legal, (b) *RGB* to colour difference transform. (c) *RGB* cube mapped into colour difference space is no longer a cube. Projection down creates conventional vectorscope display.

the colour difference signals being bipolar. If the signal resolution has to be maintained, eight-bit *RGB* should be transformed to a longer wordlength in the colour difference domain, nine bits being adequate. At this stage the colour difference transform doesn't seem efficient because twenty-four-bit *RGB* converts to twenty-six-bit Y, C_r, C_b.

In most cases the loss of colour resolution is invisible to the eye, and eight-bit resolution is retained. The results of the transform computation must be digitally dithered to avoid posterizing.

The inverse transform to obtain *RGB* again at the display is straightforward. *R* and *B* are readily obtained by adding *Y* to the two colour difference signals. *G* is obtained by re-arranging the expression for *Y* above such that:

$$G = \frac{Y - 0.3R - 0.11B}{0.59}$$

If a monochrome source having only a *Y* output is supplied to a colour display, C_r and C_b will be zero. It is reasonably obvious that if there are no colour difference signals the colour signals cannot be different from one another and $R = G = B$. As a result the colour display can only produce a neutral picture.

The use of colour difference signals is essential for compatibility in both directions between colour and monochrome, but it has a further advantage which follows from the way in which the eye works. In order to produce the highest resolution in the fovea, the eye will use signals from all types of cone, regardless of colour. In order to determine colour the stimuli from three cones must be compared.

There is evidence that the nervous system uses some form of colour difference processing to make this possible. As a result the full acuity of the human eye is only available in monochrome. Detail in colour changes cannot be resolved so well. A further factor is that the lens in the human eye is not achromatic and this means that the ends of the spectrum are not well focused. This is particularly noticeable on blue.

In this case there is no point is expending valuable bandwidth sending high-resolution colour signals. Colour difference working allows the luminance to be sent separately at full bandwidth. This determines the subjective sharpness of the picture. The colour difference information can be sent with considerably reduced resolution, as little as one quarter that of luminance, and the human eye is unable to tell.

The acuity of human vision is axisymmetric. In other words detail can be resolved equally at all angles. When the human visual system assesses the sharpness of a TV picture, it will measure the quality of the worst axis and the extra information on the better axis is wasted. Consequently the most efficient row-and-column image sampling arrangement is the so-called 'square pixel'. Now pixels are dimensionless and so this is meaningless. However, it is understood to mean that the horizontal and vertical spacing between pixels is the same. Thus it is the sampling grid which is square, rather than the pixel.

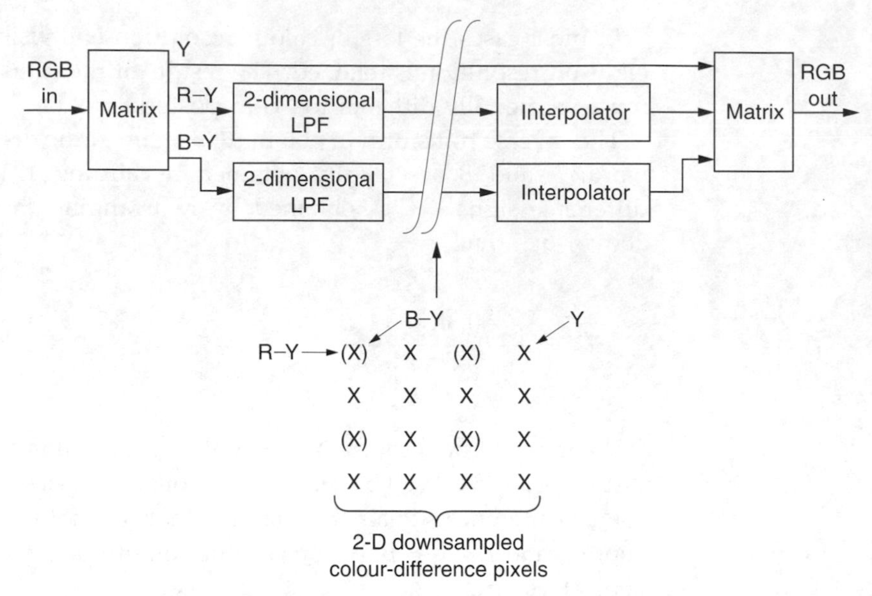

Figure 4.39 Ideal two-dimensionally downsampled colour-difference system. Colour resolution is half of luma resolution, but the eye cannot tell the difference.

The square pixel is optimal for luminance and also for colour difference signals. Figure 4.39(a) shows the ideal. The colour sampling is co-sited with the luminance sampling but the colour sample spacing is twice that of luminance. The colour difference signals after matrixing from *RGB* have to be low-pass filtered in two dimensions prior to downsampling in order to prevent aliasing of HF detail. At the display, the downsampled colour data have to be interpolated in two dimensions to produce colour information in every pixel. In an oversampling display the colour interpolation can be combined with the display upsampling stage.

Co-siting the colour and luminance pixels means that the transmitted colour values are displayed unchanged. Only the interpolated values need to be calculated. This minimizes generation loss in the filtering. Downsampling the colour by a factor of two in both axes means that the colour data is reduced to one quarter of its original amount. When viewed by a human this is essentially a lossless process.

References

1. Shannon, C.E., A mathematical theory of communication. *Bell Syst. Tech. J.*, **27**, 379 (1948)
2. Jerri, A.J., The Shannon sampling theorem – its various extensions and applications: a tutorial review. *Proc. IEEE*, **65**, 1565–1596 (1977)
3. Betts, J.A., *Signal Processing Modulation and Noise*, Chapter 6. Sevenoaks: Hodder and Stoughton (1970)

4. Harris, S., The effects of sampling clock jitter on Nyquist sampling analog to digital convertors and on oversampling delta-sigma ADCs. *J. Audio Eng. Soc.*, **38**, 537–542 (1990)
5. Bennett, W.R., Spectra of quantized signals. *Bell System Tech. J.*, **27**, 446–472 (1948)
6. Widrow, B., Statistical analysis of amplitude quantized sampled-data systems. *Trans. AIEE*, Part II, **79**, 555–568 (1961)
7. Lipshitz, S.P., Wannamaker, R.A. and Vanderkooy, J., Quantization and dither: a theoretical survey. *J. Audio Eng. Soc.*, **40**, 355–375 (1992)
8. Maher, R.C., On the nature of granulation noise in uniform quantization systems. *J. Audio Eng. Soc.*, **40**, 12–20 (1992)
9. Goodall, W. M., Television by pulse code modulation. *Bell System Tech. Journal*, **30**, 33–49 (1951)
10. Roberts, L.G., Picture coding using pseudo-random noise. *IRE Trans. Inform. Theory*, **IT-8**, 145–154 (1962)
11. Schuchman, L., Dither signals and their effect on quantizing noise. *Trans. Commun. Technol.*, **COM-12**, 162–165 (1964)
12. Sherwood, Some theorems on quantization and an example using dither. In *Conf. Rec., 19th Asilomar Conf. on circuits, systems and computers*, Pacific Grove, CA (1985)
13. Lipshitz, S.P., Wannamaker, R.A. and Vanderkooy, J., Quantization and dither: a theoretical survey. *J. Audio Eng. Soc.*, **40**, 355–375 (1992)
14. Gerzon, M. and Craven, P.G., Optimal noise shaping and dither of digital signals. Presented at 87th Audio Eng. Soc. Conv. New York (1989), Preprint No. 2822 (J-1)
15. Vanderkooy, J. and Lipshitz, S.P., Resolution below the least significant bit in digital systems with dither. *J. Audio Eng. Soc.*, **32**, 106–113 (1984)
16. Vanderkooy, J. and Lipshitz, S.P., Digital dither. Presented at 81st Audio Eng. Soc. Conv. Los Angeles (1986), Preprint 2412 (C-8)
17. Vanderkooy, J. and Lipshitz, S.P., Digital dither. In *Audio in Digital Times*, New York: AES (1989)
18. Hauser, M.W., Principles of oversampling A/D conversion. *J. Audio Eng. Soc.*, **39**, 3–26 (1991)

5

Digital video processing

5.1 A simple digital vision mixer

The luminance path of a simple component digital mixer is shown in Figure 5.1. The CCIR-601 digital input is offset binary in that it has a nominal black level of 16_{10} in an eight-bit system (64 in a ten-bit system), and a subtraction has to be made in order that fading will take place with respect to black. On a perfect signal, subtracting 16 (or 64) would achieve this, but on a slightly out-of-range signal, it would not. Since the digital active line is slightly longer than the analog active line, the first sample should be blanking level, and this will be the value to subtract to obtain pure binary luminance with respect to black. This is the digital equivalent of black-level clamping. The two inputs are then multiplied by their respective coefficients, and added together to achieve the mix. Peak limiting will be required as in section 2.7, and then, if the output is to be to CCIR-601, 16_{10} (or 64) must be added to each sample value to establish the correct offset. In some video applications, a cross-fade will be needed, and a rearrangement of the crossfading equation allows one multiplier to be used instead of two, as shown in Figure 5.2.

The colour difference signals are offset binary with an offset of 128_{10} in eight-bit systems (512 in ten-bit systems), and again it is necessary to normalize these with respect to blanking level so that proper fading can be carried out. Since colour difference signals can be positive or negative, this process results in two's complement samples. Figure 5.3 shows some examples. In this form, the samples can be added with respect to blanking level.

Following addition, a limiting stage is used as before, and then, if it is desired to return to CCIR-601 standard, the MSB must be inverted once more in order to convert from two's complement to offset binary.

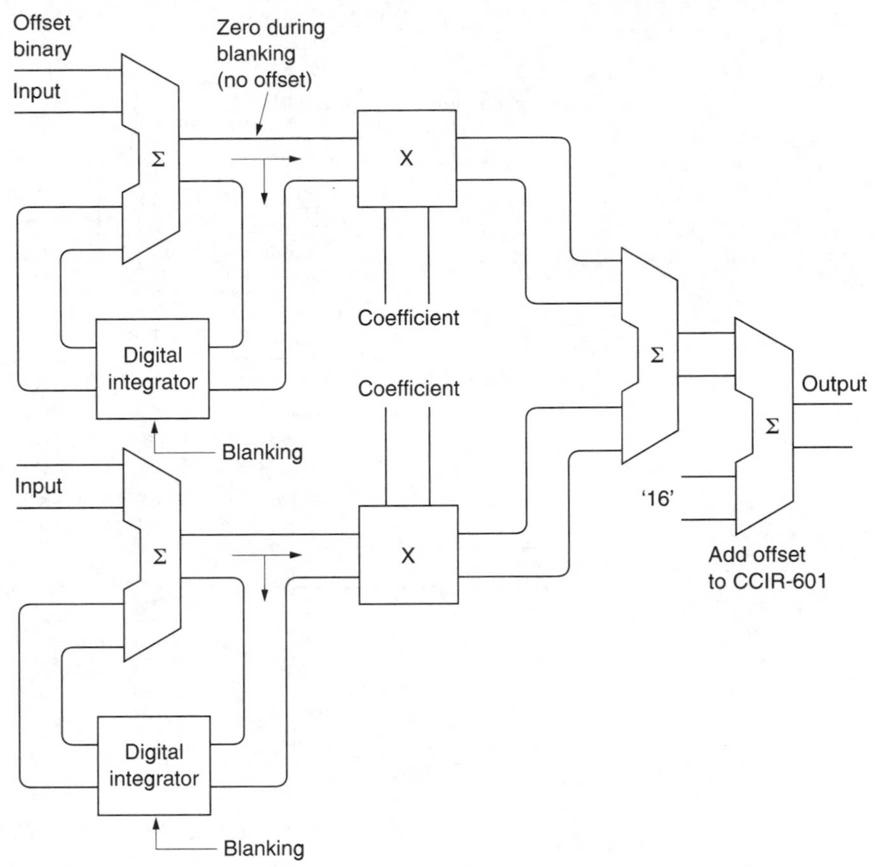

Figure 5.1 A simple digital mixer. Offset binary inputs must have the offset removed. A digital integrator will produce a counter-offset which is subtracted from every input sample. This will increase or reduce until the output of the subtractor is zero during blanking. The offset must be added back after processing if a CCIR-601 output is required.

In practice the same multiplier can be used to process luminance and colour difference signals. Since these will be arriving time multiplexed at 27 MHz, it is only necessary to ensure that the correct coefficients are provided at the right time. Figure 5.4 shows an example of part of a slow fade. As the co-sited samples C_B, Y and C_R enter, all are multiplied by the same coefficient K_n, but the next sample will be luminance only, so this will be multiplied by K_{n+1}. The next set of co-sited samples will be multiplied by K_{n+2} and so on. Clearly coefficients must be provided which change at 13.5 MHz. The sampling rate of the two inputs must be exactly the same, and in the same phase, or the circuit will not be able to add on a sample-by-sample basis. If the two inputs have come from different sources, they must be synchronized by the same master clock, and/or timebase correction must be provided on the inputs.

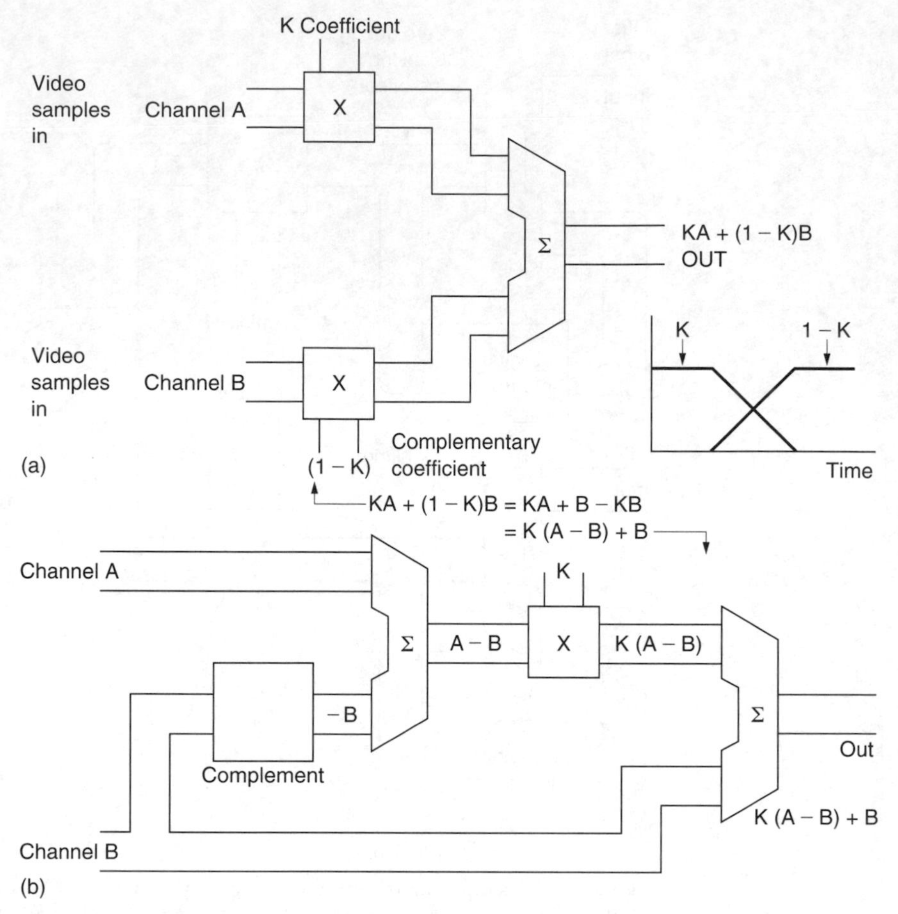

Figure 5.2 Crossfade at (a) requires two multipliers. Reconfiguration at (b) requires only one multiplier.

Some thought must be given to the wordlength of the system. If a sample is attenuated, it will develop bits which are below the radix point. For example, if an eight-bit sample is attenuated by 24 dB, the sample value will be shifted four places down. Extra bits must be available within the mixer to accommodate this shift. Digital vision mixers may have an internal wordlength of sixteen bits or more. When several attenuated sources are added together to produce the final mix, the result will be a sixteen-bit sample stream. As the output will generally need to be of the same format as the input, the wordlength must be shortened. Shortening the wordlength of samples effectively makes the quantizing intervals larger and can thus be called requantizing. This must be done very carefully to avoid artifacts and the necessary processes were shown in section 4.11.

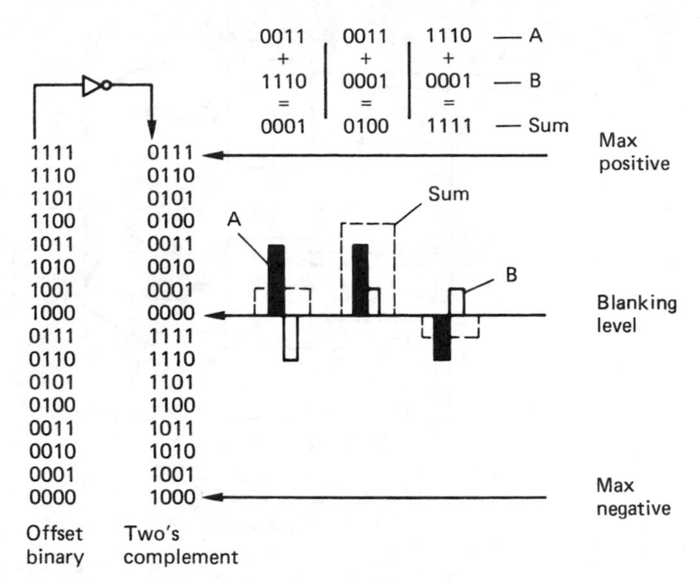

Figure 5.3 Offset binary colour difference values are converted to two's complement by reversing the state of the first bit. Two's complement values A and B will then add around blanking level.

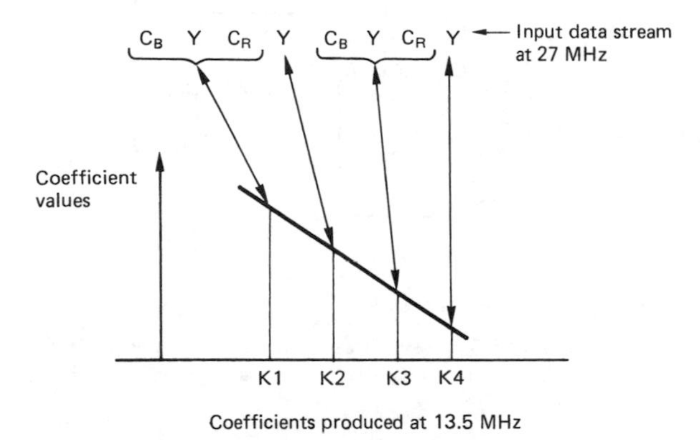

Figure 5.4 When using one multiplier to fade both luminance and colour difference in a 27 MHz multiplex 4:2:2 system, one coefficient will be used three times on the co-sited samples, whereas the next coefficient will only be used for a single luminance sample.

5.2 Concentrators/combiners

A concentrator or combiner is the name given to a special kind of digital switcher which is driven by a number of perspective effects machines. The illusion of a solid object can be created, where each video input represents one face, often of a cube, but other shapes are possible. Using the principle of transparency, images can be placed on the inside of a solid, and those at the rear can be seen through those at the front.

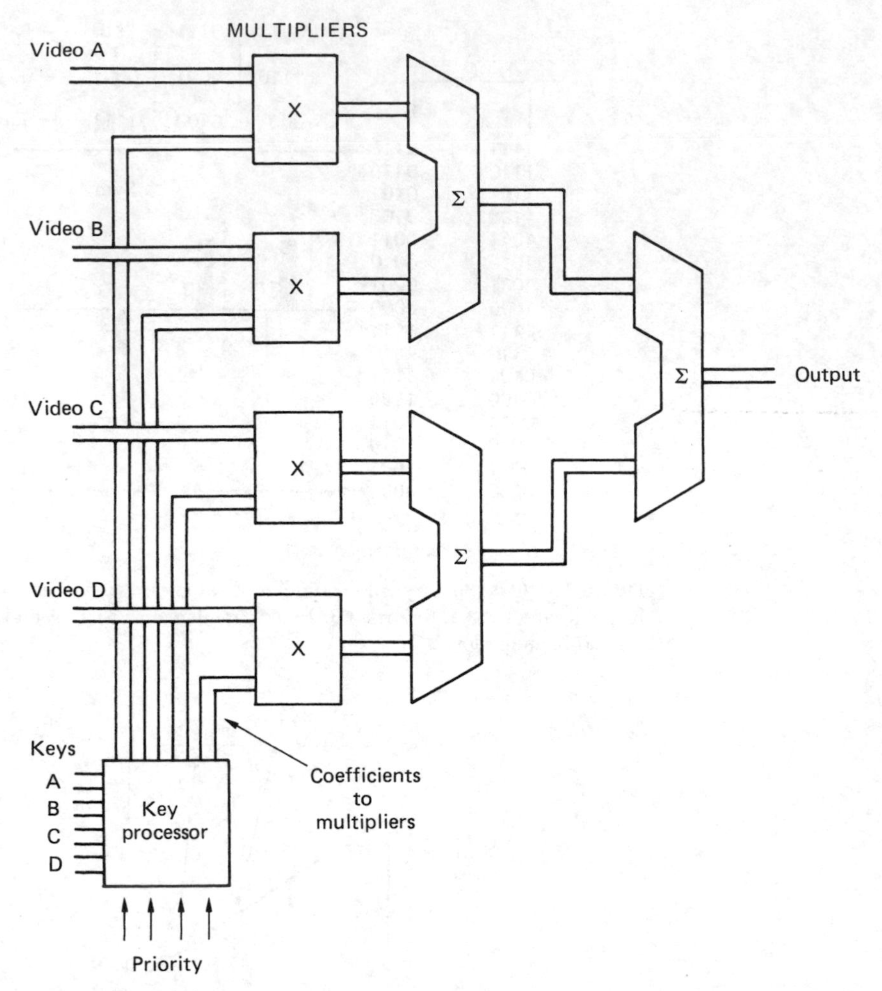

Figure 5.5 Transparency mixing requires one multiplication per input. The transparency of the highest-priority input determines the gain given to lower-priority inputs and so on.

Transparency requires control of the amplitude of keying signals. Figure 5.5 shows that a digital mixer is used, where the key signals are processed to become the coefficients for the mixer. The key processing is the most important aspect of the system, because it determines the realism of the final result. It is necessary to specify the priority of the images, that is, which lies closest to the viewer, in the order A, B, C, D for a four-channel system. The priority determines the order in which the incoming key signals are processed. The goal of the key processor is to produce four coefficients which never sum to more than unity. In this way the output can be prevented from clipping.

On a pixel-by-pixel basis, the key signals are used to compute new coefficients. Suppose that the highest priority, or front, picture A has been given a transparency of 33 per cent. This will result from a two-thirds amplitude key signal accompanying picture A. The key processor produces the A coefficient and also passes down to lower-priority circuits the fact that two thirds of the total permitted sum of coefficients has been used up. The transparency of A is 33 per cent, so only 33 per cent of the brightness of the B picture can be added where A obscures B. Suppose that B also has a transparency of 33 per cent, then light from picture C will

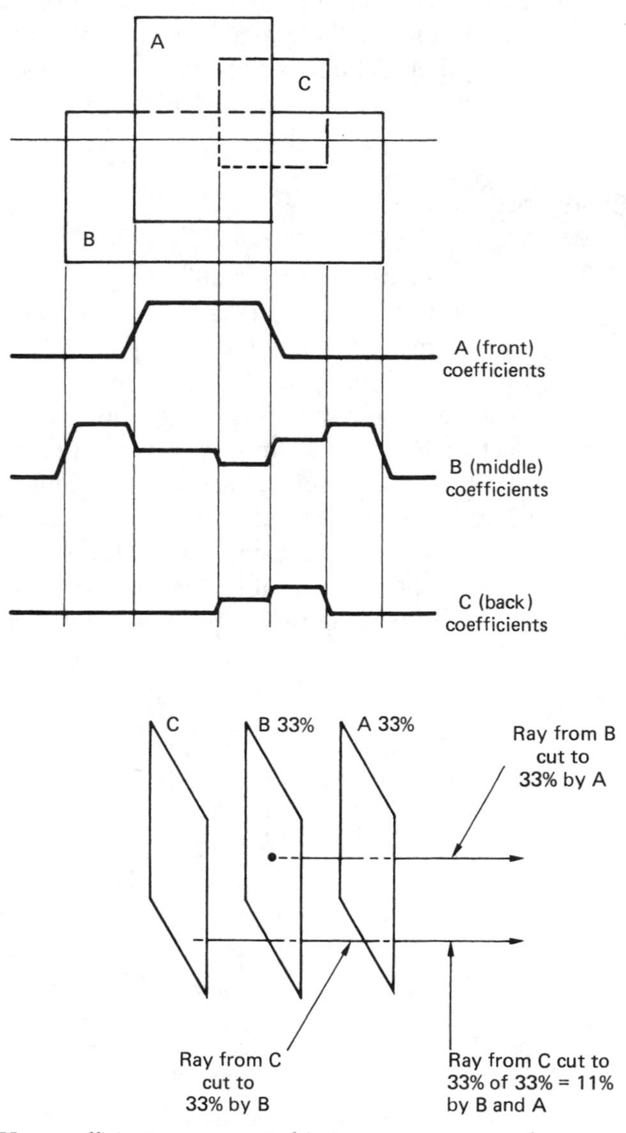

Figure 5.6 How coefficients are computed in transparency processing.

be reduced to 33 per cent of its original brightness passing through B, and reduced again by passing through A, so the coefficient needed will be about 11 per cent. The process continues for as many channels of input are available, plus a background value.

The coefficients then multiply their respective channels, and the products are summed to produce a single digital video stream carrying the final effect. Figure 5.6 shows the effect on the key coefficients during a single raster line scan across a multi-layer effect with transparency.

In some machines it is possible to make pictures in different planes intersect realistically. The position of the two planes in the Z-axis is compared, and the picture with the lowest Z at a given pixel position will get priority in the combiner. At the intersection point, priority will pass to the other picture. The transparencies allocated to both pictures will then be used to produce coefficients as before.

5.3 Blanking

It is often necessary to blank the ends of active lines smoothly to prevent out-of-band signals being generated. This is usually the case where an effects machine has cropped the picture to fit inside a coloured border. The border will be generated by supplying constant luminance and colour difference values to the data stream. Blanking consists of sloping off the active line by multiplying the sample values by successively smaller coefficients until blanking is reached. This is easy where the sample values have been normalized so that zero represents black, but where the usual offset of 16_{10} is present, multiplication by descending coefficients will cause a black-level shift. The solution is to use a correction PROM which can be seen in Figure 5.7. This is addressed by the multiplier coefficient and

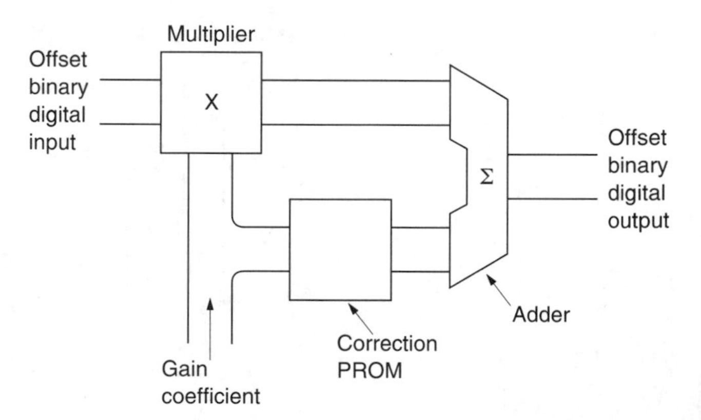

Figure 5.7 In order to fade an offset binary signal, a correction term from a PROM can be added to remove the level shift caused by fading.

adds a suitable constant to the multiplier output. If the multiplier were to have a gain of one half, this would shift the black level by eight quantizing intervals, and so the correction PROM would add eight to the output. Where the multiplier has fully blanked, the output will be zero, and the correction PROM has to add 16_{10} to the output.

5.4 Keying

Keying is the process where one video signal can be cut into another to replace part of the picture with a different image. One application of keying is where a switcher can wipe from one input to another using one of a variety of different patterns. Figure 5.8 shows that an analog switcher performs such an effect by generating a binary switching waveform in a pattern generator. Video switching between inputs actually takes place during the active line. In most analog switchers, the switching waveform is digitally generated, then fed to a D-A convertor, whereas in a digital switcher, the pattern generator outputs become the coefficients supplied to the crossfader, which is sometimes referred to as a cutter. The switching edge must be positioned to an accuracy of a few nanoseconds, much less than the spacing of the pixels, otherwise slow wipes will not appear to move smoothly, and diagonal wipes will have stepped edges, a phenomenon known as *ratcheting*.

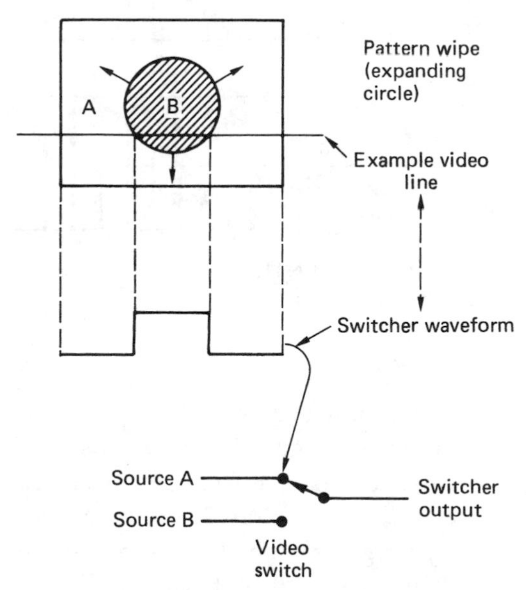

Figure 5.8 In a video switcher a pattern generator produces a switching waveform which changes from line to line and from frame to frame to allow moving pattern wipes between sources.

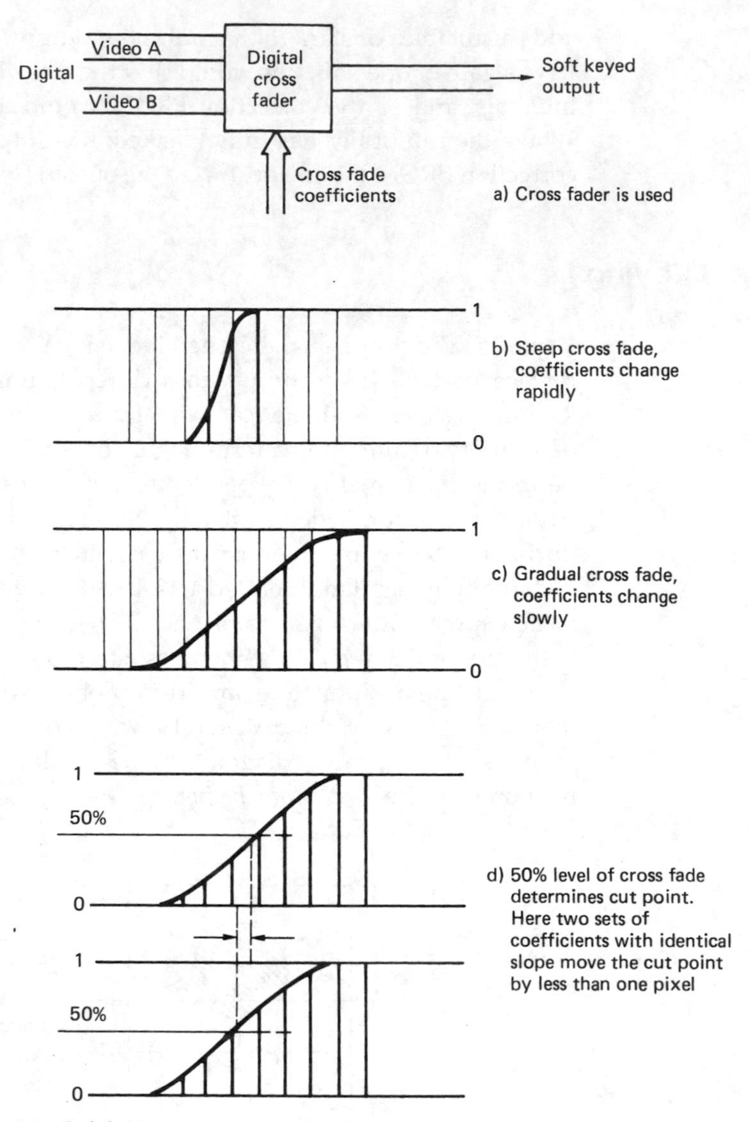

Figure 5.9 Soft keying.

Positioning the switch point to sub-pixel accuracy is not particularly difficult, as Figure 5.9 shows. A suitable series of coefficients can position the effective cross-over point anywhere. The finite slope of the coefficients results in a brief crossfade from one video signal to the other. This soft-keying gives a much more realistic effect than binary switchers, which often give a 'cut out with scissors' appearance. In some machines the slope of the crossfade can be adjusted to achieve the desired degre of softness.

5.5 Chroma keying

Another application of keying is to derive the switching signal by processing video from a camera in some way. By analysing colour difference signals, it is possible to determine where in a picture a particular colour occurs. When a key signal is generated in this way, the process is known as chroma keying, which is the electronic equivalent of matting in film.

In a 4:2:2 component system, it will be necessary to provide coefficients to the luminance crossfader at 13.5 MHz. Chroma samples only occur at half this frequency, so it is necessary to provide a chroma interpolator artificially to raise the chroma sampling rate. For chroma keying a simple linear interpolator is perfectly adequate. Intermediate chroma samples are simply the average of two adjacent samples. Figure 5.10 shows how a multiplexed C_r, C_b signal can be averaged using a delay of two clocks.

As with analog switchers, chroma keying is also possible with composite digital inputs, but decoding must take place before it is possible to obtain the key signals. The video signals which are being keyed will, however, remain in the composite digital format.

In switcher/keyers, it is necessary to obtain a switching signal which ramps between two states from an input signal which can be any

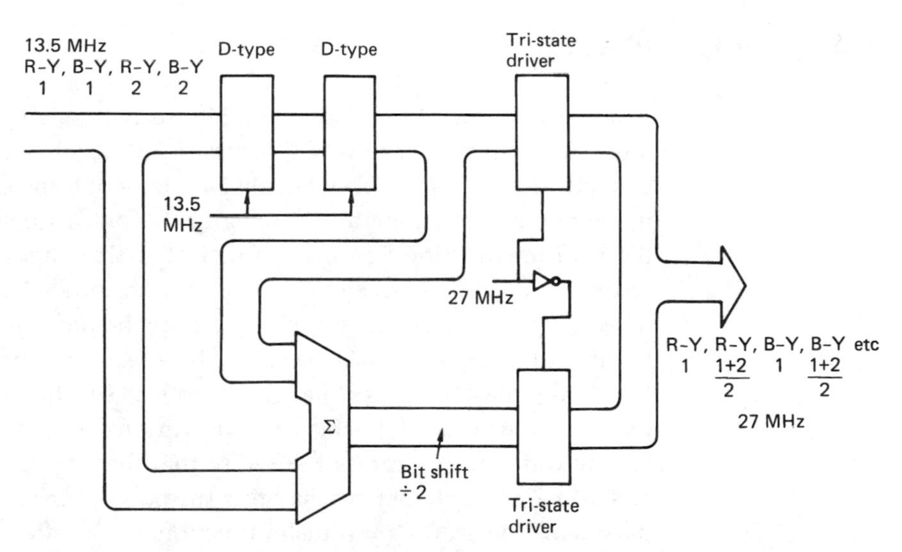

Figure 5.10 Alternate *R–Y*, *B–Y* samples can be averaged by a *two*-sample delay and an adder. Output will then alternate between sample values and averaged sample values at 27 MHz. Demultiplexing the output will give two colour difference signals each at 13.5 MHz so that they can be used to produce coefficients at that rate to key luminance.

allowable video waveform. Manual controls are provided so that the operator can set thresholds and gains to obtain the desired effect. In the analog domain, these controls distort the transfer function of a video amplifier so that it is no longer linear. A digital keyer will perform the same functions using logic circuits.

Figure 5.11(a) shows the effect of a non-linear transfer function is to switch when the input signal passes through a particular level. The transfer function is implemented in a memory in digital systems. The incoming video sample value acts as the memory address, so that the selected memory location is proportional to the video level. At each memory location, the appropriate output level code is stored. If, for example, each memory location stored its own address, the output would equal the input, and the device would be transparent. In practice, switching is obtained by distorting the transfer function to obtain more gain in one particular range of input levels at the expense of less gain at other input levels. With the transfer function shown in Figure 5.11(b), an input level change from a to b causes a smaller output change, whereas the same level change between c and d causes a considerable output change. If the memory is RAM, different transfer functions can be loaded in by the control system, and this requires multiplexers in both data and address lines as shown in Figure 5.11(c). In practice such a RAM will be installed in Y, C_r and C_b channels, and the results will be combined to obtain the final switching coefficients.

5.6 Simple effects

If a RAM of the type shown in Figure 5.11 is inserted in a digital luminance path, the result will be *solarizing* which is a form of contrast enhancement. Figure 5.12 shows that a family of transfer functions can be implemented which control the degree of contrast enhancement. When the transfer function becomes so distorted that the slope reverses, the result is *luminance reversal*, where black and white are effectively interchanged. Solarizing can also be implemented in colour difference channels to obtain *chroma solarizing*. In effects machines, the degree of solarizing may need to change smoothly so that the effect can be gradually introduced. In this case the various transfer functions will be kept in different pages of a PROM, so that the degree of solarization can be selected immediately by changing the page address of the PROM. One page will have a straight transfer function, so the effect can be turned off by selecting that page.

In the digital domain it is easy to introduce various forms of quantizing distortion to obtain special effects. Figure 5.13 shows that eight-bit luminance allows 256 different brightnesses, which to the naked eye

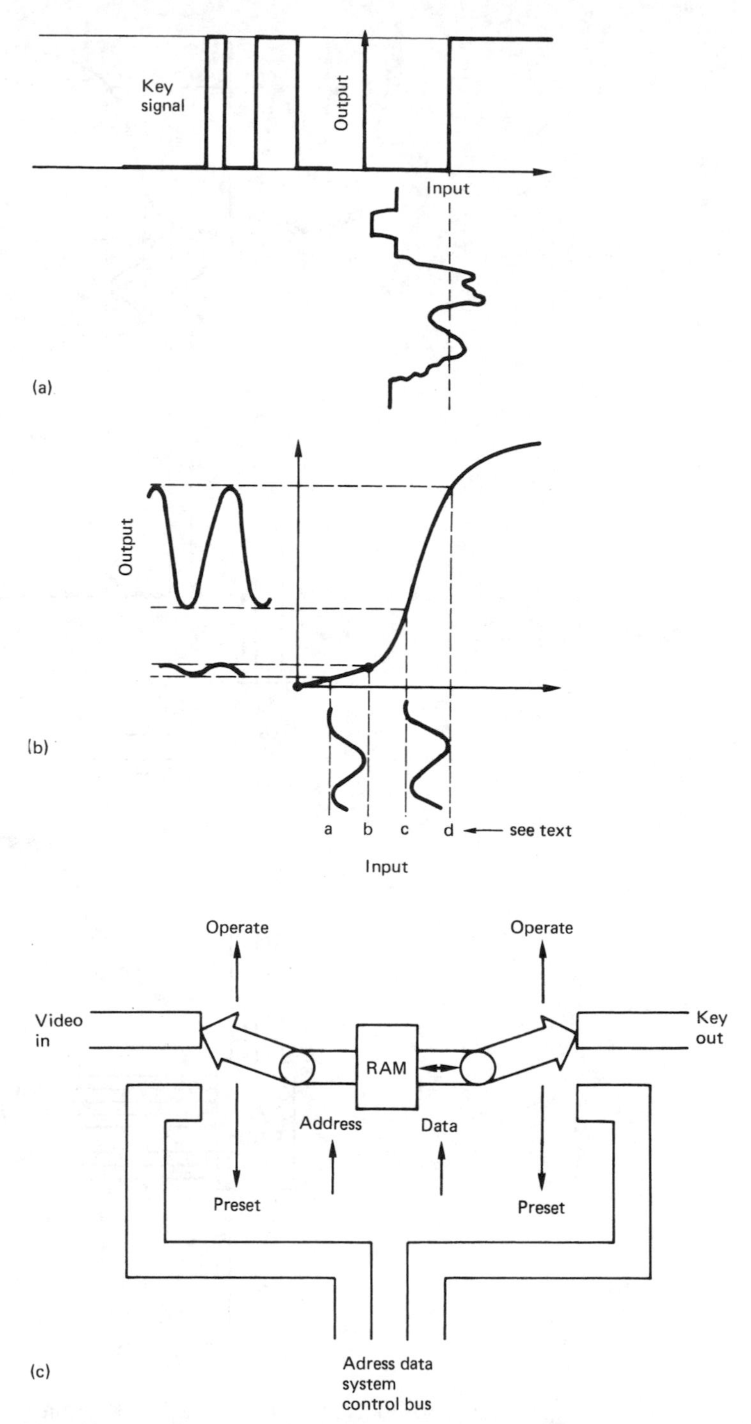

Figure 5.11 (a) A non-linear transfer function can be used to produce a keying signal. (b) The non-linear transfer function emphasizes contrast in part of the range but reduces it at other parts. (c) If a RAM is used as a flexible transfer function, it will be necessary to provide multiplexers so that the RAM can be preset with the desired values from the control system.

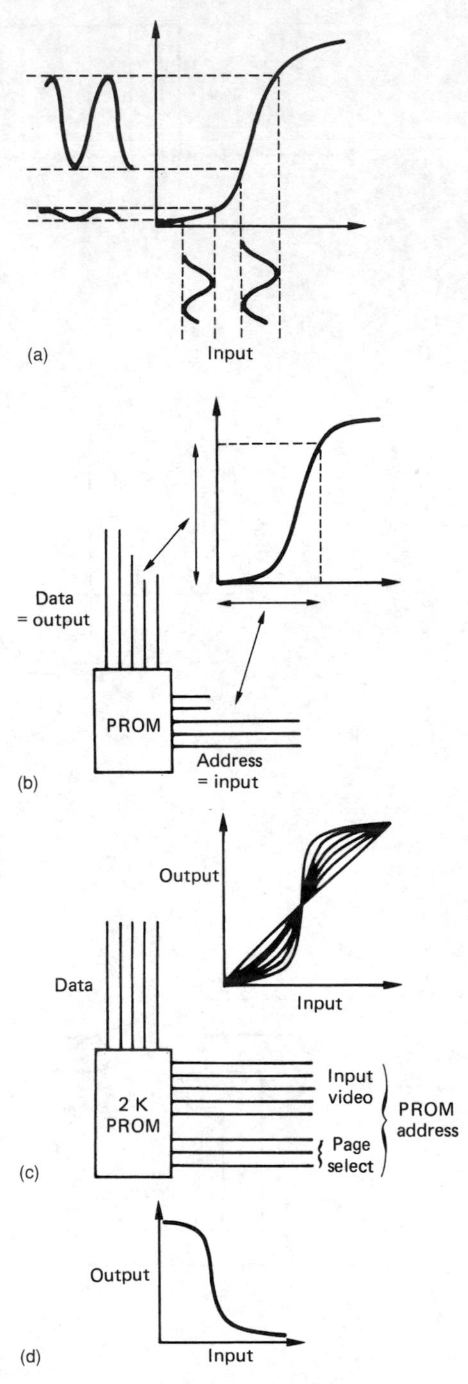

Figure 5.12 Solarization. (a) The non-linear transfer function emphasizes contrast in part of the range but reduces it at other parts. (b) The desired transfer function is implemented in a PROM. Each input sample value is used as the address to select a corresponding output value stored in the PROM. (c) A family of transfer functions can be accommodated in a larger PROM. Page select affects the high-order address bits. (d) Transfer function for luminance reversal.

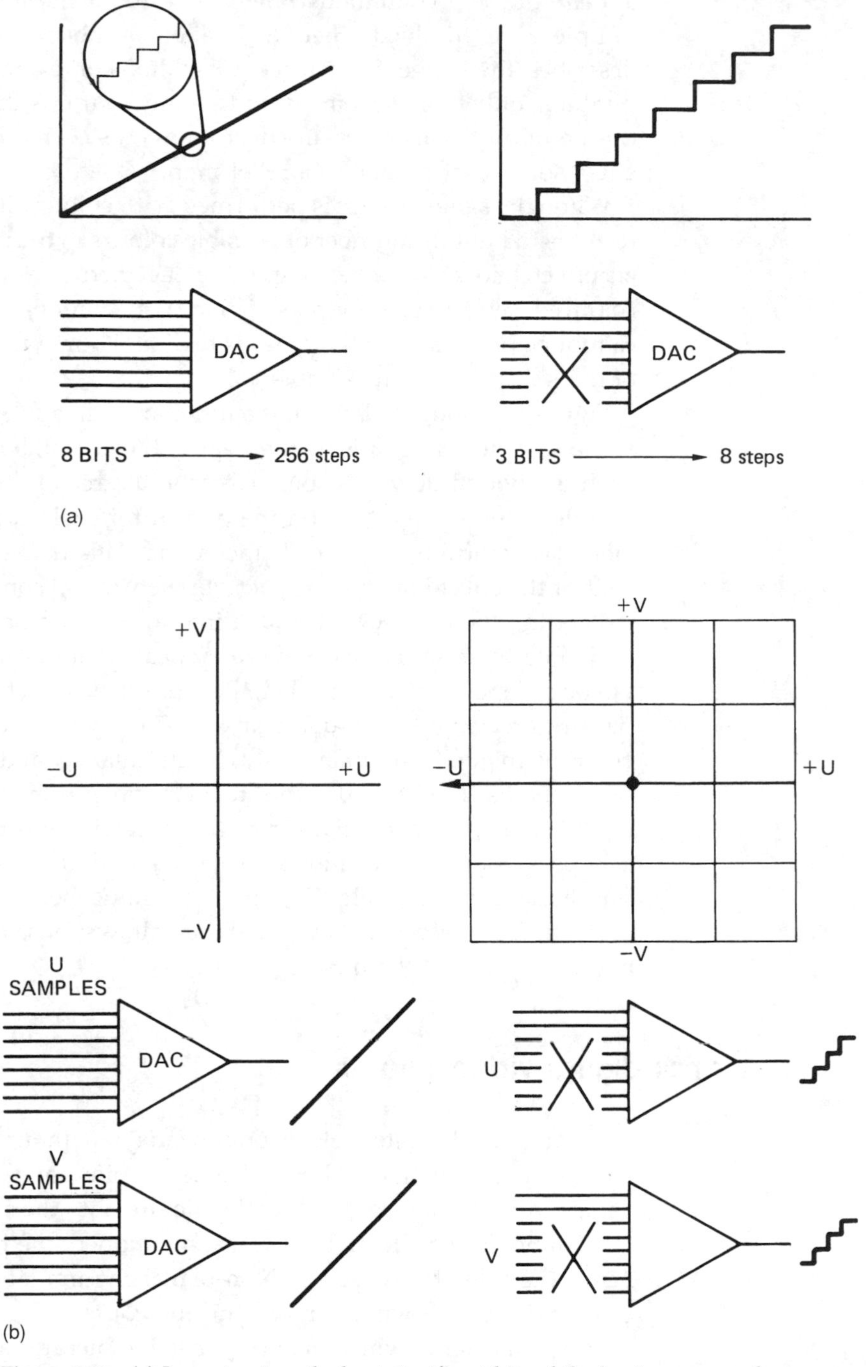

Figure 5.13 (a) In contouring, the least significant bits of the luminance samples are discarded, which reduces the number of possible output levels. (b) At left, the eight-bit colour difference signals allow 2^{16} different colours. At right, eliminating all but two bits of each colour difference signals allows only 2^4 different colours.

appears to be a continuous range. If some of the low-order bits of the samples are disabled, then a smaller number of brightness values describes the range from black to white. For example, if six bits are disabled, only two bits remain, and so only four possible brightness levels can be output. This gives an effect known as *contouring* since the visual effect somewhat resembles a relief map.

When the same process is performed with colour difference signals, the result is to limit the number of possible colours in the picture, which gives an effect known as *posterizing*, since the picture appears to have been coloured by paint from pots. Solarizing, contouring and posterizing cannot be performed in the composite digital domain, due to the presence of the subcarrier in the sample values.

Figure 5.14 shows a latch in the luminance data which is being clocked at the sampling rate. It is transparent to the signal, but if the clock to the latch is divided down by some factor n, the result will be that the same sample value will be held on the output for n clock periods, giving the video waveform a staircase characteristic. This is the horizontal component of the effect known as *mosaicing*. The vertical component is obtained by feeding the output of the latch into a line memory, which stores one horizontally mosaiced line and then repeats that line m times. As n and m can be independently controlled, the mosaic tiles can be made to be of any size, and rectangular or square at will. Clearly the mosaic circuitry must be implemented simultaneously in luminance and colour difference signal paths. It is not possible to perform mosaicing on a composite digital signal, since it will destroy the subcarrier. It is common to provide a bypass route which allows mosaiced and unmosaiced video to be simultaneously available. Dynamic switching between the two sources controlled by a seperate key signal then allows mosaicing to be restricted to certain parts of the picture.

5.7 Planar digital video effects

One can scarcely watch television nowadays without becoming aware of picture manipulation. Flips, tumbles, spins and page-turn effects, perspective rotation, and rolling the picture onto the surface of a solid are all commonly seen. In all but the last mentioned, the picture remains flat, hence the title of this section. Non-planar manipulation requires further complexity which will be treated in due course.

Effects machines which manipulate video pictures are close relatives of the machines which produce computer-generated images.[1] Computer-generated images require enormous processing power, and even with state-of-the-art CPUs the time needed to compute a single frame is several orders of magnitude too long to work with real-time video. Video

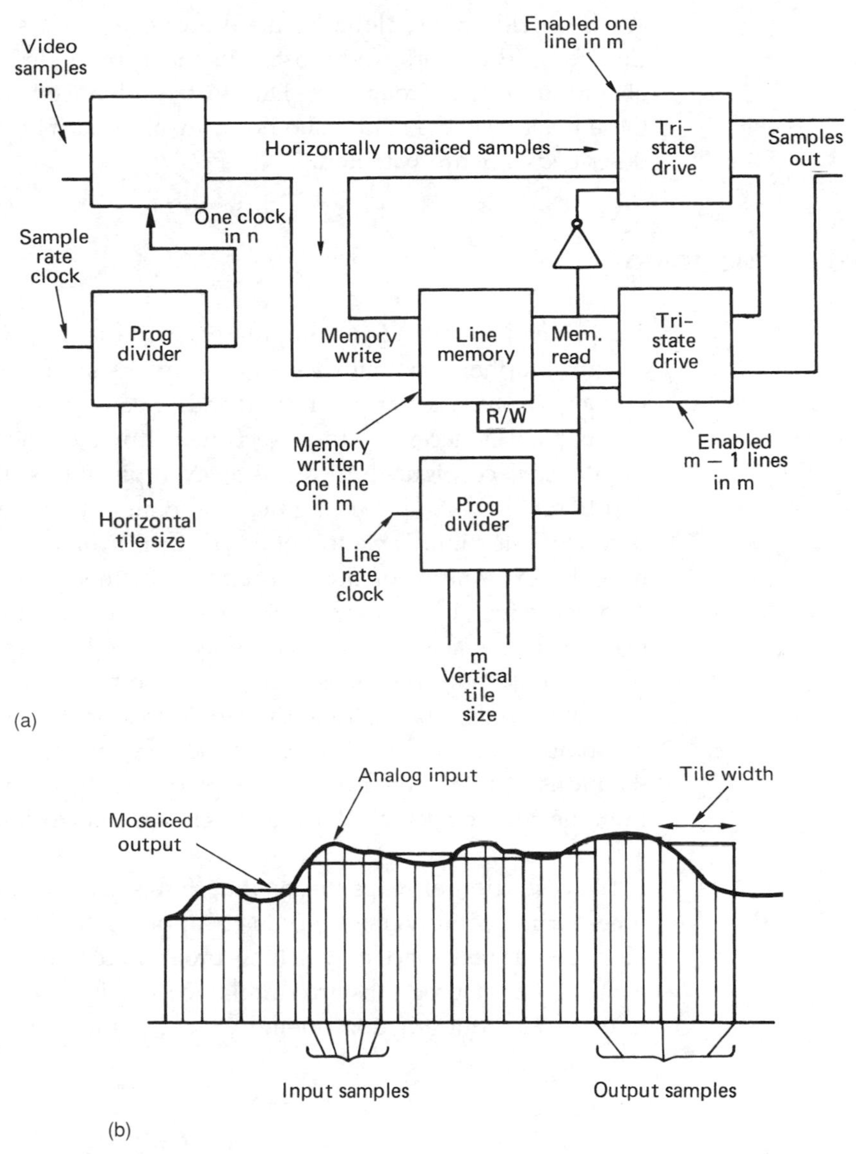

Figure 5.14 (a) Simplified diagram of mosaicing system. At the left-hand side, horizontal mosaicing is done by intercepting sample clocks. On one line in *m*, the horizontally mosaiced line becomes the output, and is simultaneously written into a one-line memory. On the remaining (*m*-1) lines the memory is read to produce several identical successive lines to give the vertical dimensions of the tile. (b) In mosaicing, input samples are neglected, and the output is held constant by failing to clock a latch in the data stream for several sample periods. Heavy vertical lines here correspond to the clock signal occurring. Heavy horizontal line is resultant waveform.

effects machines, then, represent a significant technical achievement, because they have only a field period to complete the processing before another field comes along. General-purpose processors are usually too slow for effects work, and most units incorporate dedicated hardware to obtain sufficient throughput. Due to the advanced technology used in these machines, the reader should be aware that many of the techniques described here are patented.

5.8 Mapping

The principle of all video manipulators is the same as the technique used by cartographers for centuries. Cartographers are faced with a continual problem in that the earth is round, and paper is flat. In order to produce flat maps, it is necessary to project the features of the round original on to a flat surface. Figure 5.15 shows an example of this. There are a number of different ways of projecting maps, and all of them must, by definition, produce distortion. The effect of this distortion is that distances measured near the extremities of the map appear further than they actually are. Another effect is that *great circle routes* (the shortest or longest path between two places on a planet) appear curved on a projected map. The type of projection used is usually printed somewhere on the map, a very common system being that due to Mercator. Clearly the process of mapping involves some three-dimensional geometry in order to simulate the paths of light rays from the map so that they appear to have come from the curved surface. Video effects machines work in exactly the same way.

The distortion of maps means that things are not where they seem. In timesharing computers, every user appears to have his or her own identical address space in which his program resides, despite the fact that many different programs are simultaneously in the memory. In order to resolve this contradiction, memory management units are constructed

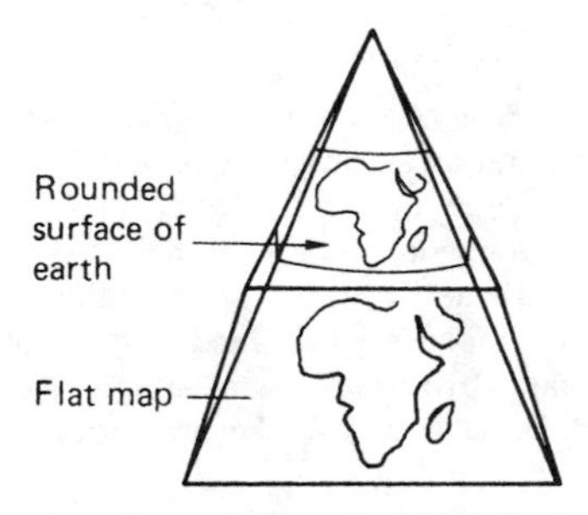

Figure 5.15 Map projection is a close relative of video effects units which manipulate the shape of pictures.

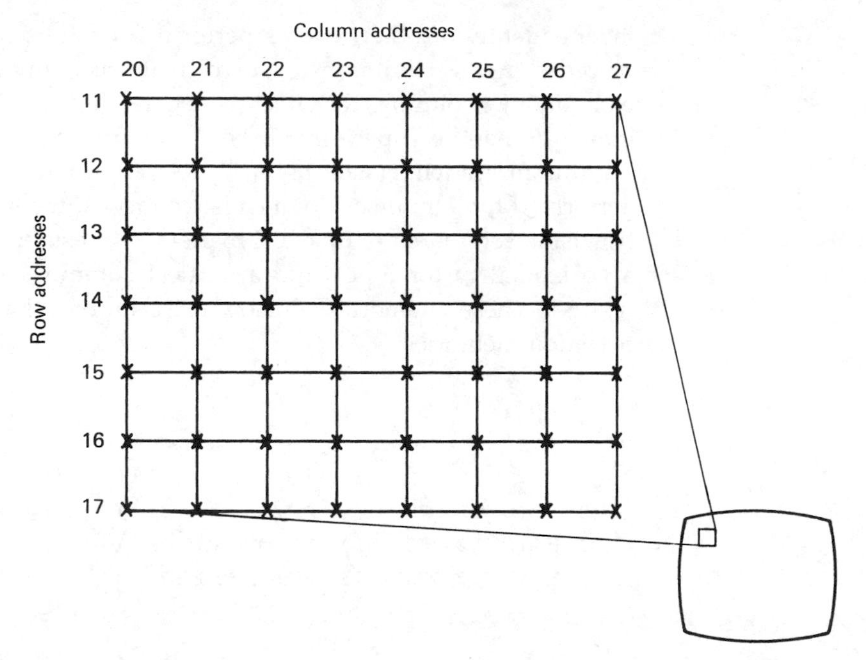

Figure 5.16 The entire TV picture can be broken down into uniquely addressable pixels.

which add a constant value to the address which the user thinks he has (the *virtual address*) in order to produce the *physical address*. As long as the unit gives each user a different constant, they can all program in the same virtual address space without one corrupting another's programs. Because the program is no longer where it seems to be, the term of mapping was introduced. The address space of a computer is one-dimensional, but a video frame expressed as rows and columns of pixels can be considered to have a two-dimensional address as in Figure 5.16. Video manipulators work by mapping the pixel addresses in two dimensions.

5.9 Separability and transposition

Every pixel in the frame array has an address. The address is two-dimensional, because in order to uniquely specify one pixel, the column address and the row address must be supplied. It is possible to transform a picture by simultaneously addressing in rows and columns, but this is complicated, and very difficult to do in real time. It was discovered some time ago in connection with computer graphics that the two-dimensional problem can be converted with care into two one-dimensional problems.[2] Essentially if a horizontal transform affecting whole rows of pixels independently of other rows is performed on the array, followed by or preceded by a vertical transform which affects entire columns independently of other columns, the effect will be the same as if a two-dimensional transform had been performed. This is the principle of separability.

From an academic standpoint, it does not matter which transform is performed first. In a world which is wedded to the horizontally scanned television set, there are practical matters to consider. In order to convert a horizontal raster input into a vertical column format, a memory is needed. These memories already exist in the shape of the motion-detection memories.

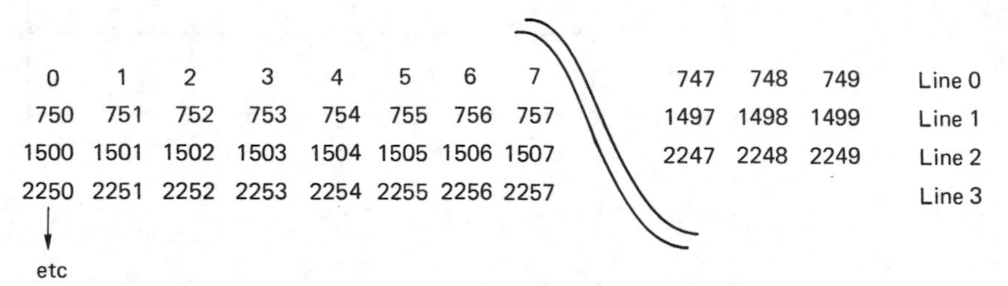

Figure 5.17 Pixels are written into a memory in the order of address counting a line at a time. If the read address is given by $(n \times 750) + k$, for each value of k, a different column will be read by increasing n. For example k=4, addresses will be 4, 754, 1504, 2254, etc. This transposes the incoming raster to a vertical scan format.

Figure 5.17 shows a simplified TV standard where there are exactly 750 samples per line. As the first line of a field enters, it will be put into addresses 0 to 749. As the second line of the frame appears, it will be put into addresses 750 to 1499 and so on. If subsequently a small accumulator is arranged to provide addresses which are given by $n \times 750 + k$, where n goes from zero upwards, if $k = 0$, the addresses 0, 750, 1500, 2250 etc. will be generated, which correspond to the first column of pixels. If k is increased to 1, the addresses 1, 751, 1501, 2251 etc will be generated, which correspond to the second column of pixels. In order to de-interlace two fields A and B, the two address accumulators would need to be synchronized so that they alternated. The effect would then be to provide a single column of pixels in the sequence A0, B0, A750, B750, A1500, B1500 etc. Where there is motion, an alternate pixel stream would be selected, which would be A0, A0/2 + A750/2, A750, A750/2 + A1500/2, A2250 etc., which contains no B field data. The average of two successive pixels can be obtained using an adder which has its two inputs connected one at each side of a pixel delay latch.

The process of writing rows and reading columns in a memory is called transposition. Clearly two stages of transposition are necessary to return to a horizontal raster output, as shown in Figure 5.18. The vertical

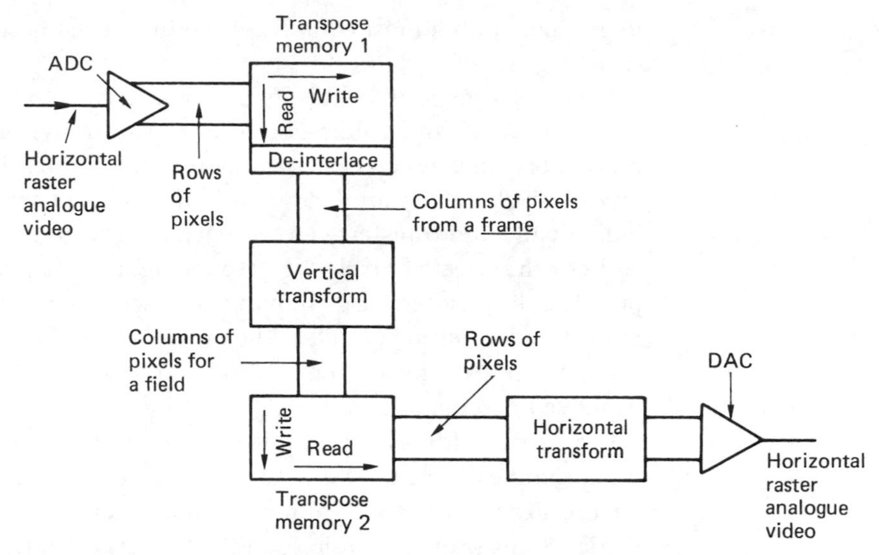

Figure 5.18 Two transposing memories are necessary, one before and one after the vertical transform.

transform must take place between the two transposes, but the horizontal transform could take place before the first transpose or after the second. In practice the horizontal transform cannot be placed before the first transpose, because it would interfere with the de-interlace and motion-sensing process. The horizontal transform is placed after the second transpose, and reads rows from the second transpose memory. As the output of the machine must be a horizontal raster, the horizontal transform can be made to work in synchronism with reference H-sync so that the digital output samples from the H-transform can be taken direct to a DAC to become analog video with a standard structure again. A further advantage is that the real-time horizontal output is one field at a time. The preceding vertical transform need only compute array values which lie in the next field to be required at the output.

It is not possible to take a complete column from a memory until all the rows have entered. The presence of the two transposes in a DVE results in an unavoidable delay of one frame in the output image. Some DVEs can manage greater delay. The effect on lip-sync may have to be considered. It is not advisable to cut from the input of a DVE to the output, since a frame will be lost, and on cutting back a frame will be repeated.

5.10 Address generation and interpolation

There are many different manipulations possible, and the approach here will be to begin with the simplest, which require the least processing, and to graduate to the most complex, introducing the necessary processes at each stage.

It has been stated that address mapping is used to perform transforms. Now that rows and columns are processed individually, the mapping process becomes much easier to understand. Figure 5.19 shows a single row of pixels which are held in a buffer where each can be addressed individually and transferred to another. If a constant is added to the read address, the selected pixel will be to the right of the place where it will be put. This has the effect of moving the picture to the left. If the buffer represented a column of pixels, the picture would be moved vertically. As these two transforms can be controlled independently, the picture could be moved diagonally.

If the read address is multiplied by a constant, say 2, the effect is to bring samples from the input closer together on the output, so that the picture size is reduced. Again independent control of the horizontal and vertical transforms is possible, so that the aspect ratio of the picture can be modified. This is very useful for telecine work when CinemaScope films are to be broadcast. Clearly the secret of these manipulations is in the constants fed to the address generators. The added constant represents displacement, and the multiplied constant represents magnification. A multiplier constant of less than one will result in the picture getting larger. Figure 5.19 also shows, however, that there is a problem. If a constant of 0.5 is used, to make the picture twice as big, half of the addresses generated are not integers. A memory does not understand an address of two and a half! If an arbitrary magnification is used, nearly all the addresses generated are non-integer. A similar problem crops up if a constant of less than one is added to the address in an attempt to move the picture less than the pixel spacing. The solution to the problem is interpolation. Because the input image is spatially sampled, those samples contain enough information to represent the brightness and colour all over the screen. When the address generator comes up with an address of 2.5, it actually means that what is wanted is the value of the signal interpolated half-way between pixel two and pixel three. The output of the address generator will thus be split into two parts. The integer part will become the memory address, and the fractional part is the phase of the necessary interpolation. In order to interpolate pixel values a digital filter is necessary.

Figure 5.20 shows that the input and output of an effects machine must be at standard sampling rates to allow digital interchange with other equipment. When the size of a picture is changed, this causes the pixels in the picture to fail to register with output pixel spacing. The problem is exactly the same as sampling rate conversion, which produces a differently spaced set of samples which still represent the original waveform. One pixel value actually represents the peak brightness of a

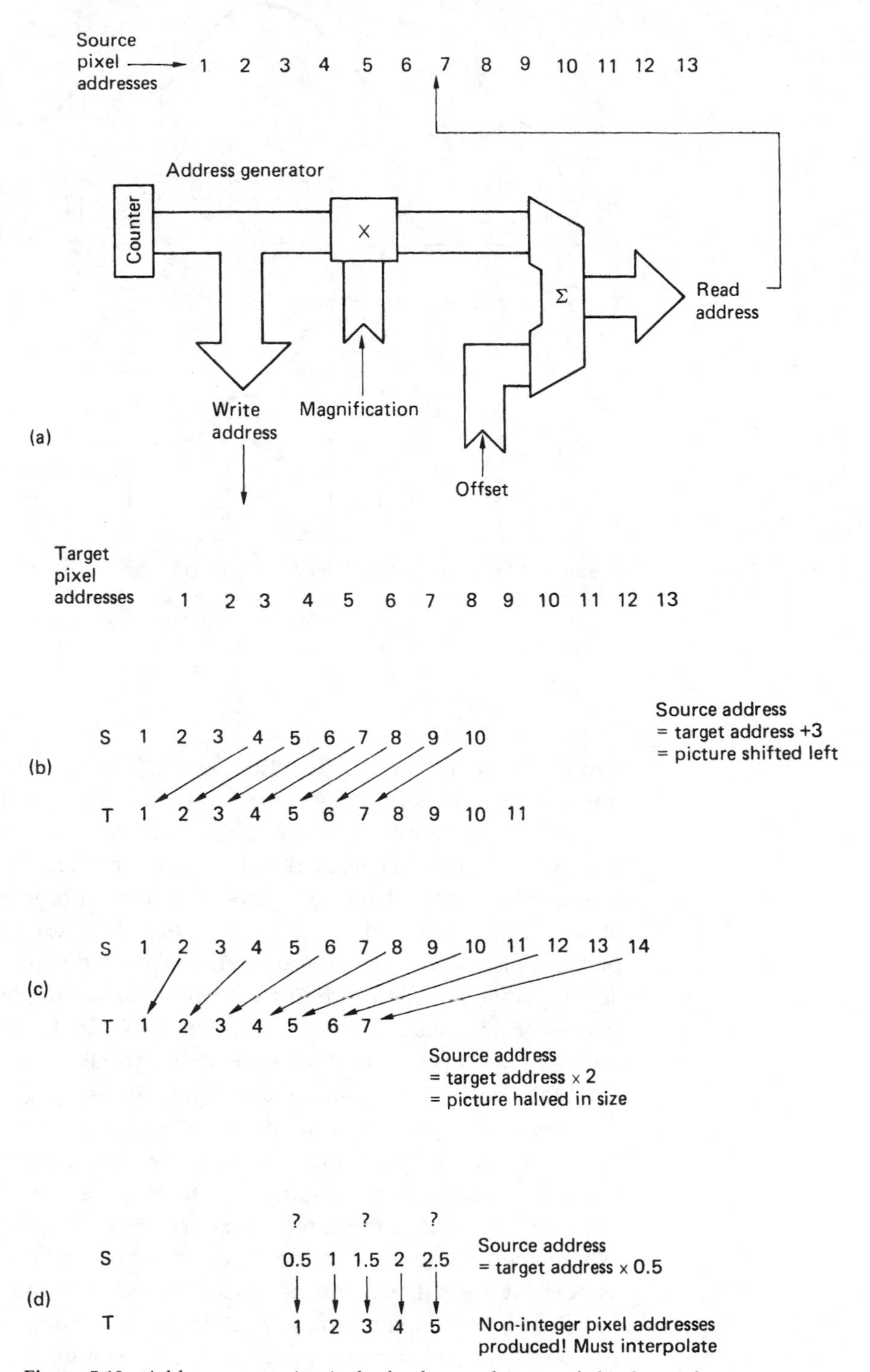

Figure 5.19 Address generation is the fundamental process behind transforms.

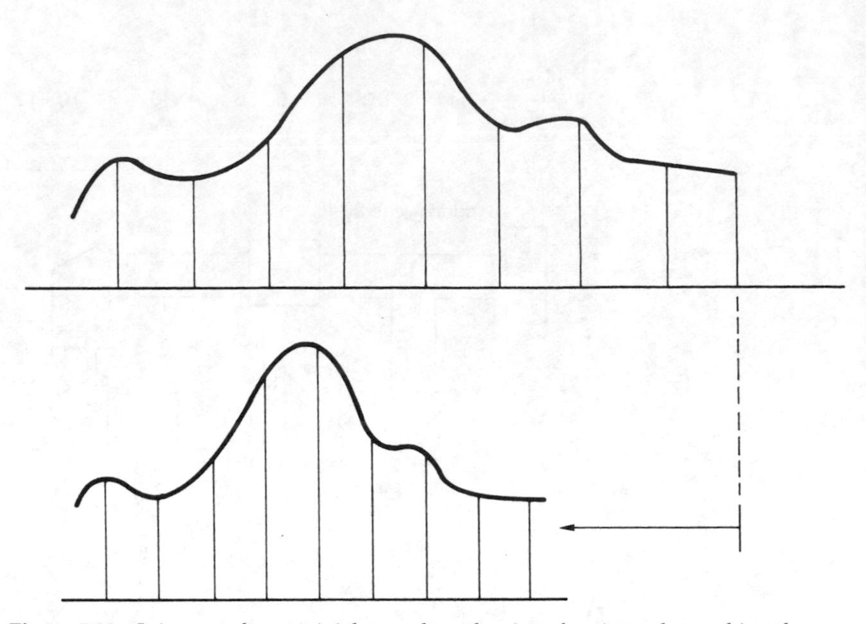

Figure 5.20 It is easy, almost trivial, to reduce the size of a picture by pushing the samples closer together, but this is not often of use, because it changes the sampling rate in proportion to the compression. Where a standard sampling-rate output is needed, interpolation must be used.

two-dimensional intensity function, which is the effect of the modulation transfer function of the system on an infinitely small point. As each dimension can be treated separately, the equivalent in one axis is that the pixel value represents the peak value of an infinitely short impulse which has been low-pass filtered to the system bandwidth. The waveform is that of a $\sin x/x$ curve, which has value everywhere except at the centre of other pixels. In order to compute an interpolated value, it is necessary to add together the contribution from all relevant samples, at the point of interest. Each contribution can be obtained by looking up the value of a unity $\sin x/x$ curve at the distance from the input pixel to the output pixel to obtain a coefficient, and multiplying the input pixel value by that coefficient. The process of taking several pixel values, multiplying each by a different coefficient and summing the products can be performed by the FIR (finite-impulse response) configuration described earlier. The impulse response of the filter necessary depends on the magnification. Where the picture is being enlarged, the impulse response can be the same as at normal size, but as the size is reduced, the impulse response has to become broader (corresponding to a reduced spatial frequency response) so that more input samples are averaged together to prevent aliasing. The coefficient store will need a two-dimensional structure, such that the magnification and the interpolation phase must both be supplied to obtain a set of coefficients. The magnification can easily be obtained by comparing successive outputs from the address generator.

As was seen in Chapter 3, the number of points in the filter is a compromise between cost and performance, eight being a typical number for high quality. As there are two transform processes in series, every output pixel will be the result of sixteen multiplications, so there will be 216 million multiplications per second taking place in the luminance channel alone for a 13.5 MHz sampling rate unit. The quality of the output video also depends on the number of different interpolation phases available between pixels. The address generator may compute fractional addresses to any accuracy, but these will be rounded off to the nearest available phase in the digital filter. The effect is that the output pixel value provided is actually the value a tiny distance away, and has the same result as sampling clock jitter, which is to produce program-modulated noise. The greater the number of phases provided, the larger will be the size of the coefficient store needed. As the coefficient store is two-dimensional, an increase in the number of filter points and phases causes an exponential growth in size and cost. The filter itself can be implemented readily with fast multiplier chips, but one problem is accessing the memory to provide input samples. What the memory must do is to take the integer part of the address generator output and provide simultaneously as many adjacent pixels as there are points in the filter. This problem is usually solved by making the memory from several smaller memories with an interleaved address structure, so that several pixel values can be provided from one address.

Figure 5.21(a) shows one way in which this can be done. The incoming memory write address is split into two sections. The three low-order bits are decoded to successively enable one out of eight RAMs, which are fed the remainder of the address bits in parallel. The result of this arrangement is that samples 0 through 7 are written into location 0 of successive RAMs, samples 8 through 15 are written in location 1 of the RAMs and so on.

In order to read the memories, a different addressing configuration is used. The read address comes from the address generator, and is split into two sections as before. The most significant bits are fed in parallel to all RAMs, but each RAM is furnished with an adder which allows the address to be increased by one. The least significant bits are also treated in a different manner. They decide *how many* of the addresses will be incremented. Figure 5.21(b) lists the result for a series of input addresses. In each case, providing one *source address* produces eight adjacent sample values simultaneously. The correct sample numbers are produced, but they suffer an end-around rotation of position. As each one is to be fed to one point of a parallel FIR filter, this does not matter, provided that the coefficients fed to the filter are given the same rotation.

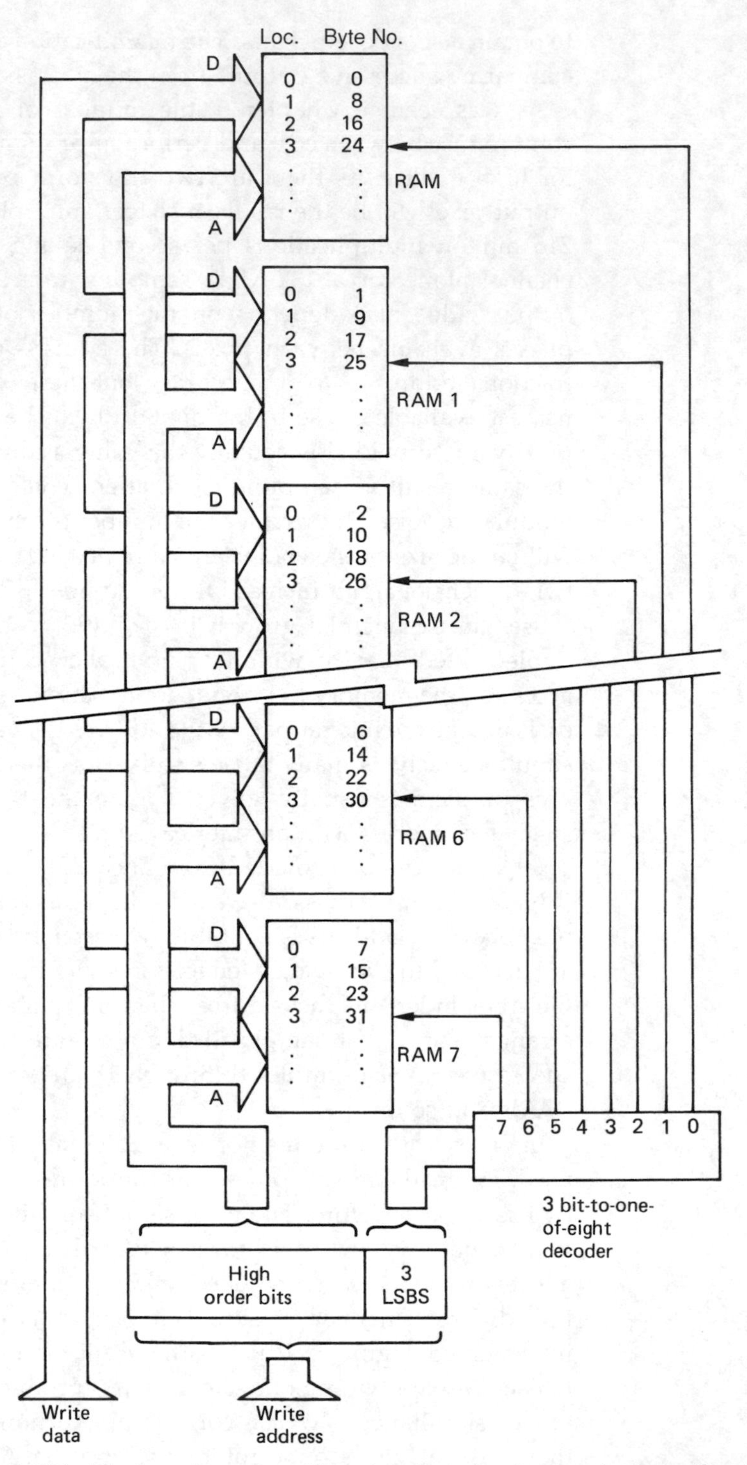

Figure 5.21(a) The incoming write address is split so that the three least significant bits sequentially enable one of eight parallel RAMs. Then every eighth byte will be written into the same RAM.

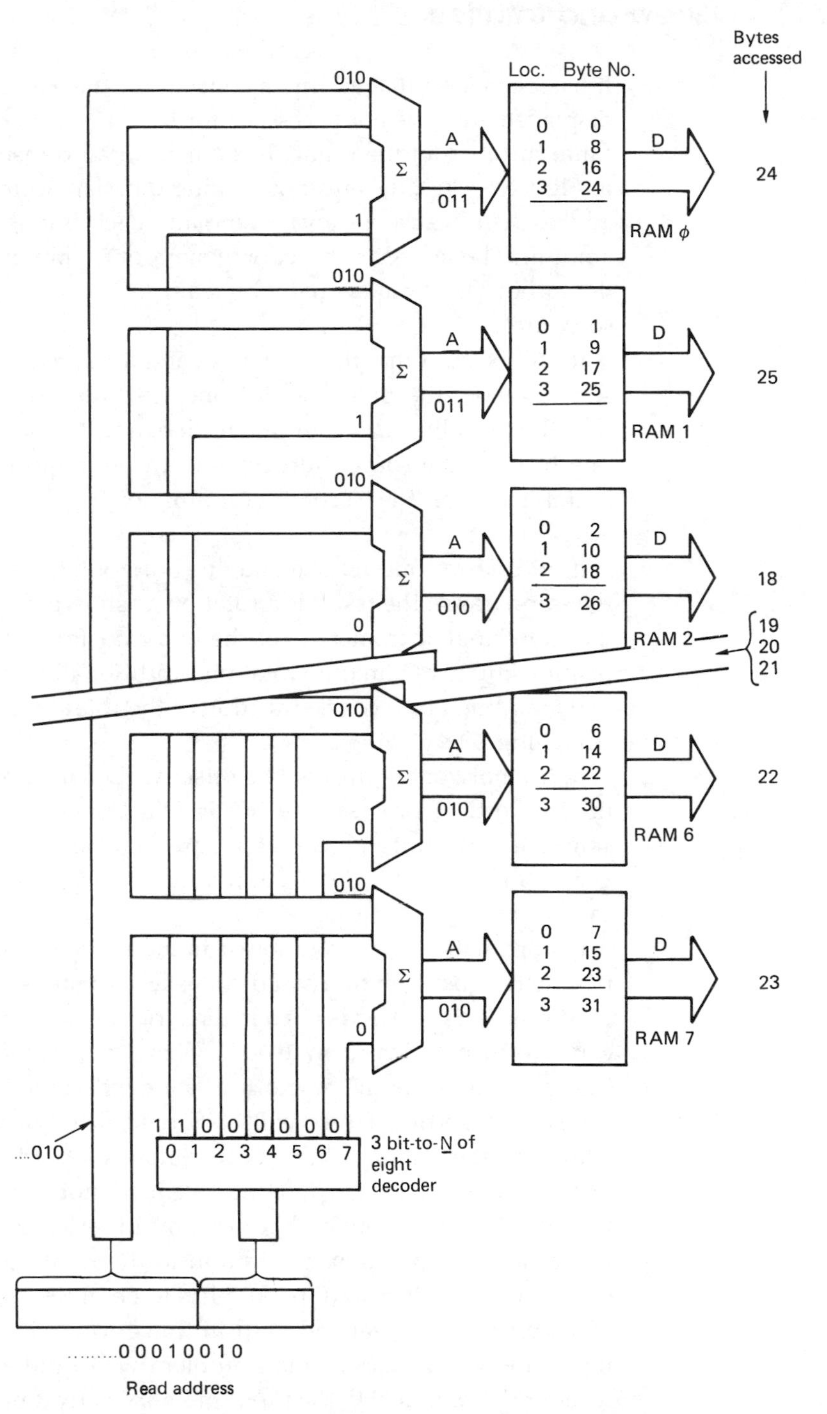

Figure 5.21(b) The incoming read address is split as before, but now the three least significant bits determine the *number* of carries which will be added to the remainder of the address. In this example two carries are added, so the top two RAMs access an address 8 bytes higher, and the system accesses bytes 18, 19, 20, 21, 22, 23, 24, 25.

5.11 Skew and rotation

It has been seen that adding a constant to the source address produces a displacement. It is not necessary for the displacement constant to be the same throughout the frame. If the horizontal transform is considered, as in Figure 5.22(a), the effect of making the displacement a function of line address is to cause a skew. Essentially each line is displaced a different amount. The necessary function generator is shown in simplified form in (b), although it could equally be realized in a fast CPU with appropriate software.

It will be seen that the address generator is really two accumulators in series, where the first operates once per line to calculate a new offset which grows linearly from line to line, and the second operates at pixel rate to calculate source addresses from the required magnification. The initial state of the second accumulator is the offset from the first accumulator.

If two skews, one vertical and one horizontal, are performed in turn on the same frame, the result is a rotation as shown in Figure 5.22(c). Clearly the skew angle parameters for the two transforms must be in the correct relationship to obtain pure rotation. Additionally the magnification needs to be modified by a cosine function of the rotation angle to counteract the stretching effect of the skews.

In the horizontal process, the offset will change once per line, whereas in the vertical process, the offset will change once per column. For simplicity, the offset generators are referred to as the *slow* address generators, whereas the accumulators which operate at pixel rate are called the *fast* address generators.

Unfortunately skew rotations cannot approach 90°, because the skew parameter goes to infinity, and so a skew rotate is generally restricted to rotations of ±45°. This is not a real restriction, since the apparatus already exists to turn a picture on its side. This can be done readily by failing to transpose in one of the memories. There will no longer be two cancelling transposes, so the picture will be turned through 90° when it emerges. Although the failure to transpose could be at either of the memories if only this effect is required, in practice it cannot be at the second memory, because this would put both transforms in series on the same axis, so that any other manipulation in addition to the rotate could only be done in one axis. The failure to transpose has to be in the input memory, in order to keep the horizontal and vertical transforms at right angles. Using the input memory causes a small problem when motion is detected. When turning the picture 90°, the input memory is read in rows, as it is written, but the readout is treated as a column. In order to de-interlace under these conditions, whole lines must be read from the A and B field memories alternately, and treated as columns by the vertical transform.

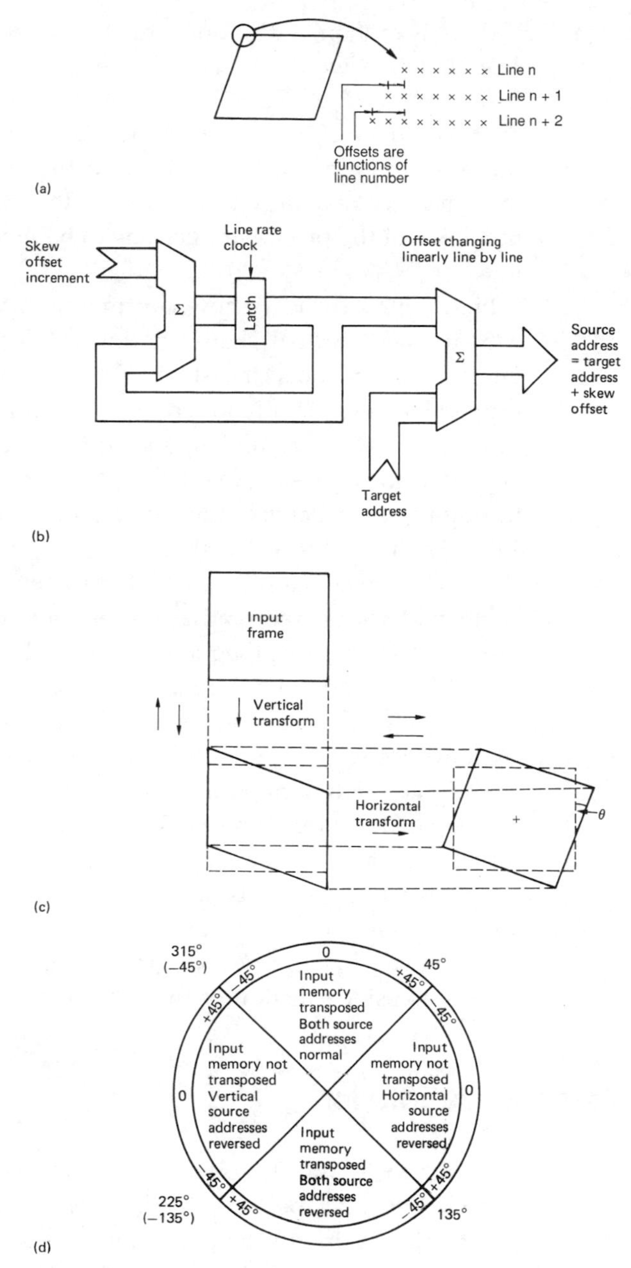

Figure 5.22 (a) Skew is achieved by subjecting each line of pixels to a different offset. (b) The hardware necessary to perform a skew where the left-hand accumulator produces the offset which increases every line, and the right-hand accumulator adds it to the address. (c) A z-axis rotate is performed using a pair of skews in succession. The magnification of each transform must also change from unity to cos θ because horizontal and vertical components of distances on the frame reduce as the frame turns. (d) The four modes necessary for a complete z-axis rotation using skews. Switching between modes at the vertical interval allows a skew range of ±45° (outer ring) to embrace a complete revolution in conjunction with memory transposes which exchange rows and columns to give 90° changes.

If there is motion, entire lines must be interpolated from the A memory. The first line would be A0, A1, A2 etc., but the next line would be A0/2 + A750/2, A1/2 + A751/2, A2/2 + A752/2 etc., and the line after that would be A750, A751, A752 etc. This is a different interpolation process from that shown earlier, and the de-interlacer must be told that there has been a failure to transpose. To obtain the correct interpolation, the delay in the pixel averager has to be increased from one pixel to one line.

Figure 5.22(d) shows how continuous rotation can be obtained. From –45° to +45°, normal skew rotation is used. At 45° during the vertical interval, the memory transpose is turned off, causing the picture to be flipped 90° and laterally inverted. Reversing the source address sequence cancels the lateral inversion, and at the same time the skew parameters are changed from +45° to –45°. In this way the picture passes smoothly through the 45° barrier, and skew parameters continue to change until 135° (90° transpose + 45° skew) degrees is reached. At this point, three things happen, again during the vertical interval. The transpose is switched back on, reorienting the picture, the source addresses are both reversed, which turns the picture upside down, and a skew rotate of –45° is applied, returning the picture to 135° of rotation, from which point motion can continue. The remainder of the rotation takes place along similar lines which can be followed in the diagram.

The rotation described is in the Z axis, i.e. the axis coming out of the source picture at right angles. Rotation about the other axes is rather more difficult, because to perform the effect properly, perspective is needed. In simple machines, there is no perspective, and the effect of rotation is as if viewed from a long way away. These non-perspective pseudo-rotations are achieved by simply changing the magnification in the appropriate axis as a cosine function of the rotation angle.

5.12 Perspective rotation

In order to follow the operation of a true perspective machine, some knowledge of perspective is necessary. Stated briefly, the phenomenon of perspective is due to the angle subtended to the eye by objects being a function not only of their size but also of their distance. Figure 5.23 shows that the size of an image on the rear wall of a pinhole camera can be increased either by making the object larger or bringing it closer. In the absence of stereoscopic vision, it is not possible to tell which has happened. The pinhole camera is very useful for study of perspective, and has indeed been used by artists for that purpose. The clinically precise perspective of Canaletto paintings was achieved through the use of the camera obscura ('darkened room' in Italian).[3]

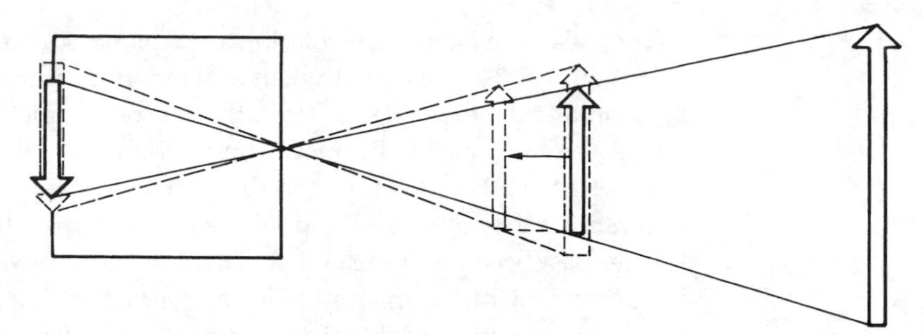

Figure 5.23 The image on the rear of the pinhole camera is identical for the two solid objects shown because the size of the object is proportional to distance, and the subtended angle remains the same. The image can be made larger (dotted) by making the object larger or moving it closer.

It is sometimes claimed that the focal length of the lens used on a camera changes the perspective of a picture. This is not true, perspective is only a function of the relative positions of the camera and the subject. Fitting a wide-angle lens simply allows the camera to come near enough to keep dramatic perspective within the frame, whereas fitting a long-focus lens allows the camera to be far enough away to display a reasonably sized image with flat perspective.[4]

Since a single eye cannot tell distance unaided, all current effects machines work by simply producing the correct subtended angles which the brain perceives as a three-dimensional effect. Figure 5.24 shows that to a single eye, there is no difference between a three-dimensional scene and a two-dimensional image formed where rays traced from features to the eye intersect an imaginary plane. This is exactly the reverse of the map projection shown in Figure 5.15, and is the principle of all perspective manipulators.

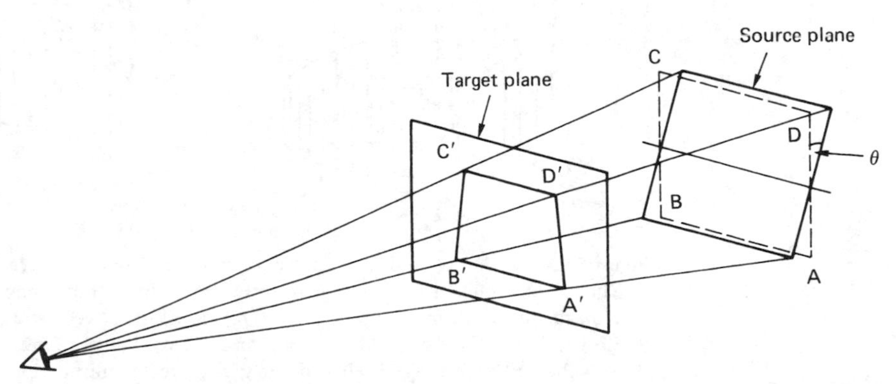

Figure 5.24 In a planar rotation effect the source plane ABCD is the rectangular input picture. If it is rotated through the angle θ, ray tracing to a single eye at left will produce a trapezoidal image A′B′C′D′ on the target. Magnification will now vary with position on the picture.

The case of perspective rotation of a plane source will be discussed first. Figure 5.24 shows that when a plane input frame is rotated about a horizontal axis, the distance from the top of the picture to the eye is no longer the same as the distance from the bottom of the picture to the eye. The result is that the top and bottom edges of the picture subtend different angles to the eye, and where the rays cross the target plane, the image has become trapezoidal. There is now no such thing as the magnification of the picture. The magnification changes continuously from top to bottom of the picture, and if a uniform grid is input, after a perspective rotation it will appear non-linear as the diagram shows.

As the two axes of the picture are transformed consecutively, it can be seen in Figure 5.25(a) that the first process will be a vertical transform

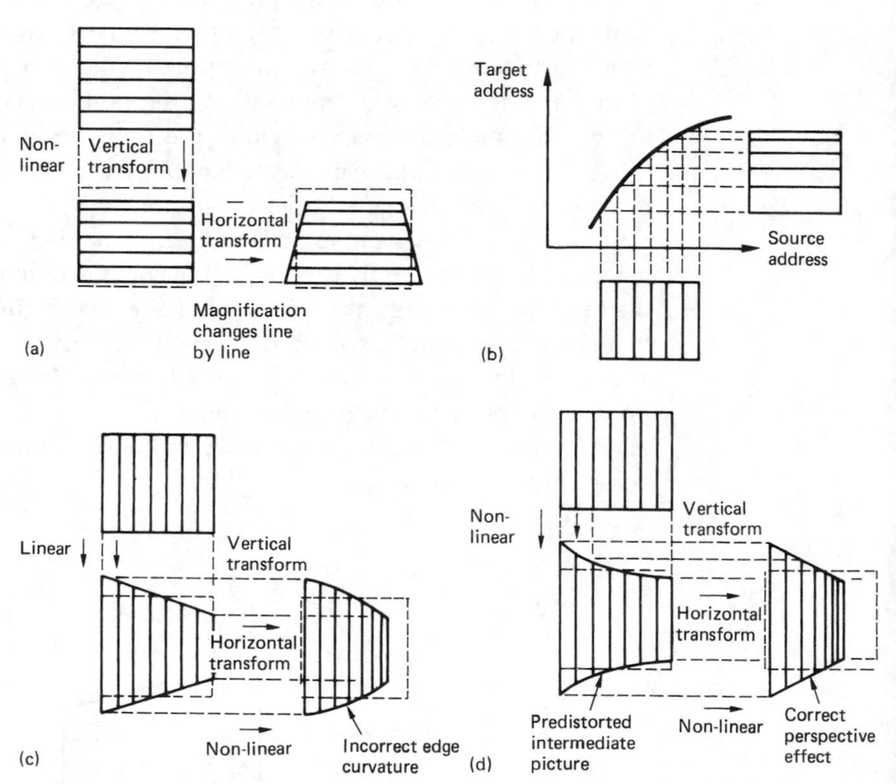

Figure 5.25 (a) A perspective rotation consist of a non-linear vertical transform followed by a horizontal transform where the magnification is linear but changes line by line. (b) Non-linear transforms require a curved relationship between source and target address. (c) How not to perform a rotation about a vertical axis. The non-linear horizontal process distorts the vertical process. (d) In order to give correct manipulation, the vertical process must produce columns which will be the correct length for the position the horizontal process will put them in. This implies that the vertical transform knows what the horizontal transform will do and predistorts accordingly. The non-linear horizontal transform then cancels the distortion.

which is non-linear. The address generator produces a function or series of addresses which is curved, which will be seen in (b) to produce the desired distortion. The vertical process is followed by the horizontal process where each line or row of pixels has the same magnification, but the magnification changes from one line to the next. The result is the desired trapezoidal shape with the correct non-linearity, which the eye takes to be a rotation of the source, even though the image never leaves the plane of the monitor. Rotation about a vertical axis will now be considered. Again this results in a trapezoidal picture, but now the non-linearity appears in the horizontal transform. Figure 5.25(c) shows that if the vertical and horizontal transforms are performed in that order, the non-linear horizontal transform causes an unwanted distortion – the edges of the picture become curved. This is a fundamental problem of separating the axes, and some thought is necessary to provide a solution. Figure 5.25(d) shows that the problem is caused by the vertical process moving a given pixel vertically by what appears to be the correct distance, whereas it should move the pixel so that it will be at the correct vertical position *after* it has been horizontally transformed. In other words, the vertical process must pre-distort the intermediate image in such a way that the non-linear horizontal process will cancel the pre-distortion to give the correct overall transform.

The above process implies that the vertical transform must be exactly aware of what the horizontal transform will do. This is no problem because both transforms are computed from high-level commands put

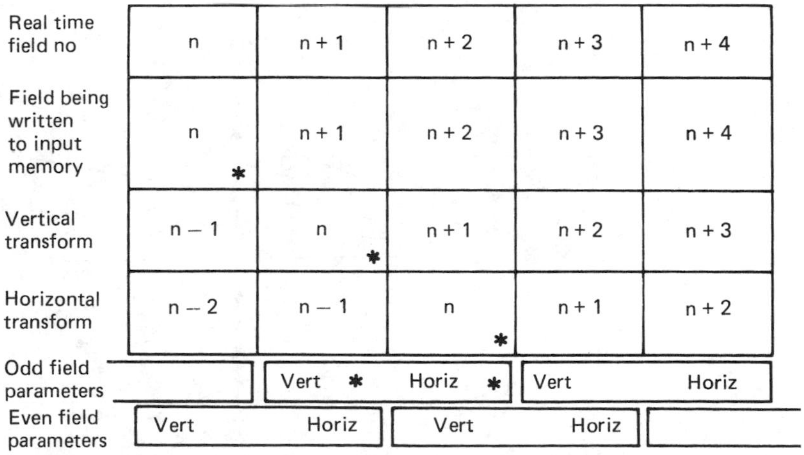

Real time field no	n	n + 1	n + 2	n + 3	n + 4
Field being written to input memory	n *	n + 1	n + 2	n + 3	n + 4
Vertical transform	n – 1	n *	n + 1	n + 2	n + 3
Horizontal transform	n – 2	n – 1	n *	n + 1	n + 2

Odd field parameters: | | Vert * Horiz * | Vert Horiz |

Even field parameters: | Vert Horiz | Vert Horiz | |

Figure 5.26 Timing chart showing how the transform parameters are supplied at the correct time. For a field which arrives at time n it can be read from input memory at field $n + 1$ where the vertical transform will take place. The horizontal transform will take place in the next field. In order to have matching transform parameters, the vertical parameters must be supplied one field ahead of the horizontal parameters. (Follow asterisk for progress of field n.)

into the system by the operator, but it does require some care to supply the transform parameters to the processors at the correct time, because the horizontal transform cannot begin until the vertical transform is completed, due to the presence of the transpose memory. This means that the transform parameters for the horizontal processor will need to be supplied one field later than the parameters for the vertical process. Figure 5.26 shows how the field shift in the parameters relates to the progress of fields through a machine.

The address generators for perspective operation are necessarily complex, and a careful approach is necessary to produce the complex calculations at the necessary speed.[5] Figure 5.27 shows a section through a transform where the source plane has been rotated about an axis perpendicular to the page. The mapping or ray tracing process must produce a straight line from every target pixel (corresponding to where an output value is needed) to locate a source address (corresponding to where an input value is available). Moving the source value to the target performs the necessary transform.

Clearly it is possible to compute any source address from the target address and a knowledge of the geometry of the system. This will take several minutes with a pocket calculator, making liberal use of trigonometric functions. The standard-definition DVE must compute the source address in 75 ns, which is non-trivial. One solution is to split the problem

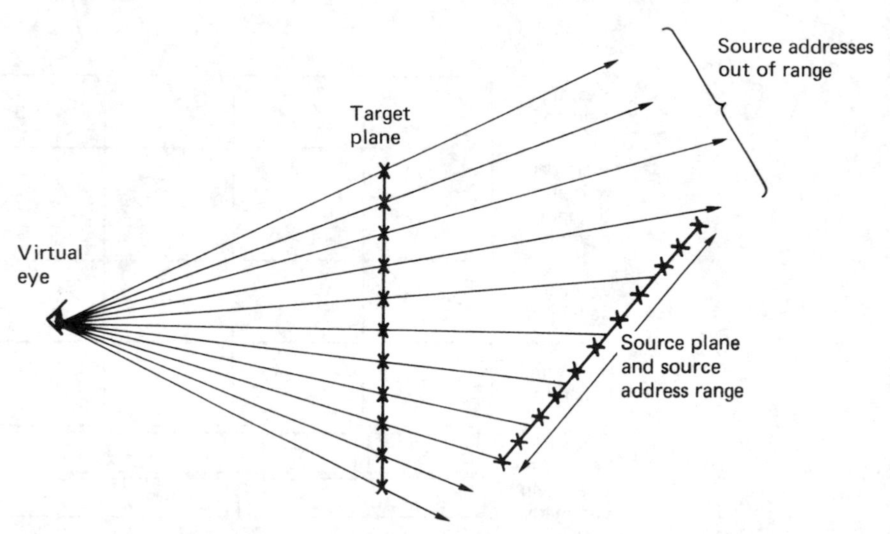

Figure 5.27 A rotation of the source plane along with a movement away from the observer is shown here. The system has to produce pixel values at the spacing demanded by the output. Thus a ray from the eye to each target pixel is produced to locate a source pixel. Since the chances of exactly hitting a source pixel are small, the need for interpolation is clear. If the source plane is missed, this will result in an out-of-range source address, and a background value will be substituted.

into two parts. For a given position of the source plane in space, all the angles except that of the ray being traced become constants, and therefore the trigonometrical functions of those angles also become constants. A general-purpose computer can work out what the constants are for a given source orientation in less than a field period, and these constants are then used by the slow and fast address generators which execute much simpler equations at line/column and pixel rate.

The derivation of the address generator equations is in highly complex mathematics which can be found in Newman and Sproull[2] and in the ADO patent.[5] For the purposes of this chapter, an understanding of the result can be had by considering the example of Figure 5.28.

In this figure, successive integer values of x have been used in a simple source address calculation equation which contains a division stage. The addresses produced form a non-linear sequence, and will be seen to lie on a rotated source plane. The actual equations necesary for a planar transform are shown in Figure 5.29 along with the block diagram of one possible implementation. Note that the vertical and horizontal equations are different because of the use of separability. As Figure 5.25 showed, the vertical transform must pre-distort if the subsequent horizontal transform is non-linear.

The accumulation stages of the address generator are quite straightforward, as these are simply repeated adders or digital integrators. The difficulty arises with the division. There is no parallel with the fast multiplier. Fast division is impossible, as all divisions are in the form of an experiment, to see how many times one number will fit into another. Division can be made easier by calculating the reciprocal of the divisor and multiplying. The calculation of a reciprocal is, admittedly, still a division, but it is a division into a constant, i.e. one, and advantage can be taken of PROMs. For short wordlengths, the reciprocal PROM will be of moderate size, but address wordlengths in DVEs are very large because of the enormous virtual space in which the picture can move, and a more complex solution is necessary. In this approach, the transfer function of the reciprocal is approximated by a piecewise linear graph, as shown in Figure 5.30. The high-order bits of the number to be inverted now address two PROMs, one to give a coarse reciprocal, and one to give the slope of the curve. The low-order bits are multiplied by the slope to give a correction term which is added to the coarse reciprocal.

Where really long wordlengths must be multiplied or divided, the only practicable solution is to use floating-point notation. In floating point, the numbers are expressed as a mantissa, and an exponent. Only the mantissae need to be multiplied or divided, the exponents are simply added or subtracted. Following the floating-point division stage, the addresses must be returned to fixed point. The source address then performs a number of functions as can be seen in Figure 5.29.

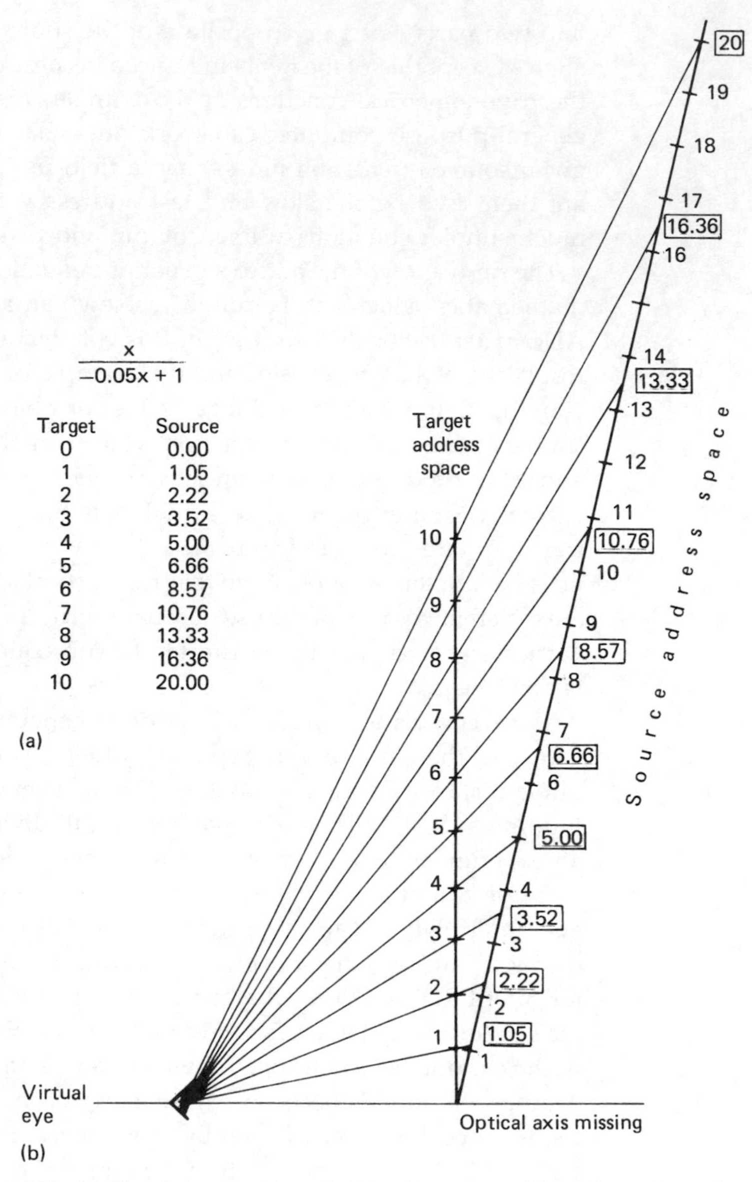

$$\frac{x}{-0.05x + 1}$$

Target	Source
0	0.00
1	1.05
2	2.22
3	3.52
4	5.00
5	6.66
6	8.57
7	10.76
8	13.33
9	16.36
10	20.00

Figure 5.28 (a) The above equation calculates the source address for each evenly spaced target address from 0 to 10. All numbers are kept positive for simplicity, so only one side of the picture is represented here. (b) The ray tracing diagram corresponding to the calculations of (a). Following a ray from the virtual eye through any target pixel address will locate the source addresses calculated.

The source address will be compared with limits to see if a ray from target has hit the source picture or missed it. If it has missed, a border or background value must be substituted. The source address also passes to the interpolator memory where it will access a contiguous group of

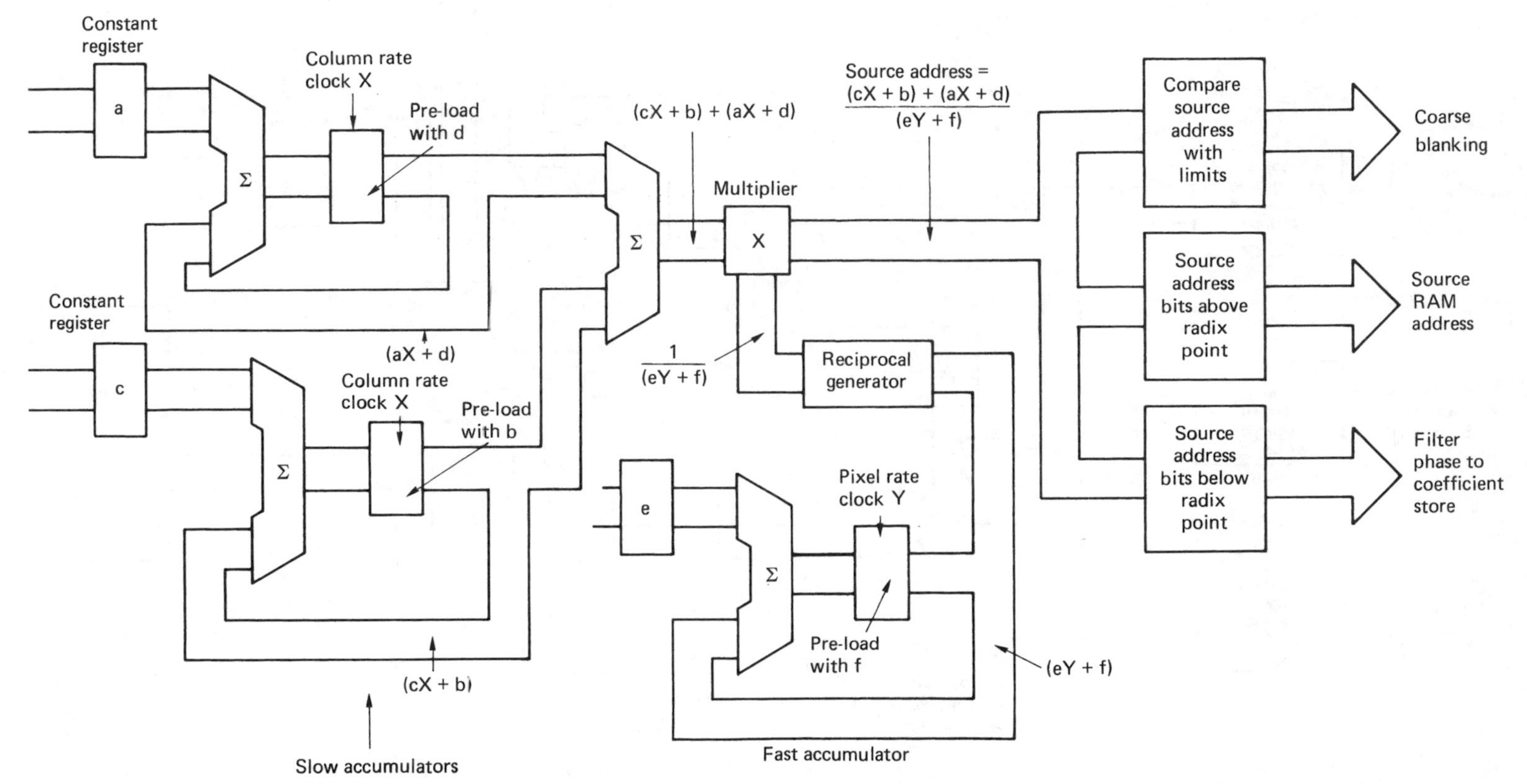

Figure 5.29(a) Hardware block diagram of vertical address generator. Accumulators on left operate once per column. When latch is clocked, preloaded value d (or b) is added to constant a (or c). Each clock X adds a (or c) once more, hence (aX + d) or (cX + b) is calculated. Fast accumulator produces a new denominator by clocking at pixel rate Y. Reciprocal of denominator is multiplied by numerator to give final source address.

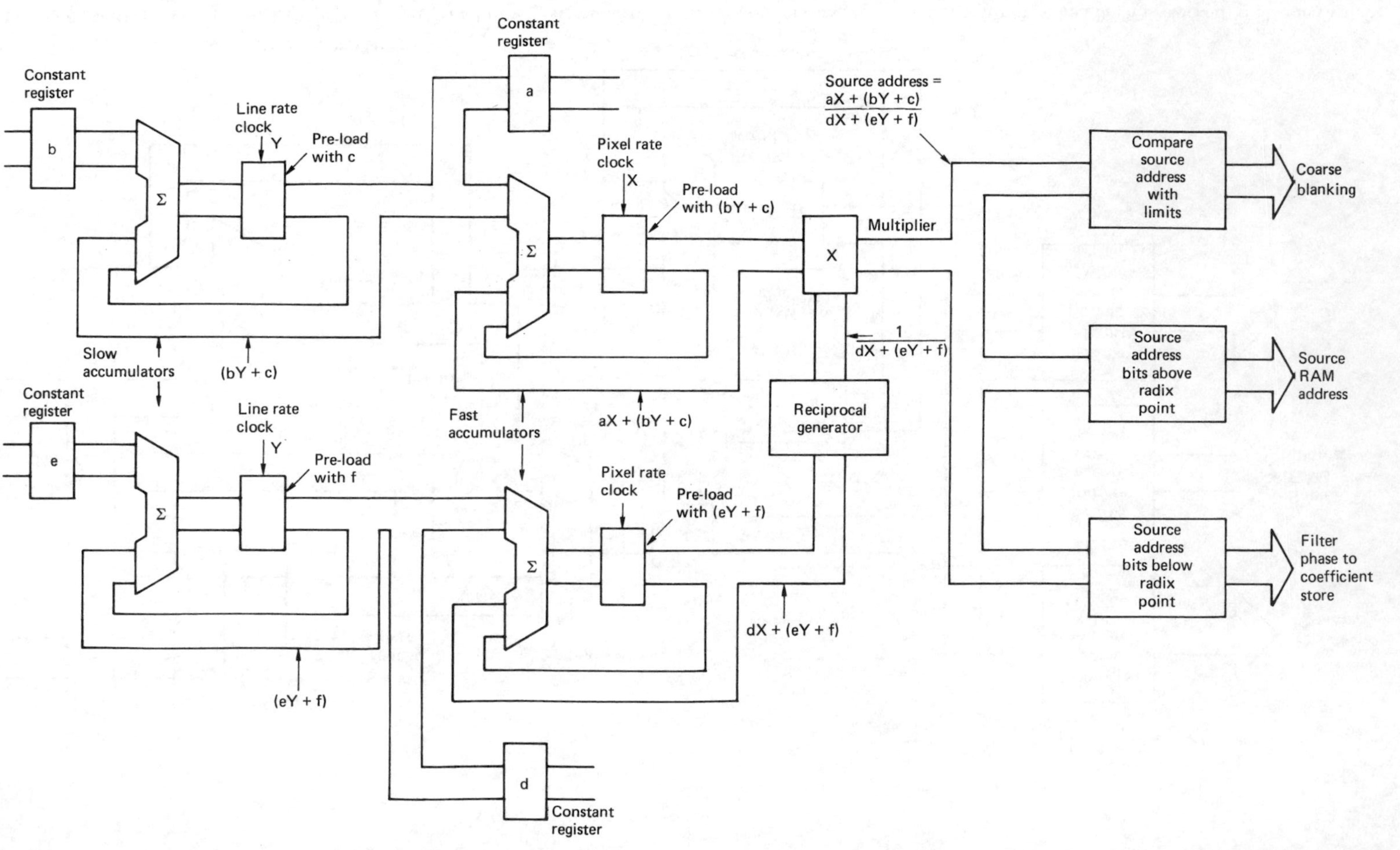

Figure 5.29(b) Hardware block diagram to implement horizontal equation. Slow accumulators are preloaded with constants c and f at the beginning of each field and calculate a new value (bY + c) or (eY + f) every line. Fast accumulators are preloaded with the values from the slow accumulators at the start of every line, then revert to constant registers a, d.

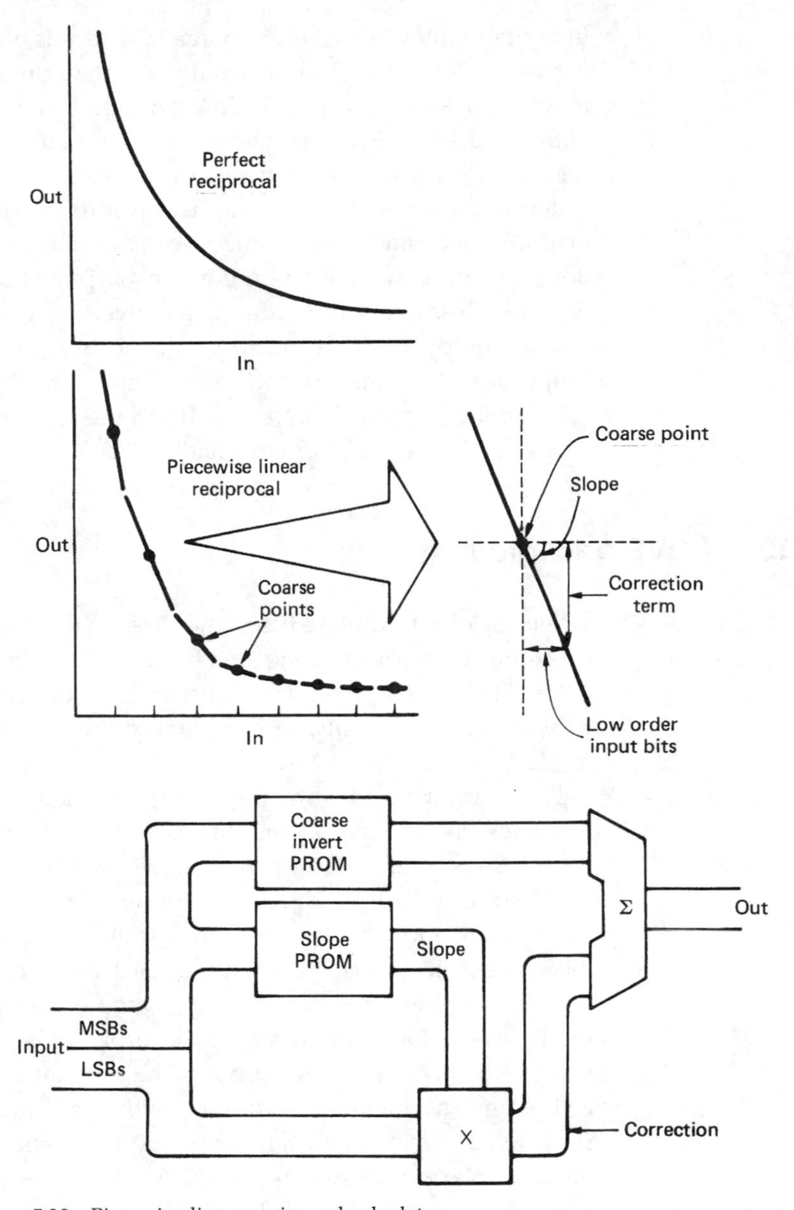

Figure 5.30 Piecewise linear reciprocal calculator.

(typically) eight pixels. The low-order bits of the source address below the radix point (the fractional address) determine the position between source samples of the interpolated target pixel, and so these bits control the phase of the interpolator.

All perspective machines must work with dynamic changes in magnification throughout the frame. The situation often arises where at one end of a pixel row the magnification is greater than unity, and the FIR filter has to interpolate between available pixels, whereas at the other end of the row the magnification will be less than unity and the FIR filter has to adopt a low pass and decimate mode to eliminate excessive pixels without aliasing. The characteristics of the filter are changed at will by selecting different coefficient sets from pages of memory according to the instantaneous magnification at the centre of the filter window. The magnification can be determined by computing the address slope. This is done by digitally differentiating the output of the address generator, which is to say that the difference between one source address and the next is computed. This produces the address slope, which is inversely proportional to the magnification, and can be used to select the appropriate impulse response width in the interpolator. The interpolator output is then a transformed image.

5.13 DVE backgrounds

When an effects unit reduces the size of the picture, it is necesary to control what happens in the area between the edge of the picture and the edge of the screen. At least, the system must ensure that this area defaults to black, but it is usually possible to insert some kind of plain background instead.

There are two conflicting requirements. The transition between picture and background has to pass through the interpolator so that a smooth edge results. This will be achieved by switching to background values (luminance and colour differences in the respective processing channels) in source space before the interpolator. Figure 5.31(a) shows the mechanism necessary. When the source memory of the interpolator is loaded from the transpose memory, sample values must pass through a multiplexer which can switch between picture samples and background values from a control system port. The switching points are obtained by comparing the memory addresses with limit addresses which are also provided by a system port. If the memory address falls outside the limits, the multiplexer will switch to background values, and the result will be a column or row of samples in the interpolator memory which has background values at both ends. If the picture is cropped, the limit addresses are simply brought closer, so that more of the source picture is replaced by background.

The transition to background will now be manipulated with the source picture, and can be positioned to sub-pixel accuracy according to the phase of the interpolator.

This is not, however, the complete solution. Figure 5.31(b) shows that when the magnification becomes very small, the background area in source space must tend towards infinite size if the screen is to be filled.

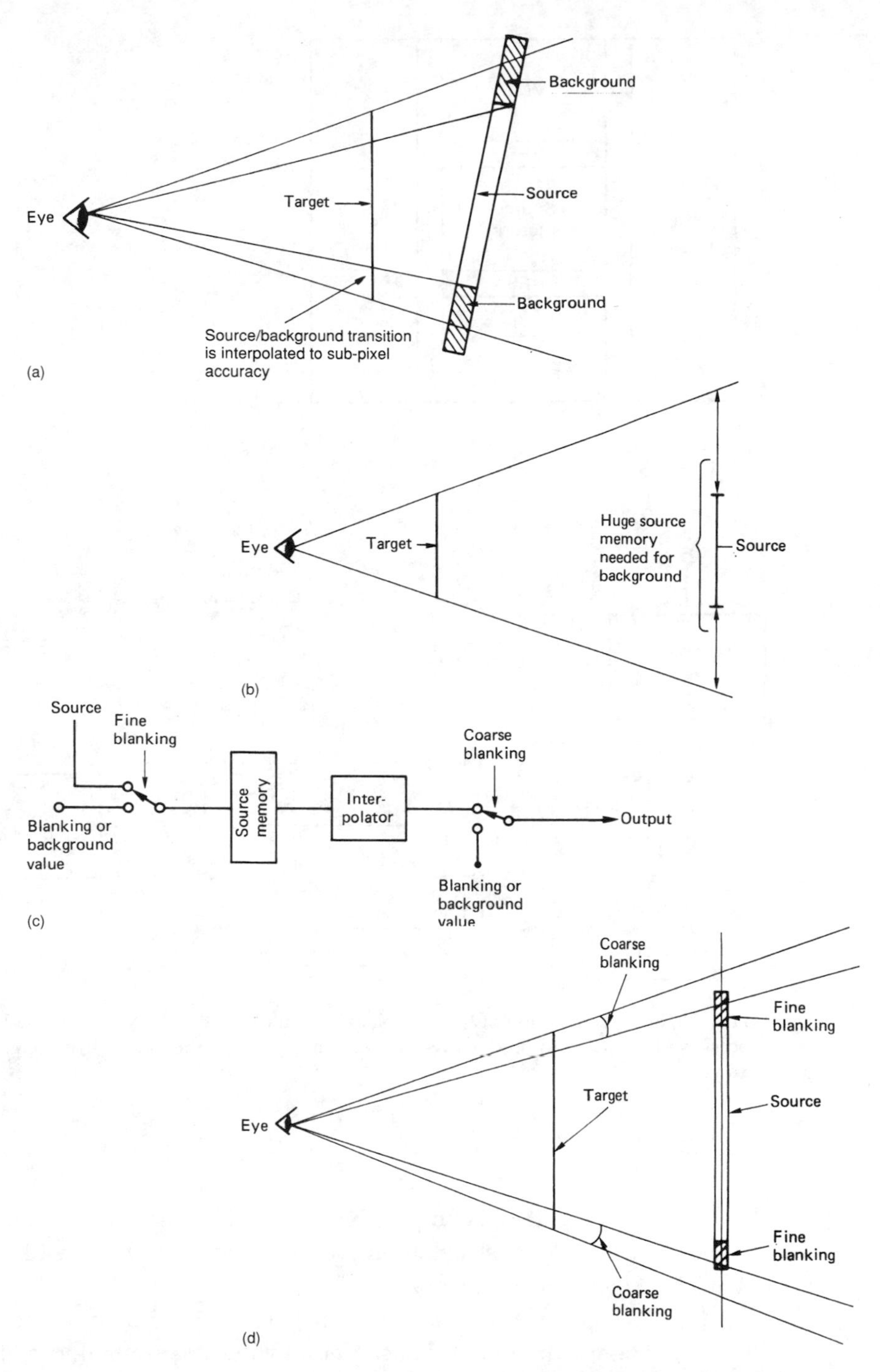

Figure 5.31 (a) Source/background transition is interpolated to subpixel accuracy. (b) At small magnifications the memory needed in source space to hold background values becomes very large. (c) Two multiplexers are necessary to insert background before and after the interpolator. (d) Coarse blanking multiplexer is switched in before the fine blanking in source space runs out at the ends of the address sequence.

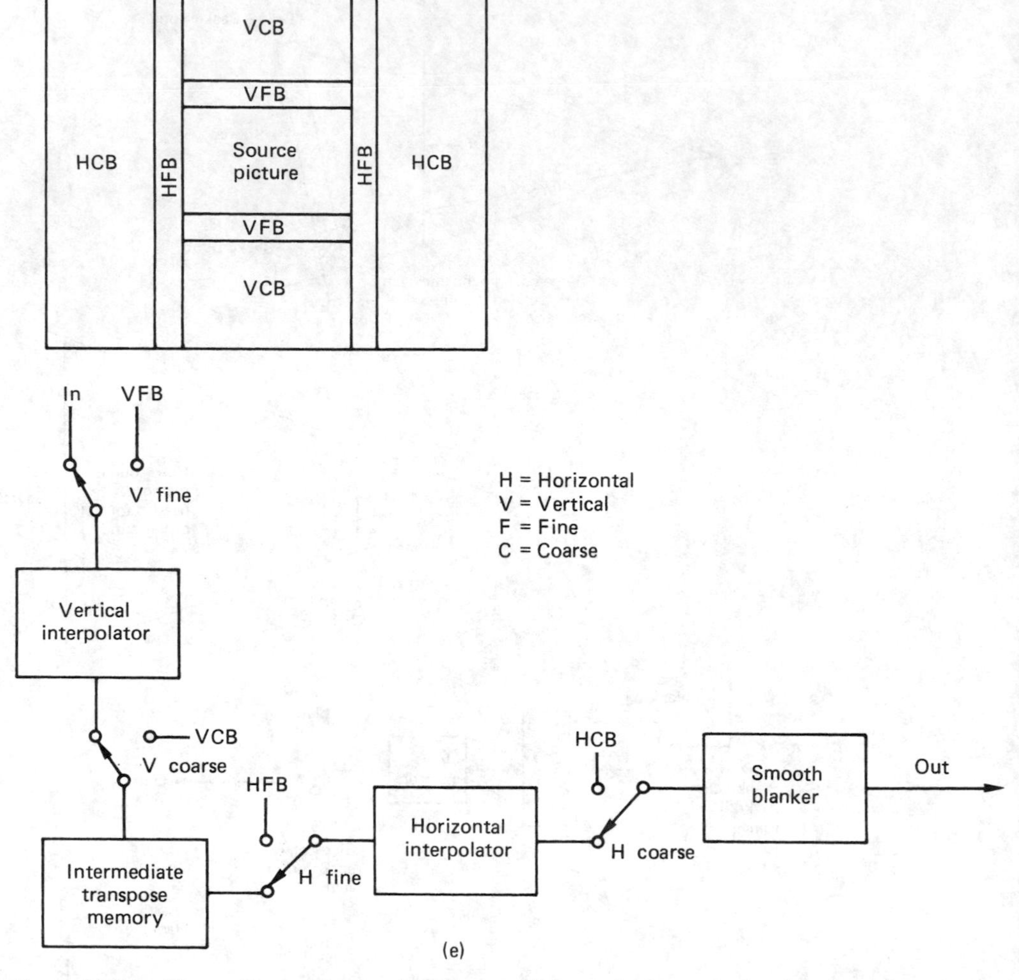

Figure 5.31(e) The position of the four blanking multiplexers in the vertical and horizontal process, and the position on the screen of the contributions from each when the picture size is greatly reduced.

Clearly this is impossible. In order to allow background generation without a restriction on the magnification which will be used, a further stage is necessary.

The arrangement is shown in Figure 5.31(c). A further multiplexer inserts an identical background value *after* the interpolator. This device is working in target space, and so can only switch at target pixel spacing, but this is of no consequence provided it operates at the correct time. Several pixels away from the source picture, the interpolator will be outputting background values, and as the result of interpolating a continuous plain area is the same as sampling it, the output of the interpolator will be the same as the background value, and the switch can operate invisibly. The two processes before and after the interpolator are called fine and coarse blanking respectively, and should not be confused with the blanking which is necessary to return background values to black level smoothly to prevent out-of-band signals with bright backgrounds. Clearly this is only necessary in the horizontal domain. It is sometimes called smooth blanking to avoid confusion.

Figure 5.31(d) shows the sequence of events in a horizontal line output from a DVE. At the beginning of the line the smooth blanker slowly releases so that a gradual transition to background takes place. The background values have come from the coarse blanker after the interpolator. As the line proceeds, the output of the transform address generator will fall within the range of the interpolator memory, and meaningful samples will begin to come from the interpolator. These will be numerically identical to the coarse blanking value, because they have come from the fine-blanking multiplexer via the interpolator.

It is now possible to switch invisibly to the interpolator output. The picture/background transition will thus have come through the interpolator. At the opposite edge of the picture, the return transition will be similarly treated. Shortly after this, the interpolator output will again be identical to the coarse blanking value, and a second invisible switch will be made to coarse background. At the end of active line, the smooth blanker will gradually fade the background value to black.

With a reduced picture it is possible to see the origin of sample values in different parts of the screen (Figure 5.31(e)), and this is one of the most powerful tools available for locating the position of a failure in a DVE.

5.14 Non-planar effects

The basic approach to perspective rotation of plane pictures has been described, and this can be extended to embrace transforms which make the source picture appear non-planar. Effects in this category include rolling the picture onto the surface of an imaginary solid such as a cylinder or a cone. Figure 5.32 shows that the ray tracing principle is still used, but that the relationship between source and target addresses has become much more complex. The problem is that when a source picture can be curved, it may be put in such an attitude that one part of the source can be seen through another part. This results in two difficulties. First, the source address function needs to be of higher order, and second, the target needs to be able to accept and accumulate pixel data from two different source addresses, with weighting given to the one nearer the viewer according to the transparency allocated to the picture.

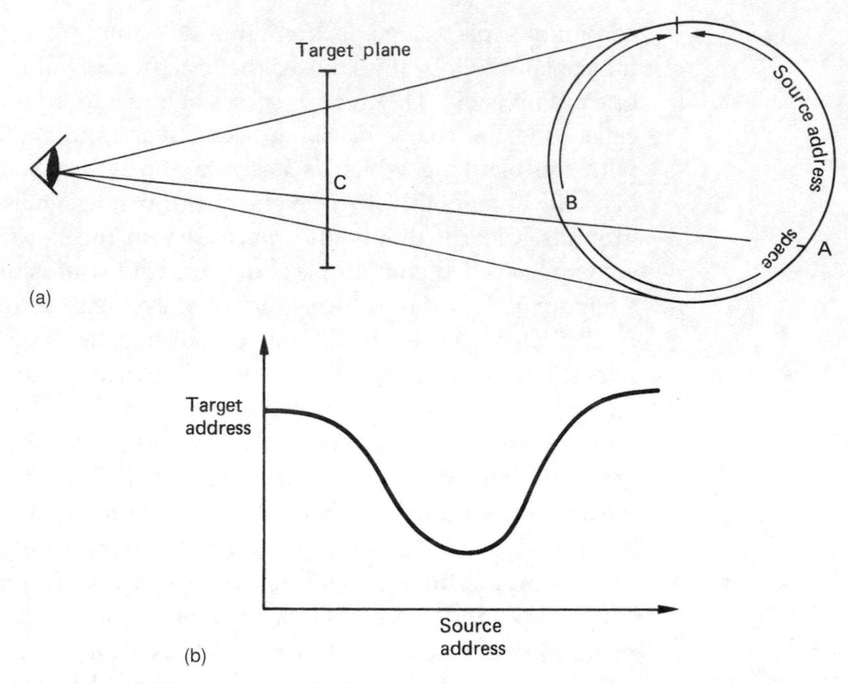

Figure 5.32 (a) To produce a rolled-up image for a given target pixel address C, there will be two source addresses A and B. Pixel data from A and B must be added with weighting dependent on the transparency of the nearer pixel to produce the pixel value to be put on the target plane at C. (b) Transfer function for a rolling-up transform. There are two source addresses for every target address; hence the need for target accumulation.

The high-order address functions needed for non-planar images cannot be readily implemented in fast hard-wired accumulators, because too much hardware would be needed. A fast CPU will be needed to generate these address sequences in real time. At one time this restricted non-planar effects to very expensive machines, but the falling cost of digital processing means that such effects are now commonplace.

5.15 Controlling effects

The basic mechanism of the transform process has been described, but this is only half of the story, because these transforms have to be controlled. There is a lot of complex geometrical calculation necessary to perform even the simplest effect, and the operator cannot be expected to calculate directly the parameters required for the transforms. All effects machines require a computer of some kind, with which the operator communicates using keyboard entry or joystick/trackball movements at high level. These high-level commands will specify such things as the position of the axis of rotation of the picture relative to the viewer, the position of the axis of rotation relative to the source picture, and the angle of rotation in the three axes.

An essential feature of this kind of effects machines is fluid movement of the source picture as the effect proceeds. If the source picture is to be made to move smoothly, then clearly the transform parameters will be different in each field. The operator cannot be expected to input the source position for every field, because this would be an enormous task. Additionally, storing the effect would require a lot of space. The solution is for the operator to specify the picture position at strategic points during the effect, and then digital filters are used to compute the intermediate positions so that every field will have different parameters.

The specified positions are referred to as knots, nodes or keyframes, the first being the computer graphics term. The operator is free to enter knots anywhere in the effect, and so they will not necessarily be evenly spaced in time, i.e. there may well be different numbers of fields between each knot. In this environment it is not possible to use conventional FIR-type digital filtering, because a fixed-impulse response is inappropriate for irregularly spaced samples.

Interpolation of various orders is used ranging from zero-order hold for special jerky effects through linear interpolation to cubic interpolation for very smooth motion. The algorithms used to perform the interpolation are known as splines, a term which has come down from shipbuilding via computer graphics.[6] When a ship is designed, the draughtsman produces hull cross-sections at intervals along the keel, whereas the shipyard needs to recreate a continuous structure. The solution is a lead-filled bar, known as a spline, which can be formed to join up each cross-section in a smooth curve, and then used as a template to form the hull plating.

The filter which does not ring cannot be made, and so the use of spline algorithms for smooth motion sometimes results in unintentional overshoots of the picture position. This can be overcome by modifying the filtering algorithm. Spline algorithms usually look ahead beyond the next knot in order to compute the degree of curvature in the graph of the parameter against time. If a break is put in that parameter at a given knot, the spline algorithm is prevented from looking ahead, and no overshoot will occur. In practice the effect is created and run without breaks, and then breaks are added later where they are subjectively thought necessary.

It will be seen that there are several levels of control in an effects machine. At the highest level, the operator can create, store and edit knots, and specify the times which elapse between them. The next level is for the knots to be interpolated by spline algorithms to produce parameters for every field in the effect. The field frequency parameters are then used as the inputs to the geometrical computation of transform parameters which the lowest level of the machine will use as micro-instructions to act upon the pixel data. Each of these layers will often have a separate processor, not just for speed, but also to allow software to be updated at certain levels without disturbing others.

5.16 Graphics

Although there is no easy definition of a video graphics system which distinguishes it from a graphic art system, for the purposes of discussion it can be said that graphics consists of generating alphanumerics on the screen, whereas graphic art is concerned with generating more general images. The simplest form of screen presentation of alphanumerics is the ubiquitous visual display unit (VDU) which is frequently necessary to control computer-based systems. The mechanism used for character generation in such devices is very simple, and thus makes a good introduction to the subject.

In VDUs, there is no grey scale, and the characters are formed by changing the video signal between two levels at the appropriate place in the line. Figure 5.33 shows how a character is built up in this way, and also illustrates how easy it is to obtain the reversed video used in some word processor displays to simulate dark characters on white paper. Also shown is the method of highlighting single characters or words by using localized reverse video.

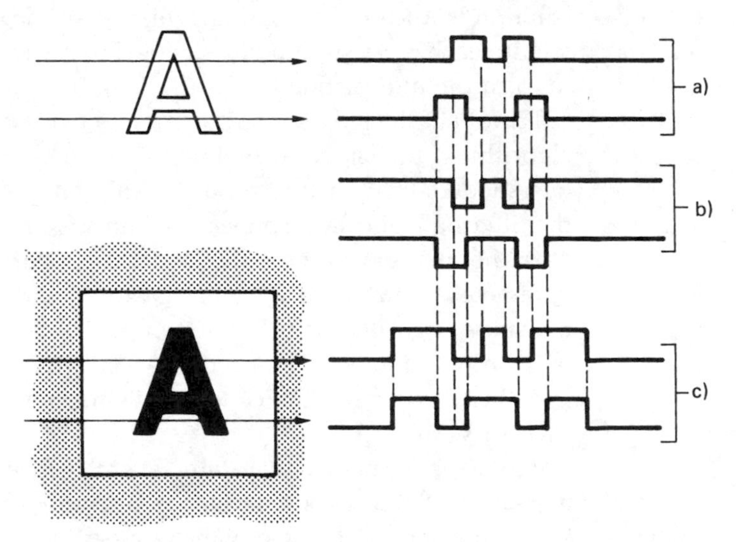

Figure 5.33 Elementary character generation. At (a), white on black waveform for two raster lines passing through letter A. At (b), black on white is simple inversion. At (c), reverse video highlight waveforms.

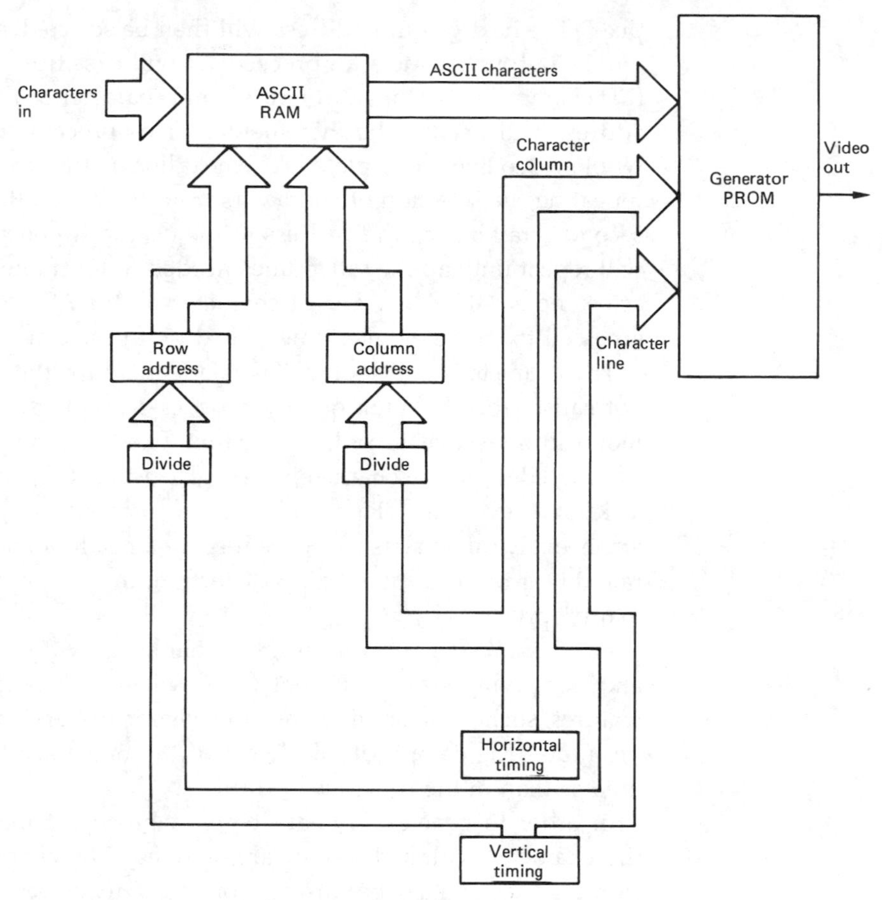

Figure 5.34 Simple character generator produces characters as rows and columns of pixels. See text for details.

Figure 5.34 is a representative character generator, as might be used in a VDU. The characters to be displayed are stored as ASCII symbols in a RAM, which has one location for each character position on each available text line on the screen. Each character must be used to generate a series of dots on the screen which will extend over several lines. Typically the characters are formed by an array five dots by nine. In order to convert from the ASCII code to a dot pattern, a ROM is programmed with a conversion. This will be addressed by the ASCII character, and the column and row addresses in the character array, and will output a high or low (bright or dark) output.

As the VDU screen is a raster-scanned device, the display scan will begin at the left-hand end of the top line. The first character in the ASCII RAM will be selected, and this and the first row and column addresses will be sent to the character generator, which ouputs the video level for the first pixel. The next column address will then be selected, and the next pixel will be output. As the scan proceeds, it will pass from the top line of the first character to the top line of the second character, so that the ASCII RAM address will need to be incremented. This process continues until the whole video line is completed. The next line on the screen is generated by repeating the selection of characters from the ASCII RAM, but using the second array line as the address to the character generator. This process will repeat until all the video lines needed to form one row of characters are complete. The next row of characters in the ASCII RAM can then be accessed to create the next line of text on the screen and so on.

The character quality of VDUs is adequate for the application, but is not satisfactory for high-quality broadcast graphics. The characters are monochrome, have a fixed, simple font, fixed size, no grey scale, and the sloping edges of the characters have the usual stepped edges due to the lack of grey scale. This stepping of diagonal edges is sometimes erroneously called aliasing. Since it is not a result of inadequate sampling rate, but a result of quantizing distortion, the use of the term is wholly inappropriate.

In a broadcast graphics unit, the characters will be needed in colour, and in varying sizes. Different fonts will be necessary, and additional features such as solid lines around characters and drop shadows are desirable. The complexity and cost of the necessary hardware is much greater than in the previous example.

In order to generate a character in a broadcast machine, a font and the character within that font are selected. The characters are actually stored as key signals, because the only difference between one character and another in the same font is the shape. A character is generated by specifying a constant background colour and luminance, and a constant character colour and luminance, and using the key signal to cut a hole in the background and insert the character colour. This is illustrated in Figure 5.35. The problem of stepped diagonal edges is overcome by giving the key signal a grey scale. The grey scale eliminates the quantizing distortion responsible for the stepped edges. The edge of the character now takes the form of a ramp, which has the desirable characteristic of limiting the bandwidth of the character generator output. Early character generators were notorious for producing out-of-band frequencies which drove equipment further down the line to distraction and in some cases would interfere with the sound channel on being broadcast. Figure 5.36 illustrates how in a system with grey scale and sloped edges, the edge of a character can be positioned to sub-pixel resolution, which completely removes the stepped effect on diagonals.

In a powerful system, the number of fonts available will be large, and all the necessary characters will be stored on disk drives. Some systems

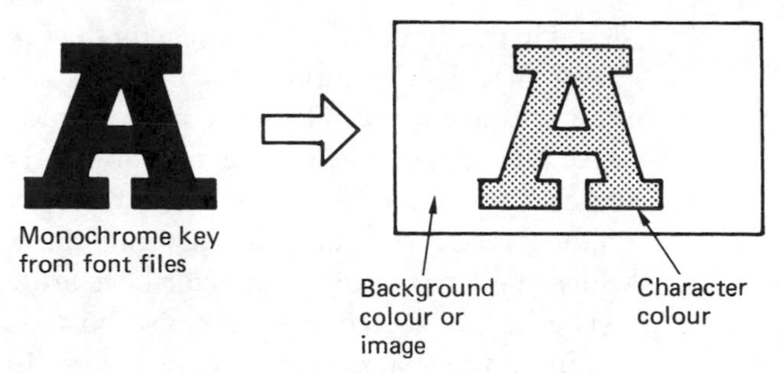

Figure 5.35 Font characters only store the shape of the character. This can be used to key any coloured character into a background.

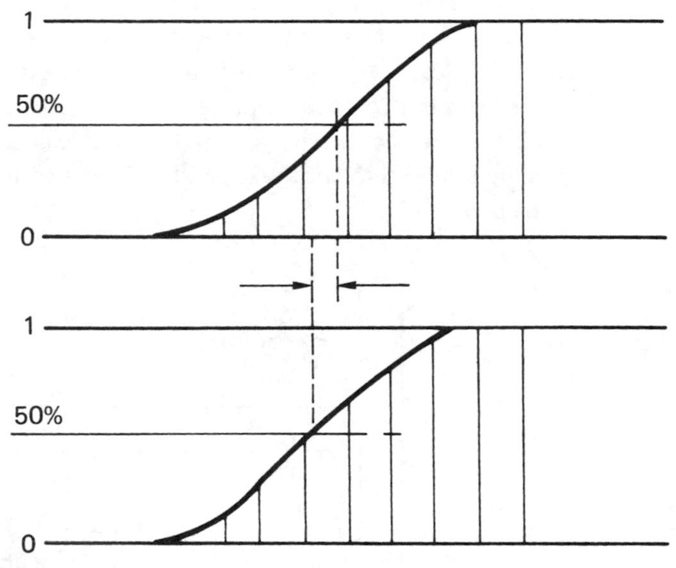

Figure 5.36 When a character has a ramped edge, the edge position can be moved in subpixel steps by changing the pixel values in the ramp.

allow users to enter their own fonts using a rostrum camera. A frame grab is performed, but the system can be told to file the image as a font character key signal rather than as a still frame. This approach allows infinite flexibility if it is desired to work in Kanji or Cyrillic, and allows European graphics to be done with all necessary umlauts, tildes and cedillas.

In order to create a character string on the screen, it is necessary to produce a key signal which has been assembled from all the individual character keys. The keys are usually stored in a large format to give highest quality, and it will be necessary to reduce the size of the characters to fit the available screen area. The size reduction of a key signal in the digital domain is exactly the same as the zoom function of an effects machine, requiring FIR filtering and interpolation, but again, it is not necessary for it to be done in real time, and so less hardware can be used. The key source for the generation of the final video output is a RAM which has one location for every screen pixel. Position of the characters on the screen is controlled by changing the addresses in the key RAM into which the size-reduced character keys are written.

The keying system necessary is shown in Figure 5.37. The character colour and the background colour are produced by latches on the control system bus, which output continuous digital parameters. The grey scale key signal obtained by scanning the key memory is used to provide coefficients for the digital cross-fader which cuts between background and character colour to assemble the video signal in real time.

If characters with contrasting edges are required, an extra stage of keying can be used. The steps described above take place, but the background colour is replaced by the desired character edge colour. The size of each character key is then increased slightly, and the new key signal is used to cut the characters and a contrasting border into the final background.

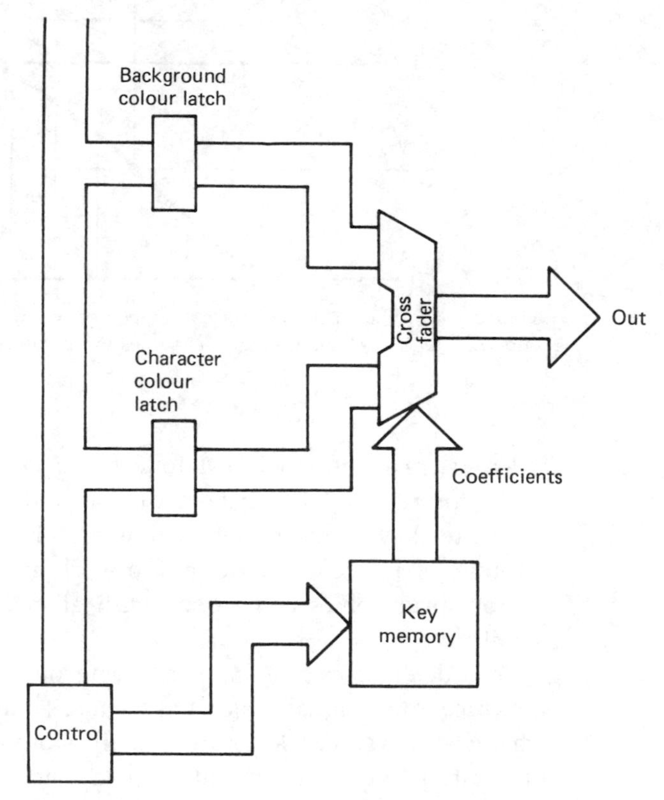

Figure 5.37 Simple character generator using keying. See text for details.

Early character generators were based on a frame store which refreshes the dynamic output video. Recent devices abandon the frame-store approach in favour of real-time synthesis. The symbols which make up a word can move on and turn with respect to the plane in which they reside as a function of time in any way individually or together. Text can also be mapped on to an arbitrarily shaped line. The angle of the characters can follow a tangent to the line, or can remain at a fixed angle regardless of the line angle.

By controlling the size of planes, characters or words can appear to zoom into view from a distance and recede again. Rotation of the character planes off the plane of the screen allows the perspective effects to be seen. Rotating a plane back about a horizontal axis by 90° will reduce it to an edge-on line, but lowering the plane to the bottom of the screen allows the top surface to be seen, receding into the distance like a road. Characters or text strings can then roll off into the distance, getting smaller as they go. In fact the planes do not rotate, but a perspective transform is performed on them.

Since characters can be moved without restriction, it will be possible to make them overlap. Either one can be declared to be at the front, so that it cuts out the other, but if desired the overlapping area can be made a different colour from either of the characters concerned. If this is done with care, the overlapping colour will give the effect of transparency. In fact colour is attributed to characters flexibly so that a character may change colour with respect to time. If this is combined with a movement, the colour will appear to change with position. The background can also be allocated a colour in this way, or the background can be input video. Instead of filling characters with colour on a video background, the characters, or only certain characters, or only the overlapping areas of characters can be filled with video.

There will be several planes on which characters can move, and these are assigned a priority sequence so that the first is essentially at the front of the screen and the last is at the back. Where no character exists, a plane is transparent, and every character on a plane has an eight-bit transparency figure allocated to it. When characters on two different planes are moved until they overlap, the priority system and the transparency parameter decide what will be seen. An opaque front character will obscure any character behind it, whereas using a more transparent parameter will allow a proportion of a character to be seen through one in front of it. Since characters have multi-bit accuracy, all transparency effects are performed without degradation, and character edges always remain smooth, even at overlaps. Clearly the character planes are not memories or frame stores, because perspective rotation of images held in a frame store in real time in this way would require staggering processing power. Instead, the contents of output fields are

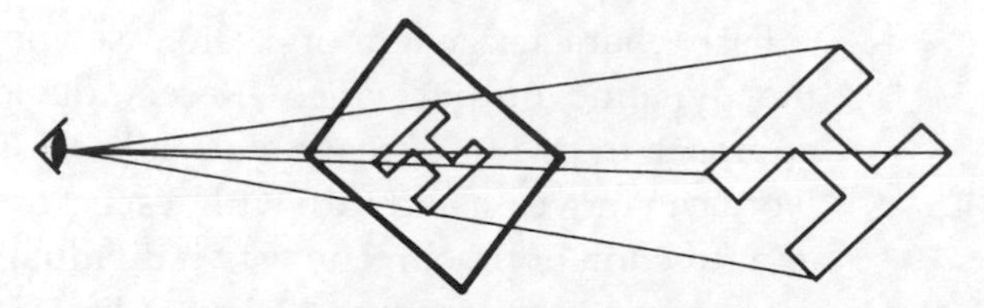

Figure 5.38 Perspective effects results from mapping or projecting the character outline on to a plane which represents the screen.

computed in real time. Every field is determined individually, so it is scarcely more difficult to animate by making successive fields different.

In a computer a symbol or character exists not as an array of pixels, but as an outline, rather as if a thin wire frame had been made to fit the edge of the symbol. The wire frame is described by a set of mathematical expressions. The operator positions the wire frame in space, and turns it to any angle. The computer then calculates what the wire frame will look like from the viewing position. Effectively the shape of the frame is projected onto a surface which will become the TV screen. The principle is shown in Figure 5.38, and will be seen to be another example of mapping. If drop or plane shadows are being used, a further projection takes place which determines how the shadow of the wire frame would fall on a second plane. The only difference here between drop and plane shadows is the angle of the plane the shadow falls on. If it is parallel to the symbol frame, the result is a drop shadow. If it is at right angles, the result is a plane shadow.

Since the wire frame can be described by a minimum amount of data, the geometrical calculations needed to project onto the screen and shadow planes are quick, and are repeated for every symbol. This process also reveals where overlaps occur within a plane, since the projected frames will cross each other. This computation takes place for all planes. The positions of the symbol edges are then converted to a pixel array in a field interlaced raster scan. Because the symbols are described by eight-bit pixels, edges can be positioned to sub-pixel accuracy. An example is shown in Figure 5.39. The pixel sequence is now effectively a digital key signal, and the priority and transparency parameters are now used to reduce the amplitude of a given key when it is behind a symbol on a plane of higher priority which has less than 100 per cent transparency.

The key signals then pass to a device known as the filler which is a fast digital mixer which has as inputs the colour of each character, and the background, whether colour or video. On a pixel-by-pixel basis, the filler cross-fades between different colours as the video line proceeds. The cross-fading is controlled by the key signals. The output is three data streams, red, green and blue. The output board takes the data and converts it to the video standard needed, and the result can then be seen.

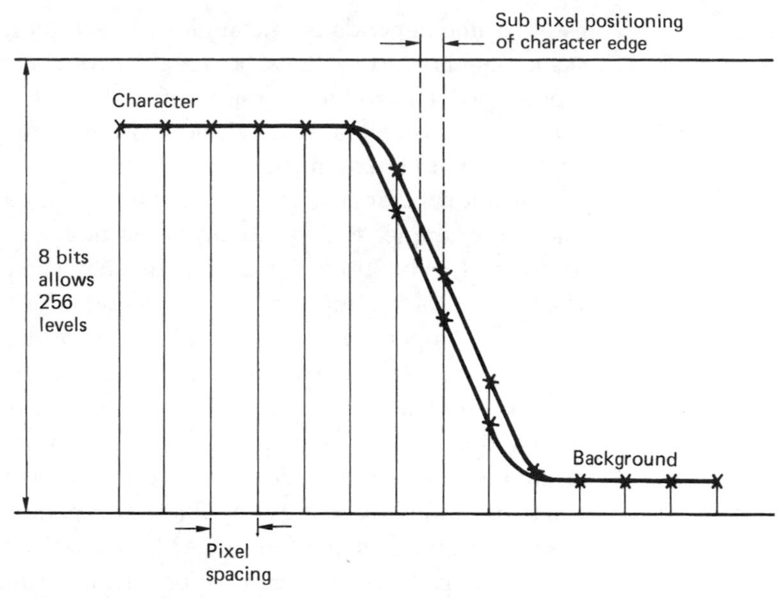

Figure 5.39 The eight-bit resolution of pixels allows the edge of characters to be placed accurately independent of the sampling points. This offers smooth character edges.

5.17 Graphic art/paint systems

In graphic art systems, there is a requirement for disk storage of the generated images, and some art machines incorporate a still store unit, whereas others can be connected to a separate one by an interface. Disk-based stores are discussed in Chapter 12. The essence of an art system is that an artist can draw images which become a video signal directly with no intermediate paper and paint. Central to the operation of most art systems is a digitizing tablet, which is a flat surface over which the operator draws a stylus. The tablet can establish the position of the stylus in vertical and horizontal axes. One way in which this can be done is to launch ultrasonic pulses down the tablet, which are detected by a transducer in the stylus. The time taken to receive the pulse is proportional to the distance to the stylus. The coordinates of the stylus are converted to addresses in the frame store which correspond to the same physical position on the screen. In order to make a simple sketch, the operator specifies a background parameter, perhaps white, which would be loaded into every location in the frame store. A different parameter is then written into every location addressed by movement of the stylus, which results in a line drawing on the screen. The art world uses pens and brushes of different shapes and sizes to obtain a variety of effects, one common example being the rectangular pen nib where the width of the resulting line depends on the angle at which the pen is moved. This can be simulated on art systems, because the address derived from the tablet is processed to produce a range of addresses within a certain screen distance of the stylus. If all these locations are updated as the stylus moves, a broad stroke results.

If the address range is larger in the horizontal axis than in the vertical axis, for example, the width of the stroke will be a function of the direction of stylus travel. Some systems have a sprung tip on the stylus which connects to a force transducer, so that the system can measure the pressure the operator uses. By making the address range a function of pressure, broader strokes can be obtained simply by pressing harder. In order to simulate a set of colours available on a palette, the operator can select a mode where small areas of each colour are displayed in boxes on the monitor screen. The desired colour is selected by moving a screen cursor over the box using the tablet. The parameter to be written into selected locations in the frame RAM now reflects the chosen colour. In more advanced systems, simulation of airbrushing is possible. In this technique, the transparency of the stroke is great at the edge, where the background can be seen showing through, but transparency reduces to the centre of the stroke. A read modify write process is necessary in the frame memory, where background values are read, mixed with paint values with the appropriate transparency, and written back. The position of the stylus effectively determines the centre of a two-dimensional transparency contour, which is convolved with the memory contents as the stylus moves.

5.18 Applications of motion compensation

Section 2.16 introduced the concept of eye tracking and the optic flow axis. The optic flow axis is the locus of some point on a moving object which will be in a different place in successive pictures. Any device which computes with respect to the optic flow axis is said to be *motion compensated*. Until recently the amount of computation required in motion compensation was too expensive, but now this is no longer the case the technology has become very important in moving-image portrayal systems.

Figure 5.40(a) shows an example of a moving object which is in a different place in each of three pictures. The optic flow axis is shown. The object is not moving with respect to the optic flow axis and if this axis can be found some very useful results are obtained. The proces of finding the optic flow axis is called *motion estimation*. Motion estimation is literally a process which analyses successive pictures and determines how objects move from one to the next. It is an important enabling technology because of the way it parallels the action of the human eye.

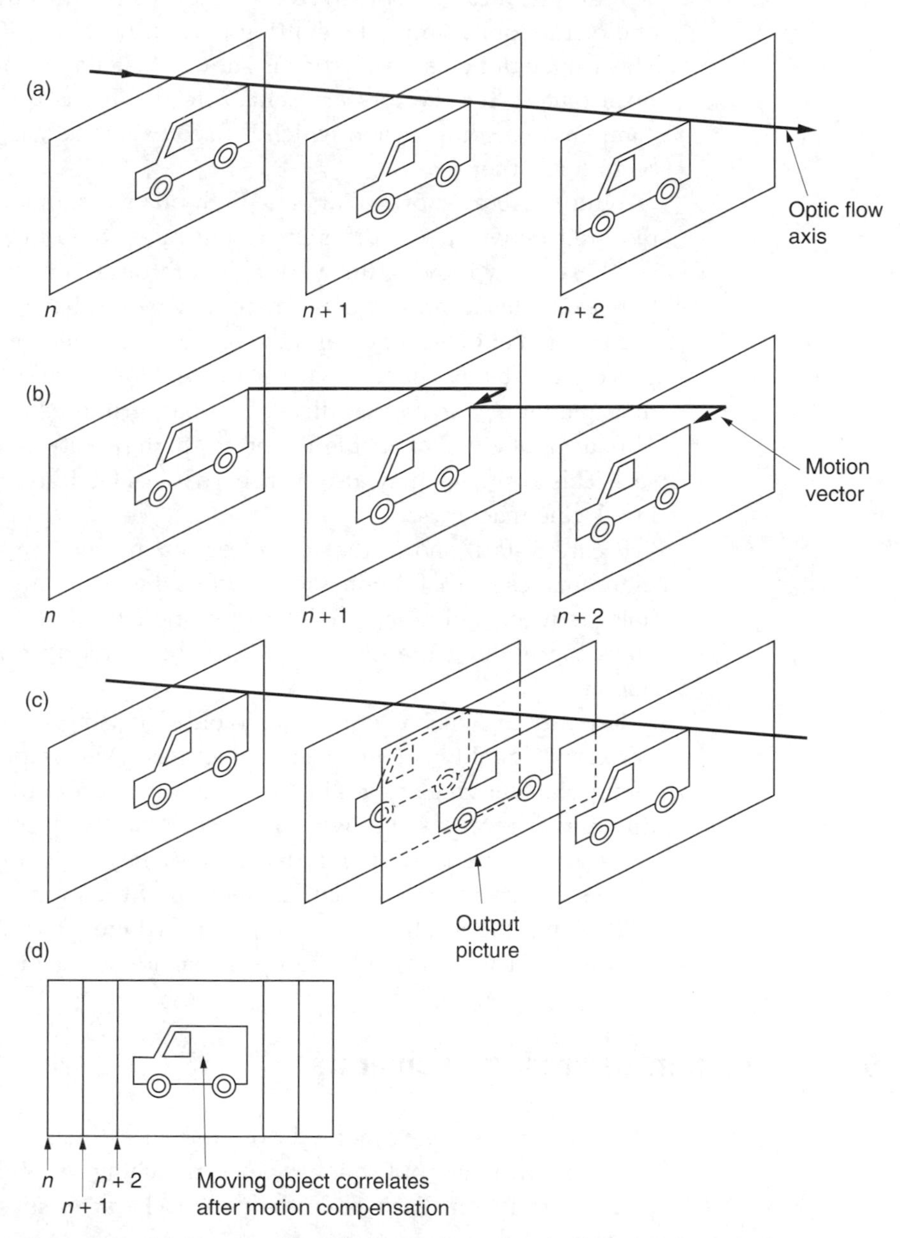

Figure 5.40 Motion compensation is an important technology. (a) The optic flow axis is found for a moving object. (b) The object in picture (n + 1) and (n + 2) can be recreated by shifting the object of picture n using motion vectors. MPEG uses this process for compression. (c) A standards convertor creates a picture on a new timebase by shifting object data along the optic flow axis. (d) With motion compensation a moving object can still correlate from one picture to the next so that noise reduction is possible.

Figure 5.40(b) shows that if the object does not change its appearance as it moves, it can be portrayed in two of the pictures by using data from one picture only, simply by shifting part of the picture to a new location. This can be done using vectors as shown. Instead of transmitting a lot of pixel data, a few vectors are sent instead. This is the basis of motion-compensated compression which is used extensively in MPEG as will be seen in Chapter 6.

Figure 5.40(c) shows that if a high-quality standards conversion is required between two different frame rates, the output frames can be synthesized by moving-image data, not through time but along the optic flow axis. This locates objects where they would have been if frames had been sensed at those times, and the result is a judder-free conversion. This process can be extended to drive image displays at a frame rate higher than the input rate so that flicker and background strobing are reduced. This technology is available in certain high-quality consumer television sets. This approach may also be used with 24 Hz film to eliminate judder in telecine machines.

Figure 5.40(d) shows that noise reduction relies on averaging two or more images so that the images add but the noise cancels. Conventional noise reducers fail in the presence of motion, but if the averaging process takes place along the optic flow axis, noise reduction can continue to operate.

The way in which eye tracking avoides aliasing is fundamental to the perceived quality of television pictures. Many processes need to manipulate moving images in the same way in order to avoid the obvious difficulty of processing with respect to a fixed frame of reference. Processes of this kind are referred to as *motion compensated* and rely on a quite separate process which has measured the motion.

Motion compensation is also important where interlaced video needs to be processed as it allows the best possible de-interlacing performance.

5.19 Motion-estimation techniques

There are three main methods of motion estimation which are to be found in various applications: block matching, gradient matching and phase correlation. Each have their own characteristics which are quite different.

5.19.1 Block matching

This is the simplest technique to follow. In a given picture, a block of pixels is selected and stored as a reference. If the selected block is part of

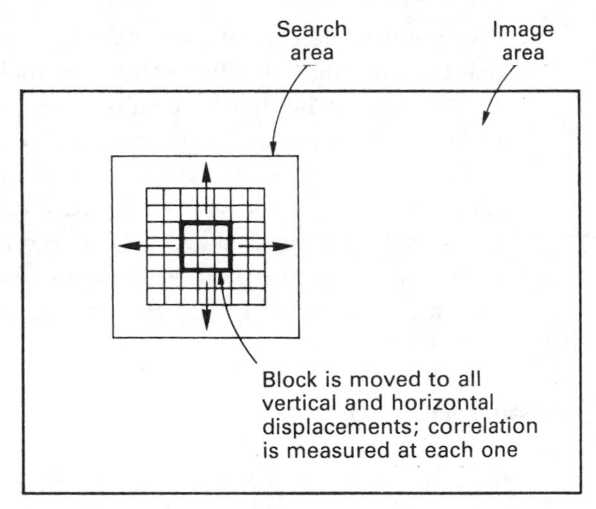

Figure 5.41 In block matching the search block has to be positioned at all possible relative motions within the search area and a correlation measured at each one.

a moving object, a similar block of pixels will exist in the next picture, but not in the same place. As Figure 5.41 shows, block matching simply moves the reference block around over the second picture looking for matching pixel values. When a match is found, the displacement needed to obtain it is used as a basis for a motion vector.

Whilst simple in concept, block matching requires an enormous amount of computation because every possible motion must be tested over the assumed range. Thus if the object is assumed to have moved over a 16-pixel range, then it will be necessary to test sixteen different horizontal displacements in each of sixteen vertical positions; in excess of 65 000 positions. At each position every pixel in the block must be compared with every pixel in the second picture. In typical video displacements of twice the figure quoted here may be found, particularly in sporting events, and the computation then required becomes enormous. If the motion is required to sub-pixel accuracy, then before any matching can be attempted the picture will need to be interpolated, requiring further computation.

One way of reducing the amount of computation is to perform the matching in stages where the first stage is inaccurate but covers a large motion range but the last stage is accurate but covers a small range (4.16). The first matching stage is performed on a heavily filtered and sub-sampled picture, which contains far fewer pixels. When a match is found, the displacement is used as a basis for a second stage which is performed with a less heavily filtered picture. Eventually the last stage takes place to any desired accuracy, including sub-pixel. This hierarchical approach does reduce the computation required, but it suffers from the problem that the filtering of the first stage may make small objects disappear and they can never be found by subsequent stages if they are moving with respect to their background. This is not a problem for compression, since a prediction error will provide the missing detail, but it is an issue for standards convertors which require more accurate motion than compressors. Many televised sports events contain small, fast-moving objects. As the matching process depends upon finding similar luminance values, this can be confused by objects moving into shade or fades.

5.19.2 Gradient matching

At some point in a picture, the function of brightness with respect to distance across the screen will have a certain slope, known as the spatial luminance gradient. If the associated picture area is moving, the slope will traverse a fixed point on the screen and the result will be that the brightness now changes with respect to time. This is a temporal luminance gradient. Figure 5.42 shows the principle. For a given spatial gradient, the temporal gradient becomes steeper as the speed of movement increases. Thus motion speed can be estimated from the ratio of the spatial and temporal gradients.[7]

In practice this is difficult because there are numerous processes which can change the luminance gradient. When an object moves so as to

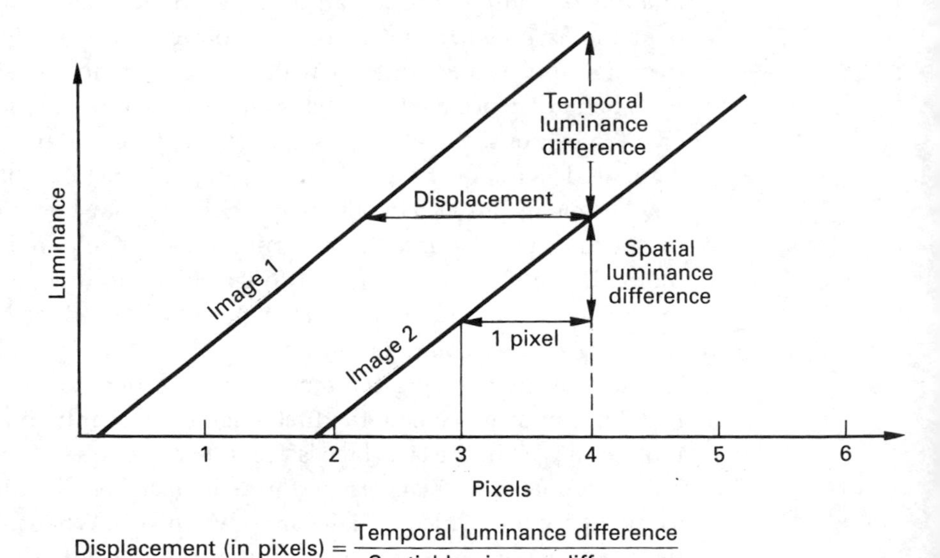

Figure 5.42 The principle of gradient matching. The luminance gradient across the screen is compared with that through time.

obscure or reveal the background, the spatial gradient will change from field to field even if the motion is constant. Variations in illumination, such as when an object moves into shade, also cause difficulty. The process can be assisted by recursion, in which the motion in a current picture is predicted by extrapolating the optic flow axis from earlier pictures, but this will result in problems at cuts.

5.19.3 Phase correlation

Phase correlation works by performing a discrete Fourier transform on two successive fields and then subtracting all the phases of the spectral components. The phase differences are then subject to a reverse transform which directly reveals peaks whose positions correspond to motions between the fields.[8,9] The nature of the transform domain means that if the distance and direction of the motion is measured accurately, the area of the screen in which it took place is not. Thus in practical systems the phase correlation stage is followed by a matching stage not dissimilar to the block-matching process. However, the matching process is steered by the motions from the phase correlation, and so there is no need to attempt to match at all possible motions. By attempting matching on measured motion the overall process is made much more efficient.

One way of considering phase correlation is that by using the Fourier transform to break the picture into its constituent spatial frequencies the hierarchical structure of block matching at various resolutions is in fact performed in parallel. In this way small objects are not missed because they will generate high-frequency components in the transform.

Although the matching process is simplified by adopting phase correlation, the Fourier transforms themselves require complex calculations. The high performance of phase correlation would remain academic if it were too complex to put into practice. However, if realistic values are used for the motion speeds which can be handled, the computation required by block matching actually exceeds that required for phase correlation. The elimination of amplitude information from the phase correlation process ensures that motion-estimation continues to work in the case of fades, objects moving into shade or flashguns firing.

The details of the Fourier transform have been described in section 3.23. A one-dimensional example of phase correlation will be given here by way of introduction. A line of luminance, which in the digital domain consists of a series of samples, is a function of brightness with respect to distance across the screen. The Fourier transform converts this function into a spectrum of spatial frequencies (units of cycles per picture width) and phases.

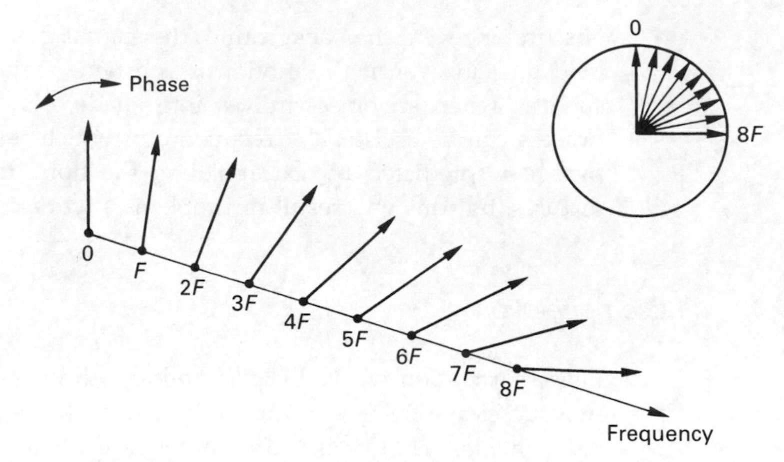

Figure 5.43 The definition of phase linearity is that phase shift is proportional to frequency. In phase-linear systems the waveform is preserved, and simply moves in time or space.

All television signals must be handled in linear-phase systems. A linear-phase system is one in which the delay experienced is the same for all frequencies. If video signals pass through a device which does not exhibit linear phase, the various frequency components of edges become displaced across the screen.

Figure 5.43 shows what phase linearity means. If the left-hand end of the frequency axis (DC) is considered to be firmly anchored, but the right-hand end can be rotated to represent a change of position across the screen, it will be seen that as the axis twists evenly the result is phase shift proportional to frequency. A system having this characteristic is said to display linear phase.

In the spatial domain, a phase shift corresponds to a physical movement. Figure 5.44 shows that if between fields a waveform moves

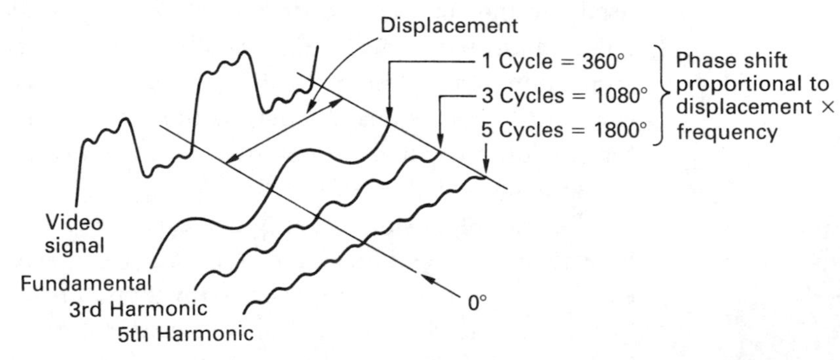

Figure 5.44 In a phase-linear system, shifting the video waveform across the screen causes phase shifts in each component proportional to frequency.

along the line, the lowest frequency in the Fourier transform will suffer a given phase shift, twice that frequency will suffer twice that phase shift and so on. Thus it is potentially possible to measure movement between two successive fields if the phase differences between the Fourier spectra are analysed. This is the basis of phase correlation.

Figure 5.45 shows how a one-dimensional phase correlator works. The Fourier transforms of two lines from successive fields are computed and expressed in polar (amplitude and phase) notation (see section 3.23). The phases of one transform are all subtracted from the phases of the same frequencies in the other transform. Any frequency component having significant amplitude is then normalized, or boosted to full amplitude.

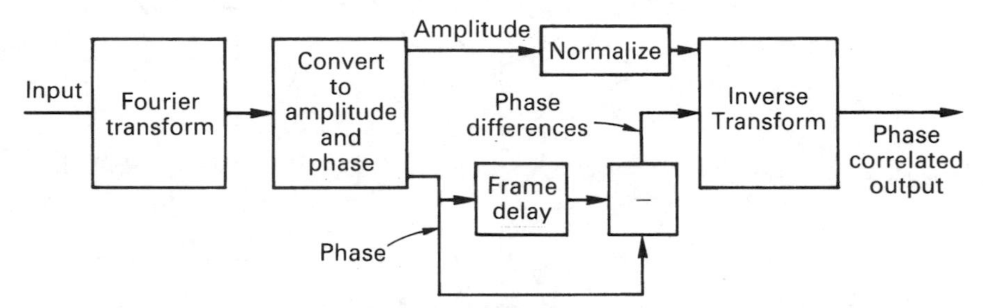

Figure 5.45 The basic components of a phase correlator.

The result is a set of frequency components which all have the same amplitude, but have phases corresponding to the difference between two fields. These coefficients form the input to an inverse transform. Figure 5.46(a) shows what happens. If the two fields are the same, there are no phase differences between the two, and so all the frequency components are added with zero-degree phase to produce a single peak in the centre of the inverse transform. If, however, there was motion between the two fields, such as a pan, all the components will have phase differences, and this results in a peak shown in Figure 5.46(b) which is displaced from the centre of the inverse transform by the distance moved. Phase correlation thus actually measures the movement between fields. In the case where the line of video in question intersects objects moving at different speeds, Figure 5.46(c) shows that the inverse transform would contain one peak corresponding to the distance moved by each object.

Whilst this explanation has used one dimension for simplicity, in practice the entire process is two-dimensional. A two-dimensional Fourier transform of each field is computed, the phases are subtracted, and an inverse two-dimensional transform is computed, the output of which is a flat plane out of which three-dimensional peaks rise. This is known as a correlation surface.

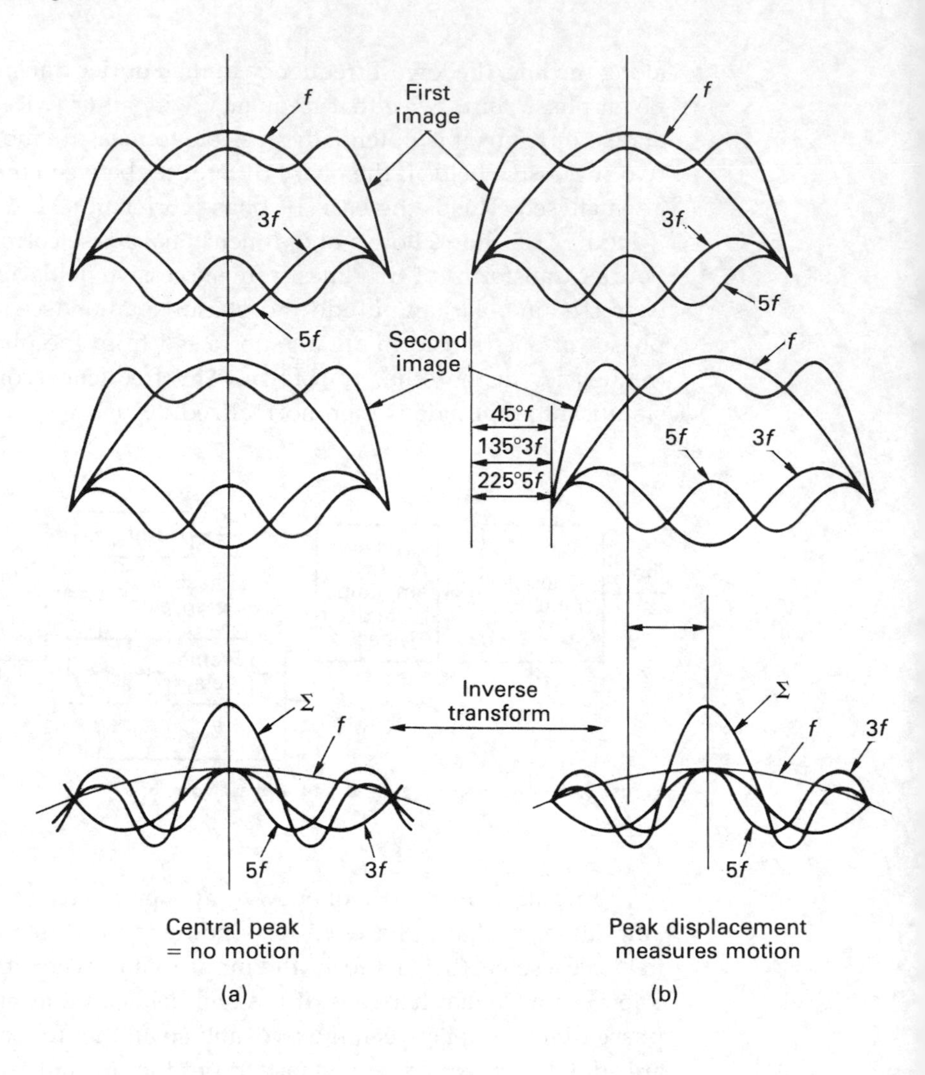

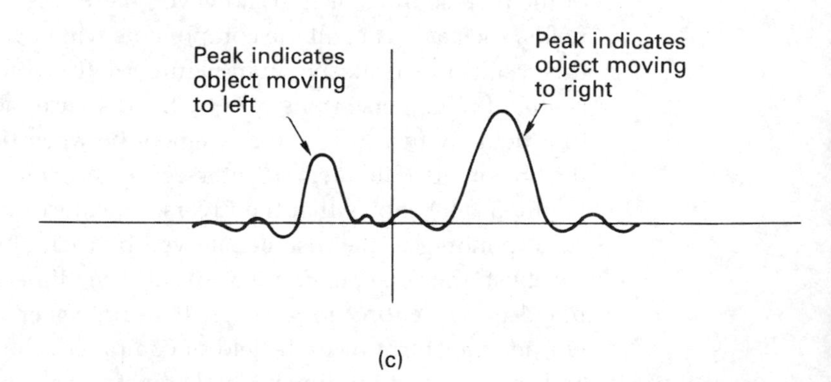

Figure 5.46 (a) The peak in the inverse transform is central for no motion. (b) In the case of motion, the peak shifts by the distance moved. (c) If there are several motions, each one results in a peak.

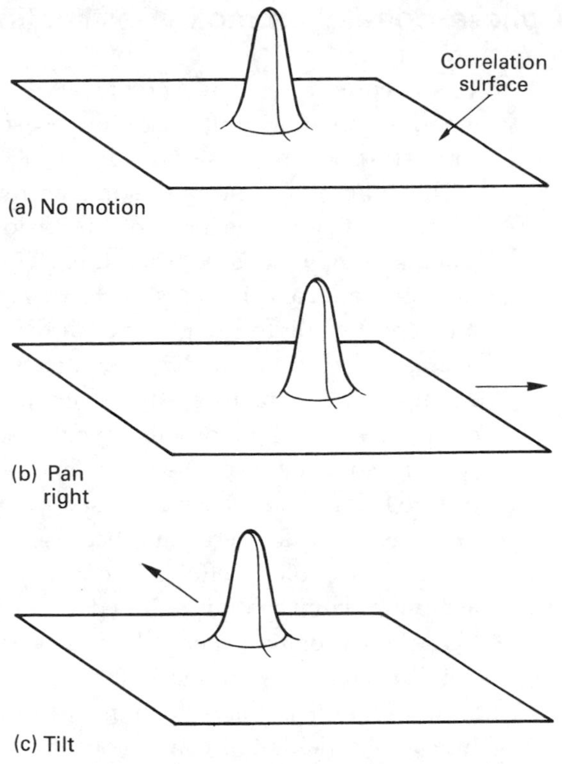

Figure 5.47 (a) A two-dimensional correlation surface has a central peak when there is no motion, (b) In the case of a pan, the peak moves laterally. (c) A camera tilt moves the peak at right angles to the pan.

Figure 5.47 shows some examples of a correlation surface. At (a) there has been no motion between fields and so there is a single central peak. At (b) there has been a pan and the peak moves across the surface. At (c) the camera has been depressed and the peak moves upwards.

Where more complex motions are involved, perhaps with several objects moving in different directions and/or at different speeds, one peak will appear in the correlation surface for each object.

It is a fundamental strength of phase correlation that it actually measures the direction and speed of moving objects rather than estimating, extrapolating or searching for them. The motion can be measured to sub-pixel accuracy. However, it should be understood that according to Heisenberg's uncertainty theorem, accuracy in the transform domain is incompatible with accuracy in the spatial domain. Although phase correlation accurately measures motion speeds and directions, it cannot specify where in the picture these motions are taking place. It is necessary to look for them in a further matching process. The efficiency of this process is dramatically improved by the inputs from the phase-correlation stage.

5.20 A phase-correlation motion-estimation system

This section gives an example of the practical implementation of a motion-estimation unit based on phase correlation. In fact the phase-correlation stage forms only part of the overall structure.

The input to a motion estimator for most applications consists of interlaced fields. The lines of one field lie between those of the next, making comparisons between them difficult. A further problem is that vertical spatial aliasing may exist in the fields. Pre-processing solves these problems by performing a two-dimensional spatial low-pass filtering operation on input fields. Alternate fields are also interpolated up or down by half a line using the techniques of section 4.13 so that interlace disappears and all fields subsequently have the same sampling grid. The spatial frequency response in 625-line systems is filtered to 72 cycles per picture height. This is half the response possible from the number of lines in a field, but is necessary because subsequent correlation causes a frequency-doubling effect. The spatial filtering also cuts down the amount of computation required.

The computation needed to perform a two-dimensional Fourier transform increases dramatically with the size of the block employed, and so no attempt is made to transform the downsampled fields directly. Instead the fields are converted into overlapping blocks by the use of window functions as shown in Figure 5.48. The size of the window controls the motion speed which can be handled, and so a window size is chosen which allows motion to be detected up to the limit of human judder visibility.

Figure 5.49 shows a block diagram of a phase-correlated motion-estimation system. Following the pre-processing, each windowed block is subject to a FFT, and the output spectrum is converted to the amplitude and phase representation. The phases are subtracted from those of the previous field in each window, and the amplitudes are normalized to eliminate any variations in illumination or the effect of fades from the

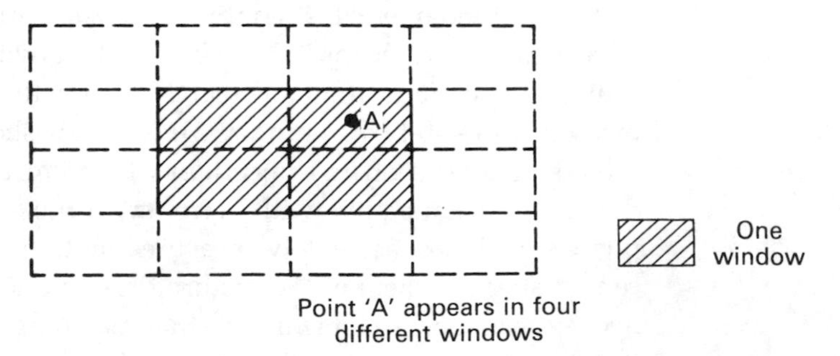

Figure 5.48 The input fields are converted into overlapping windows. Each window is individually transformed.

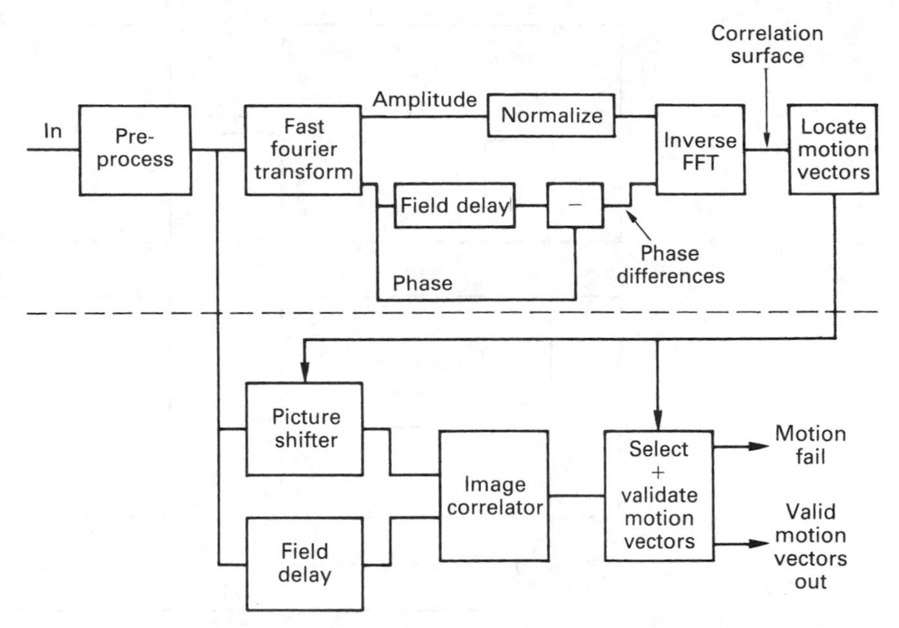

Figure 5.49 The block diagram of a phase-correlated motion estimator. See text for details.

motion sensing. A reverse transform is performed, which results in a correlation surface. The correlation surface contains peaks whose positions actually measure distances and directions moved by some feature in the window.

It is a characteristic of all transforms that the more accurately the spectrum of a signal is known, the less accurately the spatial domain is known. Thus the whereabouts within the window of the moving objects which gave rise to the correlation peaks is not known. Figure 5.50 illustrates the phenomenon. Two windowed blocks are shown in consecutive fields. Both contain the same objects, moving at the same speed, but from different starting points. The correlation surface will be the same in both cases. The phase-correlation process therefore needs to be followed by a further process called image correlation which identifies the picture areas in which the measured motion took place and establishes a level of confidence in the identification. This stage can also be seen in the block diagram of Figure 5.49.

To employ the terminology of motion estimation, the phase-correlation process produces candidate vectors, and the image-correlation process assigns the vectors to specific areas of the picture. In many ways the vector-assignment process is more difficult than the phase-correlation process as the latter is a fixed computation whereas the vector assignment has to respond to infinitely varying picture conditions.

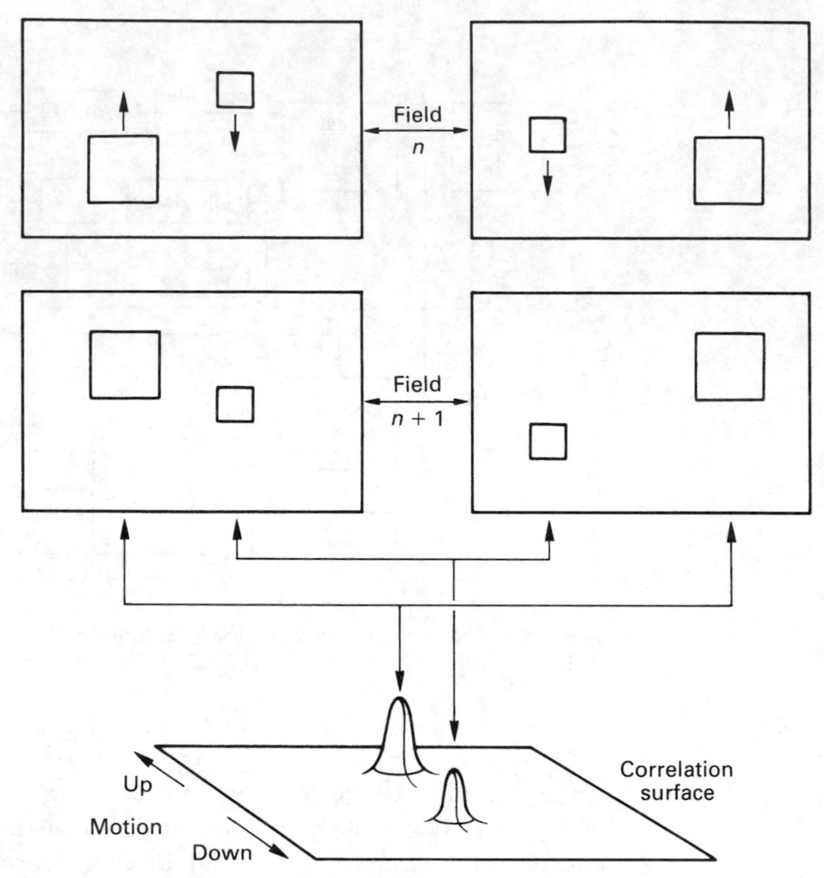

Figure 5.50 Phase correlation measures motion, not the location of moving objects. These two examples give the same correlation surface.

Figure 5.51 shows the image-correlation process which is used to link the candidate vectors from the phase correlator to the picture content. In this example, the correlation surface contains three peaks which define three possible motions between two successive fields. One downsampled field is successively shifted by each of the candidate vectors and compared with the next field a pixel at a time. Similarities or correlations between pixel values indicate that an area with the measured motion has been found. This happens for two of the candidate vectors, and these vectors are then assigned to those areas. However, shifting by the third vector does not result in a meaningful correlation. This is taken to mean that it was a spurious vector; one which was produced in error because of difficult program material. The ability to eliminate spurious vectors and establish confidence levels in those which remain is essential to artifact-free conversion.

The phase-correlation process produces candidate vectors in each window. The vectors from all windows must be combined to obtain an

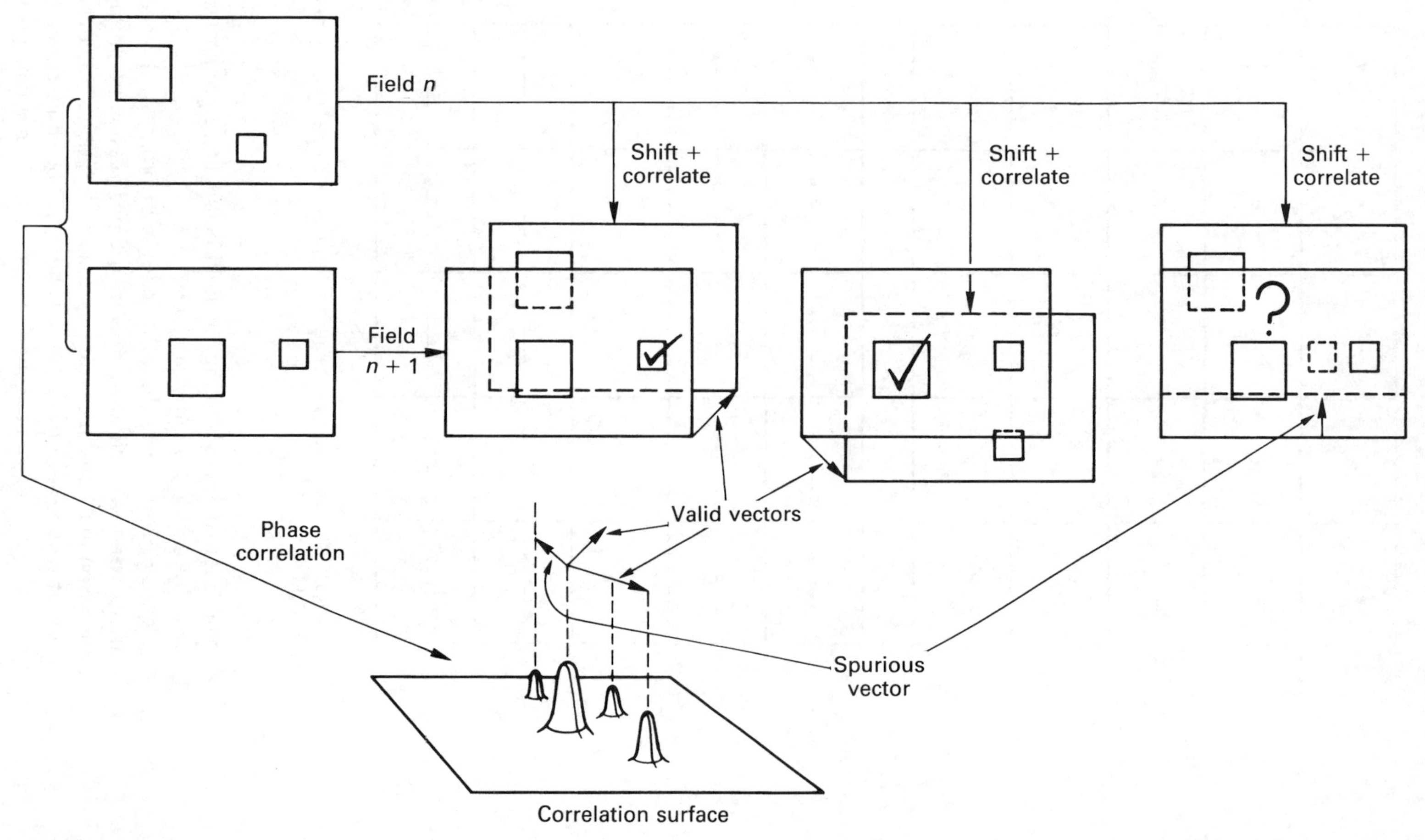

Figure 5.51 Image correlation uses candidate vectors to locate picture areas with the corresponding motion. If no image correlation is found the vector was spurious and is discounted.

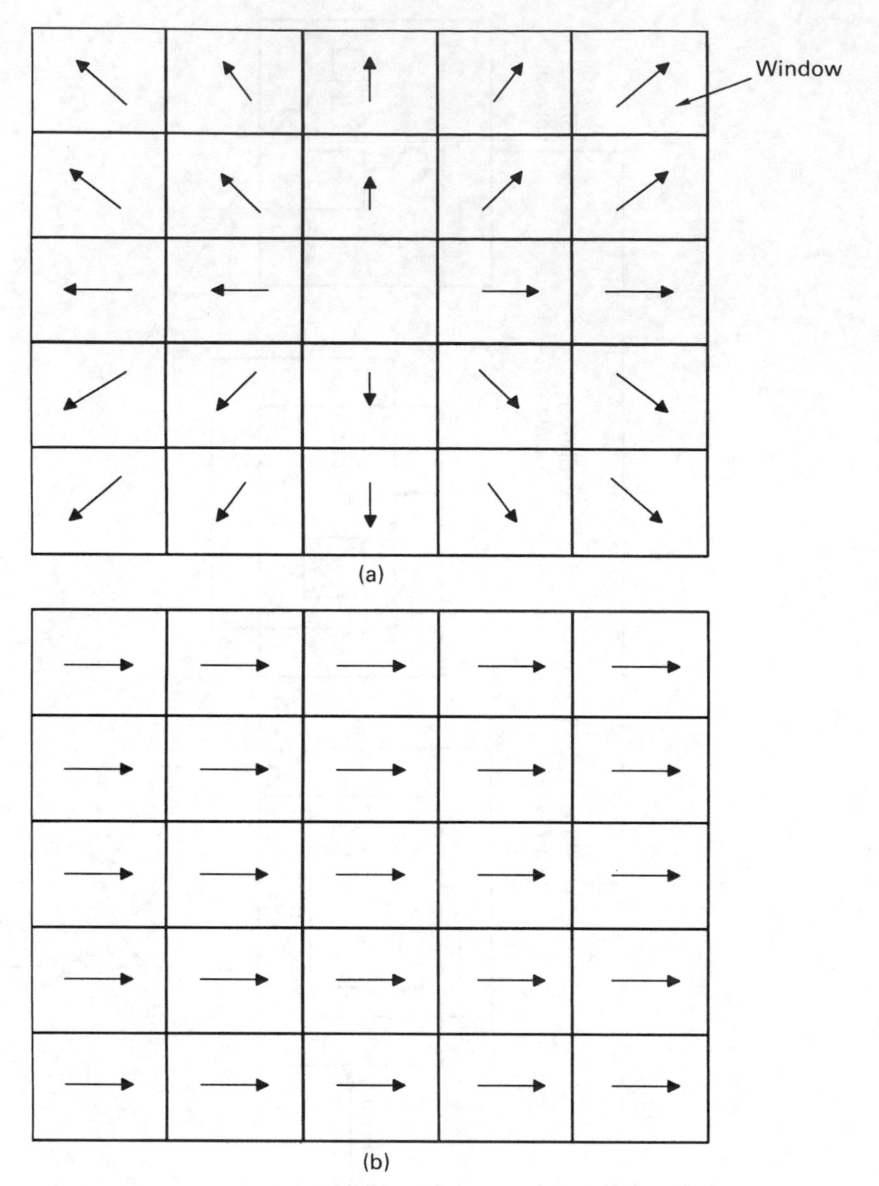

Figure 5.52 The results of (a) a zoom and (b) a pan on the vectors in various windows in the field.

overall view of the motion in the field before attempting to describe the motion of each pixel individually.

Figure 5.52(a) shows that if a zoom is in progress, the vectors in the various windows will form a geometric progression becoming longer in proportion to the distance from the axis of the zoom. However, if there is a pan, it will be seen from (b) that there will be similar vectors in all the windows. In practice both motions may occur simultaneously.

An estimate will be made of the speed of a field-wide zoom, or of the speed of picture areas which contain receding or advancing motions which give a zoom-like effect. If the effect of zooming is removed from each window by shifting the peaks by the local zoom magnitude, but in the opposite direction, the position of the peaks will reveal any component due to panning. This can be found by summing all the windows to create a histogram. Panning results in a dominant peak in the histogram where all windows contain peaks in a similar place which reinforce.

Each window is then processed in turn. Where only a small part of an object overlaps into a window, it will result in a small peak in the correlation surface which might be missed. The windows are deliberately overlapped so that a given pixel may appear in four windows. Thus a moving object will appear in more than one window. If the majority of an object lies within one window, a large peak will be produced from the motion in that window. The resulting vector will be added to the candidate vector list of all adjacent windows. When the vector assignment is performed, image correlations will result if a small overlap occurred, and the vector will be validated. If there was no overlap, the vector will be rejected.

The peaks in each window reflect the degree of correlation between the two fields for different offsets in two dimensions. The volume of the peak corresponds to the amount of the area of the window (i.e. the number of pixels) having that motion. Thus peaks should be selected starting with the largest. However, periodic structures in the input field, such as grilles and striped shirts will result in partial correlations at incorrect distances which differ from the correct distance by the period of the structure. The effect is that a large peak on the correlation surface will be flanked by smaller peaks at uniform spacing in a straight line. The characteristic pattern of subsidiary peaks can be identified and the vectors invalidated.

One way in which this can be done is to compare the position of the peaks in each window with those estimated by the pan/zoom process. The true peak due to motion will be similar; the false sub-peaks due to image periodicity will not be and can be rejected.

Correlations with candidate vectors are then performed. The image in one field is shifted in an interpolator by the amount specified by a candidate vector and the degree of correlation is measured. Note that this interpolation is to sub-pixel accuracy because phase correlation can accurately measure sub-pixel motion. High correlation results in vector assignment, low correlation results in the vector being rejected as unreliable.

If all the peaks are evaluated in this way, then most of the time valid assignments will be made for which there is acceptable confidence from the correlator. Should it not be possible to obtain any correlation with confidence in a window, then the pan/zoom values will be inserted so that that the window moves in a similar way to the overall field motion.

5.21 Motion-compensated standards conversion

A conventional standards convertor is not transparent to motion portrayal, and the effect is judder and loss of resolution. Figure 5.53 shows what happens on the time axis in a conversion between 60 Hz and 50 Hz (in either direction). Fields in the two standards appear in different planes cutting through the spatio-temporal volume, and the job of the standards convertor is to interpolate along the time axis between input planes in one standard in order to estimate what an intermediate plane in the other standard would look like. With still images, this is easy, because planes can be slid up and down the time axis with no ill effect. If an object is moving, it will be in a different place in successive fields. Interpolating between several fields results in multiple images of the object. The position of the dominant image will not move smoothly, an effect which is perceived as judder. Motion compensation is designed to eliminate this undesirable judder.

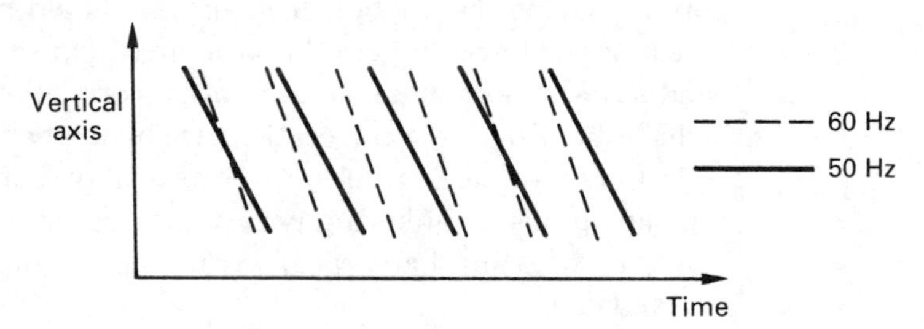

Figure 5.53 The different temporal distribution of input and output fields in a 50/60 Hz convertor.

A conventional standards convertor interpolates only along the time axis, whereas a motion-compensated standards convertor can swivel its interpolation axis off the time axis. Figure 5.54(a) shows the input fields in which three objects are moving in a different way. At (b) it will be seen that the interpolation axis is aligned with the optic flow axis of each moving object in turn.

Each object is no longer moving with respect to its own optic flow axis, and so on that axis it no longer generates temporal frequencies due to motion and temporal aliasing due to motion cannot occur.[10] Interpolation

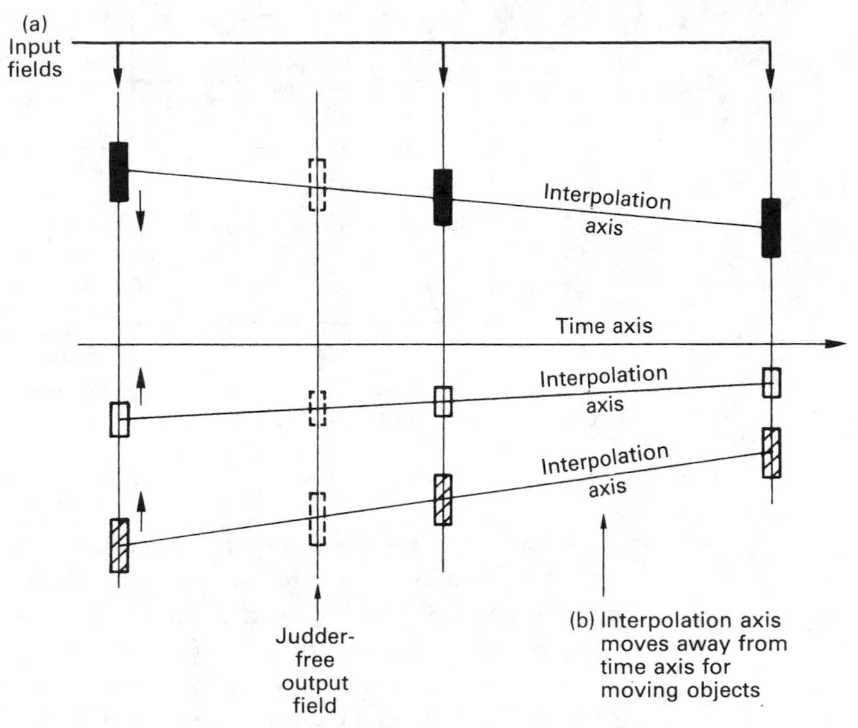

Figure 5.54 (a) Input fields with moving objects. (b) Moving the interpolation axes to make them parallel to the trajectory of each object.

along the optic flow axes will then result in a sequence of output fields in which motion is properly portrayed. The process requires a standards convertor which contains filters that are modified to allow the interpolation axis to move dynamically within each output field. The signals which move the interpolation axis are known as motion vectors. It is the job of the motion-estimation system to provide these motion vectors. The overall performance of the convertor is determined primarily by the accuracy of the motion vectors. An incorrect vector will result in unrelated pixels from several fields being superimposed and the result is unsatisfactory.

Figure 5.55 shows the sequence of events in a motion-compensated standards convertor. The motion estimator measures movements between successive fields. These motions must then be attributed to objects by creating boundaries around sets of pixels having the same motion. The result of this process is a set of motion vectors, hence the term vector assignation. The motion vectors are then input to a modified four-field standards convertor in order to deflect the inter-field interpolation axis.

The vectors from the motion estimator actually measure the distance moved by an object from one input field to another. What the standards

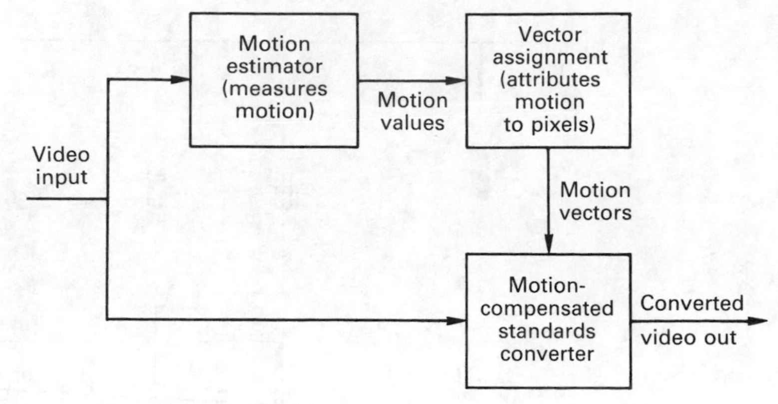

Figure 5.55 The essential stages of a motion-compensated standards convertor.

convertor requires is the value of motion vectors at an output field. A vector interpolation stage is required which computes where between the input fields A and B the current output field lies, and uses this to proportion the motion vector into two parts. Figure 5.56(a) shows that the first part is the motion between field A and the output field; the second is the motion between field B and the output field. Clearly the difference between these two vectors is the motion between input fields. These processed vectors are used to displace parts of the input fields so that the axis of interpolation lies along the optic flow axis. The moving object is stationary with respect to this axis so interpolation between fields along it will not result in any judder.

Whilst a conventional convertor only needs to interpolate vertically and temporally, a motion-compensated convertor also needs to interpolate horizontally to account for lateral movement in images. Figure 5.56(b) shows that the motion vector from the motion estimator is resolved into two components, vertical and horizontal. The spatial impulse response of the interpolator is shifted in two dimensions by these components. This shift may be different in each of the fields which contribute to the output field.

When an object in the picture moves, it will obscure its background. The vector interpolator in the standards convertor handles this automatically provided the motion estimation has produced correct vectors. Figure 5.57 shows an example of background handling. The moving object produces a finite vector associated with each pixel, whereas the stationary background produces zero vectors except in the area O–X where the background is being obscured. Vectors converge in the area where the background is being obscured, and diverge where it is being revealed. Image correlation is poor in these areas so no valid vector is assigned.

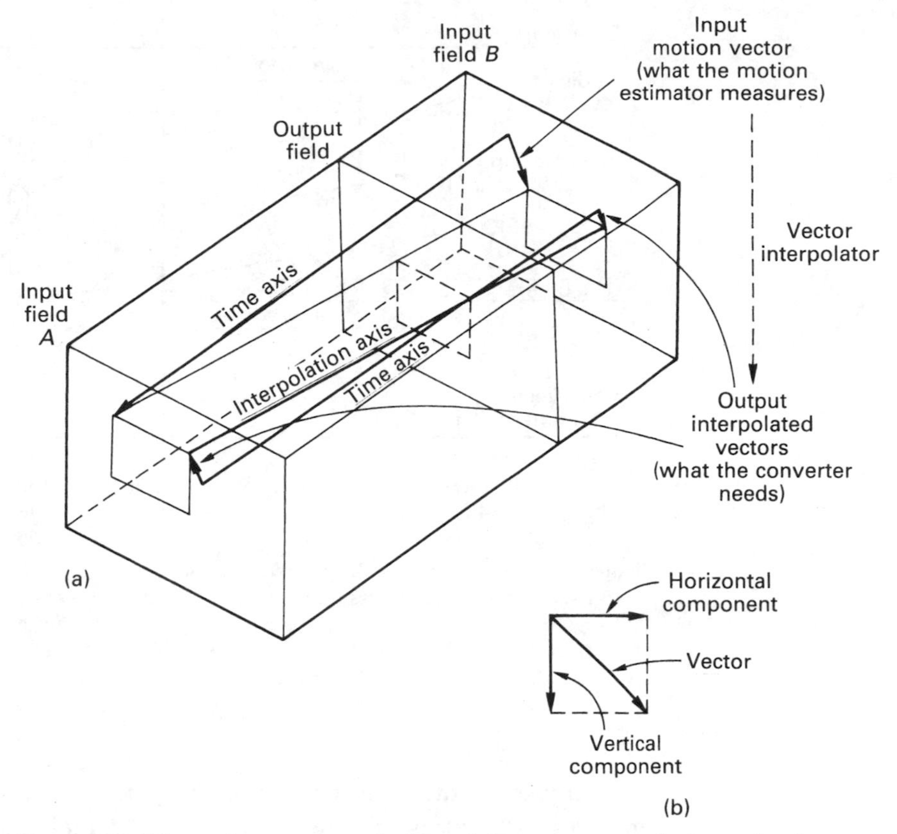

Figure 5.56 The motion vectors on the input field structure must be interpolated onto the output field structure as in (a). The field to be interpolated is positioned temporally between source fields and the motion vector between them is apportioned according to the location. Motion vectors are two dimensional, and can be transmitted as vertical and horizontal components shown at (b) which control the spatial shifting of input fields.

An output field is located between input fields, and vectors are projected through it to locate the intermediate position of moving objects. These are interpolated along an axis which is parallel to the optic flow axis. This results in address mapping which locates the moving object in the input field RAMs. However, the background is not moving and so the optic flow axis is parallel to the time axis. The pixel immediately below the leading edge of the moving object does not have a valid vector because it is in the area O–X where forward image correlation failed.

The solution is for that pixel to assume the motion vector of the background below point X, but only to interpolate in a backwards direction, taking pixel data from previous fields. In a similar way, the pixel immediately behind the trailing edge takes the motion vector for the background above point Y and interpolates only in a forward direction, taking pixel data from future fields. The result is that the moving object

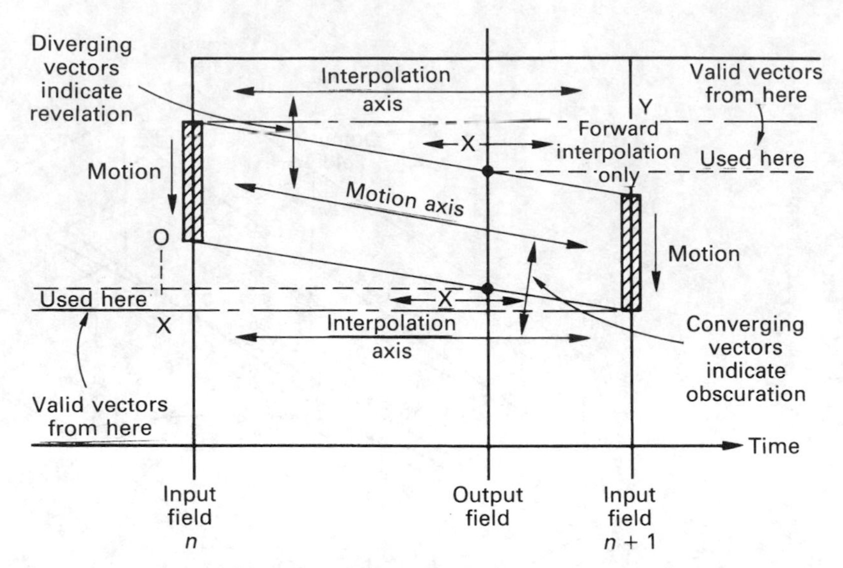

Figure 5.57 Background handling. When a vector for an output pixel near a moving object is not known, the vectors from adjacent background areas are assumed. Converging vectors imply obscuring is taking place which requires that interpolation can only use previous field data. Diverging vectors imply that the background is being revealed and interpolation can only use data from later fields.

is portrayed in the correct place on its trajectory, and the background around it is filled in only from fields which contain useful data.

The technology of the motion-compensated standards convertor can be used in other applications. When video recordings are played back in slow motion, the result is that the same picture is displayed several times, followed by a jump to the next picture. Figure 5.58 shows that a moving object would remain in the same place on the screen during picture repeats, but jump to a new position as a new picture was played. The eye attempts to track the moving object, but, as Figure 5.58 also shows, the location of the moving object wanders with respect to the trajectory of the eye, and this is visible as judder.

Motion-compensated slow-motion systems are capable of synthesizing new images which lie between the original images from a slow-motion source. Figure 5.59 shows that two successive images in the original recording (using DVE terminology, these are source fields) are fed into the unit, which then measures the distance travelled by all moving objects between those images. Using interpolation, intermediate fields (target fields) are computed in which moving objects are positioned so that they lie on the eye trajectory. Using the principles described above, background information is removed as moving objects conceal it, and replaced as the rear of an object reveals it. Judder is thus removed and motion with a fluid quality is obtained.

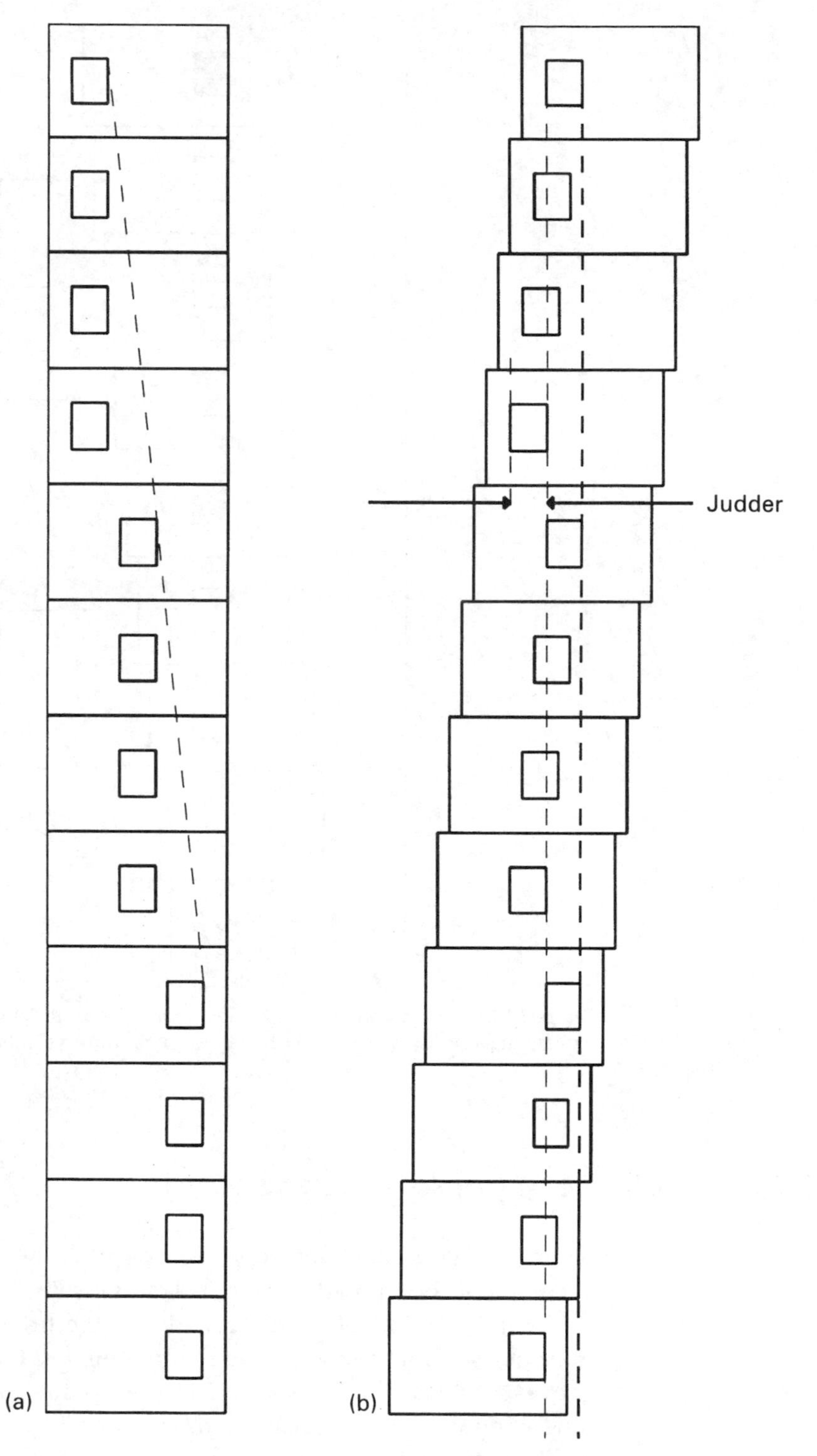

Figure 5.58 Conventional slow motion using field repeating with stationary eye shown at (a). With tracking eye at (b) the source of judder is seen.

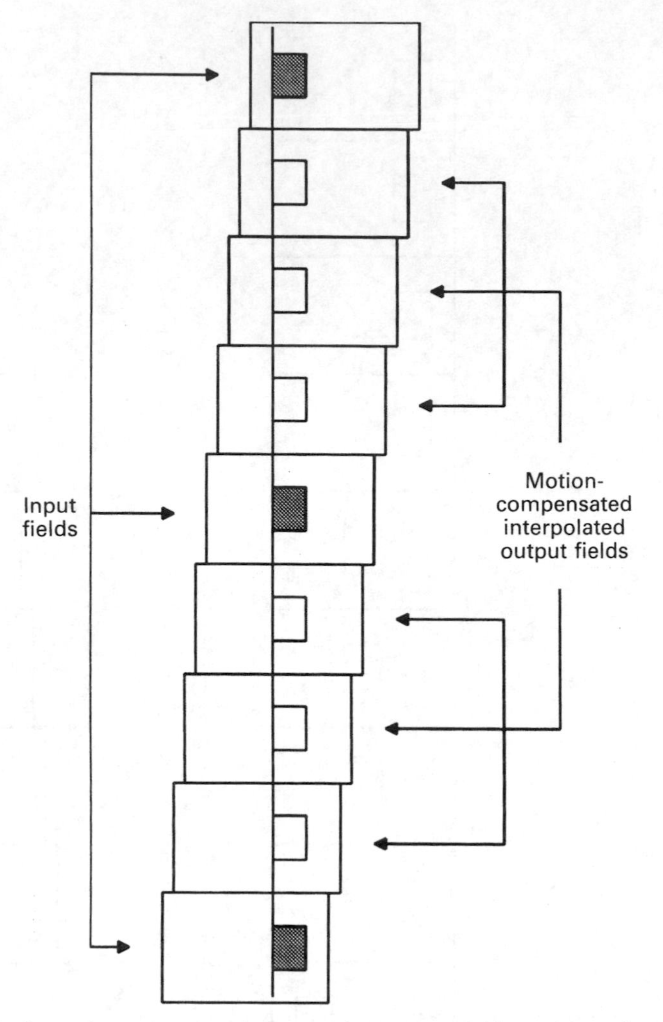

Figure 5.59 In motion-compensated slow motion, output fields are interpolated with moving objects displaying judder-free linear motion between input fields.

5.22 Motion-compensated telecine system

Figure 5.60(a) shows the time axis of film, where entire frames are simultaneously exposed, or sampled, at typically 24 Hz. The result is that the image is effectively at right angles to the time axis. During filming, some of the frame period is required to transport the film, and the shutter is closed whilst this takes place. The temporal aperture or exposure is thus somewhat shorter than the frame period.

When displayed in the cinema, each frame of a film is generally projected twice to produce a flicker frequency of 48 Hz. The result with a

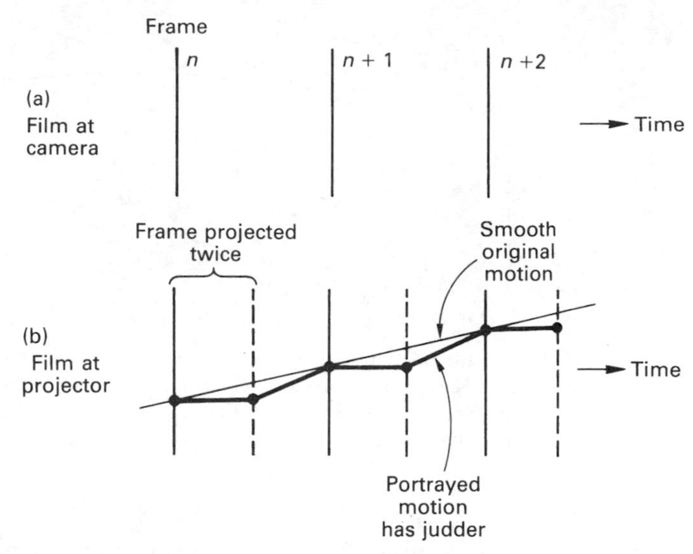

Figure 5.60 (a) The spatio-temporal characteristic of film. Note that each frame is repeated twice on projection. (b) The frame repeating results in motion judder as shown here.

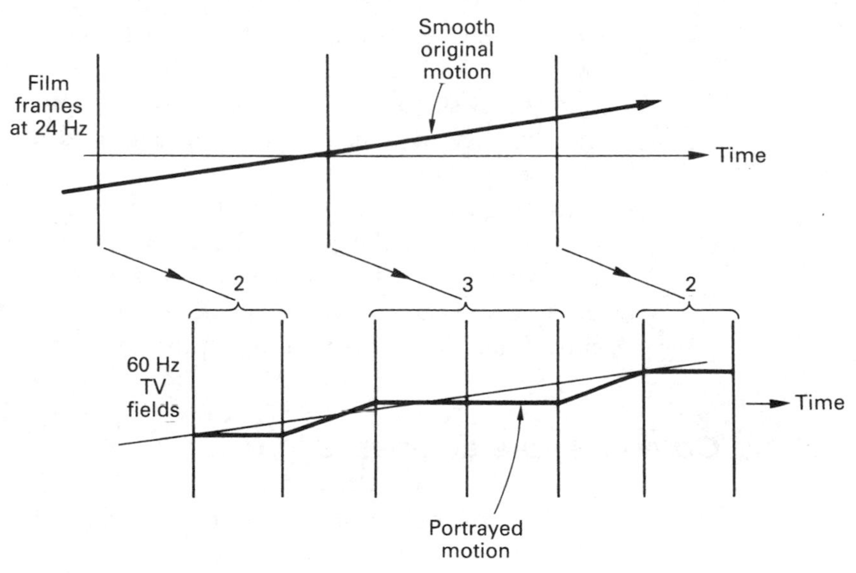

Figure 5.60(c) Telecine machines must use 3:2 pulldown to produce 60 Hz field rate video.

moving object is that the motion is not properly portrayed and there is judder. Figure 5.60(b) shows the origin of the judder.

The same effect is evident if film is displayed on a CRT via a conventional telecine machine. In telecine the film is transported at 25 fps and each frame results in two fields in 50 Hz standards and this will result in judder as well. In 60 Hz telecine the film travels at 24 fps, but odd

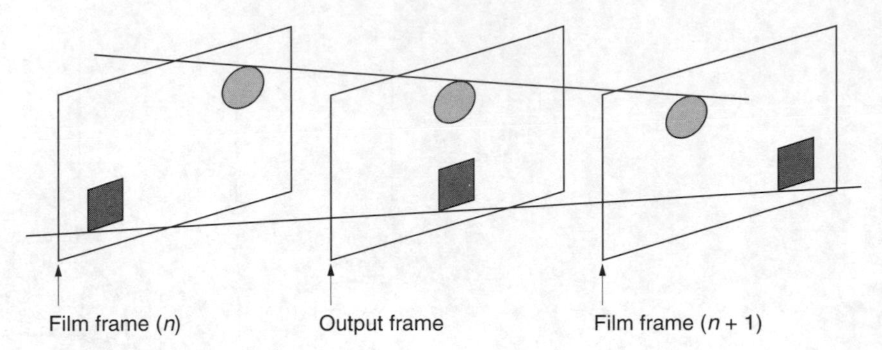

Figure 5.61 A film with a frame rate of 24 Hz cannot be displayed directly because of flicker. Using a motion-compensated standards conversion process extra frames can be synthesized in which moving objects are correctly positioned. Any television picture rate can then be obtained from film.

frames result in three fields, even frames result in two fields; the well known 3/2 pulldown. Motion portrayal (or lack of it) in this case is shown in Figure 5.60(c).

In fact the telecine machine is a perfect application for motion compensation. As Figure 5.61 shows, each film frame is converted to a progressive scan image in a telecine machine, and then a motion-compensated standards conversion process is used to output whatever frame rate is required without judder, leading to much-improved subjective quality.

If the original film is not available, 50 and 60 Hz video recording can be used. In the case of 50 Hz, pairs of fields are combined to produce progressively scanned frames. In the case of 60 Hz, the third field in the 3/2 sequence is identified and discarded, prior to de-interlacing. Motion-compensated rate conversion then proceeds as before.

5.23 Camera shake compensation

As video cameras become smaller and lighter, it becomes increasingly difficult to move them smoothly and the result is camera shake. This is irritating to watch, as well as requiring a higher bit rate in compression systems. There are two solutions to the problem, one which is contained within the camera, and one which can be used at some later time on the video data.

Figure 5.62 shows that image-stabilizing cameras contain miniature gyroscopes which produce an electrical output proportional to their rate of turn about a specified axis. A pair of these, mounted orthogonally, can produce vectors describing the camera shake. This can be used to oppose the shake by shifting the image. In one approach, the shifting is done

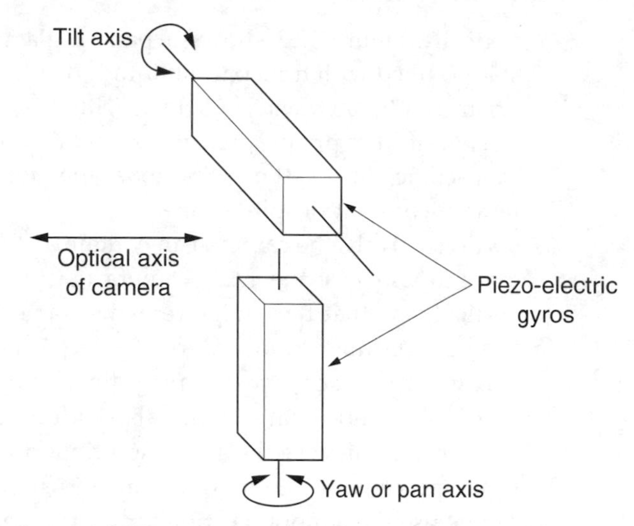

Figure 5.62 Image-stabilizing cameras sense shake using a pair of orthogonal gyros which sense movement of the optical axis.

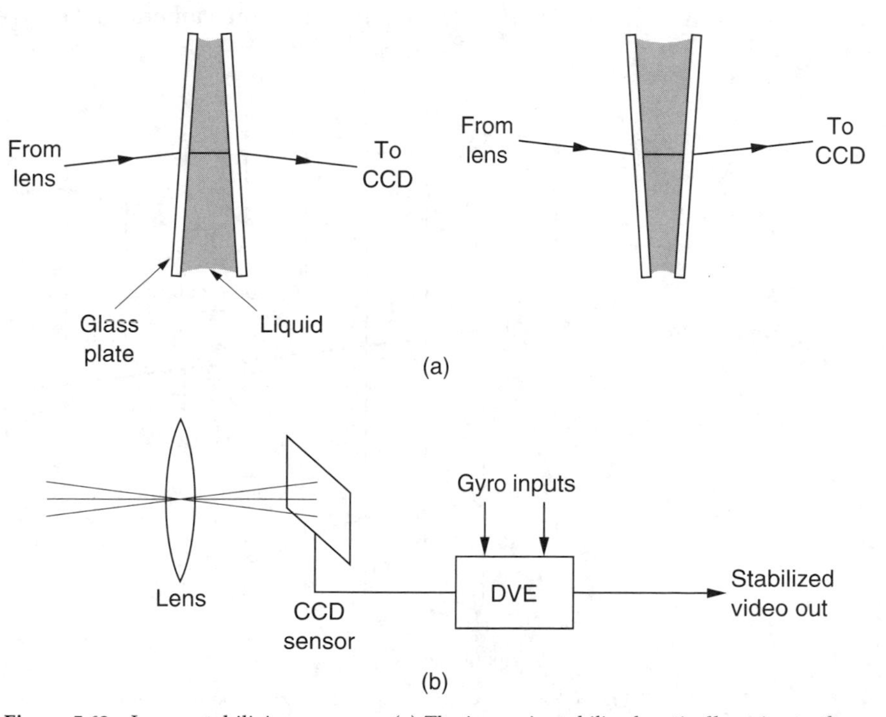

Figure 5.63 Image-stabilizing cameras. (a) The image is stabilized optically prior to the CCD sensors. (b) The CCD output contains image shake, but this is opposed by the action of a DVE configured to shift the image under control of the gyro inputs.

optically. Figure 5.63 shows a pair of glass plates with the intervening spaced filled with transparent liquid. By tilting the plates a variable-angle prism can be obtained and this is fitted in the optical system before the sensor. If the prism plates are suitably driven by servos from the gyroscopic sensors, the optical axis along which the camera is looking can remain constant despite shake.

Alternatively, the camera can contain a DVE where the vectors from the gyroscopes cause the CCD camera output to be shifted horizontally or vertically so that the image remains stable. This approach is commonly used in consumer camcorders.

A great number of video recordings and films already exist in which there is camera shake. Film also suffers from weave in the telecine machine. In this case the above solutions are inappropriate and a suitable signal processor is required. Figure 5.64 shows that motion compensation can be used. If a motion estimator is arranged to find the motion between a series of pictures, camera shake will add a fixed component in each picture to the genuine object motions. This can be used to compute the optic flow axis of the camera, independently of the objects portrayed.

Operating over several pictures, the trend in camera movement can be separated from the shake by filtering, to produce a position error for each picture. Each picture is then shifted in a DVE in order to cancel the position error. The result is that the camera shake is gone and the camera

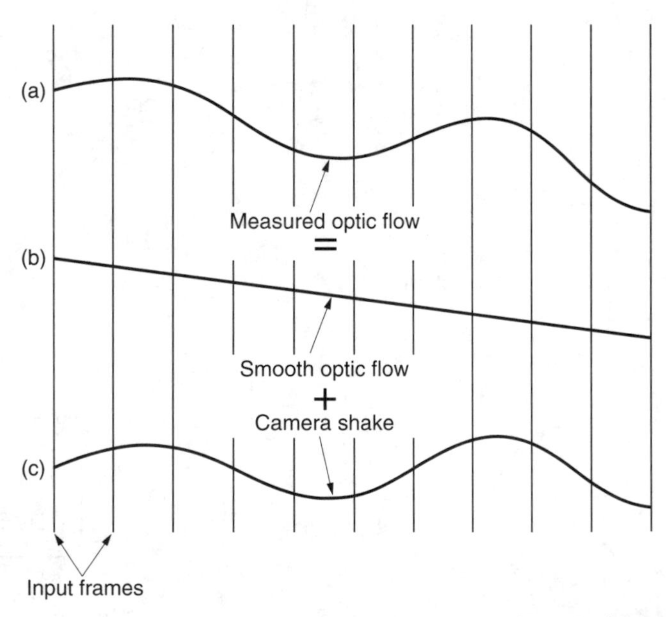

Figure 5.64 In digital image stabilizing the optic flow axis of objects in the input video is measured as in (a). This motion is smoothed to obtain a close approximation to the original motion (b). If this is subtracted from (a) the result is the camera shake motion which is used to drive the image stabilizer.

movements appear smooth. In order to prevent the edges of the frame moving visibly, the DVE also performs a slight magnification so that the edge motion is always outside the output frame.

5.24 De-interlacing

Interlace is a compression technique which sends only half of the picture lines in each field. Whilst this works reasonably well for transmission, it causes difficulty in any process which requires image manipulation. This includes DVEs, standards convertors and display convertors. All these devices give better results when working with progressively scanned data and if the source material is interlaced, a de-interlacing process will be necessary.

Interlace distributes vertical detail information over two fields and for maximum resolution all that information is necessary. Unfortunately it is not possible to use the information from two different fields directly. Figure 5.65 shows a scene in which an object is moving. When the second field of the scene leaves the camera, the object will have assumed a different position from the one it had in the first field, and the result of combining the two fields to make a de-interlaced frame will be a double image. This effect can easily be demonstrated on any video recorder which offers a choice of still field or still frame. Stationary objects before a stationary camera, however, can be de-interlaced perfectly.

In simple de-interlacers, motion sensing is used so that de-interlacing can be disabled when movement occurs, and interpolation from a single field is used instead. Motion sensing implies comparison of one picture with the next. If interpolation is only to be used in areas where there is movement, it is necessary to test for motion over the entire frame. Motion can be simply detected by comparing the luminance value of a given pixel with the value of the same pixel two fields earlier. As two fields are to be combined, and motion can occur in either, then the comparison must be made between two odd fields and two even fields. Thus four fields of memory are needed to correctly perform motion sensing. The luminance from four fields requires about a megabyte of storage.

At some point a decision must be made to abandon pixels from the previous field which are in the wrong place due to motion, and to interpolate them from adjacent lines in the current field. Switching suddenly in this way is visible, and there is a more sophisticated mechanism which can be used. In Figure 5.66, two fields are shown, separated in time. Interlace can be seen by following lines from pixels in one field, which pass between pixels in the other field. If there is no movement, the fact that the two fields are separated in time is irrelevant, and the two can be superimposed to make a frame array. When there is

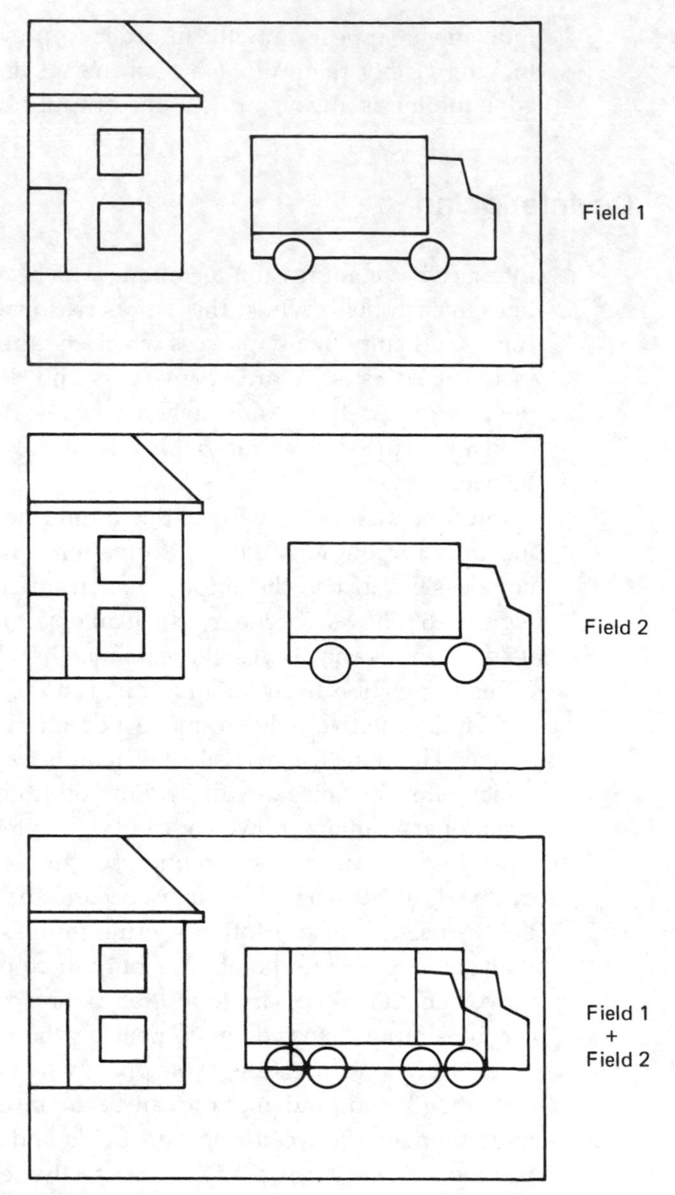

Figure 5.65 Moving object will be in a different place in two successive fields and will produce a double image.

motion, pixels from above and below the unknown pixels are added together and divided by two, to produce interpolated values. If both of these mechanisms work all the time, a better-quality picture results if a cross-fade is made between the two based on the amount of motion. A suitable digital cross-fader was shown in Chapter 3. At some motion value, or some magnitude of pixel difference, the loss of resolution due to

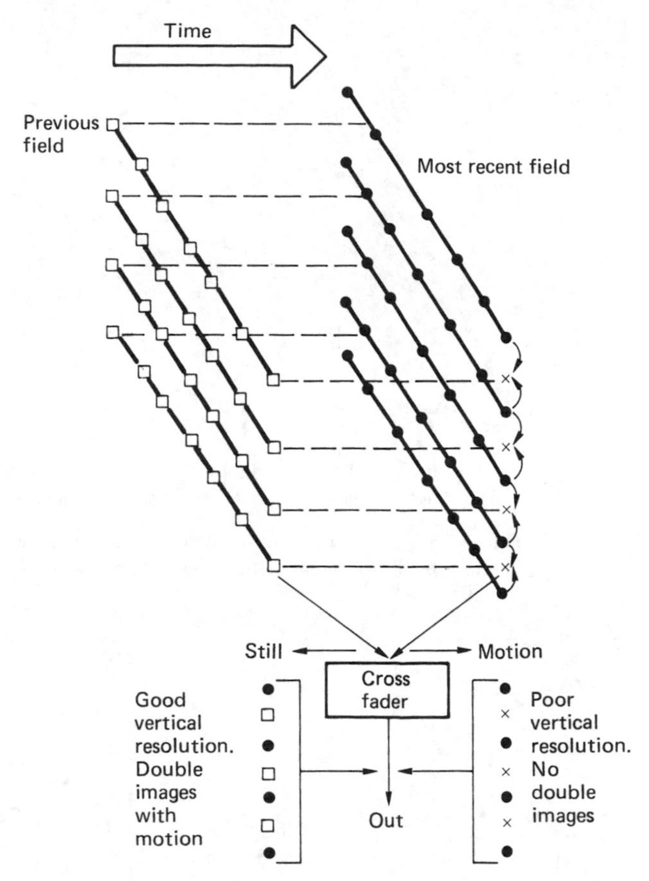

Figure 5.66 Pixels from the most recent field (●) are interpolated spatially to form low vertical resolution pixels (×) which will be used if there is excessive motion; pixels from the previous field (□) will be used to give maximum vertical resolution. The best possible de-interlaced frame results.

a double image is equal to the loss of resolution due to interpolation. That amount of motion should result in the cross-fader arriving at a 50/50 setting. Any less motion will result in a fade towards both fields, any more motion resulting in a fade towards the interpolated values.

The most efficient way of de-interlacing is to use motion compensation. Figure 5.67 shows that when an object moves in an interlaced system, the interlace breaks down with respect to the optic flow axis as was seen in section 2.17. If the motion is known, two or more fields can be shifted so that a moving object is in the same place in both. Pixels from both field can then be used to describe the object with better resolution than would be possible from one field alone. It will be seen from Figure 5.68 that the combination of two fields in this way will result in pixels having a highly irregular spacing and a special type of filter is needed to convert this back to a progressive frame with regular pixel spacing. At some critical vertical

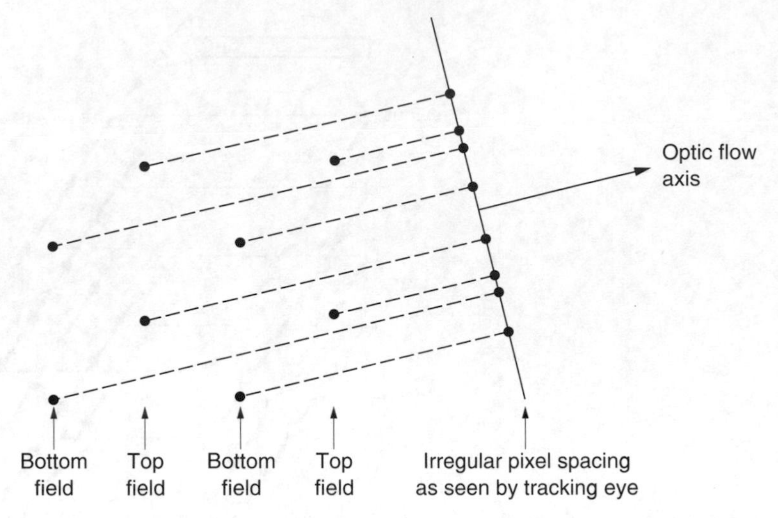

Figure 5.67 In the presence of vertical motion or motion having a vertical component, interlace breaks down and the pixel spacing with respect to the tracking eye becomes irregular.

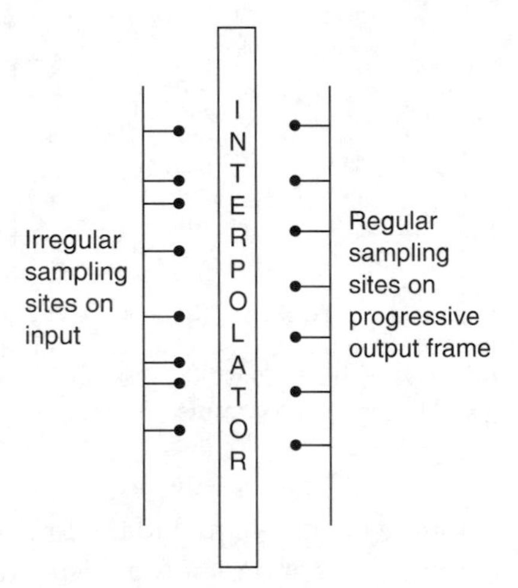

Figure 5.68 A de-interlacer needs an interpolator which can operate with input samples which are positioned arbitrarily rather than regularly.

speeds there will be alignment between pixels in adjacent fields and no improvement is possible, but at other speeds the process will always give better results.

5.25 Noise reduction

The basic principle of all video noise reducers is that there is a certain amount of correlation between the video content of successive frames, whereas there is no correlation between the noise content.

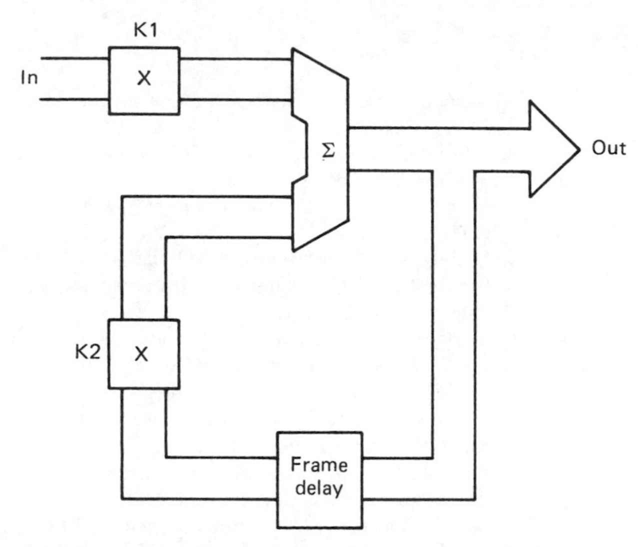

Figure 5.69 A basic recursive device feeds back the output to the input via a frame store which acts as a delay. The characteristics of the device are controlled totally by the values of the two coefficients K1 and K2 which control the multipliers.

A basic recursive device is shown in Figure 5.69. There is a frame store which acts as a delay, and the output of the delay can be fed back to the input through an attenuator, which in the digital domain will be a multiplier. In the case of a still picture, successive frames will be identical, and the recursion will be large. This means that the output video will actually be the average of many frames. If there is movement of the image, it will be necessary to reduce the amount of recursion to prevent the generation of trails or smears. Probably the most famous examples of recursion smear are the television pictures sent back of astronauts walking on the moon. The received pictures were very noisy and needed a lot of averaging to make them viewable. This was fine until the astronaut moved. The technology of the day did not permit motion sensing.

The noise reduction increases with the number of frames over which the noise is integrated, but image motion prevents simple combining of frames. If motion estimation is available, the image of a moving object in a particular frame can be integrated from the images in several frames which have been superimposed on the same part of the screen by displacements derived from the motion measurement. The result is that greater reduction of noise becomes possible.[11] In fact a motion-compensated standards convertor performs such a noise-reduction process automatically and can be used as a noise reducer, albeit an expensive one, by setting both input and output to the same standard.

References

1. Aird, B., Three dimensional picture synthesis. *Broadcast Systems Engineering*, **12** No.3, 34–40 (1986)
2. Newman, W.M. and Sproull, R.F., *Principles of Interactive Computer Graphics*. Tokyo: McGraw-Hill (1979)
3. Gernsheim, H., *A Concise History of Photography*. London: Thames and Hudson, 9–15 (1971)
4. Hedgecoe, J., *The Photographer's Handbook*. London: Ebury Press, 104–105 (1977)
5. Bennett, P., *et al*. Spatial transformation system including key signal generator. US Patent No. 4,463,372 (1984)
6. de Boor, C., *A Practical Guide to Splines*. Berlin: Springer (1978)
7. Limb, J.O. and Murphy, J.A., Measuring the speed of moving objects from television signals. *IEEE Trans. Commun.*, 474–478 (1975)
8. Thomas, G.A., Television motion measurement for DATV and other applications. *BBC Res. Dept. Rept*, RD 1987/11 (1987)
9. Pearson, J.J. *et al*. Video rate image correlation processor. *SPIE*, Vol.119, Application of digital image processing, IOCC (1977)
10. Lau, H. and Lyon, D., Motion compensated processing for enhanced slow motion and standards conversion. *IEE Conf. Publ. No. 358*, 62–66 (1992)
11. Weiss, P. and Christensson, J., Real time implementation of sub-pixel motion estimation for broadcast applications. *IEE Digest*, 1990/128

6

Video compression and MPEG

6.1 Introduction to compression

Compression, bit rate reduction and data reduction, are all terms which mean basically the same thing in this context. In essence the same (or nearly the same) information is carried using a smaller quantity or rate of data. It should be pointed out that in audio *compression* traditionally means a process in which the dynamic range of the sound is reduced. In the context of MPEG the same word means that the bit rate is reduced, ideally leaving the dynamics of the signal unchanged. Provided the context is clear, the two meanings can co-exist without a great deal of confusion.

There are several reasons why compression techniques are popular:

1 Compression extends the playing time of a given storage device.
2 Compression allows miniaturization. With less data to store, the same playing time is obtained with smaller hardware. This is useful in ENG (electronic news gathering) and consumer devices.
3 Tolerances can be relaxed. With less data to record, storage density can be reduced, making equipment which is more resistant to adverse environments and which requires less maintenance.
4 In transmission systems, compression allows a reduction in bandwidth which will generally result in a reduction in cost to make possible some process which would be impracticable without it.
5 If a given bandwidth is available to an uncompressed signal, compression allows faster than real-time transmission in the same bandwidth.
6 If a given bandwidth is available, compression allows a better quality signal in the same bandwidth.

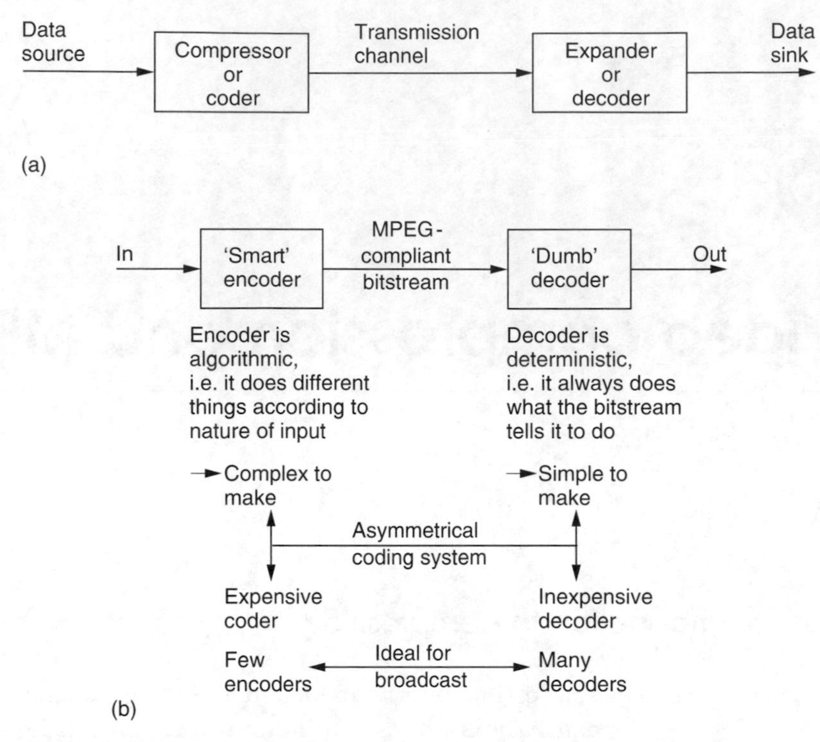

Figure 6.1 In (a) a compression system consists of compressor or coder, a transmission channel and a matching expander or decoder. The combination of coder and decoder is known as a codec. (b) MPEG is asymmetrical since the encoder is much more complex than the decoder.

Compression is summarized in Figure 6.1. It will be seen in (a) that the data rate is reduced at source by the *compressor*. The compressed data are then passed through a communication channel and returned to the original rate by the *expander*. The ratio between the source data rate and the channel data rate is called the *compression factor*. The term *coding gain* is also used. Sometimes a compressor and expander in series are referred to as a *compander*. The compressor may equally well be referred to as a *coder* and the expander a *decoder* in which case the tandem pair may be called a *codec*.

In audio and video compression, where the encoder is more complex than the decoder the system is said to be asymmetrical as in Figure 6.1(b). The encoder needs to be algorithmic or adaptive whereas the decoder is 'dumb' and carries out fixed actions. This is advantageous in applications such as broadcasting where the number of expensive complex encoders is small but the number of simple inexpensive decoders is large. In point-to-point applications the advantage of asymmetrical coding is not so great.

Although there are many different coding techniques, all of them fall into one or other of these categories. In *lossless* coding, the data from the expander are identical bit-for-bit with the original source data. The so-called 'stacker' programs which increase the apparent capacity of disk drives in personal computers use lossless codecs. Clearly with computer programs the corruption of a single bit can be catastrophic. Lossless coding is generally restricted to compression factors of around 2:1.

It is important to appreciate that a lossless coder cannot guarantee a particular compression factor and the communications link or recorder used with it must be able to function with the variable output data rate. Source data which result in poor compression factors on a given codec are described as *difficult*. It should be pointed out that the difficulty is often a function of the codec. In other words data which one codec finds difficult may not be found difficult by another. Lossless codecs can be included in bit-error-rate testing schemes. It is also possible to cascade or *concatenate* lossless codecs without any special precautions.

In *lossy* coding data from the expander are not identical bit-for-bit with the source data and as a result comparing the input with the output is bound to reveal differences. Lossy codecs are not suitable for computer data, but are used in MPEG as they allow greater compression factors than lossless codecs. Successful lossy codecs are those in which the errors are arranged so that a human viewer or listener finds them subjectively difficult to detect. Thus lossy codecs must be based on an understanding of psycho-acoustic and psycho-visual perception and are often called *perceptive* codes.

In perceptive coding, the greater the compression factor required, the more accurately must the human senses be modelled. Perceptive coders can be forced to operate at a fixed compression factor. This is convenient for practical transmission applications where a fixed data rate is easier to handle than a variable rate. The result of a fixed compression factor is that the subjective quality can vary with the 'difficulty' of the input material. Perceptive codecs should not be concatenated indiscriminately especially if they use different algorithms. As the reconstructed signal from a perceptive codec is not bit-for-bit accurate, clearly such a codec cannot be included in any bit error rate testing system as the coding differences would be indistinguishable from real errors.

Although the adoption of digital techniques is recent, compression itself is as old as television. Figure 6.2 shows some of the compression techniques used in traditional television systems.

One of the oldest techniques is interlace, which has been used in analog television from the very beginning as a primitive way of reducing bandwidth. As seen in Chapter 5, interlace is not without its problems, particularly in motion rendering. MPEG-2 supports interlace simply because legacy interlaced signals exist and there is a requirement to

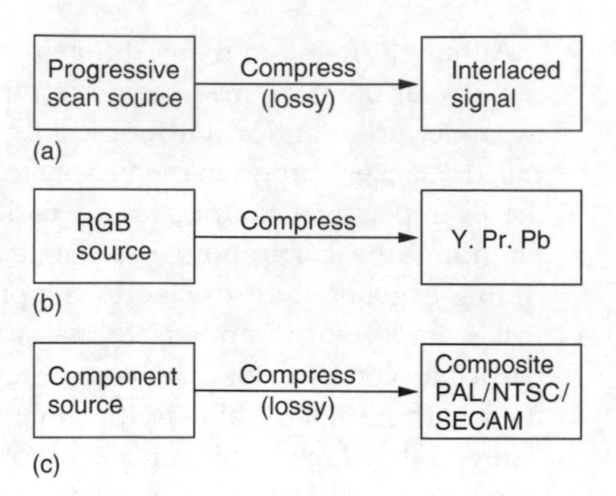

Figure 6.2 Compression is as old as television. (a) Interlace is a primitive way of halving the bandwidth. (b) Colour difference working invisibly reduces colour resolution. (c) Composite video transmits colour in the same bandwidth as monochrome.

compress them. This should not be taken to mean that it is a good idea.

The generation of colour difference signals from *RGB* in video represents an application of perceptive coding. The human visual system (HVS) sees no change in quality although the bandwidth of the colour difference signals is reduced. This is because human perception of detail in colour changes is much less than in brightness changes. This approach is sensibly retained in MPEG.

Composite video systems such as PAL, NTSC and SECAM are all analog compression schemes which embed a subcarrier in the luminance signal so that colour pictures are available in the same bandwidth as monochrome. In comparison with a progressive scan *RGB* picture, interlaced composite video has a compression factor of 6:1.

In a sense MPEG-2 can be considered to be a modern digital equivalent of analog composite video as it has most of the same attributes. For example, the eight field sequence of PAL subcarrier which makes editing diffficult has its equivalent in the GOP (group of pictures) of MPEG.[1]

In a PCM digital system the bit rate is the product of the sampling rate and the number of bits in each sample and this is generally constant. Nevertheless the *information* rate of a real signal varies. In all real signals, part of the signal is obvious from what has gone before or what may come later and a suitable receiver can predict that part so that only the true information actually has to be sent. If the characteristics of a predicting receiver are known, the transmitter can omit parts of the message in the knowledge that the receiver has the ability to recreate it. Thus all encoders must contain a model of the decoder.

One definition of information is that it is the unpredictable or surprising element of data. Newspapers are a good example of information because they only mention items which are surprising. Newspapers never carry items about individuals who have *not* been involved in an accident as this is the normal case. Consequently the phrase 'no news is good news' is remarkably true because if an information channel exists but nothing has been sent then it is most likely that nothing remarkable has happened.

The difference between the information rate and the overall bit rate is known as the redundancy. Compression systems are designed to eliminate as much of that redundancy as practicable or perhaps affordable. One way in which this can be done is to exploit statistical predictability in signals. The information content or *entropy* of a sample is a function of how different it is from the predicted value. Most signals have some degree of predictability. A sine wave is highly predictable, because all cycles look the same. According to Shannon's theory, any signal which is totally predictable carries no information. In the case of the sine wave this is clear because it represents a single frequency and so has no bandwidth.

At the opposite extreme a signal such as noise is completely unpredictable and as a result all codecs find noise *difficult*. There are two consequences of this characteristic. First, a codec which is designed using the statistics of real material should not be tested with random noise because it is not a representative test. Second, a codec which performs well with clean source material may perform badly with source material containing superimposed noise. Most practical compression units require some form of pre-processing before the compression stage proper and appropriate noise reduction should be incorporated into the pre-processing if noisy signals are anticipated. It will also be necessary to restrict the degree of compression applied to noisy signals.

All real signals fall part-way between the extremes of total predictability and total unpredictability or noisiness. If the bandwidth (set by the sampling rate) and the dynamic range (set by the wordlength) of the transmission system are used to delineate an area, this sets a limit on the information capacity of the system. Figure 6.3(a) shows that most real signals only occupy part of that area. The signal may not contain all frequencies, or it may not have full dynamics at certain frequencies.

Entropy can be thought of as a measure of the actual area occupied by the signal. This is the area that *must* be transmitted if there are to be no subjective differences or *artifacts* in the received signal. The remaining area is called the *redundancy* because it adds nothing to the information conveyed. Thus an ideal coder could be imagined which miraculously sorts out the entropy from the redundancy and only sends the former. An

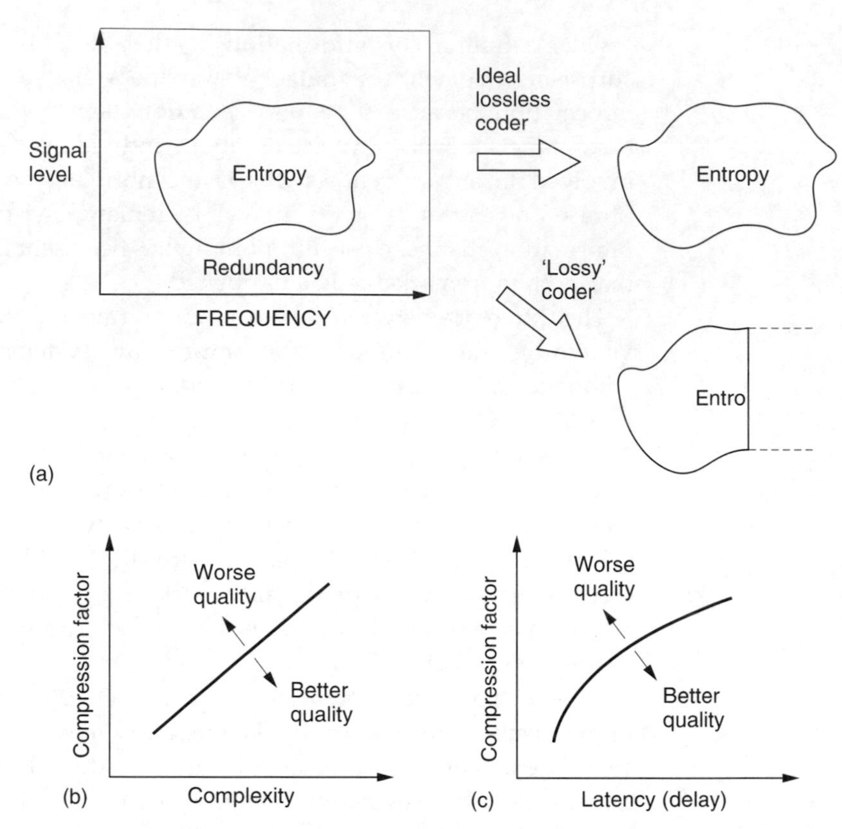

Figure 6.3 (a) A perfect coder removes only the redundancy from the input signal and results in subjectively lossless coding. If the remaining entropy is beyond the capacity of the channel some of it must be lost and the codec will then be lossy. An imperfect coder will also be lossy as it falls to keep all entropy. (b) As the compression factor rises, the complexity must also rise to maintain quality. (c) High compression factors also tend to increase latency or delay through the system.

ideal decoder would then recreate the original impression of the information quite perfectly.

As the ideal is approached, the coder complexity and the latency or delay both rise. Figure 6.3(b) shows how complexity increases with compression factor and (c) shows how increasing the codec latency can improve the compression factor. Obviously we would have to provide a channel which could accept whatever entropy the coder extracts in order to have transparent quality. As a result moderate coding gains which only remove redundancy need not cause artifacts and result in systems which are described as *subjectively lossless*.

If the channel capacity is not sufficient for that, then the coder will have to discard some of the entropy and with it useful information. Larger coding gains which remove some of the entropy must result in artifacts. It will also be seen from Figure 6.3 that an imperfect coder will fail to separate the redundancy and may discard entropy instead, resulting in artifacts at a sub-optimal compression factor.

A single variable-rate transmission or recording channel is inconvenient and unpopular with channel providers because it is difficult to police. The requirement can be overcome by combining several compressed channels into one constant rate transmission in a way which flexibly allocates data rate between the channels. Provided the material is unrelated, the probability of all channels reaching peak entropy at once is very small and so those channels which are at one instant passing easy material will free up transmission capacity for those channels which are handling difficult material. This is the principle of statistical multiplexing.

Where the same type of source material is used consistently, e.g. English text, then it is possible to perform a statistical analysis on the frequency with which particular letters are used. Variable-length coding is used in which frequently used letters are allocated short codes and letters which occur infrequently are allocated long codes. This results in a lossless code. The well-known Morse code used for telegraphy is an example of this approach. The letter e is the most frequent in English and is sent with a single dot.

An infrequent letter such as z is allocated a long complex pattern. It should be clear that codes of this kind which rely on a prior knowledge of the statistics of the signal are only effective with signals actually having those statistics. If Morse code is used with another language, the transmission becomes significantly less efficient because the statistics are quite different; the letter z, for example, is quite common in Czech.

The Huffman code[2] is one which is designed for use with a data source having known statistics and shares the same principles with the Morse code. The probability of the different code values to be transmitted is studied, and the most frequent codes are arranged to be transmitted with short wordlength symbols. As the probability of a code value falls, it will be allocated a longer wordlength. The Huffman code is used in conjunction with a number of compression techniques and is shown in Figure 6.4.

The input or *source* codes are assembled in order of descending probability. The two lowest probabilities are distinguished by a single code bit and their probabilities are combined. The process of combining probabilities is continued until unity is reached and at each stage a bit is used to distinguish the path. The bit will be a zero for the most probable path and one for the least. The compressed output is obtained by reading the bits which describe which path to take going from right to left.

In the case of computer data, there is no control over the data statistics. Data to be recorded could be instructions, images, tables, text files and so on; each having their own code value distributions. In this case a coder

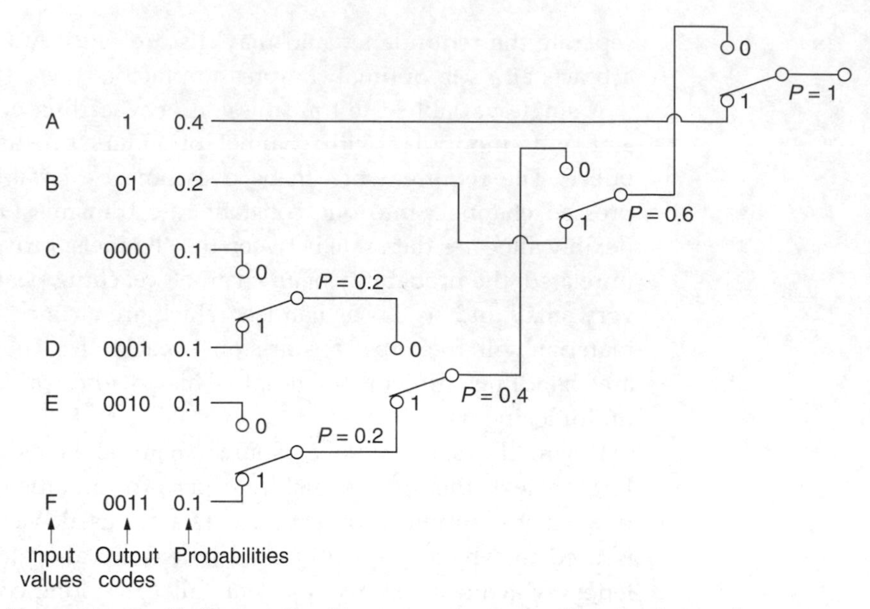

Figure 6.4 The Huffman code achieves compression by allocating short codes to frequent values. To aid deserializing the short codes are not prefixes of longer codes.

relying on fixed-source statistics will be completely inadequate. Instead a system is used which can learn the statistics as it goes along. The Lempel–Ziv–Welch (LZW) lossless codes are in this category. These codes build up a conversion table between frequent long-source data strings and short transmitted data codes at both coder and decoder, and initially their compression factor is below unity as the contents of the conversion tables are transmitted along with the data. However, once the tables are established, the coding gain more than compensates for the initial loss. In some applications, a continuous analysis of the frequency of code selection is made and if a data string in the table is no longer being used with sufficient frequency it can be deselected and a more common string substituted.

Lossless codes are less common for audio and video coding where perceptive codes are permissible. The perceptive codes often obtain a coding gain by shortening the wordlength of the data representing the signal waveform. This must increase the noise level and the trick is to ensure that the resultant noise is placed at frequencies where human senses are least able to perceive it. As a result although the received signal is measurably different from the source data, it can *appear* the same to the human listener or viewer at moderate compressions factors. As these codes rely on the characteristics of human sight and hearing, they can only fully be tested subjectively.

The compression factor of such codes can be set at will by choosing the wordlength of the compressed data. Whilst mild compression will be undetectable, with greater compression factors, artifacts become noticeable. Figure 6.3 shows that this is inevitable from entropy considerations.

6.2 What is MPEG?

MPEG is actually an acronym for the Moving Pictures Experts Group which was formed by the ISO (International Standards Organization) to set standards for audio and video compression and transmission. The first compression standard for audio and video was MPEG-1,[3,4] but this was of limited application and the subsequent MPEG-2 standard was considerably broader in scope and of wider appeal. For example, MPEG-2 supports interlace whereas MPEG-1 did not.

The approach of the ISO to standardization in MPEG is novel because it is not the encoder which is standardized. Figure 6.5(a) shows that instead the way in which a decoder shall interpret the bitstream is defined. A decoder which can successfully interpret the bitstream is said to be *compliant*. Figure 6.5(b) shows that the advantage of standardizing the decoder is that, over time, encoding algorithms can improve yet compliant decoders will continue to function with them.

Manufacturers can supply encoders using algorithms which are proprietary and their details do not need to be published. A useful result is that there can be competition between different encoder designs which means that better designs will evolve. The user will have greater choice because different levels of cost and complexity can exist in a range of coders yet a compliant decoder will operate with them all.

MPEG is, however, much more than a compression scheme as it also standardizes the protocol and syntax under which it is possible to combine or multiplex audio data with video data to produce a digital equivalent of a television program. Many such programs can be combined in a single multiplex and MPEG defines the way in which such multiplexes can be created and transported. The definitions include the metadata which decoders require to demultiplex correctly and which users will need to locate programs of interest.

As with all video systems there is a requirement for synchronizing or genlocking and this is particularly complex when a multiplex is assembled from many signals which are not necessarily synchronized to one another.

The applications of audio and video compression are limitless and the ISO has done well to provide standards which are appropriate to the wide range of possible compression products.

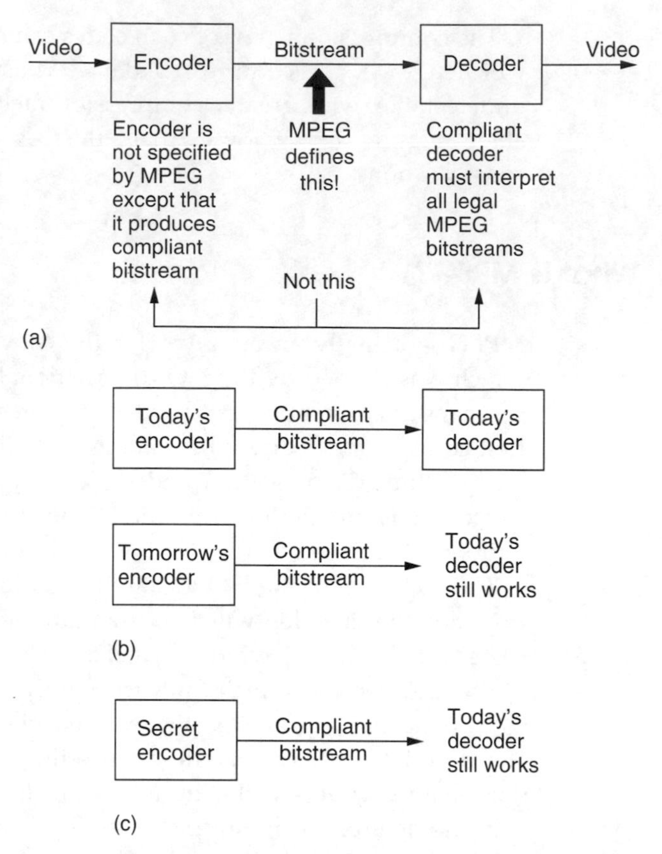

Figure 6.5 (a) MPEG defines the protocol of the bitstream between encoder and decoder. The decoder is defined by implication, the encoder is left very much to the designer. (b) This approach allows future encoders of better performance to remain compatible with existing decoders. (c) This approach also allows an encoder to produce a standard bitstream while its technical operation remains a commercial secret.

MPEG-2 embraces video pictures from the tiny screen of a videophone to the high-definition images needed for electronic cinema. Audio coding stretches from speech-grade mono to multichannel surround sound.

Figure 6.6 shows the use of a codec with a recorder. The playing time of the medium is extended in proportion to the compression factor. In the case of tapes, the access time is improved because the length of tape needed for a given recording is reduced and so it can be rewound more quickly.

In the case of DVD (Digital Video Disk aka Digital Versatile Disk) the challenge was to store an entire movie on one 12 cm disk. The storage density available with today's optical disk technology is such that recording of conventional uncompressed video would be out of the question.

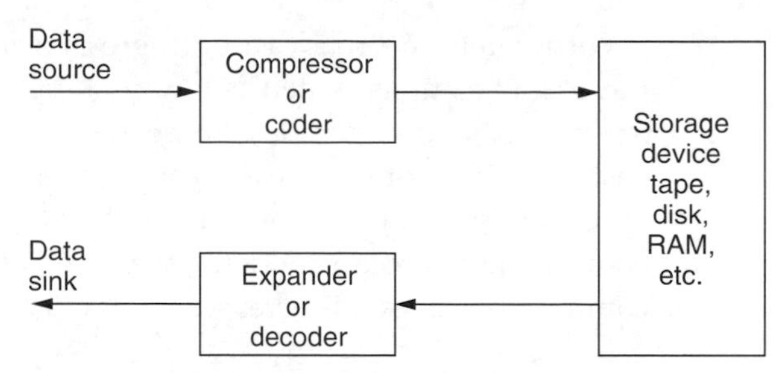

Figure 6.6 Compression can be used around a recording medium. The storage capacity may be increased or the access time reduced according to the application.

In communications, the cost of data links is often roughly proportional to the data rate and so there is simple economic pressure to use a high compression factor. However, it should be borne in mind that implementing the codec also has a cost which rises with compression factor and so a degree of compromise will be inevitable.

In the case of Video-On-Demand, technology exists to convey full bandwidth video to the home, but to do so for a single individual at the moment would be prohibitively expensive. Without compression, HDTV (high-definition television) requires too much bandwidth. With compression, HDTV can be transmitted to the home in a similar bandwidth to an existing analog SDTV channel. Compression does not make Video-On-Demand or HDTV possible, it makes them economically viable.

In workstations designed for the editing of audio and/or video, the source material is stored on hard disks for rapid access. Whilst top-grade systems may function without compression, many systems use compression to offset the high cost of disk storage. When a workstation is used for *offline* editing, a high compression factor can be used and artifacts will be visible in the picture.

This is of no consequence as the picture is only seen by the editor who uses it to make an EDL (Edit Decision List) which is no more than a list of actions and the timecodes at which they occur. The original uncompressed material is then *conformed* to the EDL to obtain a high-quality edited work. When *online* editing is being performed, the output of the workstation is the finished product and clearly a lower compression factor will have to be used.

Perhaps it is in broadcasting where the use of compression will have its greatest impact. There is only one electromagnetic spectrum and pressure from other services such as cellular telephones makes efficient use of bandwidth mandatory. Analog television broadcasting is an old technology and makes very inefficient use of bandwidth. Its replacement by a compressed digital transmission will be inevitable for the practical reason that the bandwidth is needed elsewhere.

Fortunately in broadcasting there is a mass market for decoders and these can be implemented as low-cost integrated circuits. Fewer encoders are needed and so it is less important if these are expensive. Whilst the cost of digital storage goes down year on year, the cost of electromagnetic spectrum goes up. Consequently in the future the pressure to use compression in recording will ease whereas the pressure to use it in radio communications will increase.

6.3 Spatial and temporal redundancy in MPEG

Video signals exist in four dimensions: these are the attributes of the sample, the horizontal and vertical spatial axes and the time axis. Compression can be applied in any or all of those four dimensions. MPEG-2 assumes eight-bit colour difference signals as the input, requiring rounding if the source is ten-bit. The sampling rate of the colour signals is less than that of the luminance. This is done by downsampling the colour samples horizontally and generally vertically as well. Essentially an MPEG-2 system has three parallel simultaneous channels, one for luminance and two colour difference, which after coding are multiplexed into a single bitstream.

Figure 6.7(a) shows that when individual pictures are compressed without reference to any other pictures, the time axis does not enter the process which is therefore described as *intra-coded* (intra = within) compression. The term *spatial coding* will also be found. It is an advantage of intra-coded video that there is no restriction to the editing which can be carried out on the picture sequence. As a result compressed VTRs such as Digital Betacam, DVC and D-9 use spatial coding. Cut editing may take place on the compressed data directly if necessary. As spatial coding treats each picture independently, it can employ certain techniques developed for the compression of still pictures. The ISO JPEG (Joint Photographic Experts Group) compression standards[5,6] are in this category. Where a succession of JPEG coded images are used for television, the term 'Motion JPEG' will be found.

Greater compression factors can be obtained by taking account of the redundancy from one picture to the next. This involves the time axis, as Figure 6.7(b) shows, and the process is known as *inter-coded* (inter = between) or *temporal* compression.

Temporal coding allows a higher compression factor, but has the disadvantage that an individual picture may exist only in terms of the differences from a previous picture. Clearly editing must be undertaken with caution and arbitrary cuts simply cannot be performed on the MPEG bitstream. If a previous picture is removed by an edit, the difference data will then be insufficient to recreate the current picture.

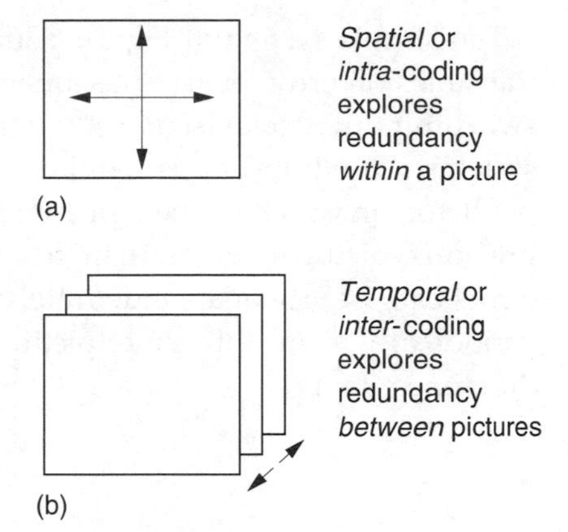

Figure 6.7 (a) Spatial or intra-coding works on individual images. (b) Temporal or inter-coding works on successive images.

Intra-coding works in three dimensions on the horizontal and vertical spatial axes and on the sample values. Analysis of typical television pictures reveals that whilst there is a high spatial frequency content due to detailed areas of the picture, there is a relatively small amount of energy at such frequencies. Often pictures contain sizeable areas in which the same or similar pixel values exist. This gives rise to low spatial frequencies. The average brightness of the picture results in a substantial zero frequency component. Simply omitting the high-frequency components is unacceptable as this causes an obvious softening of the picture.

A coding gain can be obtained by taking advantage of the fact that the amplitude of the spatial components falls with frequency. It is also possible to take advantage of the eye's reduced sensitivity to noise in high spatial frequencies. If the spatial frequency spectrum is divided into frequency bands the high-frequency bands can be described by fewer bits not only because their amplitudes are smaller but also because more noise can be tolerated. The Wavelet Transform and the Discrete Cosine Transform used in MPEG allows two-dimensional pictures to be described in the frequency domain and these were discussed in Chapter 3.

Inter-coding takes further advantage of the similarities between successive pictures in real material. Instead of sending information for each picture separately, inter-coders will send the difference between the previous picture and the current picture in a form of differential coding. Figure 6.8 shows the principle. A picture store is required at the coder to allow comparison to be made between successive pictures and a similar store is required at the decoder to make the previous picture available. The difference data may be treated as a picture itself and subjected to some form of transform-based spatial compression.

The simple system of Figure 6.8(a) is of limited use as in the case of a transmission error, every subsequent picture would be affected. Channel switching in a television set would also be impossible. In practical systems a modification is required. One approach is the so-called 'leaky predictor' in which the next picture is predicted from a limited number of previous pictures rather than from an indefinite number. As a result errors cannot propagate indefinitely. The approach used in MPEG is that periodically some absolute picture data are transmitted in place of difference data.

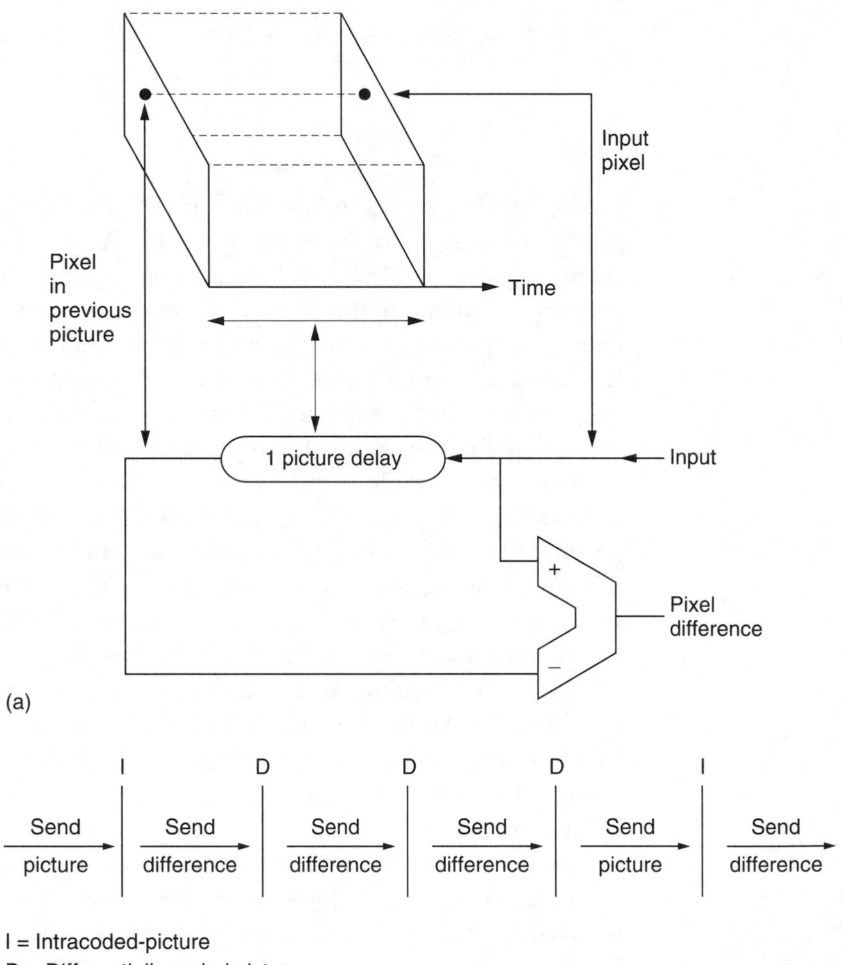

Figure 6.8 An inter-coded system (a) uses a delay to calculate the pixel differences between successive pictures. To prevent error propagation, intra-coded pictures (b) may be used periodically.

Figure 6.8(b) shows that absolute picture data, known as *I* or *intra pictures* are interleaved with pictures which are created using difference data, known as *P* or *predicted* pictures. The *I* pictures require a large amount of data, whereas the *P* pictures require less data. As a result the instantaneous data rate varies dramatically and buffering has to be used to allow a constant transmission rate. The leaky predictor needs less buffering as the compression factor does not change so much from picture to picture.

The *I* picture and all the *P* pictures prior to the next *I* picture are called a group of pictures (GOP). For a high compression factor, a large number of *P* pictures should be present between *I* pictures, making a long GOP. However, a long GOP delays recovery from a transmission error. The compressed bitstream can only be edited at *I* pictures as shown.

In the case of moving objects, although their appearance may not change greatly from picture to picture, the data representing them on a fixed sampling grid will change and so large differences will be generated between successive pictures. It is a great advantage if the effect of motion can be removed from difference data so that they only reflect the changes in appearance of a moving object since a much greater coding gain can then be obtained. This is the objective of motion compensation.

In real television program material objects move around before a fixed camera or the camera itself moves. Motion compensation is a process which effectively measures motion of objects from one picture to the next so that it can allow for that motion when looking for

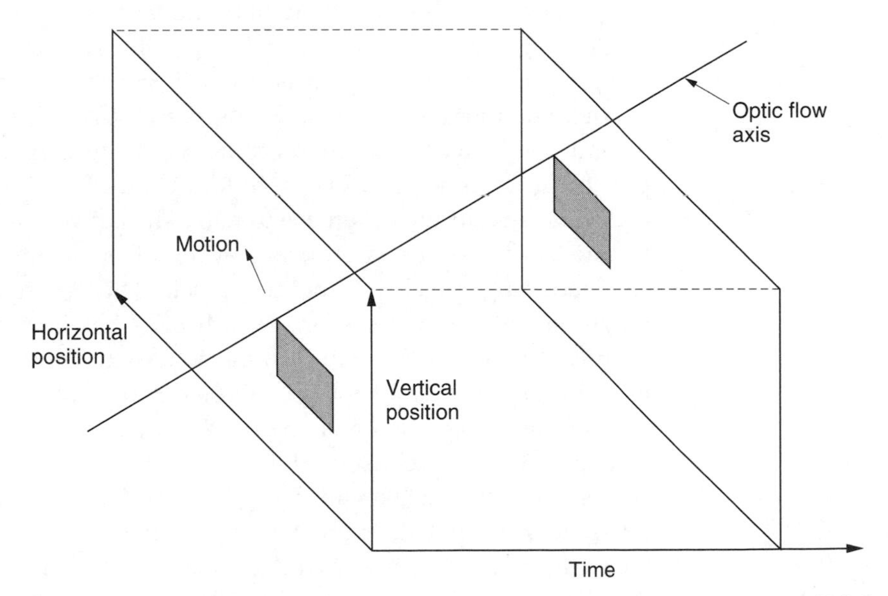

Figure 6.9 Objects travel in a three-dimensional space along the optic flow axis which is only parallel to the time axis if there is no movement.

redundancy between pictures. Figure 6.9 shows that moving pictures can be expressed in a three-dimensional space which results from the screen area moving along the time axis. In the case of still objects, the only motion is along the time axis. However, when an object moves, it does so along the *optic flow axis* which is not parallel to the time axis. The optic flow axis joins the same point on a moving object as it takes on various screen positions.

It will be clear that the data values representing a moving object change with respect to the time axis. However, looking along the optic flow axis the appearance of an object only changes if it deforms, moves into shadow or rotates. For simple translational motions the data representing an object are highly redundant with respect to the optic flow axis. Thus if the optic flow axis can be located, coding gain can be obtained in the presence of motion.

A motion-compensated coder works as follows. An *I* picture is sent, but is also locally stored so that it can be compared with the next input picture to find motion vectors for various areas of the picture. The *I* picture is then shifted according to these vectors to cancel inter-picture motion. The resultant *predicted* picture is compared with the actual picture to produce a *prediction error* also called a *residual*. The prediction error is transmitted with the motion vectors. At the receiver the original *I* picture is also held in a memory. It is shifted according to the transmitted motion vectors to create the predicted picture and then the prediction error is added to it to recreate the original. When a picture is encoded in this way MPEG calls it a *P* picture.

Figure 6.10(a) shows that spatial redundancy is redundancy within a single image, for example repeated pixel values in a large area of blue sky. Temporal redundancy (b) exists between successive images.

Where temporal compression is used, the current picture is not sent in its entirety; instead the difference between the current picture and the previous picture is sent. The decoder already has the previous picture, and so it can add the difference to make the current picture. A difference picture is created by subtracting every pixel in one picture from the corresponding pixel in another pixel. This is trivially easy in a progressively scanned system, but MPEG-2 has had to develop greater complexity so that this can also be done with interlaced pictures. The handling of interlace in MPEG will be detailed later.

A difference picture is an image of a kind, although not a viewable one, and so should contain some kind of spatial redundancy. Figure 6.10(c) shows that MPEG-2 takes advantage of both forms of redundancy. Picture differences are spatially compressed prior to transmission. At the decoder the spatial compression is decoded to recreate the difference picture, then this difference picture is added to the previous picture to complete the decoding process.

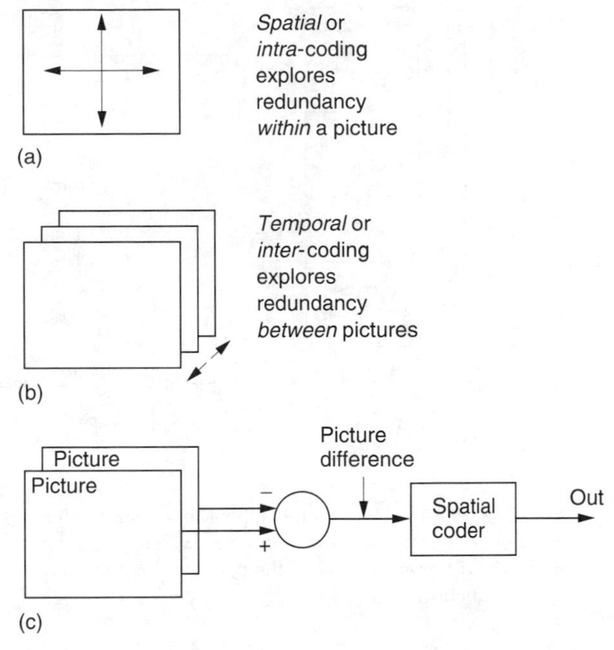

Figure 6.10 (a) Spatial or intra-coding works on individual images. (b) Temporal or inter-coding works on successive images. (c) In MPEG inter-coding is used to create difference images. These are then compressed spatially.

Whenever objects move they will be in a different place in successive pictures. This will result in large amounts of difference data. MPEG-2 overcomes the problem using motion compensation. The encoder contains a motion estimator which measures the direction and distance of motion between pictures and outputs these as vectors which are sent to the decoder. When the decoder receives the vectors it uses them to shift data in a previous picture to resemble the current picture more closely. Effectively the vectors are describing the optic flow axis of some moving screen area, along which axis the image is highly redundant. Vectors are bipolar codes which determine the amount of horizontal and vertical shift required.

In real images, moving objects do not necessarily maintain their appearance as they move. For example, objects may turn, move into shade or light, or move behind other objects. Consequently motion compensation can never be ideal and it is still necessary to send a picture difference to make up for any shortcomings in the motion compensation.

Figure 6.11 shows how this works. In addition to the motion-encoding system, the coder also contains a motion decoder. When the encoder outputs motion vectors, it also uses them locally in the same way that a real decoder will, and is able to produce a *predicted picture* based solely on

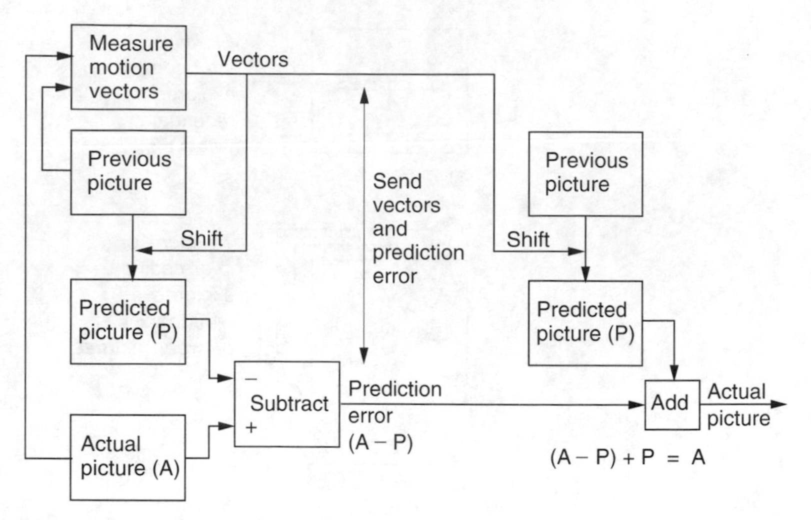

Figure 6.11 A motion-compensated compression system. The coder calculates motion vectors which are transmitted as well as being used locally to create a predicted picture. The difference between the predicted picture and the actual picture is transmitted as a prediction error.

the previous picture shifted by motion vectors. This is then subtracted from the *actual* current picture to produce a *prediction error* or *residual* which is an image of a kind that can be spatially compressed.

The decoder takes the previous picture, shifts it with the vectors to recreate the predicted picture and then decodes and adds the prediction error to produce the actual picture. Picture data sent as vectors plus prediction error are said to be *P* coded.

The concept of sending a prediction error is a useful approach because it allows both the motion estimation and compensation to be imperfect.

A good motion-compensation system will send just the right amount of vector data. With insufficient vector data, the prediction error will be large, but transmission of excess vector data will also cause the the bit rate to rise. There will be an optimum balance which minimizes the sum of the prediction error data and the vector data.

In MPEG-2 the balance is obtained by dividing the screen into areas called *macroblocks* which are 16 luminance pixels square. Each macroblock is steered by a vector. The location of the boundaries of a macroblock are fixed and so the vector does not move the macroblock. Instead the vector tells the decoder where to look in another frame to find pixel data to *fetch* to the macroblock. Figure 6.12(a) shows this concept. The shifting process is generally done by modifying the read address of a RAM using the vector. This can shift by one-pixel steps. MPEG-2 vectors have half-pixel resolution so it is necessary to interpolate between pixels from RAM to obtain half-pixel shifted values.

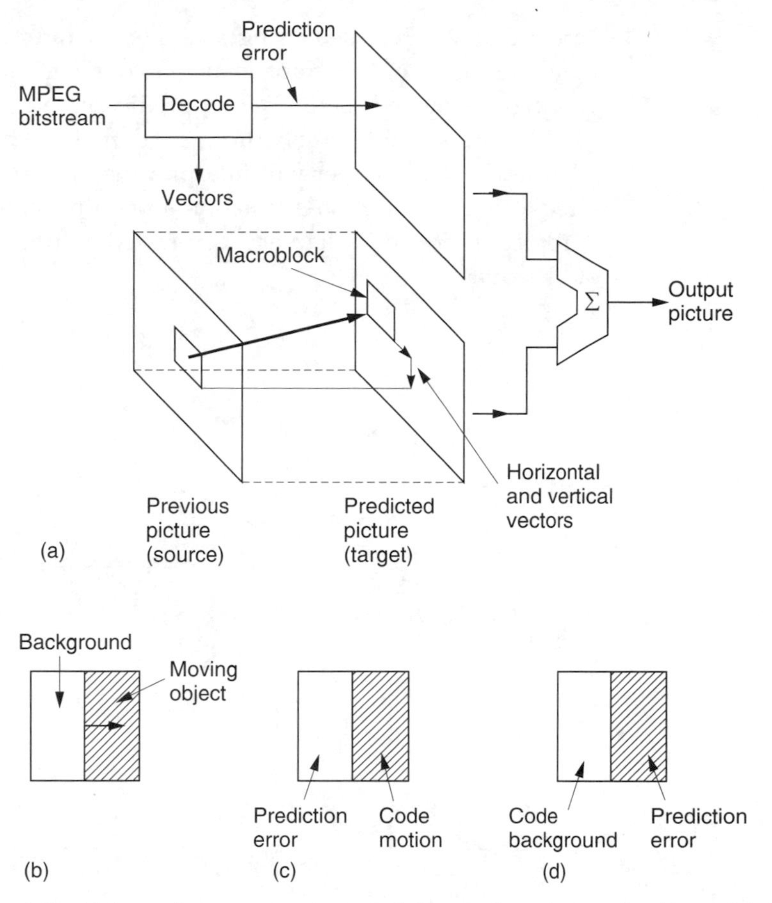

Figure 6.12 (a) In motion compensation, pixel data are brought to a fixed macroblock in the target picture from a variety of places in another picture. (b) Where only part of a macroblock is moving, motion compensation is non-ideal. The motion can be coded (c), causing a prediction error in the background, or the background can be coded (d) causing a prediction error in the moving object.

Real moving objects will not coincide with macroblocks and so the motion compensation will not be ideal but the prediction error makes up for any shortcomings. Figure 6.12(b) shows the case where the boundary of a moving object bisects a macroblock. If the system measures the moving part of the macroblock and sends a vector, the decoder will shift the entire block making the stationary part wrong. If no vector is sent, the moving part will be wrong. Both approaches are legal in MPEG-2 because the prediction error sorts out the incorrect values. An intelligent coder might try both approaches to see which required the least prediction error data.

The prediction error concept also allows the use of simple but inaccurate motion estimators in low-cost systems. The greater prediction error data are handled using a higher bit rate. On the other hand, if a precision motion estimator is available, a very high compression factor may be achieved because the prediction error data are minimized. MPEG-2 does not specify how motion is to be measured; it simply defines how a decoder will interpret the vectors. Encoder designers are free to use any motion-estimation system provided that the right vector protocol is created. Chapter 3 contrasted a number of motion estimation techniques.

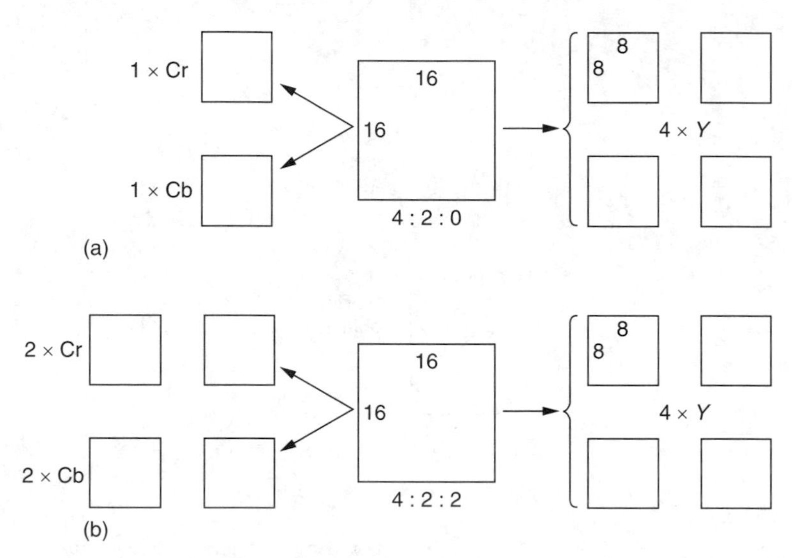

Figure 6.13 The structure of a macroblock. (A macroblock is the screen area steered by one vector.) (a) In 4:2:0, there are two chroma DCT blocks per macroblock whereas in 4:2:2 (b) there are four, 4:2:2 needs 33% more data than 4:2:0.

Figure 6.13(a) shows that a macroblock contains both luminance and colour difference data at different resolutions. Most of the MPEG-2 Profiles use a 4:2:0 structure which means that the colour is downsampled by a factor of two in both axes. Thus in a 16 × 16 pixel block, there are only 8 × 8 colour difference sampling sites. MPEG-2 is based upon the 8 × 8 DCT (see section 3.7) and so the 16 × 16 block is the screen area which contains an 8 × 8 colour difference sampling block. Thus in 4:2:0 in each macroblock there are four luminance DCT blocks, one $R-Y$ DCT block and one $B-Y$ DCT block, all steered by the same vector.

In the 4:2:2 Profile of MPEG-2, shown in Figure 6.13(b), the chroma is not downsampled vertically, and so there is twice as much chroma data in each macroblock which is otherwise substantially the same.

6.4 *I* and *P* coding

Predictive (*P*) coding cannot be used indefinitely, as it is prone to error propagation. A further problem is that it becomes impossible to decode the transmission if reception begins part-way through. In real video signals, cuts or edits can be present across which there is little redundancy and which make motion estimators throw up their hands.

In the absence of redundancy over a cut, there is nothing to be done but to send the new picture information in absolute form. This is called *I* coding where *I* is an abbreviation of *intra* coding. As *I* coding needs no previous picture for decoding, then decoding can begin at *I* coded information.

MPEG-2 is effectively a toolkit and there is no compulsion to use all the tools available. Thus an encoder may choose whether to use *I* or *P* coding, either once and for all or dynamically on a macroblock-by-macroblock basis.

For practical reasons, an entire frame may be encoded as *I* macroblocks periodically. This creates a place where the bitstream might be edited or where decoding could begin.

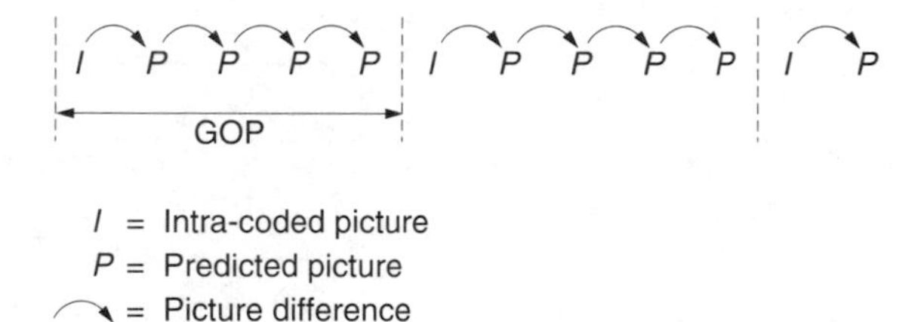

Figure 6.14 A Simple Profile MPEG-2 signal may contain periodic *I* pictures with a number of *P* pictures between.

Figure 6.14 shows a typical application of the Simple Profile of MPEG-2. Periodically an *I* picture is created. Between *I* pictures are *P* pictures which are based on the previous picture. These *P* pictures predominantly contain macroblocks having vectors and prediction errors. However it is perfectly legal for *P* pictures to contain *I* macroblocks. This might be useful where, for example, a camera pan introduces new material at the edge of the screen which cannot be created from an earlier picture.

Note that although what is sent is called a *P* picture, it is not a picture at all. It is a set of instructions to convert the previous picture into the current picture. If the previous picture is lost, decoding is impossible. An *I* picture together with all the pictures before the next *I* picture form a *Group of Pictures* (GOP).

6.5 Bidirectional coding

Motion-compensated predictive coding is a useful compression technique, but it does have the drawback that it can only take data from a previous picture. Where moving objects reveal a background this is completely unknown in previous pictures and forward prediction fails. However, more of the background is visible in later pictures. Figure 6.15 shows the concept. In the centre of the diagram, a moving object has revealed some background. The previous picture can contribute nothing, whereas the next picture contains all that is required.

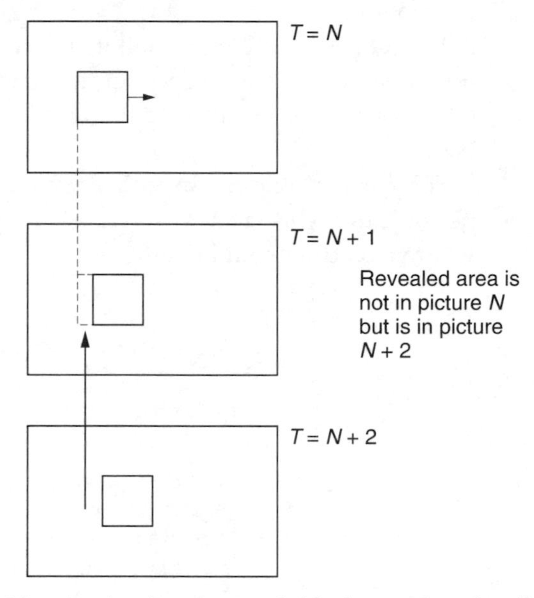

Figure 6.15 In bidirectional coding the revealed background can be efficiently coded by bringing data back from a future picture.

Bidirectional coding is shown in Figure 6.16. A bidirectional or *B* macroblock can be created using a combination of motion compensation and the addition of a prediction error. This can be done by forward prediction from a previous picture or backward prediction from a subsequent picture. It is also possible to use an average of both forward and backward prediction. On noisy material this may result in some reduction in bit rate. The technique is also a useful way of portraying a dissolve.

The averaging process in MPEG-2 is a simple linear interpolation which works well when only one *B* picture exists between the reference pictures before and after. A larger number of *B* pictures would require weighted interpolation but MPEG-2 does not support this.

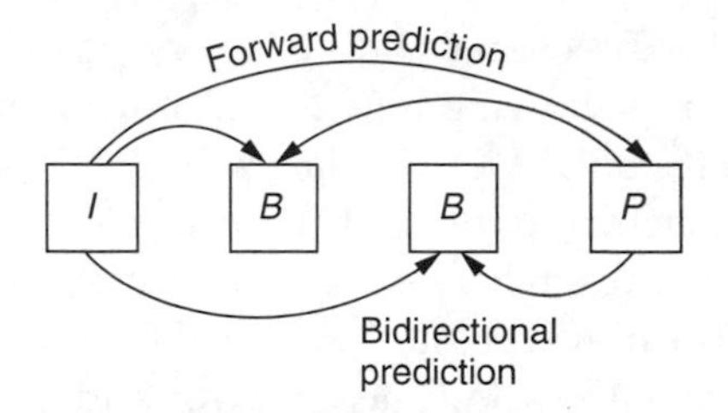

I = Intra- or spatially coded 'anchor' picture

P = Forward predicted. Coder sends difference between *I* and *P* decoder. Adds difference to create *P*

B = Bidirectionally coded picture can be coded from a previous *I* or *P* picture or a later *I* or *P* picture. *B* pictures are not coded from each other

Figure 6.16 In bidirectional coding, a number of *B* pictures can be inserted between periodic forward predicted pictures. See text.

Typically two *B* pictures are inserted between *P* pictures or between *I* and *P* pictures. As can be seen, *B* pictures are never predicted from one another, only from *I* or *P* pictures. A typical GOP for broadcasting purposes might have the structure *IBBPBBPBBPBB*. Note that the last *B* pictures in the GOP require the *I* picture in the next GOP for decoding and so the GOPs are not truly independent. Independence can be obtained by creating a *closed GOP* which may contain *B* pictures but which ends with a *P* picture. It is also legal to have a *B* picture in which every macroblock is forward predicted, needing no future picture for decoding.

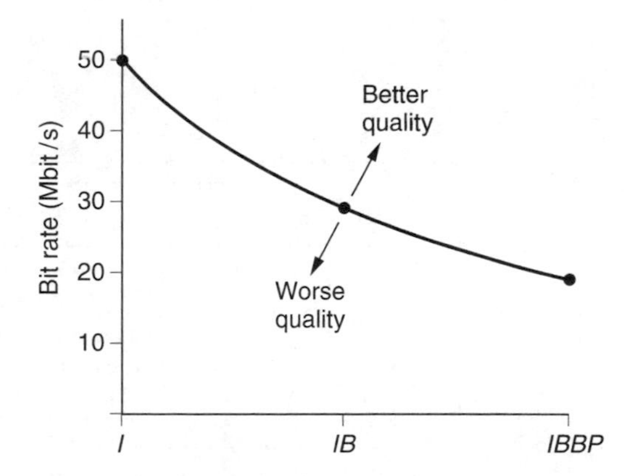

Figure 6.17 Bidirectional coding is very powerful as it allows the same quality with only 40 per cent of the bit rate of intra-coding. However, the encoding and decoding delays must increase. Coding over a longer time span is more efficient but editing is more difficult.

Bidirectional coding is very powerful. Figure 6.17 is a constant quality curve showing how the bit rate changes with the type of coding. On the left, only *I* or spatial coding is used, whereas on the right an *IBBP* structure is used. This means that there are two bidirectionally coded pictures in between a spatially coded picture (*I*) and a forward predicted picture (*P*). Note how for the same quality the system which only uses spatial coding needs two and a half times the bit rate that the bidirectionally coded system needs.

Clearly information in the future has yet to be transmitted and so is not normally available to the decoder. MPEG-2 gets around the problem by sending pictures in the wrong order. Picture reordering requires delay in the encoder and a delay in the decoder to put the order right again. Thus the overall codec delay must rise when bidirectional coding is used. This is quite consistent with Figure 6.3 which showed that as the compression factor rises the latency must also rise.

Figure 6.18 shows that although the original picture sequence is *IBBPBBPBBIBB* . . . , this is transmitted as *IPBBPBBIBB* . . . so that the future picture is already in the decoder before bidirectional decoding begins. Note that the *I* picture of the next GOP is actually sent before the last *B* pictures of the current GOP.

Figure 6.18 also shows that the amount of data required by each picture is dramatically different. *I* pictures have only spatial redundancy and so need a lot of data to describe them. *P* pictures need less data because they are created by shifting the *I* picture with vectors and then adding a prediction error picture. *B* pictures need the least data of all because they can be created from *I* or *P*.

With pictures requiring a variable length of time to transmit, arriving in the wrong order, the decoder needs some help. This takes the form of picture-type flags and time stamps.

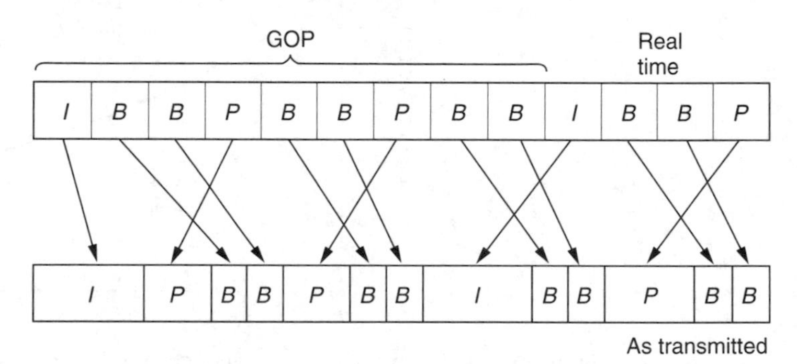

Figure 6.18 Comparison of pictures before and after compression showing sequence change and varying amount of data needed by each picture type. *I*, *P*, *B* pictures use unequal amounts of data.

6.6 Coding applications

Figure 6.19 shows a variety of GOP structures. The simplest is the *III* . . . sequence in which every picture is intra-coded. Pictures can be fully decoded without reference to any other pictures and so editing is straightforward. However, this approach requires about two-and-one-half times the bit rate of a full bidirectional system. Bidirectional coding is most useful for final delivery of post-produced material either by broadcast or on prerecorded media as there is then no editing requirement. As a compromise the *IBIB* . . . structure can be used which has some

I I I I I I ...	*I* only freely editable, needs high bit rate
I P P P P I P ...	Forward predicted only, needs less decoder memory, used in Simple Profile
I B B P B B P B ...	Forward and bidirectional, best compression factor, needs large decoder memory, hard to edit
I B I B I B ...	Lower bit rate than *I* only, editable with moderate processing

Figure 6.19 Various possible GOP structures used with MPEG. See text for details.

of the bit rate advantage of bidirectional coding but without too much latency. It is possible to edit an *IBIB* stream by performing some processing. If it is required to remove the video following a *B* picture, that *B* picture could not be decoded because it needs *I* pictures either side of it for bidirectional decoding. The solution is to decode the *B* picture first, and then re-encode it with forward prediction only from the previous *I* picture. The subsequent *I* picture can then be replaced by an edit process. Some quality loss is inevitable in this process but this is acceptable in applications such as ENG and industrial video.

6.7 Spatial compression

Spatial compression in MPEG-2 is used in *I* pictures on actual picture data and in *P* and *B* pictures on prediction error data. MPEG-2 uses the discrete cosine transform described in section 3.7. The DCT works on blocks and in MPEG-2 these are 8×8 pixels. Section 5.7 showed how the macroblocks of the motion-compensation structure are designed so they can be broken down into 8×8 DCT blocks. In a 4:2:0 macroblock there

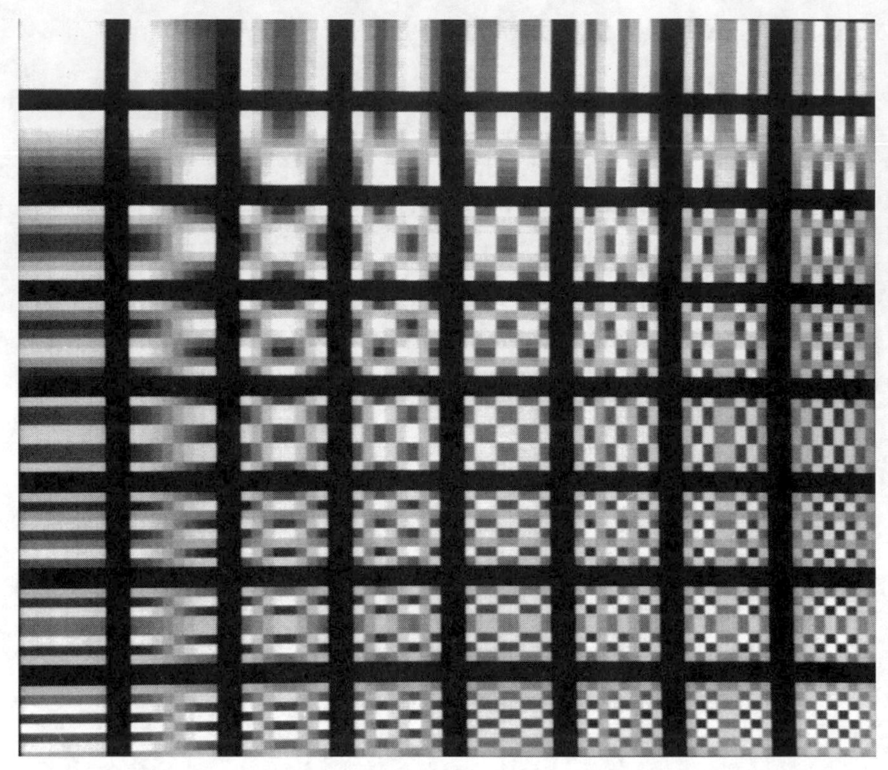

Figure 6.20 The discrete cosine transform breaks up an image area into discrete frequencies in two dimensions. The lowest frequency can be seen here at the top-left corner. Horizontal frequency increases to the right and vertical frequency increases downwards.

will be six DCT blocks whereas in a 4:2:2 macroblock there will be eight.

Figure 6.20 shows the table of basis functions or *wave table* for an 8 × 8 DCT. Adding these two-dimensional waveforms together in different proportions will give any original 8 × 8 pixel block. The coefficients of the DCT simply control the proportion of each wave which is added in the inverse transform. The top-left wave has no modulation at all because it conveys the DC component of the block. This coefficient will be a unipolar (positive only) value in the case of luminance and will typically be the largest value in the block as the spectrum of typical video signals is dominated by the DC component.

Increasing the DC coefficient adds a constant amount to every pixel. Moving to the right, the coefficients represent increasing horizontal spatial frequencies and moving downwards, the coefficients represent increasing vertical spatial frequencies. The bottom-right coefficient represents the highest diagonal frequencies in the block. All these

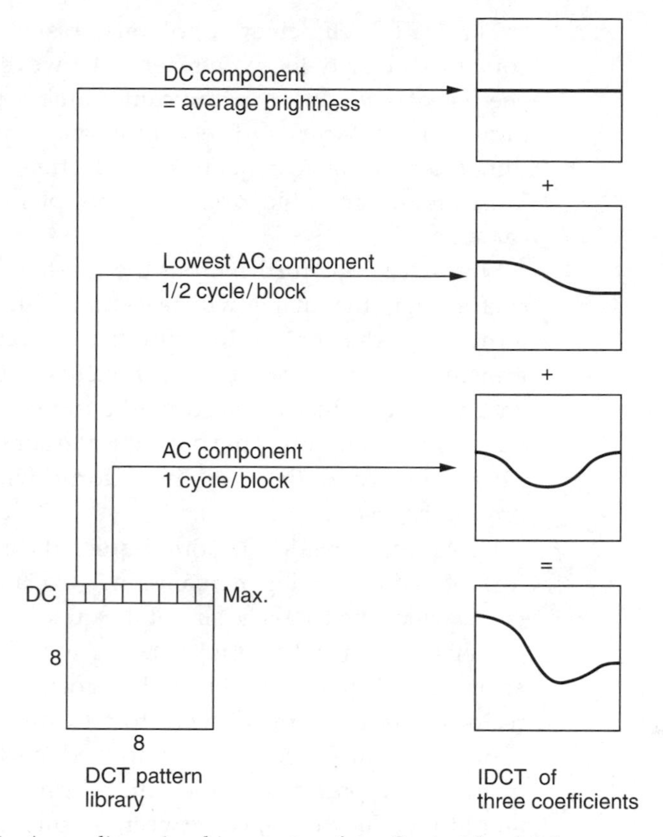

Figure 6.21 A one-dimensional inverse transform. See text for details.

coefficients are bipolar, where the polarity indicates whether the original spatial waveform at that frequency was inverted.

Figure 6.21 shows a one-dimensional example of an inverse transform. The DC coefficient produces a constant level throughout the pixel block. The remaining waves in the table are AC coefficients. A zero coefficient would result in no modulation, leaving the DC level unchanged. The wave next to the DC component represents the lowest frequency in the transform which is half a cycle per block. A positive coefficient would make the left side of the block brighter and the right side darker whereas a negative coefficient would do the opposite. The magnitude of the coefficient determines the amplitude of the wave which is added. Figure 6.21 also shows that the next wave has a frequency of one cycle per block. i.e. the block is made brighter at both sides and darker in the middle.

Consequently an inverse DCT is no more than a process of mixing various pixel patterns from the wave table where the relative amplitudes and polarity of these patterns are controlled by the coefficients. The original transform is simply a mechanism which finds the coefficient amplitudes from the original pixel block.

The DCT itself achieves no compression at all. Sixty-four pixels are converted to sixty-four coefficients. However, in typical pictures, not all coefficients will have significant values; there will often be a few dominant coefficients. The coefficients representing the higher two-dimensional spatial frequencies will often be zero or of small value in large areas, due to blurring or simply plain undetailed areas before the camera.

Statistically, the further from the top-left corner of the wave table the coefficient is, the smaller will be its magnitude. Coding gain (the technical term for reduction in the number of bits needed) is achieved by transmitting the low-valued coefficients with shorter wordlengths. The zero-valued coefficients need not be transmitted at all. Thus it is not the DCT which compresses the data, it is the subsequent processing. The DCT simply expresses the data in a form which makes the subsequent processing easier.

Higher compression factors require the coefficient wordlength to be further reduced using requantizing. Coefficients are divided by some factor which increases the size of the quantizing step. The smaller number of steps which results permits coding with fewer bits, but of course, with an increased quantizing error. The coefficients will be multiplied by a reciprocal factor in the decoder to return to the correct magnitude.

Inverse transforming a requantized coefficient means that the frequency it represents is reproduced in the output with the wrong amplitude. The difference between original and reconstructed amplitude is regarded as a noise added to the wanted data. Figure 6.22 shows that the visibility of such noise is far from uniform. The maximum sensitivity is found at DC and falls thereafter. As a result the top-left coefficient is often treated as a special case and left unchanged. It may warrant more error protection than other coefficients.

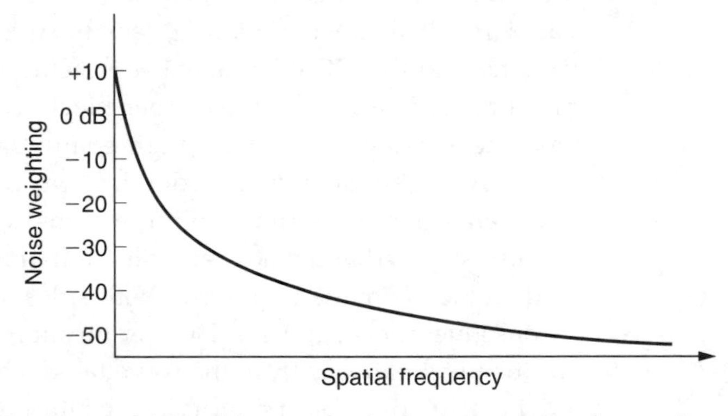

Figure 6.22 The sensitivity of the eye to noise is greatest at low frequencies and drops rapidly with increasing frequency. This can be used to mask quantizing noise caused by the compression process.

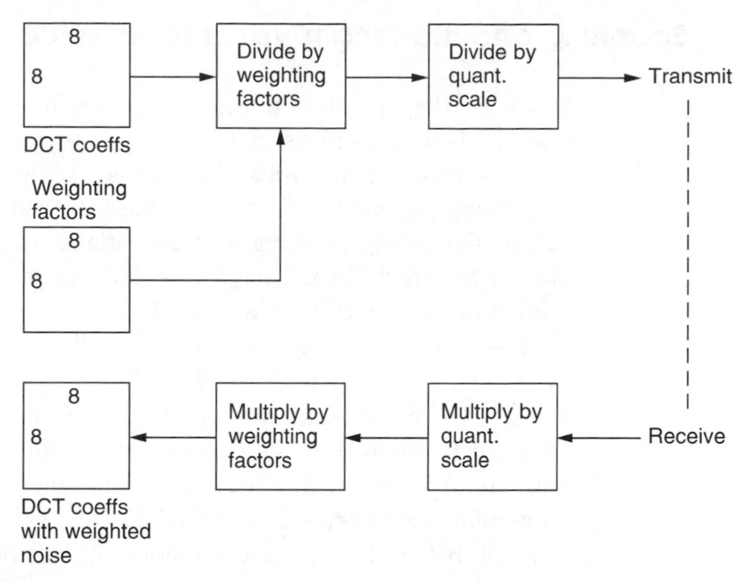

Figure 6.23 Weighting is used to make the noise caused by requantizing different at each frequency.

MPEG-2 takes advantage of the falling sensitivity to noise. Prior to requantizing, each coefficient is divided by a different weighting constant as a function of its frequency. Figure 6.23 shows a typical weighting process. Naturally the decoder must have a corresponding inverse weighting. This weighting process has the effect of reducing the magnitude of high-frequency coefficients disproportionately. Clearly, different weighting will be needed for colour difference data as colour is perceived differently.

P and *B* pictures are decoded by adding a prediction error image to a reference image. That reference image will contain weighted noise. One purpose of the prediction error is to cancel that noise to prevent tolerance build-up. If the prediction error were also to contain weighted noise this result would not be obtained. Consequently prediction error coefficients are flat weighted.

When forward prediction fails, such as in the case of new material introduced in a *P* picture by a pan, *P* coding would set the vectors to zero and encode the new data entirely as an unweighted prediction error. In this case it is better to encode that material as an *I* macroblock because then weighting can be used and this will require fewer bits.

Requantizing increases the step size of the coefficients, but the inverse weighting in the decoder results in step sizes which increase with frequency. The larger step size increases the quantizing noise at high frequencies where it is less visible. Effectively the noise floor is shaped to match the sensitivity of the eye. The quantizing table in use at the encoder can be transmitted to the decoder periodically in the bitstream.

6.8 Scanning and run-length/variable-length coding

Study of the signal statistics gained from extensive analysis of real material is used to measure the probability of a given coefficient having a given value. This probability turns out to be highly non-uniform, suggesting the possibility of a variable-length encoding for the coefficient values. On average, the higher the spatial frequency, the lower the value of a coefficient will be. This means that the value of a coefficient falls as a function of its radius from the DC coefficient.

Typical material often has many coefficients which are zero valued, especially after requantizing. The distribution of these also follows a pattern. The non-zero values tend to be found in the top-left corner of the DCT block, but as the radius increases, not only do the coefficient values fall, but it becomes increasingly likely that these small coefficients will be interspersed with zero-valued coefficients. As the radius increases further it is probable that a region where all coefficients are zero will be entered.

MPEG-2 uses all these attributes of DCT coefficients when encoding a coefficient block. By sending the coefficients in an optimum order, by describing their values with Huffman coding and by using run-length encoding for the zero-valued coefficients it is possible to achieve a significant reduction in coefficient data which remains entirely lossless. Despite the complexity of this process, it does contibute to improved picture quality because for a given bit rate lossless coding of the coefficients must be better than requantizing, which is lossy. Of course, for lower bit rates both will be required.

It is an advantage to scan in a sequence where the largest coefficient values are scanned first. Then the next coefficient is more likely to be zero than the previous one. With progressively scanned material, a regular zig-zag scan begins in the top-left corner and ends in the bottom-right corner as shown in Figure 6.24. Zig-zag scanning means that significant values are more likely to be transmitted first, followed by the zero values. Instead of coding these zeros, an unique 'end of block' (EOB) symbol is transmitted instead.

As the zig-zag scan approaches the last finite coefficient it is increasingly likely that some zero-value coefficients will be scanned. Instead of transmitting the coefficients as zeros, the *zero-run-length*, i.e. the number of zero-valued coefficients in the scan sequence, is encoded into the next non-zero coefficient which is itself variable-length coded. This combination of run-length and variable-length coding is known as RLC/VLC in MPEG-2.

The DC coefficient is handled separately because it is differentially coded and this discussion relates to the AC coefficients. Three items need to be handled for each coefficient: the zero-run-length prior to this

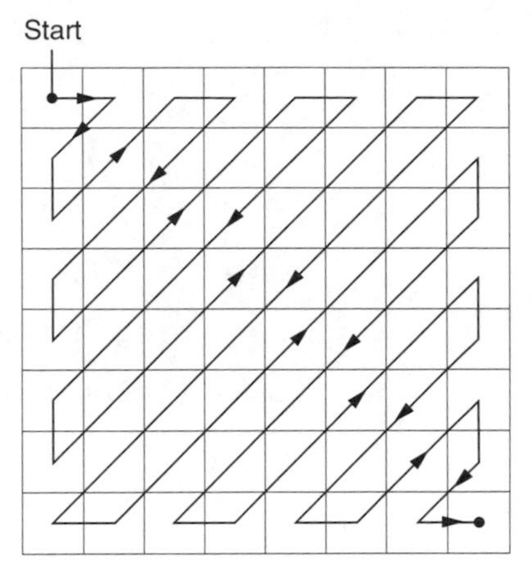

Figure 6.24 The zig-zag scan for a progressively scanned image.

coefficient, the wordlength and the coefficient value itself. The wordlength needs to be known by the decoder so that it can correctly parse the bitstream. The wordlength of the coefficient is expressed directly as an integer called the *size*.

Figure 6.25(a) shows that a two-dimensional run/size table is created. One dimension expresses the zero-run-length; the other the size. A run length of zero is obtained when adjacent coefficients are non-zero, but a code of 0/0 has no meaningful run/size interpretation and so this bit pattern is used for the EOB symbol.

In the case where the zero-run-length exceeds 14, a code of 15/0 is used, signifying that there are fifteen zero-valued coefficients. This is then followed by another run/size parameter whose run-length value is added to the previous 15.

The run/size parameters contain redundancy because some combinations are more common than others. Figure 6.25(b) shows that each run/size value is converted to a variable-length Huffman codeword for transmission. As was shown in section 1.5, the Huffman codes are designed so that short codes are never a prefix of long codes so that the decoder can deduce the parsing by testing an increasing number of bits until a match with the lookup table is found. Having parsed and decoded the Huffman run/size code, the decoder then knows what the coefficient wordlength will be and can correctly parse that.

The variable-length coefficient code has to describe a bipolar coefficient, i.e one which can be positive or negative. Figure 6.25(c) shows that for a particular size, the coding scale has a certain gap in it. For example, all values from –7 to +7 can be sent by a size 3 code, so a size 4 code only has to send the values of –15 to –8 and +8 to +15. The coefficient code is sent as a pure binary number whose value ranges from all zeros to all

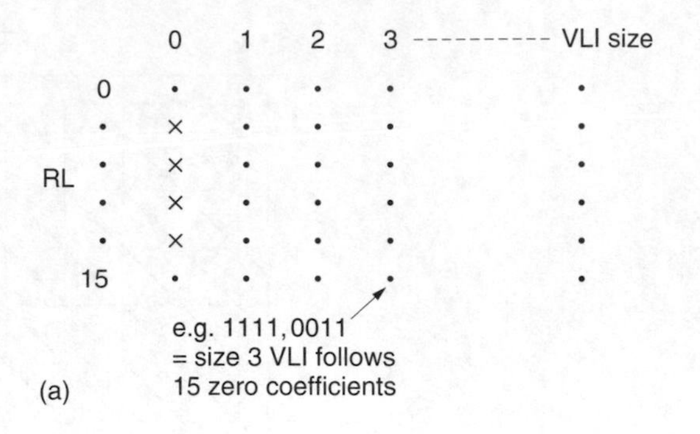

Run \ Size	0	1	2	3 → etc.
0	1010 (EOB)	00	01	100
1	–	1100	11011	
2	–	11100	11111001	
3	–	111010	111110111	
4	–	111011	1111111000	
5 ↓ etc.	–	1111010		

(b)

Figure 6.25 Run-length and variable-length coding simultaneously compresses runs of zero-valued coefficients and describes the wordlength of a non-zero coefficient.

ones where the maximum value is a function of the size. The number range is divided into two, the lower half of the codes specifying negative values and the upper half specifying positive ones.

In the case of positive numbers, the transmitted binary value is the actual coefficient value, whereas in the case of negative numbers a constant must be subtracted which is a function of the size. In the case of a size 4 code, the constant is 15_{10}. Thus a size 4 parameter of 0111_2 (7_{10}) would be interpreted as 7 – 15 = –8. A size of 5 has a constant of 31 so a transmitted coded of 01010_2 (10_2) would be interpreted as 10 – 31 = –21.

This technique saves a bit because, for example, 63 values from –31 to +31 are coded with only five bits having only 32 combinations. This is possible because that extra bit is effectively encoded into the run/size parameter.

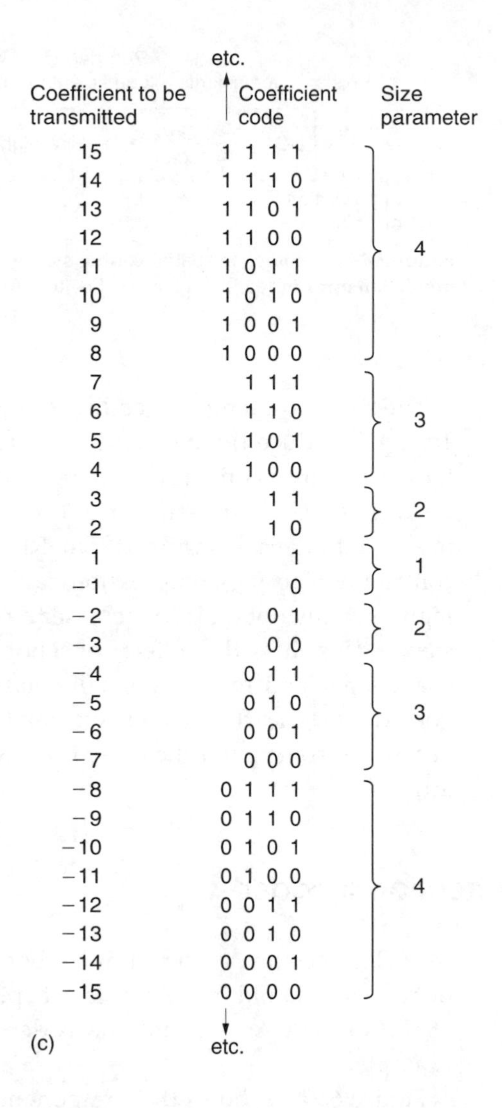

Figure 6.25(c)

Figure 6.26 shows the whole spatial coding subsystem. Macroblocks are subdivided into DCT blocks and the DCT is calculated. The resulting coefficients are multiplied by the weighting matrix and then requantized. The coefficients are then reordered by the zig-zag scan so that full advantage can be taken of run-length and variable-length coding. The last non-zero coefficient in the scan is followed by the EOB symbol.

In predictive coding, sometimes the motion-compensated prediction is nearly exact and so the prediction error will be almost zero. This can also happen on still parts of the scene. MPEG-2 takes advantage of this by sending a code to tell the decoder there is no prediction error data for the macroblock concerned.

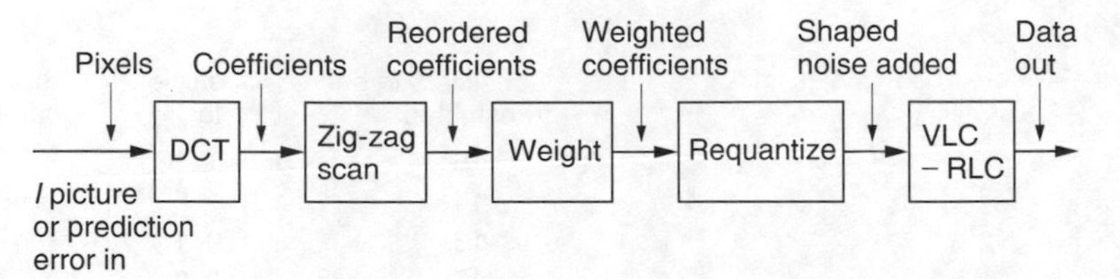

Figure 6.26 A complete spatial coding system which can compress an *I* picture or the prediction error in *P* and *B* pictures. See text for details.

The success of temporal coding depends on the accuracy of the vectors. Trying to reduce the bit rate by reducing the accuracy of the vectors is false economy as this simply increases the prediction error. Consequently for a given GOP structure it is only in the the spatial coding that the overall bit rate is determined. The RLC/VLC coding is lossless and so its contribution to the compression cannot be varied. If the bit rate is too high, the only option is to increase the size of the coefficient-requantizing steps. This has the effect of shortening the wordlength of large coefficients, and rounding small coefficients to zero, so that the bit rate goes down. Clearly if taken too far the picture quality will also suffer because at some point the noise floor will become visible as some form of artifact.

6.9 A bidirectional coder

MPEG-2 does not specify how an encoder is to be built or what coding decisions it should make. Instead it specifies the protocol of the bitstream at the output. As a result the coder shown in Figure 6.27 is only an example.

Figure 6.27(a) shows the component parts of the coder. At the input is a chain of picture stores which can be bypassed for reordering purposes. This allows a picture to be encoded ahead of its normal timing when bidirectional coding is employed.

At the centre is a dual-motion estimator which can simultaneously measure motion between the input picture, an earlier picture and a later picture. These reference pictures are held in frame stores. The vectors from the motion estimator are locally used to shift a picture in a frame store to form a predicted picture. This is subtracted from the input picture to produce a prediction error picture which is then spatially coded.

The bidirectional encoding process will now be described. A GOP begins with an *I* picture which is intra-coded. In Figure 6.27(b) the *I* picture emerges from the reordering delay. No prediction is possible on an *I* picture so the motion estimator is inactive. There is no predicted

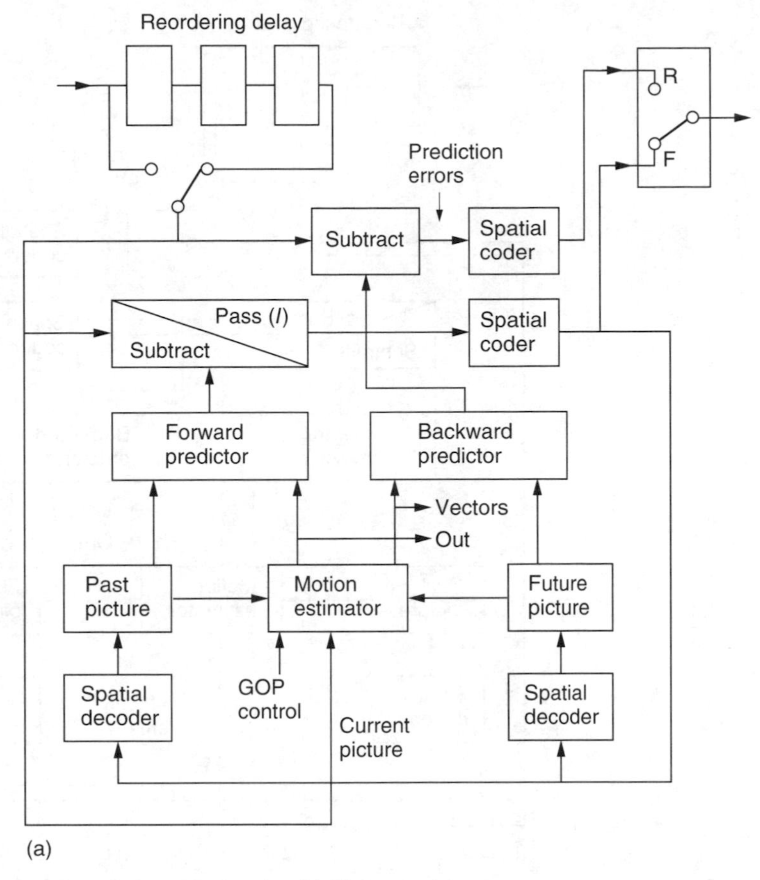

Figure 6.27 A bidirectional coder. (a) The essential components.

picture and so the prediction error subtractor is set simply to pass the input. The only processing which is active is the forward spatial coder which describes the picture with DCT coefficients. The output of the forward spatial coder is locally decoded and stored in the past picture frame store.

The reason for the spatial encode/decode is that the past picture frame store now contains exactly what the decoder frame store will contain, including the effects of any requantizing errors. When the same picture is used as a reference at both ends of a differential coding system, the errors will cancel out.

Having encoded the *I* picture, attention turns to the *P* picture. The input sequence is *IBBP*, but the transmitted sequence must be *IPBB*. Figure 6.27(c) shows that the reordering delay is bypassed to select the *P* picture. This passes to the motion estimator which compares it with the *I* picture and outputs a vector for each macroblock. The forward predictor uses these vectors to shift the *I* picture so that it more closely resembles

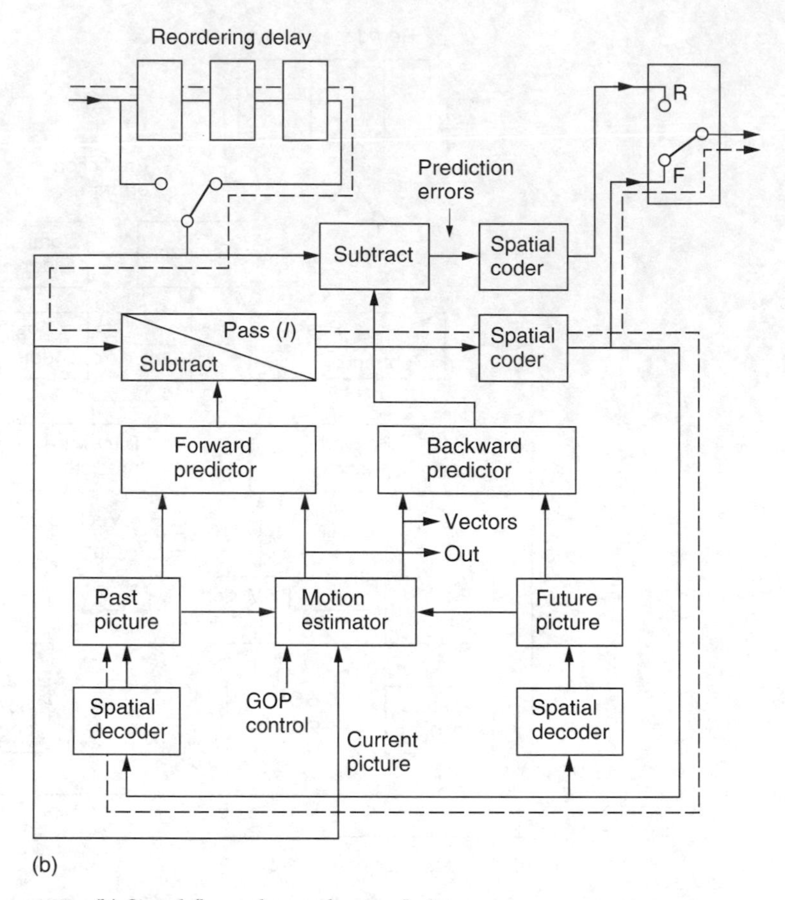

Figure 6.27 (b) Signal flow when coding an *I* picture.

the *P* picture. The predicted picture is then subtracted from the actual picture to produce a forward prediction error. This is then spatially coded. Thus the *P* picture is transmitted as a set of vectors and a prediction error image.

The *P* picture is locally decoded in the right-hand decoder. This takes the forward predicted picture and adds the decoded prediction error to obtain exactly what the decoder will obtain.

Figure 6.27(d) shows that the encoder now contains an *I* picture in the left store and a *P* picture in the right store. The reordering delay is reselected so that the first *B* picture can be input. This passes to the motion estimator where it is compared with both the *I* and *P* pictures to produce forward and backward vectors. The forward vectors go to the forward predictor to make a *B* prediction from the *I* picture. The backward vectors go to the backward predictor to make a *B* prediction from the *P* picture. These predictions are simultaneously subtracted from the actual *B* picture to produce a forward prediction error and a backward

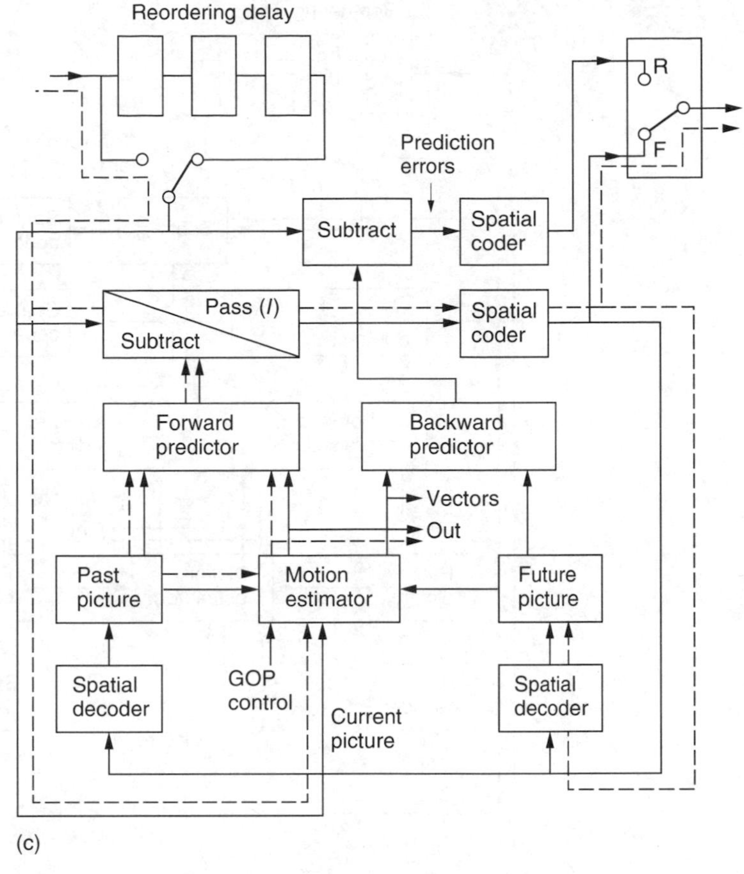

Figure 6.27 (c) Signal flow when coding a *P* picture.

prediction error. These are then spatially encoded. The encoder can then decide which direction of coding resulted in the best prediction; i.e. the smallest prediction error.

Not shown in the interests of clarity is a third signal path which creates a predicted *B* picture from the average of forward and backward predictions. This is subtracted from the input picture to produce a third prediction error. In some circumstances this prediction error may use less data than either forward or backward prediction alone.

As *B* pictures are never used to create other pictures, the decoder does not locally decode the *B* picture. After decoding and displaying the *B* picture the decoder will discard it. At the encoder the *I* and *P* pictures remain in their frame stores and the second *B* picture is input from the reordering delay.

Following the encoding of the second *B* picture, the encoder must reorder again to encode the second *P* picture in the GOP. This will locally be decoded and will replace the *I* picture in the left store. The stores and

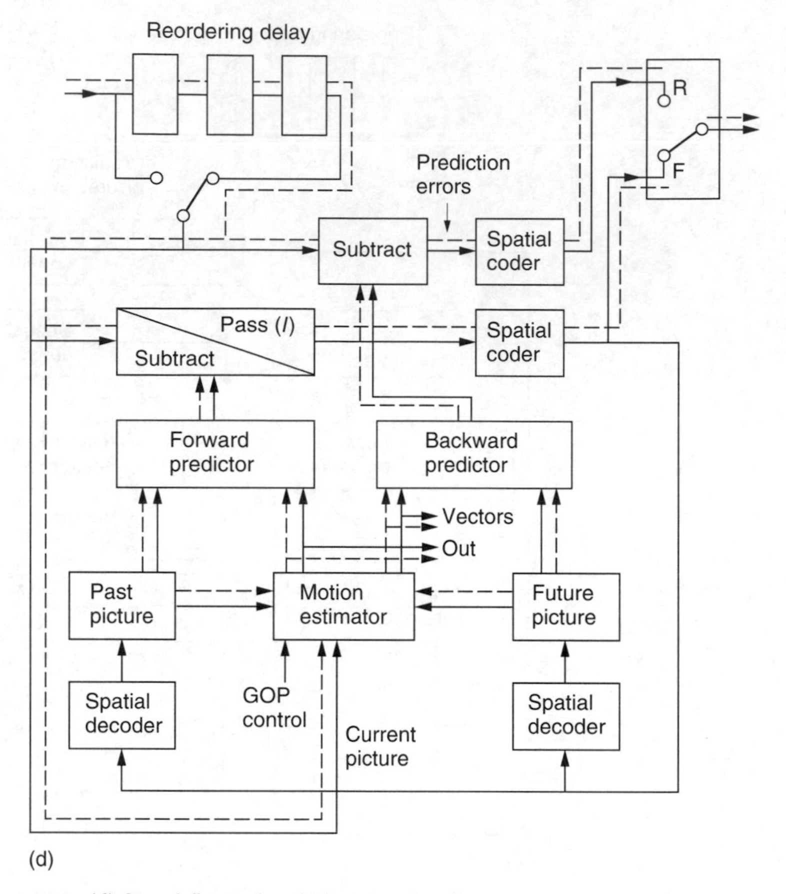

Figure 6.27 (d) Signal flow when bidirectional coding.

predictors switch designation because the left store is now a future *P* picture and the right store is now a past *P* picture. *B* pictures between them are encoded as before.

6.10 Slices

There is still some redundancy in the output of a bidirectional coder and MPEG-2 is remarkably diligent in finding it. In *I* pictures, the DC coefficient describes the average brightness of an entire DCT block. In real video the DC component of adjacent blocks will be similar much of the time. A saving in bit rate can be obtained by differentially coding the DC coefficient.

In *P* and *B* pictures this is not done because these are prediction errors, not actual images and the statistics are different. However, *P* and *B* pictures send vectors and instead the redundancy in these is explored. In a large moving object, many macroblocks will be moving at the same velocity and their vectors will be the same. Thus differential vector coding will be advantageous.

As has been seen above, differential coding cannot be used indiscriminately as it is prone to error propagation. Periodically absolute DC coefficients and vectors must be sent and the *slice* is the logical structure which supports this mechanism. In *I* pictures, the first DC coefficient in a slice is sent in absolute form, whereas the subsequent coefficients are sent differentially. In *P* or *B* pictures, the first vector in a slice is sent in absolute form, but the subsequent vectors are differential.

Slices are horizontal picture strips which are one macroblock (16 pixels) high and which proceed from left to right across the screen. The sides of the picture must coincide with the beginning or the end of a slice in MPEG-2, but otherwise the encoder is free to decide how big slices should be and where they begin.

In the case of a central dark building silhouetted against a bright sky, there would be two large changes in the DC coefficients, one at each edge of the building. It may be advantageous to the encoder to break the width of the picture into three slices, one each for the left and right areas of sky and one for the building. In the case of a large moving object, different slices may be used for the object and the background.

Each slice contains its own synchronizing pattern, so following a transmission error, correct decoding can resume at the next slice. Slice size can also be matched to the characteristics of the transmission channel. For example, in an error-free transmission system the use of a large number of slices in a packet simply wastes data capacity on surplus synchronizing patterns. However, in a non-ideal system it might be advantageous to have frequent resynchronizing.

6.11 Handling interlaced pictures

Spatial coding, predictive coding and motion compensation can still be performed using interlaced source material at the cost of considerable complexity. Despite that complexity, MPEG-2 cannot be expected to perform as well with interlaced material.

Figure 6.28 shows that in an incoming interlaced frame there are two fields each of which contain half of the lines in the frame. In MPEG-2 these are known as the *top field* and the *bottom field*. In video from a camera, these fields represent the state of the image at two different times. Where there is little image motion, this is unimportant and the fields can be combined, obtaining more effective compression. However, in the presence of motion the fields become increasingly decorrelated because of the displacement of moving objects from one field to the next.

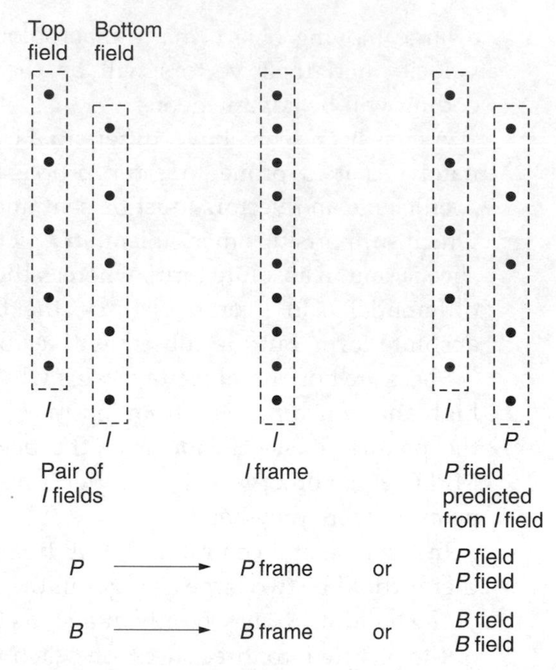

Figure 6.28 An interlaced frame consists of top and bottom fields. MPEG-2 can code a frame in the ways shown here.

This characteristic determines that MPEG-2 must be able to handle fields independently or together. This dual approach permeates all aspects of MPEG-2 and affects the definition of pictures, macroblocks, DCT blocks and zig-zag scanning.

Figure 6.28 also shows how MPEG-2 designates interlaced fields. In picture types *I*, *P* and *B*, the two fields can be superimposed to make a *frame-picture* or the two fields can be coded independently as two *field-pictures*. As a third possibility, in *I* pictures only, the bottom field-picture can be predictively coded from the top field-picture to make an *IP* frame-picture.

A frame-picture is one in which the macroblocks contain lines from both field types over a picture area 16 scan lines high. Each luminance macroblock contains the usual four DCT blocks but there are two ways in which these can be assembled. Figure 6.29(a) shows how a frame is divided into *frame DCT* blocks. This is identical to the progressive scan approach in that each DCT block contains eight contiguous picture lines. In 4:2:0, the colour difference signals have been downsampled by a factor of two and shifted as was shown in section 4.18. Figure 6.29(a) also shows how one 4:2:0 DCT block contains the chroma data from 16 lines in two fields.

Even small amounts of motion in any direction can destroy the correlation between odd and even lines and a frame DCT will result in an

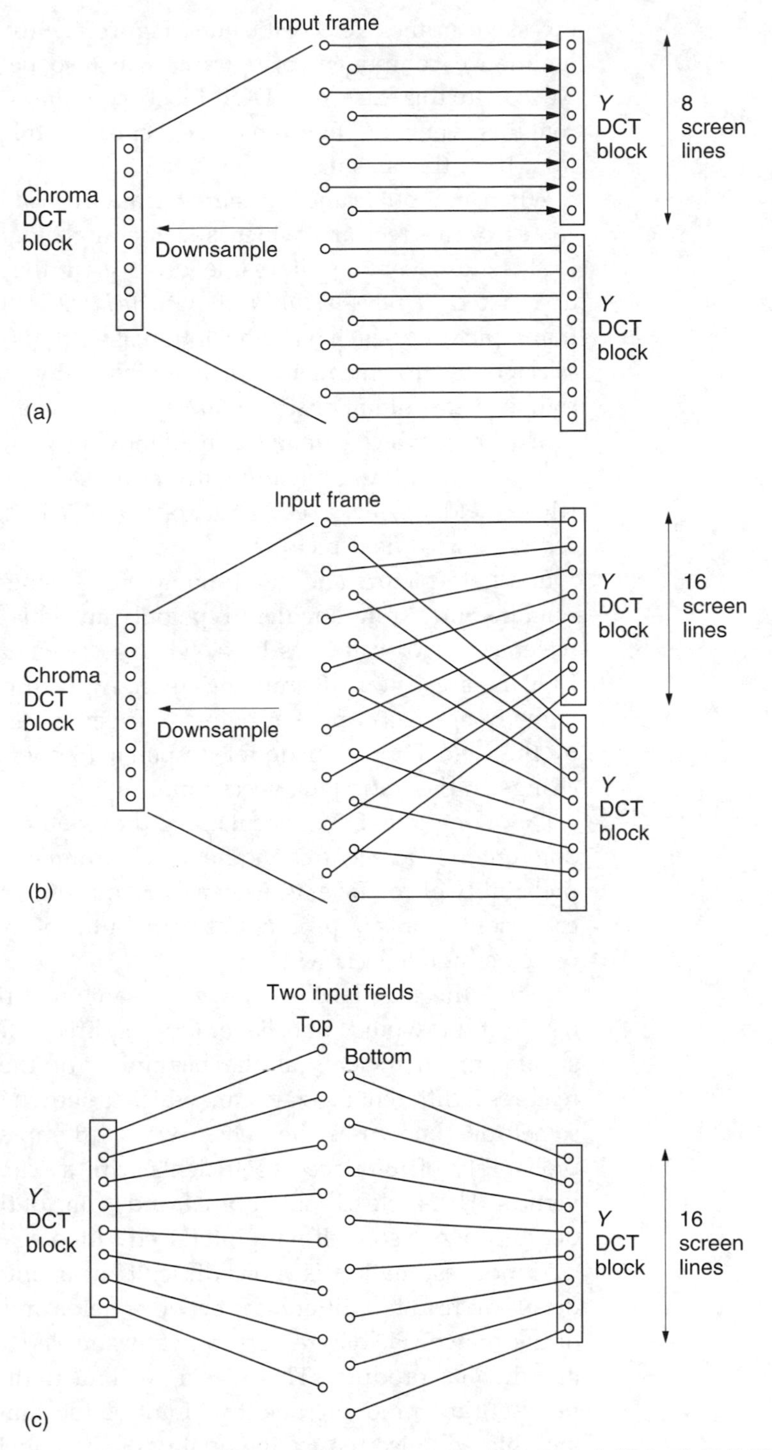

Figure 6.29 (a) In Frame-DCT, a picture is effectively de-interlaced. (b) In Field-DCT, each DCT block only contains lines from one field, but over twice the screen area. (c) The same DCT content results when field-pictures are assembled into blocks.

excessive number of coefficients. Figure 6.29(b) shows that instead the luminance component of a frame can also be divided into *field DCT* blocks. In this case one DCT block contains odd lines and the other contains even lines. In this mode the chroma still produces one DCT block from both fields as in Figure 6.29(a).

When an input frame is designated as two *field-pictures*, the macroblocks come from a screen area which is 32 lines high. Figure 6.29(c) shows that the DCT blocks contain the same data as if the input frame had been designated a *frame-picture* but with *field DCT*. Consequently it is only frame-pictures which have the option of field or frame DCT. These may be selected by the encoder on a macroblock-by-macroblock basis and, of course, the resultant bitstream must specify what has been done.

In a frame which contains a small moving area, it may be advantageous to encode as a frame-picture with frame DCT except in the moving area where field DCT is used. This approach may result in fewer bits than coding as two field-pictures.

In a field-picture and in a frame-picture using field DCT, a DCT block contains lines from one field type only and this must have come from a screen area 16 scan lines high, whereas in progressive scan and frame DCT the area is only 8 scan lines high. A given vertical spatial frequency in the image is sampled at points twice as far apart which is interpreted by the field DCT as a doubled spatial frequency, whereas there is no change in the horizontal spectrum.

Following the DCT calculation, the coefficient distribution will be different in field-pictures and field DCT frame-pictures. In these cases, the probability of coefficients is not a constant function of radius from the DC coefficient as it is in progressive scan, but is elliptical where the ellipse is twice as high as it is wide.

Using the standard 45° zig-zag scan with this different coefficient distribution would not have the required effect of putting all the significant coefficients at the beginning of the scan. To achieve this requires a different zig-zag scan, which is shown in Figure 6.30. This scan, sometimes known as the Yeltsin walk, attempts to match the elliptical probability of interlaced coefficients with a scan slanted at 67.5° to the vertical. This is clearly suboptimal, and is one of the reasons why MPEG-2 does not work so well with interlaced video.

Motion estimation is more difficult in an interlaced system. Vertical detail can result in differences between fields and this reduces the quality of the match. Fields are vertically subsampled without filtering and so contain alias products. This aliasing will mean that the vertical waveform representing a moving object will not be the same in successive pictures and this will also reduce the quality of the match.

Even when the correct vector has been found, the match may be poor so the estimator fails to recognize it. If it is recognized, a poor match

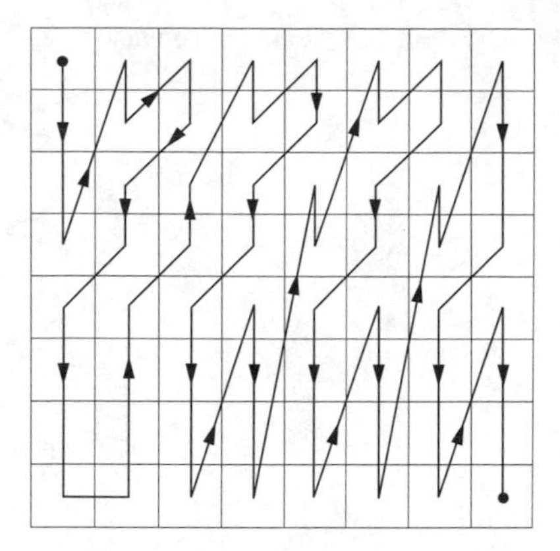

Figure 6.30 The zig-zag scan for an interlaced image has to favour vertical frequencies twice as much as horizontal.

means that the quality of the prediction in *P* and *B* pictures will be poor and so a large prediction error or residual has to be transmitted. In an attempt to reduce the residual, MPEG-2 allows field-pictures to use motion-compensated prediction from either the adjacent field or from the same field type in another frame. In this case the encoder will use the better match. This technique can also be used in areas of frame-pictures which use field DCT.

The motion compensation of MPEG-2 has half-pixel resolution and this is inherently compatible with interlace because an interpolator must be present to handle the half-pixel shifts. Figure 6.31(a) shows that in an interlaced system, each field contains half of the frame lines and so interpolating half-way between lines of one field type will actually create values lying on the sampling structure of the other field type. Thus it is equally possible for a predictive system to decode a given field type based on pixel data from the other field type or of the same type.

If when using predictive coding from the other field type the vertical motion vector contains a half-pixel component, then no interpolation is needed because the act of transferring pixels from one field to another results in such a shift.

Figure 6.31(b) shows that a macroblock in a given *P* field-picture can be encoded using a vector which shifts data from the previous field or from the field before that, irrespective of which frames these fields occupy. As noted above, field-picture macroblocks come from an area of screen 32 lines high and this means that the vector density is halved, resulting in larger prediction errors at the boundaries of moving objects.

As an option, field-pictures can restore the vector density by using 16 $\times$ 8 motion compensation where separate vectors are used for the top and bottom halves of the macroblock. Frame-pictures can also use 16 $\times$ 8

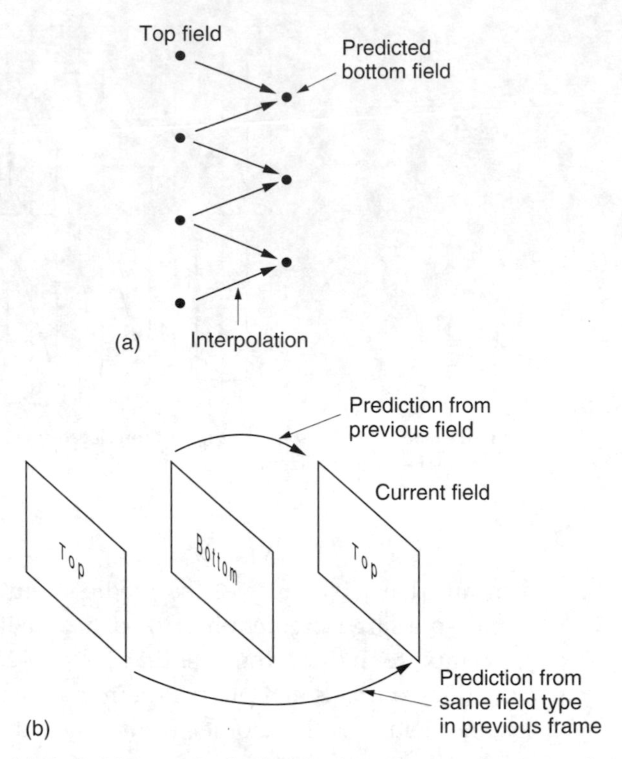

Figure 6.31 (a) Each field contains half of the frame lines and so interpolation is needed to create values lying on the sampling structure of the other field type. (b) Prediction can use data from the previous field or the one before that.

motion compensation in conjunction with field DCT. Whilst the 2×2 DCT block luminance structure of a macroblock can easily be divided vertically in two, in 4:2:0 the same screen area is represented by only one chroma macroblock of each component type. As it cannot be divided in half, this chroma is deemed to belong to the luminance DCT blocks of the upper field. In 4:2:2 no such difficulty arises.

MPEG-2 supports interlace simply because interlaced video exists in legacy systems and there is a requirement to compress it. However, where the opportunity arises to define a new system, interlace should be avoided. Legacy interlaced source material should be handled using a motion-compensated de-interlacer prior to compression in the progressive domain.

6.12 An MPEG-2 coder

Figure 6.32 shows the complete coder. The bidirectional coder outputs coefficients and vectors, and the quantizing table in use. The vectors of *P* and *B* pictures and the DC coefficients of *I* pictures are differentially

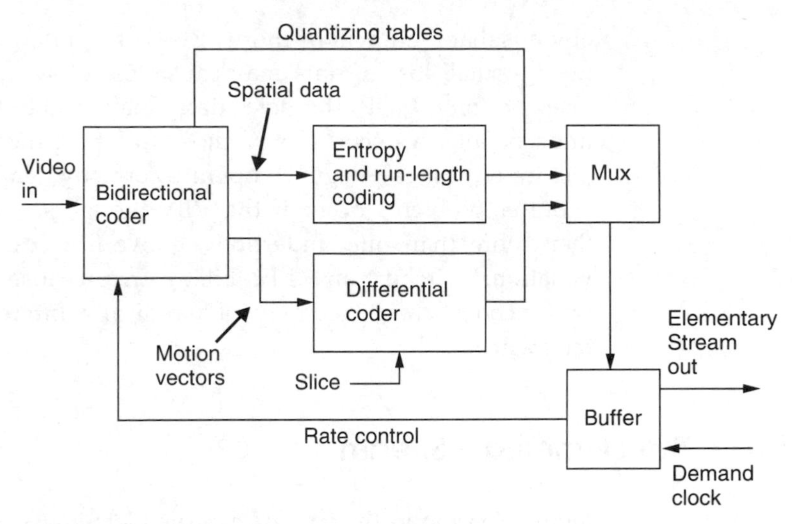

Figure 6.32 An MPEG-2 coder. See text for details.

encoded in slices and the remaining coefficients are RLC/VLC coded. The multiplexer assembles all this data into a single bitstream called an Elementary Stream. The output of the encoder is a buffer which absorbs the variations in bit rate between different picture types. The buffer output has a constant bit rate determined by the demand clock. This comes from the transmission channel or storage device. If the bit rate is low, the buffer will tend to fill up, whereas if it is high the buffer will tend to empty. The buffer content is used to control the severity of the requantizing in the spatial coders. The more the buffer fills, the bigger the requantizing steps get.

The buffer in the decoder has a finite capacity and the encoder must model the decoder's buffer occupancy so that it neither overflows nor underflows. An overflow might occur if an *I* picture is transmitted when the buffer content is already high. The buffer occupancy of the decoder depends somewhat on the memory access strategy of the decoder. Instead of defining a specific buffer size, MPEG-2 defines the size of a particular mathematical model of a hypothetical buffer. The decoder designer can use any strategy which implements the model, and the encoder can use any strategy which doesn't overflow or underflow the model. The Elementary Stream has a parameter called the video buffer verifier (VBV) which defines the minimum buffering assumptions of the encoder.

As was seen in Chapter 1, buffering is one way of ensuring constant quality when picture entropy varies. An intelligent coder may run down the buffer contents in anticipation of a difficult picture sequence so that a large amounts of data can be sent.

MPEG-2 does not define what a decoder should do if a buffer underflow or overflow occurs, but since both irrecoverably lose data it is obvious that there will be more or less of an interruption to the decoding. Even a small loss of data may cause loss of synchronization and in the case of long GOP the lost data may make the rest of the GOP undecodable. A decoder may chose to repeat the last properly decoded picture until it can begin to operate correctly again.

Buffer problems occur if the VBV model is violated. If this happens then more than one underflow or overflow can result from a single violation. Switching an MPEG bitstream can cause a violation because the two encoders concerned may have radically different buffer occupancy at the switch.

6.13 The Elementary Stream

Figure 6.33 shows the structure of the Elementary Stream from an MPEG-2 encoder. The structure begins with a set of coefficients representing a DCT block. Six or eight DCT blocks form the luminance and chroma content of one macroblock. In *P* and *B* pictures a macroblock will be associated with a vector for motion compensation. Macroblocks are associated into slices in which DC coefficients of *I* pictures and vectors in *P* and *B* pictures are differentially coded. An arbitrary number of slices forms a picture and this needs *I*/*P*/*B* flags describing the type of picture it is. The picture may also have a global vector which efficiently deals with pans.

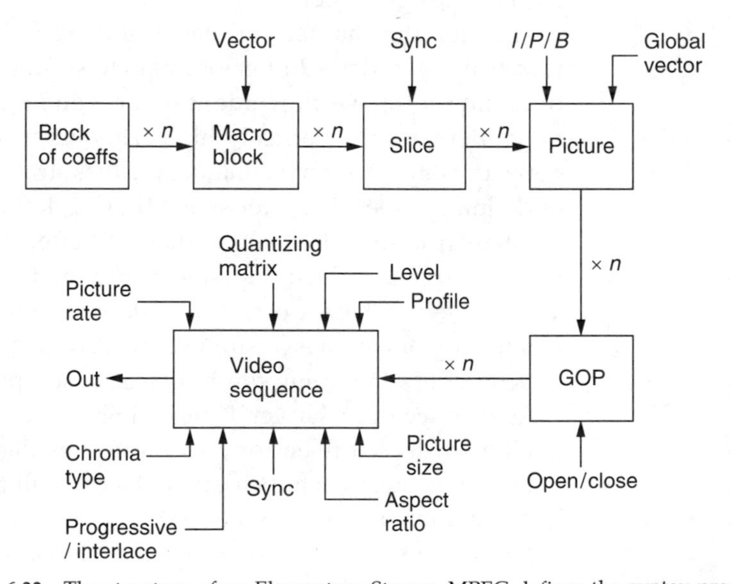

Figure 6.33 The structure of an Elementary Stream. MPEG defines the syntax precisely.

Several pictures form a Group of Pictures (GOP). The GOP begins with an *I* picture and may or may not include *P* and *B* pictures in a structure which may vary dynamically.

Several GOPs form a Sequence which begins with a Sequence header containing important data to help the decoder. It is possible to repeat the header within a sequence, and this helps lock-up in random access applications. The Sequence header describes the MPEG-2 profile and level, whether the video is progressive or interlaced, whether the chroma is 4:2:0 or 4:2:2, the size of the picture and the aspect ratio of the pixels. The quantizing matrix used in the spatial coder can also be sent. The sequence begins with a standardized bit pattern which is detected by a decoder to synchronize the deserialization.

6.14 An MPEG-2 decoder

The decoder is only defined by implication from the definitions of syntax and any decoder which can correctly interpret all combinations of syntax at a particular profile will be deemed compliant however it works.

The first problem a decoder has is that the input is an endless bitstream which contains a huge range of parameters many of which have variable length. Unique synchronizing patterns must be placed periodically throughout the bitstream so that the decoder can identify a known starting point. The pictures which can be sent under MPEG-2 are so flexible that the decoder must first find a Sequence header so that it can establish the size of the picture, the frame rate, the colour coding used etc.

The decoder must also be supplied with a 27 MHz system clock. In a DVD player, this would come from a crystal, but in a transmission system this would be provided by a numerically locked loop running from clock reference parameter in the bitstream (see Chapter 2). Until this loop has achieved lock the decoder cannot function properly.

Figure 6.34 shows a bidirectional decoder. The decoder can only begin decoding with an *I* picture and as this only uses intra-coding there will be no vectors. An *I* picture is transmitted as a series of slices. These slices begin with subsidiary synchronizing patterns. The first macroblock in the slice contains an absolute DC coefficient, but the remaining macroblocks code the DC coefficient differentially so the decoder must subtract the differential values from the previous one to obtain the absolute value.

The AC coefficients are sent as Huffman coded run/size parameters followed by coefficient value codes. The variable-length Huffman codes are decoded by using a lookup table and extending the number of bits considered until a match is obtained. This allows the zero-run-length and the coefficient size to be established. The right number of bits is taken

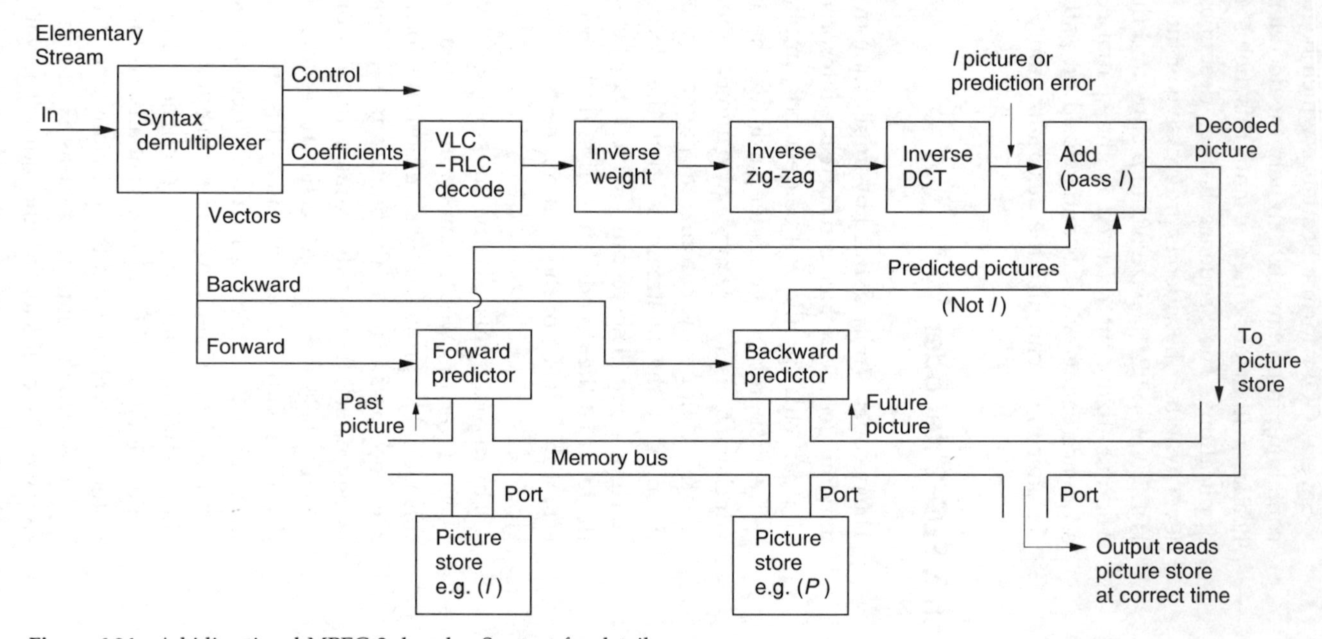

Figure 6.34 A bidirectional MPEG-2 decoder. See text for details.

from the bitstream corresponding to the coefficient code and this is decoded to the actual coefficient using the size parameter.

If the correct number of bits has been taken from the stream, the next bit must be the beginning of the next run/size code and so on until the EOB symbol is reached. The decoder uses the coefficient values and the zero-run-lengths to populate a DCT coefficient block following the appropriate zig-zag scanning sequence. Following EOB, the bitstream then continues with the next DCT block. Clearly this Huffman decoding will work perfectly or not at all. A single bit slippage in synchronism or a single corrupted data bit can cause a spectacular failure.

Once a complete DCT coefficient block has been received, the coefficients need to be inverse quantized and inverse weighted. Then an inverse DCT can be performed and this will result in an 8×8 pixel block. A series of DCT blocks will allow the luminance and colour information for an entire macroblock to be decoded and this can be placed in a frame store. Decoding continues in this way until the end of the slice when an absolute DC coefficient will once again be sent. Once all the slices have been decoded, an entire picture will be resident in the frame store.

The amount of data needed to decode the picture is variable and the decoder just keeps going until the last macroblock is found. It will obtain data from the input buffer. In a constant bit rate transmission system, the decoder will remove more data to decode an *I* picture than has been received in one picture period, leaving the buffer emptier than it began. Subsequent *P* and *B* pictures need much less data and allow the buffer to fill again. The picture will be output when the time stamp sent with the picture matches the state of the decoder's time count.

Following the *I* picture may be another *I* picture or a *P* picture. Assuming a *P* picture, this will be predictively coded from the *I* picture. The *P* picture will be divided into slices as before. The first vector in a slice is absolute, but subsequent vectors are sent differentially. However, the DC coefficients are not differential.

Each macroblock may contain a forward vector. The decoder uses this to shift pixels from the *I* picture into the correct position for the predicted *P* picture. The vectors have half-pixel resolution and where a half-pixel shift is required, an interpolator will be used.

The DCT data is sent much as for an *I* picture, it will require inverse quantizing, but not inverse weighting because *P* and *B* coefficients are flat-weighted. When decoded this represents an error-cancelling picture which is added pixel-by-pixel to the motion predicted picture. This results in the output picture.

If bidirectional coding is being used, the *P* picture may be stored until one or more *B* pictures have been decoded. The *B* pictures are sent essentially as a *P* picture might be, except that the vectors can be forward, backward or bidirectional. The decoder must take pixels from the *I* picture, the *P* picture, or both, and shift them according to the vectors to make a predicted picture. The DCT data decodes to produce an error-cancelling image as before.

In an interlaced system, the prediction mechanism may alternatively obtain pixel data from the previous field or the field before that. Vectors may relate to macroblocks or to 16 × 8 pixel areas. DCT blocks after decoding may represent frame lines or field lines. This adds up to a lot of different possibilities for a decoder handling an interlaced input.

6.15 Coding artifacts

This section describes the visible results of imperfect coding. Imperfect coding may be where the coding algorithm is sub-optimal, where the coder latency is too short or where the compression factor in use is simply too great for the material.

In motion-compensated systems such as MPEG, the use of periodic intra-fields means that the coding noise varies from picture to picture and this may be visible as noise pumping. Noise pumping may also be visible where the amount of motion changes. If a pan is observed, as the pan speed increases the motion vectors may become less accurate and reduce the quality of the prediction processes. The prediction errors will get larger and will have to be more coarsely quantized. Thus the picture gets noisier as the pan accelerates and the noise reduces as the pan slows down. The same result may be apparent at the edges of a picture during zooming. The problem is worse if the picture contains fine detail. Panning on grass or trees waving in the wind taxes most coders severely. Camera shake from a hand-held camera also increases the motion vector data and results in more noise as does film weave.

Input video noise or film grain degrades inter-coding as there is less redundancy between pictures and the difference data become larger, requiring coarse quantizing and adding to the existing noise.

Where a codec is really fighting the quantizing may become very coarse and as a result the video level at the edge of one DCT block may not match that of its neighbour. Therefore the DCT block structure becomes visible as a mosaicing or tiling effect. Coarse quantizing also causes some coefficients to be rounded up and appear larger than they should be. High-frequency coefficients may be eliminated by heavy quantizing and this forces the DCT to act as a steep-cut low-pass filter. This causes fringeing or ringing around sharp edges and extra shadowy edges which were not in the original. This is most noticeable on text.

Excess compression may also result in colour bleed where fringeing has taken place in the chroma or where high-frequency chroma coefficients have been discarded. Graduated colour areas may reveal banding or posterizing as the colour range is restricted by requantizing. These artifacts are almost impossible to measure with conventional test gear.

Neither noise pumping nor blocking are visible on analog video recorders and so it is nonsense to liken the performance of a codec to the quality of a VCR. In fact noise pumping is extremely objectionable because, unlike steady noise, it attracts attention in peripheral vision and may result in viewing fatigue.

In addition to highly detailed pictures with complex motion, certain types of video signal are difficult for MPEG-2 to handle and will usually result in a higher level of artifacts than usual. Noise has already been mentioned as a source of problems. Timebase error from, for example, VCRs is undesirable because this puts succesive lines in different horizontal positions. A straight vertical line becomes jagged and this results in high spatial frequencies in the DCT process. Spurious coefficients are created which need to be coded.

Much archive video is in composite form and MPEG-2 can only handle this after it has been decoded to components. Unfortunately many general-purpose composite decoders have a high level of residual subcarrier in the outputs. This is normally not a problem because the subcarrier is designed to be invisible to the naked eye. Figure 6.35 shows that in PAL and NTSC the subcarrier frequency is selected so that a phase reversal is achieved between successive lines and frames.

Whilst this makes the subcarrier invisible to the eye, it is not invisible to an MPEG decoder. The subcarrier waveform is interpreted as a horizontal frequency, the vertical phase reversals are interpreted as a vertical spatial frequency and the picture-to-picture reversals increase the

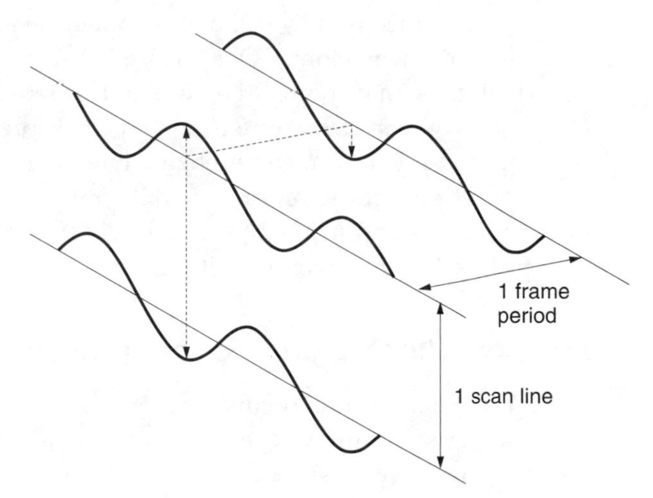

Figure 6.35 In composite video the subcarrier frequency is arranged so that inversions occur between adjacent lines and pictures to help reduce the visibility of the chroma.

magnitude of the prediction errors. The subcarrier level may be low but it can be present over the whole screen and may require an excess of coefficients to describe it.

Composite video should not in general be used as a source for MPEG-2 encoding, but where this is inevitable the standard of the decoder must be much higher than average, especially in the residual subcarrier specification. Some MPEG preprocessors support high-grade composite decoding options.

Judder from conventional linear standards convertors degrades the performance of MPEG-2. The optic flow axis is corrupted and linear filtering causes multiple images which confuse motion estimators and result in larger prediction errors. If standards conversion is necessary, the MPEG-2 system must be used to encode the signal in its original format and the standards convertor should be installed after the decoder. If a standards convertor has to be used before the encoder, then it must be a type which has effective motion compensation.

Film weave causes movement of one picture with respect to the next and this results in more vector activity and larger prediction errors. Movement of the centre of the film frame along the optical axis causes magnification changes which also result in excess prediction error data. Film grain has the same effect as noise: it is random and so cannot be compressed.

Perhaps because it is relatively uncommon, MPEG-2 cannot handle image rotation well because the motion-compensation system is only designed for translational motion. Where a rotating object is highly detailed, such as in certain fairground rides, the motion-compensation failure requires a significant amount of prediction error data and if a suitable bit rate is not available the level of artifacts will rise.

Flash guns used by still photographers are a serious hazard to MPEG-2 especially when long GOPs are used. At a press conference where a series of flashes may occur, the resultant video contains intermittent white frames which defeat prediction. A huge prediction error is required to turn the previous picture into a white picture, followed by another huge prediction error to return the white frame to the next picture. The output buffer fills and heavy requantizing is employed. After a few flashes the picture has generally gone to tiles.

6.16 Processing MPEG-2 and concatenation

Concatenation loss occurs when the losses introduced by one codec are compounded by a second codec. All practical compressers, MPEG-2 included, are lossy because what comes out of the decoder is not bit-identical to what went into the encoder. The bit differences are controlled so that they have minimum visibility to a human viewer.

MPEG-2 is a toolbox which allows a variety of manipulations to be performed in both the spatial and the temporal domain. There is a limit to the compression which can be used on a single frame, and if higher compression factors are needed, temporal coding will have to be used. The longer the run of pictures considered, the lower the bit rate needed, but the harder it becomes to edit.

The most editable form of MPEG-2 is to use *I* pictures only. As there is no temporal coding, pure-cut edits can be made between pictures. The next best thing is to use a repeating *IB* structure which is locked to the odd/even field structure. Cut edits cannot be made as the *B* pictures are bidirectionally coded and need data from both adjacent *I* pictures for decoding. The *B* picture has to be decoded prior to the edit and re-encoded after the edit. This will cause a small concatenation loss.

Beyond the *IB* structure processing gets harder. If a long GOP is used for the best compression factor, an *IBBPBBP* . . . structure results. Editing this is very difficult because the pictures are sent out of order so that bidirectional decoding can be used. MPEG allows closed GOPs where the last *B* picture is coded wholly from the previous pictures and does not need the *I* picture in the next GOP. The bitstream can be switched at this point but only if the GOP structures in the two source video signals are synchronized (makes colour framing seem easy). Consequently in practice a long GOP bitstream will need to be decoded prior to any production step. Afterwards it will need to be re-encoded.

This is known as *naive* concatenation and an enormous pitfall awaits. Unless the GOP structure of the output is identical to and synchronized with the input the results will be disappointing. The worst case is where an *I* picture is encoded from a picture which was formerly a *B* picture. It is easy enough to lock the GOP structure of a coder to a single input, but if an edit is made between two inputs, the GOP timings could well be different.

As there are so many structures allowed in MPEG, there will be a need to convert between them. If this has to be done, it should only be in the direction which increases the GOP length and reduces the bit rate. Going the other way is inadvisable. The ideal way of converting from, say, the *IB* structure of a news system to the *IBBP* structure of an emission system is to use a recompressor. This is a kind of standards convertor which will give better results than a decode followed by an encode.

The DCT part of MPEG-2 itself is lossless. If all the coefficients are preserved intact an inverse transform yields the same pixel data. Unfortunately this does not give enough compression for many applications. In practice the coefficients are made less accurate by removing bits starting at the least significant end and working upwards. This process is weighted, or made progressively more aggressive as spatial frequency increases.

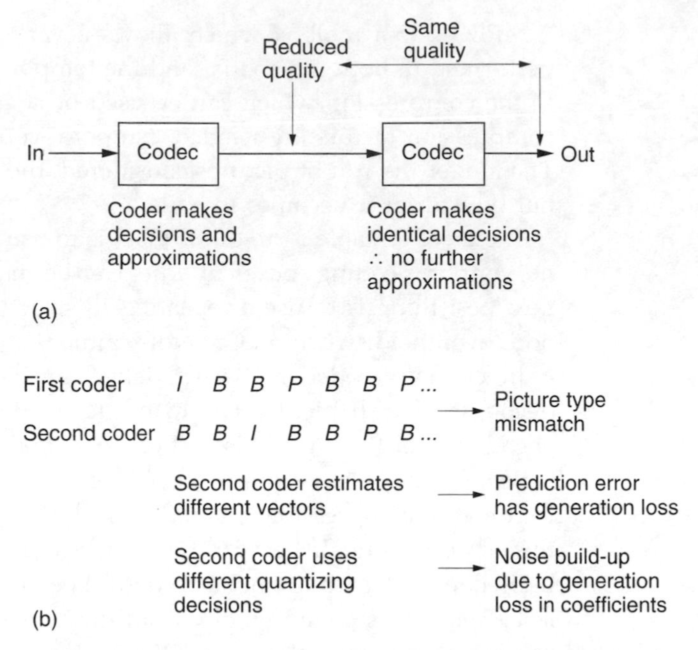

Figure 6.36 (a) Two identical coders in tandem which are synchronized make similar coding decisions and cause little loss. (b) There are various ways in which concatenated coders can produce non-ideal performance.

Small-value coefficients may be truncated to zero and large-value coefficients are most coarsely truncated at high spatial frequencies where the effect is least visible.

Figure 6.36 shows what happens in the ideal case where two *identical* coders are put in tandem and synchronized. The first coder quantizes the coefficients to finite accuracy and causes a loss on decoding. However, when the second coder performs the DCT calculation, the coefficients obtained will be identical to the quantized coefficients in the first coder and so if the second weighting and requantising step is identical the same truncated coefficient data will result and there will be no further loss of quality.[7]

In practice this ideal situation is elusive. If the two DCTs become non-identical for any reason, the second requantizing step will introduce further error in the coefficients and the artifact level goes up. Figure 6.36(b) shows that non-identical concatenation can result from a large number of real-world effects.

An intermediate processing step such as a fade will change the pixel values and thereby the coefficients. A DVE resize or shift will move pixels from one DCT block to another. Even if there is no processing step, this effect will also occur if the two codecs disagree on where the MPEG picture boundaries are within the picture. If the boundaries are correct there will still be concatenation loss if the two codecs use different weighting.

One problem with MPEG is that the compressor design is unspecified. Whilst this has advantages, it does mean that the chances of finding identical coders is minute because each manufacturer will have his or her own views on the best compression algorithm. In a large system it may be worth obtaining the coders from a single supplier.

It is now increasingly accepted that concatenation of compression techniques is potentially damaging, and results are worse if the codecs are different. Clearly, feeding a digital coder such as MPEG-2 with a signal which has been subject to analog compression comes into the category of worse. Using interlaced video as a source for MPEG coding is sub-optimal and using decoded composite video is even worse.

One way of avoiding concatenation is to stay in the compressed data domain. If the goal is just to move pictures from one place to another, decoding to traditional video so an existing router can be used is not ideal, although substantially better than going through the analog domain.

Figure 6.37 shows some possibilities for picture transport. Clearly, if the pictures exist as a compressed file on a server, a file transfer is the right way to do it as there is no possibility of loss because there has been no concatenation. File transfer is also quite indifferent to the picture format.

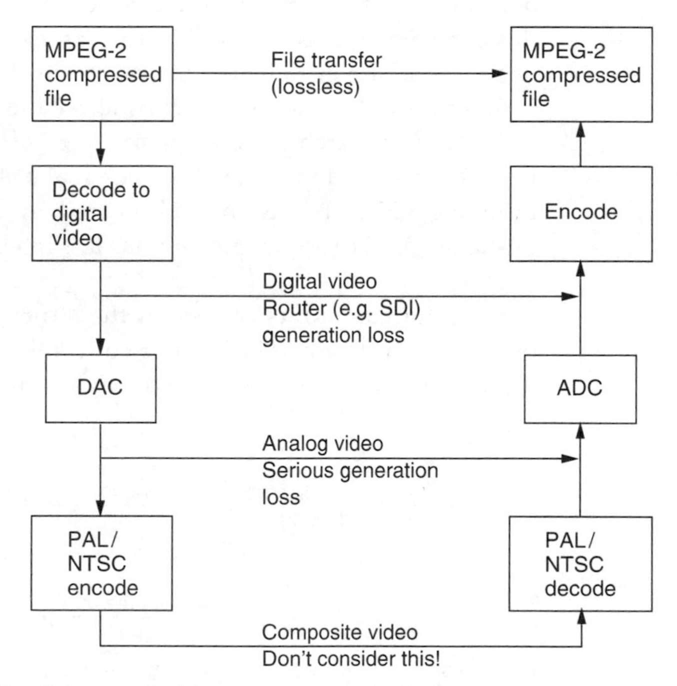

Figure 6.37 Compressed picture transport mechanisms contrasted.

It doesn't care whether the pictures are interlaced or not, whether the colour is 4:2:0 or 4:2:2.

Decoding to SDI (serial digital interface) standard is sometimes done so that existing serial digital routing can be used. This is concatenation and has to be done carefully. The compressed video can only use interlace with non-square pixels and the colour coding has to be 4:2:2 because SDI only allows that. If a compressed file has 4:2:0 the chroma has to be interpolated up to 4:2:2 for SDI transfer and then subsampled back to 4:2:0 at the second coder and this will cause generation loss. An SDI transfer also can only be performed in real time, thus negating one of the advantages of compression. In short, traditional SDI is not really at home with compression.

As 4:2:0 progressive scan gains popularity and video production moves steadily towards non-format-specific hardware using computers and data networks, use of the serial digital interface will eventually decline. In the short term, if an existing SDI router has to be used, one solution is to produce a bitstream which is sufficiently similar to SDI that a router will pass it. In other words the signal level, frequency and impedance is pure SDI, but the data protocol is different so that a bit-accurate file transfer can be performed. This has two advantages over SDI. First, the compressed data format can be anything appropriate and non-interlaced and/or 4:2:0 can be handled in any picture size, aspect ratio or frame rate. Second, a faster than real-time transfer can be used depending on the compression factor of the file. Equipment which allows this is becoming available and its use can mean that the full economic life of a SDI-routing installation can be obtained.

An improved way of reducing concatenation loss has emerged from the ATLANTIC research project.[8] Figure 6.38 shows that the second encoder in a concatenated scheme does not make its own decisions from the incoming video, but is instead steered by information from the first bitstream. As the second encoder has less intelligence, it is known as a *dim* encoder.

The information bus carries all the structure of the original MPEG-2 bitstream which would be lost in a conventional decoder. The ATLANTIC decoder does more than decode the pictures. It also places on the

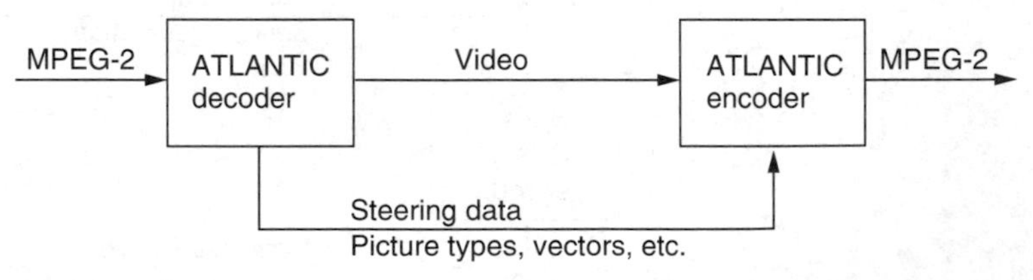

Figure 6.38 In an ATLANTIC system, the second encoder is steered by information from the decoder.

information bus all parameters needed to make the dim encoder re-enact what the initial MPEG-2 encode did as closely as possible.

The GOP structure is passed on so that pictures are re-encoded as the same type. Positions of macroblock boundaries become identical so that DCT blocks contain the same pixels and motion vectors relate to the same screen data. The weighting and quantizing tables are passed so that coefficient truncation is identical. Motion vectors from the original bitsream are passed on so that the dim encoder does not need to perform motion estimation. In this way predicted pictures will be identical to the original prediction and the prediction error data will be the same.

One application of this approach is in recompression, where an MPEG-2 bitstream has to have its bit rate reduced. This has to be done by heavier requantizing of coefficients, but if as many other parameters as possible can be kept the same, such as motion vectors, the degradation will be minimized. In a simple recompressor just requantizing the coefficients means that the predictive coding will be impaired. In a proper encode, the quantizing error due to coding say an *I* picture is removed from the *P* picture by the prediction process. The prediction error of *P* is obtained by subtracting the decoded *I* picture rather than the original *I* picture.

In simple recompression this does not happen and there may be a tolerance build-up known as drift.[9] A more sophisticated recompressor will need to repeat the prediction process using the decoded output pictures as the prediction reference.

MPEG-2 bitstreams will often be decoded for the purpose of switching. Local insertion of commercial breaks into a centrally originated bitstream is one obvious requirement. If the decoded video signal is switched, the information bus must also be switched. At the switch point identical re-encoding becomes impossible because prior pictures required for predictive coding will have disappeared. At this point the dim encoder has to become bright again because it has to create an MPEG-2 bitstream without assistance.

It is possible to encode the information bus into a form which allows it to be invisibly carried in the serial digital interface. Where a production process such as a vision mixer or DVE performs no manipulation, i.e. becomes bit transparent, the subsequent encoder can extract the information bus and operate in 'dim' mode. Where a manipulation is performed, the information bus signal will be corrupted and the encoder has to work in 'bright' mode. The encoded information signal is known as a 'mole'[10] because it burrows through the processing equipment!

There will be a generation loss at the switch point because the re-encode will be making different decisions in bright mode. This may be difficult to detect because the human visual system is slow to react to a vision cut and defects in the first few pictures after a cut are masked.

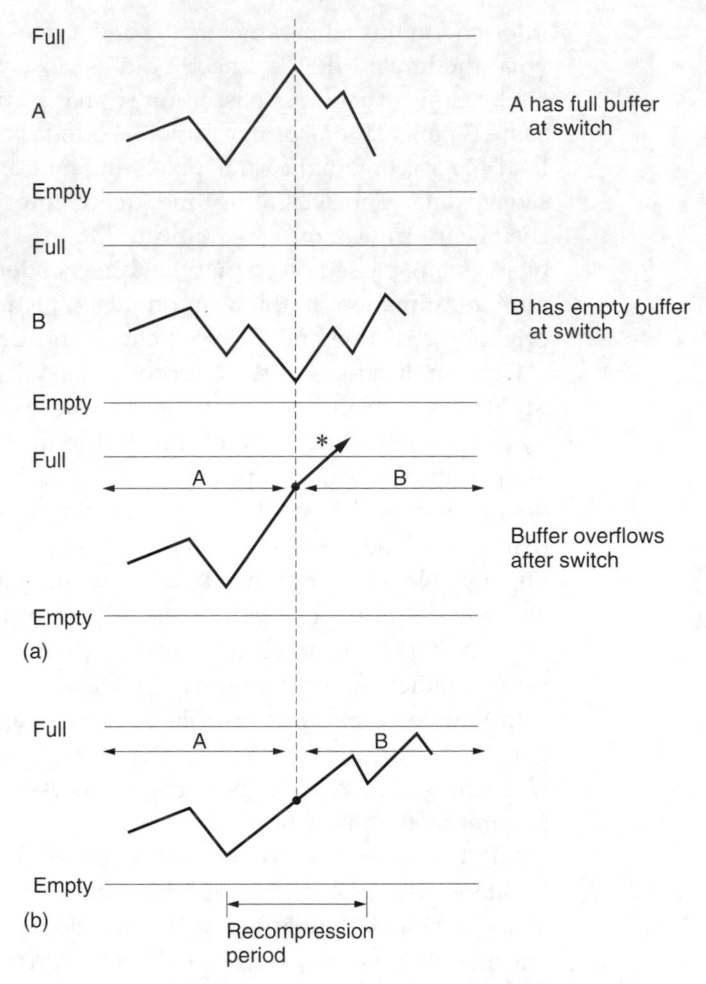

Figure 6.39 (a) A bitstream switch at a different level of buffer occupancy can cause a decoder overflow. (b) Recompression after a switch to return to correct buffer occupancy.

In addition to the video computation required to perform a cut, the process has to consider the buffer occupancy of the decoder. A downstream decoder has finite buffer memory, and individual encoders model the decoder buffer occupancy to ensure that it neither overflows nor underflows. At any instant the decoder buffer can be nearly full or nearly empty without a problem provided there is a subsequent correction. An encoder which is approaching a complex *I* picture may run down the buffer so it can send a lot of data to describe that picture. Figure 6.39(a) shows that if a decoder with a nearly full buffer is suddenly switched to an encoder which has been running down its buffer occupancy, the decoder buffer will overflow when the second encoder sends a lot of data.

An MPEG-2 switcher will need to monitor the buffer occupancy of its own output to avoid overflow of downstream decoders. Where this is a possibility the second encoder will have to recompress to reduce the output bit rate temporarily. In practice there will be a recovery period where the buffer occupancy of the newly selected signal is matched to that of the previous signal. This is shown in Figure 6.39(b).

References

1. MPEG Video Standard: ISO/IEC 13818–2: Information technology – generic coding of moving pictures and associated audio information: Video (1996) (aka ITU-T Rec. H-262 (1996))
2. Huffman, D.A., A method for the construction of minimum redundancy codes. *Proc. IRE*, **40** 1098–1101 (1952)
3. LeGall, D., MPEG: a video compression standard for multimedia applications. *Communications of the ACM*, **34**, No.4, 46–58 (1991)
4. ISO/IEC JTC1/SC29/WG11 MPEG, International standard ISO 11172 'Coding of moving pictures and associated audio for digital storage media up to 1.5 Mbits/s' (1992)
5. ISO Joint Photographic Experts Group standard JPEG-8-R8
6. Wallace, G.K., Overview of the JPEG (ISO/CCITT) still image compression standard. ISO/JTC1/SC2/WG8 N932 (1989)
7. Stone, J. and Wilkinson, J., Concatenation of video compression systems. Presented at 137th SMPTE Tech. Conf. New Orleans (1995)
8. Wells, N.D., The ATLANTIC project: Models for programme production and distribution. *Proc. Euro. Conf. Multimedia Applications Services and Techniques (ECMAST)*, 243–253 (1996)
9. Werner, O., Drift analysis and drift reduction for multiresolution hybrid video coding. *Image Communication*, **8** 387–409 (1996)
10. Knee, M.J. and Wells, N.D., Seamless concatenation – a 21st century dream. Presented at Int. Television. Symp. Montreux (1997)

7

Digital audio in video

7.1 What is sound?

Physics can tell us the mechanism by which disturbances propagate through the air. If this is our definition of sound, we have the problem that in physics there are no limits to the frequencies and levels which must be considered. Biology can tell that the ear only responds to a certain range of frequencies provided a threshold level is exceeded. This is a better definition of sound; reproduction is easier because it is only necessary to reproduce that range of levels and frequencies which the ear can detect.

Psychoacoustics can describe how our hearing has finite resolution in both time and frequency domains such that what we perceive is an inexact impression. Some aspects of the original disturbance are inaudible to us and are said to be masked. If our goal is the highest quality, we can design our imperfect equipment so that the shortcomings are masked. Conversely if our goal is economy we can use compression and hope that masking will disguise the inaccuracies it causes.

By definition, the sound quality of a perceptive coder can only be assessed by human hearing. Equally, a useful perceptive coder can only be designed with a good knowledge of the human hearing mechanism.[1] The acuity of the human ear is astonishing. The frequency range is extremely wide, covering some ten octaves (an octave is a doubling of pitch or frequency) without interruption. It can detect tiny amounts of distortion, and will accept an enormous dynamic range. If the ear detects a different degree of impairment between two codecs having the same bit rate in properly conducted tests, we can say that one of them is superior. Thus quality is completely subjective and can only be checked by listening tests. However, any characteristic of a signal which can be heard can also be measured by a suitable instrument. The subjective tests can tell us how sensitive the instrument should be. Then the objective readings from the instrument give an indication of how acceptable a signal is in respect of that characteristic. Instruments for assessing the performance of codecs are currently extremely rare and there remains much work to be done.

7.2 The ear

The sense we call hearing results from acoustic, mechanical, hydraulic, nervous and mental processes in the ear/brain combination, leading to the term psychoacoustics. It is only possible briefly to introduce the subject here. The interested reader is referred to Moore[2] for an excellent treatment.

Figure 7.1 shows that the structure of the ear is traditionally divided into the outer, middle and inner ears. The outer ear works at low impedance, the inner ear works at high impedance, and the middle ear is an impedance-matching device. The visible part of the outer ear is called the pinna which plays a subtle role in determining the direction of arrival of sound at high frequencies. It is too small to have any effect at low frequencies. Incident sound enters the auditory canal or meatus. The pipe-like meatus causes a small resonance at around 4 kHz. Sound vibrates the eardrum or tympanic membrane which seals the outer ear from the middle ear. The inner ear or cochlea works by sound travelling though a fluid. Sound enters the cochlea via a membrane called the oval window. If airborne sound were to be incident on the oval window directly, the serious impedance mismatch would cause most of the sound to be reflected. The middle ear remedies that mismatch by providing a mechanical advantage. The tympanic membrane is linked to the oval window by three bones known as ossicles which act as a lever system

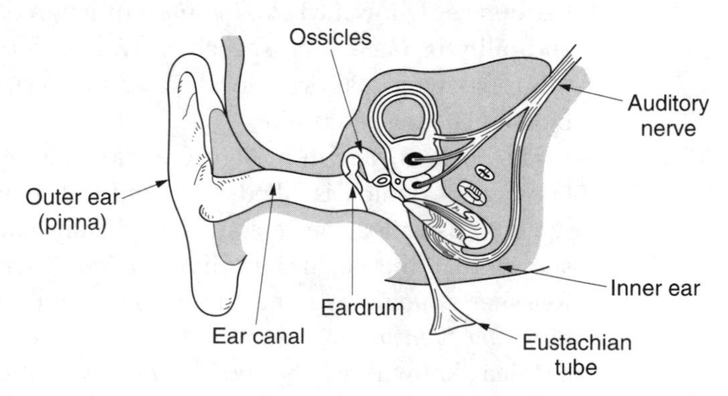

Figure 7.1 The structure of the human ear. See text for details.

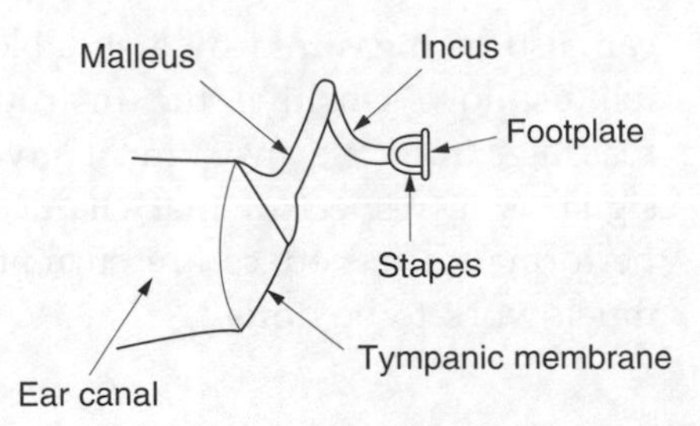

Figure 7.2 The malleus tensions the tympanic membrane into a conical shape. The ossicles provide an impedance-transforming lever system between the tympanic membrane and the oval window.

such that a large displacement of the tympanic membrane results in a smaller displacement of the oval window but with greater force. Figure 7.2 shows that the malleus applies a tension to the tympanic membrane rendering it conical in shape. The malleus and the incus are firmly joined together to form a lever. The incus acts upon the stapes through a spherical joint. As the area of the tympanic membrane is greater than that of the oval window, there is a further multiplication of the available force. Consequently small pressures over the large area of the tympanic membrane are converted to high pressures over the small area of the oval window.

The middle ear is normally sealed, but ambient pressure changes will cause static pressure on the tympanic membrane, which is painful. The pressure is relieved by the Eustachian tube which opens involuntarily while swallowing. The Eustachian tubes open into the cavities of the head and must normally be closed to avoid one's own speech appearing deafeningly loud.

The ossicles are located by minute muscles which are normally relaxed. However, the middle ear reflex is an involuntary tightening of the *tensor tympani* and *stapedius* muscles which heavily damp the ability of the tympanic membrane and the stapes to transmit sound by about 12 dB at frequencies below 1 KHz. The main function of this reflex is to reduce the audibility of one's own speech. However, loud sounds will also trigger this reflex which takes some 60–120 ms to occur, too late to protect against transients such as gunfire.

The cochlea, shown in Figure 7.3(a), is a tapering spiral cavity within bony walls which is filled with fluid. The widest part, near the oval window, is called the *base* and the distant end is the *apex*. Figure 7.3(b) shows that the cochlea is divided lengthwise into three volumes by Reissner's membrane and the basilar membrane. The *scala vestibuli* and the *scala tympani* are connected by a small aperture at the apex of the cochlea known as the *helicotrema*. Vibrations from the stapes are transferred to the oval window and become fluid pressure variations

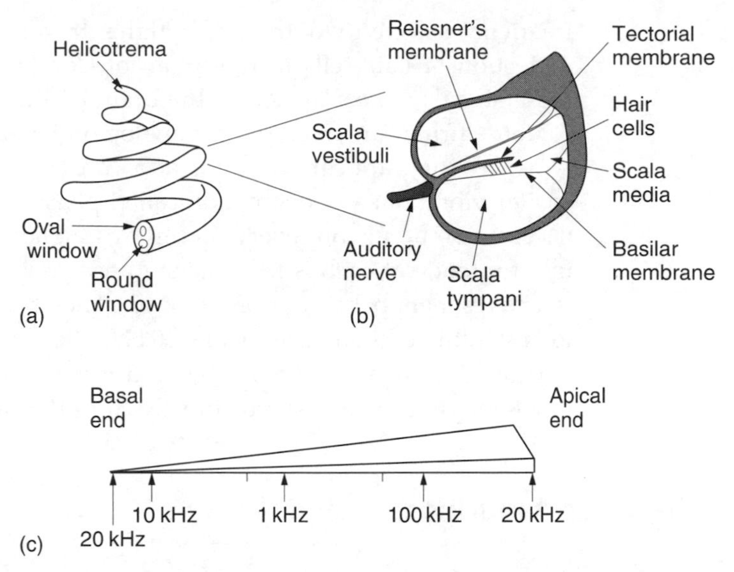

Figure 7.3 (a) The cochlea is a tapering spiral cavity. (b) The cross-section of the cavity is divided by Reissner's membrane and the basilar membrane. (c) The basilar membrane tapers so its resonant frequency changes along its length.

which are relieved by the flexing of the round window. Effectively the basilar membrane is in series with the fluid motion and is driven by it except at very low frequencies where the fluid flows through the helicotrema, bypassing the basilar membrane.

Figure 7.3(c) shows that the basilar membrane is not uniform, but tapers in width and varies in thickness in the opposite sense to the taper of the cochlea. The part of the basilar membrane which resonates as a result of an applied sound is a function of the frequency. High frequencies cause resonance near to the oval window, whereas low frequencies cause resonances further away. More precisely, the distance from the apex where the maximum resonance occurs is a logarithmic function of the frequency. Consequently tones spaced apart in octave steps will excite evenly spaced resonances in the basilar membrane. The prediction of resonance at a particular location on the membrane is called *place theory.* Essentially the basilar membrane is a mechanical frequency analyser. A knowledge of the way it operates is essential to an understanding of musical phenomena such as pitch discrimination, timbre, consonance and dissonance and to auditory phenomena such as critical bands, masking and the precedence effect.

The vibration of the basilar membrane is sensed by the organ of Corti which runs along the centre of the cochlea. The organ of Corti is active in that it contains elements which can generate vibration as well as sense it. These are connected in a regenerative fashion so that the Q factor, or frequency selectivity of the ear, is higher than it would otherwise be. The deflection of hair cells in the organ of Corti triggers nerve firings and these signals are conducted to the brain by the auditory nerve.

Nerve firings are not a perfect analog of the basilar membrane motion. A nerve firing appears to occur at a constant phase relationship to the basilar vibration; a phenomenon called phase locking, but firings do not necessarily occur on every cycle. At higher frequencies firings are intermittent, yet each is in the same phase relationship.

The resonant behaviour of the basilar membrane is not observed at the lowest audible frequencies below 50 Hz. The pattern of vibration does not appear to change with frequency and it is possible that the frequency is low enough to be measured directly from the rate of nerve firings.

7.3 Level and loudness

At its best, the ear can detect a sound pressure variation of only 2×10^{-5} Pascals r.m.s. and so this figure is used as the reference against which sound pressure level (SPL) is measured. The sensation of loudness is a logarithmic function of SPL and consequently a logarithmic unit, the deciBel, is used in audio measurement.

The dynamic range of the ear exceeds 130 dB, but at the extremes of this range, the ear is either straining to hear or is in pain. Neither of these cases can be described as pleasurable or entertaining, and it is hardly necessary to produce audio of this dynamic range since, among other things, the consumer is unlikely to have anywhere sufficiently quiet to listen to it. On the other hand, extended listening to music whose dynamic range has been excessively compressed is fatiguing.

The frequency response of the ear is not at all uniform and it also changes with SPL. The subjective response to level is called loudness and is measured in *phons*. The phon scale and the SPL scale coincide at 1 kHz, but at other frequencies the phon scale deviates because it displays the actual SPLs judged by a human subject to be equally loud as a given level at 1 kHz. Figure 7.4 shows the so-called equal loudness contours which were originally measured by Fletcher and Munson and subsequently by Robinson and Dadson. Note the irregularities caused by resonances in the meatus at about 4 kHz and 13 kHz.

Usually, people's ears are at their most sensitive between about 2 kHz and 5 kHz, and although some people can detect 20 kHz at high level, there is much evidence to suggest that most listeners cannot tell if the upper frequency limit of sound is 20 kHz or 16 kHz.[3,4] For a long time it was thought that frequencies below about 40 Hz were unimportant, but it is now clear that reproduction of frequencies down to 20 Hz improves reality and ambience.[5] The generally accepted frequency range for high-

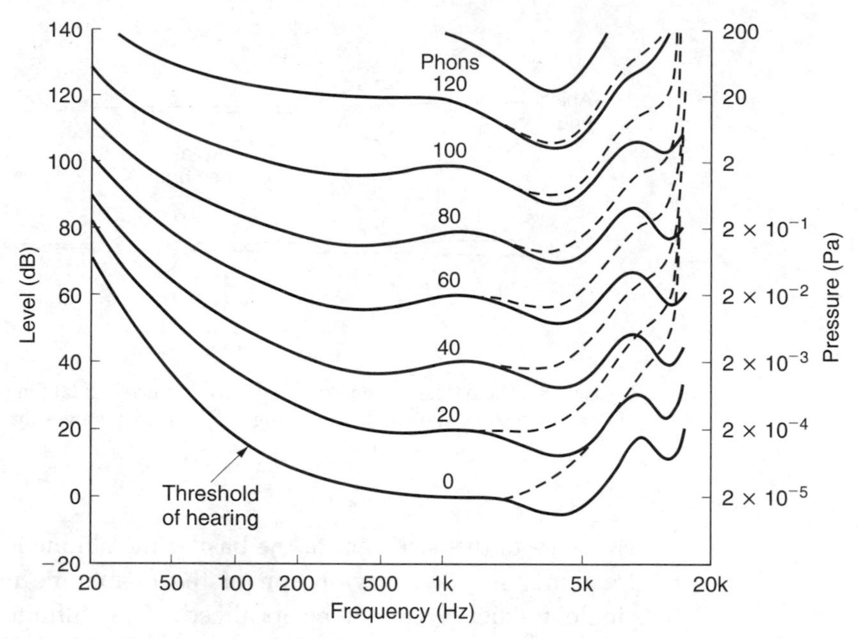

Figure 7.4 Contours of equal loudness showing that the frequency response of the ear is highly level dependent (solid line, age 20; dashed line, age 60).

quality audio is 20–20 000 Hz, although for broadcasting an upper limit of 15 000 Hz is often applied.

The most dramatic effect of the curves of Figure 7.4 is that the bass content of reproduced sound is disproportionately reduced as the level is turned down.

Loudness is a subjective reaction and is almost impossible to measure. In addition to the level-dependent frequency response problem, the listener uses the sound not for its own sake but to draw some conclusion about the source. For example, most people hearing a distant motorcycle will describe it as being loud. Clearly at the source, it *is* loud, but the listener has compensated for the distance.

The best that can be done is to make some compensation for the level-dependent response using *weighting curves*. Ideally there should be many, but in practice the A, B and C weightings were chosen where the A curve is based on the 40-phon response. The measured level after such a filter is in units of dBA. The A curve is almost always used because it most nearly relates to the annoyance factor of distant noise sources.

7.4 Critical bands

Figure 7.5 shows an uncoiled basilar membrane with the apex on the left so that the usual logarithmic frequency scale can be applied. The

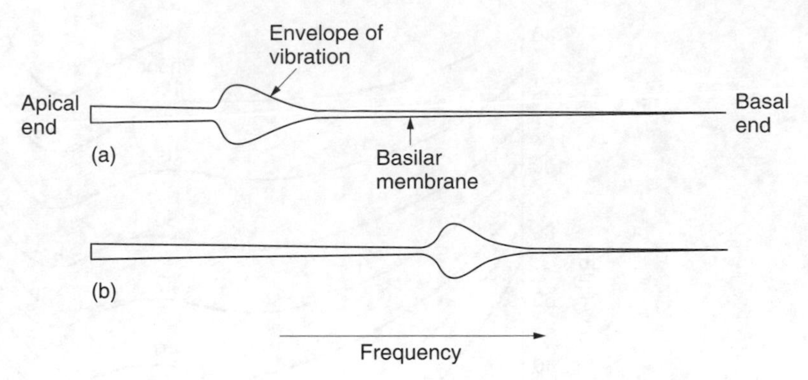

Figure 7.5 The basilar membrane symbolically uncoiled. (a) Single frequency causes the vibration envelope shown. (b) Changing the frequency moves the peak of the envelope.

envelope of displacement of the basilar membrane is shown for a single frequency in (a). The vibration of the membrane in sympathy with a single frequency cannot be localized to an infinitely small area, and nearby areas are forced to vibrate at the same frequency with an amplitude that decreases with distance. Note that the envelope is asymmetrical because the membrane is tapering and because of frequency-dependent losses in the propagation of vibrational energy down the cochlea. If the frequency is changed, as in (b), the position of maximum displacement will also change. As the basilar membrane is continuous, the position of maximum displacement is infinitely variable allowing extremely good pitch discrimination of about one twelfth of a semitone which is determined by the spacing of hair cells.

In the presence of a complex spectrum, the finite width of the vibration envelope means that the ear fails to register energy in some bands when there is more energy in a nearby band. Within those areas, other frequencies are mechanically excluded because their amplitude is insufficient to dominate the local vibration of the membrane. Thus the Q factor of the membrane is responsible for the degree of auditory masking, defined as the decreased audibility of one sound in the presence of another.

The term used in psychoacoustics to describe the finite width of the vibration envelope is *critical bandwidth*. Critical bands were first described by Fletcher.[6] The envelope of basilar vibration is a complicated function. It is clear from the mechanism that the area of the membrane involved will increase as the sound level rises. Figure 7.6 shows the bandwidth as a function of level.

As was shown in Chapter 3, transform theory teaches that the higher the frequency resolution of a transform, the worse the time accuracy. As the basilar membrane has finite frequency resolution measured in the

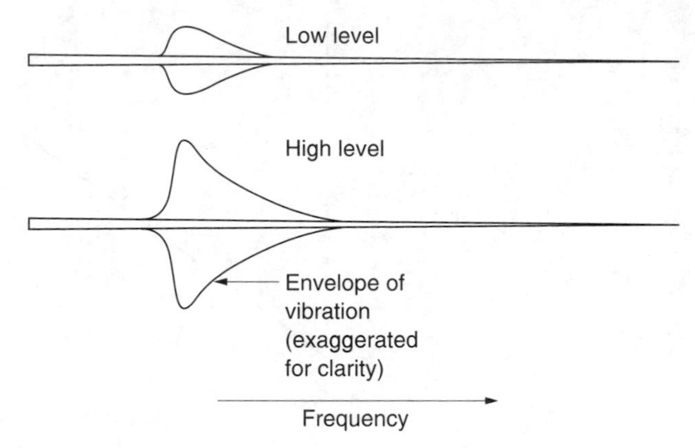

Figure 7.6 The critical bandwidth changes with SPL.

width of a critical band, it follows that it must have finite time resolution. This also follows from the fact that the membrane is resonant, taking time to start and stop vibrating in response to a stimulus. There are many examples of this. Figure 7.7 shows the impulse response. Figure 7.8 shows the perceived loudness of a tone burst increases with duration up to about 200 ms due to the finite response time.

The ear has evolved to offer intelligibility in reverberant environments which it does by averaging all received energy over a period of about 30 ms. Reflected sound which arrives within this time is integrated to produce a louder sensation, whereas reflected sound which arrives after that time can be temporally discriminated and is perceived as an echo. Our simple microphones have no such ability which is why we often need to have acoustic treatment in areas where microphones are used.

A further example of the finite time discrimination of the ear is the fact that short interruptions to a continuous tone are difficult to detect. Finite time resolution means that masking can take place even when the masking tone begins after and ceases before the masked sound. This is referred to as forward and backward masking.[7]

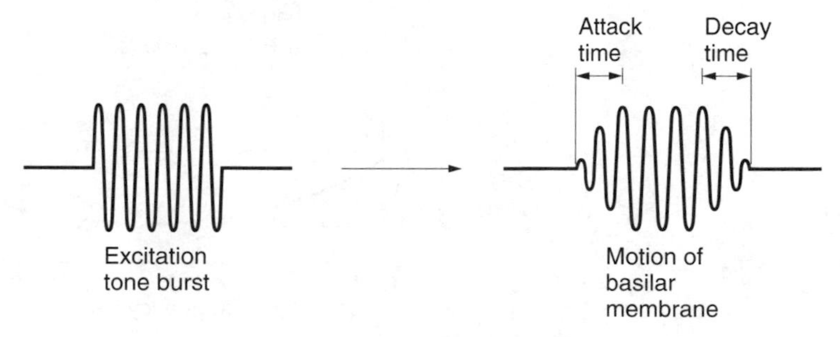

Figure 7.7 Impulse response of the ear showing slow attack and decay due to resonant behaviour.

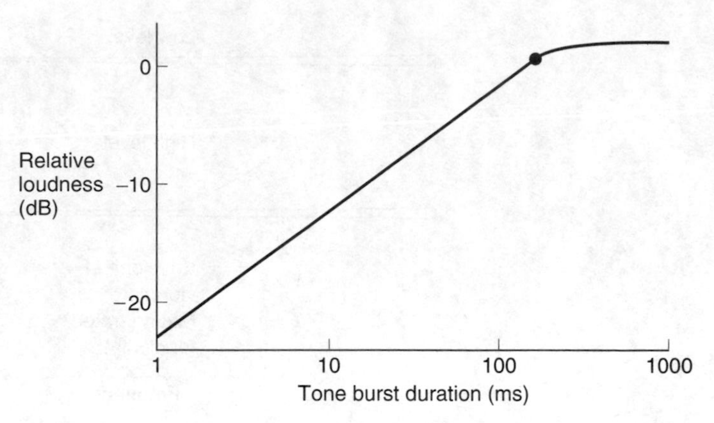

Figure 7.8 Perceived level of tone burst rises with duration as resonance builds up.

As the vibration envelope is such a complicated shape, Moore and Glasberg[8] have proposed the concept of equivalent rectangular bandwidth to simplify matters. The ERB is the bandwidth of a rectangular filter which passes the same power as a critical band. Figure 7.9(a) shows the expression they have derived linking the ERB with frequency. This is plotted in (b) where it will be seen that one third of an octave is a good approximation. This is about thirty times broader than the pitch discrimination also shown in (b).

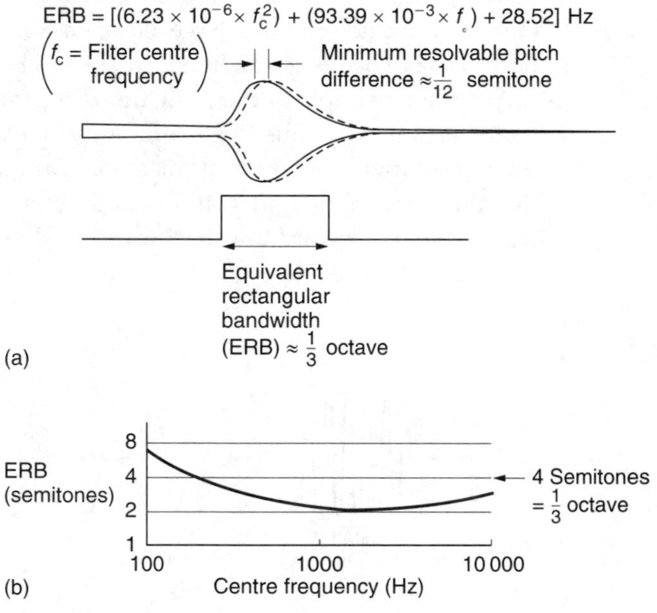

Figure 7.9 Effective rectangular bandwidth of critical band is much wider than the resolution of the pitch discrimination mechanism.

7.5 Choice of sampling rate for audio

Sampling theory is only the beginning of the process which must be followed to arrive at a suitable sampling rate. The finite slope of realizable filters will compel designers to raise the sampling rate. For consumer products, the lower the sampling rate, the better, since the cost of the medium is directly proportional to the sampling rate: thus sampling rates near to twice 20 kHz are to be expected. For professional products, there is a need to operate at variable speed for pitch correction. When the speed of a digital recorder is reduced, the offtape sampling rate falls, and Figure 7.10 shows that with a minimal sampling rate the first

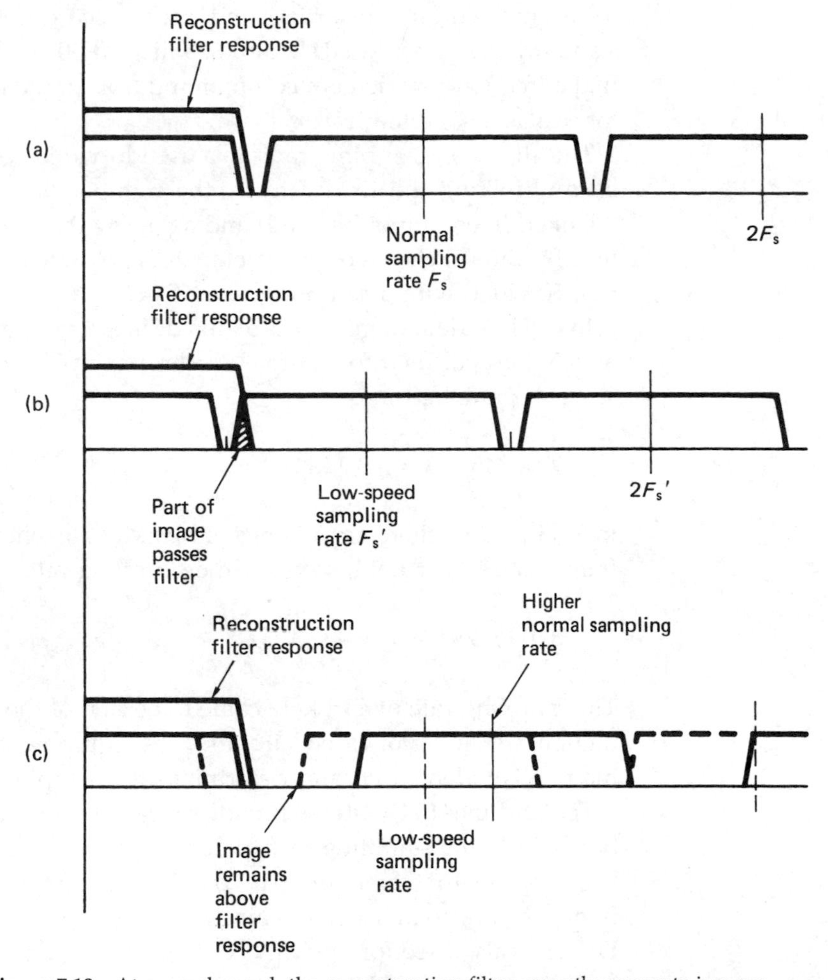

Figure 7.10 At normal speed, the reconstruction filter correctly prevents images entering the baseband, as at (a). When speed is reduced, the sampling rate falls, and a fixed filter will allow part of the lower sideband of the sampling frequency to pass. If the sampling rate of the machine is raised, but the filter characteristic remains the same, the problem can be avoided, as at (c).

image frequency can become low enough to pass the reconstruction filter. If the sampling frequency is raised without changing the response of the filters, the speed can be reduced without this problem.

In the early days of digital audio research, the necessary bandwidth of about 1 megabit per second per audio channel was difficult to store. Disk drives had the bandwidth but not the capacity for long recording time, so attention turned to video recorders. These were adapted to store audio samples by creating a pseudo-video waveform which could convey binary as black and white levels. The sampling rate of such a system is constrained to relate simply to the field rate and field structure of the television standard used, so that an integer number of samples can be stored on each usable TV line in the field. Such a recording can be made on a monochrome recorder, and these recordings are made in two standards, 525 lines at 60 Hz and 625 lines at 50 Hz. Thus it is possible to find a frequency which is a common multiple of the two and also suitable for use as a sampling rate.

The allowable sampling rates in a pseudo-video system can be deduced by multiplying the field rate by the number of active lines in a field (blanked lines cannot be used) and again by the number of samples in a line. By careful choice of parameters it is possible to use either 525/60 or 625/50 video with a sampling rate of 44.1 kHz.

In 60 Hz video, there are 35 blanked lines, leaving 490 lines per frame, or 245 lines per field for samples. If three samples are stored per line, the sampling rate becomes

$$60 \times 245 \times 3 = 44.1 \text{ kHz}$$

In 50 Hz video, there are 37 lines of blanking, leaving 588 active lines per frame, or 294 per field, so the same sampling rate is given by

$$50.00 \times 294 \times 3 = 44.1 \text{ kHz}$$

The sampling rate of 44.1 kHz came to be that of the Compact Disc. Even though CD has no video circuitry, the equipment used to make CD masters is video based and determines the sampling rate.

For landlines to FM stereo broadcast transmitters having a 15 kHz audio bandwidth, the sampling rate of 32 kHz is more than adequate, and has been in use for some time in the United Kingdom and Japan. This frequency is also in use in the NICAM 728 stereo TV sound system and in DVB. It is also used for the Sony NT format mini-cassette. The professional sampling rate of 48 kHz was proposed as having a simple relationship to 32 kHz, being far enough above 40 kHz for variable-speed operation.

Although in a perfect world the adoption of a single sampling rate might have had virtues, for practical and economic reasons digital audio now has essentially three rates to support: 32 kHz for broadcast, 44.1 kHz for CD and its mastering equipment, and 48 kHz for 'professional' use.[9] A rate of 48 kHz is used extensively in television production where it can be synchronized to both US and European line standards relatively easily. The currently available DVTR formats offer 48 kHz audio sampling. A number of formats can operate at more than one sampling rate. Both DAT and DASH formats are specified for all three rates, although not all available hardware implements every possibility. Most hard disk recorders will operate at a range of rates.

7.6 Basic digital-to-analog conversion

This direction of conversion will be discussed first, since ADCs often use embedded DACs in feedback loops.

The purpose of a digital-to-analog convertor is to take numerical values and reproduce the continuous waveform that they represent. Figure 7.11 shows the major elements of a conventional conversion subsystem, i.e. one in which oversampling is not employed. The jitter in the clock needs to be removed with a VCO or VCXO. Sample values are buffered in a latch and fed to the convertor element which operates on each cycle of the clean clock. The output is then a voltage proportional to the number for at least a part of the sample period. A resampling stage may be found next, in order to remove switching transients, reduce the aperture ratio or allow the use of a convertor which takes a substantial part of the sample period to operate. The resampled waveform is then presented to a reconstruction filter which rejects frequencies above the audio band.

This section is primarily concerned with the implementation of the convertor element. There are two main ways of obtaining an analog

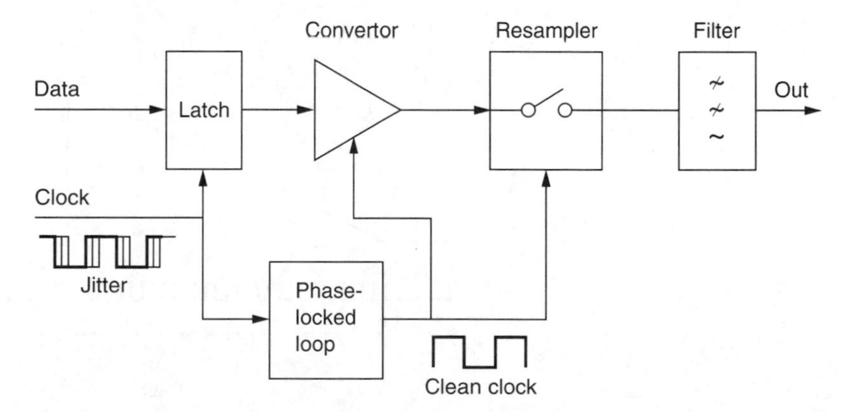

Figure 7.11 The components of a conventional convertor. A jitter-free clock drives the voltage conversion, whose output may be resampled prior to reconstruction.

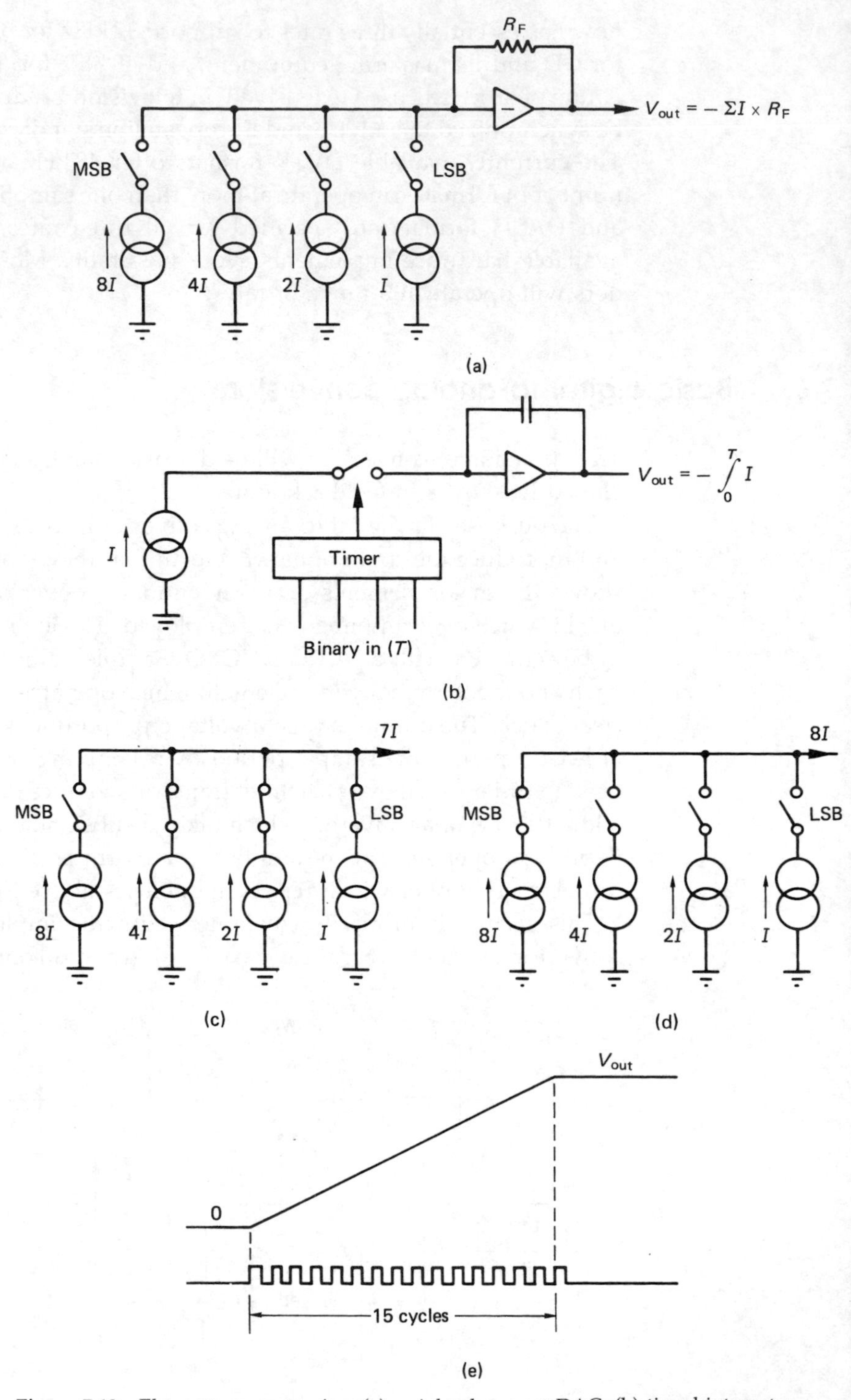

Figure 7.12 Elementary conversion: (a) weighted current DAC; (b) timed integrator DAC; (c) current flow with 0111 input; (d) current flow with 1000 input; (e) integrator ramps up for 15 cycles of clock for input 1111.

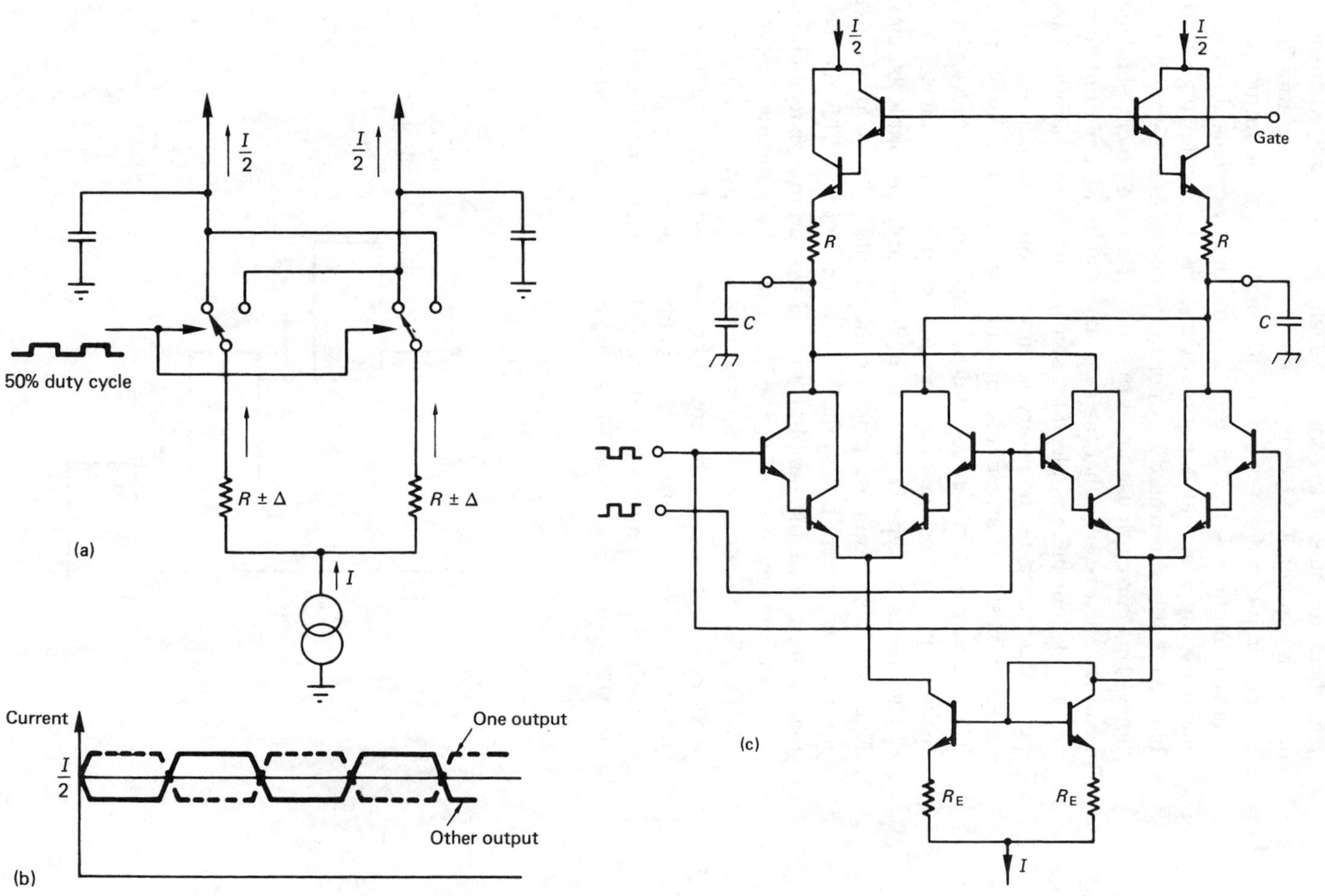

Figure 7.13 Dynamic element matching. (a) Each resistor spends half its time in each current path. (b) Average current of both paths will be identical if duty cycle is accurately 50 per cent (c) Typical monolithic implementation. Note clock frequency is arbitrary.

signal from PCM data. One is to control binary-weighted currents and sum them; the other is to control the length of time a fixed current flows into an integrator. The two methods are contrasted in Figure 7.12. They appear simple, but are of no use for audio in these forms because of practical limitations. In Figure 7.12(c), the binary code is about to have a major overflow, and all the low-order currents are flowing. In (d), the binary input has increased by one, and only the most significant current flows. This current must equal the sum of all the others plus one. The accuracy must be such that the step size is within the required limits. In this simple four-bit example, if the step size needs to be a rather casual 10 per cent accurate, the necessary accuracy is only one part in 160, but for a sixteen-bit system it would become one part in 655 360, or about 2 ppm. This degree of accuracy is almost impossible to achieve, let alone maintain in the presence of ageing and temperature change.

The integrator-type convertor in this four-bit example is shown in Figure 7.12(c); it requires a clock for the counter which allows it to count up to the maximum in less than one sample period. This will be more than sixteen times the sampling rate. However, in a sixteen-bit system, the clock rate would need to be 65 536 times the sampling rate, or about 3 GHz. Whilst there may be a market for a CD player which can defrost a chicken, clearly some refinements are necessary to allow either of these convertor types to be used in audio applications.

One method of producing currents of high relative accuracy is *dynamic element matching*.[10,11] Figure 7.13 shows a current source feeding a pair of

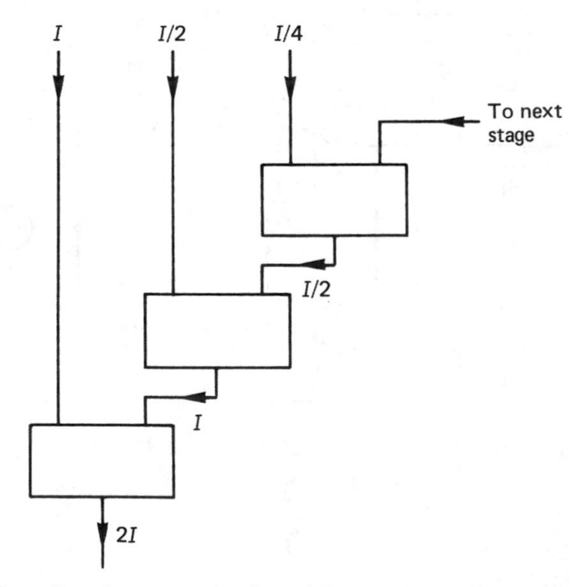

Figure 7.14 Cascading the current dividers of Figure 4.22 produces a binary-weighted series of currents.

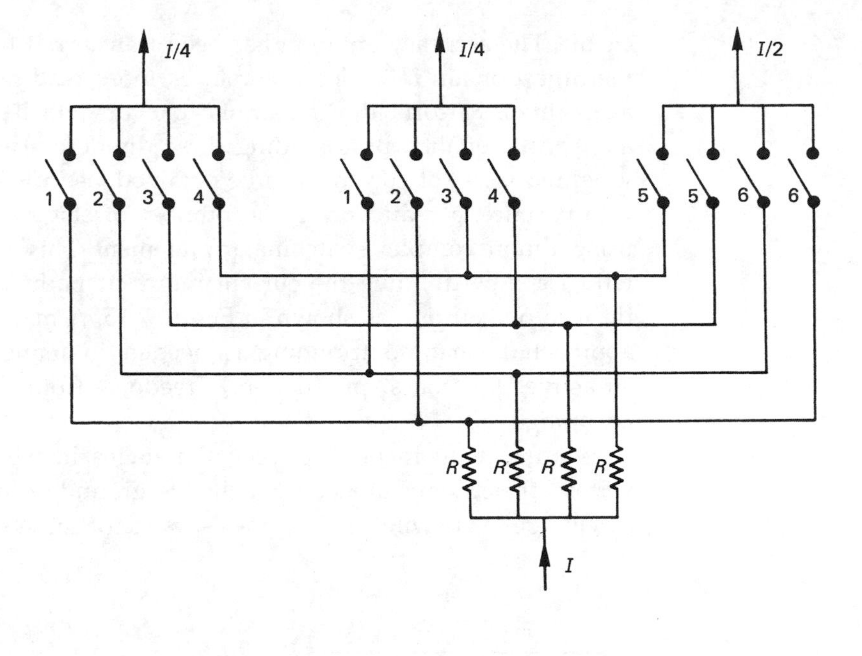

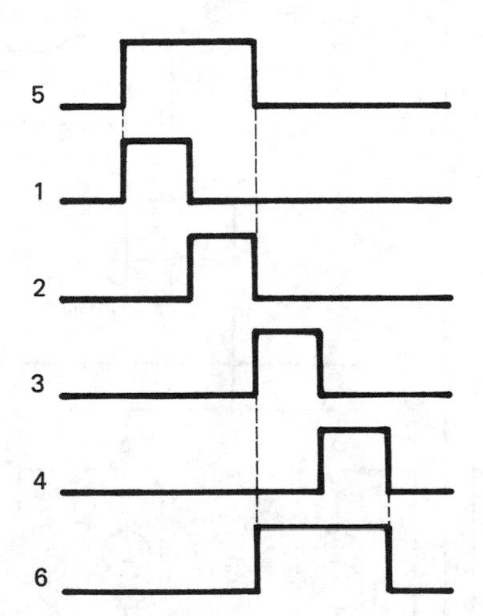

Figure 7.15 More complex dynamic element-matching system. Four drive signals (1, 2, 3, 4) of 25% duty cycle close switches of corresponding number. Two signals (5, 6) have 50% duty cycle, resulting in two current shares going to right-hand output. Division is thus into 1:1:2.

nominally equal resistors. The two will not be the same due to manufacturing tolerances and drift, and thus the current is only approximately divided between them. A pair of change-over switches places each resistor in series with each output. The average current in each output will then be identical, provided that the duty cycle of the switches is exactly 50 per cent. This is readily achieved in a divide-by-two circuit. The accuracy criterion has been transferred from the resistors to the time domain in which accuracy is more readily achieved. Current averaging is performed by a pair of capacitors which do not need to be of any special quality. By cascading these divide-by-two stages, a binary-weighted series of currents can be obtained, as in Figure 7.14.

In practice, a reduction in the number of stages can be obtained by using a more complex switching arrangement. This generates currents of ratio 1:1:2 by dividing the current into four paths and feeding two of them to one output, as shown in Figure 7.15. A major advantage of this approach is that no trimming is needed in manufacture, making it attractive for mass production. Freedom from drift is a further advantage.

To prevent interaction between the stages in weighted-current convertors, the currents must be switched to ground or into the virtual earth by change-over switches. The on-resistance of these switches is a source

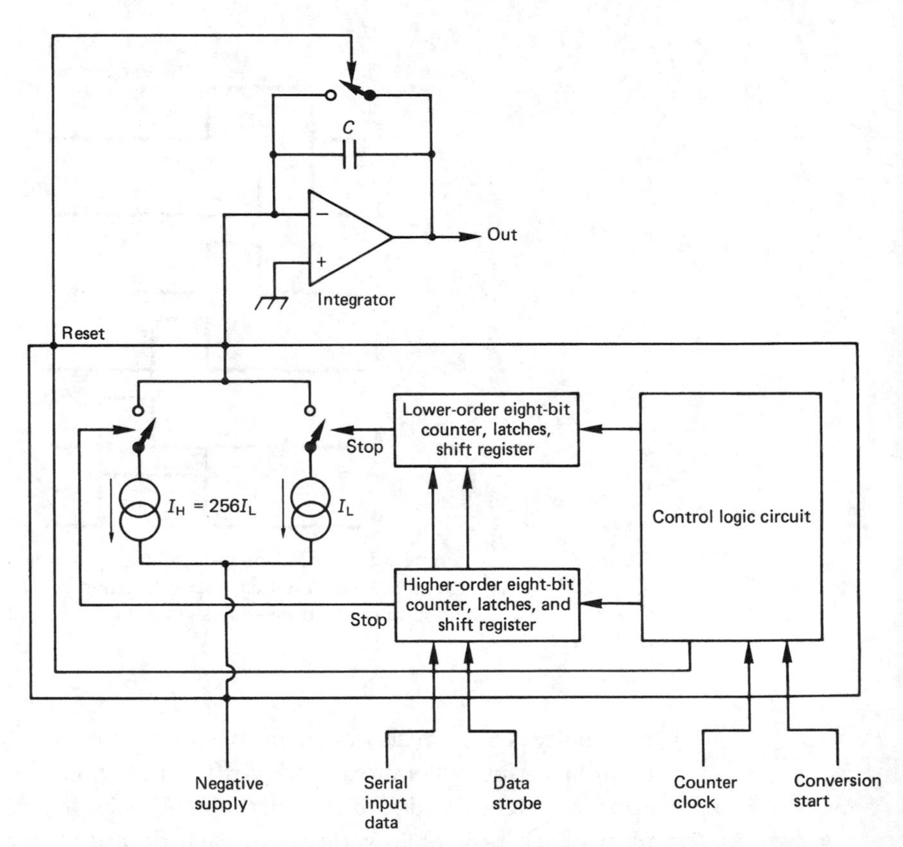

Figure 7.16 Simplified diagram of Sony CX-20017. The high-order and low-order current sources (I_H and I_L) and associated timing circuits can be seen. The necessary integrator is external.

of error, particularly the MSB, which passes most current. A solution in monolithic convertors is to fabricate switches whose area is proportional to the weighted current, so that the voltage drops of all the switches are the same. The error can then be removed with a suitable offset. The layout of such a device is dominated by the MSB switch since, by definition, it is as big as all the others put together.

The practical approach to the integrator convertor is shown in Figures 7.16 and 7.17 where two current sources whose ratio is 256:1 are used; the larger is timed by the high byte of the sample and the smaller is timed by the low byte. The necessary clock frequency is reduced by a factor of 256. Any inaccuracy in the current ratio will cause one quantizing step in

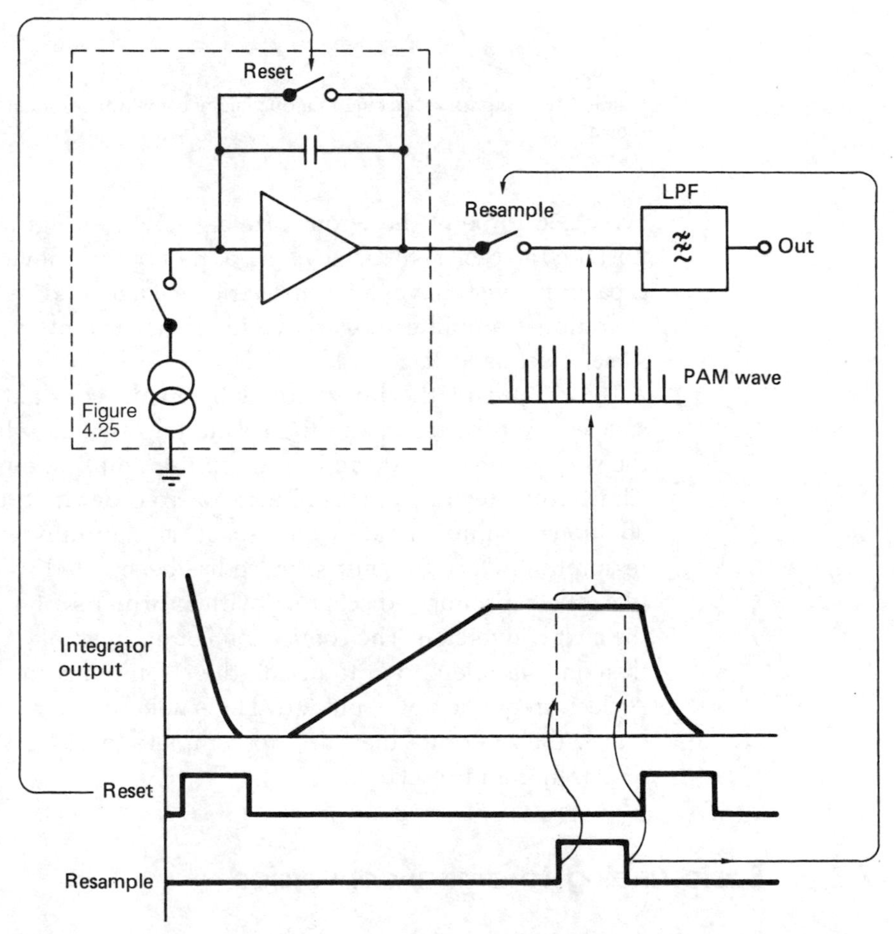

Figure 7.17 In an integrator convertor, the output level is only stable when the ramp finishes. An analog switch is necessary to isolate the ramp from subsequent circuits. The switch can also be used to produce a PAM (pulse amplitude modulated) signal which has a flatter frequency response than a zero-order hold (staircase) signal.

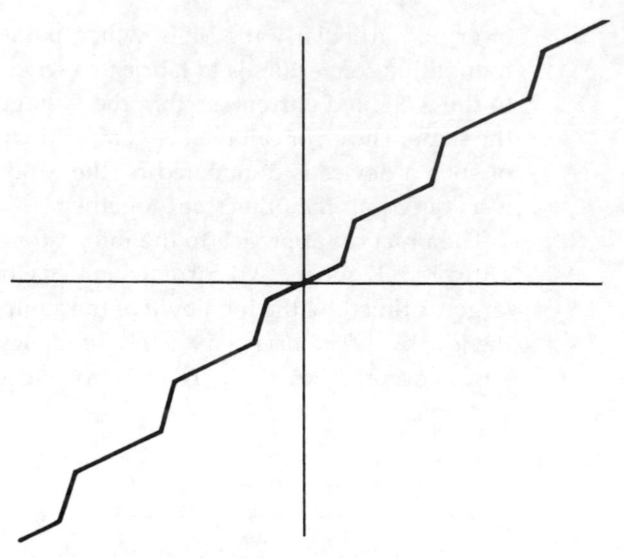

Figure 7.18 Imprecise tracking in a dual-slope convertor results in the transfer function shown here.

every 256 to be of the wrong size as shown in Figure 7.18, but current tracking is easier to achieve in a monolithic device. The integrator capacitor must have low dielectric leakage and relaxation, and the operational amplifier must have low bias current as this will have the same effect as leakage.

The output of the integrator will remain constant once the current sources are turned off, and the resampling switch will be closed during the voltage plateau to produce the pulse amplitude-modulated output. Clearly this device cannot produce a zero-order hold output without an additional sample-hold stage, so it is naturally complemented by resampling. Once the output pulse has been gated to the reconstruction filter, the capacitor is discharged with a further switch in preparation for the next conversion. The conversion count must take place in rather less than one sample period to permit the resampling and discharge phases. A clock frequency of about 20 MHz is adequate for a sixteen-bit 48 kHz unit, which permits the ramp to complete in 12.8 ms, leaving 8 ms for resampling and reset.

7.7 Basic analog-to-digital conversion

A conventional analog-to-digital subsystem is shown in Figure 7.19. Following the anti-aliasing filter there will be a sampling process. Many of the ADCs described here will need a finite time to operate, whereas an instantaneous sample must be taken from the input. The solution is to use

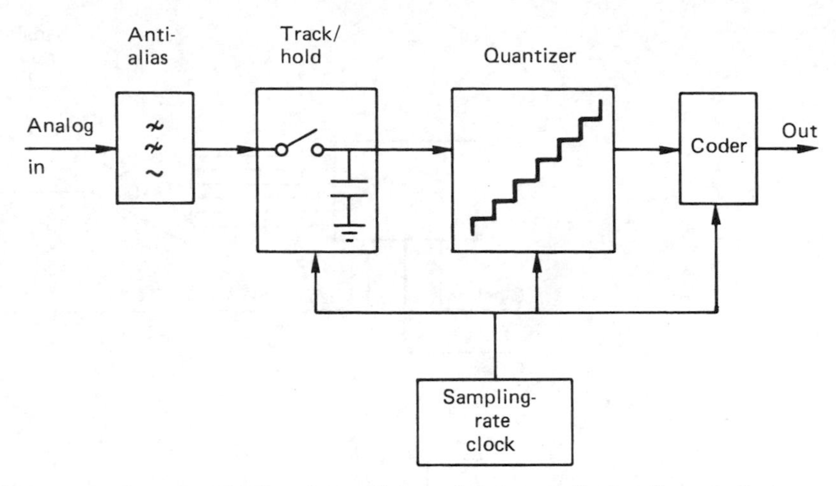

Figure 7.19 A conventional analog-to-digital subsystem. Following the anti-aliasing filter there will be a sampling process, which may include a track-hold circuit. Following quantizing, the number of the quantized level is then converted to a binary code, typically two's complement.

a track/hold circuit. Following sampling the sample voltage is quantized. The number of the quantized level is then converted in to a binary code, typically two's complement. This section is concerned primarily with the implementation of the quantizing step.

The flash convertor is probably the simplest technique available for PCM and DPCM conversion. The principle was shown in Chapter 4. Although the device is simple in principle, it contains a lot of circuitry and can only be practicably implemented on a chip. A sixteen-bit device would need a ridiculous 65 535 comparators, and thus these convertors are not practicable for direct audio conversion, although they will be used to advantage in the DPCM and oversampling convertors described later in this chapter. The analog signal has to drive a lot of inputs which results in a significant parallel capacitance, and a low-impedance driver is essential to avoid restricting the slewing rate of the input. The extreme speed of a flash convertor is a distinct advantage in oversampling. Because computation of all bits is performed simultaneously, no track/hold circuit is required, and droop is eliminated.

Reduction in component complexity can be achieved by quantizing serially. The most primitive method of generating different quantized voltages is to connect a counter to a DAC as in Figure 7.20. The resulting staircase voltage is compared with the input and used to stop the clock to the counter when the DAC output has just exceeded the input. This method is painfully slow, and is not used, as a much faster method exists which is only slightly more complex. Using successive approximation, each bit is tested in turn, starting with the MSB. If the input is greater than

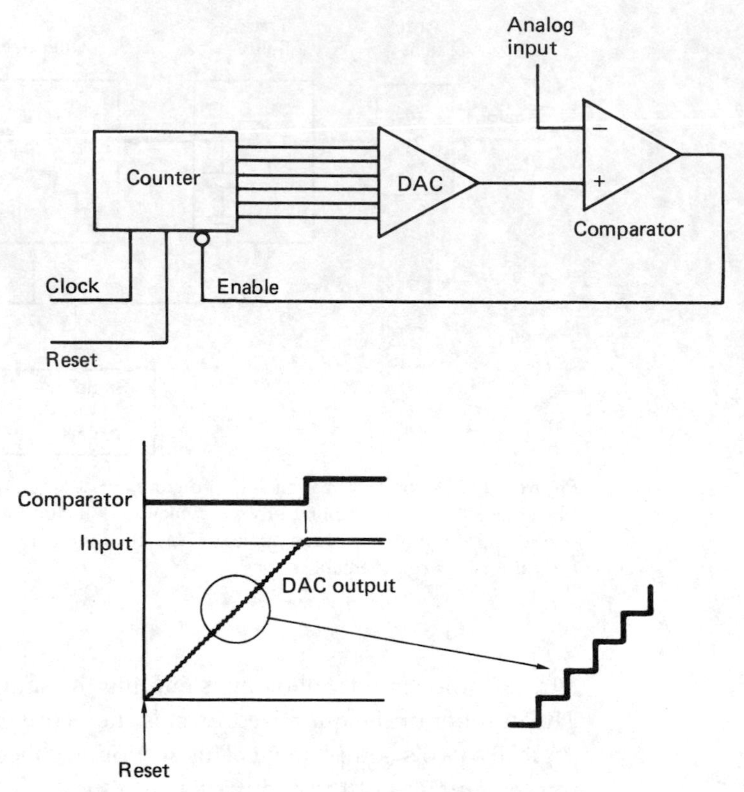

Figure 7.20 Simple ramp ADC compares output of DAC with input. Count is stopped when DAC output just exceeds input. This method, although potentially accurate, is much too slow for digital audio.

half-range, the MSB will be retained and used as a base to test the next bit, which will be retained if the input exceeds three-quarters range and so on. The number of decisions is equal to the number of bits in the word, in contrast to the number of quantizing intervals which was the case in the previous example. A drawback of the successive approximation convertor is that the least significant bits are computed last, when droop is at its worst. Figures 7.21 and 7.22 show that droop can cause a successive approximation convertor to make a significant error under certain circumstances.

Analog-to-digital conversion can also be performed using the dual-current-source type DAC principle in a feedback system; the major difference is that the two current sources must work sequentially rather than concurrently. Figure 7.23 shows a sixteen-bit application in which the capacitor of the track/hold circuit is also used as the ramp integrator. The system operates as follows. When the track/hold FET switches off, the capacitor C will be holding the sample voltage. Two currents of ratio 128:1 are capable of discharging the capacitor. As a result of this ratio, the

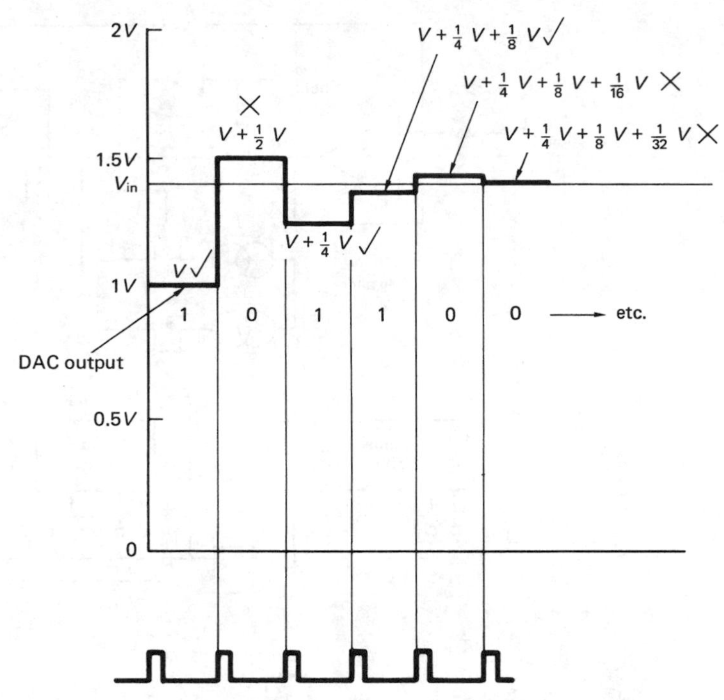

Figure 7.21 Successive approximation tests each bit in turn, starting with the most significant. The DAC output is compared with the input. If the DAC output is below the input (✓) the bit is made 1; if the DAC output is above the input (×) the bit is made zero.

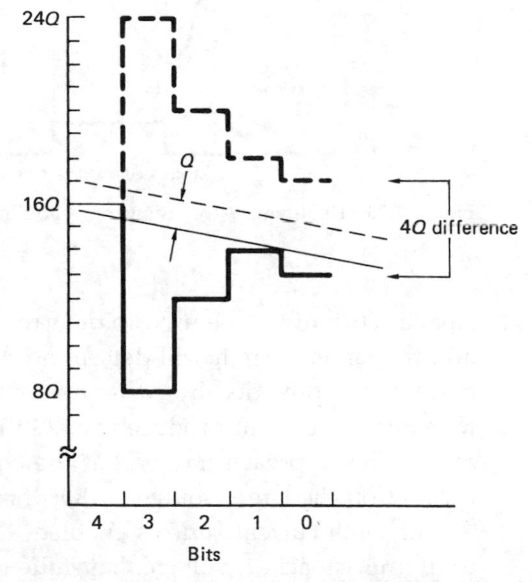

Figure 7.22 Two drooping track–hold signals (solid and dashed lines) which differ by one quantizing interval Q are shown here to result in conversions which are $4Q$ apart. Thus droop can destroy the monotonicity of a convertor. Low-level signals (near the midrange of the number system) are especially vulnerable.

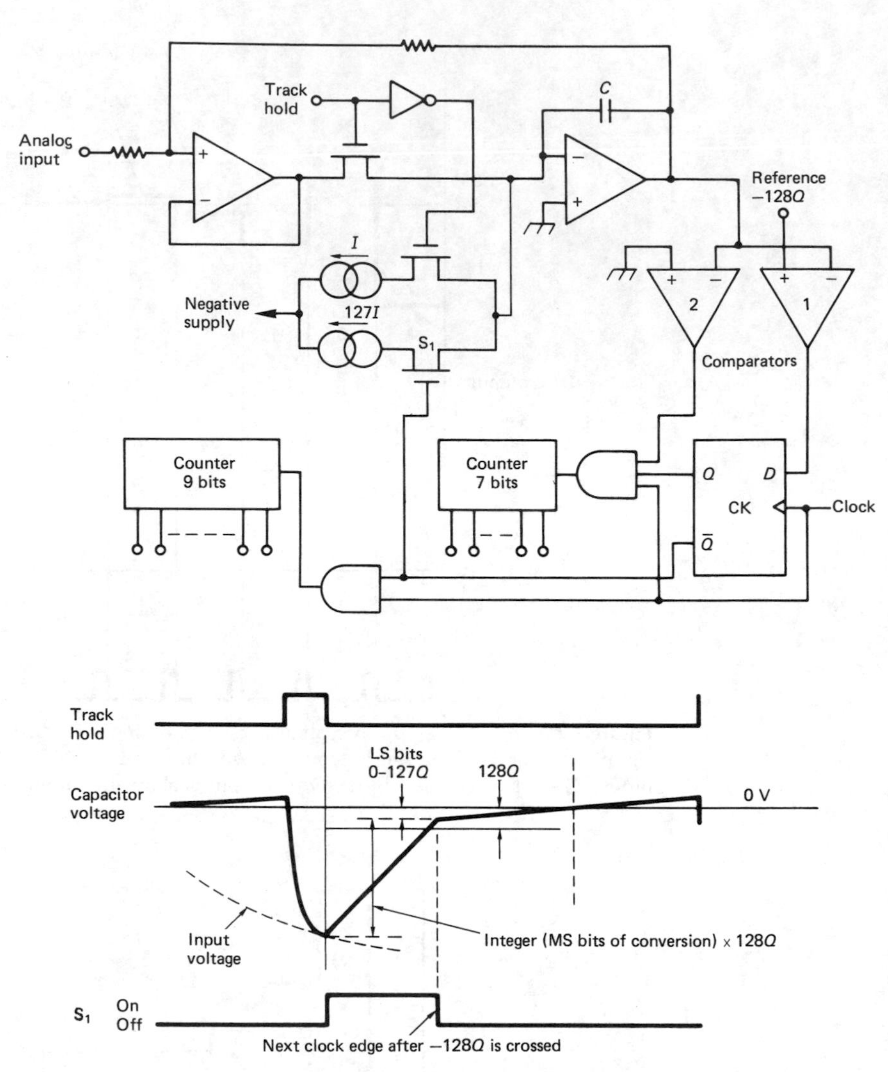

Figure 7.23 Dual-ramp ADC using track–hold capacitor as integrator.

smaller current will be used to determine the seven least significant bits, and the larger current will determine the nine most significant bits. The currents are provided by current sources of ratio 127:1. When both run together, the current produced is 128 times that from the smaller source alone. This approach means that the current can be changed simply by turning off the larger source, rather than by attempting a change-over.

With both current sources enabled, the high-order counter counts up until the capacitor voltage has fallen below the reference of –128*Q* supplied to comparator 1. At the next clock edge, the larger current source is turned off. Waiting for the next clock edge is important, because it ensures that the larger source can only run for entire clock periods, which will discharge the integrator by integer multiples of 128*Q*. The integrator voltage will overshoot the 128*Q* reference, and the remaining voltage on the integrator will be less than 128*Q* and will be measured by counting the number of clocks for which the smaller current source runs before the integrator voltage reaches zero. This process is termed residual expansion. The break in the slope of the integrator voltage gives rise to the alternative title of gear-change convertor. Following ramping to ground in the conversion process, the track/hold circuit must settle in time for the next conversion. In this sixteen-bit example, the high-order conversion needs a maximum count of 512, and the low order needs 128: a total of 640. Allowing 25 per cent of the sample period for the track/hold circuit to operate, a 48 kHz convertor would need to be clocked at some 40 MHz. This is rather faster than the clock needed for the DAC using the same technology.

7.8 Alternative convertors

Although PCM audio is universal because of the ease with which it can be recorded and processed numerically, there are several alternative related methods of converting an analog waveform to a bitstream. The output of these convertor types is not Nyquist rate PCM, but this can be obtained from them by appropriate digital processing. In advanced conversion systems it is possible to adopt an alternative convertor technique specifically to take advantage of a particular characteristic. The output is then digitally converted to Nyquist rate PCM in order to obtain the advantages of both.

Conventional PCM has already been introduced. In PCM, the amplitude of the signal depends only on the number range of the quantizer, and is independent of the frequency of the input. Similarly, the amplitude of the unwanted signals introduced by the quantizing process is also largely independent of input frequency.

Figure 7.24 introduces the alternative convertor structures. The top half of the diagram shows convertors which are differential. In differential coding the value of the output code represents the difference between the current sample voltage and that of the previous sample. The lower half of the diagram shows convertors which are PCM. In addition, the left side of the diagram shows single-bit convertors, whereas the right side shows multi-bit convertors.

In differential pulse code modulation (DPCM), shown at top-right, the difference between the previous absolute sample value and the current one is quantized into a multi-bit binary code. It is possible to produce a DPCM signal from a PCM signal simply by subtracting successive samples; this is digital differentiation. Similarly the reverse process is

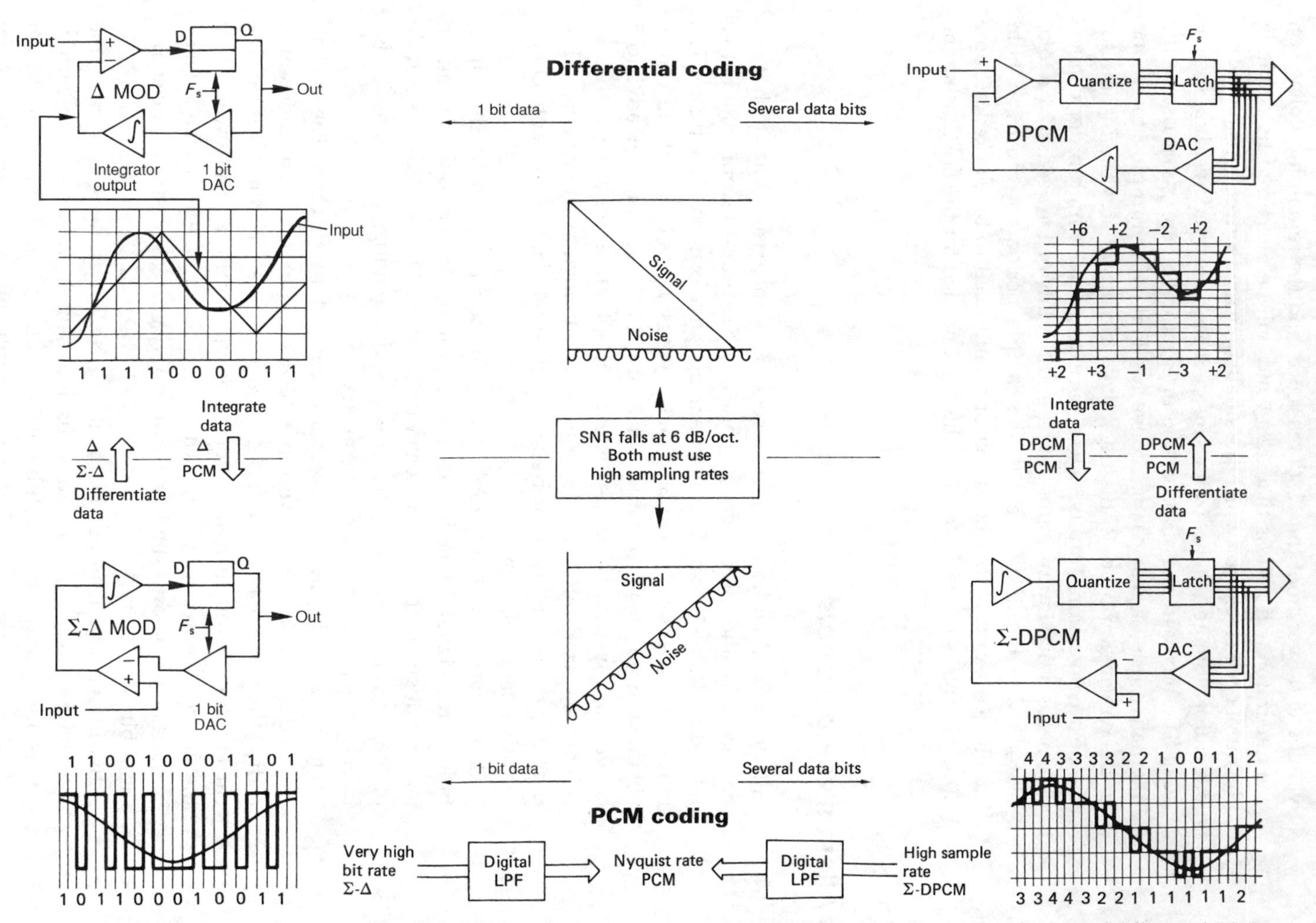

Figure 7.24 The four main alternatives to simple PCM conversion are compared here. Delta modulation is a one-bit case of differential PCM, and conveys the slope of the signal. The digital output of both can be integrated to give PCM. Σ-Δ (sigma–delta) is a one-bit case of Σ-DPCM. The application of integrator before differentiator makes the output ture PCM, but tilts the noise floor; hence these can be referred to as 'noise-shaping' convertors.

possible by using an accumulator or digital integrator (see Chapter 3) to compute sample values from the differences received. The problem with this approach is that it is very easy to lose the baseline of the signal if it commences at some arbitrary time. A digital high-pass filter can be used to prevent unwanted offsets.

Differential convertors do not have an absolute amplitude limit. Instead there is a limit to the maximum rate at which the input signal voltage can change. They are said to be slew rate limited, and thus the permissible signal amplitude falls at 6 dB per octave. As the quantizing steps are still uniform, the quantizing error amplitude has the same limits as PCM. As input frequency rises, ultimately the signal amplitude available will fall down to it.

If DPCM is taken to the extreme case where only a binary output signal is available then the process is described as delta modulation (top-left in Figure 7.24). The meaning of the binary output signal is that the current analog input is above or below the accumulation of all previous bits. The characteristics of the system show the same trends as DPCM, except that there is severe limiting of the rate of change of the input signal. A DPCM decoder must accumulate all the difference bits to provide a PCM output for conversion to analog, but with a one-bit signal the function of the accumulator can be performed by an analog integrator.

If an integrator is placed in the input to a delta modulator, the integrator's amplitude response loss of 6 dB per octave parallels the convertor's amplitude limit of 6 dB per octave; thus the system amplitude limit becomes independent of frequency. This integration is responsible for the term sigma-delta modulation, since in mathematics sigma is used to denote summation. The input integrator can be combined with the integrator already present in a delta-modulator by a slight rearrangement of the components (bottom-left in Figure 7.24). The transmitted signal is now the amplitude of the input, not the slope; thus the receiving integrator can be dispensed with, and all that is necessary to after the DAC is an LPF to smooth the bits. The removal of the integration stage at the decoder now means that the quantizing error amplitude rises at 6 dB per octave, ultimately meeting the level of the wanted signal.

The principle of using an input integrator can also be applied to a true DPCM system and the result should perhaps be called sigma DPCM (bottom-right in Figure 7.24. The dynamic range improvement over delta sigma modulation is 6 dB for every extra bit in the code. Because the level of the quantizing error signal rises at 6 dB per octave in both delta sigma modulation and sigma DPCM, these systems are sometimes referred to as 'noise-shaping' convertors, although the word 'noise' must be used with some caution. The output of a sigma DPCM system is again PCM, and a DAC will be needed to receive it, because it is a binary code.

As the differential group of systems suffer from a wanted signal that converges with the unwanted signal as frequency rises, they must all use very high sampling rates.[12] It is possible to convert from sigma DPCM to conventional PCM by reducing the sampling rate digitally. When the sampling rate is reduced in this way, the reduction of bandwidth excludes a disproportionate amount of noise because the noise shaping concentrated it at frequencies beyond the audio band. The use of noise shaping and oversampling is the key to the high resolution obtained in advanced convertors.

7.9 Oversampling and noise shaping

It was seen in Chapter 4 that oversampling has a number of advantages for video conversion and the same is true for audio. Although it can be used alone, the advantages of oversampling in audio are better realized when it is used in conjunction with noise shaping. Thus in practice the two processes are generally used together and the terms are often seen used in the loose sense as if they were synonymous. For a detailed and quantitative analysis of audio oversampling having exhaustive references the serious reader is referred to Hauser.[13]

In section 7.6, where dynamic element matching was described, it was seen that component accuracy was traded for accuracy in the time domain. Oversampling is another example of the same principle. Oversampling permits the use of a convertor element of shorter wordlength, making it possible to use a flash convertor for audio conversion. The flash convertor is capable of working at very high frequency and so large oversampling factors are easily realized. The flash convertor needs no track/hold system as it works instantaneously. The drawbacks of track/hold set out in section 7.7 are thus eliminated.

If the sigma-DPCM convertor structure of Figure 7.24 is realized with a flash convertor element, it can be used with a high oversampling factor. This class of convertor has a rising noise floor. If the highly oversampled output is fed to a digital low-pass filter which has the same frequency response as an analog anti-aliasing filter used for Nyquist rate sampling, the result is a disproportionate reduction in noise because the majority of the noise was outside the audio band. A high-resolution convertor can be obtained using this technology without requiring unattainable component tolerances.

Noise shaping dates from the work of Cutler[14] in the 1950s. It is a feedback technique applicable to quantizers and requantizers in which the quantizing process of the current sample is modified in some way by the quantizing error of the previous sample.

When used with requantizing, noise shaping is an entirely digital process which is used, for example, following word extension due to the arithmetic in digital mixers or filters in order to return to the required wordlength. It will be found in this form in oversampling DACs. When used with quantizing, part of the noise-shaping circuitry will be analog. As the feedback loop is placed around an ADC it must contain a DAC. When used in convertors, noise shaping is primarily an implementation technology. It allows processes which are conveniently available in integrated circuits to be put to use in audio conversion. Once integrated circuits can be employed, complexity ceases to be a drawback and low-cost mass production is possible.

It has been stressed throughout this chapter that a series of numerical values or samples is just another analog of an audio waveform. Chapter 3 showed that all analog processes such as mixing, attenuation or integration have exact numerical parallels. It has been demonstrated that digitally dithered requantizing is no more than a digital simulation of analog quantizing. It should be no surprise that in this section noise shaping will be treated in the same way. Noise shaping can be performed by manipulating analog voltages or numbers representing them or both. If the reader is content to make a conceptual switch between the two, many obstacles to understanding fall, not just in this topic, but in digital audio and video in general.

The term noise shaping is idiomatic and in some respects unsatisfactory because not all devices which are called noise shapers produce true noise. The caution which was given when treating quantizing error as noise is also relevant in this context. Whilst 'quantizing-error-spectrum shaping' is a bit of a mouthful, it is useful to keep in mind that noise shaping means just that in order to avoid some pitfalls. Some noise shaper architectures do not produce a signal-decorrelated quantizing error and need to be dithered.

Figure 7.25(a) shows a requantizer using a simple form of noise shaping. The low-order bits which are lost in requantizing are the quantizing error. If the value of these bits is added to the next sample before it is requantized, the quantizing error will be reduced. The process is somewhat like the use of negative feedback in an operational amplifier except that it is not instantaneous, but encounters a one-sample delay. With a constant input, the mean or average quantizing error will be brought to zero over a number of samples, achieving one of the goals of additive dither. The more rapidly the input changes, the greater the effect of the delay and the less effective the error feedback will be. Figure 7.25(b) shows the equivalent circuit seen by the quantizing error, which is created at the requantizer and subtracted from itself one sample period later. As a result the quantizing error spectrum is not uniform, but has the shape of a raised sinewave shown in (c), hence the term noise shaping. The

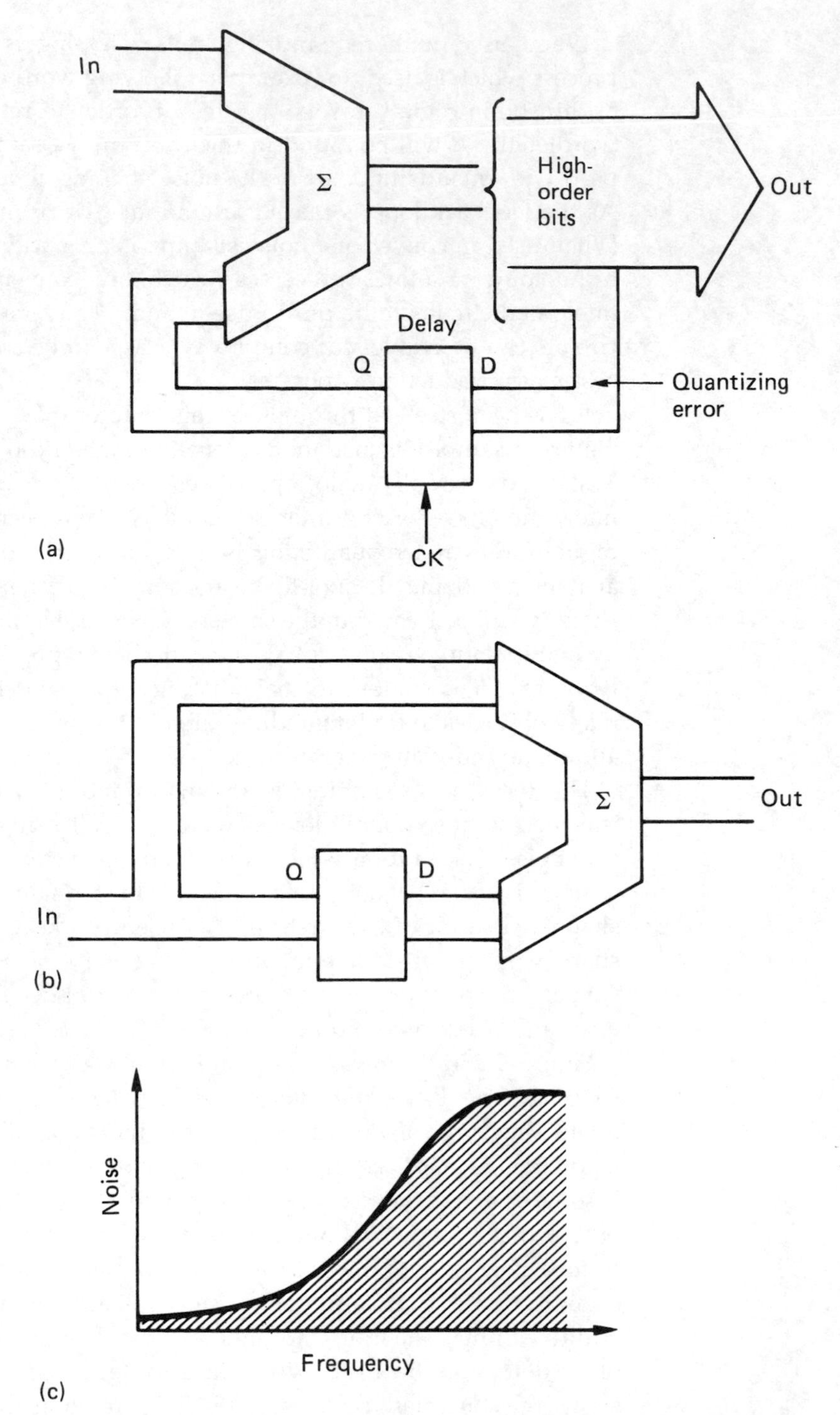

Figure 7.25 (a) A simple requantizer which feeds back the quantizing error to reduce the error of subsequent samples. The one-sample delay causes the quantizing error to see the equivalent circuit shown in (b) which results in a sinusoidal quantizing error spectrum shown in (c).

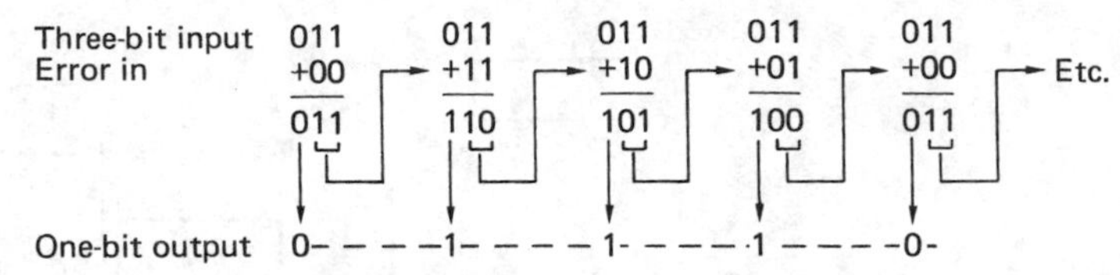

Figure 7.26 By adding the error caused by truncation to the next value, the resolution of the lost bits is maintained in the duty cycle of the output. Here, truncation of 011 by two bits would give continuous zeros, but the system repeats 0111, 0111, which, after filtering, will produce a level of three-quarters of a bit.

noise is very small at DC and rises with frequency, peaking at the Nyquist frequency at a level determined by the size of the quantizing step. If used with oversampling, the noise peak can be moved outside the audio band.

Figure 7.26 shows a simple example in which two low-order bits need to be removed from each sample. The accumulated error is controlled by using the bits which were neglected in the truncation, and adding them to the next sample. In this example, with a steady input, the roundoff mechanism will produce an output of 01110111 . . . If this is low-pass filtered, the three ones and one zero result in a level of three-quarters of a quantizing interval, which is precisely the level which would have been obtained by direct conversion of the full digital input. Thus the resolution is maintained even though two bits have been removed.

Noise shaping can also be used without oversampling. In this case the noise cannot be pushed outside the audio band. Instead the noise floor is shaped or weighted to complement the unequal spectral sensitivity of the ear to noise.[15–17] Unless we wish to violate Shannon's theory, this psychoacoustically optimal noise shaping can only reduce the noise power at certain frequencies by increasing it at others. Thus the average log psd over the audio band remains the same, although it may be raised slightly by noise induced by imperfect processing.

Figure 7.27 shows noise shaping applied to a digitally dithered requantizer. Such a device might be used when, for example, making a CD master from a twenty-bit recording format. The input to the dithered requantizer is subtracted from the output to give the error due to requantizing. This error is filtered (and inevitably delayed) before being subtracted from the system input. The filter is not designed to be the exact inverse of the perceptual weighting curve because this would cause extreme noise levels at the ends of the band. Instead the perceptual curve is levelled off[18] such that it cannot fall more than e.g. 40 dB below the peak.

Psychoacoustically optimal noise shaping can offer nearly three bits of increased dynamic range when compared with optimal spectrally flat dither. Enhanced Compact Discs recorded using these techniques are now available.

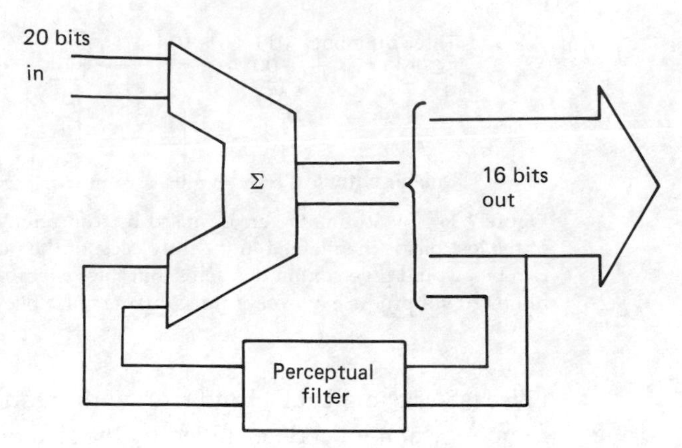

Figure 7.27 Perceptual filtering in a requantizer gives a subjectively improved SNR.

The sigma DPCM convertor introduced in Figure 7.24 has a natural application here and is shown in more detail in Figure 7.28. The current digital sample from the quantizer is converted back to analog in the embedded DAC. The DAC output differs from the ADC input by the quantizing error. It is subtracted from the analog input to produce an error which is integrated to drive the quantizer in such a way that the error is reduced. With a constant input voltage the average error will be zero because the loop gain is infinite at DC. If the average error is zero, the mean or average of the DAC outputs must be equal to the analog input. The instantaneous output will deviate from the average in what is

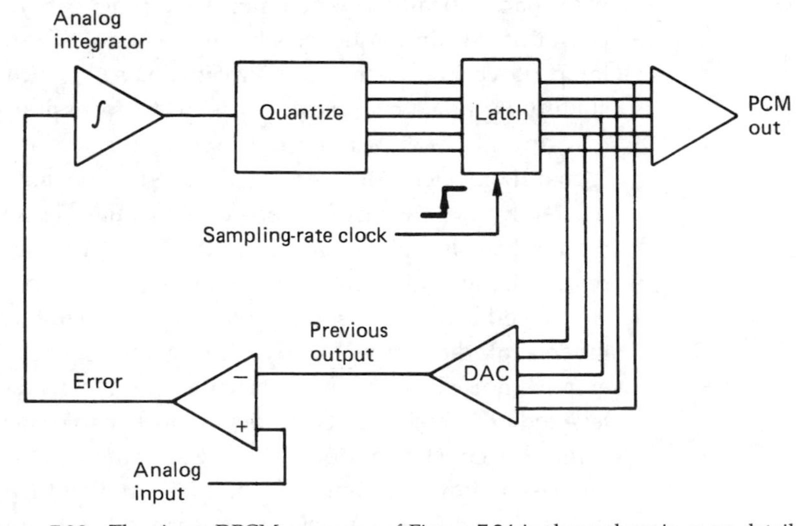

Figure 7.28 The sigma DPCM convertor of Figure 7.24 is shown here in more detail.

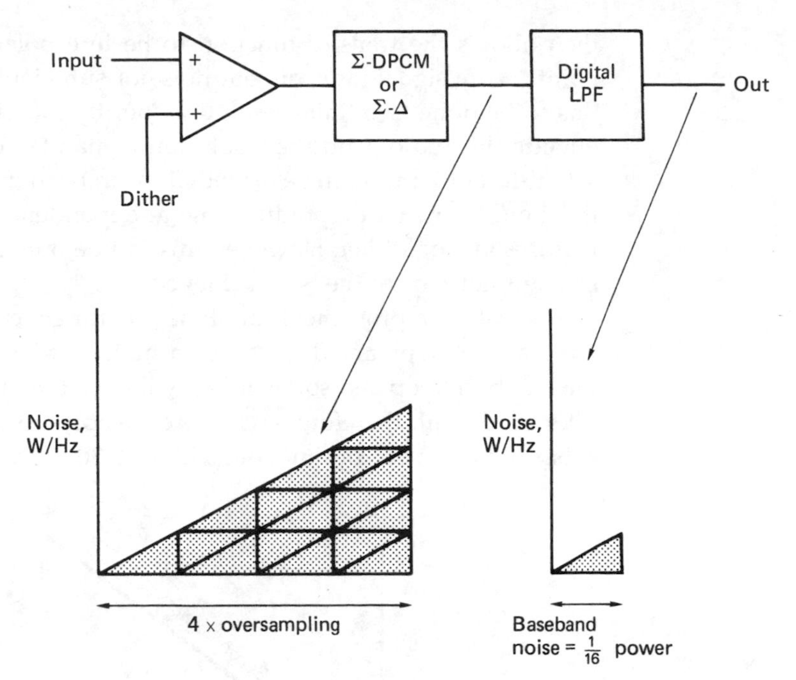

Figure 7.29 In a sigma-DPCM or Σ–Δ convertor, noise amplitude increases by 6 dB/octave, noise power by 12 dB/octave. In this 4 × oversampling convertor, the digital filter reduces bandwidth by four, but noise power is reduced by a factor of 16. Noise voltage falls by a factor of four or 12 dB.

called an idling pattern. The presence of the integrator in the error feedback loop makes the loop gain fall with rising frequency. With the feedback falling at 6 dB per octave, the noise floor will rise at the same rate.

Figure 7.29 shows a simple oversampling system using a sigma DPCM convertor and an oversampling factor of only four. The sampling spectrum shows that the noise is concentrated at frequencies outside the audio part of the oversampling baseband. Since the scale used here means that noise power is represented by the area under the graph, the area left under the graph after the filter shows the noise-power reduction. Using the relative areas of similar triangles shows that the reduction has been by a factor of sixteen. The corresponding noise-voltage reduction would be a factor of four, or 12 dB which corresponds to an additional two bits in wordlength. These bits will be available in the wordlength extension which takes place in the decimating filter. Due to the rise of 6 dB per octave in the PSD of the noise, the SNR will be 3 dB worse at the edge of the audio band.

One way in which the operation of the system can be understood is to consider that the coarse DAC in the loop defines fixed points in the audio transfer function. The time averaging which takes place in the decimator then allows the transfer function to be interpolated between the fixed points. True signal-independent noise of sufficient amplitude will allow this to be done to infinite resolution, but by making the noise primarily outside the audio band the resolution is maintained but the audio band signal-to-noise ratio can be extended. A first-order noise-shaping ADC of the kind shown can produce signal-dependent quantizing error and requires analog dither. However, this can be outside the audio band and so need not reduce the SNR achieved.

A greater improvement in dynamic range can be obtained if the integrator is supplanted to realize a higher-order filter.[19] The filter is in the feedback loop and so the noise will have the opposite response to the filter and will therefore rise more steeply to allow a greater SNR enhancement after decimation. Figure 7.30 shows the theoretical SNR

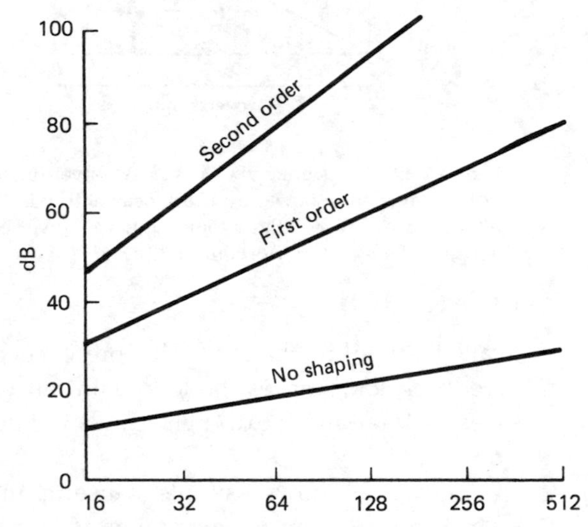

Figure 7.30 The enhancement of SNR possible with various filter orders and oversampling factors in noise-shaping convertors.

enhancement possible for various loop filter orders and oversampling factors. A further advantage of high-order loop filters is that the quantizing noise can be decorrelated from the signal, making dither unnecessary. High-order loop filters were at one time thought to be impossible to stabilize, but this is no longer the case, although care is necessary. One technique which may be used is to include some feedforward paths as shown in Figure 7.31.

An ADC with high-order noise shaping was disclosed by Adams[20] and a simplified diagram is shown in Figure 7.32. The comparator outputs of the 128 times oversampled four-bit flash ADC are directly fed to the DAC which consists of fifteen equal resistors fed by CMOS switches. As with

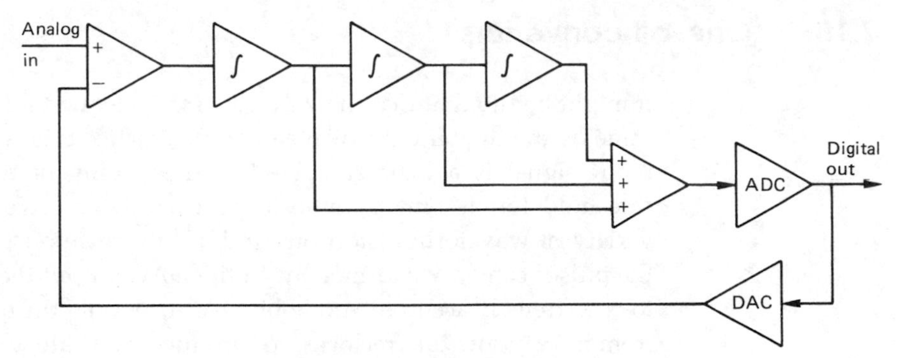

Figure 7.31 Stabilizing the loop filter in a noise-shaping convertor can be assisted by the incorporation of feedforward paths as shown here.

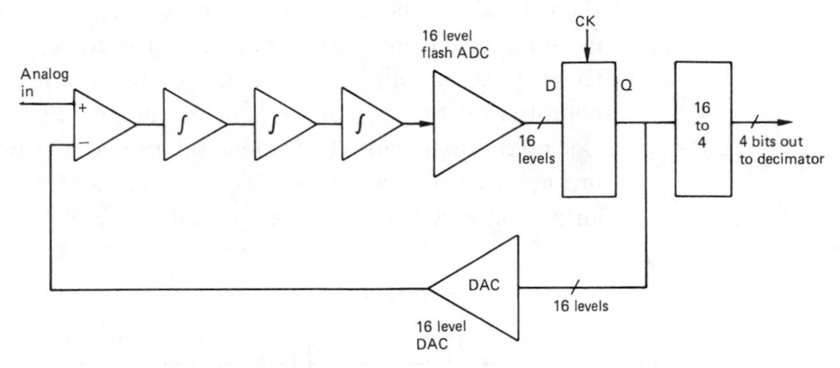

Figure 7.32 An example of a high-order noise-shaping ADC. See text for details.

all feedback loops, the transfer characteristic cannot be more accurate than the feedback, and in this case the feedback accuracy is determined by the precision of the DAC.[21] Driving the DAC directly from the ADC comparators is more accurate because each input has equal weighting. The stringent MSB tolerance of the conventional binary-weighted DAC is then avoided. The comparators also drive a 16 to 4 priority encoder to provide the four-bit PCM output to the decimator. The DAC output is subtracted from the analog input at the integrator. The integrator is followed by a pair of conventional analog operational amplifiers having frequency-dependent feedback and a passive network which gives the loop a fourth-order response overall. The noise floor is thus shaped to rise at 24 dB per octave beyond the audio band. The time constants of the loop filter are optimized to minimize the amplitude of the idling pattern as this is an indicator of the loop stability. The four-bit PCM output is low-pass filtered and decimated to the Nyquist frequency. The high oversampling factor and high-order noise shaping extend the dynamic range of the four-bit flash ADC to 108 dB at the output.

7.10 One-bit convertors

It might be thought that the waveform from a one-bit DAC is simply the same as the digital input waveform. In practice this is not the case. The input signal is a logic signal which need only be above or below a threshold for its binary value to be correctly received. It may have a variety of waveform distortions and a duty cycle offset. The area under the pulses can vary enormously. In the DAC output the amplitude needs to be extremely accurate. A one-bit DAC uses only the binary information from the input, but reclocks to produce accurate timing and uses a reference voltage to produce accurate levels. The area of pulses produced is then constant. One-bit DACs will be found in noise-shaping ADCs as well as in the more obvious application of producing analog audio.

Figure 7.33(a) shows a one-bit DAC which is implemented with MOS field-effect switches and a pair of capacitors. Quanta of charge are driven into or out of a virtual earth amplifier configured as an integrator by the switched capacitor action. Figure 7.33(b) shows the associated waveforms. Each data bit period is divided into two equal portions; that for which the clock is high, and that for which it is low. During the first half of the bit period, pulse P+ is generated if the data

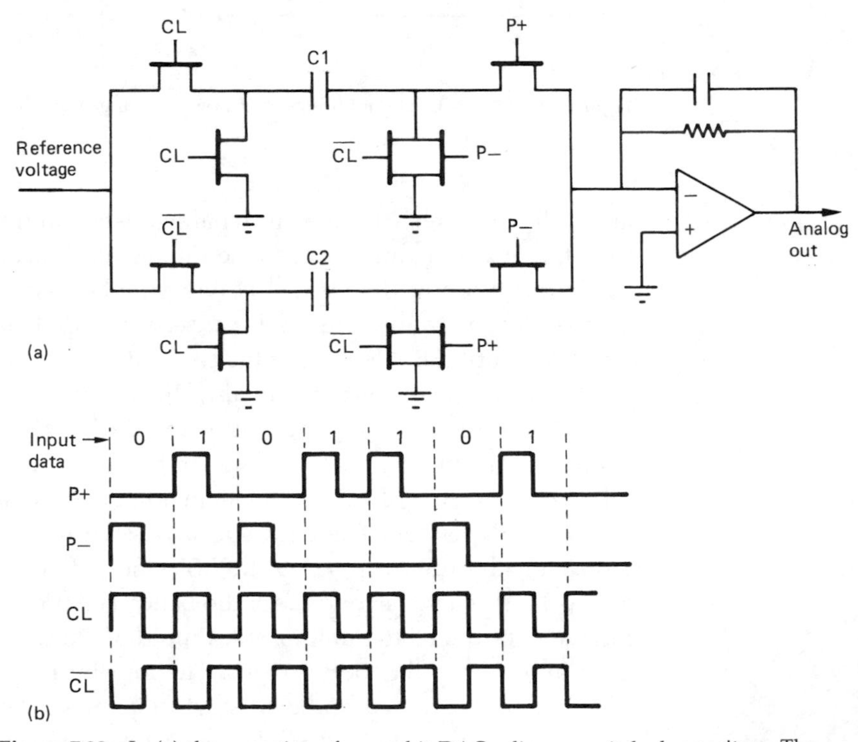

Figure 7.33 In (a) the operation of a one-bit DAC relies on switched capacitors. The switching waveforms are shown in (b).

bit is a 1, or pulse P– is generated if the data bit is a 0. The reference input is a clean voltage corresponding to the gain required.

C1 is *discharged* during the second half of every cycle by the switches driven from the complemented clock. If the next bit is a 1, during the next high period of the clock the capacitor will be connected between the reference and the virtual earth. Current will flow into the virtual earth until the capacitor is charged. If the next bit is not a 1, the current through C1 will flow to ground.

C2 is *charged* to reference voltage during the second half of every cycle by the switches driven from the complemented clock. On the next high period of the clock, the reference end of C2 will be grounded, and so the op-amp end wil assume a negative reference voltage. If the next bit is a 0, this negative reference will be switched into the virtual earth. If not, the capacitor will be discharged.

Thus on every cycle of the clock, a quantum of charge is either pumped into the integrator by C1 or pumped out by C2. The analog output therefore precisely reflects the ratio of ones to zeros.

In order to overcome the DAC accuracy constraint of the sigma DPCM convertor, the sigma delta convertor can be used as it has only one-bit internal resolution. A one-bit DAC cannot be non-linear by definition as it defines only two points on a transfer function. It can, however, suffer from other deficiencies such as DC offset and gain error although these are less offensive in audio. The one-bit ADC is a comparator.

As the sigma delta convertor is only a one-bit device, clearly it must use a high oversampling factor and high-order noise shaping in order to have sufficiently good SNR for audio.[22] In practice the oversampling factor is limited not so much by the convertor technology as by the difficulty of computation in the decimator. A sigma delta convertor has the advantage that the filter input 'words' are one bit long and this simplifies the filter design as multiplications can be replaced by selection of constants.

Conventional analysis of loops falls down heavily in the one-bit case. In particular the gain of a comparator is difficult to quantify, and the loop is highly non-linear so that considering the quantizing error as additive white noise in order to use a linear loop model gives rather optimistic results. In the absence of an accurate mathematical model, progress has been made empirically, with listening tests and by using simulation.

Single-bit sigma delta convertors are prone to long idling patterns because the low resolution in the voltage domain requires more bits in the time domain to be integrated to cancel the error. Clearly the longer the period of an idling pattern, the more likely it is to enter the audio band as an objectional whistle or 'birdie'. They also exhibit threshold effects or deadbands where the output fails to react to an input change at certain levels. The problem is reduced by the order of the filter and the wordlength of the embedded DAC. Second- and third-order feedback loops are still prone to audible idling patterns and threshold effect.[23] The traditional approach to linearizing sigma delta convertors is to use dither. Unlike conventional quantizers, the dither used was of a frequency outside the audio band and of considerable level. Squarewave dither has been used and it is advantageous to choose a frequency which is a multiple of the final output sampling rate as then the harmonics will coincide with the troughs in the stopband ripple of the decimator. Unfortunately the level of dither needed to linearize the convertor is high enough to cause premature clipping of high-level signals, reducing the dynamic range. This problem is overcome by using in-band white-noise dither at low level.[24]

An advantage of the one-bit approach is that in the one-bit DAC, precision components are replaced by precise timing in switched capacitor networks. The same approach can be used to implement the loop filter in an ADC. Figure 7.34 shows a third-order sigma delta modulator incorporating a DAC based on the principle of Figure 7.33. The loop filter is also implemented with switched capacitors.

7.11 Operating levels in digital audio

Analog tape recorders use operating levels which are some way below saturation. The range between the operating level and saturation is called the headroom. In this range, distortion becomes progressively worse and sustained recording in the headroom is avoided. However, transients may be recorded in the headroom as the ear cannot respond to distortion products unless they are sustained. The PPM level meter has an attack time constant which simulates the temporal distortion sensitivity of the ear. If a transient is too brief to deflect a PPM into the headroom, it will not be heard either.

Operating levels are used in two ways. On making a recording from a microphone, the gain is increased until distortion is just avoided, thereby obtaining a recording having the best SNR. In post-production the gain will be set to whatever level is required to obtain the desired subjective effect in the context of the program material. This is particularly important to broadcasters who require the relative loudness of different material to be controlled so that the listener does not need to make continuous adjustments to the volume control.

In order to maintain level accuracy, analog recordings are traditionally preceded by line-up tones at standard operating level. These are used to adjust the gain in various stages of dubbing and transfer along land lines so that no level changes occur to the program material.

Unlike analog recorders, digital recorders do not have headroom, as there is no progressive onset of distortion until convertor clipping, the

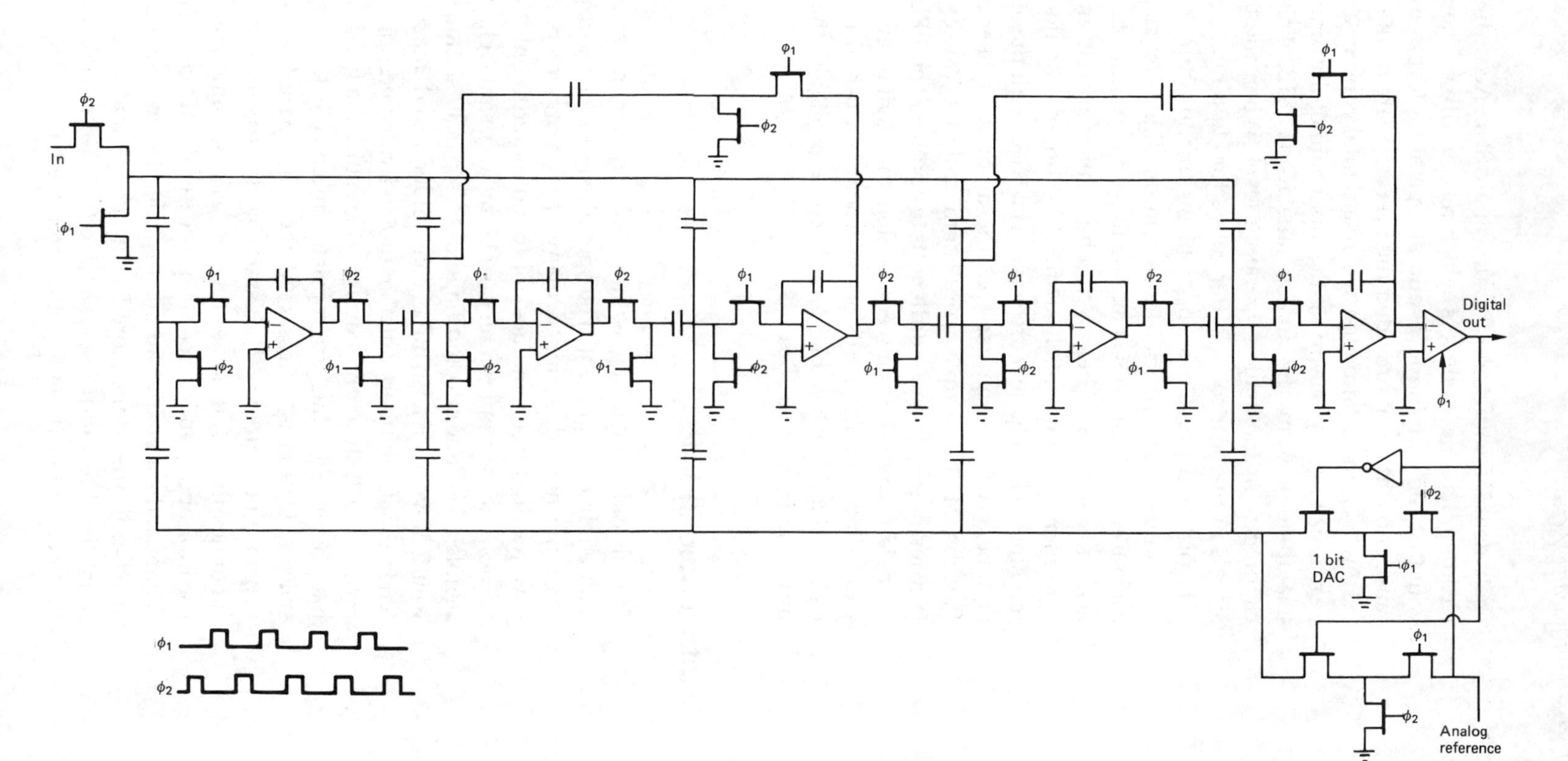

Figure 7.34 A third-order Σ–Δ modulator using a switched capacitor loop filter.

equivalent of saturation, occurs at 0 dBFs. Accordingly many digital recorders have level meters which read in dBFs. The scales are marked with 0 at the clipping level and all operating levels are below that. This causes no difficulty provided the user is aware of the consequences.

However, in the situation where a digital copy of an analog tape is to be made, it is very easy to set the input gain of the digital recorder so that line-up tone from the analog tape reads 0 dB. This lines up digital clipping with the analog operating level. When the tape is dubbed, all signals in the headroom suffer convertor clipping.

In order to prevent such problems, manufacturers and broadcasters have introduced artificial headroom on digital level meters, simply by calibrating the scale and changing the analog input sensitivity so that 0 dB analog is some way below clipping. Unfortunately there has been little agreement on how much artificial headroom should be provided, and machines which have it are seldom labelled with the amount. There is an argument which suggests that the amount of headroom should be a function of the sample wordlength, but this causes difficulties when transferring from one wordlength to another. The EBU[25] concluded that a single relationship between analog and digital level was desirable. In sixteen-bit working, 12 dB of headroom is a useful figure, but now that eighteen- and twenty-bit convertors are available, the new EBU recommendation specifies 18 dB.

7.12 Digital audio in VTRs

Digital audio recording with video is rather more difficult than in an audio-only environment. The special requirements of professional video recording have determined many of the parameters of the format, and have resulted in techniques not found in audio-only recorders.

The audio samples are carried by the same channel as the video samples. The audio could have used separate stationary heads, but this would have increased tape consumption and machine complexity. In order to permit independent audio and video editing, the tape tracks are given a block structure. Editing will require the heads momentarily to go into record as the appropriate audio block is reached. Accurate synchronization is necessary if the other parts of the recording are to remain uncorrupted. The sync block structure of the video sector continues in the audio sectors because the same read/write circuitry is used for audio and video data. Clearly the ID code structure must also continue through the audio. In order to prevent audio samples from arriving in the frame store in shuttle, the audio addresses are different from the video addresses. Despite the additional complexity of sharing the medium with video, the professional DVTR must support such functions as track bouncing, synchronous recording and split audio edits with variable crossfade times.

The audio samples in a DVTR are binary numbers just like the video samples, and although there is an obvious difference in sampling rate and wordlength, the use of time compression means that this affects only the relative areas of tape devoted to the audio and video samples. The most important difference between audio and video samples is the tolerance to errors. The acuity of the ear means that uncorrected audio samples must not occur more than once every few hours. There is little redundancy in sound when compared to video, and concealment of errors is not desirable on a routine basis. In video, the samples are highly redundant, and concealment can be effected using samples from previous or subsequent lines or, with care, from the previous frame.

Whilst subjective considerations require greater data reliability in the audio samples, audio data form a small fraction of the overall data and it is difficult to protect them with an extensive interleave whilst still permitting independent editing. For these reasons major differences can be expected between the ways that audio and video samples are handled in a digital video recorder. One such difference is that the error-correction strategy for audio samples uses a greater amount of redundancy. Whilst this would cause a serious playing time penalty in an audio recorder, even doubling the audio data rate in a video recorder raises the overall data rate by only a few per cent. The arrangement of the audio blocks is also designed to maximize data integrity in the presence of tape defects and head clogs. The audio blocks are at the ends of the head sweeps in D-2 and D-3, but are placed in the middle of the segment in D-1, D-5, Digital Betacam and DCT.

7.13 MPEG audio compression

The ISO (International Standards Organization) and the IEC (International Electrotechnical Commission) recognized that compression would have an important part to play in future digital video products and in 1988 established the ISO/IEC/MPEG (Moving Picture Experts Group) to compare and assess various coding schemes in order to arrive at an international standard. The terms of reference were extended the same year to include audio and the MPEG/Audio group was formed.

As part of the Eureka 147 project, a system known as MUSICAM[26] (Masking pattern adapted Universal Sub-band Integrated Coding And Multiplexing) was developed jointly by CCETT in France, IRT in Germany and Philips in the Netherlands. MUSICAM was designed to be suitable for DAB (digital audio broadcasting).

As a parallel development, the ASPEC[27] (Adaptive Spectral Perceptual Entropy Coding) system was developed from a number of earlier systems as a joint proposal by AT&T Bell Labs, Thomson, the Fraunhofer Society and CNET. ASPEC was designed for high degrees of compression to allow audio transmission on ISDN.

These two systems were both fully implemented by July 1990 when comprehensive subjective testing took place at the Swedish Broadcasting Corporation.[28,29] As a result of these tests, the MPEG/Audio group combined the attributes of both ASPEC and MUSICAM into a draft standard[30] having three levels of complexity and performance.

These three different levels are needed because of the number of possible applications. Audio coders can be operated at various compression factors with different quality expectations. Stereophonic classical music requires different quality criteria to monophonic speech. As was seen in Chapter 6, the complexity of the coder will be reduced with a smaller compression factor. For moderate compression, a simple codec will be more cost effective. On the other hand, as the compression factor is increased, it will be necessary to employ a more complex coder to maintain quality.

At each level, MPEG coding allows input sampling rates of 32, 44.1 and 48 kHz and supports output bit rates of 32, 48, 56, 64, 96, 112, 128, 192, 256 and 384 kbits/s. The transmission can be mono, dual channel (e.g. bilingual), stereo and joint stereo which is where advantage is taken of redundancy between the two audio channels.

MPEG Layer 1 is a simplified version of MUSICAM which is appropriate for mild compression applications at low cost. It is very similar to PASC. Layer II is identical to MUSICAM and is very likely to be used for DAB. Layer III is a combination of the best features of ASPEC and MUSICAM and is mainly applicable to telecommunications where high compression factors are required.

The earlier MPEG-1 standard compresses audio and video into about 1.5 Mbits/s. The audio content of MPEG-1 may be used on its own to encode one or two channels at bit rates up to 448 kbits/s. MPEG-2 allows the number of channels to increase to five: Left, Right, Centre, Left surround, Right surround and Subwoofer. In order to retain reverse compatibility with MPEG-1, the MPEG-2 coding converts the five-channel input to a compatible two-channel signal, L_o,R_o, by matrixing.[31] The data from these two channels are encoded in a standard MPEG-1 audio frame, and this is followed in MPEG-2 by an ancillary data frame which an MPEG-1 decoder will ignore. The ancillary frame contains data for three further audio channels. An MPEG-2 decoder will extract those three channels in addition to the MPEG-1 frame and then recover all five original channels by an inverse matrix.

In various countries, it has been proposed to use an alternative compression technique for the audio content of DVB broadcasts. This is the AC-3 system developed by Dolby Laboratories. The MPEG transport stream structure has also been standardized to allow it to carry AC-3 coded audio. The Digital Video Disk can also carry AC-3 or MPEG audio coding.

There are many different approaches to audio compression, each having advantages and disadvantages. MPEG audio coding combines these tools in various ways in the three different coding levels. The approach of this section will be to examine the tools separately before seeing how they are used in MPEG and AC-3.

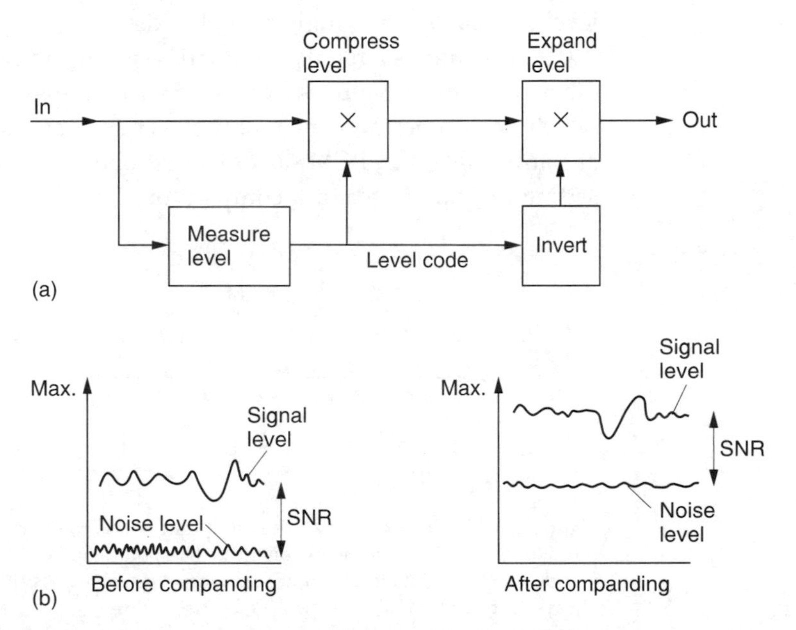

Figure 7.35 Digital companding. In (a) the encoder amplifies the input to maximum level and the decoder attenuates by the same amount. (b) In a companded system, the signal is kept as far as possible above the noise caused by shortening the sample wordlength.

The simplest coding tool is companding which is a digital parallel of the analog noise reducers used in tape recording. Figure 7.35(a) shows that in companding the input signal level is monitored. Whenever the input level falls below maximum, it is amplified at the coder. The gain which was applied at the coder is added to the data stream so that the decoder can apply an equal attenuation. The advantage of companding is that the signal is kept as far away from the noise floor as possible. In analog noise reduction this is used to maximize the SNR of a tape recorder, whereas in digital compression it is used to keep the signal level as far as possible above the noises and artifacts introduced by various coding steps.

One common way of obtaining coding gain is to shorten the wordlength of samples so that fewer bits need to be transmitted. Figure 7.35(b) shows that when this is done, the noise floor will rise by 6 dB for every bit removed. This is because removing a bit halves the number of quantizing intervals which then must be twice as large, doubling the noise level. Clearly if this step follows the compander in (a), the audibility of the noise will be minimized. As an alternative to shortening the wordlength, the uniform quantized PCM signal can be converted to a non-uniform format. In non-uniform coding, shown in (c), the size of the quantizing step rises with the magnitude of the sample so that the noise level is greater when higher levels exist.

Companding is a relative of floating-point coding shown in Figure 7.36 where the sample value is expressed as a mantissa and a binary exponent which determines how the mantissa needs to be shifted to have its correct absolute value on a PCM scale. The exponent is the equivalent of the gain setting or scale factor of a compandor.

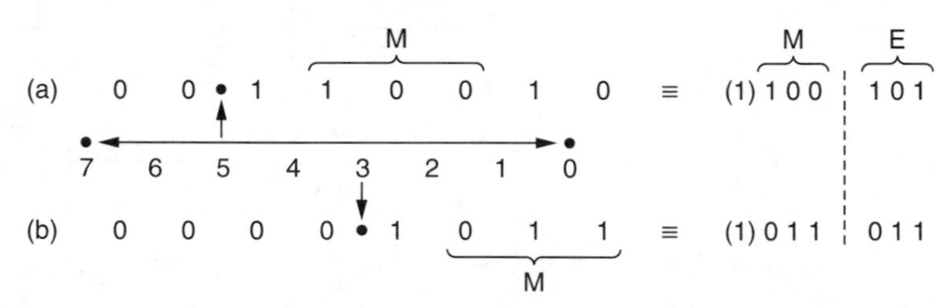

Figure 7.36 In this example of floating-point notation, the radix point can have eight positions determined by the exponent E. The point is placed to the left of the first '1', and the next four bits to the right form the mantissa M. As the MSB of the mantissa is always 1, it need not always be stored.

Clearly in floating point the signal-to-noise ratio is defined by the number of bits in the mantissa, and as shown in Figure 7.37, this will vary as a sawtooth function of signal level, as the best value, obtained when the mantissa is near overflow, is replaced by the worst value when the mantissa overflows and the exponent is incremented. Floating-point notation is used within DSP chips as it eases the computational problems involved in handling long wordlengths. For example, when multiplying floating-point numbers, only the mantissae need to be multiplied. The exponents are simply added.

A floating-point system requires one exponent to be carried with each mantissa and this is wasteful because in real audio material the level does not change so rapidly and there is redundancy in the exponents. A better alternative is floating-point block coding, also known as near-instantaneous companding, where the magnitude of the largest sample in a block

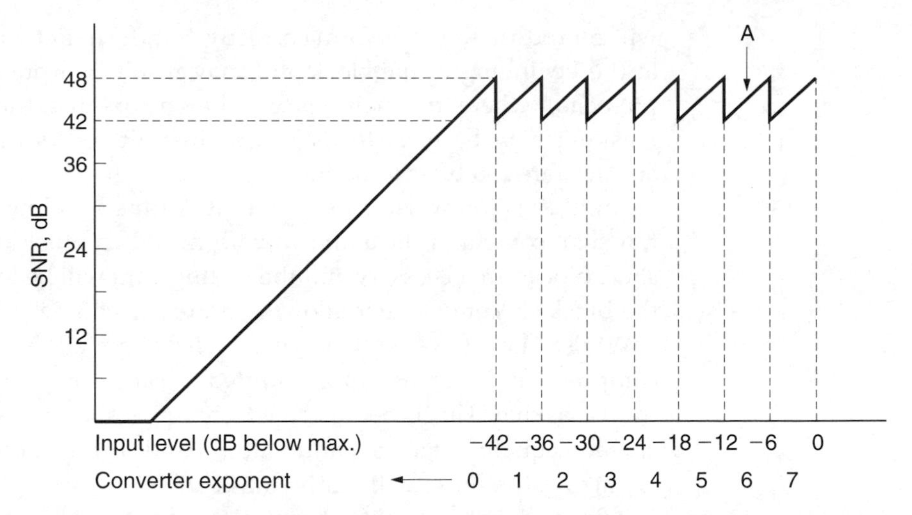

Figure 7.37 In this example of an eight-bit mantissa, three-bit exponent system, the maximum SNR is 6 dB × 8 = 48 dB with maximum input of 0 dB. As input level falls by 6 dB, the convertor noise remains the same, so SNR falls to 42 dB. Further reduction in signal level causes the convertor to shift range (point A in the diagram) by increasing the input analog gain by 6 dB. The SNR is restored, and the exponent changes from 7 to 6 in order to cause the same gain change at the receiver. The noise modulation would be audible in this simple system. A longer mantissa word is needed in practice.

is used to determine the value of an exponent which is valid for the whole block. Sending one exponent per block requires a lower data rate in than true floating point.[32]

In block coding the requantizing in the coder raises the quantizing noise, but it does so over the entire duration of the block. Figure 7.38 shows that if a transient occurs towards the end of a block, the decoder

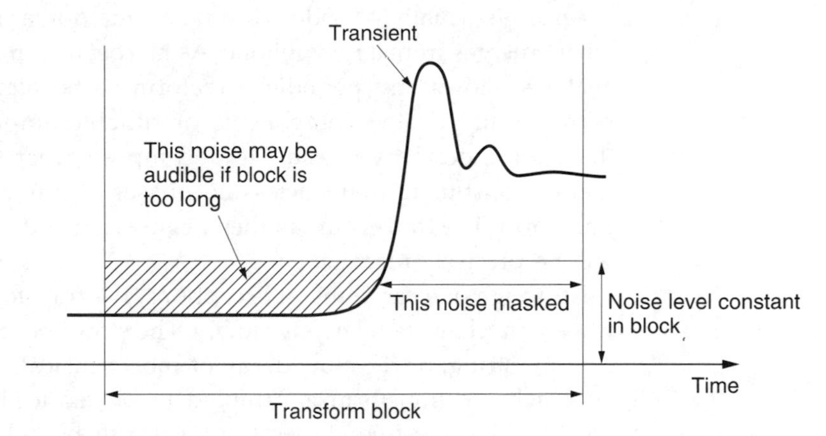

Figure 7.38 If a transient occurs towards the end of a transform block, the quantizing noise will still be present at the beginning of the block and may result in a pre-echo where the noise is audible before the transient.

will reproduce the waveform correctly, but the quantizing noise will start at the beginning of the block and may result in a pre-noise (also called pre-echo) where the noise is audible before the transient. Temporal masking may be used to make this inaudible. With a 1 ms block, the artifacts are too brief to be heard.

Another solution is to use a variable time window according to the transient content of the audio waveform. When musical transients occur, short blocks are necessary and the coding gain will be low. At other times the blocks become longer allowing a greater coding gain.

Whilst the above systems used alone do allow coding gain, the compression factor has to be limited because little benefit is obtained from masking. This is because the techniques above produce noise which spreads equally over the entire audio band. If the audio input spectrum is narrow, the noise will not be masked.

Sub-band coding splits the audio spectrum up into many different frequency bands. Once this has been done, each band can individually be processed. In real audio signals most bands will contain lower-level signals than the loudest one. Individual companding of each band will be more effective than broadband companding. Sub-band coding also allows the noise floor to be raised selectively so that noise is added only at frequencies where spectral masking will be effective.

There is little conceptual difference between a sub-band coder with a large number of bands and a transform coder. In transform coding, a Fourier or DCT transform of the waveform is computed periodically. Since the transform of an audio signal changes slowly, it need be sent much less often than audio samples. The receiver performs an inverse transform. Transform coding was considered in Chapter 3. Finally the data may be subject to a lossless binary compression using, for example, a Huffman code.

Audio is usually considered to be a time domain waveform as this is what emerges from a microphone. As has been seen in Chapter 3, spectral analysis allows any periodic waveform to be represented by a set of harmonically related components of suitable amplitude and phase. In theory it is perfectly possible to decompose a periodic input waveform into its constituent frequencies and phases, and to record or transmit the transform. The transform can then be inverted and the original waveform will be precisely recreated.

Although one can think of exceptions, the transform of a typical audio waveform changes relatively slowly. The slow speech of an organ pipe or a violin string, or the slow decay of most musical sounds allow the rate at which the transform is sampled to be reduced, and a coding gain results. At some frequencies the level will be below maximum and a shorter wordlength can be use to describe the coefficient. Further coding gain will be achieved if the coefficients describing frequencies which will experience masking are quantized more coarsely. The transform of an audio signal is computed in the main signal path in a transform coder, and has sufficient frequency resolution to drive the masking model directly.

In practice there are some difficulties. Real sounds are not periodic, but contain transients which transformation cannot accurately locate in time. The solution to this difficulty is to cut the waveform into short segments and then to transform each individually. The delay is reduced, as is the computational task, but there is a possibility of artifacts arising because of the truncation of the waveform into rectangular time windows. A solution is to use window functions (see Chapter 3) and to overlap the segments as shown in Figure 7.39. Thus every input sample appears in just two transforms, but with variable weighting depending upon its position along the time axis.

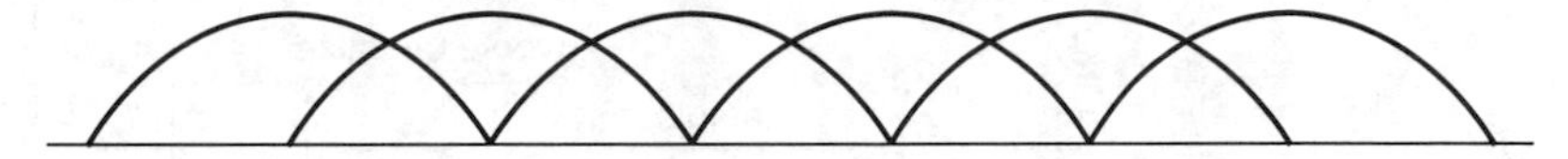

Figure 7.39 Transform coding can only be practically performed on short blocks. These are overlapped using window functions in order to handle continuous waveforms.

The DFT (discrete frequency transform) does not produce a continuous spectrum, but instead produces coefficients at discrete frequencies. The frequency resolution (i.e. the number of different frequency coefficients) is equal to the number of samples in the window. If overlapped windows are used, twice as many coefficients are produced as are theoretically necessary. In addition, the DFT requires intensive computation, due to the requirement to use complex arithmetic to render the phase of the components as well as the amplitude. An alternative is to use discrete cosine transforms (DCT).

Figure 7.40 shows a block diagram of a Layer I coder which is a simplified version of that used in the MUSICAM system. A polyphase quadrature mirror filter network divides the audio spectrum into 32 equal sub-bands. The output data rate of the filter bank is no higher than the input rate because each band has been heterodyned to a frequency range from DC upwards.

Sub-band compression takes advantage of the fact that real sounds do not have uniform spectral energy. The wordlength of PCM audio is based on the dynamic range required and this is generally constant with frequency although any pre-emphasis will affect the situation. When a signal with an uneven spectrum is conveyed by PCM, the whole dynamic range is occupied only by the loudest spectral component, and all the

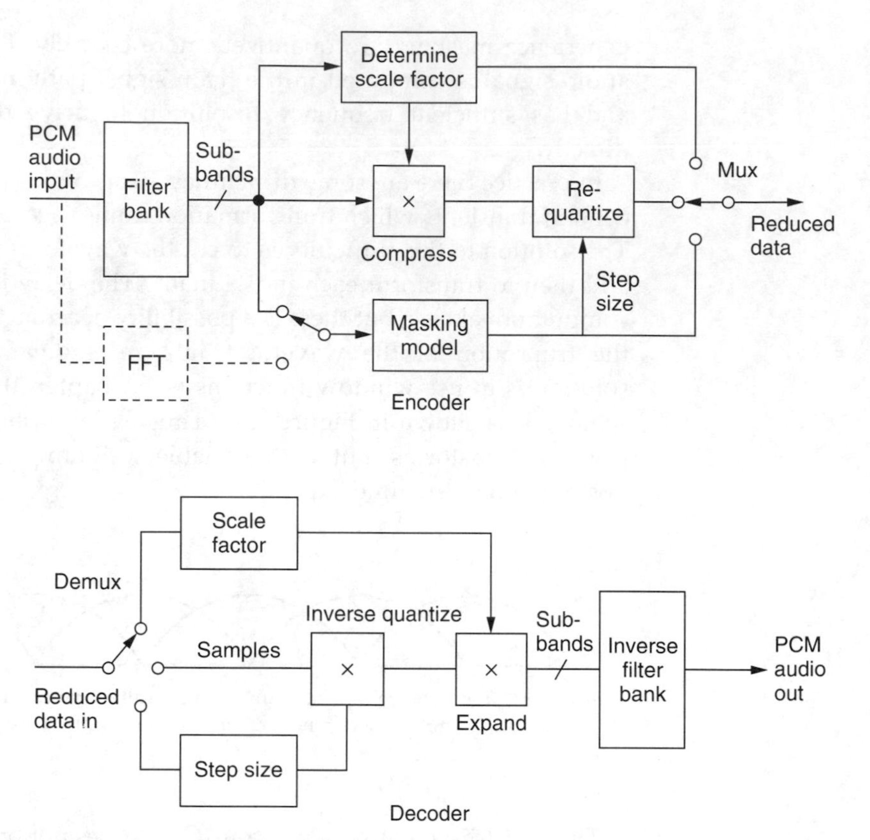

Figure 7.40 A simple sub-band coder. The bit allocation may come from analysis of the sub-band energy, or, for greater reduction, from a spectral analysis in a side chain.

other components are coded with excessive headroom. In its simplest form, sub-band coding[33] works by splitting the audio signal into a number of frequency bands and companding each band according to its own level. Bands in which there is little energy result in small amplitudes which can be transmitted with short wordlength. Thus each band results in variable-length samples, but the sum of all the sample wordlengths is less than that of PCM and so a coding gain can be obtained.

As MPEG audio coding relies on auditory masking, the sub-bands should preferably be narrower than the critical bands of the ear, hence the large number required. Figure 7.41 shows the critical condition where the masking tone is at the top edge of the sub-band. It will be seen that the narrower the sub-band, the higher the requantizing noise that can be masked. The use of an excessive number of sub-bands will, however, raise complexity and the coding delay.

Constant size input blocks are used, containing 384 samples. At 48 kHz, 384 samples corresponds to a period of 8 ms. After the sub-band filter each

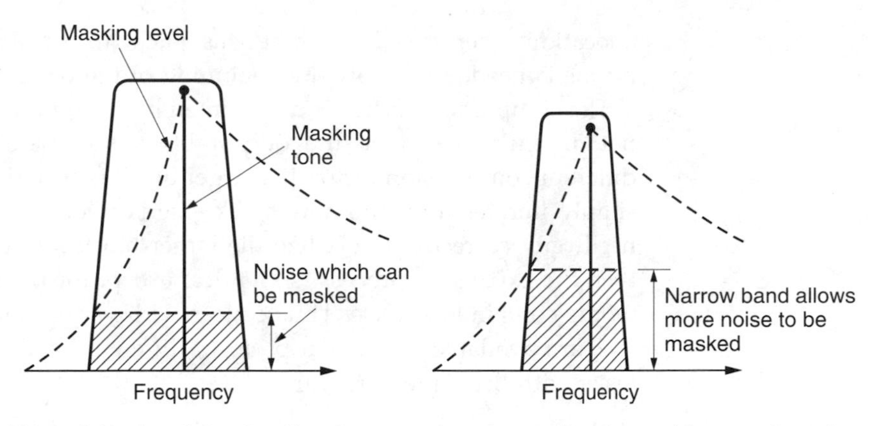

Figure 7.41 In sub-band coding the worst case occurs when the masking tone is at the top edge of the sub-band. The narrower the band, the higher the noise level which can be masked.

band contains 12 samples per block. The block size was based on the pre-masking phenomenon of Figure 7.38. The samples in each sub-band block or *bin* are companded according to the peak value in the bin. A six-bit scale factor is used for each sub-band which applies to all 12 samples.

If a fixed compression factor is employed, the size of the coded output block will be fixed. The wordlengths in each bin will have to be such that the sum of the bits from all of the sub-band equals the size of the coded block. Thus some sub-bands can have long wordlength coding if others have short wordlength coding. The process of determining the requantization step size, and hence the wordlength in each sub-band is known as bit allocation.

For simplicity, in Layer I the levels in the 32 sub-bands themselves are used as a crude spectral analysis of the input in order to drive the masking model. The masking model uses the input spectrum to determine a new threshold of hearing which in turn determines how much the noise floor can be raised in each sub-band. Where masking takes place, the signal is quantized more coarsely until the quantizing noise is raised to just below the masking level. The coarse quantization requires shorter wordlengths and allows a coding gain. The bit allocation may be iterative as adjustments are made to obtain the best NMR within the allowable data rate.

The samples of differing wordlength in each bin are then assembled into the output coded block. Unlike a PCM block, which contains samples of fixed wordlength, a coded block contains many different wordlengths and these can vary from one block to the next. In order to deserialize the block into samples of various wordlengths and demultiplex the samples into the appropriate frequency bins, the decoder has to be told what bit allocations were used when it was packed, and some synchronizing means is needed to allow the beginning of the block to be identified.

The compression factor is determined by the bit-allocation system. It is not difficult to change the output block size parameter to obtain a different compression factor. If a larger block is specified, the bit allocator simply iterates until the new block size is filled. Similarly the decoder need only correctly deserialize the larger block into coded samples and then the expansion process is identical except for the fact that expanded words contain less noise. Thus codecs with varying degrees of compression are available which can perform different bandwidth/performance tasks with the same hardware.

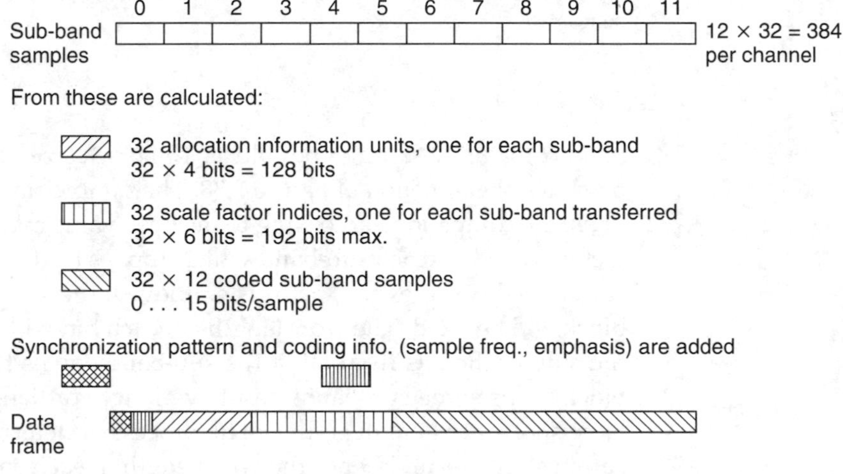

Figure 7.42 The Layer I data frame showing the allocation codes, the scale factors and the sub-band samples.

Figure 7.42 shows the format of the Layer I data stream. The frame begins with a sync pattern to reset the phase of deserialization, and a header which describes the sampling rate and any use of pre-emphasis. Following this is a block of 32 four-bit allocation codes. These specify the wordlength used in each sub-band and allow the decoder to deserialize the sub-band sample block. This is followed by a block of 32 six-bit scale factor indices, which specify the gain given to each band during companding. The last block contains 32 sets of 12 samples. These samples vary in wordlength from one block to the next, and can be from zero to fifteen bits long. The deserializer has to use the 32 allocation information codes to work out how to deserialize the sample block into individual samples of variable length.

The Layer I MPEG decoder is shown in Figure 7.43. The Elementary Stream is deserialized using the sync pattern and the variable-length

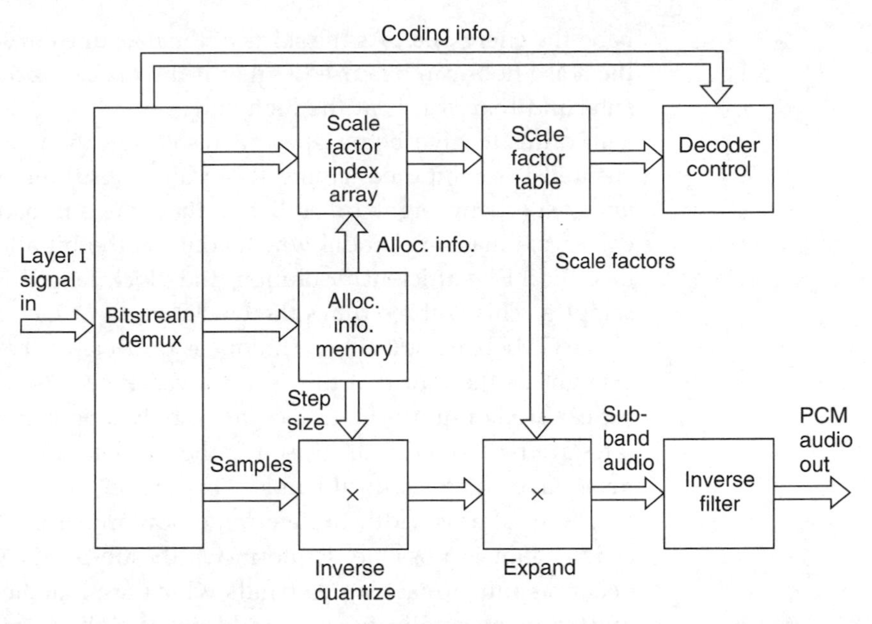

Figure 7.43 The Layer I decoder. See text for details.

samples are asembled using the allocation codes. The variable-length samples are returned to fifteen-bit wordlength by adding zeros. The scale factor indices are then used to determine multiplication factors used to return the waveform in each sub-band to its original level. The 32 sub-band signals are then merged into one spectrum by the synthesis filter. This is a set of bandpass filters which returns every sub-band to the correct place in the audio spectrum and then adds them to produce the audio output.

MPEG Layer II audio coding is identical to MUSICAM. The same 32-band filterbank and the same block companding scheme as Layer I is used. Figure 7.44 shows that using the level in a sub-band to drive the masking model is sub-optimal because it is not known where in the sub-

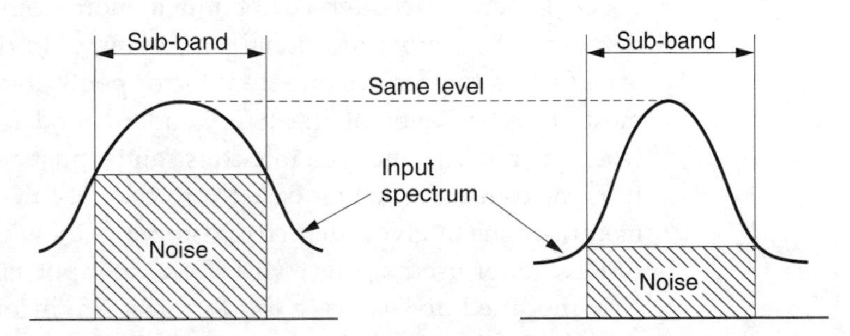

Figure 7.44 Accurate knowledge of the spectrum allows the noise floor to be raised higher while remaining masked.

band the energy lies. As the skirts of the masking curve are asymmetrical, the noise floor can be raised higher if the masker is at the low end of the sub-band than if it is at the high end.

In order to give better spectral resolution than the filterbank, a side chain FFT is computed having 1024 points, resulting in an analysis of the audio spectrum eight times better than the sub-band width. The FFT drives the masking model which controls the bit allocation. In order to give the FFT sufficient resolution, the block length is increased to 1152 samples. This is three times the block length of Layer I.

The QMF band splitting technique is restricted to bands of equal width. It might be thought that this is a drawback because the critical bands of the ear are non-uniform. In fact this is only a problem when very low bit rates are required. In all cases it is the masking model of hearing which must have correct critical bands. This model can then be superimposed on bands of any width to determine how much masking and therefore coding gain is possible. Uniform-width sub-bands will not be able to obtain as much masking as bands which are matched to critical bands, but for many applications the additional coding gain is not worth the added filter complexity.

The block-companding scheme of Layer II is the same as in Layer I because the 1152-sample block is divided into three 384-sample blocks. However, not all the scale factors are transmitted, because they contain a degree of redundancy on real program material. The difference between scale factors in successive blocks in the same band exceeds 2 dB less than 10 per cent of the time. Layer II analyses the set of three successive scale factors in each sub-band. On a stationary program, these will be the same and only one scale factor out of three is sent. As the transient content increases in a given sub-band, two or three scale factors will be sent. A scale factor select code must be sent to allow the decoder to determine what has been sent in each sub-band. This technique effectively halves the scale factor bit rate. The requantized samples in each sub-band, bit-allocation data, scale factors and scale factor select codes are multiplexed into the output bitstream.

The Layer II decoder is not much more complex than the Layer I decoder as the only additional processing is to decode the compressed scale factors to produce one scale factor per 384-sample block. This is the most complex layer of the ISO standard, and is only really necessary when the most severe data rate constraints must be met with high quality. It is a transform code based on the ASPEC system with certain modifications to give a degree of commonality with Layer II. The original ASPEC coder used a direct MDCT on the input samples. In Layer III this was modified to use a hybrid transform incorporating the existing polyphase 32-band QMF of Layers I and II. In Layer 3, the 32 sub-bands from the QMF are each processed by a 12 band MDCT to obtain 384 output coefficients. Two window sizes are used to avoid pre-echo on transients. The window switching is performed by the psychoacoustic model. It has been found that pre-echo is associated with the entropy in the audio rising above the average value.

A highly accurate perceptive model is used to take advantage of the high-frequency resolution available. Non-uniform quantizing is used, along with Huffman coding. This is a technique where the most common code values are allocated the shortest wordlength.

7.14 Dolby AC-3

Dolby AC-3 is in fact a family of transform coders based on time domain aliasing cancellation (TDAC) which allow various compromises between coding delay and bit rate to be used. In the modified discrete cosine transform (MDCT),[34] windows with 50 per cent overlap are used. Thus twice as many coefficients as necessary are produced. These are sub-sampled by a factor of two to give a critically sampled transform, which results in potential aliasing in the frequency domain. However, by making a slight change to the transform, the alias products in the second half of a given window are equal in size but of opposite polarity to the alias products in the first half of the next window, and so will be cancelled on reconstruction. This is the principle of TDAC.

Figure 7.45 shows the generic block diagram of the AC-3 coder. Input audio is divided into 50 per cent overlapped blocks of 512 samples. These are subject to a TDAC transform which uses alternate modified sine and cosine transforms. The transforms produce 512 coefficients per block, but these are redundant and after the redundancy has been removed there are 256 coefficients per block. The input

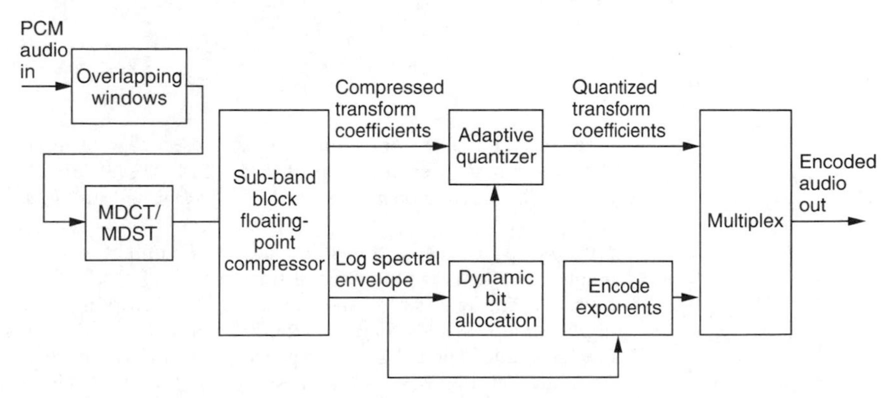

Figure 7.45 Block diagram of the Dolby AC-3 coder. See text for details.

waveform is constantly analysed for the presence of transients and if these are present the block length will be halved to prevent pre-noise. This halves the frequency resolution but doubles the temporal resolution.

The coefficients have high frequency resolution and are selectively combined in sub-bands which approximate the critical bands. Coefficients in each sub-band are normalized and expressed in floating-point block notation with common exponents. The exponents in fact represent the logarithmic spectral envelope of the signal and can be used to drive the perceptive model which operates the bit allocation. The mantissae of the transform coefficients are then requantized according to the bit allocation.

The output bitstream consists of the requantized coefficients and the log spectral envelope in the shape of the exponents. There is a great deal of redundancy in the exponents. In any block, only the first exponent, corresponding to the lowest frequency, is transmitted absolutely. Remaining coefficients are transmitted differentially. Where the input has a smooth spectrum the exponents in several bands will be the same and the differences will then be zero. In this case exponents can be grouped using flags.

Further use is made of temporal redundancy. An AC-3 sync frame contains six blocks. The first block of the frame contains absolute exponent data, but where stationary audio is encountered, successive blocks in the frame can use the same exponents.

The receiver uses the log spectral envelope to deserialize the mantissae of the coefficients into the correct wordlengths. The highly redundant exponents are decoded starting with the lowest-frequency coefficient in the first block of the frame and adding differences to create the remainder. The exponents are then used to return the coefficients to fixed-point notation. Inverse transforms are then computed, followed by a weighted overlapping of the windows to obtain PCM data.

References

1. Johnston, J.D., Transform coding of audio signals using perceptual noise criteria. *IEEE J. Selected Areas in Comms.*, **JSAC-6**, 314–323 (1988)
2. Moore, B.C.J., *An introduction to the psychology of hearing*, London: Academic Press (1989)
3. Muraoka, T., Iwahara, M. and Yamada, Y., Examination of audio bandwidth requirements for optimum sound signal transmission. *J. Audio Eng. Soc.*, **29**, 2–9 (1982)
4. Muraoka, T., Yamada, Y. and Yamazaki, M., Sampling frequency considerations in digital audio. *J. Audio Eng. Soc.*, **26**, 252–256 (1978)
5. Fincham, L.R., The subjective importance of uniform group delay at low frequencies. Presented at the 74th Audio Engineering Society Convention (New York, 1983), Preprint 2056(H-1)

6. Fletcher, H., Auditory patterns. *Rev. Modern Physics*, **12**, 47–65 (1940)
7. Zwicker, E., Subdivision of the audible frequency range into critical bands. *J. Acoust. Soc. Amer.*, **33**, 248 (1961)
8. Moore, B. and Glasberg, B., Formulae describing frequency selectivity as a function of frequency and level, and their use in calculating excitation patterns. *Hearing Research*, **28**, 209–225 (1987)
9. Anon., AES recommended practice for professional digital audio applications employing pulse code modulation: preferred sampling frequencies. AES5–1984 (ANSI S4.28–1984), *J. Audio Eng. Soc.*, **32**, 781–785 (1984)
10. v.d. Plassche, R.J., Dynamic element matching puts trimless convertors on chip.*Electronics*, 16 June 1983
11. v.d. Plassche, R.J. and Goedhart, D., A monolithic 14 bit D/A convertor. *IEEE J. Solid-State Circuits*, **SC-14**, 552–556 (1979)
12. Adams, R.W., Companded predictive delta modulation: a low-cost technique for digital recording. *J. Audio Eng. Soc.*, **32**, 659–672 (1984)
13. Hauser, M.W., Principles of oversampling A/D conversion. *J. Audio Eng. Soc.*, **39**, 3–26 (1991)
14. Cutler, C.C., Transmission systems employing quantization. US Patent No. 2.927,962 (1960)
15. Fielder, L.D., Human auditory capabilities and their consequences in digital audio convertor design. In *Audio in Digital Times*, New York: Audio Engineering Society (1989)
16. Gerzon, M. and Craven, P.G., Optimal noise shaping and dither of digital signals. Presented at the 87th Audio Engineering Society Convention (New York, 1989), Preprint No. 2822 (J-1)
17. Wannamaker, R.A., Psychoacoustically optimal noise shaping. *J. Audio Eng. Soc.*, **40**, 611–620 (1992)
18. Lipshitz, S.P., Wannamaker, R.A. and Vanderkooy, J., Minimally audible noise shaping. *J. Audio Eng. Soc.*, **39**, 836–852 (1991)
19. Adams, R.W., Design and implementation of an audio 18-bit A/D convertor using oversampling techniques. Presented at the 77th Audio Engineering Society Convention (Hamburg, 1985), Preprint 2182
20. Adams, R.W., An IC chip set for 20 bit A/D conversion. In *Audio in Digital Times*, New York: Audio Engineering Society (1989)
21. Richards, M., Improvements in oversampling analogue to digital convertors. Presented at the 84th Audio Engineering Society Convention (Paris, 1988), Preprint 2588 (D-8)
22. Inose, H. and Yasuda, Y., A unity bit coding method by negative feedback. *Proc. IEEE*, **51**, 1524–1535 (1963)
23. Naus, P.J. *et al.*, Low signal level distortion in sigma-delta modulators. Presented at the 84th Audio Engineering Society Convention (Paris 1988), Preprint 2584
24. Stikvoort, E., High order one bit coder for audio applications. Presented at the 84th Audio Engineering Society Convention (Paris, 1988), Preprint 2583(D-3)
25. Moller, L., Signal levels across the EBU/AES digital audio interface. In *Proc. 1st NAB Radio Montreux Symp.*, Montreux (1992) 16–28
26. Wiese, D., MUSICAM: flexible bitrate reduction standard for high quality audio. Presented at the Digital Audio Broadcasting Conference (London, March 1992)
27. Brandenburg, K., ASPEC coding. *Proc. 10th. Audio Eng. Soc. Int. Conf.*, 81–90, New York: Audio Engineering Society (1991)
28. ISO/IEC JTC1/SC2/WG11 N0030: MPEG/AUDIO test report, Stockholm (1990)
29. ISO/IEC JTC1/SC2/WG11 MPEG 91/010 The SR report on: The MPEG/AUDIO subjective listening test. Stockholm (1991)
30. ISO/IEC JTC1/SC2/WG11 Committee draft 11172
31. Bonicel, P. *et al.*, A real time ISO/MPEG2 Multichannel decoder. Presented at the 96th Audio Engineering Society Convention (1994) Preprint No. 3798 (P3.7)4.30; Wiese, D., MUSICAM: flexible bitrate reduction standard for high quality audio. Presented at the Digital Audio Broadcasting Conference, London (March 1992)

32. Caine, C.R., English, A.R. and O'Clarey, J.W.H. NICAM-3: near-instantaneous companded digital transmission for high-quality sound programmes. *J. IERE*, **50**, 519–530 (1980)
33. Crochiere, R.E., Sub-band coding. *Bell System Tech. J.*, **60**, 1633–1653 (1981)
34. Princen, J.P., Johnson, A. and Bradley, A.B., Sub-band/transform coding using filter bank designs based on time domain aliasing cancellation. *Proc. ICASSP*, 2161–2164 (1987)

8

Digital recording principles

8.1 Introduction to the channel

Data can be recorded on many different media and conveyed using many forms of transmission. The generic term for the path down which the information is sent is the *channel*. In analog systems, the characteristics of the channel affect the signal directly. It is a fundamental strength of digital video that by using pulse-code modulation the quality can be made independent of the channel.

In a transmission application, the channel may be a length of cable or optical fibre, or a radio link. In a recording application the channel will include the record head, the medium and the replay head. This chapter concentrates on recording channels, and transmission channels are considered in Chapters 10 and 14.

In digital circuitry there is a great deal of noise immunity because the signal has only two states, which are widely separated compared with the amplitude of noise. In both digital recording and transmission this is not always the case. In magnetic recording, noise immunity is a function of track width and reduction of the working SNR of a digital track allows the same information to be carried in a smaller area of the medium, improving economy of operation. In broadcasting, the noise immunity is a function of the transmitter power and reduction of working SNR allows lower power to be used with consequent economy. These reductions also increase the random error rate, but, as was seen in Chapter 1, an error-correction system may already be necessary in a practical system and it is simply made to work harder.

In real channels, the signal may *originate* with discrete states which change at discrete times, but the channel will treat it as an analog waveform and so it will not be *received* in the same form. Various loss mechanisms will reduce the amplitude of the signal. These attenuations will not be the same at all frequencies. Noise will be picked up in the channel as a result of stray electric fields or magnetic induction. As a result the voltage received at the end of the channel will have an infinitely varying state along with a degree of uncertainty due to the noise. Different frequencies can propagate at different speeds in the channel; this is the phenomenon of group delay. An alternative way of considering group delay is that there will be frequency-dependent phase shifts in the signal and these will result in uncertainty in the timing of pulses.

Thus it is not the channel which is digital; instead the term describes the way in which the received signals are *interpreted*. When the receiver makes discrete decisions from the input waveform it attempts to reject the uncertainties in voltage and time. The technique of channel coding is one where transmitted waveforms are restricted to those which still allow the receiver to make discrete decisions despite the degradations caused by the analog nature of the channel.

In digital circuitry, the signals are generally accompanied by a separate clock signal which reclocks the data to remove jitter as was shown in Chapter 1. In contrast, it is generally not feasible to provide a separate clock in recording and transmission applications.

The solution is to use a self-clocking waveform and the generation of this is a further essential function of the coding process. Clearly if data bits are simply clocked serially from a shift register in so-called direct recording or transmission this characteristic will not be obtained. If all the data bits are the same, for example all zeros, there is no clock when they are serialized.

8.2 Types of recording medium

There is considerably more freedom of choice for digital media than was the case for analog signals. Once converted to the digital domain, video and audio are no more than data and can take advantage of the research expended in computer data recording.

Digital media do not need to have linear transfer functions, nor do they need to be noise-free, error-free or continuous. All they need to do is to allow the player to be able to distinguish the presence or absence of replay events, such as the generation of pulses, with reasonable (rather than perfect) reliability. In a magnetic medium, the event will be a flux change from one direction of magnetization to another. In an optical medium, the event may cause the pickup to perceive a change in the intensity or polarization of the light falling on the sensor. In optical disks, the apparent contrast may be obtained by interference or through selective absorption of light by dyes. In magneto-optical disks the recording itself is magnetic, but it is made and read using light.

8.3 Magnetic recording

Magnetic recording relies on the hysteresis of certain magnetic materials. After an applied magnetic field is removed, the material remains magnetized in the same direction. By definition the process is non-linear, and analog magnetic recorders have to use bias to linearize it. Digital recorders are not concerned with the non-linearity, and HF bias is unnecessary.

Figure 8.1 shows the construction of a typical digital record head, which is not dissimilar to an analog record head. A magnetic circuit carries a coil through which the record current passes and generates flux. A non-magnetic gap forces the flux to leave the magnetic circuit of the

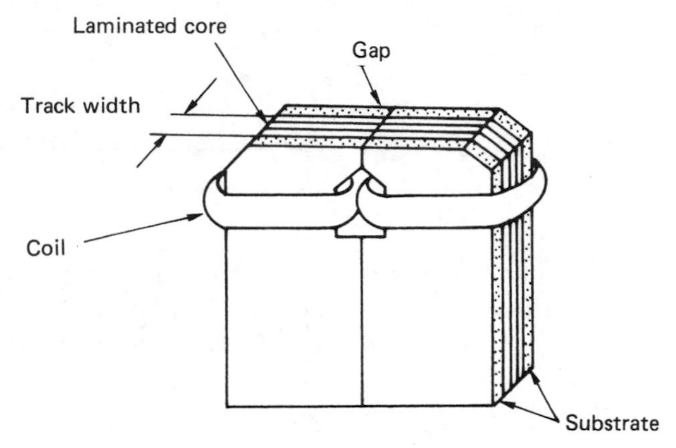

Figure 8.1 A digital record head is similar in principle to an analog head but uses much narrower tracks.

head and penetrate the medium. The current through the head must be set to suit the coercivity of the tape, and is arranged almost to saturate the track. The amplitude of the current is constant, and recording is performed by reversing the direction of the current with respect to time. As the track passes the head, this is converted to the reversal of the magnetic field left on the tape with respect to distance. The magnetic recording is therefore bipolar. Figure 8.2 shows that the recording is actually made just after the trailing pole of the record head where the flux strength from the gap is falling. As in analog recorders, the width of the gap is generally made quite large to ensure that the full thickness of the magnetic coating is recorded, although this cannot be done if the same head is intended to replay.

Figure 8.3 shows what happens when a conventional inductive head, i.e. one having a normal winding, is used to replay the bipolar track made by reversing the record current. The head output is proportional to the

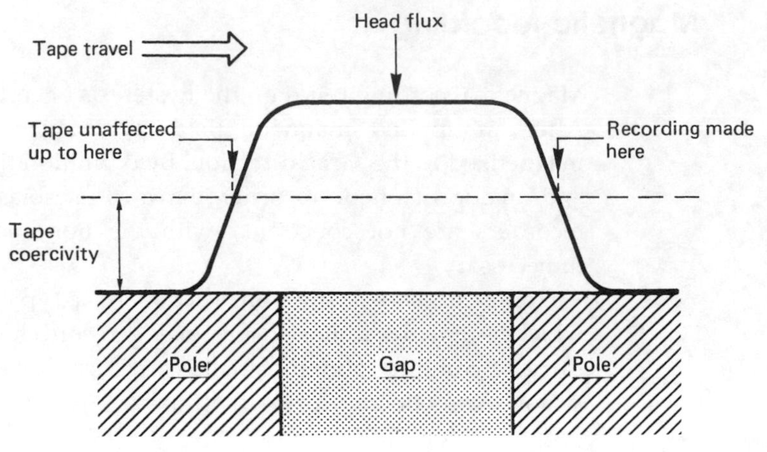

Figure 8.2 The recording is actually made near the trailing pole of the head where the head flux falls below the coercivity of the tape.

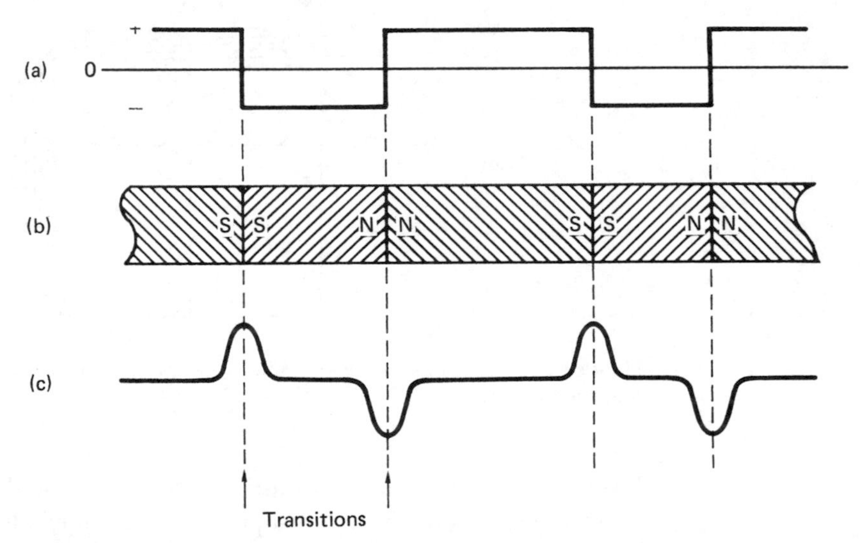

Figure 8.3 Basic digital recording. At (a) the write current in the head is reversed from time to time, leaving a binary magnetization pattern shown at (b). When replayed, the waveform at (c) results because an output is only produced when flux in the head changes. Changes are referred to as transitions.

rate of change of flux and so only occurs at flux reversals. In other words, the replay head differentiates the flux on the track. The polarity of the resultant pulses alternates as the flux changes and changes back. A circuit is necessary which locates the peaks of the pulses and outputs a signal corresponding to the original record current waveform. There are two ways in which this can be done.

The amplitude of the replay signal is of no consequence and often an AGC system is used to keep the replay signal constant in amplitude.

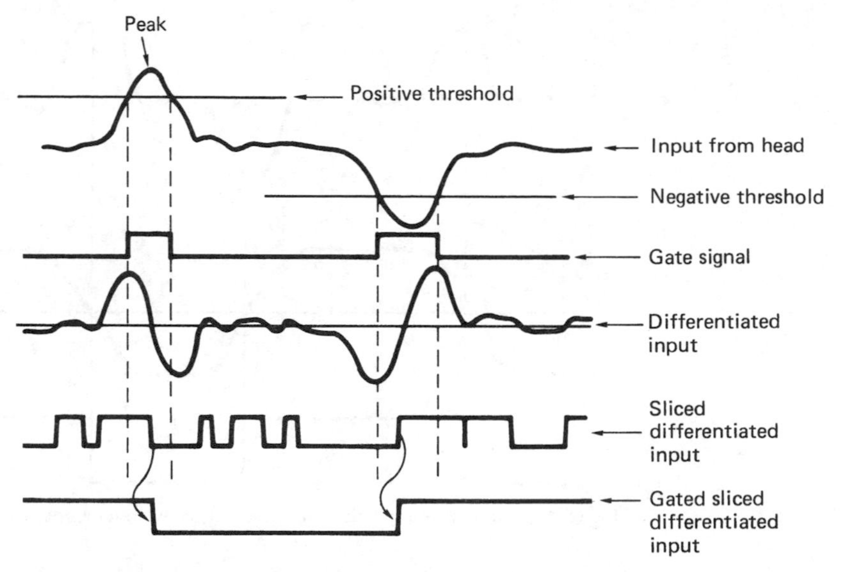

Figure 8.4 Gated peak detection rejects noise by disabling the differentiated output between transitions.

What matters is the time at which the write current, and hence the flux stored on the medium, reverses. This can be determined by locating the peaks of the replay impulses, which can conveniently be done by differentiating the signal and looking for zero crossings. Figure 8.4 shows that this results in noise between the peaks. This problem is overcome by the gated peak detector, where only zero crossings from a pulse which exceeds the threshold will be counted. The AGC system allows the thresholds to be fixed. As an alternative, the record waveform can also be restored by integration, which opposes the differentiation of the head as in Figure 8.5.[1]

The head shown in Figure 8.1 has a frequency response shown in Figure 8.6. At DC there is no change of flux and no output. As a result inductive heads are at a disadvantage at very low speeds. The output rises with frequency until the rise is halted by the onset of thickness loss. As the frequency rises, the recorded wavelength falls and flux from the shorter magnetic patterns cannot be picked up so far away. At some point, the wavelength becomes so short that flux from the back of the tape coating cannot reach the head and a decreasing thickness of tape contributes to the replay signal.[2]

In digital recorders using short wavelengths to obtain high density, there is no point in using thick coatings. As wavelength further reduces, the familiar gap loss occurs, where the head gap is too big to resolve detail on the track. The construction of the head results in the same action as that of a two-point transversal filter, as the two poles of the head see

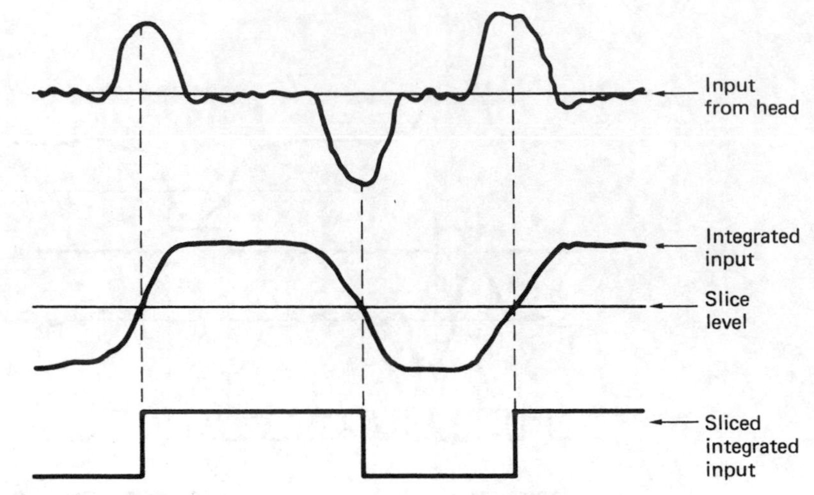

Figure 8.5 Integration method for re-creating write-current waveform.

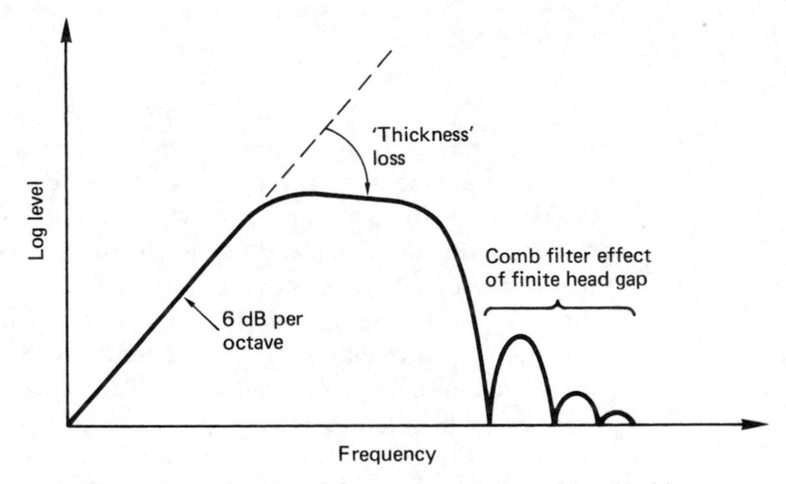

Figure 8.6 The major mechanisms defining magnetic channel bandwidth.

the tape with a small delay interposed due to the finite gap. As expected, the head response is like a comb filter with the well-known nulls where flux cancellation takes place across the gap.

Clearly the smaller the gap, the shorter the wavelength of the first null. This contradicts the requirement of the record head to have a large gap. In quality analog audio recorders, it is the norm to have different record and replay heads for this reason, and the same will be true in digital machines which have separate record and playback heads. Clearly where the same pair of heads are used for record and play, the head gap size will be determined by the playback requirement. As can be seen, the frequency response is far from ideal, and steps must be taken to ensure that recorded data waveforms do not contain frequencies which suffer excessive losses.

A more recent development is the magneto-resistive (M-R) head. This is a head which measures the flux on the tape rather than using it to generate a signal directly. Flux measurement works down to DC and so offers advantages at low tape speeds. Unfortunately flux-measuring heads are not polarity conscious but sense the modulus of the flux and if used directly they respond to positive and negative flux equally, as shown in Figure 8.7. This is overcome by using a small extra winding in

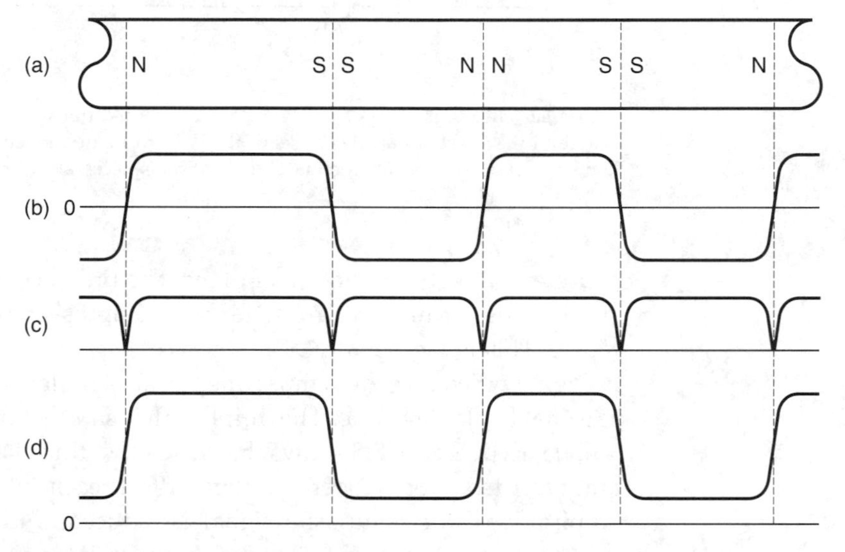

Figure 8.7 The sensing element in a magneto-resistive head is not sensitive to the polarity of the flux, only the magnitude. At (a) the track magnetization is shown and this causes a bidirectional flux variation in the head as at (b), resulting in the magnitude output at (c). However, if the flux in the head due to the track is biased by an additional field, it can be made unipolar as at (d) and the correct waveform is obtained.

the head carrying a constant current. This creates a steady bias field which adds to the flux from the tape. The flux seen by the head is now unipolar and changes between two levels and a more useful output waveform results. M-R heads are advantageous for recorders which have low head-to-medium speed, and so they are not generally found in digital VTRs or magnetic disk drives which use inductive heads.

Heads designed for use with tape work in actual contact with the magnetic coating. The tape is tensioned to pull it against the head. There will be a wear mechanism and need for periodic cleaning.

In the hard disk, the rotational speed is high in order to reduce access time, and the drive must be capable of staying on line for extended periods. In this case the heads do not contact the disk surface, but are

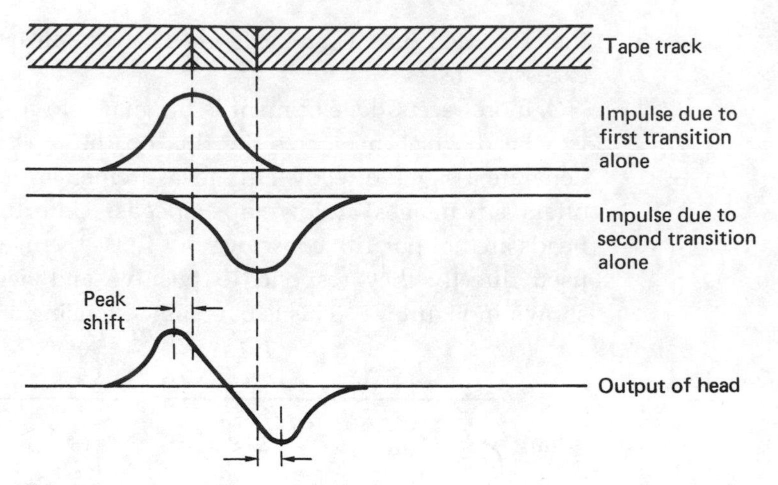

Figure 8.8 Readout pulses from two closely recorded transitions are summed in the head and the effect is that the peaks of the waveform are moved outwards. This is known as peak-shift distortion and equalization is necessary to reduce the effect.

supported on a boundary layer of air. The presence of the air film causes spacing loss, which restricts the wavelengths at which the head can replay. This is the penalty of rapid access.

Digital video recorders must operate at high density in order to offer a reasonable playing time. This implies that shortest possible wavelengths will be used. Figure 8.8 shows that when two flux changes, or transitions, are recorded close together, they affect each other on replay. The amplitude of the composite signal is reduced, and the position of the peaks is pushed outwards. This is known as inter-symbol interference, or peak-shift distortion and it occurs in all magnetic media. The effect is primarily due to high-frequency loss and it can be reduced by equalization on replay, as is done in most tapes, or by pre-compensation on record as is done in hard disks.

8.4 Azimuth recording and rotary heads

Figure 8.9(a) shows that in azimuth recording, the transitions are laid down at an angle to the track by using a head which is tilted. Machines using azimuth recording must always have an even number of heads, so that adjacent tracks can be recorded with opposite azimuth angle. The two track types are usually referred to as A and B. Figure 8.9(b) shows the effect of playing a track with the wrong type of head. The playback process suffers from an enormous azimuth error. The effect of azimuth error can be understood by imagining the tape track to be made from many identical parallel strips. In the presence of azimuth error, the strips

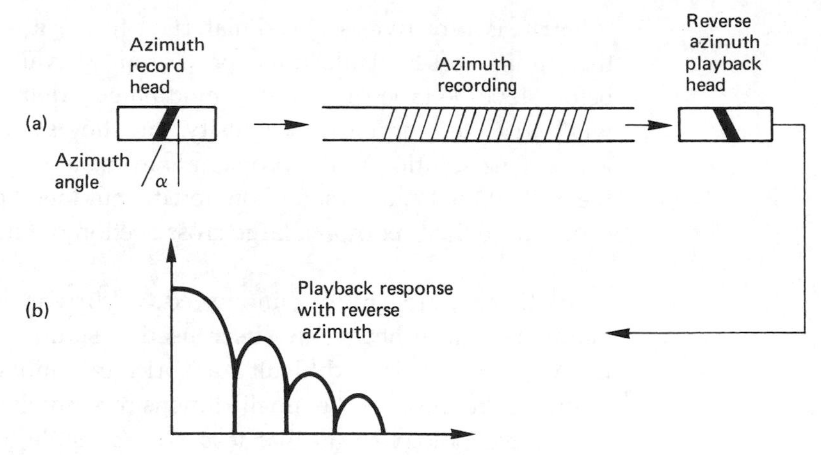

Figure 8.9 In azimuth recording (a), the head gap is tilted. If the track is played with the same head, playback is normal, but the response of the reverse azimuth head is attenuated (b).

at one edge of the track are played back with a phase shift relative to strips at the other side. At some wavelengths, the phase shift will be 180°, and there will be no output; at other wavelengths, especially long wavelengths, some output will reappear. The effect is rather like that of a comb filter, and serves to attenuate crosstalk due to adjacent tracks so that no guard bands are required. Since no tape is wasted between the tracks, more efficient use is made of the tape. The term guard-bandless recording is often used instead of, or in addition to, the term azimuth recording. The failure of the azimuth effect at long wavelengths is a characteristic of azimuth recording, and it is necessary to ensure that the spectrum of the signal to be recorded has a small low-frequency content. The signal will need to pass through a rotary transformer to reach the heads, and cannot therefore contain a DC component.

In machines such as early D-2 recorders and some digital camcorders there is no separate erase process, and erasure is achieved by overwriting with a new waveform. Overwriting is only successful when there are no long wavelengths in the earlier recording, since these penetrate deeper into the tape, and the short wavelengths in a new recording will not be able to erase them. In this case the ratio between the shortest and longest wavelengths recorded on tape should be limited. Restricting the spectrum of the code to allow erasure by overwrite also eases the design of the rotary transformer.

Design of a head for DVTRs is not easy, since there are a number of conflicting requirements. The high coercivity of the tape used requires that the head should be able to pass high flux densities into the tape without saturating. The bit rate of digital video is such that eddy current losses in the head are significant.

Ferrite is attractive as a head material because it is a non-conductor and this limits losses. Unfortunately currently available ferrites saturate before 1500 Oe tape can be fully modulated. Metals are available which will carry the required flux density, but they suffer from eddy current losses. One solution to the problem is to make a composite head, where the bulk of the head is made from ferrite, but metal poles are fitted which concentrate the flux from a large cross-section of ferrite into a small track width.

Another approach to limiting eddy current losses is lamination. Laminated metal heads have been used in stationary head recorders for many years, but it is difficult to fabricate laminated heads for digital recorders because of the small dimensions involved. Developments in plating technology mean that it is now possible to fabricate laminated heads on an extremely small scale by plating layers of magnetic material alternately with thinly deposited insulating layers. The construction of such a head is shown in Figure 8.1.

In the amorphous head, particles of metal are sintered under pressure to form the magnetic circuit. The metal allows high flux density, but the magnetic circuit has much higher electrical resistance than solid metal because the metal particles are only joined at their corners. In this way eddy current losses are reduced.

8.5 Optical disks

Optical recorders have the advantage that light can be focused at a distance whereas magnetism cannot. This means that there need be no physical contact between the pickup and the medium and no wear mechanism.

In the same way that the recorded wavelength of a magnetic recording is limited by the gap in the replay head, the density of optical recording is limited by the size of light spot which can be focused on the medium. This is controlled by the wavelength of the light used and by the aperture of the lens. When the light spot is as small as these limits allow, it is said to be diffraction limited. The recorded details on the disk are minute, and could easily be obscured by dust particles. In practice the information layer needs to be protected by a thick transparent coating. Light enters the coating well out of focus over a large area so that it can pass around dust particles, and comes to a focus within the thickness of the coating. Although the number of bits per unit area is high in optical recorders the number of bits per unit volume is not as high as that of tape because of the thickness of the coating.

Figure 8.10 shows the principle of readout of the CD and DVD which are read-only disks manufactured by pressing. The track consists of

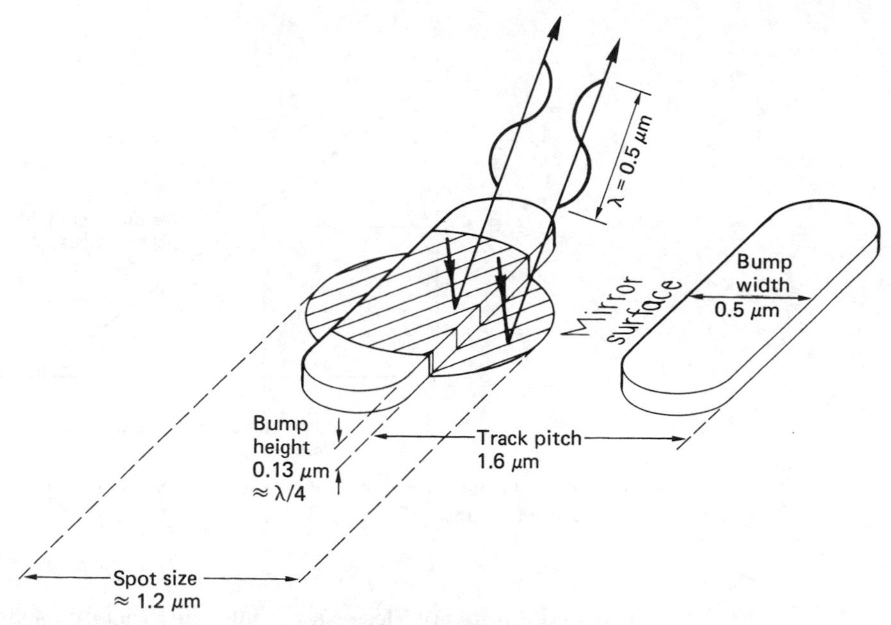

Figure 8.10 CD readout principle and dimensions. The presence of a bump causes destructive interference in the reflected light.

raised bumps separated by flat areas. The entire surface of the disk is metallized, and the bumps are one quarter of a wavelength in height. The player spot is arranged so that half of its light falls on top of a bump, and half on the surrounding surface. Light returning from the flat surface has travelled half a wavelength further than light returning from the top of the bump, and so there is a phase reversal between the two components of the reflection. This causes destructive interference, and light cannot return to the pickup. It must reflect at angles which are outside the aperture of the lens and be lost. Conversely, when light falls on the flat surface between bumps, the majority of it is reflected back to the pickup. The pickup thus sees a disk *apparently* having alternately good or poor reflectivity. The sensor in the pickup responds to the incident intensity and so the replay signal is unipolar and varies between two levels in a manner similar to the output of a M-R head.

Some disks can be recorded once, but not subsequently erased or re-recorded. These are known as WORM (Write Once Read Mostly) disks. One type of WORM disk uses a thin metal layer which has holes punched in it on recording by heat from a laser. Others rely on the heat raising blisters in a thin metallic layer by decomposing the plastic material beneath. Yet another alternative is a layer of photochemical dye which darkens when struck by the high-powered recording beam. Whatever the recording principle, light from the pickup is reflected more or less, or

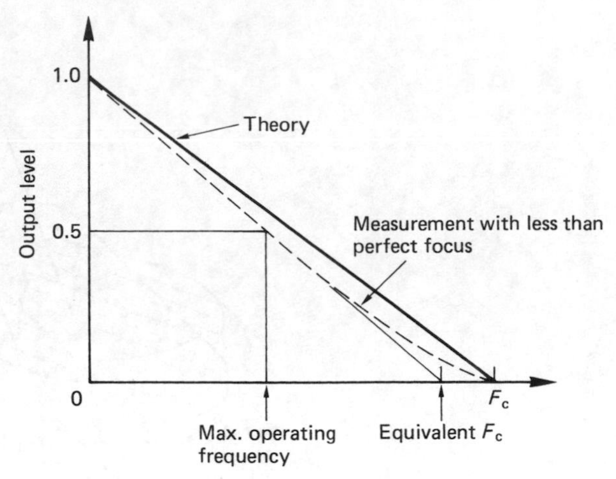

Figure 8.11 Frequency response of laser pickup. Maximum operating frequency is about half of cut-off frequency F_c.

absorbed more or less, so that the pickup senses a change in reflectivity.

All optical disks need mechanisms to keep the pickup following the track and sharply focused on it, and these will be discussed in Chapter 12 and need not be treated here.

The frequency response of an optical disk is shown in Figure 8.11. The response is best at DC and falls steadily to the optical cut-off frequency. Although the optics work down to DC, this cannot be used for the data recording. DC and low frequencies in the data would interfere with the focus and tracking servos and, as will be seen, difficulties arise when attempting to demodulate a unipolar signal. In practice the signal from the pickup is split by a filter. Low frequencies go to the servos, and higher frequencies go to the data circuitry. As a result the optical disk channel has the same inability to handle DC as does a magnetic recorder, and the same techniques are needed to overcome it.

8.6 Magneto-optical disks

When a magnetic material is heated above its Curie temperature, it becomes demagnetized, and on cooling will assume the magnetization of an applied field which would be too weak to influence it normally. This is the principle of magneto-optical recording. The heat is supplied by a finely focused laser, the field is supplied by a coil which is much larger.

Figure 8.12 shows that the medium is initially magnetized in one direction only. In order to record, the coil is energized with a current in

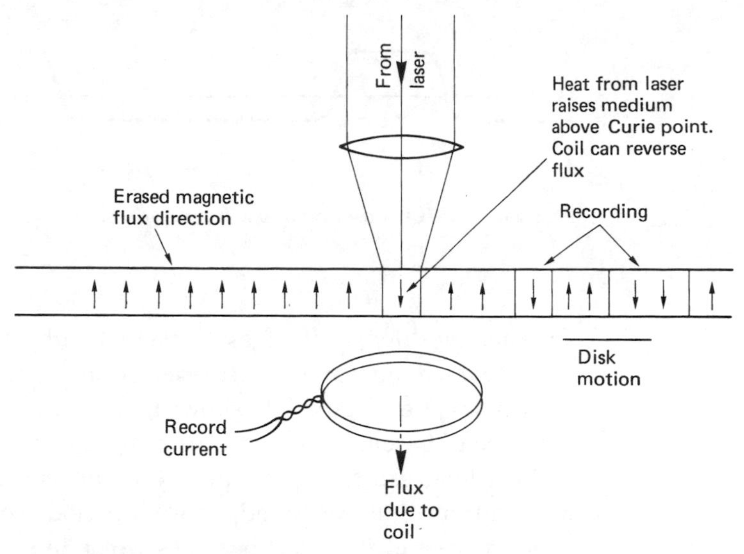

Figure 8.12 The thermomagneto-optical disk uses the heat from a laser to allow magnetic field to record on the disk.

the opposite direction. This is too weak to influence the medium in its normal state, but when it is heated by the recording laser beam the heated area will take on the magnetism from the coil when it cools. Thus a magnetic recording with very small dimensions can be made even though the magnetic circuit involved is quite large in comparison.

Readout is obtained using the Kerr effect or the Faraday effect, which are phenomena whereby the plane of polarization of light can be rotated by a magnetic field. The angle of rotation is very small and needs a sensitive pickup. The pickup contains a polarizing filter before the sensor. Changes in polarization change the ability of the light to get through the polarizing filter and results in an intensity change which once more produces a unipolar output.

The magneto-optic recording can be erased by reversing the current in the coil and operating the laser continuously as it passes along the track. A new recording can then be made on the erased track.

A disadvantage of magneto-optical recording is that all materials having a Curie point low enough to be useful are highly corrodible by air and need to be kept under an effectively sealed protective layer. The magneto-optical channel has the same frequency response as that shown in Figure 8.12.

8.7 Equalization

The characteristics of most channels are that signal loss occurs which increases with frequency. This has the effect of slowing down rise times

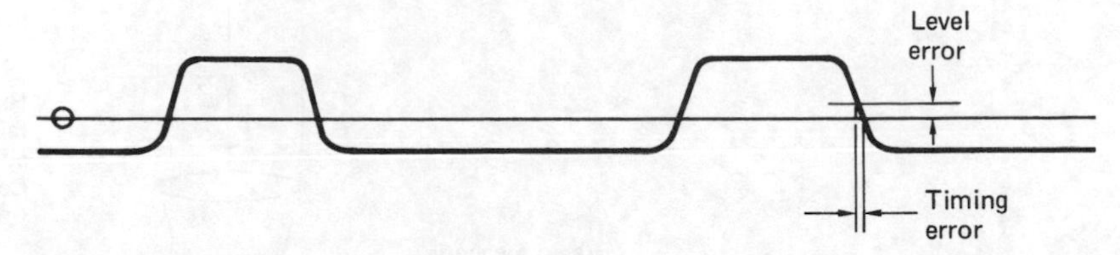

Figure 8.13 A DC offset can cause timing errors.

and thereby sloping off edges. If a signal with sloping edges is sliced, the time at which the waveform crosses the slicing level will be changed, and this causes jitter. Figure 8.13 shows that slicing a sloping waveform in the presence of baseline wander causes more jitter.

On a long cable, high-frequency rolloff can cause sufficient jitter to move a transition into an adjacent bit period. This is called inter-symbol interference and the effect becomes worse in signals which have greater asymmetry, i.e. short pulses alternating with long ones. The effect can be reduced by the application of equalization, which is typically a high-frequency boost, and by choosing a channel code which has restricted asymmetry.

Compensation for peak shift distortion in recording requires equalization of the channel,[3] and this can be done by a network after the replay head, termed an equalizer or pulse sharpener,[4] as in Figure 8.14(a). This technique uses transversal filtering to oppose the inherent transversal effect of the head. As an alternative, pre-compensation in the record stage can be used as shown in (b). Transitions are written in such a way that the anticipated peak shift will move the readout peaks to the desired timing.

8.8 Data separation

The important step of information recovery at the receiver or replay circuit is known as data separation. The data separator is rather like an analog-to-digital convertor because the two processes of sampling and quantizing are both present. In the time domain, the sampling clock is derived from the clock content of the channel waveform. In the voltage domain, the process of *slicing* converts the analog waveform from the channel back into a binary representation. The slicer is thus a form of quantizer which has only one-bit resolution. The slicing process makes a discrete decision about the voltage of the incoming signal in order to reject noise. The sampler makes discrete decisions along the time axis in order to reject jitter. These two processes will be described in detail.

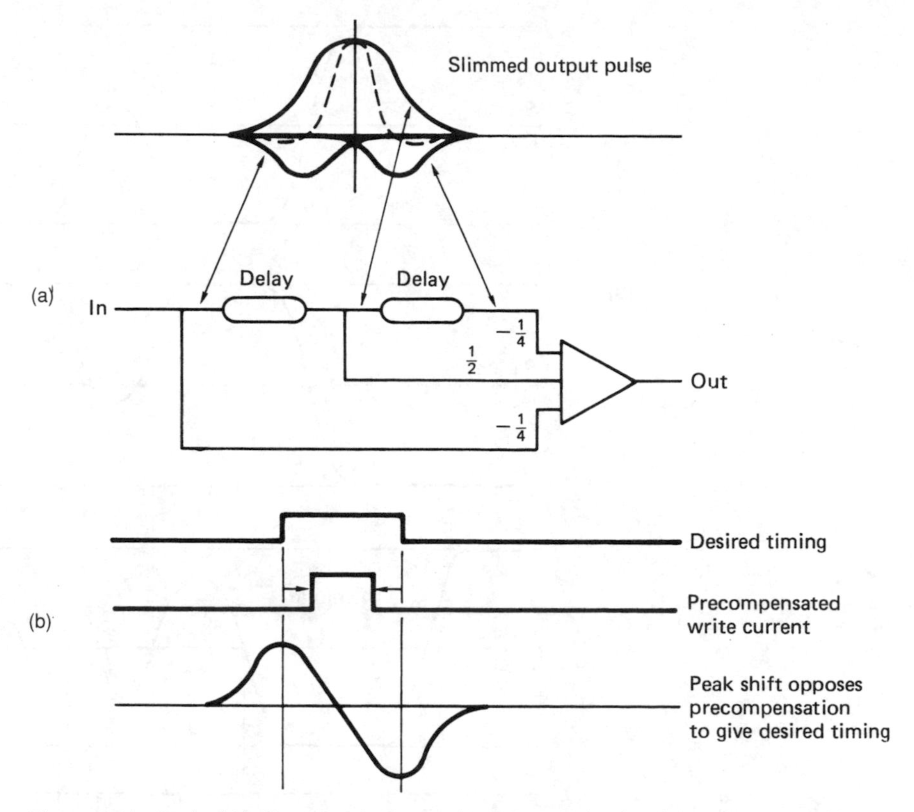

Figure 8.14 Peak-shift distortion is due to the finite width of replay pulses. The effect can be reduced by the pulse slimmer shown in (a) which is basically a transversal filter. The use of a linear operational amplifier emphasizes the analog nature of channels. Instead of replay pulse slimming, transitions can be written with a displacement equal and opposite to the anticipated peak shift as shown in (b).

8.9 Slicing

The slicer is implemented with a comparator which has analog inputs but a binary output. In a cable receiver, the input waveform can be sliced directly. In an inductive magnetic replay system, the replay waveform is differentiated and must first pass through a peak detector (Figure 8.4) or an integrator (Figure 8.5). The signal voltage is compared with the midway voltage, known as the threshold, baseline or slicing level by the comparator. If the signal voltage is above the threshold, the comparator outputs a high level, if below, a low level results.

Figure 8.15 shows some waveforms associated with a slicer. In (a) the transmitted waveform has an uneven duty cycle. The DC component, or average level, of the signal is received with high amplitude, but the pulse amplitude falls as the pulse gets shorter. Eventually the waveform cannot

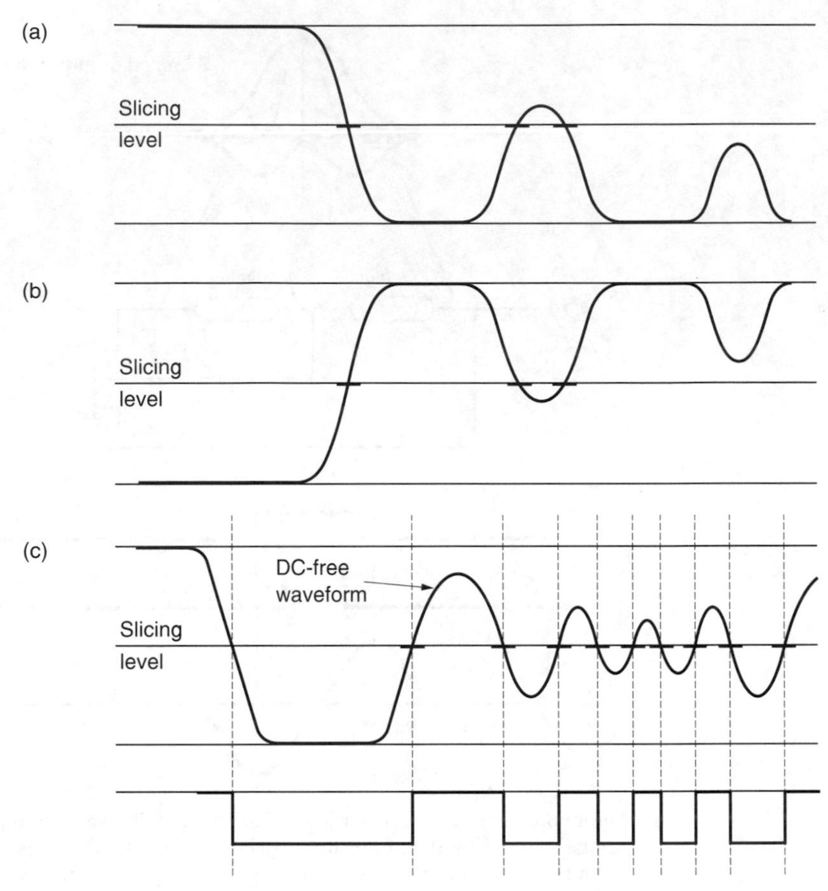

Figure 8.15 Slicing a signal which has suffered losses works well if the duty cycle is even. If the duty cycle is uneven, as at (a), timing errors will become worse until slicing fails. With the opposite duty cycle, the slicing fails in the opposite direction as at (b). If, however, the signal is DC free, correct slicing can continue even in the presence of serious losses, as (c) shows.

be sliced. In (b) the opposite duty cycle is shown. The signal level drifts to the opposite polarity and once more slicing is impossible. The phenomenon is called baseline wander and will be observed with any signal whose average voltage is not the same as the slicing level. In (c) it will be seen that if the transmitted waveform has a relatively constant average voltage, slicing remains possible up to high frequencies even in the presence of serious amplitude loss, because the received waveform remains symmetrical about the baseline.

It is clearly not possible simply to serialize data in a shift register for so-called direct transmission, because successful slicing can only be obtained if the number of ones is equal to the number of zeros; there is little chance of this happening consistently with real data. Instead, a modulation code or channel code is necessary. This converts the data into a waveform which is DC-free or nearly so for the purpose of transmission.

The slicing threshold level is naturally zero in a bipolar system such as magnetic inductive replay or a cable. When the amplitude falls it does so symmetrically and slicing continues. The same is not true of M-R heads and optical pickups, which both respond to intensity and therefore produce a unipolar output. If the replay signal is sliced directly, the threshold cannot be zero, but must be some level approximately half the amplitude of the signal as shown in Figure 8.16(a). Unfortunately when the signal level falls it falls toward zero and not towards the slicing level. The threshold will no longer be appropriate for the signal as can be seen in (b). This can be overcome by using a DC-free coded waveform. If a

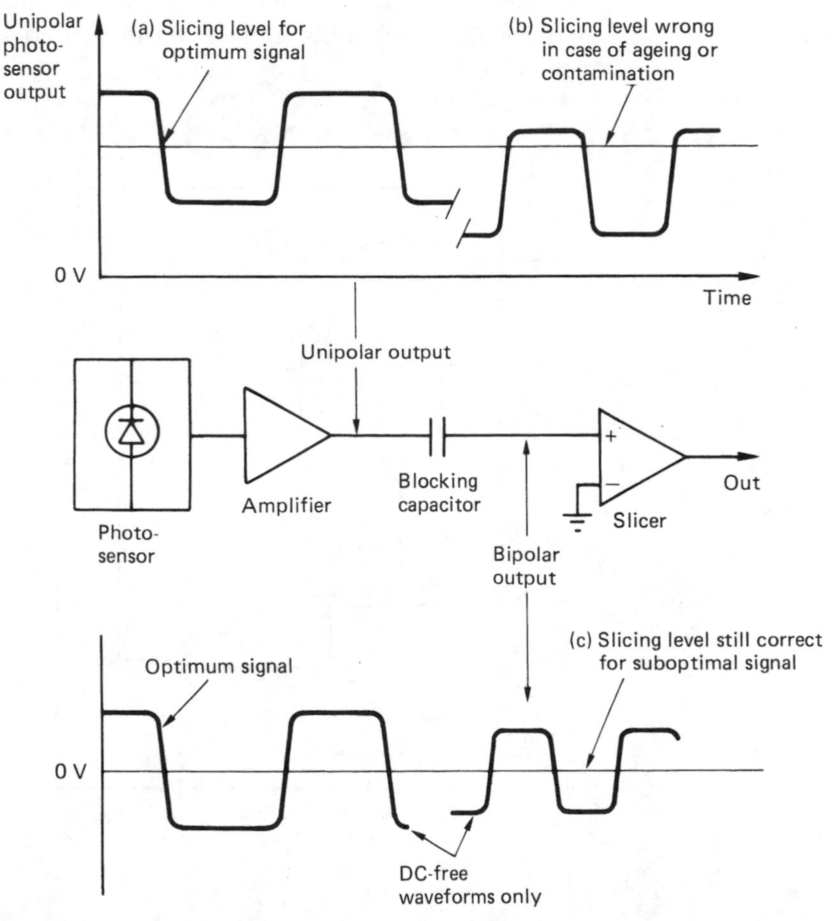

Figure 8.16 (a) Slicing a unipolar signal requires a non-zero threshold. (b) If the signal amplitude changes, the threshold will then be incorrect. (c) If a DC-free code is used, a unipolar waveform can be converted to a bipolar waveform using a series capacitor. A zero threshold can be used and slicing continues with amplitude variations.

series capacitor is connected to the unipolar signal from an optical pickup, the waveform is rendered bipolar because the capacitor blocks any DC component in the signal. The DC-free channel waveform passes through unaltered. If an amplitude loss is suffered, (c) shows that the resultant bipolar signal now reduces in amplitude about the slicing level and slicing can continue.

Whilst cables and optical recording channels need to be DC-free, some channel waveforms used in magnetic recording have a reduced DC component, but are not completely DC-free. As a result the received waveform will suffer from baseline wander. If this is moderate, an adaptive slicer which can move its threshold can be used. As Figure 8.17 shows, the adaptive slicer consists of a pair of delays. If the input and output signals are linearly added together with equal weighting, when a transition passes, the resultant waveform has a plateau which is at the half-amplitude level of the signal and can be used as a threshold voltage for the slicer.

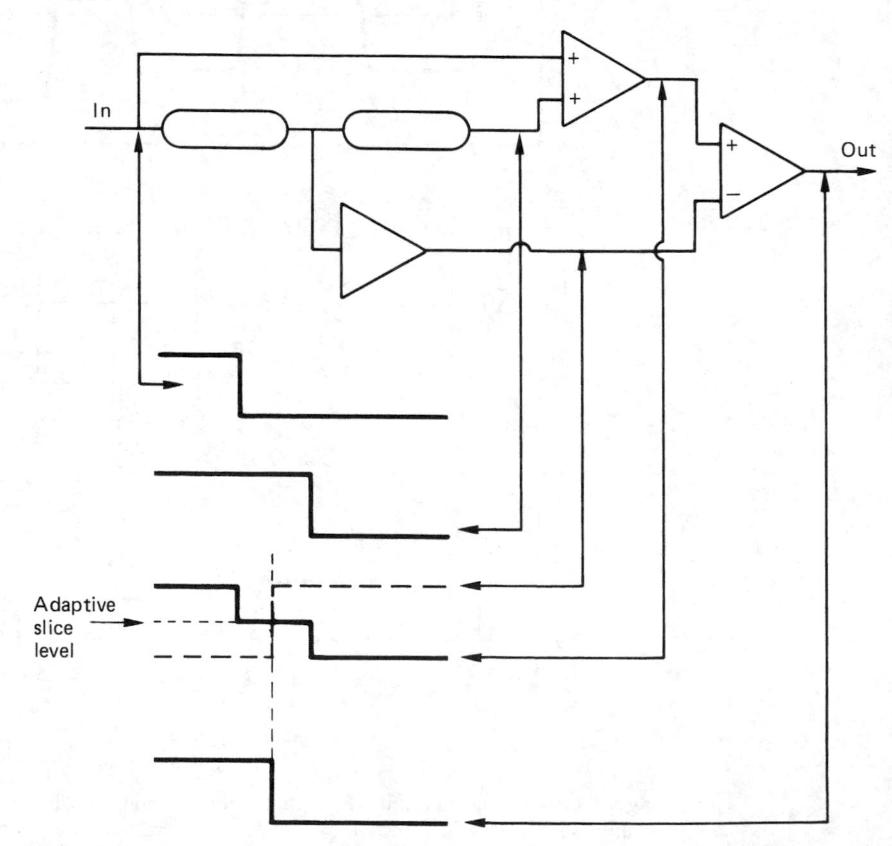

Figure 8.17 An adaptive slicer uses delay lines to produce a threshold from the waveform itself. Correct slicing will then be possible in the presence of baseline wander. Such a slicer can be used with codes which are not DC free.

8.10 Jitter rejection

The binary waveform at the output of the slicer will be a replica of the transmitted waveform, except for the addition of jitter or time uncertainty in the position of the edges due to noise, baseline wander, intersymbol interference and imperfect equalization.

Binary circuits reject noise by using discrete voltage levels which are spaced further apart than the uncertainty due to noise. In a similar manner, digital coding combats time uncertainty by making the time axis discrete using events, known as transitions, spaced apart at integer multiples of some basic time period, called a detent, which is larger than the typical time uncertainty. Figure 8.18 shows how this jitter-rejection mechanism works. All that matters is to identify the detent in which the transition occurred. Exactly where it occurred within the detent is of no consequence.

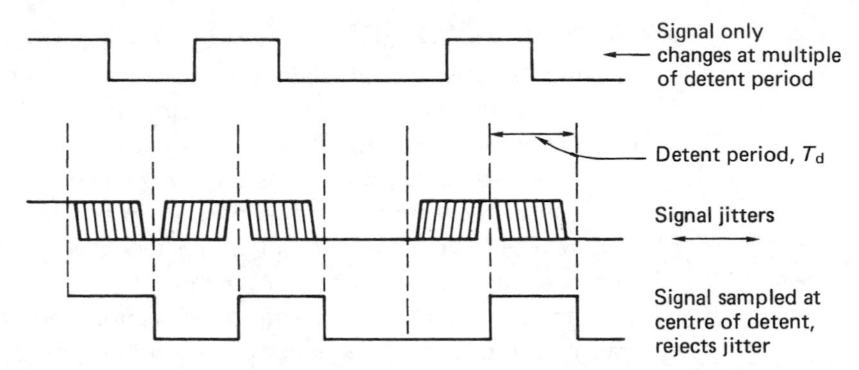

Figure 8.18 A certain amount of jitter can be rejected by changing the signal at multiples of the basic detent period T_d.

As ideal transitions occur at multiples of a basic period, an oscilloscope, which is repeatedly triggered on a channel-coded signal carrying random data, will show an eye pattern if connected to the output of the equalizer. Study of the eye pattern reveals how well the coding used suits the channel. In the case of transmission, with a short cable, the losses will be small, and the eye opening will be virtually square except for some edge sloping due to cable capacitance. As cable length increases, the harmonics are lost and the remaining fundamental gives the eyes a diamond shape. The same eye pattern will be obtained with a recording channel where it is uneconomic to provide bandwidth much beyond the fundamental.

Noise closes the eyes in a vertical direction, and jitter closes the eyes in a horizontal direction, as in Figure 8.19. If the eyes remain sensibly open, data separation will be possible. Clearly more jitter can be tolerated if

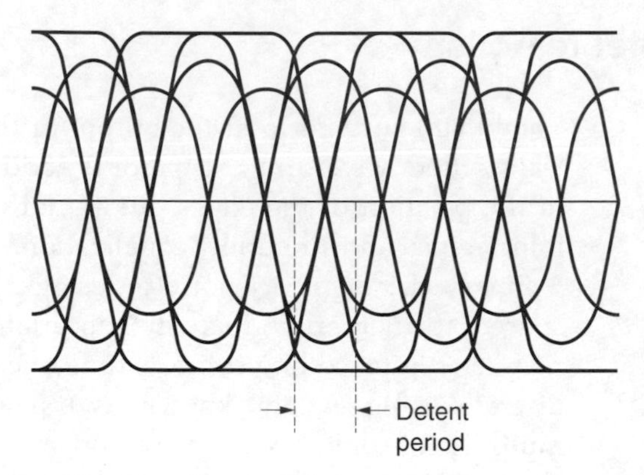

Figure 8.19 A transmitted waveform which is generated according to the principle of Figure 8.17 will appear like this on an oscilloscope as successive parts of the waveform are superimposed on the tube. When the waveform is rounded off by losses, diamond-shaped eyes are left in the centre, spaced apart by the detent period.

there is less noise, and vice versa. If the equalizer is adjustable, the optimum setting will be where the greatest eye opening is obtained.

In the centre of the eyes, the receiver must make binary decisions at the channel bit rate about the state of the signal, high or low, using the slicer output. As stated, the receiver is sampling the output of the slicer, and it needs to have a sampling clock in order to do that. In order to give the best rejection of noise and jitter, the clock edges which operate the sampler must be in the centre of the eyes.

As mentioned above, a separate clock is not practicable in recording or transmission. A fixed-frequency clock at the receiver is of no use as even if it was sufficiently stable, it would not know what phase to run at.

The only way in which the sampling clock can be obtained is to use a phase-locked loop to regenerate it from the clock content of the self-clocking channel-coded waveform. In phase-locked loops, the voltage-controlled oscillator is driven by a phase error measured between the output and some reference, as described in Chapter 3, such that the output eventually has the same frequency as the reference. If a divider is placed between the VCO and the phase comparator, the VCO frequency can be made to be a multiple of the reference. This also has the effect of making the loop more heavily damped. If a channel-coded waveform is used as a reference to a PLL, the loop will be able to make a phase comparison whenever a transition arrives and will run at the channel bit rate. When there are several detents between transitions, the loop will *flywheel* at the last known frequency and phase until it can rephase at a subsequent transition. Thus a continuous clock is recreated from the clock content of the channel waveform. In a recorder, if the speed of the medium should change, the PLL will change frequency to follow. Once the loop is locked, clock edges will be phased with the average phase of the jittering edges of the input waveform. If, for example, rising edges of the clock are phased to input transitions, then falling edges will be in the centre of the eyes. If these edges are used to clock the sampling process, the maximum jitter and noise can be rejected. The output of the slicer when sampled by the PLL edge at the centre of an eye is the value of a channel bit. Figure 8.20 shows the complete clocking system of a channel code from encoder to data separator.

Clearly data cannot be separated if the PLL is not locked, but it cannot be locked until it has seen transitions for a reasonable period. In recorders, which have discontinuous recorded blocks to allow editing, the solution is to precede each data block with a pattern of transitions whose sole purpose is to provide a timing reference for synchronizing the phase-locked loop. This pattern is known as a preamble. In interfaces, the transmission can be continuous and there is no difficulty remaining in lock indefinitely. There will simply be a short delay on first applying the signal before the receiver locks to it.

One potential problem area which is frequently overlooked is to ensure that the VCO in the receiving PLL is correctly centred. If it is not, it will be running with a static phase error and will not sample the received waveform at the centre of the eyes. The sampled bits will be more prone to noise and jitter errors. VCO centring can simply be checked by displaying the control voltage. This should not change significantly when the input is momentarily interrupted.

8.11 Channel coding

In summary, it is not practicable simply to serialize raw data in a shift register for the purpose of recording or for transmission except over relatively short distances. Practical systems require the use of a modulation scheme, known as a channel code, which expresses the data as waveforms which are self-clocking in order to reject jitter, separate the received bits and to avoid skew on separate clock lines. The coded waveforms should further be DC-free or nearly so to enable slicing in the presence of losses and have a narrower spectrum than the raw data to make equalization possible.

Jitter causes uncertainty about the time at which a particular event occurred. The frequency response of the channel then places an overall limit on the spacing of events in the channel. Particular emphasis must be placed on the interplay of bandwidth, jitter and noise, which will be shown here to be the key to the design of a successful channel code.

Figure 8.21 shows that a channel coder is necessary prior to the record stage, and that a decoder, known as a data separator, is necessary after the

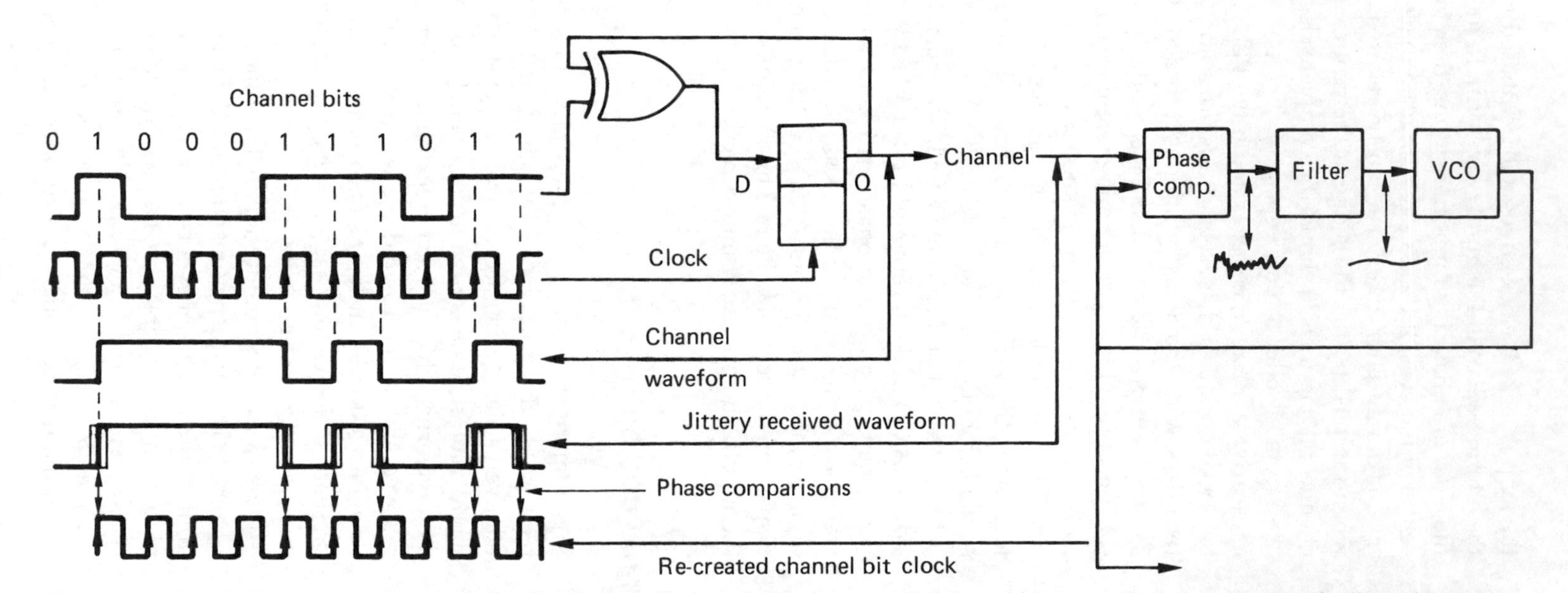

Figure 8.20 The clocking system when channel coding is used. The encoder clock runs at the channel bit rate, and any transitions in the channel must coincide with encoder clock edges. The reason for doing this is that, at the data separator, the PLL can lock to the edges of the channel signal. which represent an intermittent clock, and turn it into a continuous clock. The jitter in the edges of the channel signal causes noise in the phase error of the PLL, but the damping acts as a filter and the PLL runs at the average phase of the channel bits, rejecting the jitter.

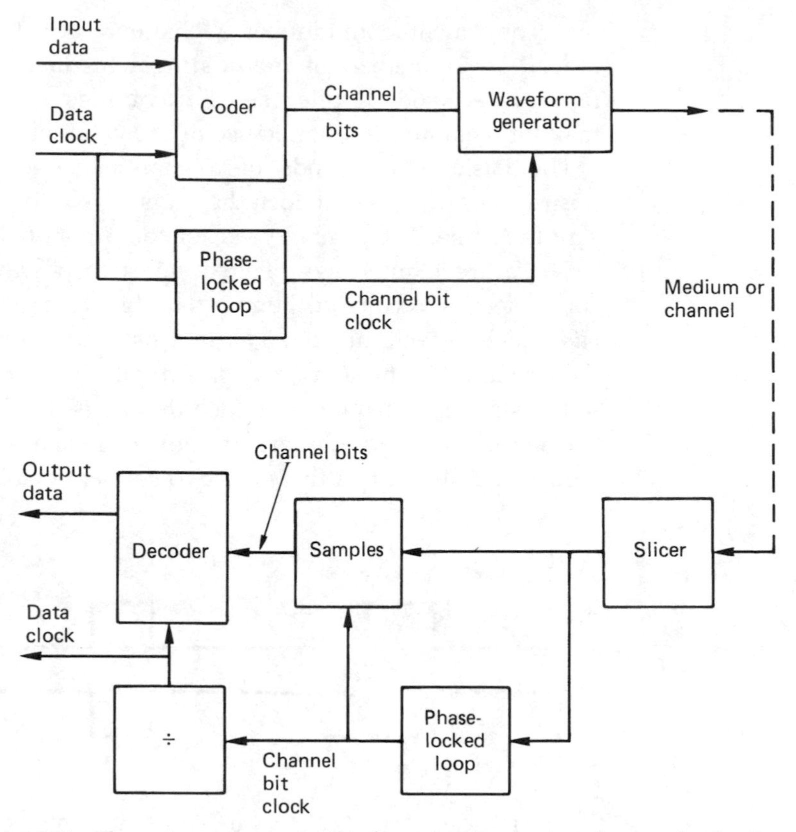

Figure 8.21 The major components of a channel coding system. See text for details.

replay stage. The output of the channel coder is generally a logic level signal which contains a 'high' state when a transition is to be generated. The waveform generator produces the transitions in a signal whose level and impedance is suitable for driving the medium or channel. The signal may be bipolar or unipolar as appropriate.

Some codes eliminate DC entirely, which is advantageous for optical media and for rotary head recording. Other codes can reduce the channel bandwidth needed by lowering the upper spectral limit. This permits higher linear density, usually at the expense of jitter rejection. Yet other codes narrow the spectrum by raising the lower limit. A code with a narrow spectrum has a number of advantages. The reduction in asymmetry will reduce peak shift and data separators can lock more readily because the range of frequencies in the code is smaller. In theory the narrower the spectrum, the less noise will be suffered, but this is only achieved if filtering is employed. Filters can easily cause phase errors which will nullify any gain.

A convenient definition of a channel code (for there are certainly others) is 'A method of modulating real data such that they can be reliably received despite the shortcomings of a real channel, while making maximum economic use of the channel capacity'.

The basic time periods of a channel-coded waveform are called positions or detents, in which the transmitted voltage will be reversed or stay the same. The symbol used for the units of channel time is T_d.

There are many ways of creating such a waveform, but the most convenient is to convert the raw data bits to a larger number of *channel bits* which are output from a shift register to the waveform generator at the detent rate. The coded waveform will then be high or low according to the state of a channel bit which describes the detent.

Channel coding is the art of converting real data into channel bits. Figure 8.22 shows that there are two conventions in use relating channel

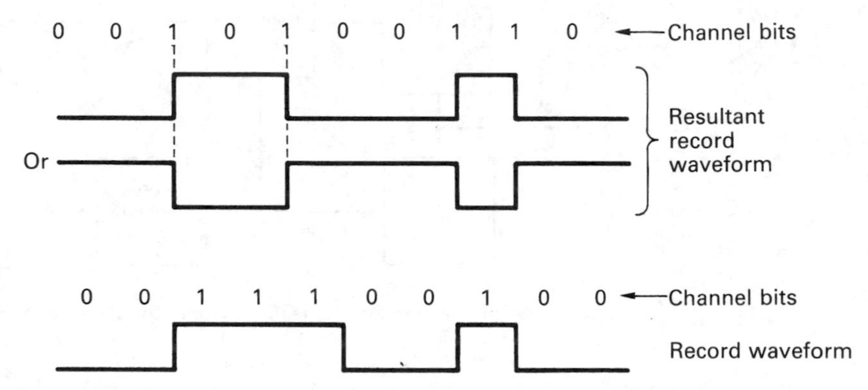

Figure 8.22 The two conventions for channel bit notation. A channel bit one most commonly represents a transition, but some codes use the bit status to describe the direction of the record current.

bits to the recorded waveform. The most commonly used convention is one in which a channel-bit one represents a voltage (or flux) change, whereas a zero represents no change. This convention is used because it is possible to assemble sequential groups of channel bits together without worrying about whether the polarity of the end of the last group matches the beginning of the next. The polarity is unimportant in most codes and all that matters is the length of time between transitions. Less common is the convention where the channel bit state directly represents the direction of the recording current. Clearly steps then need to be taken to ensure that the boundary between two groups is properly handled. Such an approach is used in the D-3/D-5 formats and will be explained in section 5.20.

It should be stressed that channel bits are not recorded. They exist only in a circuit technique used to control the waveform generator. In many media, for example D-3/D-5, the channel bit rate is beyond the frequency response of the channel and so it *cannot* be recorded.

One of the fundamental parameters of a channel code is the density ratio (DR). One definition of density ratio is that it is the worst-case ratio of the number of data bits recorded to the number of transitions in the channel. It can also be thought of as the ratio between the Nyquist rate of the data (one-half the bit rate) and the frequency response required in the channel. The storage density of data recorders has steadily increased due to improvements in medium and transducer technology, but modern storage densities are also a function of improvements in channel coding. Figure 8.23(a) shows how density ratio has improved as more sophisticated codes have been developed.

As jitter is such an important issue in digital recording and transmission, a parameter has been introduced to quantify the ability of a channel code to reject time instability. This parameter, the jitter margin, also known as the window margin or phase margin (T_w), is defined as the permitted range of time over which a transition can still be received correctly, divided by the data bit-cell period (T).

Since equalization is often difficult in practice, a code which has a large jitter margin will sometimes be used because it resists the effects of

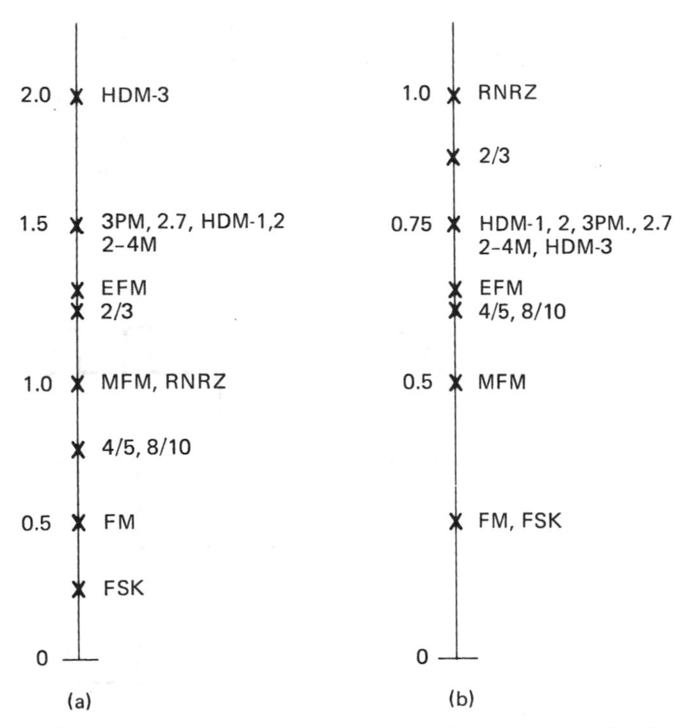

Figure 8.23 (a) Comparison of codes by density ratio; (b) comparison of codes by figure of merit. Note how 4/5, 2/3, 8/10 + RNRZ move up because of good jitter performance; HDM-3 moves down because of jitter sensitivity.

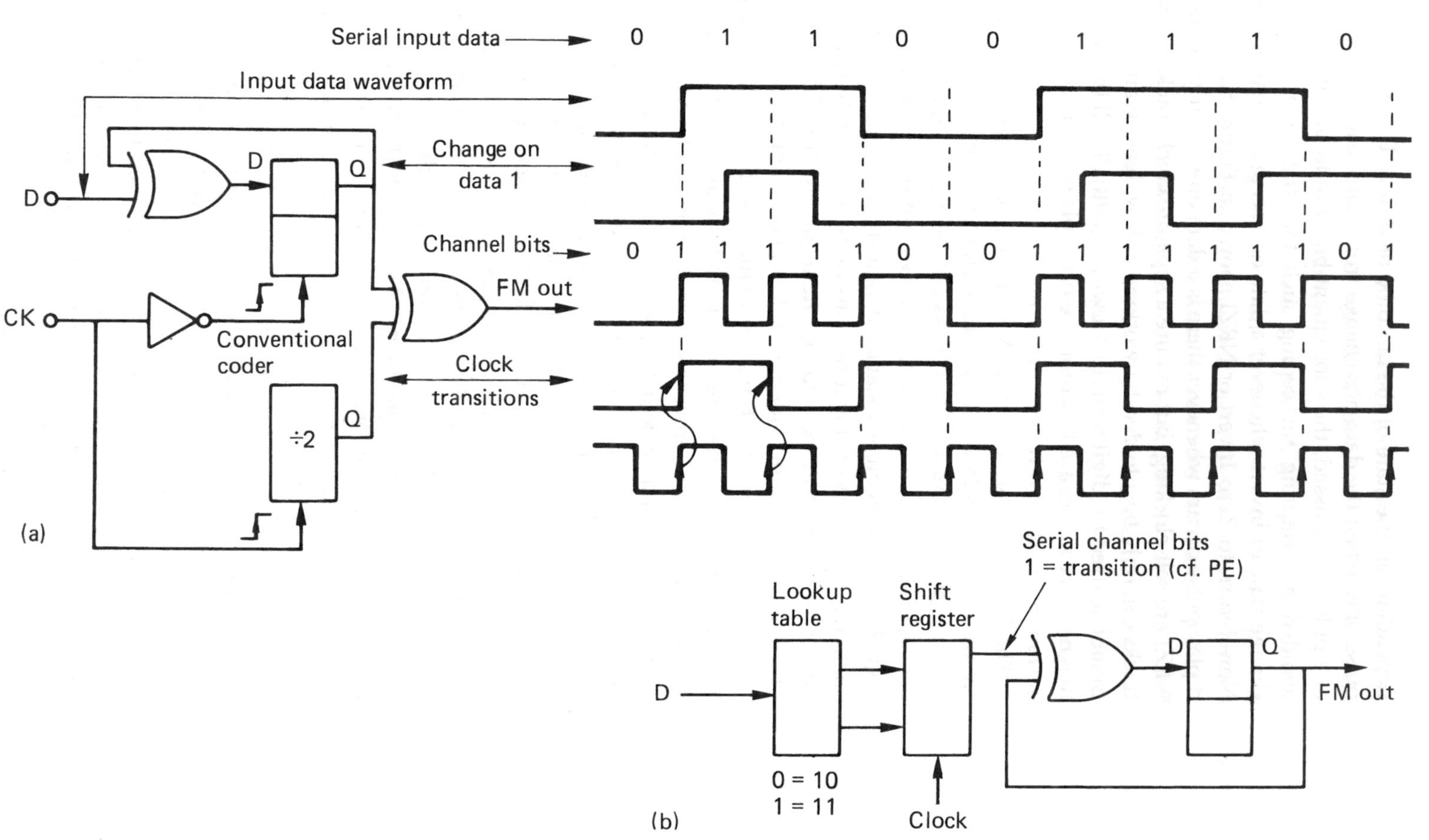

Figure 8.25 FM encoding. At (a) are the FM waveform and the channel bits which may be used to describe transitions in it. The FM coder is shown at (b).

clocked at the channel bit rate, and an exclusive-OR (XOR) gate. This changes state when a channel bit one is input. The result is a coded FM waveform where there is always a transition at the beginning of the data bit period, and a second optional transition whose presence indicates a one.

In modified frequency modulation (MFM) also known as Miller code,[5] the highly redundant clock content of FM was reduced by the use of a phase-locked loop in the receiver which could flywheel over missing clock transitions. This technique is implicit in all the more advanced codes. Figure 8.26(a) shows that the bit-cell centre transition on a data one was retained, but the bit-cell boundary transition is now only required between successive zeros. There are still two channel bits for every data bit, but adjacent channel bits will never be one, doubling the minimum time between transitions, and giving a DR of 1. Clearly the coding of the current bit is now influenced by the preceding bit. The maximum number of prior bits which affect the current bit is known as the constraint length L_c, measured in data-bit periods. For MFM $L_c = T$. Another way of considering the constraint length is that it assesses the number of data bits which may be corrupted if the receiver misplaces one transition. If L_c is long, all errors will be burst errors.

MFM doubled the density ratio compared to FM and PE without changing the jitter performance; thus the FoM also doubles, becoming 0.5. It was adopted for many hard disks at the time of its development, and remains in use on double-density floppy disks. It is not, however, DC-free. Figure 8.26(b) shows how MFM can have DC content under certain conditions.

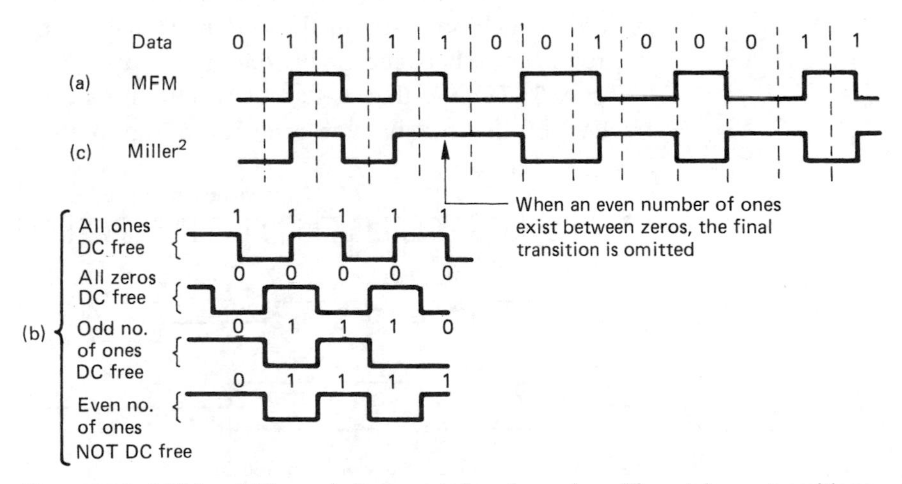

Figure 8.26 MFM or Miller code is generated as shown here. The minimum transition spacing is twice that of FM or PE. MFM is not always DC-free as shown at (b). This can be overcome by the modification of (c) which results in the Miller^2 code.

The Miller^2 code is derived from MFM, and Figure 8.26(c) shows that the DC content is eliminated by a slight increase in complexity.[6,7] Wherever an even number of ones occurs between zeros, the transition at the last one is omitted. This creates two additional, longer run lengths and increases the T_{max} of the code. The decoder can detect these longer run lengths in order to re-insert the suppressed ones. The FoM of Miller^2 is 0.5 as for MFM. Miller^2 is used in high rate instrumentation recorders and in the D-2 and DCT DVTR formats.

8.12 Group codes

Further improvements in coding rely on converting patterns of real data to patterns of channel bits with more desirable characteristics using a conversion table known as a codebook. If a data symbol of m bits is considered, it can have 2^m different combinations. As it is intended to discard undesirable patterns to improve the code, it follows that the number of channel bits n must be greater than m. The number of patterns which can be discarded is:

$$2^n - 2^m$$

One name for the principle is group code recording (GCR), and an important parameter is the code rate, defined as:

$$R = \frac{m}{n}$$

It will be evident that the jitter margin T_w is numerically equal to the code rate, and so a code rate near to unity is desirable. The choice of patterns which are used in the codebook will be those which give the desired balance between clock content, bandwidth and DC content.

Figure 8.27 shows that the upper spectral limit can be made to be some fraction of the channel bit rate according to the minimum distance

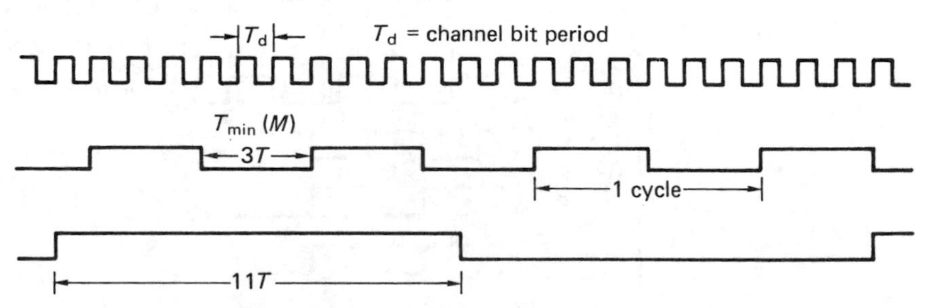

Figure 8.27 A channel code can control its spectrum by placing limits on T_{min} (M) and T_{max} which define upper and lower frequencies. The ratio of T_{max}/T_{min} determines the asymmetry of waveform and predicts DC content and peak shift. Example shown is EFM.

between ones in the channel bits. This is known as T_{min}, also referred to as the minimum transition parameter M and in both cases is measured in data bits T. It can be obtained by multiplying the number of channel detent periods between transitions by the code rate. Unfortunately, codes are measured by the number of consecutive zeros in the channel bits, given the symbol d, which is always one less than the number of detent periods. In fact T_{min} is numerically equal to the density ratio.

$$T_{min} = M = DR = \frac{(d + 1) \times m}{n}$$

It will be evident that choosing a low code rate could increase the density ratio, but it will impair the jitter margin. The figure of merit is:

$$FoM = DR \times T_w = \frac{(d + 1) \times m^2}{n^2}$$

since $T_w = m/n$

Figure 8.28 also shows that the lower spectral limit is influenced by the maximum distance between transitions T_{max}. This is also obtained by multiplying the maximum number of detent periods between transitions by the code rate. Again, codes are measured by the maximum number of zeros between channel ones, k, and so:

$$T_{max} = \frac{(k + 1) \times m}{n}$$

and the maximum/minimum ratio P is:

$$P = \frac{(k + 1)}{(d + 1)}$$

The length of time between channel transitions is known as the *run length*. Another name for this class is the run-length-limited (RLL) codes.[8] Since m data bits are considered as one symbol, the constraint length L_c will be increased in RLL codes to at least m. It is, however, possible for a code to have run-length limits without it being a group code.

In practice, the junction of two adjacent channel symbols may violate run-length limits, and it may be necessary to create a further codebook of symbol size $2n$ which converts violating code pairs to acceptable patterns. This is known as merging and follows the golden rule that the substitute $2n$ symbol must finish with a pattern which eliminates the possibility of a subsequent violation. These patterns must also differ from all other symbols.

Substitution may also be used to different degrees in the same nominal code in order to allow a choice of maximum run length, e.g. 3PM. The maximum number of symbols involved in a substitution is denoted by *r*.[10,11] There are many RLL codes and the parameters *d,k,m,n,* and *r* are a way of comparing them.

Sometimes the code rate forms the name of the code, as in 2/3, 8/10 and EFM; at other times the code may be named after the *d,k* parameters, as in 2,7 code. Various examples of group codes will be given to illustrate the principles involved.

Figure 8.28(a) shows the codebook of an optimized code which illustrates one merging technique. This is a 1,7,2,3,2 code known as 2/3. It is designed to have a good jitter window in order to resist peak shift distortion in disk drives, but it also has a good density ratio.[12] In 2/3

Data	Code
0 0	1 0 1
0 1	1 0 0
1 0	0 0 1
1 1	0 1 0

(a)

Data	Illegal code	Substitution
0 0 0 0	1 0 1 1 0 1	1 0 1 0 0 0
0 0 0 1	1 0 1 1 0 0	1 0 0 0 0 0
1 0 0 0	0 0 1 1 0 1	0 0 1 0 0 0
1 0 0 1	0 0 1 1 0 0	0 1 0 0 0 0

(b)

Figure 8.28 2/3 code. At (a) two data bits (m) are expressed as three channel bits (n) without adjacent transitions ($d = 1$). Violations are dealt with by substitution

$$\mathrm{DR} = \frac{(d+1)m}{n} = \frac{2 \times 2}{3} = 1.33$$

Adjacent data pairs can break the encoding rule; in these cases substitutions are made, as shown in (b)

code, pairs of data bits create symbols of three channel bits. For bandwidth reduction, codes having adjacent ones are eliminated so that $d = 1$. This halves the upper spectral limit and the DR is improved accordingly:

$$\text{DR} = \frac{(d + 1) \times m}{n} = \frac{2 \times 2}{3} = 1.33$$

In Figure 8.28(b) it will be seen that some group combinations cause violations. To avoid this, pairs of three channel-bit symbols are replaced with a new six channel-bit symbol. L_c is thus 4T, the same as for the 4/5 code. The jitter window is given by:

$$T_w = \frac{m}{n} = \frac{2}{3}\,T$$

and the FoM is:

$$\frac{2}{3} \times \frac{4}{3} = \frac{8}{9}$$

This is an extremely good figure for an RLL code and is some 10 per cent better than the FoM of 3PM[13] and 2,7 and as a result 2/3 has been highly successful in Winchester disk drives.

8.13 EFM code in D-3/D-5

The $\frac{1}{2}$-inch D-3 and D-5 formats use a group code in which $m = 8$ and $n = 14$ so the code rate is 0.57. The code is called 8,14 after the main parameters. It will be evident that the jitter margin T_w is numerically equal to the code rate, and so for jitter resistance a code rate close to unity is preferable. The choice of patterns which are used in the codebook will be those which give the desired balance between clock content, bandwidth and DC content.

The code used in D-3/D-5 uses the convention in which a channel bit one represents a high in the recorded waveform. In this convention a flux reversal or transition will be written when the channel bits change.

In Figure 8.29 it is shown that the upper spectral limit can be made to be some fraction of the channel bit rate according to the minimum distance between transitions in the channel bits, which in 8,14 is two-channel bits. This is known as T_{min}, also referred to as the minimum transition parameter M, and in both cases is measured in data bits T. It can be obtained by multiplying the number of channel detent periods

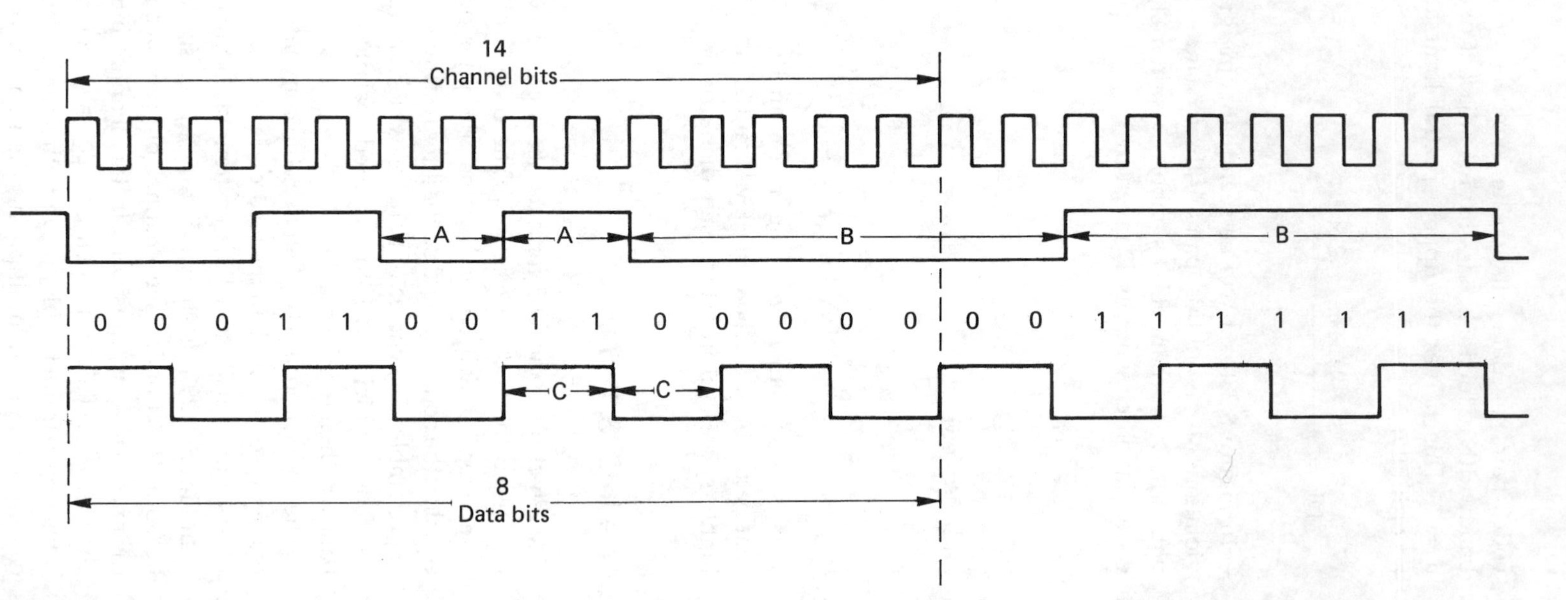

Figure 8.29 In the 8.14 code of D-3, eight data bit periods are divided into fourteen channel bit periods. However, the shortest run length allowed in the channel is two bits, shown at A. This restriction is obtained by selecting fourteen bit patterns from those available. The longest run length in the code is seven channel bits. Note that the shortest run length A is 14% longer than the shortest run length in the raw data C. Thus density ratio DR is 1.14. Using this code 14% more data can be recorded with the same wavelength on tape as a simple code.

between transitions by the code rate. In fact T_{min} is numerically equal to the density ratio,

$$T_{min} = M = DR = \frac{2 \times 8}{14} = 1.14$$

This DR is a little better than the figure of 1 for the codes used in D-1 and D-2.

The figure of merit is:

$$FoM = DR \times T_w = \frac{2 \times 8^2}{14^2} = 0.65$$

$$\text{since} \quad T_w = \frac{m}{n} = \frac{8}{14}$$

Figure 8.29 also shows that the lower spectral limit is influenced by the maximum distance between transitions T_{max}, which also determines the minimum clock content. This is also obtained by multiplying the maximum number of detent periods between transitions by the code rate. In 8,14 code this is seven-channel bits, and so:

$$T_{max} = \frac{7 \times 8}{14} = 4$$

and the maximum/minimum ratio P is:

$$P = \frac{4}{1.14} = 3.51$$

The length of time between channel transitions is known as the run length. Another name for this class is the run-length-limited (RLL) codes.[14] Since eight data bits are considered as one symbol, the constraint length L_c will be increased in this code to at least eight bits.

In practice, the junction of some adjacent channel symbols may violate coding rules, and it is necessary to extend the codebook so that the original data can be represented by a number of alternative codes at least one of which will be acceptable. This is known as substitution. Due to the coding convention used, which generates a transition when the channel bits change, transitions will also be generated at the junction of two channel symbols if the adjacent bits are different. It will be seen in Figure 8.29 that the presence of a junction transition can be controlled by selectively inverting all the bits in the second channel symbol. Thus for every channel bit pattern in the code, an inverted version also exists.

Data	Code A (begins with 0)	CDS	Code B (begins with 1)	CDS	Code C (begins with 0)	CDS	Code D (begins with 1)	CDS
00								
↑								
42	01100011000111	0	10011100111000	0	01100011000111	0	10011100111000	0
43	01100001111100	0	10011110000011	0	01100001111100	0	10011110000011	0
44	01100001111001	0	10011110000110	0	01100001111001	0	10011110000110	0
45	01100001110011	0	10011110001100	0	01100001110011	0	10011110001100	0
46	01100001100111	0	10011110011000	0	01100001100111	0	10011110011000	0
47	01100000111110	0	10011111000001	0	01100000111110	0	10011111000001	0
48	01100000011111	0	10011111100000	0	01100000011111	0	10011111100000	0
49	01111100000001	−2	10000000110011	−4	01111111001100	4	10000011111110	2
50	01111001100000	−2	10000000111001	−4	01111111000110	4	10000110011111	2
51	01111000110000	−2	10000000111100	−4	01111111000011	4	10000111001111	2
52	01111000011000	−2	10000001100011	−4	01111110011100	4	10000111100111	2
53	01111000001100	−2	10000001100110	−4	01111110011001	4	10000111110011	2
54	01111000000110	−2	10000001110001	−4	01111110001110	4	10000111111001	2
55	01111000000011	−2	10000001111000	−4	01111110000111	4	10000111111100	2
56	01110011100000	−2	10000011000011	−4	01111100111100	4	10001100011111	2
57	01110011000001	−2	10000011000110	−4	01111100111001	4	10001100111110	2
↓								
255								

Figure 8.30 Part of the codebook for D-3 EFM code. For every eight-bit data word (left) there are four possible channel bit patterns, two of which begin with 0 and two of which begin with 1. This allows successive symbols to avoid violating the minimum run-length limit (A in Figure 8.36) at the junction of the two symbols. See Figures 8.31 and 8.32 for further details.

Figure 8.30 shows part of the 8,14 code (EFM) used in the D-3 and D-5 formats. As stated, eight-bit data symbols are represented by fourteen-bit channel symbols. There are 256 combinations of eight data bits, whereas fourteen bits have 2^{14} or 16 384 combinations. The initial range of possible codes is reduced by the requirements of the maximum and minimum run length limits and by a requirement that there shall not be more than five identical bits in the first six bits of the code and not more than six in the last seven bits as shown in Figure 8.31.

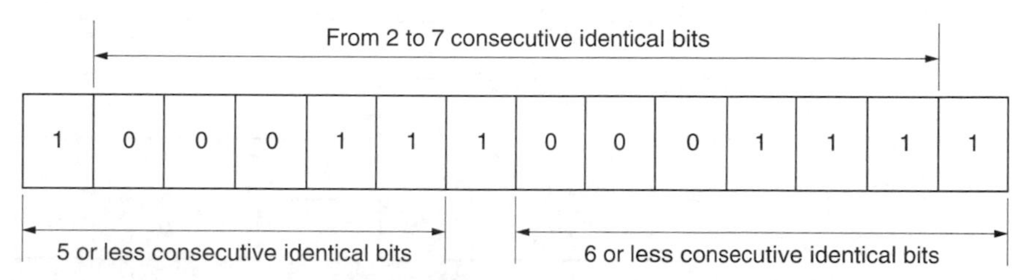

Figure 8.31 The selected fourteen-bit patterns follow the rules that there cannot be more than five consecutive identical bits at the beginning or more than six at the end. In addition the code digital sum (CDS) cannot exceed 4. See Figure 8.32 for derivation of CDS.

One of the most important parameters of a channel pattern is the code digital sum (CDS) shown in Figure 8.32. This is the number of channel ones minus the number of channel zeros. As the CDS represents the DC content or average voltage of the channel pattern it is to be kept to a minimum.

There are only 118 codes (and 118 inverse codes) which are DC-free (CDS = 0) as well as meeting the other constraints. In order to obtain 256 data combinations, codes with non-zero CDS have to be accepted. The actual codes used in 8,14 have CDS of 0, ±2 or ±4.

The CDS is a special case of a more general parameter called the Digital Sum Value (DSV). DSV is a useful way of predicting how an analog channel such as a tape/head system will handle a binary waveform. In a stream of bits, a one causes one to be added to the DSV whereas a zero causes one to be subtracted. Thus DSV is a form of running discrete integration. Figure 8.33(a) shows how the DSV varies along the time axis. The End DSV is the DSV at the end of a string of channel symbols. Clearly CDS is the End DSV of one code pattern in isolation. The next End DSV is the current one plus the CDS of the next symbol as shown in Figure 8.33(b). The Absolute DSV of a symbol is shown in Figure 8.33(c) This is obtained by finding the peak DSV. Large Absolute DSVs are associated with low frequencies and low clock content.

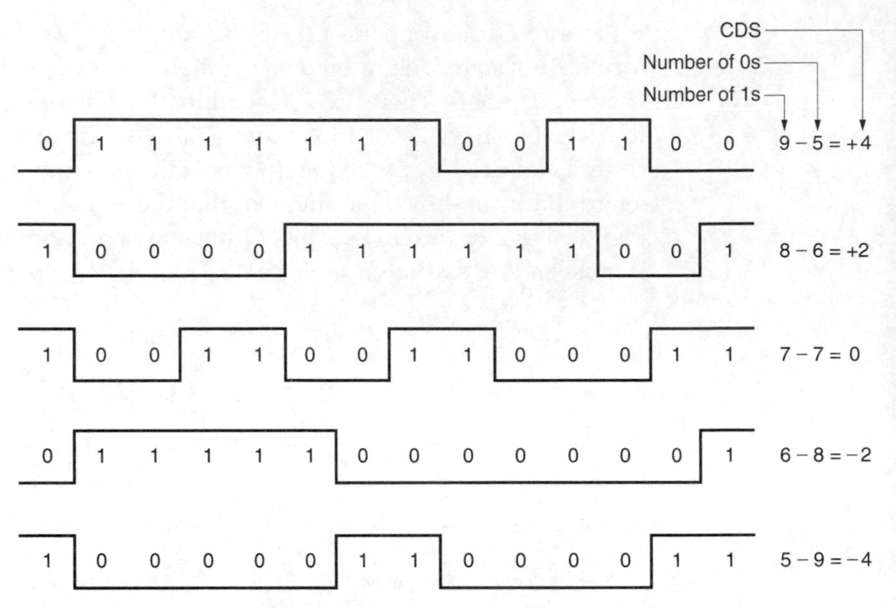

Figure 8.32 Code digital sum (CDS) is the measure of the DC content of a given channel symbol. It is obtained by subtracting the number of zeros in the symbol from the number of ones. Shown above are five actual fourteen-bit symbols from the D-3 code showing the range of CDS allowed from –4 to +4. Although CDS of zero is optimum, there are not enough patterns with zero CDS to represent all combinations of eight data bits.

When encoding is performed, each eight-bit data symbol selects four locations in the lookup table each of which contains a fourteen-bit pattern. The recording is made by selecting the most appropriate one of four candidate channel bit patterns. The decoding process is such that any of the four channel patterns will decode to the same data.

The four possible channel symbols for each data byte are classified according to Figure 8.31 into types A, B, C or D. Since 8,14 coding requires alternative inverse symbols, two of the candidates are simply bit inversions of the other two. Where codes are not DC-free there may be a pair of +2 CDS candidates and their –2 CDS inverses or a +2 CDS and a +4 CDS candidate and their inverses which will, of course, have –2 and –4 CDS. In the case of all but two of the DC-free codes the two candidates are one and the same code and the four lookup table locations contain two identical codes and two identical inverse codes. As a result of some of the codes being identical, althought there are 1024 locations in the table, only 770 different fourteen-bit patterns are used, representing 4.7 per cent of the total.

Code classes are further subdivided into Class Numbers, which go from 1 to 5 and specify the number of identical channel bits at the

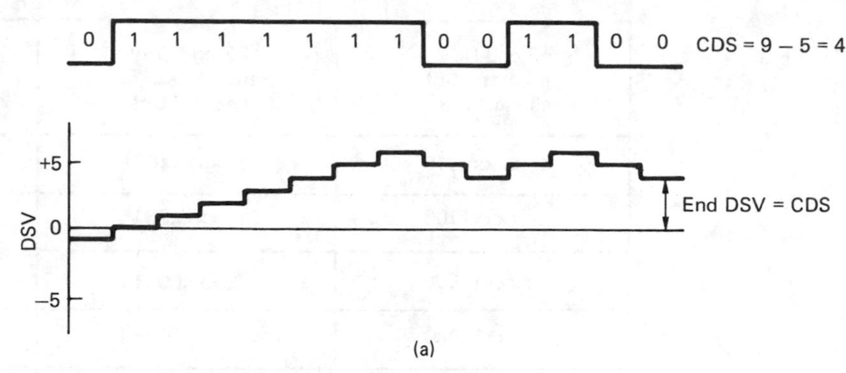

Figure 8.33(a) DSV is obtained by subtracting one for every zero, and adding one for every one which passes during the integration time. Thus DSV changes at every channel bit. Note that end DSV = CDS.

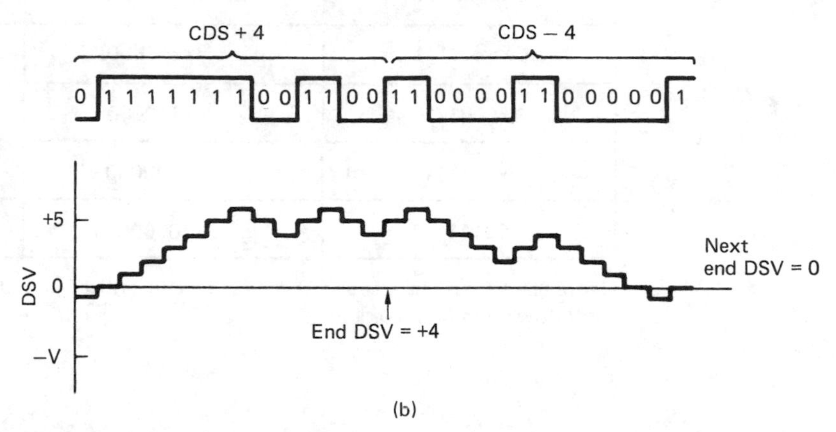

Figure 8.33(b) The next end DSV is the current end DSV plus the CDS of the next symbol. Note how the choice of a CDS –4 symbol after a +4 symbol brings DSV back to zero.

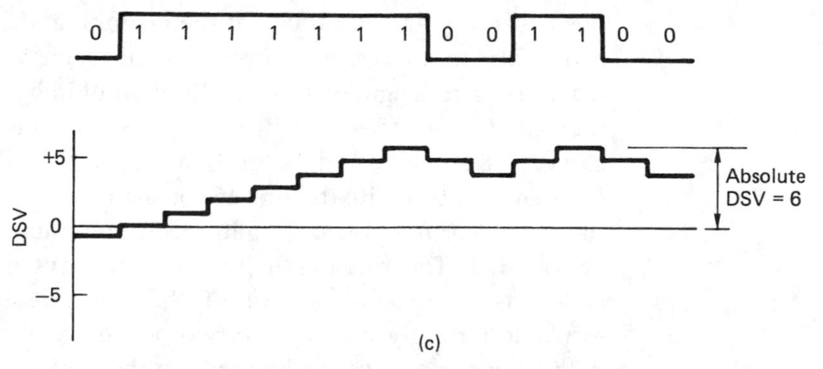

Figure 8.33(c) The absolute or peak DSV is the greatest value the DSV can have during the symbol. As shown here, it can be greater than CDS.

+2 end DSV end pattern of channel bits	−2 end DSV end pattern of channel bits	Priority
...xxxxx110	...xxxxx001	4
...xxxx1100	...xxxx0011	1
...xxx11000	...xxx00111	2
...xx110000	...xx001111	3
...x1100000	...x0011111	8
...xxxxx001	...xxxxx110	10
...xxxx0011	...xxxx1100	5
...xxx00111	...xxx11000	6
...xx001111	...xx110000	7
...x0011111	...x1100000	9
...00111111	...11000000	11

Figure 8.34 Codes having end DSV of ±2 are prioritized according to the above table as part of the selection process.

beginning of the symbol and a Priority Number which only applies to codes of CDS ±2 and is obtained from the channel bit pattern at the end of the code according to Figure 8.34.

Clearly the junction of two channel patterns cannot be allowed to violate the run-length limits. As the next code cannot have more than five consecutive identical bits at the beginning, the previous code could end with up to two bits in the same state without exceeding the maximum run length of 7. If this limit would be exceeded by one of the four candidate codes, it will be rejected and one of those having an earlier transition would be chosen instead. Similarly, if the previous code ends in 01 or 10, the first bit of the next code must be the same as the last bit of the previous code or the minimum run-length limit will be violated. The run-length limits can always be met because every code has an inverse, so out of the four possible channel symbols available for a given data byte two of them will begin with 0 and two with 1. In some cases, such as where the first code ends in 1100, up to four of the candidates for the next code could meet the run-length limits. In this case the best candidate will be chosen to optimize some other parameter.

In order to follow how the encoder selects the best channel pattern it is necessary to discuss the criteria that are used. There are a number of these, some of which are compulsory such as the run-length limits, and others which will be met if possible on a decreasing scale of importance. If a higher criterion cannot be met the decision will still attempt to meet as many of the lower ones as possible.

The overall goal is to meet the run-length limits with a sequence of symbols which has the highest clock content, lowest LF and DC content and the least asymmetry to reduce peak shift. Some channel symbols will be better than others, a phenomenon known as pattern sensitivity. The least optimal patterns are not so bad that they will *cause* errors, but they will be more prone to errors due to other causes. Minimizing the use of sensitive waveforms will enhance the data reliability and is as good as an improvement in the signal-to-noise ratio.

As there are 2^{10} different patterns, there will be 2^{20} different combinations of two patterns. Clearly it is out of the question to create a lookup table to determine how best to merge two patterns. It has to be done algorithmically. The flowchart of the algorithm is shown in Figure 8.35.

The process begins when a data byte selects four candidate channel symbols. Initially tests are made to eliminate unsuitable patterns, and on some occasions, only one code will emerge from these tests. There are so many combinations, however, that it is possible for several candidate codes to pass the initial tests, and so the flowchart continues to select the best code by further criteria.

The first selection is based on maintaining the run-length limits at the junction between the previous code and the current one and keeping the End DSV as small as possible. If two or more codes give an equal minimum End DSV, the code with the smallest Absolute DSV will be selected. If this criterion still results in more than one code being available, a further selection is made to optimize the spectrum due to the junction of the previous and current symbols. The codes are tested to see if any permit six or fewer identical channel bits across the junction. If none of the codes has this characteristic there will be a large run length at the junction and the best code will be one that has a run length of less than six bits later. This criterion will also be applied if more than one code passes the junction run-length test.

If a decision still cannot be made, the remaining candidates could have ±2 End DSV or 0 End DSV. In the former case the priority tables are invoked, and the code(s) with the highest priority is selected. In the latter case, or if two codes emerge equal from the priority test, any code(s) having a run-length of less than six bits at the end is selected. This makes

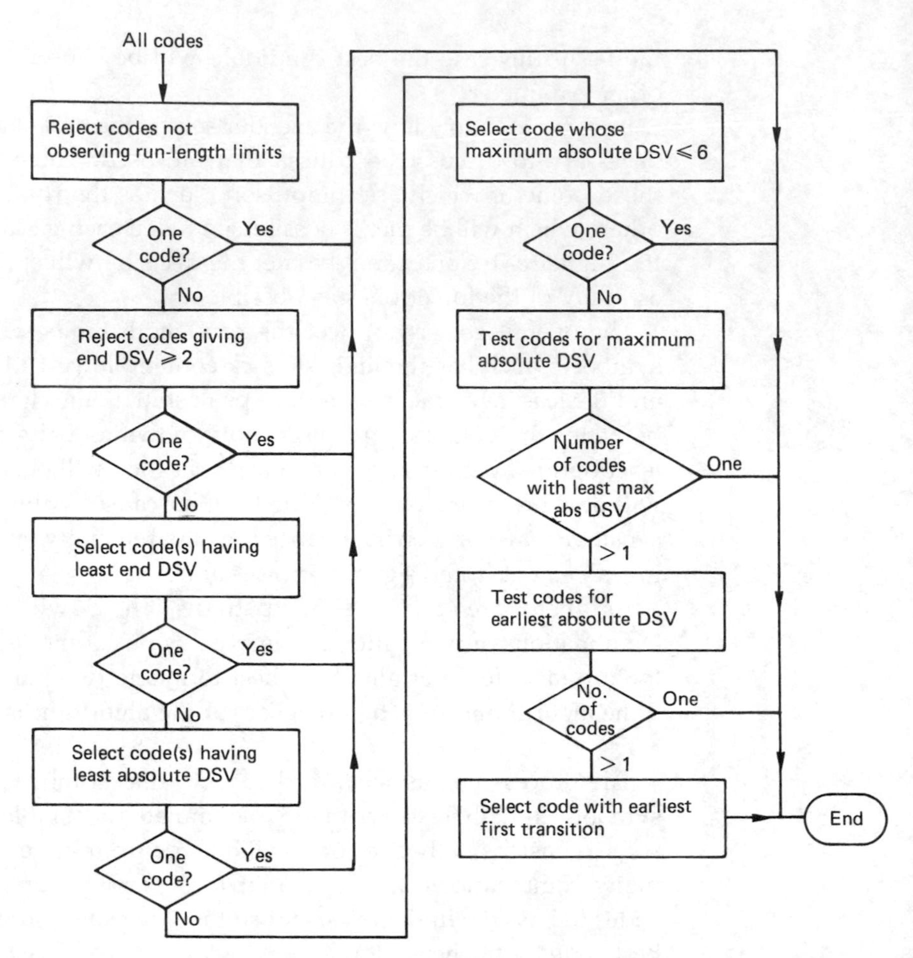

Figure 8.35 The four candidate codes, A,B,C,D, of Figure 8.30 are compared according to this flow chart to locate the single best code which will represent the data byte. Any of the codes will result in the same decode on replay.

merging with the next symbol easier. It is possible that no code will pass the end run-length test, or that two or more codes are still equal. In this case some of the criteria cannot be met or are equally met, and the final choice reverts to a further selection of the code with minimum Absolute DSV. This test was made much earlier, but some codes with equal minimum values could have ben rejected in intermediate tests. If two codes still remain, the one whose Absolute DSV appears earliest in the bitstream will be selected. If two codes still remain, the one with the earliest transition is selected.

The complexity of the coding rules in 8,14 is such that it could not have been economically implemented until recently. This illustrates the dependence of advanced recorders on LSI technology.

8.14 EFM code of CD and DVD

The original EFM code of CD is described here. Slight modifications were made to the code to improve its error resistance for the recording of MPEG data in DVD. This resulted in the EFM Plus code.

Figure 8.36 shows that in the 8,14 code (EFM), eight-bit symbols are represented by fourteen-bit channel symbols.[15] There are 256 combina-

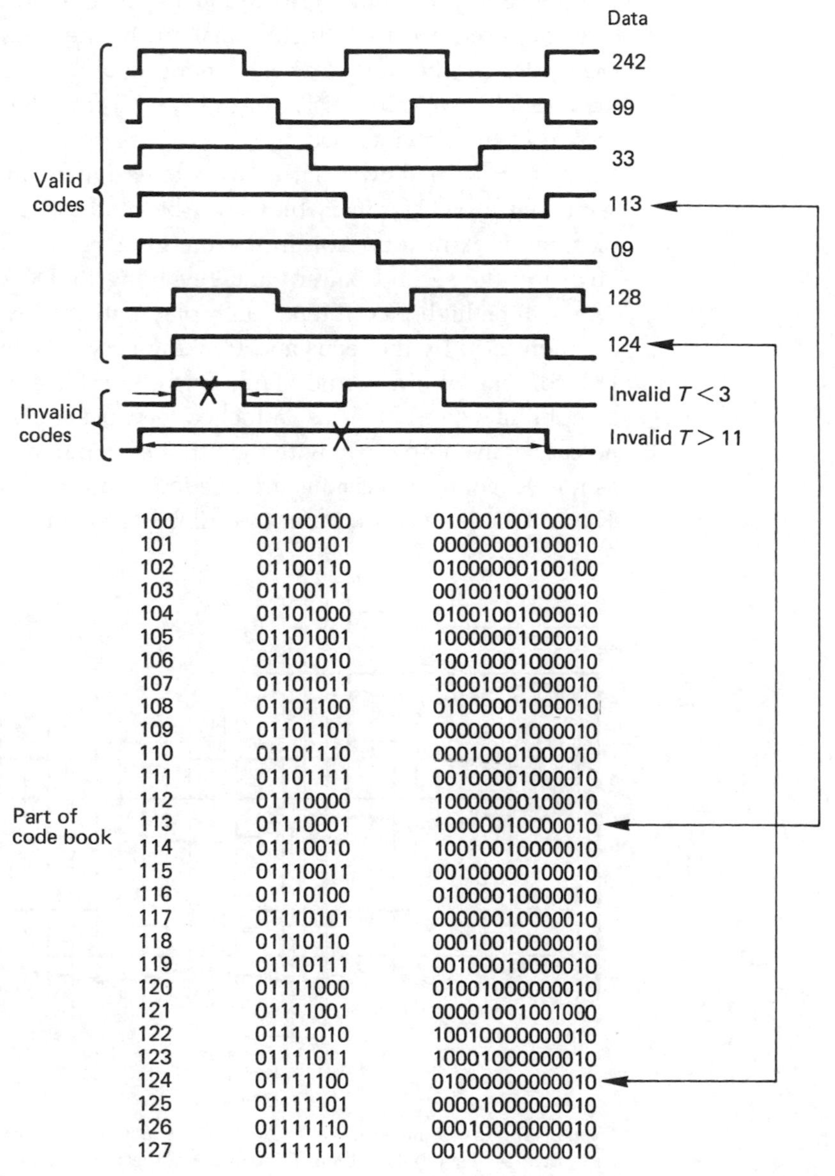

Figure 8.36 EFM code: $d = 2$, $k = 10$. Eight data bits produce fourteen channel bits plus three packing bits. Code rate is 8/17. DR = $(3 \times 8)/17 = 1.41$.

tions of eight data bits, whereas fourteen bits have 16 K combinations. Of these only 267 satisfy the criteria that the maximum run-length shall not exceed eleven channel bits (k = 10) nor be less than three channel bits (d = 2). A section of the codebook is shown in the figure. In fact 258 of the the 267 possible codes are used because two unique patterns are used to synchronize the subcode blocks (see Chapter 12). It is not possible to prevent violations between adjacent symbols by substitution, and extra merging bits having no data meaning are placed between the symbols. Two merging bits would be adequate to prevent violations, but in practice three are used because a further task of the merging bits is to control the DC content of the waveform. The merging bits are selected by computing the Digital Sum Value (DSV) of the waveform. The DSV is computed as shown in the previous section.

Figure 8.37(b) shows that if two successive channel symbols have the same sense of DC offset, these can be made to cancel one another by placing an extra transition in the merging period. This has the effect of inverting the second pattern and reversing its DC content. The DC-free code can be high-pass filtered on replay and the lower-frequency signals are then used by the focus and tracking servos without noise due to the DC content of the audio data. Encoding EFM is complex, but was acceptable when CD was launched because only a few encoders are necessary in comparison with the number of players. Decoding is simpler as no DC content decisions are needed and a lookup table can be used. The codebook was computer optimized to permit the implementation of

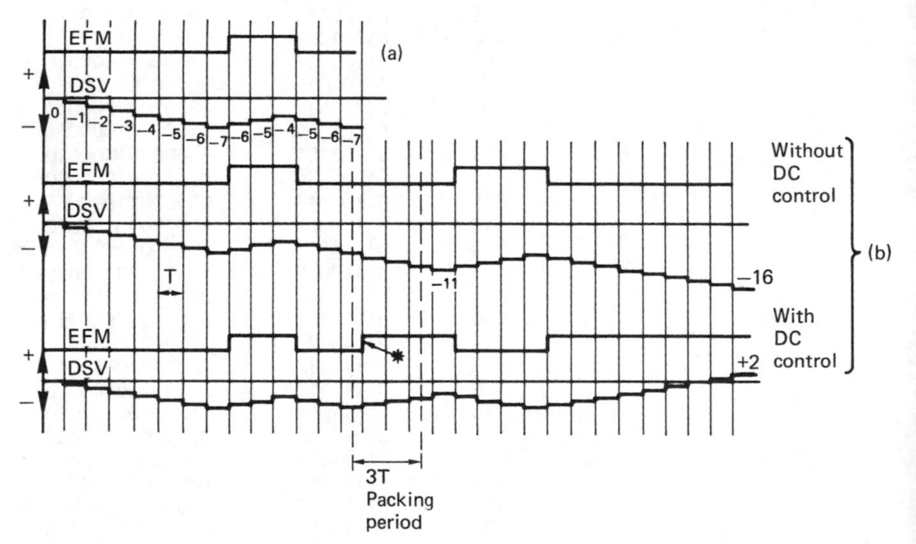

Figure 8.37 (a) Digital sum value example calculated from EFM waveform. (b) Two successive 14 T symbols without DC control (upper) give DSV of –16. Additional transition (*) results in DSV of +2, anticipating negative content of next symbol.

a programmable logic array (PLA) decoder with the minimum complexity.

Due to the inclusion of merging bits, the code rate becomes 8/17, and the density ratio becomes:

$$\frac{3 \times 8}{17} = 1.41$$

and the FoM is:

$$\frac{3 \times 8^2}{17^2} = 0.66$$

The code is thus a 2, 10, 8, 17, r system where r has meaning only in the context of DC control.[16] The constraints d and k can still be met with $r = 1$ because of the merging bits. The figure of merit is less useful for optical media because the straight-line frequency response does not produce peak shift and the rigid, non-contact medium has good speed stability. The density ratio and the freedom from DC are the most important factors.

8.15 Error detection in group codes

In the 8, 14 code only 4.7 per cent of the fourteen-bit patterns are actually recorded, since the others have undesirable characteristics such as excessive DC content or insufficient clock content. In practice, reading errors will occur which can corrupt some or all of the channel bits in a symbol. Random noise effects or peak shift could result in a tape transition being shifted along the time axis, or in the wrong number of transitions being detected in a symbol. This will change the channel-bit pattern determined by the data separator.

As there are many more channel-bit combinations than those which are actually used, it is probable that a random error will convert a channel symbol into one of the patterns which are not used. Thus it is possible for the lookup table which decodes fourteen channel bits back to eight data bits to perform an error-detecting function. When an illegal fourteen-bit code is detected, the lookup table will output an error flag. Clearly this method is not infallible, as it is possible for an error to convert one valid code into a different valid code, which this scheme would not detect. However, the additional detection capability can be used to enhance the power of the error-correction systems which can make use of the error flags produced by the 14,8 decoder. This subject will be discussed in more detail in Chapter 9.

8.16 Tracking signals

Many recorders use track-following systems to help keep the head(s) aligned with the narrow tracks used in digital media. These can operate by sensing low-frequency tones which are recorded along with the data. Whilst this can be done by linearly adding the tones to the coder output, this requires a linear record amplifier. An alternative is to use the DC content of group codes. A code is devised where for each data pattern, several code patterns exist having a range of DC components. By choosing groups with a suitable sequence of DC offsets, a low frequency can be added to the coded signal. This can be filtered from the data waveform on replay.

8.17 Randomizing

Randomizing is not a channel code, but a technique which can be used in conjunction with a channel code for various purposes. Many channel codes have the characteristic that certain data bit patterns suffer from a higher probability of error than others. If such a pattern occurs frequently in the data, the error rate will rise. Randomizing ensures that this cannot happen so that the error rate is largely independent of the data. Figure 8.38 shows that the randomizing system is arranged outside the channel coder.

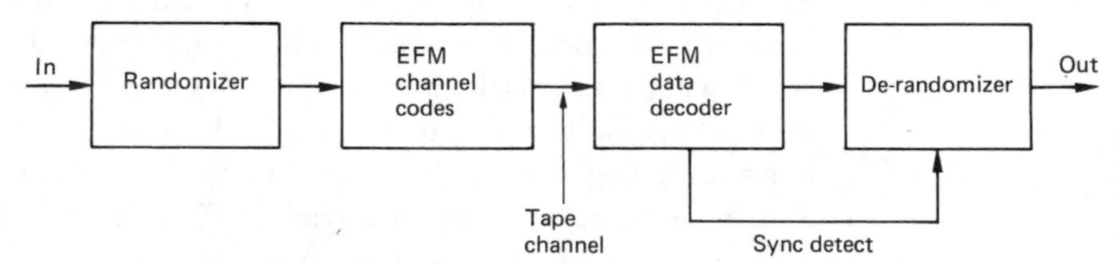

Figure 8.38 The randomizing system of D-3 is arranged outside the EFM channel coder. The pseudo-random sequence added during recording must be generated in a synchronized manner during replay so it can be subtracted. This is done by the EFM sync detector. Note that the sync pattern is not randomized!

In digital transmission over a radio path randomizing can be used to ensure that the transmitted spectrum remains uniform and independent of the data. This technique is called *energy dispersal* and will be considered in Chapter 14.

It will be appreciated that if the data were truly randomized it could not be recovered. Strictly speaking, the term should be pseudo-randomizing. A pseudo-random process is one which is sufficiently random to have the desired effect on the data spectrum, whilst being sufficiently deterministic that the process can be reversed to recover the data.

In a DVTR, the randomizing process would be standardized as part of the format and all players would be able to de-randomize any compliant tape. However, if the randomizing process were made variable, only those in possession of the specific code in use could recover the data. This forms the basis of encryption. In a pay-per-view system, only subscribers who are entitled to view the data would be given the code.

The NRZ code has a DR of 1 and a jitter window of 1 and so has a FoM of 1 which is better than the group codes. It does, however, suffer from an unconstrained spectrum and poor clock content. This can be overcome using randomizing. Figure 8.39 shows that, at the encoder, a pseudo-random sequence (see Chapter 3) is added modulo-2 to the serial data and the resulting ones generate transitions in the channel. This process drastically reduces T_{max} and reduces DC content. At the receiver the transitions are converted back to a serial bitstream to which the same pseudo-random sequence is again added modulo-2. As a result the random signal cancels itself out to leave only the serial data, provided that the two pseudo-random sequences are synchronized to bit accuracy.

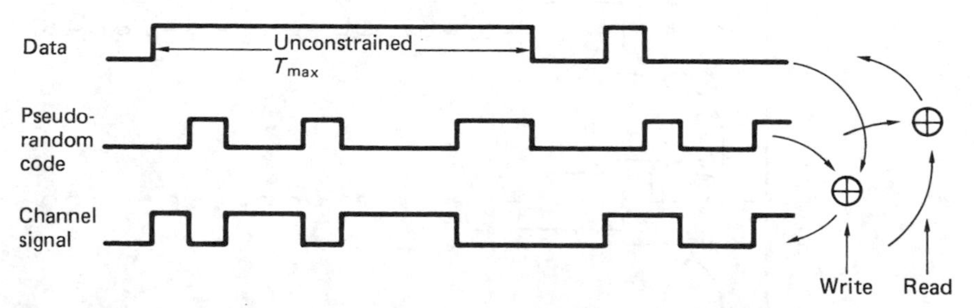

Figure 8.39 Modulo-2 addition with a pseudo-random code removes unconstrained runs in real data. Identical process must be provided on replay.

The 8, 14 code displays pattern sensitivity because some waveforms are more sensitive to peak shift distortion than others. Pattern sensitivity is only a problem if a sustained series of sensitive symbols needs to be recorded. Randomizing ensures that this cannot happen because it breaks up any regularity or repetition in the data. The data randomizing is performed by using the exclusive-OR function of the data and a pseudo-random sequence as the input to the channel coder. On replay the same sequence is generated, synchronized to bit accuracy, and the exclusive-OR of the replay bitstream and the sequence is the original data.

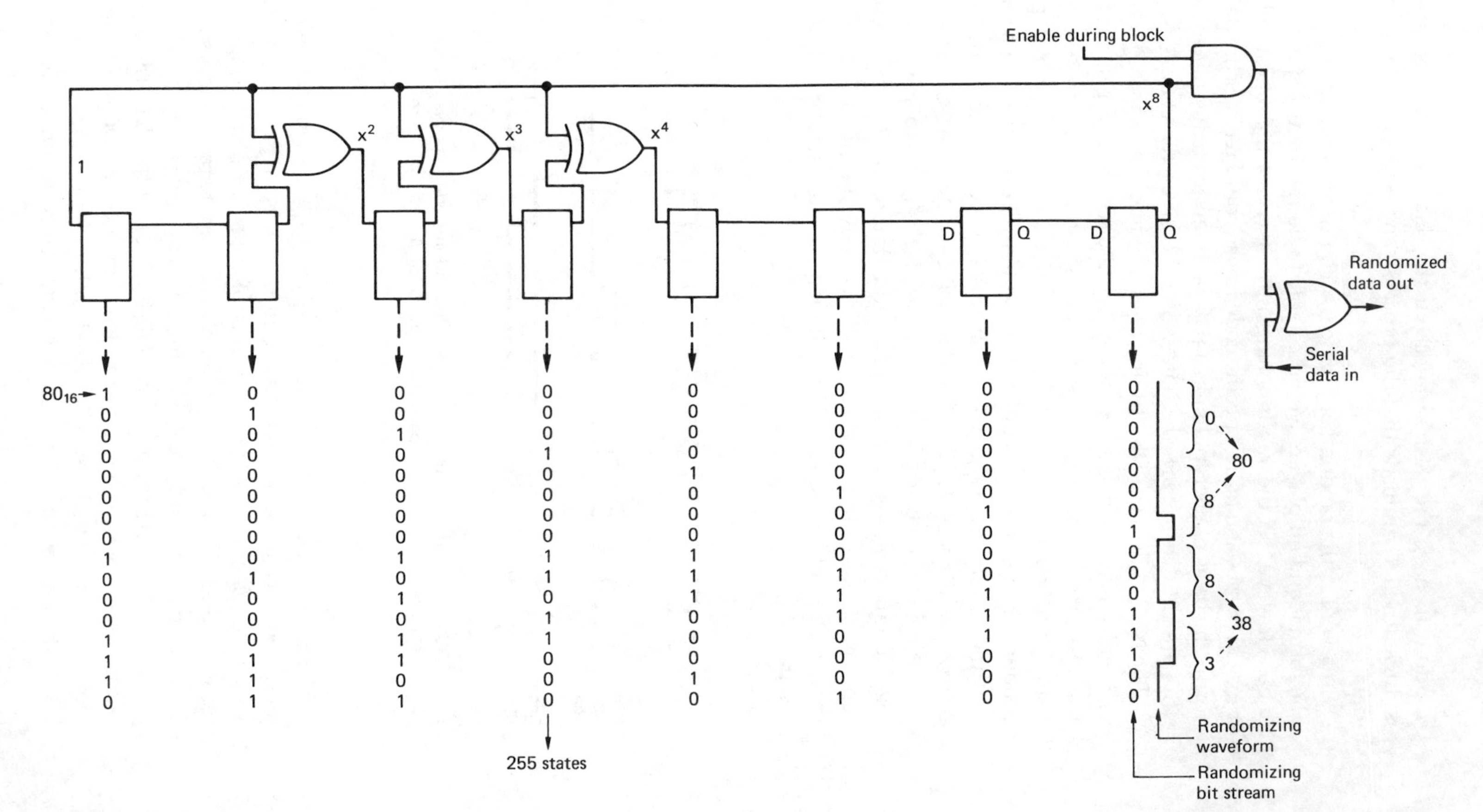

Figure 8.40 The polynomial generator circuit shown here calculates $x^8 + x^4 + x^3 + x^2 + 1$ and is preset to 80_{16} at the beginning of every sync block. When the generator is clocked it will produce a Galois field having 255 states (see Chapter 3). The right-hand bit of each field element becomes the randomizing bitstream and is fed to an exclusive OR gate in the data stream. Randomizing is disabled during preambles and sync patterns. It is also possible to randomize using a counter working at byte rate which addresses a PROM. Eight exclusive OR gates will then be needed to randomize one byte at a time.

80	38	D2	81	49	76	82	DA	9A	86	6F	AF	8B	B0	F1	9C
D1	12	A5	72	37	EF	97	59	31	B8	EA	53	C8	3F	F4	58
40	1C	E9	C0	24	38	41	6D	4D	C3	B7	D7	45	D8	78	CE
68	89	52	B9	9B	F7	CB	AC	18	5C	F5	29	E4	1F	7A	2C
20	8E	74	60	92	9D	A0	B6	A6	E1	DB	EB	22	6C	3C	67
B4	44	A9	DC	CD	FB	65	56	0C	AE	FA	14	F2	0F	3D	16
10	47	3A	30	C9	4E	50	5B	D3	F0	ED	75	11	36	9E	33
5A	A2	54	EE	E6	FD	32	2B	06	57	7D	0A	F9	87	1E	0B
88	23	1D	98	64	27	A8	AD	69	F8	F6	BA	08	1B	CF	19
2D	51	2A	77	F3	7E	99	15	83	AB	3E	85	FC	43	8F	05
C4	91	0E	4C	B2	13	D4	D6	34	7C	7B	5D	84	8D	E7	8C
96	28	95	BB	79	BF	CC	8A	C1	55	9F	42	FE	A1	C7	02
E2	48	07	26	D9	09	6A	6B	1A	BE	BD	2E	C2	C6	73	46
4B	94	CA	DD	BC	5F	66	C5	E0	AA	4F	21	FF	D0	63	01
71	A4	03	93	EC	04	B5	35	0D	DF	5E	17	61	E3	39	A3
25	4A	E5	6E	DE	2F	B3	62	70	D5	A7	90	7F	E8	B1	

Figure 8.41 The sequence which results when the randomizer of Figure 8.40 is allowed to run.

The randomizing polynomial and one way in which it can be implemented are shown in Figure 8.40. As the polynomial generates a maximum length sequence from an eight-bit wordlength, the sequence length is given by $2^8 - 1 = 255$. The sequence would repeat endlessly but for the fact that it is pre-set to 80_{16} at the beginning of each sync block immediately after the sync pattern is detected. Figure 8.41 shows the randomizing sequence starting from 80_{16}.

Clearly the sync pattern cannot be randomized, since this causes a Catch-22 situation where it is not possible to synchronize the sequence for replay until the sync pattern is read, but it is not possible to read the sync pattern until the sequence is synchronized!

The randomizing in D-3 is clearly block based, since this matches the block structure on tape. Where there is no obvious block structure, convolutional, or endless randomizing can be used. This is the approach used in the SDI digital video interconnect described in Chapter 10, which allows composite or component video of up to ten-bit wordlength to be sent serially.

8.18 Partial response

It has been stated that a magnetic head acts as a transversal filter, because it has two poles. In addition the output is differentiated, so that the head may be thought of as a $(1 - D)$ impulse response system, where D is the delay which is a function of the tape speed and gap size. It is this delay which results in intersymbol interference. Conventional equalizers attempt to oppose this effect, and succeed in raising the noise level in the process of making the frequency response linear. Figure 8.42 shows that the frequency response necessary to pass data with insignificant peak

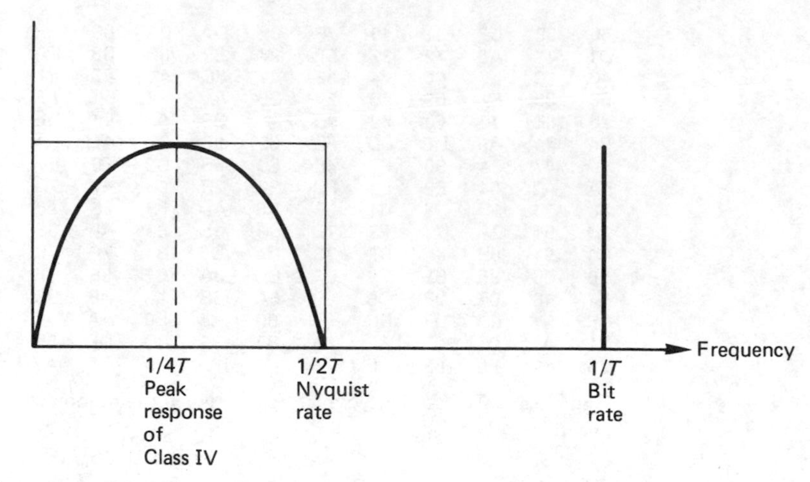

Figure 8.42 Class IV response has spectral nulls at DC and the Nyquist rate, giving a noise advantage, since magnetic replay signal is weak at both frequencies in a high-density channel.

shift is a bandwidth of half the bit rate, which is the Nyquist rate. In Class IV partial response, the frequency response of the system is made to have nulls at DC and at the Nyquist rate. Such a frequency response is particularly advantageous for rotary head recorders as it is DC-free and the low-frequency content is minimal, hence the use in Digital Betacam. The required response is achieved by an overall impulse response of $(1 - D^2)$ where D is now the bit period. There are a number of ways in which this can be done.

If the head gap is made equal to one bit, the $(1 - D)$ head response may be converted to the desired response by the use of a $(1 + D)$ filter, as in Figure 8.43(a).[17] Alternatively, a head of unspecified gapwidth may be connected to an integrator, and equalized flat to reproduce the record current waveform before being fed to a $(1 - D^2)$ filter as in Figure 8.43(b).[18]

The result of both of these techniques is a ternary signal. The eye pattern has two sets of eyes as in Figure 8.43(c).[19] When slicing such a signal, a smaller amount of noise will cause an error than in the binary case.

The treatment of the signal thus far represents an equalization technique, and not a channel code. However, to take full advantage of Class IV partial response, suitable precoding is necessary prior to recording, which does then constitute a channel-coding technique. This precoding is shown in Figure 8.44(a). Data are added modulo-2 to themselves with a two-bit delay. The effect of this precoding is that the outer levels of the ternary signals, which represent data ones, alternate in polarity on all odd bits and on all even bits. This is because the precoder

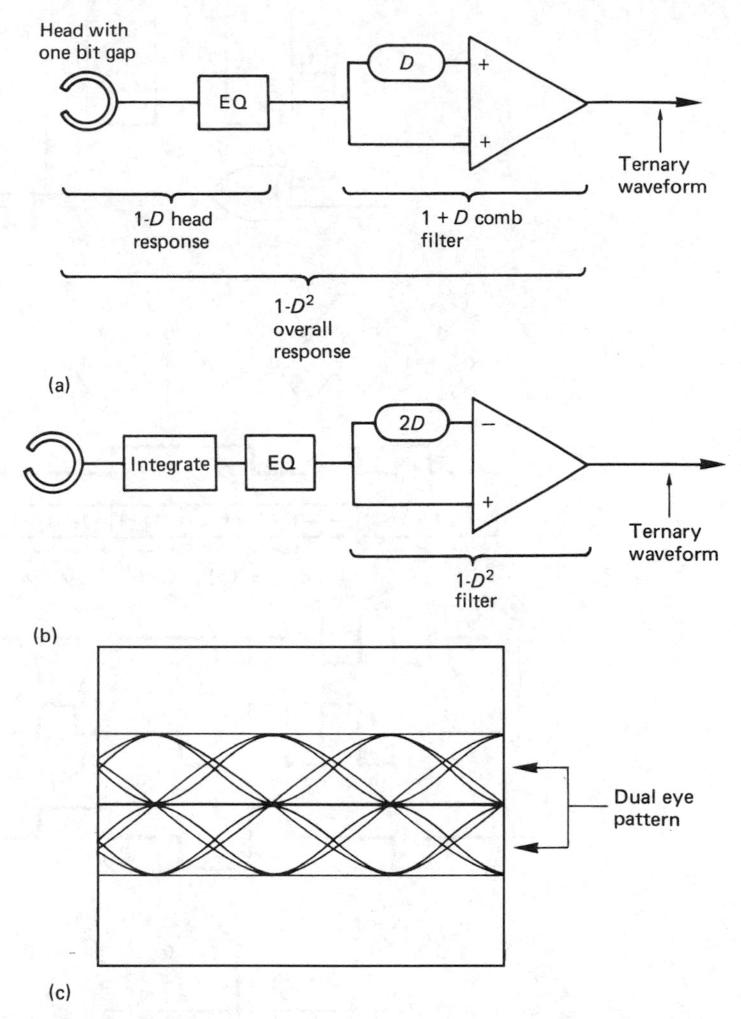

Figure 8.43 (a), (b) Two ways of obtaining partial response. (c) Characteristic eye pattern of ternary signal.

acts like two interleaved one-bit delay circuits, as in (b). As this alternation of polarity is a form of redundancy, it can be used to recover the 3-dB SNR loss encountered in slicing a ternary eye pattern. Viterbi decoding[20] can be used for this purpose. In Viterbi decoding, each channel bit is not sliced individually; the slicing decision is made in the context of adjacent decisions. Figure 8.45 shows a replay waveform which is so noisy that, at the decision point, the signal voltage crosses the centre of the eye, and the slicer alone cannot tell whether the correct decision is an inner or an outer level. In this case, the decoder essentially allows both decisions to stand, in order to see what happens. A symbol representing indecision is output. It will be seen from the figure that as subsequent bits are received, one of these decisions will result in an absurd situation,

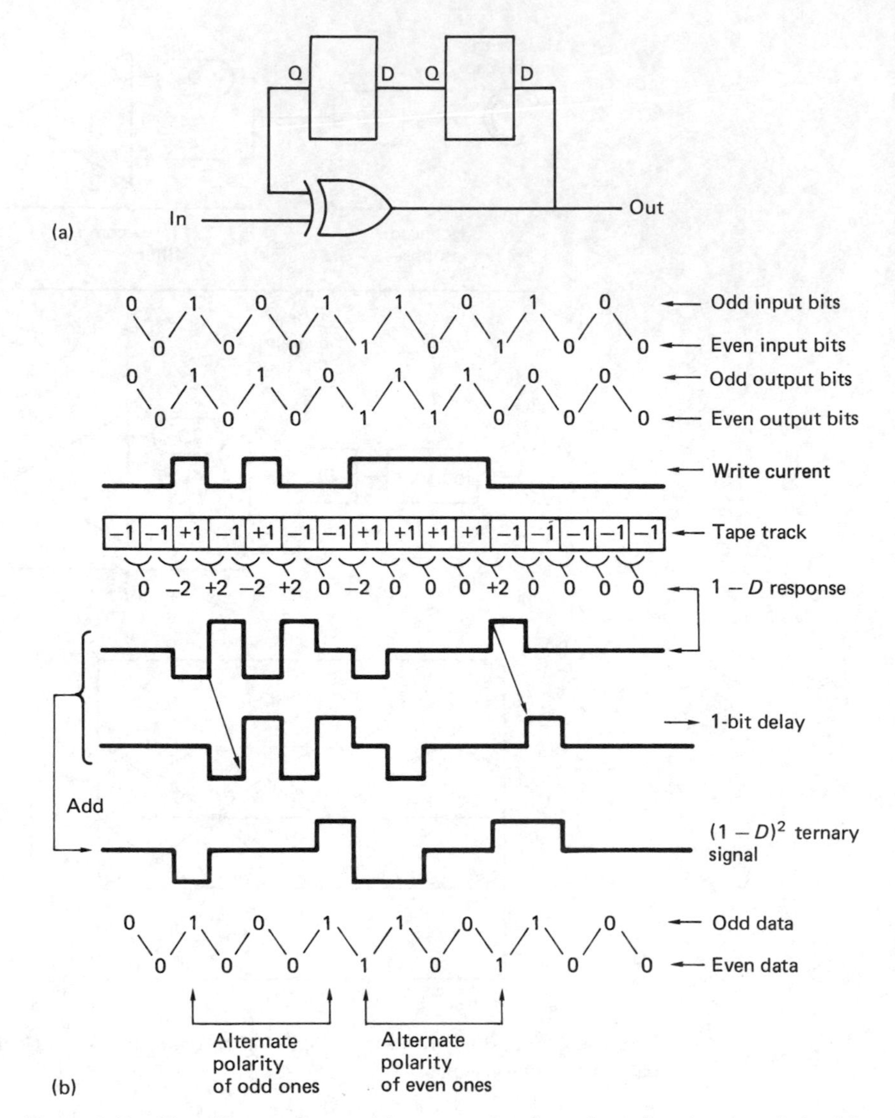

Figure 8.44 Class IV precoding at (a) causes redundancy in replay signal as derived in (b).

which indicates that the other decision was the right one. The decoder can then locate the undecided symbol and set it to the correct value.

Viterbi decoding requires more information about the signal voltage than a simple binary slicer can discern. Figure 8.46 shows that the replay waveform is sampled and quantized so that it can be processed in digital logic. The sampling rate is obtained from the embedded clock content of the replay waveform. The digital Viterbi processing logic must be able to operate at high speed to handle serial signals from a DVTR head. Its application in Digital Betacam is eased somewhat by

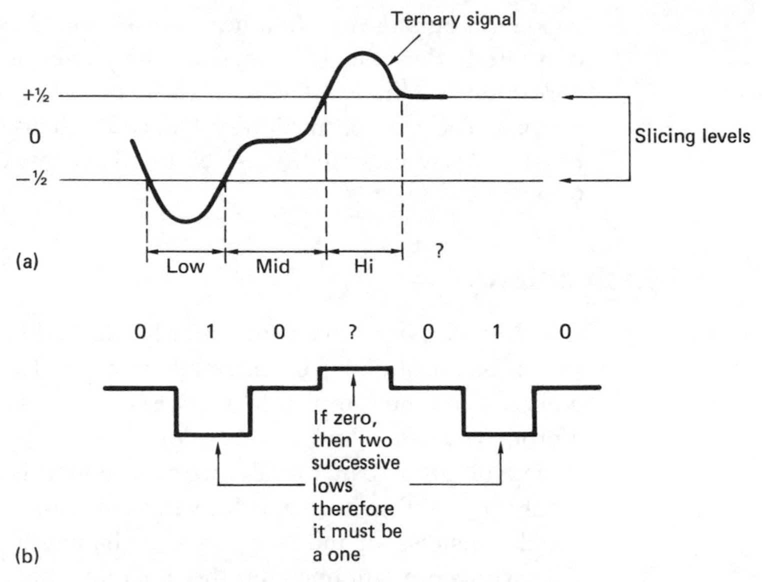

Figure 8.45 (a) A ternary signal suffers a noise penalty because there are two slicing levels. (b) The redundancy is used to determine the bit value in the presence of noise. Here the pulse height has been reduced to make it ambiguous 1/0, but only 1 is valid as zero violates the redundancy rules.

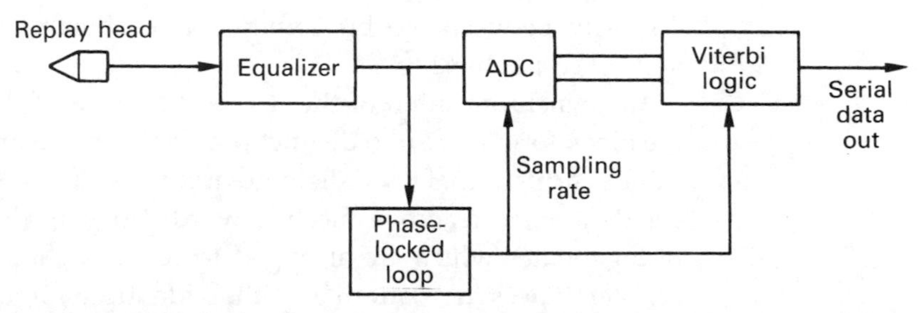

Figure 8.46 A Viterbi decoder is implemented in the digital domain by sampling the replay waveform with a clock locked to the embedded clock of the channel code.

the adoption of compression which reduces the data rate at the heads by a factor of two.

Clearly a ternary signal having a dual eye pattern is more sensitive than a binary signal, and it is important to keep the maximum run length T_{max} small in order to have accurate AGC. The use of pseudo-random coding along with partial response equalization and precoding is a logical combination.[21]

There is then considerable overlap between the channel code and the error-correction system. Viterbi decoding is primarily applicable to channels with random errors due to Gaussian statistics, and they cannot cope with burst errors. In a head-noise-limited system, however, the use of a Viterbi detector could increase the power of an separate burst error-correction system by relieving it of the need to correct random errors due to noise. The error-correction system could then concentrate on correcting burst errors unimpaired. This point will become clearer upon referring to Chapter 9.

8.19 Synchronizing

Once the PLL in the data separator has locked to the clock content of the transmission, a serial channel bitstream and a channel bit clock will emerge from the sampler. In a group code, it is essential to know where a group of channel bits begins in order to assemble groups for decoding to data bit groups. In a randomizing system it is equally vital to know at what point in the serial data stream the words or samples commence. In serial transmission and in recording, channel bit groups or randomized data words are sent one after the other, one bit at a time, with no spaces in between, so that although the designer knows that a data block contains, say, 128 bytes, the receiver simply finds 1024 bits in a row. If the exact position of the first bit is not known, then it is not possible to put all the bits in the right places in the right bytes; a process known as deserializing. The effect of sync slippage is devastating, because a one-bit disparity between the bit count and the bitstream will corrupt every symbol in the block.[22]

The synchronization of the data separator and the synchronization to the block format are two distinct problems, which are often solved by the same sync pattern. Deserializing requires a shift register which is fed with serial data and read out once per word. The sync detector is simply a set of logic gates which are arranged to recognize a specific pattern in the register. The sync pattern is either identical for every block or has a restricted number of versions and it will be recognized by the replay circuitry and used to reset the bit count through the block. Then by counting channel bits and dividing by the group size, groups can be deserialized and decoded to data groups. In a randomized system, the pseudo-random sequence generator is also reset. Then counting derandomized bits from the sync pattern and dividing by the wordlength enables the replay circuitry to deserialize the data words.

In digital audio the two's complement coding scheme is universal and traditionally no codes have been reserved for synchronizing; they are all available for sample values. It would in any case be impossible to reserve all ones or all zeros as these are in the centre of the range in two's complement. Even if a specific code were excluded from the recorded data so that it could be used for synchronizing, this cannot ensure that the same pattern cannot be falsely created at the junction between two allowable

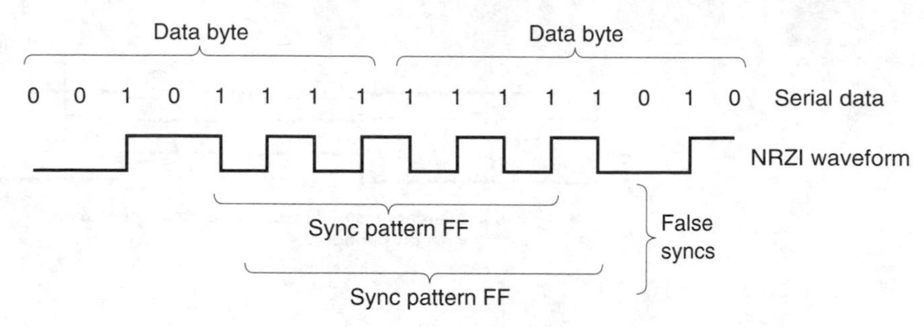

Figure 8.47 Concatenation of two words can result in the accidental generation of a word which is reserved for synchronizing.

data words. Figure 8.47 shows how false synchronizing can occur due to concatenation. It is thus not practical to use a bit pattern which is a data code value in a simple synchronizing recognizer. The problem is overcome in NICAM 728 by using the fact that sync patterns occur exactly once per millisecond or 728 bits. The sync pattern of NICAM 728 is just a bit pattern and no steps are taken to prevent it from appearing in the randomized data. If the pattern is seen by the recognizer, the recognizer is disabled for the rest of the frame and only enabled when the next sync pattern is expected. If the same pattern recurs every millisecond, a genuine sync condition exists. If it does not, there is a false sync and the recognizer will be enabled again. As a result it will take a few milliseconds before sync is achieved, but once achieved it should not be lost unless the transmission is interrupted. This is fine for the application and no-one objects to the short mute of the NICAM sound during a channel switch. The principle cannot, however, be used for recording because channel interruptions are more frequent due to head switches and dropouts and loss of several blocks of data due to a single dropout is unacceptable.

In run-length-limited codes this is not a problem. The sync pattern is no longer a data bit pattern but is a specific waveform. If the sync waveform contains run lengths which violate the normal coding limits, there is no way that these run lengths can occur in encoded data, nor any possibility that they will be interpreted as data. They can, however, be readily detected by the replay circuitry. The sync patterns of the AES/EBU interface are shown in Figure 8.48. It was seen from Figure 8.25 that the maximum run length in FM coded data is one bit. The sync pattern begins with a run length of one and a half bits which is unique. There are three types of sync pattern in the AES/EBU interface, as will be seen in Chapter 10. These are distinguished by the position of a second pulse after the run-length violation. Note that the sync patterns are also DC-free like the FM code.

In a group code there are many more combinations of channel bits than there are combinations of data bits. Thus after all data bit patterns have

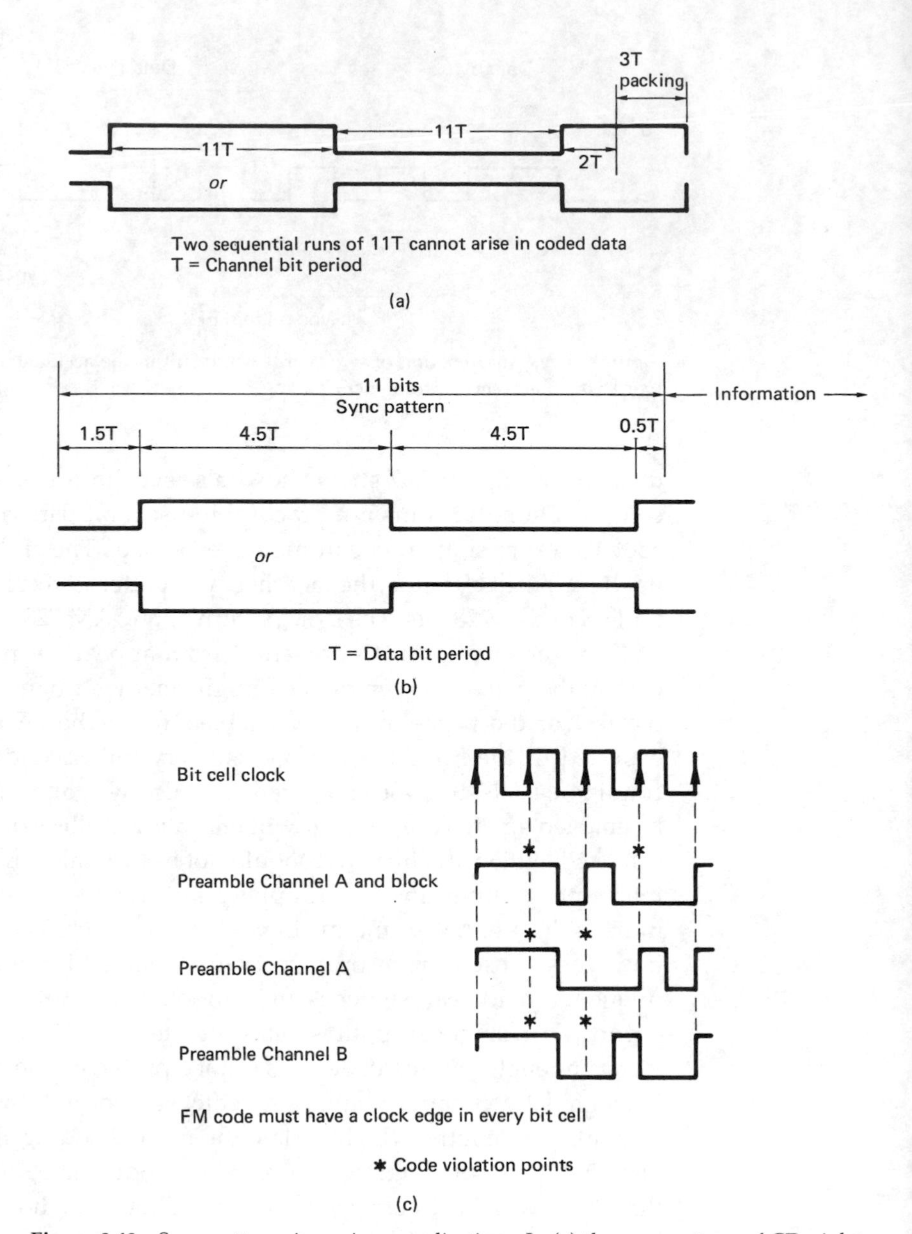

Figure 8.48 Sync patterns in various applications. In (a) the sync pattern of CD violates EFM coding rules, and is uniquely identifiable. In (b) the sync pattern of DASH stays within the run length of HDM-1. In (c) the sync patterns of AES/EBU interconnect.

been allocated group patterns, there are still many unused group patterns which cannot occur in the data. With care, group patterns can be found which cannot occur due to the concatenation of any pair of groups representing data. These are then unique and can be used for synchronizing.

References

1. Deeley, E.M., Integrating and differentiating channels in digital tape recording. *Radio Electron. Eng.*, **56**, 169–173 (1986)
2. Mee, C.D., *The Physics of Magnetic Recording*, Amsterdam and New York: Elsevier–North Holland Publishing (1978)
3. Jacoby, G.V., Signal equalization in digital magnetic recording. *IEEE Trans. Magn.*, **MAG-11**, 302–305 (1975)
4. Schneider, R.C., An improved pulse-slimming method for magnetic recording. *IEEE Trans. Magn.*, **MAG-11**, 1240–1241 (1975)
5. Miller, A., US Patent No.3, 108,261
6. Mallinson, J.C. and Miller, J.W., Optimum codes for digital magnetic recording. *Radio and Electron. Eng.*, **47**, 172–176 (1977)
7. Miller, J.W., DC-free encoding for data transmission system. US Patent No. 4,027,335 (1977)
8. Tang, D.T., Run-length-limited codes. IEEE International Symposium on Information Theory (1969)
9. Cohn, M. and Jacoby, G., Run-length reduction of 3PM code via lookahead technique. *IEEE Trans. Magn.*, **18**, 1253–1255 (1982)
10. Horiguchi, T. and Morita, K., On optimisation of modulation codes in digital recording. *IEEE Trans. Magn.*, **12**, 740–742 (1976)
11. Franaszek, P.A., Sequence state methods for run-length linited coding. *IBM J. Res. Dev.*, **14**, 376–383 (1970)
12. Jacoby, G.V. and Kost, R., Binary two-thirds-rate code with full word lookahead. *IEEE Trans. Magn.*, **20**, 709–714 (1984)
13. Jacoby, G.V., A new lookahead code for increased data density. *IEEE Trans. Magn.*, **13**, 1202–1204 (1977)
14. Uehara, T., Nakayama, T., Minaguchi, H., Shibaya, H., Sekiguchi, T., Oba, Y., A new 8–14 modulation and its application to small format VTR. SMPTE Tech. Conf. (1989)
15. Ogawa, H. and Schouhamer Immink, K.A., EFM – the modulation method for the Compact Disc digital audio system. In B. Blesser, B. Locanthi and T.G. Stockham Jr (eds), *Digital Audio*, pp. 117–124. New York: Audio Engineering Society (1983)
16. Schouhamer Immink, K.A. and Gross, U., Optimisation of low frequency properties of eight-to-fourteen modulation. *Radio Electron. Eng.*, **53**, 63–66 (1983)
17. Yokoyama, K., Digital video tape recorder. *NHK Technical Monograph*, No.31 (March 1982)
18. Coleman, C.H. *et al.*, High data rate magnetic recording in a single channel. *J. IERE*, **55**, 229–236 (1985)
19. Kobayashi, H., Application of partial response channel coding to magnetic recording systems. *IBM J. Res. Dev.*, **14**, 368–375 (1970)
20. Forney, G.D. Jr, The Viterbi algorithm, *Proc. IEEE*, **61**, 268–278 (1973)
21. Wood, R.W. and Petersen, D.A., Viterbi detection of Class IV partial response on a magnetic recording channel. *IEEE Trans. Commun.*, **34**, 454–461 (1968)
22. Griffiths, F.A., A digital audio recording system. Presented at 65th Audio Engineering Society Convention (London 1980), Preprint 1580(C1)

9

Error correction

9.1 Sensitivity of message to error

Before attempting to specify an error-correction system, the causes of errors must be studied to quantify the problem, and the sensitivity of the destination to errors must be assessed. In video and audio the sensitivity to errors must be subjective. Figure 9.1 shows the relative sensitivity of different types of data to error.

Whilst the exact BER (bit error rate) which can be tolerated will depend on the application, digital audio is less tolerant of errors than digital video and more tolerant than computer data. The use of compression changes the picture if redundancy has been removed from a signal, and it becomes less tolerant of errors.

In PCM audio and video, the effect of a single bit in error depends upon the significance of the bit. If the least significant bit of a sample is wrong,

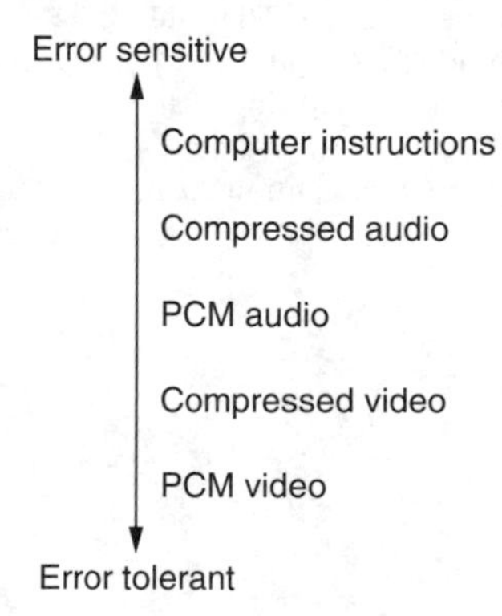

Figure 9.1 Data vary in their tolerance to error. Compressed data are more sensitive to error.

the chances are that the effect will be lost in the noise. Conversely, if a high-order bit is in error, a massive transient will be added to the waveform.

MPEG video compression uses variable-length coding. If a bit error occurs it may cause the length of a symbol incorrectly to be assessed and the decoder will lose synchronism with the bitstream. This may cause an error affecting a significant area of the picture which might propagate from one picture to the next.

If the maximum error rate which the destination can tolerate is likely to be exceeded by the unaided channel, some form of error handling will be necessary.

There are a number of terms which have idiomatic meanings in error correction. The raw BER is the error rate of the medium, whereas the residual or uncorrected BER is the rate at which the error-correction system fails to detect or miscorrects errors. In practical digital systems, the residual BER is negligibly small. If the error correction is turned off, the two figures become the same.

It is paramount in all error-correction systems that the protection used should be appropriate for the probability of errors to be encountered. An inadequate error-correction system is actually worse than not having any correction. Error correction works by trading probabilities. Error-free performance with a certain error rate is achieved at the expense of performance at higher error rates. Figure 9.2 shows the effect of an error-correction system on the residual BER for a given raw BER. It will be seen that there is a characteristic knee in the graph. If the expected raw BER has been misjudged, the consequences can be disastrous. Another result demonstrated by the example is that we can only guarantee to detect the same number of bits in error as there are redundant bits.

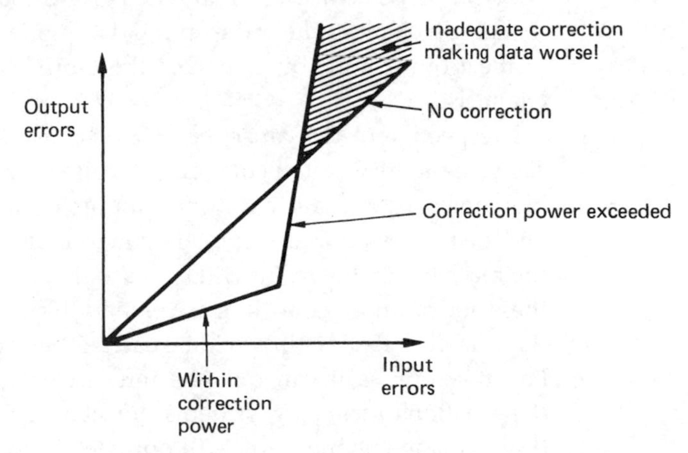

Figure 9.2 An error-correction system can only reduce errors at normal error rates at the expense of increasing errors at higher rates. It is most important to keep a system working to the left of the knee in the graph.

9.2 Error mechanisms

There are many different types of recording and transmission channel and consequently there will be many different mechanisms which may result in errors. As was the case for channel coding, although there are many different applications, the basic principles remain the same.

In magnetic recording, data can be corrupted by mechanical problems such as media dropout and poor tracking or head contact, or Gaussian thermal noise in replay circuits and heads. In optical recording, contamination of the medium interrupts the light beam. Warped disks and birefringent pressings cause defocusing. Inside equipment, data are conveyed on short wires and the noise environment is under the designer's control. With suitable design techniques, errors can be made effectively negligible. In communication systems, there is considerably less control of the electromagnetic environment.

In cables, crosstalk and electromagnetic interference occur and can corrupt data, although optical fibres are resistant to interference of this kind. In long-distance cable transmission the effects of lightning and exchange switching noise must be considered. In digital television broadcasting, multipath reception causes notches in the received spectrum where signal cancellation takes place. In MOS memories the datum is stored in a tiny charge well which acts as a capacitor (see Chapter 3) and natural radioactive decay can cause alpha particles which have enough energy to discharge a well, resulting in a single-bit error. This happens only once every few decades in a single chip, but when large numbers of chips are assembled in computer memories the probability of error rises to one every few minutes.

In Chapter 8 it was seen that when group codes are used, a single defect in a group changes the group symbol and may cause errors up to the size of the group. Single-bit errors are therefore less common in group-coded channels.

Irrespective of the cause, error mechanisms cause one of two effects. There are large isolated corruptions, called error bursts, where numerous bits are corrupted all together in an area which is otherwise error-free, and there are random errors affecting single bits or symbols. Whatever the mechanism, the result will be that the received data will not be exactly the same as those sent. It is a tremendous advantage of digital systems that the discrete data bits will be each either right or wrong. A bit cannot be off-colour as it can only be interpreted as 0 or 1. Thus the subtle degradations of analog systems are absent from digital recording and transmission channels and will only be found in convertors. Equally if a binary digit is known to be wrong, it is only necessary to invert its state and then it must be right and indistinguishable from its original value!

Thus error correction itself is trivial; the hard part is working out *which* bits need correcting.

In Chapter 3 the Gaussian nature of noise probability was discussed. Some conclusions can be drawn from the Gaussian distribution of noise.[1] First, it is not possible to make error-free digital recordings, because however high the signal-to-noise ratio of the recording, there is still a small but finite chance that the noise can exceed the signal. Measuring the signal-to-noise ratio of a channel establishes the noise power, which determines the width of the noise-distribution curve relative to the signal amplitude.

When in a binary system the noise amplitude exceeds the signal amplitude, but with the opposite polarity, a bit error will occur. Knowledge of the shape of the Gaussian curve allows the conversion of signal-to-noise ratio into bit error rate (BER). It can be predicted how many bits will fail due to noise in a given recording, but it is not possible to say *which* bits will be affected. Increasing the SNR of the channel will not eliminate errors, it just reduces their probability. The logical solution is to incorporate an error-correction system.

9.3 Basic error correction

Figure 9.3 shows that error correction works by adding some bits to the data which are calculated from the data. This creates an entity called a codeword which spans a greater length of time than one bit alone. The statistics of noise mean that whilst one bit may be lost in a codeword, the loss of the rest of the codeword because of noise is highly improbable. As will be described later in this chapter, codewords are designed to be able

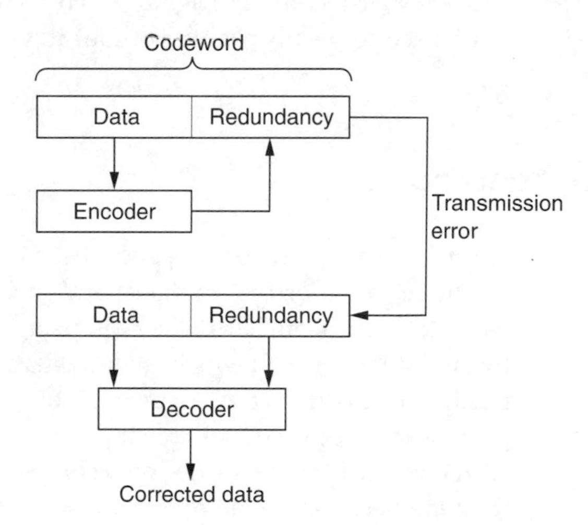

Figure 9.3 Error correction works by adding redundancy.

totally to correct a finite number of corrupted bits. The greater the timespan over which the coding is performed, or on a recording medium, the greater the area over which the coding is performed, the greater will be the reliability achieved, although this does mean a greater encoding and decoding delay.

Shannon[2] disclosed that a message can be sent to any desired degree of accuracy provided that it is spread over a sufficient timespan. Engineers have to compromise, because an infinite coding delay in the recovery of an error-free signal is not acceptable. Most short digital interfaces do not employ error correction because the build-up of coding delays in production systems is unacceptable.

If error correction is necessary as a practical matter, it is then only a small step to put it to maximum use. All error correction depends on adding bits to the original message, and this, of course, increases the number of bits to be recorded, although it does not increase the information recorded. It might be imagined that error correction is going to reduce storage capacity, because space has to be found for all the extra bits. Nothing could be further from the truth. Once an error-correction system is used, the signal-to-noise ratio of the channel can be reduced, because the raised BER of the channel will be overcome by the error-correction system.

In a digital television broadcast system the use of error correction allows a lower powered transmitter to be used. In a magnetic track, reduction of the SNR by 3 dB can be achieved by halving the track width, provided that the system is not dominated by head or preamplifier noise. This doubles the recording density, making the storage of the additional bits needed for error correction a trivial matter. In short, error correction is not a nuisance to be tolerated; it is a vital tool needed to maximize the efficiency of recorders. Digital video recording and broadcasting would not be economically viable without it.

9.4 Error handling

Figure 9.4 shows the broad subdivisions of error handling. The first stage might be called error avoidance and includes such measures as creating bad block files on hard disks or using verified media. The data pass through the channel, which causes whatever corruptions it feels like. On receipt of the data the occurrence of errors is first detected, and this process must be extremely reliable, as it does not matter how effective the correction or how good the concealment algorithm if it is not known that they are necessary! The detection of an error then results in a course of action being decided.

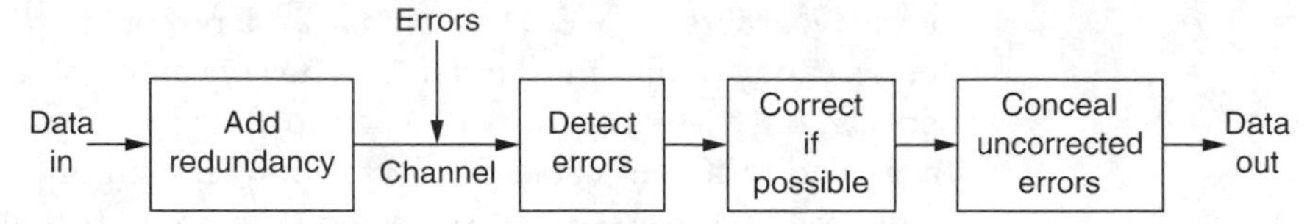

Figure 9.4 The basic stages of an error-correction system. Of these the most critical is the detection stage, since this controls the subsequent actions.

In most cases of digital video or audio replay a retry is not possible because the data are required in real time. However, if a disk-based system is transferring to tape for the purpose of back-up, real-time operation is not required. If the disk drive detects an error, a retry is easy as the disk is turning at several thousand rpm and will quickly re-present the data. An error due to a dust particle may not occur on the next revolution. Many magnetic tape systems have *read after write*. During recording, offtape data are immediately checked for errors. If an error is detected, the tape will abort the recording, reverse to the beginning of the current block and erase it. The data from that block are then recorded further down the tape. This is the recording equivalent of a retransmission in a communications system.

9.5 Concealment by interpolation

There are some practical differences between data recording for video and the computer data recording application. Although video or audio recorders seldom have time for retries, they have the advantage that there is a certain amount of redundancy in the information conveyed. Thus if an error cannot be corrected, then it can be concealed. If a sample is lost, it is possible to obtain an approximation to it by interpolating between the samples before and after the missing one. Concealment is more difficult with compressed video data. Clearly, concealment of any kind cannot be used with computer data.

If there is too much corruption for concealment, the only course in video is repeat the previous field or frame in a freeze as it is unlikely that the corrupt picture is watchable.

In general, if use is to be made of concealment on replay, the data must generally be reordered or shuffled prior to recording. To take a simple example, odd-numbered samples are recorded in a different area of the medium from even-numbered samples. On playback, if a gross error occurs on the tape, depending on its position, the result will be either corrupted odd samples or corrupted even samples, but it is most unlikely that both will be lost. Interpolation is then possible if the power of the correction system is exceeded. In practice the shuffle employed in digital video recorders is two-dimensional and rather more complex. Further details can be found in Chapter 11. The concealment technique described here is only suitable for PCM recording. If compression has been employed, different concealment techniques will be needed.

It should be stressed that corrected data are indistinguishable from the original and thus there can be no visible or audible artifacts. In contrast, concealment is only an approximation to the original information and could be detectable. In practical video equipment, concealment occurs infrequently unless there is a defect requiring attention, and its presence is difficult to see.

9.6 Parity

The error-detection and error-correction processes are closely related and will be dealt with together here. The actual correction of an error is simplified tremendously by the adoption of binary. As there are only two symbols, 0 and 1, it is enough to know that a symbol is wrong, and the correct value is obvious. Figure 9.5(a) shows a minimal circuit required for correction once the bit in error has been identified. The XOR (exclusive-OR) gate shows up extensively in error correction and the figure also shows the truth table. One way of remembering the characteristics of this useful device is that there will be an output when the inputs are different. Inspection of the truth table will show that there is an even number of ones in each row (zero is an even number) and so the device could also be called an even parity gate. The XOR gate is also an adder in modulo-2 (see Chapter 3).

Parity is a fundamental concept in error detection. In Figure 5.9(b) the example is given of a four-bit data word which is to be protected. If an extra bit is added to the word which is calculated in such a way that the total number of ones in the five-bit word is even, this property can be tested on receipt. The generation of the parity bit in (b) can be performed by a number of the ubiquitous XOR gates configured into what is known as a parity tree. In the figure, if a bit is corrupted, the received message will be seen no longer to have an even number of ones.

If two bits are corrupted, the failure will be undetected. This example can be used to introduce much of the terminology of error correction. The extra bit added to the message carries no information of its own, since it is calculated from the other bits. It is therefore called a *redundant* bit. The addition of the redundant bit gives the message a special property, i.e. the number of ones is even. A message having some special property *irrespective of the actual data content* is called a *codeword*. All error correction relies on adding redundancy to real data to form codewords for transmission. If any corruption occurs, the intention is that the received

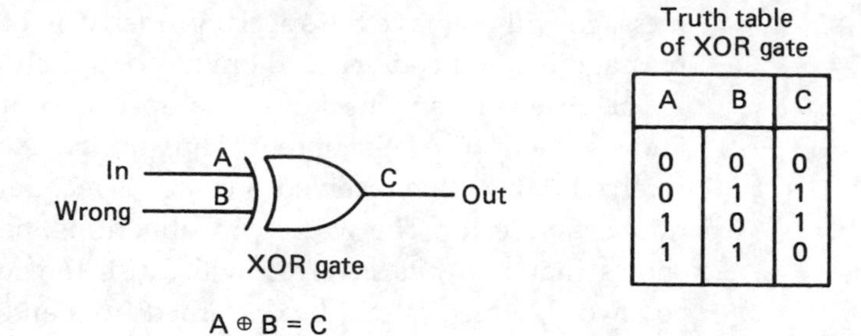

Figure 9.5(a) Once the position of the error is identified, the correction process in binary is easy.

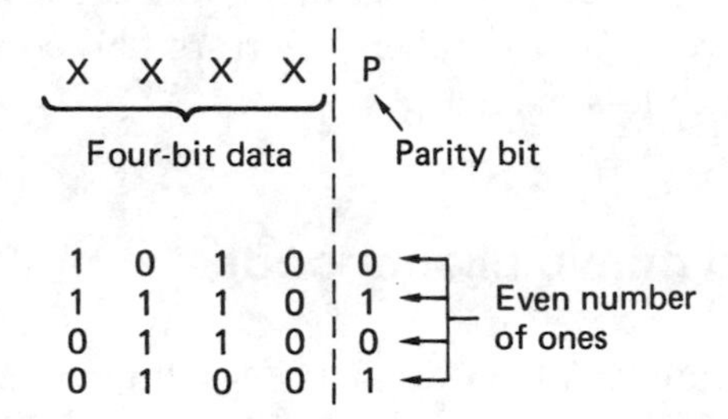

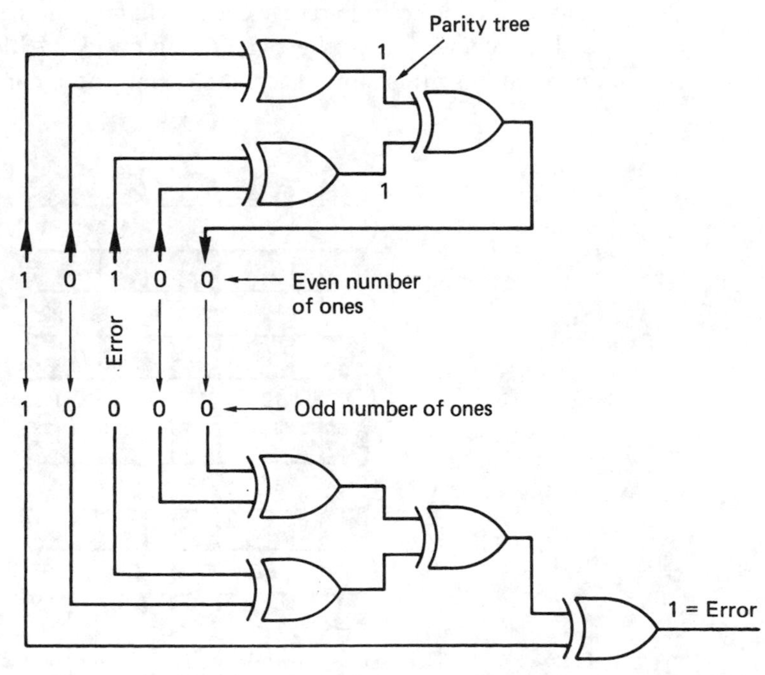

Figure 9.5(b) Parity checking adds up the number of ones in a word using, in this example, parity trees. One error bit and odd numbers of errors are detected. Even numbers of errors cannot be detected.

message will not have the special property; in other words if the received message is not a codeword there has definitely been an error.

The receiver can check for the special property without any prior knowledge of the data content. Thus the same check can be made on all received data. If the received message is a codeword, there probably has not been an error. The word 'probably' must be used because the figure shows that two bits in error will cause the received message to be a codeword, which cannot be discerned from an error-free message. If it is known that generally the only failure mechanism in the channel in question is loss of a single bit, it is *assumed* that receipt of a codeword means that there has been no error. If there is a probability of two error bits, that becomes very nearly the probability of failing to detect an error, since all odd numbers of errors will be detected, and a four-bit error is much less likely.

9.7 Block and convolutional codes

Figure 9.6(a) shows a strategy known as a crossword code, or product code. The data are formed into a two-dimensional array, in which each location can be a single bit or a multi-bit symbol. Parity is then generated on both rows and columns. If a single bit or symbol fails, one row parity check and one column parity check will fail, and the failure can be located at the intersection of the two failing checks. Although two symbols in error confuse this simple scheme, using more complex coding in a two-

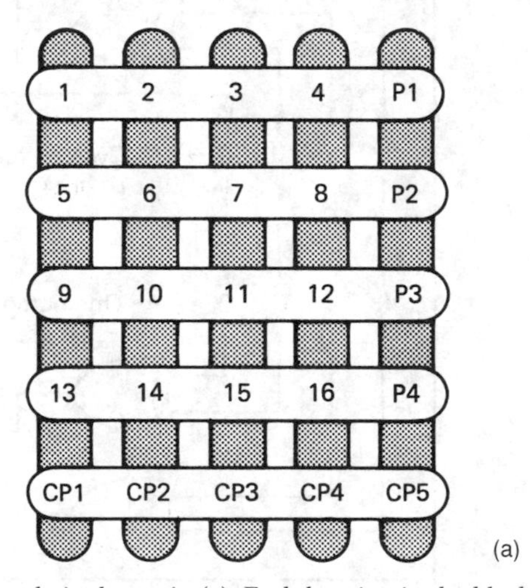

Figure 9.6 A block code is shown in (a). Each location in the block can be a bit or a word. Horizontal parity checks are made by adding P1, P2, etc., and cross-parity or vertical checks are made by adding CP1, CP2, etc. Any symbol in error will be at the intersection of the two failing codewords.

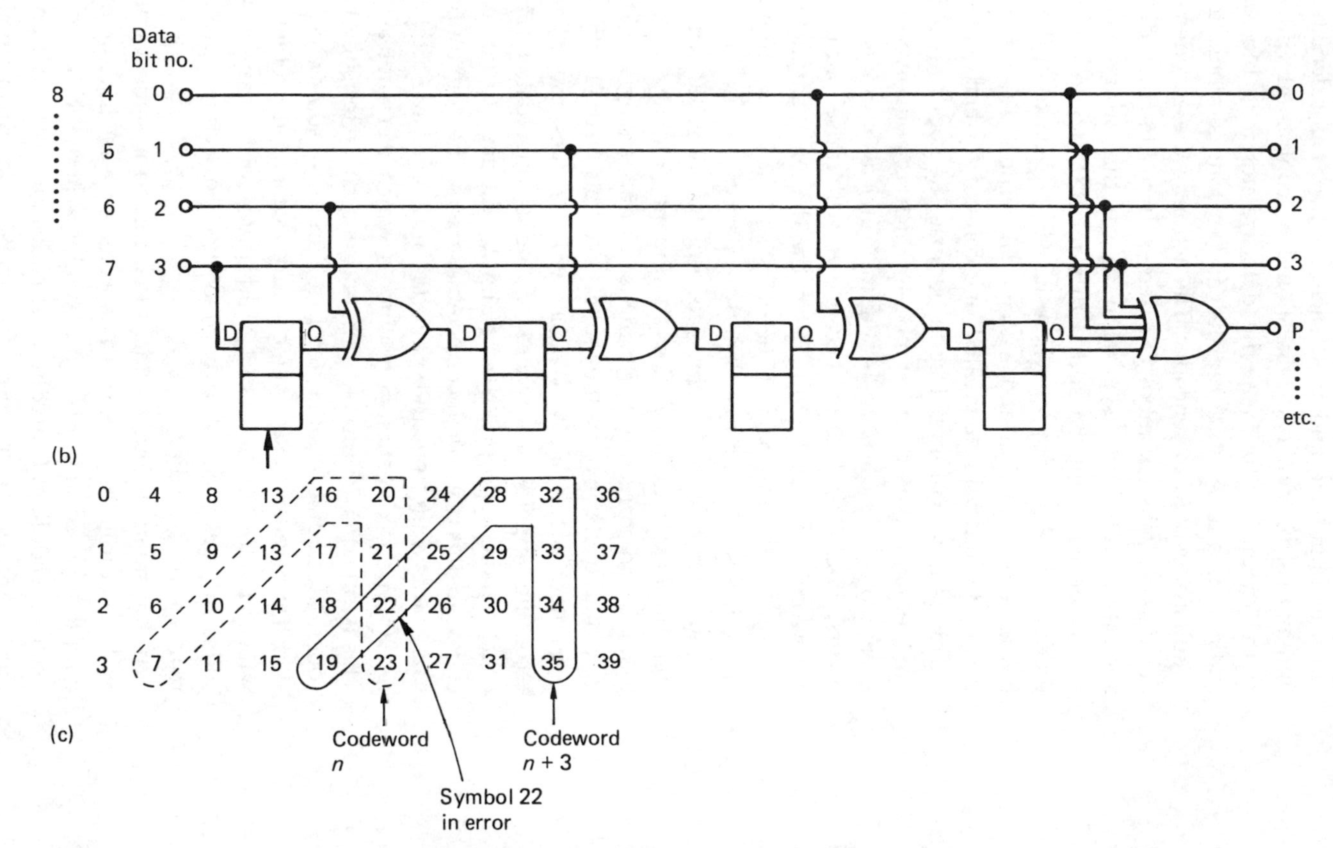

Figure 9.6 continued In (b) a convolutional coder is shown. Symbols entering are subject to different delays which result in the codewords in (c) being calculated. These have a vertical part and a diagonal part. A symbol in error will be at the intersection of the diagonal part of one code and the vertical part of another.

dimensional structure is very powerful, and further examples will be given throughout this chapter.

The example of Figure 9.6(a) assembles the data to be coded into a block of finite size and then each codeword is calculated by taking a different set of symbols. This should be contrasted with the operation of the circuit of (b). Here the data are not in a block, but form an endless stream. A shift register allows four symbols to be available simultaneously to the encoder. The action of the encoder depends upon the delays. When symbol 3 emerges from the first delay, it will be added (modulo-2) to symbol 6. When this sum emerges from the second delay, it will be added to symbol 9 and so on. The codeword produced is shown in (c) where it will be seen to be bent such that it has a vertical section and a diagonal section. Four symbols later the next codeword will be created one column further over in the data.

This is a convolutional code because the coder always takes parity on the same pattern of symbols which is convolved with the data stream on an endless basis. Figure 9.6(c) also shows that if an error occurs, it can be located because it will cause parity errors in two codewords. The error will be on the diagonal part of one codeword and on the vertical part of the other so that it can uniquely be located at the intersection and corrected by parity.

Comparison with the block code of Figure 9.6(a) will show that the convolutional code needs less redundancy for the same single symbol location and correction performance as only a single redundant symbol is required for every four data symbols. Convolutional codes are computed on an endless basis which makes them inconvenient in recording applications where editing is anticipated. Here the block code is more appropriate as it allows edit gaps to be created between codes. In the case of uncorrectable errors, the convolutional principle causes the syndromes to be affected for some time afterwards and this results in miscorrections of symbols which were not actually in error. This is called error propagation and is a characteristic of convolutional codes. Recording media tend to produce somewhat variant error statistics because media defects and mechanical problems cause errors which do not fit the classical additive noise channel. Convolutional codes can easily be taken beyond their correcting power if used with real recording media.

In transmission and broadcasting, the error statistics are more stable and the editing requirement is absent. As a result, convolutional codes tend to be used in digital broadcasting whereas block codes tend to be used in recording. Convolutional codes are not restricted to the simple parity example given here, but can be used in conjunction with more sophisticated redundancy techniques such as the Reed–Solomon codes. Examples of this will be given in Chapter 14.

9.8 Hamming code

In a one-dimensional code, the position of the failing bit can be determined by using more parity checks. In Figure 9.7(a), the four data bits have been used to compute three redundancy bits, making a seven-bit codeword. The four data bits are examined in turn, and each bit which is a one will cause the corresponding row of a generator matrix to be added to an exclusive-OR sum. For example, if the data were 1001, the top and bottom rows of the matrix would be XORed. The matrix used is known as an identity matrix, because the data bits in the codeword are identical to the data bits to be conveyed. This is useful because the original data can be stored unmodified, and the check bits are simply attached to the end to make a so-called systematic codeword. Almost all digital recording equipment uses systematic codes. The way in which the redundancy bits are calculated is simply that they do not all use every data bit. If a data bit has not been included in a parity check, it can fail without affecting the outcome of that check. The position of the error is deduced from the pattern of successful and unsuccessful checks in the check matrix. This pattern is known as a syndrome.

In the figure the example of a failing bit is given. Bit 3 fails, and because this bit is included in only two of the checks, there are two ones in the failure pattern, 011. As some care was taken in designing the matrix pattern for the generation of the check bits, the syndrome, 011, is the address of the failing bit. This is the fundamental feature of the Hamming codes due to Richard Hamming.[3] The performance of this seven-bit codeword can be assessed. In seven bits there can be 128 combinations, but in four data bits there are only sixteen combinations. Thus out of 128 possible received messages, only sixteen will be codewords, so if the message is completely trashed by a gross corruption, it will still be possible to detect that this has happened 112 times out of 127, as in these cases the syndrome will be non-zero (the 128th case is the correct data). There is thus only a probability of detecting that all of the message is corrupt.

In an idle moment it is possible to work out, in a similar way, the number of false codewords which can result from different numbers of bits being assumed to have failed. For less than three bits, the failure will always be detected, because there are three check bits. Returning to the example, if two bits fail, there will be a non-zero syndrome, but if this is used to point to a bit in error, a miscorrection will result. From these results can be deduced another important feature of error codes. The power of detection is always greater than the power of correction, which is also fortunate, since if the correcting power is exceeded by an error it will at least be a known problem, and steps can be taken to prevent any undesirable consequences.

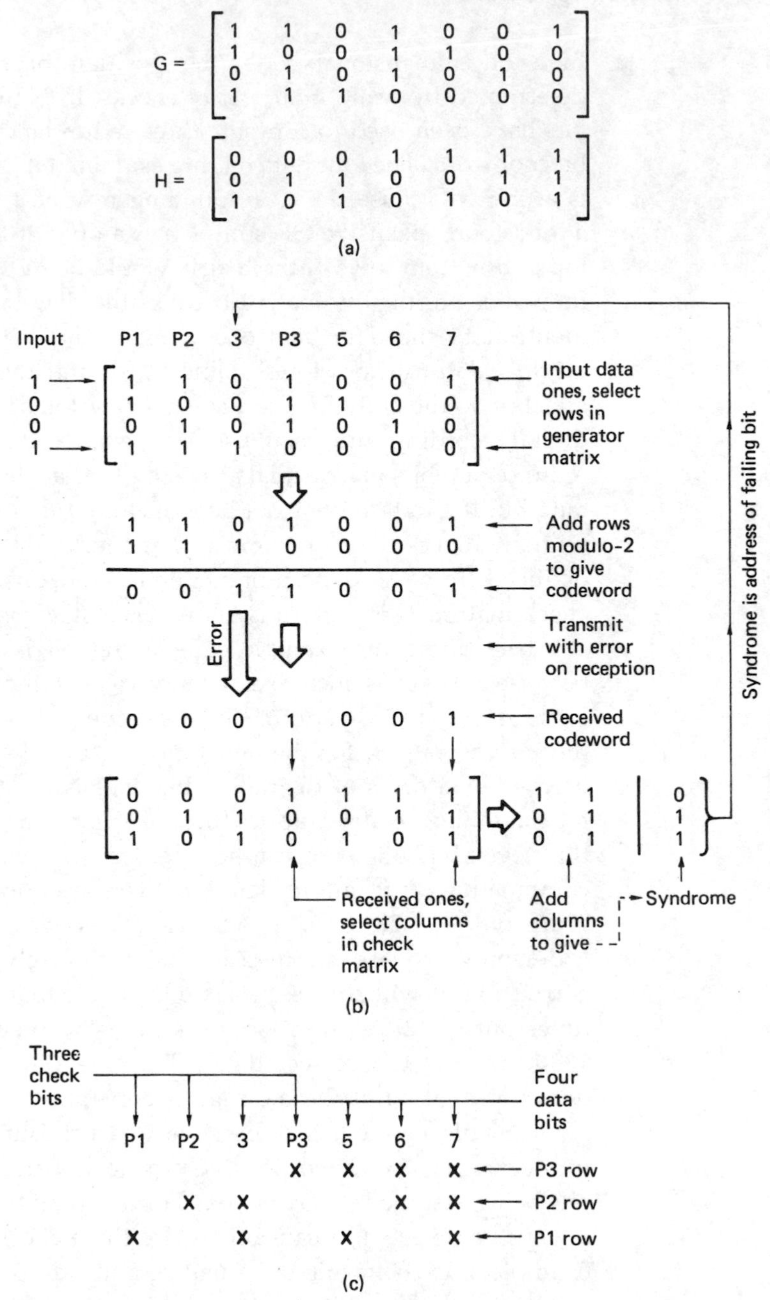

Figure 9.7 (a) The generator and check matrices of a Hamming code. The data and check bits are arranged as shown because this causes the syndrome to be the binary address of the failing bit. (b) An example of Hamming code generation and error correction. (c) Another way of looking at Hamming code is to say that the rows of crosses in this chart are calculated to have even parity. If bit 3 fails, parity check P3 is not affected, but parity checks P1 and P2 both include bit 3 and will fail.

The efficiency of the example given is not very high because three check bits are needed for every four data bits. Since the failing bit is located with a binary-split mechanism, it is possible to double the code length by adding a single extra check bit. Thus with four-bit syndromes there are fifteen non-zero codes and so the codeword will be fifteen bits long. Four bits are redundant and eleven are data. Using five bits of redundancy, the code can be thirty-one bits long and contain twenty-six data bits. Thus provided that the number of errors to be detected stays the same, it is more efficient to use long codewords. Error-correcting memories use typically four or eight data bytes plus redundancy. A drawback of long codes is that if it is desired to change a single memory byte it is necessary to read the entire codeword, modify the desired data byte and re-encode, the so-called read–modify–write process.

The Hamming code shown is limited to single-bit correction, but by addition of another bit of redundancy can be made to correct one-bit and detect two-bit errors. This is ideal for error correcting MOS memories where the SECDED (single error correcting double error detecting) characteristic matches the type of failures experienced.

The correction of one bit is of little use in the presence of burst errors, but a Hamming code can be made to correct burst errors by using interleaving. Figure 9.7(b) shows that if several codewords are calculated beforehand and woven together as shown before they are sent down the channel, then a burst of errors which corrupts several bits will become a number of single-bit errors in separate codewords upon de-interleaving.

Interleaving is used extensively in digital recording and transmission, and will be discussed in greater detail later in this chapter.

9.9 Hamming distance

It is useful at this point to introduce the concept of Hamming distance. This is not a physical distance but is a specific measure of the difference between two binary numbers. Hamming distance is defined in the general case as the number of bit positions in which a pair of words differ. The Hamming distance of a code is defined as the minimum number of bits that must be changed in any codeword in order to turn it into another codeword. This is an important yardstick because if errors convert one codeword into another, it will have the special characteristic of the code and so the corruption will not even be detected.

Figure 9.8 shows Hamming distance diagrammatically. A three-bit codeword is used with two data bits and one parity bit. With three bits, a received code could have eight combinations, but only four of these will be codewords. The valid codewords are shown in the centre of each of the

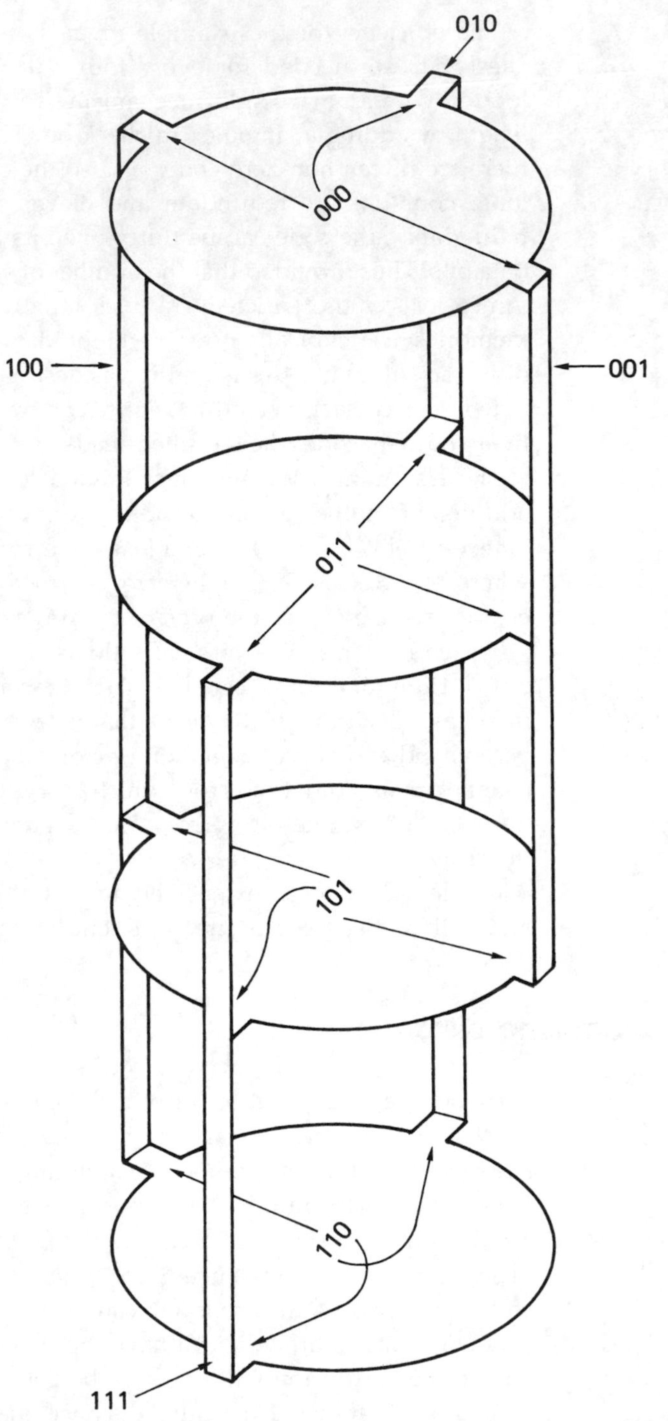

Figure 9.8 Hamming distance of two. The disk centres contain codewords. Corrupting each bit in turn produces the distance 1 values on the vertical members. In order to change one codeword to another, two bits must be changed, so the code has a Hamming distance of two.

disks, and these will be seen to be identical to the rows of the truth table in Figure 9.2. At the perimeter of the disks are shown the received words which would result from a single-bit error, i.e. they have a Hamming distance of one from codewords. It will be seen that the same received word (on the vertical bars) can be obtained from a different single-bit corruption of any three codewords. It is thus not possible to tell which codeword was corrupted, so although all single-bit errors can be detected, correction is not possible.

Figure 9.8 should be compared with that of Figure 9.9, which is a Venn diagram where there is a set in which the MSB is 1 (upper circle), a set in which the middle bit is 1 (lower-left circle) and a set in which the LSB is 1 (lower-right circle). Note that in crossing any boundary only one bit changes, and so each boundary represents a Hamming distance change of one. The four codewords of Figure 9.8 are repeated here, and it will be seen that single-bit errors in any codeword produce a non-codeword, and so single-bit errors are always detectable.

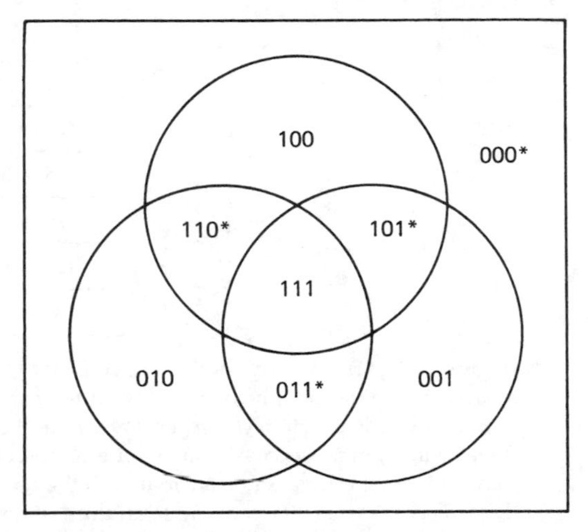

Figure 9.9 Venn diagram shows a one-bit change crossing any boundary which is a Hamming distance of one. Compare with Figure 9.11. Codewords marked*.

Correction is possible if the number of non-codewords is increased by increasing the number of redundant bits. This means that it is possible to spread out the actual codewords in Hamming distance terms.

Figure 9.10(a) shows a distance 2 code, where there is only one redundancy bit, and so half of the possible words will be codewords. There will be non-codewords at distance 1 which can be produced by altering a single bit in either of two codewords. In this case it is not possible to tell what the original codeword was in the case of a single-bit error.

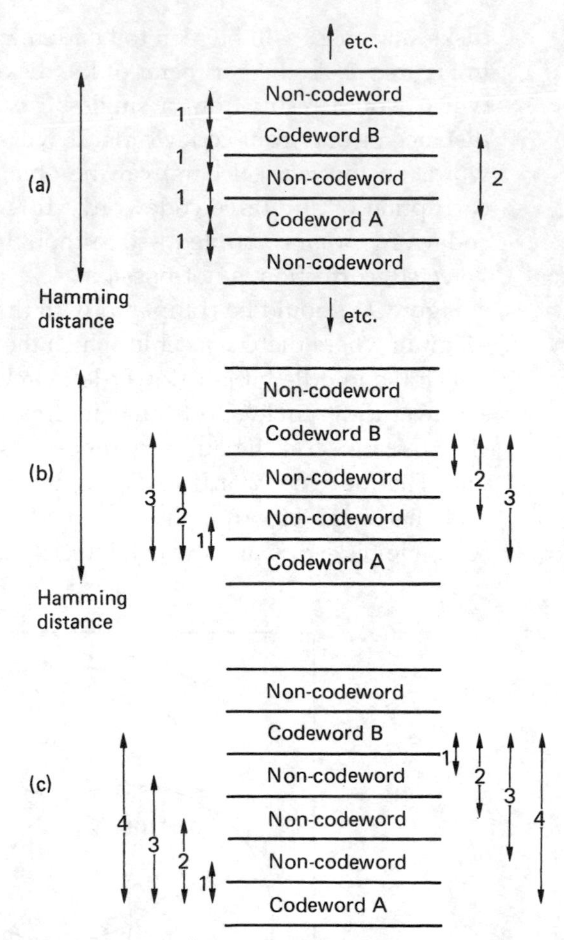

Figure 9.10 (a) Distance 2 code; non-codewords are at distance 1 from two possible codewords so it cannot be deduced what the correct one is, (b) Distance 3 code; non-codewords which have *single-bit errors* can be attributed to the nearest codeword. Breaks down in presence of double-bit errors. (c) Distance 4 code; non-codewords which have single-bit errors can be attributed to the nearest codeword, *and* double-bit errors form *different* non-codewords, and can thus be detected but not corrected.

Figure 9.10(b) shows a distance 3 code, where there will now be at least two non-codewords between codewords. If a single-bit error occurs in a codeword, the resulting non-codeword will be at distance 1 from the original codeword. This same non-codeword could also have been produced by changing *two* bits in a different codeword. If it is known that the failure mechanism is a single bit, it can be *assumed* that the original codeword was the one which is closest in Hamming distance to the received bit pattern, and so correction is possible. If, however, our assumption about the error mechanism proved to be wrong, and in fact a two-bit error had occurred, this assumption would take us to the wrong codeword, turning the event into a three-bit error. This is an illustration of the knee in the graph of Figure 9.2, where if the power of the code is exceeded it makes things worse.

Figure 9.10(c) shows a distance 4 code. There are now three non-codewords between codewords, and clearly single-bit errors can still be corrected by choosing the nearest codeword. Double-bit errors will be detected, because they result in non-codewords equidistant in Hamming terms from codewords, but it is not possible uniquely to determine what the original codeword was.

9.10 Cyclic codes

The parallel implementation of a Hamming code can be made very fast using parity trees, which is ideal for memory applications where access time is increased by the correction process. However, in digital recording applications, the data are stored serially on a track, and it is desirable to use relatively large data blocks to reduce the amount of the medium devoted to preambles, addressing and synchronizing. Where large data blocks are to be handled, the use of a look-up table or tree has to be abandoned because it would become impossibly large. The principle of codewords having a special characteristic will still be employed, but they will be generated and checked algorithmically by equations. The syndrome will then be converted to the bit(s) in error not by looking them up, but by solving an equation.

Where data can be accessed serially, simpler circuitry can be used because the same gate will be used for many XOR operations. Unfortunately the reduction in component count is only paralleled by an increase in the difficulty of explaining what takes place.

The circuit of Figure 9.11 is a kind of shift register, but with a particular feedback arrangement which leads it to be known as a twisted-ring counter. If seven message bits A–G are applied serially to this circuit, and each one of them is clocked, the outcome can be followed in the diagram. As bit A is presented and the system is clocked, bit A will enter the left-hand latch. When bits B and C are presented, A moves across to the right. Both XOR gates will have A on the upper input from the right-hand latch, the left one has D on the lower input and the right one has B on the lower input. When clocked, the left latch will thus be loaded with the XOR of A and D, and the right one with the XOR of A and B.

The remainder of the sequence can be followed, bearing in mind that when the same term appears on both inputs of an XOR gate, it goes out, as the exclusive-OR of something with itself is nothing. At the end of the process, the latches contain three different expressions. Essentially, the circuit makes three parity checks through the message, leaving the result

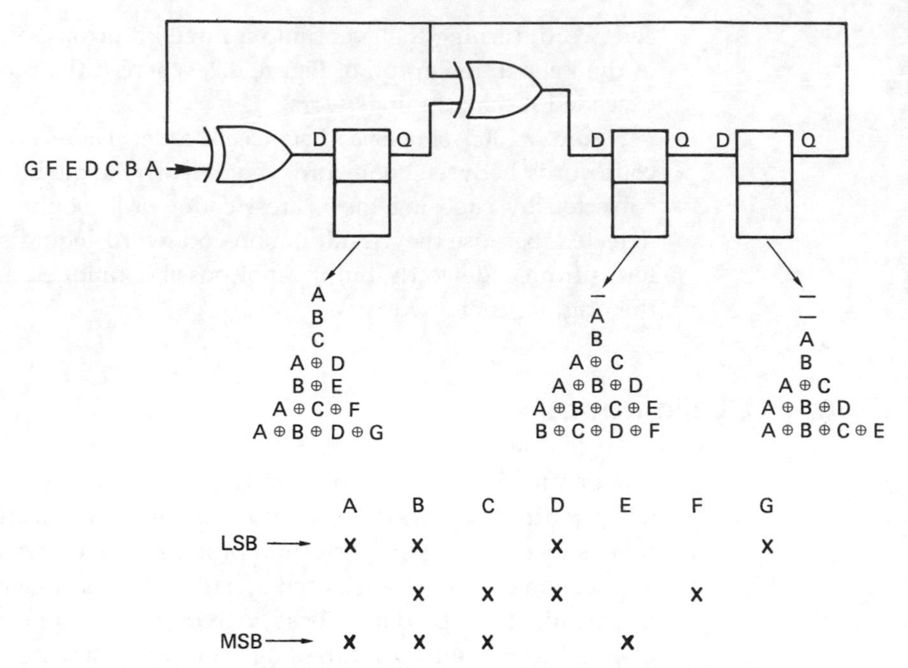

Figure 9.11 When seven successive bits A–G are clocked into this circuit, the contents of the three latches are shown for each clock. The final result is a parity-check matrix.

of each in the three stages of the register. In the figure, these expressions have been used to draw up a check matrix. The significance of these steps can now be explained. The bits A B C and D are four data bits, and the bits E F and G are redundancy.

When the redundancy is calculated, bit E is chosen so that there are an even number of ones in bits A B C and E; bit F is chosen such that the same applies to bits B C D and F, and similarly for bit G. Thus the four data bits and the three check bits form a seven-bit codeword.

If there is no error in the codeword, when it is fed into the circuit shown, the result of each of the three parity checks will be zero and every stage of the shift register will be cleared. As the register has eight possible states, and one of them is the error-free condition, then there are seven remaining states, hence the seven-bit codeword. If a bit in the codeword is corrupted, there will be a non-zero result. For example, if bit D fails, the check on bits A B D and G will fail, and a one will appear in the left-hand latch. The check on bits B C D F will also fail, and the centre latch will set. The check on bits A B C E will not fail, because D is not involved in it, making the right-hand bit zero. There will be a syndrome of 110 in the register, and this will be seen from the check matrix to correspond to an error in bit D.

Whichever bit fails, there will be a different three-bit syndrome which uniquely identifies the failed bit. As there are only three latches, there can be eight different syndromes. One of these is zero, which is the error-free condition, and so there are seven remaining error syndromes. The length of the codeword cannot exceed seven bits, or there would not be enough syndromes to correct all the bits. This can also be made to tie in with the generation of the check matrix. If fourteen bits, A to N, were fed into the circuit shown, the result would be that the check matrix repeated twice, and if a syndrome of 101 were to result, it could not be determined whether bit D or bit K failed. Because the check repeats every seven bits, the code is said to be a cyclic redundancy check (CRC) code.

In Figure 9.7 an example of a Hamming code was given. Comparison of the check matrix of Figure 9.11 with that of Figure 9.7 will show that the only difference is the order of the matrix columns. The two different processes have thus achieved exactly the same results, and the performance of both must be identical. This is not true in general, but a very small cyclic code has been used for simplicity and to allow parallels to be seen. In practice, CRC code blocks will be much longer than the blocks used in Hamming codes.

It has been seen that the circuit shown makes a matrix check on a received word to determine if there has been an error, but the same circuit can also be used to generate the check bits. To visualize how this is done, examine what happens if only the data bits A B C and D are known, and the check bits E F and G are set to zero. If this message, ABCD000, is fed into the circuit, the left-hand latch will afterwards contain the XOR of A B C and zero, which is, of course, what E should be. The centre latch will contain the XOR of B C D and zero, which is what F should be and so on. This process is not quite ideal, however, because it is necessary to wait for three clock periods after entering the data before the check bits are available. Where the data are simultaneously being recorded and fed into the encoder, the delay would prevent the check bits being easily added to the end of the data stream.

This problem can be overcome by slightly modifying the encoder circuit as shown in Figure 9.12. By moving the position of the input to the right, the operation of the circuit is advanced so that the check bits are ready after only four clocks. The process can be followed in the diagram for the four data bits A B C and D. On the first clock, bit A enters the left two latches, whereas on the second clock, bit B will appear on the upper input of the left XOR gate, with bit A on the lower input, causing the centre latch to load the XOR of A and B and so on.

The way in which the cyclic codes work has been described in engineering terms, but it can be described mathematically if analysis is contemplated.

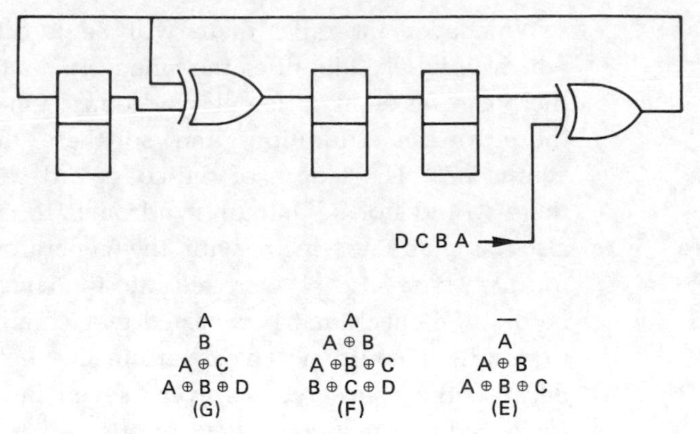

Figure 9.12 By moving the insertion point three places to the right, the calculation of the check bits is completed in only four clock periods and they can follow the data immediately. This is equivalent to premultiplying the data by x^3.

Just as the position of a decimal digit in a number determines the power of ten (whether that digit means one, ten or a hundred), the position of a binary digit determines the power of two (whether it means one, two or four). It is possible to rewrite a binary number so that it is expressed as a list of powers of two. For example, the binary number 1101 means 8 + 4 + 1, and can be written:

$$2^3 + 2^2 + 2^0$$

In fact, much of the theory of error correction applies to symbols in number bases other than 2, so that the number can also be written more generally as

$$x^3 + x^2 + 1 \ (2^0 = 1)$$

which also looks much more impressive. This expression, containing as it does various powers, is, of course, a polynomial, and the circuit of Figure 9.11 which has been seen to construct a parity-check matrix on a codeword can also be described as calculating the remainder due to dividing the input by a polynomial using modulo-2 arithmetic. In modulo-2 there are no borrows or carries, and addition and subtraction are replaced by the XOR function, which makes hardware implementation very easy. In Figure 9.13 it will be seen that the circuit of Figure 9.11 actually divides the codeword by a polynomial which is

$$x^3 + x + 1 \text{ or } 1011$$

This can be deduced from the fact that the right-hand bit is fed into two lower-order stages of the register at once. Once all the bits of the message

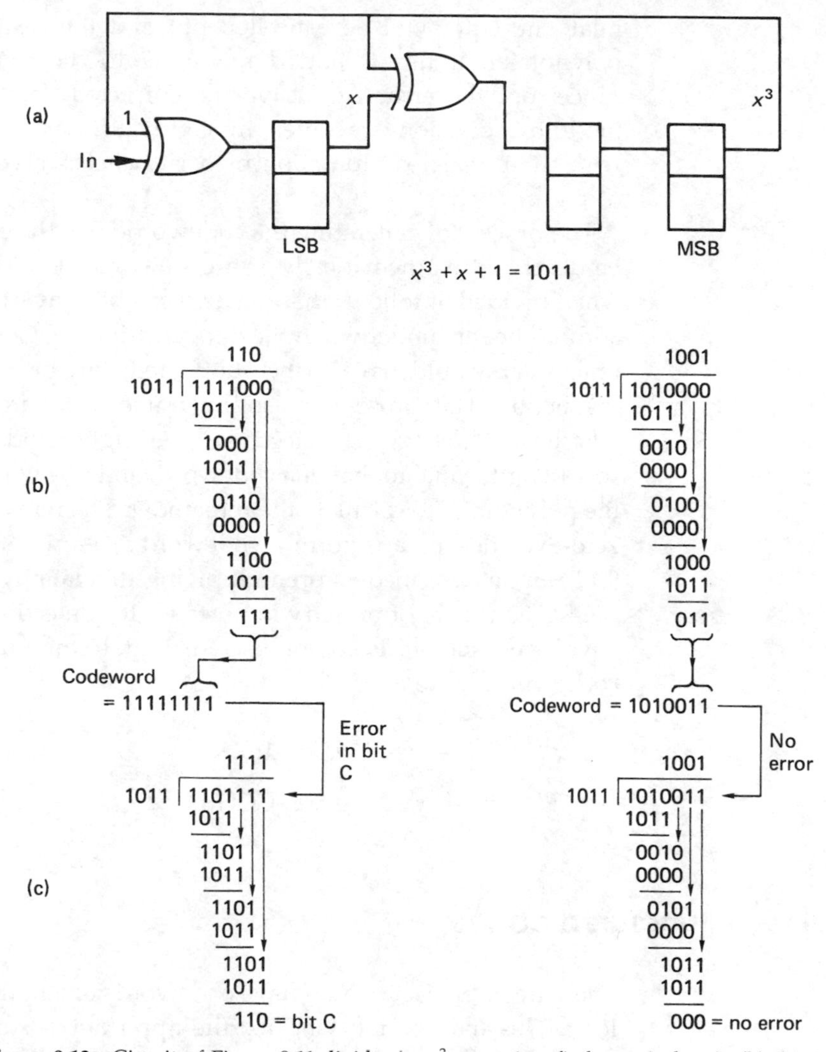

Figure 9.13 Circuit of Figure 9.11 divides by $x^3 + x + 1$ to find remainder. At (b) this is used to calculate check bits. At (c) right, zero syndrome, no error.

have been clocked in, the circuit contains the remainder. In mathematical terms, the special property of a codeword is that it is a polynomial which yields a remainder of zero when divided by the generating polynomial. The receiver will make this division, and the result should be zero in the error-free case. Thus the codeword itself disappears from the division. If an error has occurred it is considered that this is due to an error polynomial which has been added to the codeword polynomial.

If a codeword divided by the check polynomial is zero, a non-zero syndrome must represent the error polynomial divided by the check polynomial. Thus if the syndrome is multiplied by the check polynomial, the latter will be cancelled out and the result will be the error polynomial. If this is added modulo-2 to the received word, it will cancel out the error and leave the corrected data. Some examples of modulo-2 division are given in Figure 9.13 which can be compared with the parallel computation of parity checks according to the matrix of Figure 9.11.

The process of generating the codeword from the original data can also be described mathematically. If a codeword has to give zero remainder when divided, it follows that the data can be converted to a codeword by adding the remainder when the data are divided. Generally speaking, the remainder would have to be subtracted, but in modulo-2 there is no distinction. This process is also illustrated in Figure 9.13. The four data bits have three zeros placed on the right-hand end, to make the wordlength equal to that of a codeword, and this word is then divided by the polynomial to calculate the remainder. The remainder is added to the zero-extended data to form a codeword. The modified circuit of Figure 9.12 can be described as premultiplying the data by x^3 before dividing.

CRC codes are of primary importance for detecting errors, and several have been standardized for use in digital communications. The most common of these are:

$$x^{16} + x^{15} + x^2 + 1 \text{ (CRC-16)}$$

$$x^{16} + x^{12} + x^5 + 1 \text{ (CRC-CCITT)}$$

9.11 Punctured codes

The sixteen-bit cyclic codes have codewords of length $2^{16} - 1$ or 65 535 bits long. This may be too long for the application. Another problem with very long codes is that with a given raw BER, the longer the code, the more errors will occur in it. There may be enough errors to exceed the power of the code. The solution in both cases is to shorten or *puncture* the code. Figure 9.14 shows that in a punctured code, only the end of the codeword is used, and the data and redundancy are preceded by a string of zeros. It is not necessary to record these zeros, and, of course, errors cannot occur in them. Implementing a punctured code is easy. If a CRC generator starts with the register cleared and is fed with serial zeros, it will not change its state. Thus it is not necessary to provide the zeros, encoding can begin with the first data bit. In the same way, the leading zeros need not be provided during playback. The only precaution needed is that if a syndrome calculates the location of an error, this will be from the beginning of the codeword not from the beginning of the data. Where codes are used for detection only, this is of no consequence.

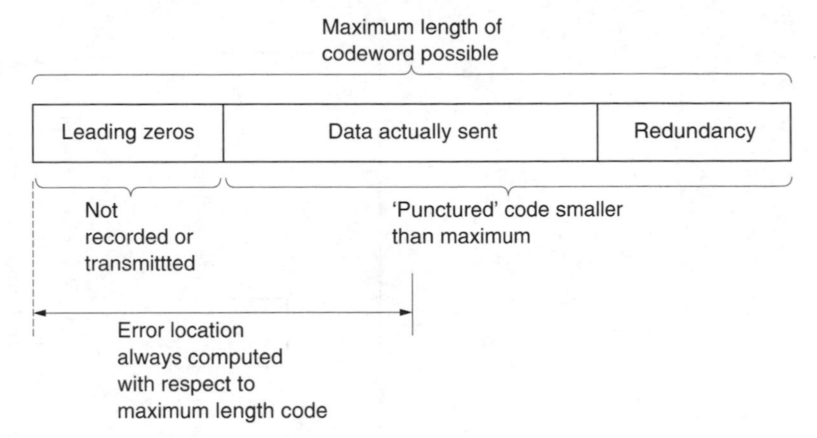

Figure 9.14 Codewords are often shortened, or punctured, which means that only the end of the codeword is actually transmitted. The only precaution to be taken when puncturing codes is that the computed position of an error will be from the beginning of the codeword, not from the beginning of the message.

9.12 Applications of cyclic codes

The AES/EBU digital audio interface described in Chapter 10 uses an eight-bit cyclic code to protect the channel-status data. The polynomial used and a typical circuit for generating it can be seen in Figure 9.15. The full codeword length is 255 bits but it is punctured to 192 bits, or 24 bytes which is the length of the AES/EBU channel status block. The CRCC is placed in the last byte.

Cyclic codes are also used for the EDH (error detection and handling) option of SDI (the serial digital interface) described in Chapter 10. Figure 9.16 shows that the entire data content of a field is divided by a polynomial. The remainder is transmitted to the receiver which also makes the same division. If the same remainder is obtained then the transmission system is error-free. It is a characteristic of cyclic codes that a single bit in error in a lengthy message will change the signature so that it can be detected. In practice, the amount of data in a digital field will exceed the length of the codeword. The cyclic code may repeat several times within the block. Thus location of errors is impossible, but this is not a goal.

9.13 Burst correction

Figure 9.17 lists all the possible codewords in the code of Figure 9.11. Examination will show that it is necessary to change at least three bits in one codeword before it can be made into another. Thus the code has a

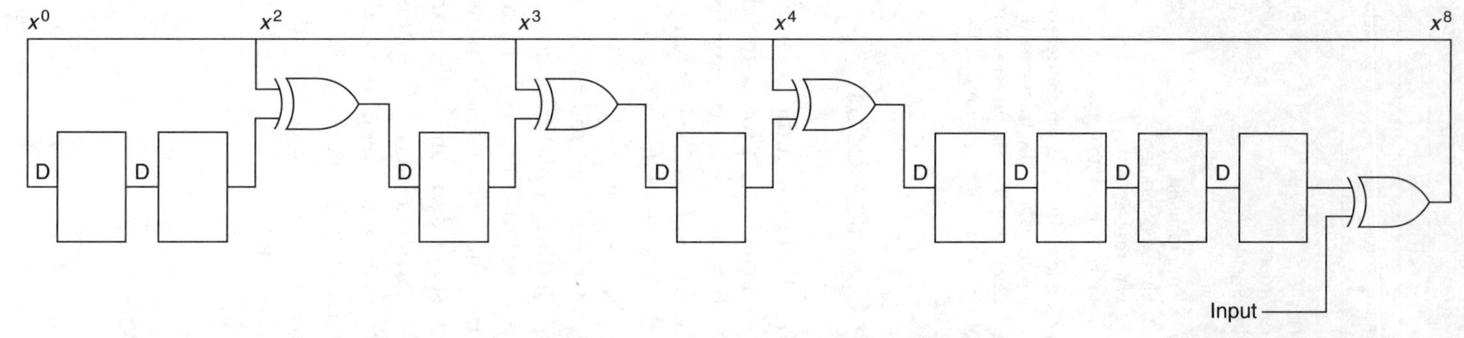

Figure 9.15 The CRCC in the AES/EBU interface is generated by premultiplying the data by x^8 and dividing by $x^8 + x^4 + x^3 + x^2 + 1$. The process can be performed on a serial input by the circuit shown. Premultiplication is achieved by connecting the input at the most significant end of the system. If the output of the right-hand XOR gate is 1 then a 1 is fed back to all of the powers shown, and the polynomial process required is performed. At the end of 23 data bytes, the CRCC will be in the eight latches. At the end of an error-free 24 byte message, the latches will be all zero.

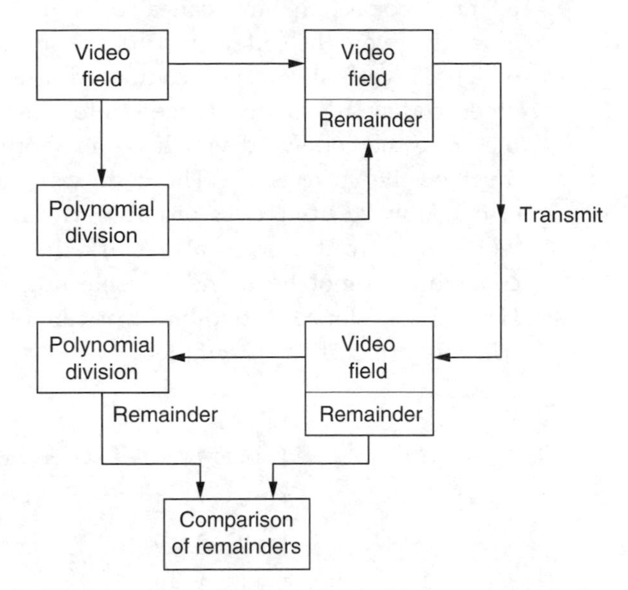

Figure 9.16 In EDH the video field is sent along with a remainder. At the receiver an identical division is made and the remainders are compared. Errors will cause the remainders to be different.

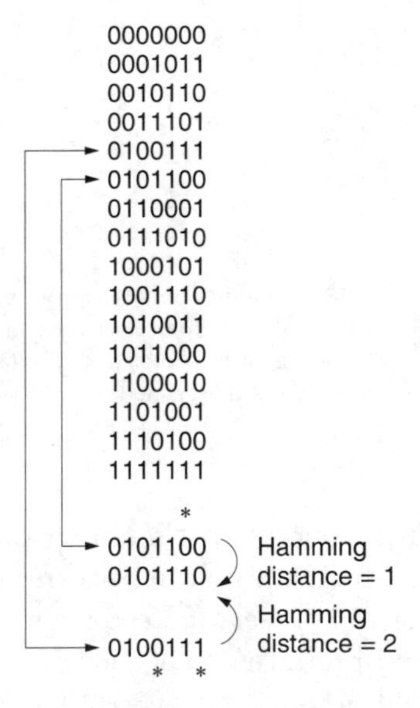

Figure 9.17 All possible codewords of $x^3 + x + 1$ are shown, and the fact that a double error in one codeword can produce the same pattern as a single error in another. Thus double errors cannot be corrected.

Hamming distance of three and cannot detect three-bit errors. The single-bit error-correction limit can also be deduced from the figure. In the example given, the codeword 0101100 suffers a single-bit error marked * which converts it to a non-codeword at a Hamming distance of 1. No other codeword can be turned into this word by a single-bit error; therefore the codeword which is the shortest Hamming distance away must be the correct one. The code can thus reliably correct single-bit errors. However, the codeword 0100111 can be made into the same failure word by a two-bit error, also marked *, and in this case the original codeword cannot be found by selecting the one which is nearest in Hamming distance. A two-bit error cannot be corrected and the system will miscorrect if it is attempted.

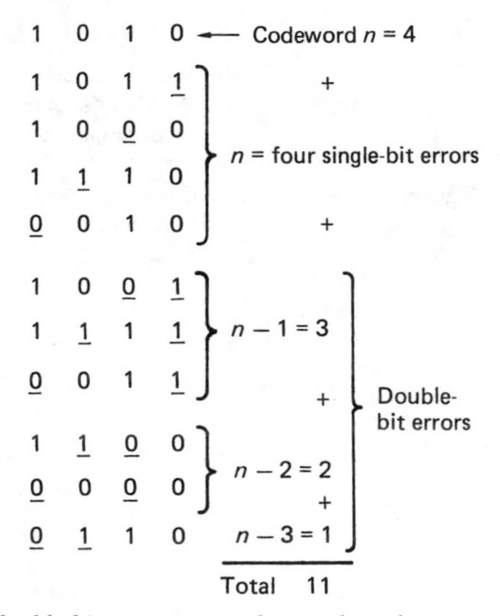

Figure 9.18 Where double-bit errors occur, the number of patterns necessary is $(n-1) + (n-2) + (n-3) + \ldots$ Total necessary is $1 + n + (n-1) + (n-2) + (n-3) + \ldots$ etc. Example here is of four bits, and all possible patterns up to a Hamming distance of two are shown (errors underlined).

The concept of Hamming distance can be extended to explain how more than one bit can be corrected. In Figure 9.18 the example of two bits in error is given. If a codeword four bits long suffers a single-bit error, it could produce one of four different words. If it suffers a two-bit error, it could produce one of 3 + 2 + 1 different words as shown in the figure (the error bits are underlined). The total number of possible words of Hamming distance 1 or 2 from a four-bit codeword is thus:

$$4 + 3 + 2 + 1 = 10$$

If the two-bit error is to be correctable, no other codeword can be allowed to become one of this number of error patterns because of a two-bit error of its own. Thus every codeword requires space for itself plus all possible error patterns of Hamming distance 2 or 1, which is eleven patterns in this example. Clearly there are only sixteen patterns available in a four-bit code, and thus no data can be conveyed if two-bit protection is necessary.

The number of different patterns possible in a word of n bits is

$$1 + n + (n - 1) + (n - 2) + (n - 3) + \ldots$$

and this pattern range has to be shared between the ranges of each codeword without overlap. For example, an eight-bit codeword could result in 1 + 8 + 7 + 6 + 5 + 4 + 3 + 2 + 1 = 37 patterns. As there are only 256 patterns in eight bits, it follows that only 256/37 pieces of information can be conveyed. The nearest integer below is six, and the nearest power of two below is four, which corresponds to two data bits and six check bits in the eight-bit word. The amount of redundancy necessary to correct *any* two bits in error is large, and as the number of bits to be corrected grows, the redundancy necessary becomes enormous and impractical. A further problem is that the more redundancy is added, the greater the probability of an error in a codeword. Fortunately, in practice errors occur in bursts, as has already been described, and it is a happy consequence that the number of patterns that result from the corruption of a codeword by *adjacent* two-bit errors is much smaller.

It can be deduced that the number of redundant bits necessary to correct a burst error is twice the number of bits in the burst for a perfect code. This is done by working out the number of received messages which could result from corruption of the codeword by bursts of from one bit up to the largest burst size allowed, and then making sure that there are enough redundant bits to allow that number of combinations in the received message.

Some codes, such as the Fire code due to Philip Fire,[4] are designed to correct single bursts, whereas later codes such as the B-adjacent code due to Bossen[5] could correct two bursts. The Reed–Solomon codes[6] have the advantage that an arbitrary number of bursts can be corrected by choosing the appropriate amount of redundancy at the design stage.

Whilst the Fire code was discovered at about the same time as the superior Reed–Solomon codes, it was dominant in disk drives for a long time because it was so much easier to implement. Now that LSI technology has advanced, the complexity of Reed–Solomon codes is no longer an issue and the Fire code is seldom used.

9.14 Introduction to the Reed–Solomon codes

The Reed–Solomon codes (Irving Reed and Gustave Solomon) are inherently burst correcting[6] because they work on multi-bit symbols rather than individual bits. The R–S codes are also extremely flexible in use. One code may be used both to detect and correct errors and the number of bursts which are correctable can be chosen at the design stage by the amount of redundancy. A further advantage of the R–S codes is that they can be used in conjunction with a separate error-detection mechanism in which case they perform only the correction by erasure. R–S codes operate at the theoretical limit of correcting efficiency. In other words, no more efficient code can be found.

In the simple CRC system described in section 9.10, the effect of the error is detected by ensuring that the codeword can be divided by a polynomial. The CRC codeword was created by adding a redundant symbol to the data. In the Reed–Solomon codes, several errors can be isolated by ensuring that the codeword will divide by a number of polynomials. Clearly if the codeword must divide by, say, two polynomials, it must have two redundant symbols. This is the minimum case of an R–S code. On receiving an R–S coded message there will be two syndromes following the division. In the error-free case, these will both be zero. If both are not zero, there is an error.

The effect of an error is to add an error polynomial to the message polynomial. The number of terms in the error polynomial is the same as the number of errors in the codeword. The codeword divides to zero and the syndromes are a function of the error only. There are two syndromes and two equations. By solving these simultaneous equations it is possible to obtain two unknowns. One of these is the position of the error, known as the *locator* and the other is the error bit pattern, known as the *corrector*. As the locator is the same size as the code symbol, the length of the codeword is determined by the size of the symbol. A symbol size of eight bits is commonly used because it fits in conveniently with both sixteen-bit audio samples and byte-oriented computers. An eight-bit syndrome results in a locator of the same wordlength. Eight bits have 2^8 combinations, but one of these is the error-free condition, and so the locator can specify one of only 255 symbols. As each symbol contains eight bits, the codeword will be $255 \times 8 = 2040$ bits long.

As further examples, five-bit symbols could be used to form a codeword 31 symbols long, and three-bit symbols would form a codeword seven symbols long. This latter size is small enough to permit some worked examples, and will be used further here. Figure 9.19 shows that in the seven-symbol codeword, five symbols of three bits each, A–E, are the data, and P and Q are the two redundant symbols. This simple

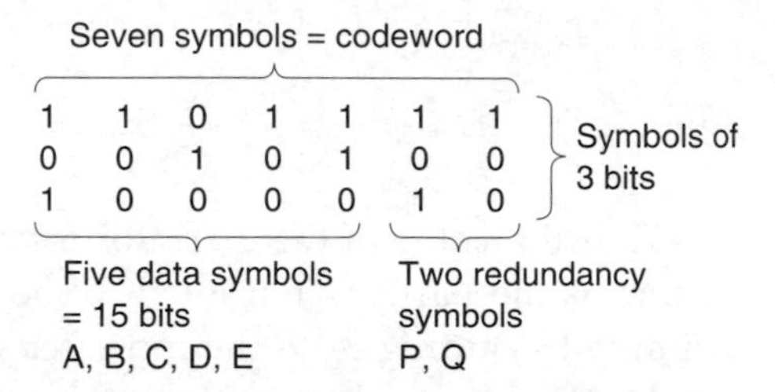

Figure 9.19 A Reed–Solomon codeword. As the symbols are of three bits, there can only be eight possible syndrome values. One of these is all zeros, the error-free case, and so it is only possible to point to seven errors; hence the codeword length of seven symbols. Two of these are redundant, leaving five data symbols.

example will locate and correct a single symbol in error. It does not matter, however, how many bits in the symbol are in error.

The two check symbols are solutions to the following equations:

$$A \oplus B \oplus C \oplus D \oplus E \oplus P \oplus Q = 0 \; [\oplus = \text{XOR symbol}]$$

$$a^7A \oplus a^6B \oplus a^5C \oplus a^4D \oplus a^3E \oplus a^2P \oplus aQ = 0$$

where a is a constant. The original data A–E followed by the redundancy P and Q pass through the channel.

The receiver makes two checks on the message to see if it is a codeword. This is done by calculating syndromes using the following expressions, where the (′) implies the received symbol which is not necessarily correct:

$$S_0 = A' \oplus B' \oplus C' \oplus D' \oplus E' \oplus P' \oplus Q'$$

(This is in fact a simple parity check.)

$$S_1 = a^7A' \oplus a^6B' \oplus a^5C' \oplus a^4D' \oplus a^3E' \oplus a^2P' \oplus aQ'$$

If two syndromes of all zeros are not obtained, there has been an error. The information carried in the syndromes will be used to correct the error. For the purpose of illustration, let it be considered that D′ has been corrupted before moving to the general case. D′ can be considered to be the result of adding an error of value X to the original value D such that $D' = D \oplus X$.

As $\quad A \oplus B \oplus C \oplus D \oplus E \oplus P \oplus Q = 0$

then $\quad A \oplus B \oplus C \oplus (D \oplus X) \oplus E \oplus P \oplus Q = E = S_0$

As $D' = D \oplus X$

then $D = D' \oplus X = D' \oplus S_0$

Thus the value of the corrector is known immediately because it is the same as the parity syndrome S_0. The corrected data symbol is obtained simply by adding S_0 to the incorrect symbol.

At this stage, however, the corrupted symbol has not yet been identified, but this is equally straightforward:

As $a^7A \oplus a^6B \oplus a^5C \oplus a^4D \oplus a^3E \oplus a^2P \oplus aQ = 0$

then:

$$a^7A \oplus a^6B \oplus a^5C \oplus a^4(D \oplus X) \oplus a^3E \oplus a^2P \oplus aQ = a^4 X = S_1$$

Thus the syndrome S_1 is the error bit pattern X, but it has been raised to a power of a which is a function of the position of the error symbol in the block. If the position of the error is in symbol k, then k is the locator value and:

$$S_0 \times a^k = S_1$$

Hence:

$$a^k = \frac{S_1}{S_0}$$

The value of k can be found by multiplying S_0 by various powers of a until the product is the same as S_1. Then the power of a necessary is equal to k. The use of the descending powers of a in the codeword calculation is now clear because the error is then multiplied by a different power of a dependent upon its position, known as the locator, because it gives the position of the error. The process of finding the error position by experiment is known as a Chien search.[7]

9.15 R–S calculations

Whilst the expressions above show that the values of P and Q are such that the two syndrome expressions sum to zero, it is not yet clear how P and Q are calculated from the data. Expressions for P and Q can be found by solving the two R–S equations simultaneously. This has been done in

Appendix 9.1. The following expressions must be used to calculate P and Q from the data in order to satisfy the codeword equations. These are:

$$P = a^6A \oplus aB \oplus a^2C \oplus a^5D \oplus a^3E$$

$$Q = a^2A \oplus a^3B \oplus a^6C \oplus a^4D \oplus aE$$

In both the calculation of the redundancy shown here and the calculation of the corrector and the locator it is necessary to perform numerous multiplications and raising to powers. This appears to present a formidable calculation problem at both the encoder and the decoder. This would be the case if the calculations involved were conventionally executed. However, the calculations can be simplified by using logarithms. Instead of multiplying two numbers, their logarithms are added. In order to find the cube of a number, its logarithm is added three times. Division is performed by subtracting the logarithms. Thus all of the manipulations necessary can be achieved with addition or subtraction, which is straightforward in logic circuits.

The success of this approach depends upon simple implementation of log tables. As was seen in Chapter 2, raising a constant, a, known as the *primitive element* to successively higher powers in modulo-2 gives rise to a Galois field. Each element of the field represents a different power n of a. It is a fundamental of the R–S codes that all the symbols used for data, redundancy and syndromes are considered to be elements of a Galois field. The number of bits in the symbol determines the size of the Galois field, and hence the number of symbols in the codeword.

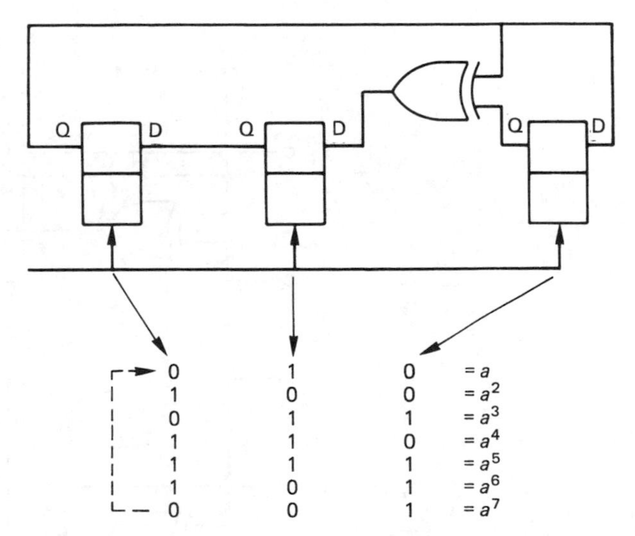

Figure 9.20 The bit patterns of a Galois field expressed as powers of the primitive element a. This diagram can be used as a form of log table in order to multiply binary numbers. Instead of an actual multiplication, the appropriate powers of a are simply added.

Figure 9.20 repeats a Galois field deduced in Chapter 3. The binary values of the elements are shown alongside the power of a they represent. In the R–S codes, symbols are no longer considered simply as binary numbers, but also as equivalent powers of a. In Reed–Solomon coding and decoding, each symbol will be multiplied by some power of a. Thus if the symbol is also known as a power of a it is only necessary to add the two powers. For example, if it is necessary to multiply the data symbol 100 by a^3, the calculation proceeds as follows, referring to Figure 9.20:

$$100 = a^2 \text{ so } 100 \times a^3 = a^{(2+3)} = a^5 = 111$$

Note that the results of a Galois multiplication are quite different from binary multiplication. Because all products must be elements of the field, sums of powers which exceed seven wrap around by having seven subtracted. For example:

$$a^5 \times a^6 = a^{11} = a^4 = 110$$

Figure 9.21 shows some examples of circuits which will perform this kind of multiplication. Note that they require a minimum amount of logic.

Figure 9.22 shows an example of the Reed–Solomon encoding process. The Galois field shown in Figure 9.20 has been used, having the primitive element a = 010. At the beginning of the calculation of P, the symbol A is multiplied by a^6. This is done by converting A to a power of a. According to Figure 9.20, 101 = a^6 and so the product will be $a^{(6+6)} = a^{12} = a^5 = 111$.

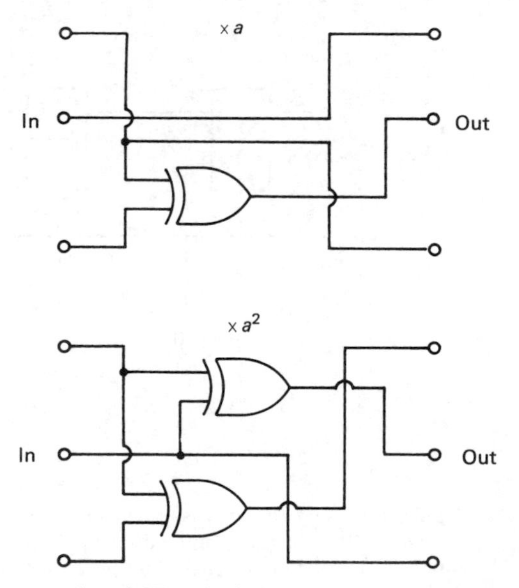

Figure 9.21 Some examples of GF multiplier circuits.

Input data	A	101	$a^6A = 111$	$a^2A = 010$
	B	100	$a\ B = 011$	$a^3B = 111$
	C	010	$a^2C = 011$	$a^6C = 001$
	D	100	$a^5D = 001$	$a^4D = 101$
	E	111	$a^3E = 010$	$a\ E = 101$
Check symbols	P	100 ←	100	
	Q	100 ←		100

Codeword	A	101	$a^7A = 101$
	B	100	$a^6B = 010$
	C	010	$a^5C = 101$
	D	100	$a^4D = 101$
	E	111	$a^3E = 010$
	P	100	$a^2P = 110$
	Q	100	$a\ Q = 011$
	S_0	= 000	$S_1 = 000$ ← Both syndromes zero

Figure 9.22 Five data symbols A–E are used as terms in the generator polynomials derived in Appendix 9.1 to calculate two redundant symbols P and Q. An example is shown at the top. Below is the result of using the codeword symbols A–Q as terms in the checking polynomials. As there is no error, both syndromes are zero.

In the same way, B is multiplied by a, and so on, and the products are added modulo-2. A similar process is used to calculate Q.

Figure 9.23 shows a circuit which can calculate P or Q. The symbols A–E are presented in succession, and the circuit is clocked for each one. On the first clock, a^6A is stored in the left-hand latch. If B is now provided at the input, the second GF multiplier produces aB and this is added to

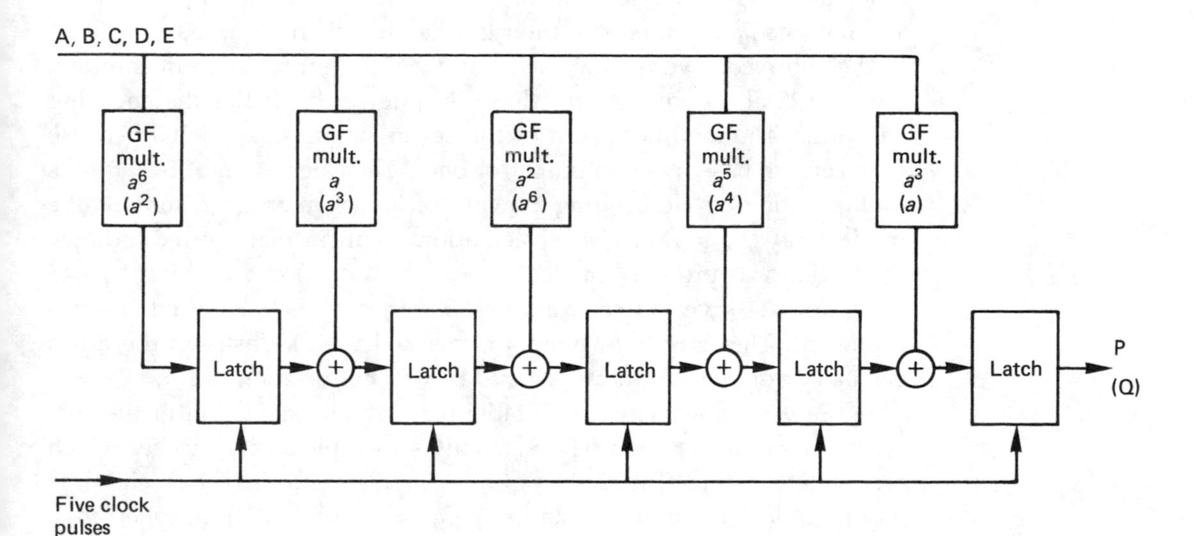

Figure 9.23 If the five data symbols of Figure 9.22 are supplied to this circuit in sequence, after five clocks, one of the check symbols will appear at the output. Terms without brackets will calculate P, bracketed terms calculate Q.

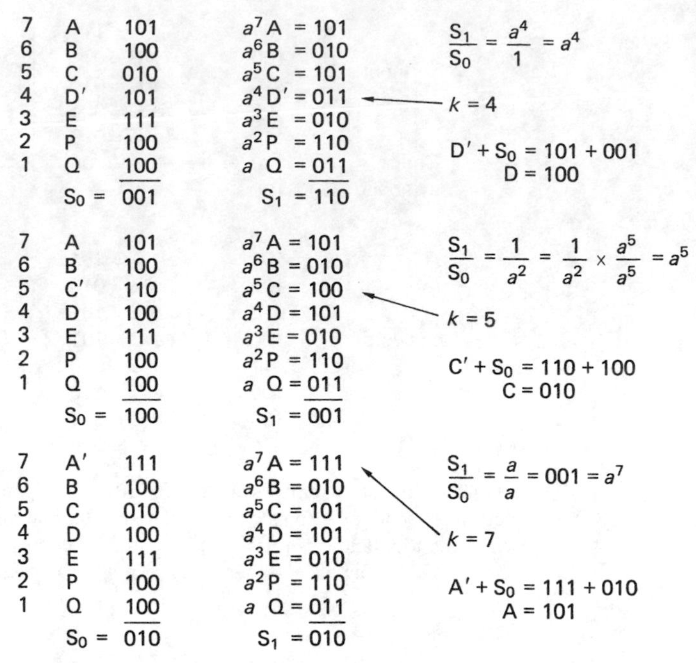

Figure 9.24 Three examples of error location and correction. The number of bits in error in a symbol is irrelevant; if all three were wrong, S_0 would be 111, but correction is still possible.

the output of the first latch and when clocked will be stored in the second latch which now contains a^6A + aB. The process continues in this fashion until the complete expression for P is available in the right-hand latch. The intermediate contents of the right-hand latch are ignored.

The entire codeword now exists, and can be recorded or transmitted. Figure 9.22 also demonstrates that the codeword satisfies the checking equations. The modulo-2 sum of the seven symbols, S_0, is 000 because each column has an even number of ones. The calculation of S_1 requires multiplication by descending powers of a. The modulo-2 sum of the products is again zero. These calculations confirm that the redundancy calculation was properly carried out.

Figure 9.24 gives three examples of error correction based on this codeword. The erroneous symbol is marked with a dash. As there has been an error, the syndromes S_0 and S_1 will not be zero.

Figure 9.25 shows circuits suitable for parallel calculation of the two syndromes at the receiver. The S_0 circuit is a simple parity checker which accumulates the modulo-2 sum of all symbols fed to it. The S_1 circuit is more subtle, because it contains a Galois field (GF) multiplier in a feedback loop, such that early symbols fed in are raised to higher powers than later symbols because they have been recirculated through the GF multiplier more often. It is possible to compare the operation of these

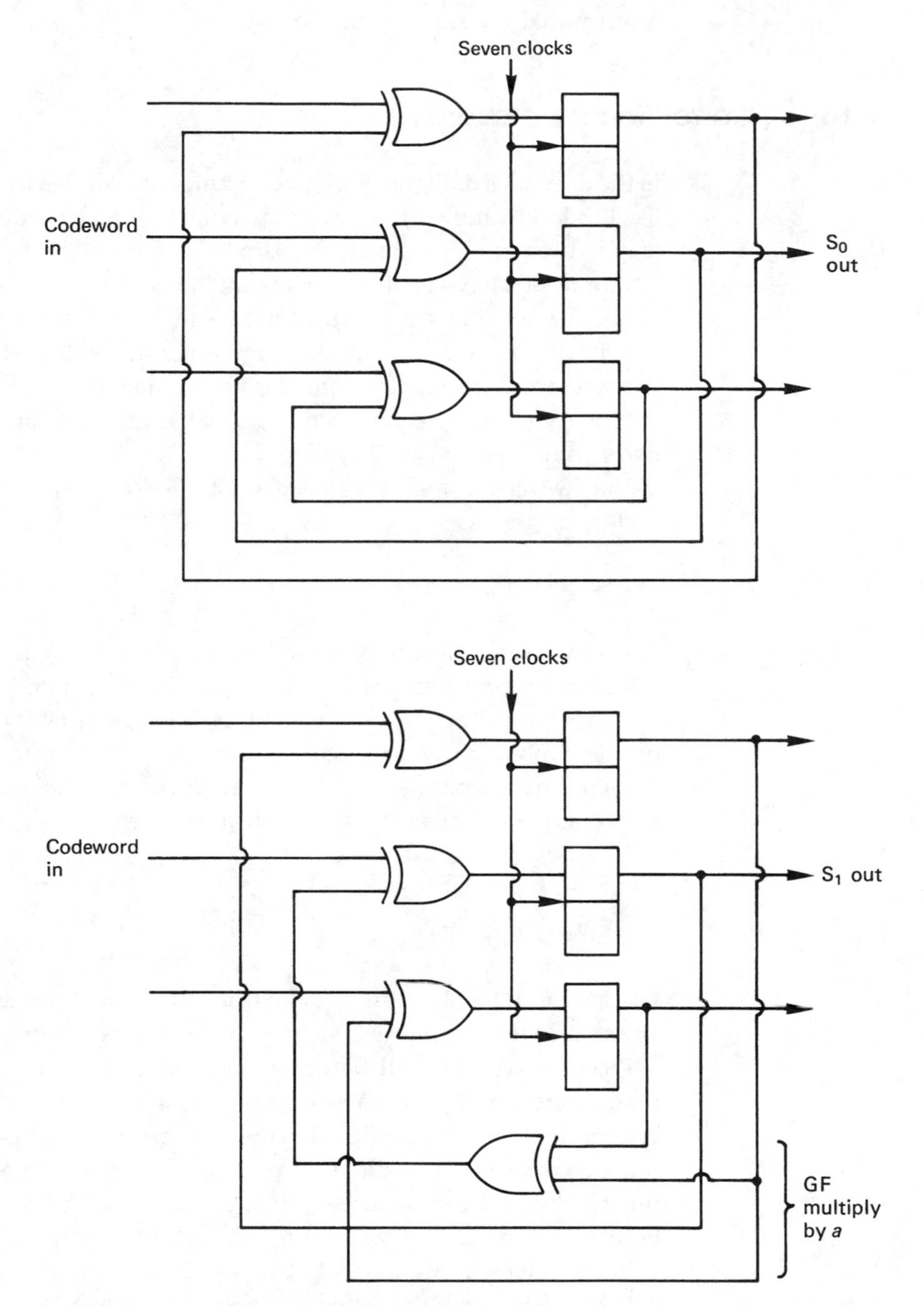

Figure 9.25 Circuits for parallel calculation of syndromes S_0, S_1, S_0 is a simple parity check. S_1 has a GF multiplication by a in the feedback, so that A is multiplied by a^7, B is multiplied by a^6, etc., and all are summed to give S_1.

circuits with the example of Figure 9.24 and with subsequent examples to confirm that the same results are obtained.

9.16 Correction by erasure

In the examples of Figure 9.24, two redundant symbols P and Q have been used to locate and correct one error symbol. If the positions of errors are known by some separate mechanism (see product codes, section 9.18) the locator need not be calculated. The simultaneous equations may instead be solved for two correctors. In this case the number of symbols which can be corrected is equal to the number of redundant symbols. In Figure 9.26(a) two errors have taken place, and it is known that they are in symbols C and D. Since S_0 is a simple parity check, it will reflect the modulo-2 sum of the two errors. Hence $S_0 = EC \oplus ED$.

The two errors will have been multiplied by different powers in S_1, such that:

$$S_1 = a^5\, EC \oplus a^4\, ED$$

These two equations can be solved, as shown in the figure, to find *EC* and *ED*, and the correct value of the symbols will be obtained by adding these correctors to the erroneous values. It is, however, easier to set the values of the symbols in error to zero. In this way the nature of the error is rendered irrelevant and it does not enter the calculation. This setting of symbols to zero gives rise to the term erasure. In this case,

$$S_0 = C \oplus D$$

$$S_1 = a^5C + a^4D$$

Erasing the symbols in error makes the errors equal to the correct symbol values and these are found more simply as shown in Figure 9.26(b).

Practical systems will be designed to correct more symbols in error than in the simple examples given here. If it is proposed to correct by erasure an arbitrary number of symbols in error given by t, the codeword must be divisible by t different polynomials. Alternatively if the errors must be located and corrected, $2t$ polynomials will be needed. These will be of the form $(x + a^n)$ where n takes all values up to t or $2t$. a is the primitive element discussed in Chapter 2.

Where four symbols are to be corrected by erasure, or two symbols are to be located and corrected, four redundant symbols are necessary, and the codeword polynomial must then be divisible by

$$(x + a^0)\,(x + a^1)\,(x + a^2)\,(x + a^3)$$

A	101	$a^7A =$	101
B	100	$a^6B =$	010
$(C \oplus E_C)$	001	$a^5(C \oplus E_C)$	111
$(D \oplus E_D)$	010	$a^4(D \oplus E_D)$	111
E	111	$a^3E =$	010
P	100	$a^2P =$	110
Q	100	$a\,Q =$	011
$S_1 =$	101	$S_1 =$	000

$$S_0 = E_C \oplus E_D \qquad S_1 = a^5 E_C \oplus a^4 E_D$$

$$S_1 = a^5 E_C \oplus a^4 (S_0 \oplus E_C)$$

$$= a^5 E_C \oplus a^4 S_0 \oplus a^4 E_C$$

$$\therefore E_C = \frac{S_1 \oplus a^4 S_0}{a^5 \oplus a^4} = \frac{000 \oplus 011}{001} = 011$$

$$C = (C \oplus E_C) \oplus E_C = 001 \oplus 011 = \underline{010}$$

$$S_1 = a^5 (S_0 \ominus E_D) \oplus a^4 E_D$$

$$= a^5 S_0 \oplus a^5 E_D \oplus a^4 E_D$$

$$\therefore E_D = \frac{S_1 \oplus a^5 S_0}{a^5 \oplus a^4} = \frac{000 \oplus 110}{001} = 110$$

$$D = (D \oplus E_D) + E_D = 010 \oplus 110 = \underline{100}$$

(a)

A	101	$a^7A =$	101	
B	100	$a^6B =$	010	$S_0 = C \oplus D$
C	000	$a^5C =$	000	$S_1 = a^5C \oplus a^4D$
D	000	$a^4D =$	000	
E	111	$a^3E =$	010	
P	100	$a^2P =$	110	
Q	100	$a\,Q =$	011	
S_0	= 100	$S_1 =$	000	

$$S_1 = a^5 S_0 \oplus a^5 D \oplus a^4 D = a^5 S_0 \oplus D$$

$$\therefore D = S_1 \oplus a^5 S_0 = 000 \oplus 100 = \underline{100}$$

$$S_1 = a^5 C \oplus a^4 C \oplus a^4 S_0 = C \oplus a^4 S_0$$

$$\therefore C = S_1 \oplus a^4 S_0 = 000 \oplus 010 = \underline{010}$$

(b)

Figure 9.26 If the location of errors is known, then the syndromes are a known function of the two errors as shown in (a). It is, however, much simpler to set the incorrect symbols to zero, i.e. to *erase* them as in (b). Then the syndromes are a function of the wanted symbols and correction is easier.

Upon receipt of the message, four syndromes must be calculated, and the four correctors or the two error patterns and their positions are determined by solving four simultaneous equations. This generally requires an iterative procedure, and a number of algorithms have been developed for the purpose.[8–10] Modern DVTR formats use eight-bit R–S codes and erasure extensively. The primitive polynomial commonly used with GF(256) is

$$x^8 + x^4 + x^3 + x^2 + 1$$

The codeword will be 255 bytes long but will often be shortened by puncturing. The larger Galois fields require less redundancy, but the computational problem increases. LSI chips have been developed specifically for R–S decoding in many high-volume formats.[11]

9.17 Interleaving

The concept of bit interleaving was introduced in connection with a single-bit correcting code to allow it to correct small bursts. With burst-correcting codes such as Reed–Solomon, bit interleave is unnecessary. In most channels, particularly high-density recording channels used for digital video or audio, the burst size may be many bytes rather than bits, and to rely on a code alone to correct such errors would require a lot of redundancy. The solution in this case is to employ symbol interleaving, as shown in Figure 9.27. Several codewords are encoded from input data, but

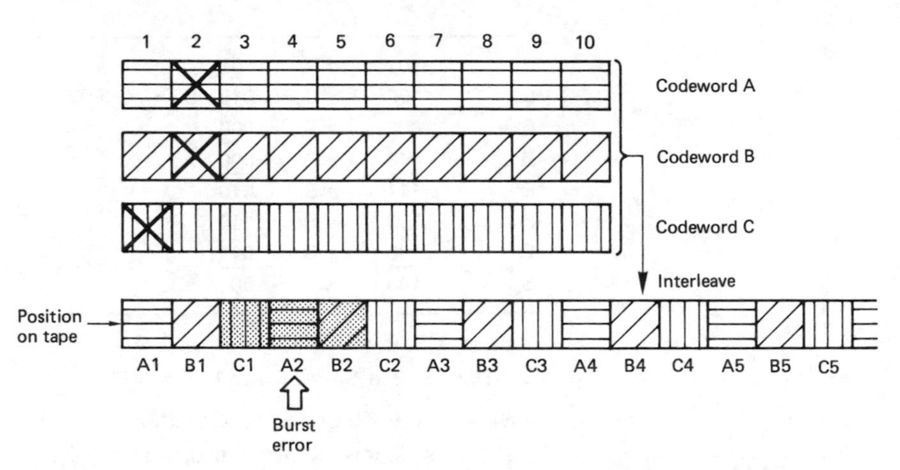

Figure 9.27 The interleave controls the size of burst errors in individual codewords.

these are not recorded in the order they were input, but are physically reordered in the channel, so that a real burst error is split into smaller bursts in several codewords. The size of the burst seen by each codeword is now determined primarily by the parameters of the interleave, and Figure 9.28 shows that the probability of occurrence of bursts with respect to the burst length in a given codeword is modified. The number of bits in the interleave word can be made equal to the burst-correcting ability of the code in the knowledge that it will be exceeded only very infrequently.

There are a number of different ways in which interleaving can be performed. Figure 9.29 shows that in block interleaving, words are reordered within blocks which are themselves in the correct order. This approach is attractive for rotary-head recorders, because the scanning

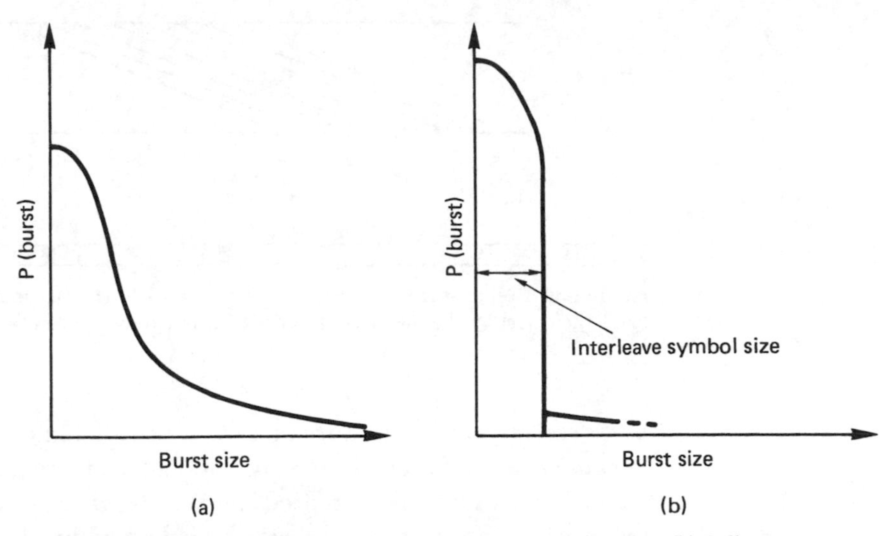

Figure 9.28 (a) The distribution of burst sizes might look like this. (b) Following interleave, the burst size within a codeword is controlled to that of the interleave symbol size, except for gross errors which have low probability.

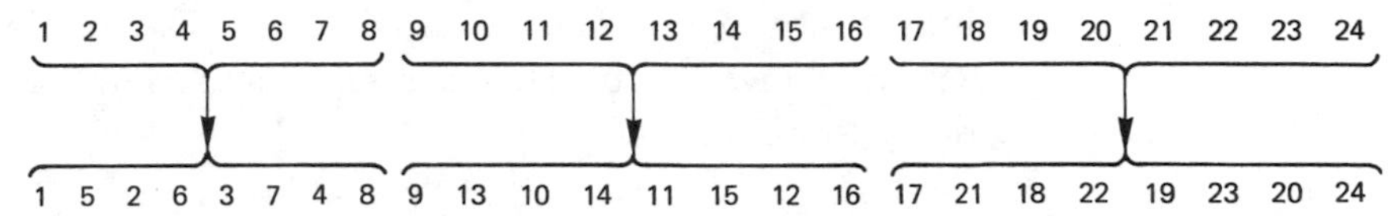

Figure 9.29 In block interleaving, data are scrambled within blocks which are themselves in the correct order.

process naturally divides the tape up into blocks. The block interleave is achieved by writing samples into a memory in sequential address locations from a counter, and reading the memory with non-sequential addresses from a sequencer. The effect is to convert a one-dimensional sequence of samples into a two-dimensional structure having rows and columns.

Rotary-head recorders naturally spatially interleave on the tape. Figure 9.30 shows that a single large tape defect becomes a series of small defects due to the geometry of helical scanning.

The alternative to block interleaving is convolutional interleaving where the interleave process is endless. In Figure 9.31 symbols are assembled into short blocks and then delayed by an amount proportional to the position in the block. It will be seen from the figure that the delays have the effect of shearing the symbols so that columns on the left side of the diagram become diagonals on the right. When the columns on the right are read, the convolutional interleave will be obtained. Convolutional interleave works well in transmission applications where there is no natural track break. Convolutional interleave has the advantage of requiring less memory to implement than a block code. This is because a

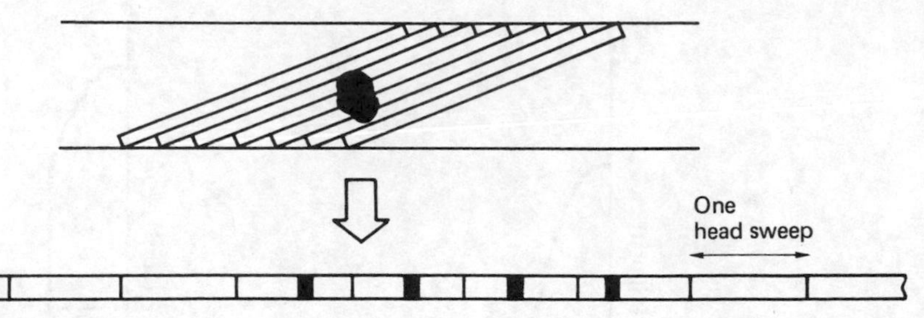

Figure 9.30 Helical-scan recorders produce a form of mechanical interleaving, because one large defect on the medium becomes distributed over several head sweeps.

block code requires the entire block to be written into the memory before it can be read, whereas a convolutional code requires only enough memory to cause the required delays. Now that RAM is relatively inexpensive, convolutional interleave is less popular.

It is possible to make a convolutional code of finite size by making a loop. Figure 9.32(a) shows that symbols are written in columns on the outside of a cylinder. The cylinder is then sheared or twisted, and the columns are read. The result is a block-completed convolutional interleave shown in (b). This technique is used in the audio blocks of the Video-8 format.

9.18 Product codes

In the presence of burst errors alone, the system of interleaving works very well, but it is known that in most practical channels there are also uncorrelated errors of a few bits due to noise. Figure 9.33 shows an interleaving system where a dropout-induced burst error has occurred which is at the maximum correctable size. All three codewords involved are working at their limit of one symbol. A random error due to noise in the vicinity of a burst error will cause the correction power of the code to be exceeded. Thus a random error of a single bit causes a further entire symbol to fail. This is a weakness of an interleave solely designed to handle dropout-induced bursts. Practical equipment must address the problem of random errors and burst errors occurring at the same time. This is done by forming codewords both before and after the interleave process. In block interleaving, this results in a *product code*, whereas in the case of convolutional interleave the result is called *cross-interleaving*.

Figure 9.34 shows that in a product code the redundancy calculated first and checked last is called the outer code, and the redundancy calculated second and checked first is called the inner code. The inner code is formed along tracks on the medium. Random errors due to noise are corrected by the inner code and do not impair the burst-correcting

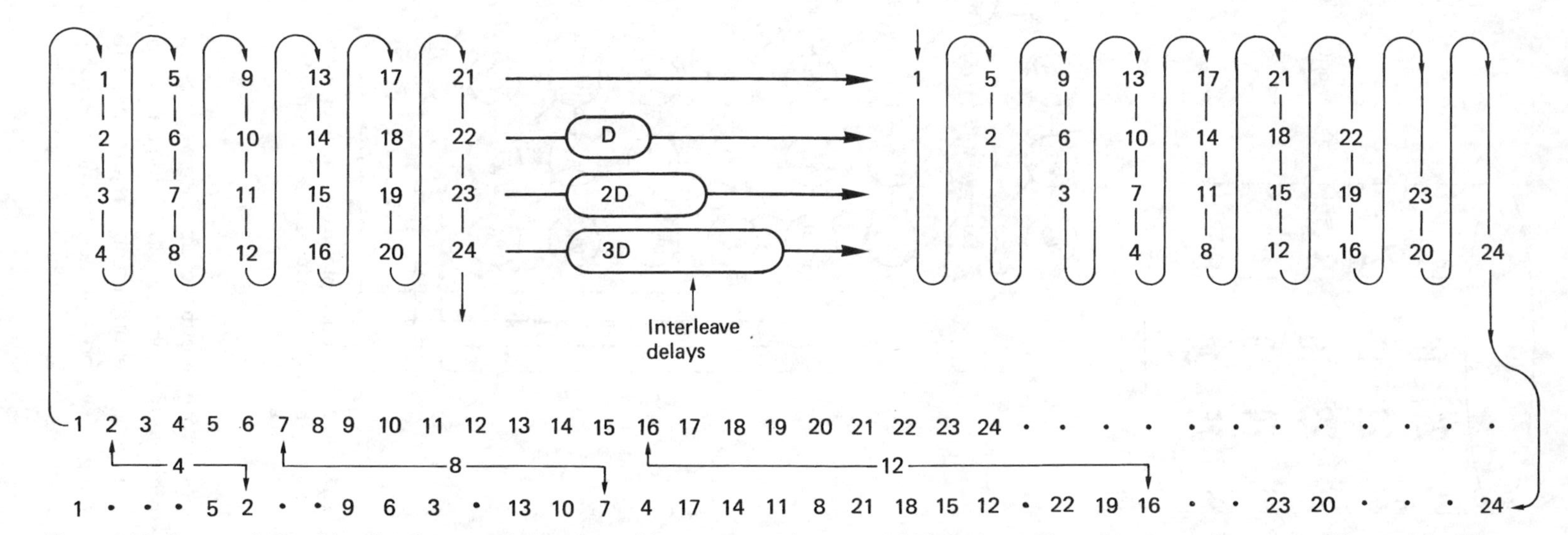

Figure 9.31 In convolutional interleaving, samples are formed into a rectangular array, which is sheared by subjecting each row to a different delay. The sheared array is read in vertical columns to provide the interleaved output. In this example, samples will be found at 4, 8 and 12 places away from their original order.

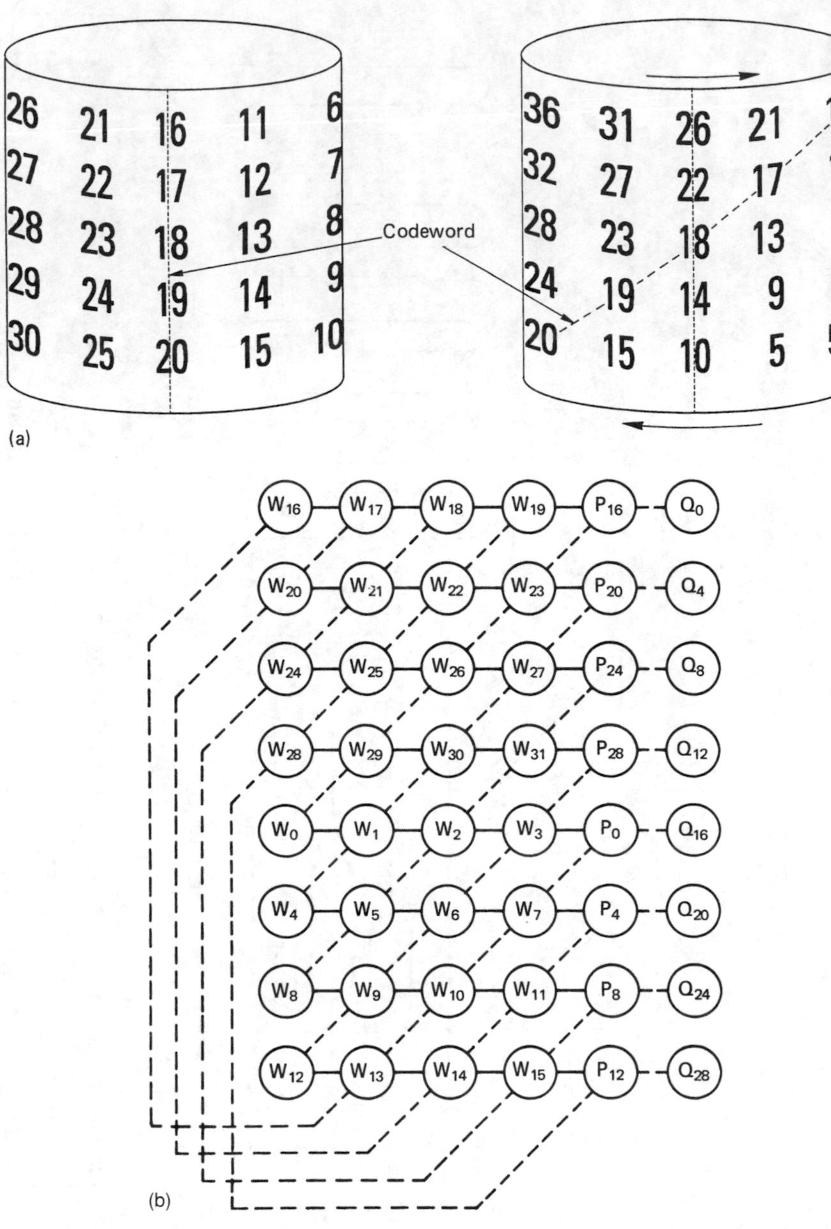

Figure 9.32 A block completed convolutional interleave can be considered to be the result of shearing a cylinder as in (a). This results in horizontal and diagonal codewords as shown in (b).

power of the outer code. Burst errors are declared uncorrectable by the inner code which flags the bad samples on the way into the de-interleave memory. The outer code reads the error flags in order to correct the flagged symbols by erasure. The error flags are also known as erasure flags. As it does not have to compute the error locations, the outer code

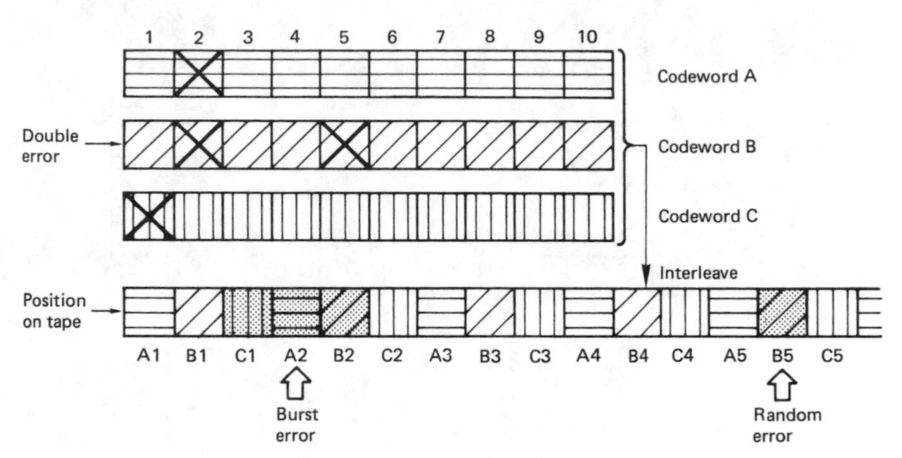

Figure 9.33 The interleave system falls down when a random error occurs adjacent to a burst.

needs half as much redundancy for the same correction power. Thus the inner code redundancy does not raise the code overhead. The combination of codewords with interleaving in several dimensions yields an error-protection strategy which is truly synergistic, in that the end result is more powerful than the sum of the parts. Needless to say, the technique is used extensively in modern DVTR formats.

9.19 Editing interleaved recordings

The interleave, de-interleave, time-compression and timebase-correction processes cause delay and any compression stages will further increase that delay. Editing must be undertaken with care. In a block-based interleave, edits can be made at block boundaries so that coded blocks are not damaged, but these blocks may be as big as a field. This is adequate for normal editing using vision cuts, but such blocks are usually too large for accurate audio editing, and editing within the video field to perform crossfades or wipes is impossible.

The only way in which accurate editing can be performed in the presence of interleave is to use a read–modify–write approach, where an entire field is read into memory and de-interleaved to the real-time sample sequence. Any desired part of the field can be replaced with new material before it is re-interleaved and re-recorded. In recorders which can only record or play at one time, an edit of this kind would take a long time because of all the tape repositioning needed. With extra heads read–modify–write editing can be performed dynamically as shown in Figure 9.35. The replay head plays back the existing

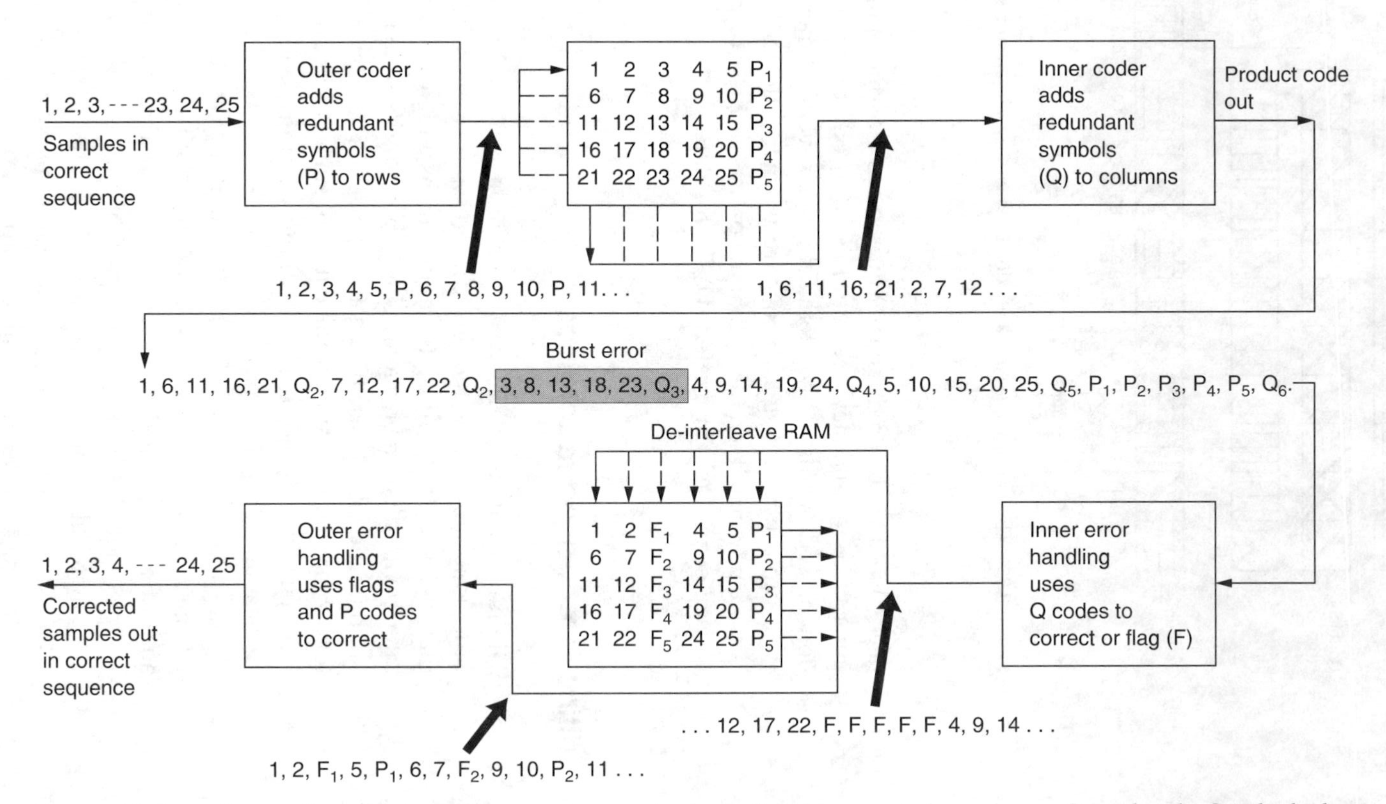

Figure 9.34 In addition to the redundancy P on rows, inner redundancy Q is also generated on columns. On replay, the Q code checker will pass on flags F if it finds an error too large to handle itself. The flags pass through the de-interleave process and are used by the outer error correction to identify which symbol in the row needs correcting with P redundancy. The concept of crossing two codes in this way is called a product code.

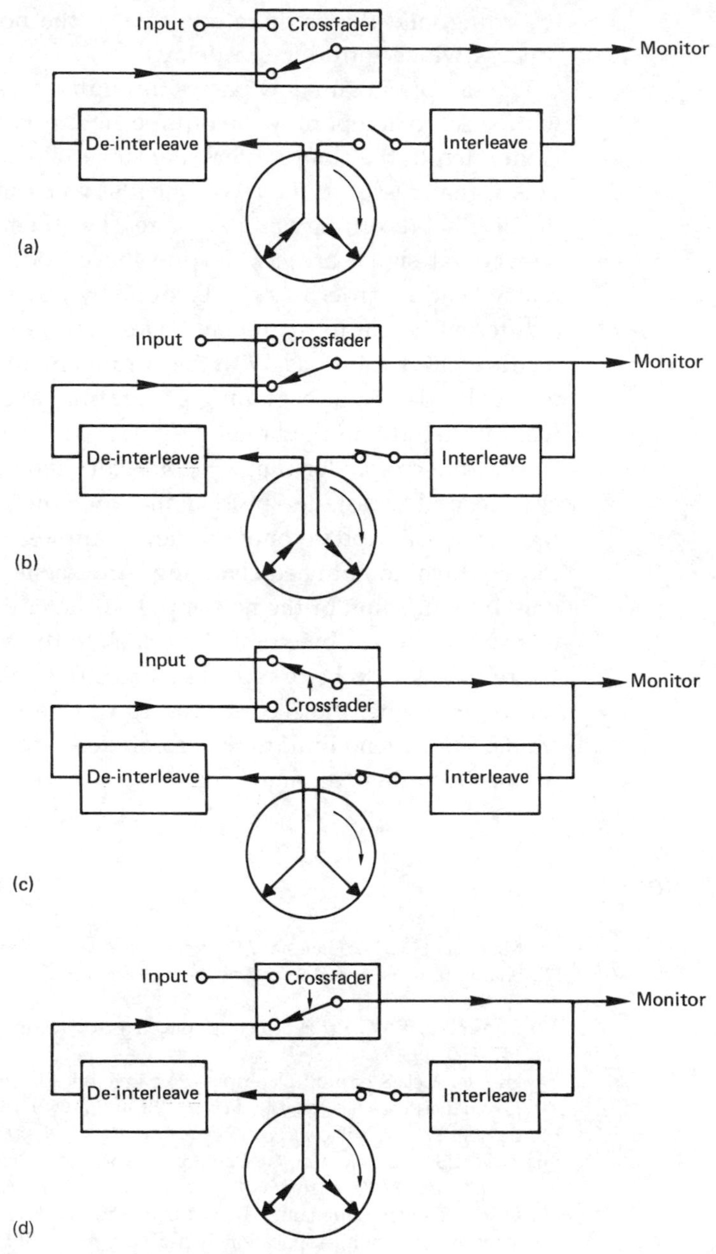

Figure 9.35 In read–modify–write or pre-read editing, there are advanced replay heads on the scanner, which allow editing to be performed on de-interleaved data. An insert sequence is shown. At (a) the replay-head signal is decoded, and fed to the encoder which, after some time, will produce an output representing what is already on the tape. In (b), at a sector boundary, the write circuits are turned on, and the machine begins to re-record. In (c) the crossfade is made to the insert material. At (d) the insert ends with a crossfade back to the signal from the advanced replay heads. After this, the write heads will once again be recording what is already on the tape, and the write circuits can be disabled at a sector boundary. An assemble edit consists of the first three of these steps only.

recording, and this is de-interleaved to the normal sample sequence, a process which introduces a delay.

The sample stream now passes through a crossfader which at this stage will be set to accept only the offtape signal. The output of the crossfader is then fed to the record interleave stage which introduces further delay. This signal passes to the record heads which must be positioned so that the original recording on the tape reaches them at the same time that the re-encoded signal arrives, despite the encode and decode delays. In a rotary head recorder this can be done by positioning the record heads at a different height to the replay heads so that they reach the same tracks on different revolutions. With this arrangement it is possible to enable the record heads at the beginning of a frame, and they will then re-record what is already on the tape.

Next the crossfader can be operated to fade across to new material, at any desired crossfade speed. Following the interleave stage, the new recording will update only the new samples in the frame and re-record those which do not need changing. After a short time, the recording will only be a function of the new input. If the edit is an insert, it is possible to end the process by crossfading back to the replay signal and allowing the replay data to be re-recorded. Once this re-recording has taken place for a short time, the record process can be terminated at the end of a frame. There is no limit to the crossfade periods which can be employed in this operating technique.

References

1. Michaels, S.R., Is it Gaussian? *Electronics World and Wireless World*, January, 72–73 (1993)
2. Shannon, C.E., A mathematical theory of communication. *Bell System Tech. J.*, **27** 379 (1948)
3. Hamming, R.W., Error-detecting and error-correcting codes. *Bell System Tech. J.*, **26**, 147–160 (1950)
4. Fire, P., A class of multiple-error correcting codes for non-independent errors. *Sylvania Receconnaissance Systems Lab. Report*, RSL-E-2 (1959)
5. Bossen, D.C., B-adjacent error correction. *IBM J. Res. Dev.*, **14**, 402–408 (1970)
6. Reed, I.S. and Solomon, G., Polynomial codes over certain finite fields. *J. Soc. Indust. Appl. Math.*, **8**, 300–304 (1960)
7. Chien, R.T.; Cunningham, B.D. and Oldham, I.B., Hybrid methods for finding roots of a polynomial – with application to BCH decoding. *IEEE Trans. Inf. Theory,* **IT-15**, 329–334 (1969)
8. Berlekamp, E.R., *Algebraic Coding Theory,* New York: McGraw-Hill (1967). Reprint edition: Laguna Hills, CA: Aegean Park Press (1983)
9. Sugiyama, Y. *et al.*, An erasures and errors decoding algorithm for Goppa codes. *IEEE Trans. Inf. Theory,* **IT-22** (1976)
10. Peterson, W.W. and Weldon, E.J., *Error Correcting Codes*, 2nd edn, Cambridge, MA: MIT Press (1972)
11. Onishi, K., Sugiyama, K., Ishida, Y., Kusonoki, Y. and Yamaguchi, T., An LSI for Reed–Solomon encoder/decoder. Presented at 80th Audio Engineering Society Convention (Montreux, 1986), Preprint 2316(A-4)

Appendix 9.1 Calculation of Reed–Solomon generator polynomials

For a Reed–Solomon codeword over GF(2^3), there will be seven three-bit symbols. For location and correction of one symbol, there must be two redundant symbols P and Q, leaving A-E for data.

The following expressions must be true, where a is the primitive element of $x^3 \oplus x \oplus 1$ and $\oplus$ is XOR throughout:

$$A \oplus B \oplus C \oplus D \oplus E \oplus P \oplus Q = 0 \quad (1)$$

$$a^7A \oplus a^6B \oplus a^5C \oplus a^4D \oplus a^3E \oplus a^2P \oplus aQ = 0 \quad (2)$$

Dividing equation (2) by a:

$$a^6A \oplus a^5B \oplus a^4C \oplus a^3D \oplus a^2E \oplus aP \oplus Q = 0$$

$$= A \oplus B \oplus C \oplus D \oplus E \oplus P \oplus Q$$

Cancelling Q, and collecting terms:

$$(a^6 \oplus 1)A \oplus (a^5 \oplus 1)B \oplus (a^4 \oplus 1)C \oplus (a^3 \oplus 1)D \oplus (a^2 \oplus 1)E$$

$$= (a + 1)P$$

Using section 3.27 to calculate $(a^n + 1)$, e.g. $a^6 + 1 = 101 + 001 = 100 = a^2$:

$$a^2A \oplus a^4B \oplus a^5C \oplus aD \oplus a^6E = a^3P$$

$$a^6A \oplus aB \oplus a^2C \oplus a^5D \oplus a^3E = P$$

Multiply equation (1) by a^2 and equating to equation (2):

$$a^2A \oplus a^2B \oplus a^2C \oplus a^2D \oplus a^2E \oplus a^2P \oplus a^2Q = 0$$

$$= a^7A \oplus a^6B \oplus a^5C \oplus a^4D \oplus a^3E \oplus a^2P \oplus aQ$$

Cancelling terms a^2P and collecting terms (remember $a^2 \oplus a^2 = 0$):

$$(a^7 \oplus a^2)A \oplus (a^6 \oplus a^2)B \oplus (a^5 \oplus a^2)C \oplus (a^4 \oplus a^2)D \oplus$$

$$(a^3 \oplus a^2)E = (a^2 \oplus a)Q$$

Adding powers according to section 3.27, e.g.

$$a^7 \oplus a^2 = 001 \oplus 100 = 101 = a^6:$$

$$a^6A \oplus B \oplus a^3C \oplus aD \oplus a^5E = a^4Q$$

$$a^2A \oplus a^3B \oplus a^6C \oplus a^4D \oplus aE = Q$$

10

Digital interfaces

10.1 Introduction

Of all the advantages of digital video the most important of these for production work is the ability to pass through multiple generations without quality loss. Effects machines perform transforms on images in the digital domain which remain impossible in the analog domain. For the highest quality post-production work, digital interconnection between such items as switchers, recorders and effects machines is highly desirable to avoid the degradation due to repeated conversion and filtering stages.

In 4:2:2 digital colour difference sampling according to CCIR 601, the luminance is sampled at 13.5 MHz, which is line synchronous to both broadcast line rates, and the two colour difference signals are sampled at one half that frequency using eight- or ten-bit resolution. This is the format most commonly used in broadcast production systems.

Early digital interfaces used parallel connection which had numerous drawbacks, not least that a router will be extremely complex. The answer to these problems is the serial connection. All the digital samples are multiplexed into a serial bitstream, and this is encoded to form a self-clocking channel code which can be sent down a single channel. The bit rate necessary is very high, but this is easily accommodated by coaxial cable. A distinct advantage of serial transmission is that a matrix distribution unit or router is more easily realized. Where numerous pieces of video equipment need to be interconnected in various ways for different purposes, a cross-point matrix is an obvious solution. With serial signals, only one switching element per signal is needed.

10.2 Interface principles

Interfacing can be by electrical conductors or optical fibre. Although these appear to be completely different, they are in fact just different examples of electromagnetic energy travelling from one place to another. If the energy is made to vary in some way, information can be carried.

Electromagnetism is not fully understood, but sufficiently good models, based on experimental results, exist so that practical equipment can be made. Electromagnetic energy propagates in a manner which is a function of frequency, and our partial understanding requires it to be considered as electrons, waves or photons so that we can predict its behaviour in given circumstances.

At DC and at the low frequencies used for power distribution, electromagnetic energy is called electricity and it needs to be guided by conductors. It has to have a complete circuit to flow in, and the resistance to current flow is determined by the cross-sectional area of the conductor. The insulation around the conductor and the spacing between the conductors have no effect on the ability of the conductor to pass current. At DC an inductor appears to be a short circuit, and a capacitor appears to be an open circuit.

As frequency rises, resistance is exchanged for impedance. Inductors display increasing impedance with frequency, capacitors show falling impedance. Electromagnetic energy tends increasingly to leave the conductor. The first symptom is that the current flows only in the outside layer of the conductor effectively causing the resistance to rise. This is the skin effect and gives rise to such techniques as Litz wire which has as much surface area as possible per unit cross-section, and to silver-plated conductors in which the surface has lower resistivity than the interior.

As the energy is starting to leave the conductors, the characteristics of the space between them become important. This determines the impedance. A change of impedance causes reflections in the energy flow and some of it heads back towards the source. At frequencies used in serial digital video, constant-impedance cables with fixed conductor spacing are necessary, and these must be suitably terminated to prevent reflections. The most important characteristic of the insulation is its thickness as this determines the spacing between the conductors.

At high frequencies the energy travels less in the conductors and more in the insulation between them, and their composition becomes important and they begin to be called dielectrics. A poor dielectric like PVC absorbs high-frequency energy and attenuates the signal. So-called low-loss dielectrics such as PTFE are used, and one way of achieving low loss is to incorporate as much air in the dielectric as possible by making it in the form of a foam or extruding it with voids.

This frequency-dependent behaviour is the most important factor in deciding how best to harness electromagnetic energy flow for information transmission. It is obvious that the higher the frequency, the greater the possible information rate, but in general, losses increase with frequency, and flat frequency response is elusive. The best that can be managed is that over a narrow band of frequencies, the response can be made reasonably constant with the help of equalization. Unfortunately raw data when serialized have an unconstrained spectrum. Runs of identical bits can produce frequencies much lower than the bit rate would suggest. One of the essential steps in a transmission system is to modify the spectrum of the data into something more suitable.

At moderate bit rates, say a few megabits per second, and with moderate cable lengths, say a few metres, the dominant effect will be the capacitance of the cable due to the geometry of the space between the conductors and the dielectric between. The capacitance behaves under these conditions as if it were a single capacitor connected across the signal. Figure 10.1 shows the equivalent circuit.

The effect of the series source resistance and the parallel capacitance is that signal edges or transitions are turned into exponential curves as the capacitance is effectively being charged and discharged through the source impedance. This effect can be observed on the AES/EBU interface with short cables. Although the position where the edges cross the centreline is displaced, the signal eventually reaches the same amplitude as it would at DC.

As cable length increases, the capacitance can no longer be lumped as if it were a single unit; it has to be regarded as being distributed along the cable. With rising frequency, the cable inductance also becomes significant, and it too is distributed.

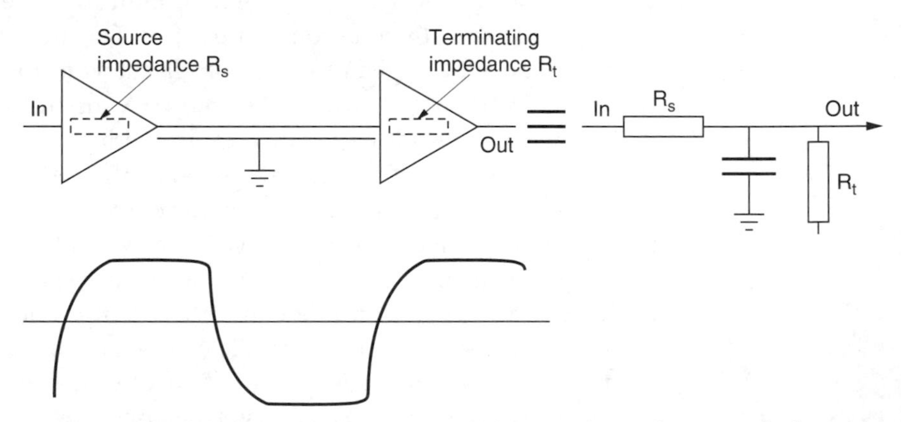

Figure 10.1 With a short cable, the capacitance between the conductors can be lumped as if it were a discrete component. The effect of the parallel capacitor is to slope off the edges of the signal.

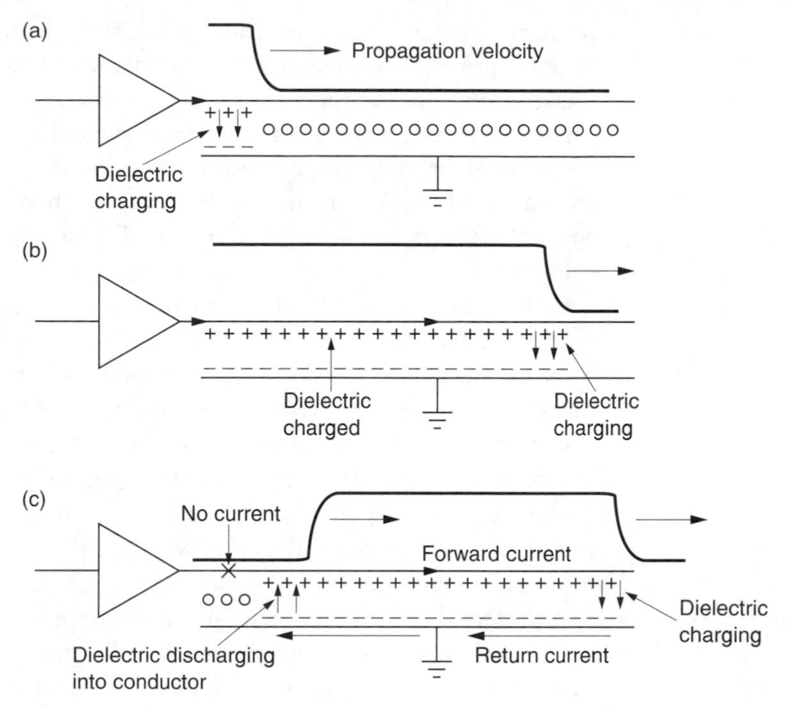

Figure 10.2 A transmission line conveys energy packets which appear with respect to the dielectric. In (a) the driver launches a pulse which charges the dielectric at the beginning of the line. As it propagates the dielectric is charged further along as in (b). When the driver ends the pulse, the charged dielectric discharges into the line. A current loop is formed where the current in the return loop flows in the opposite direction to the current in the 'hot' wire.

The cable is now a transmission line and pulses travel down it as current loops which roll along as shown in Figure 10.2. If the pulse is positive, as it is launched along the line, it will charge the dielectric locally as in (a). As the pulse moves along, it will continue to charge the local dielectric as in (b). When the driver finishes the pulse, the trailing edge of the pulse follows the leading edge along the line. The voltage of the dielectric charged by the leading edge of the pulse is now higher than the voltage on the line, and so the dielectric discharges into the line as in (c). The current flows forward as it is in fact the same current which is flowing into the dielectric at the leading edge. There is thus a loop of current rolling down the line flowing forward in the 'hot' wire and backwards in the return. The analogy with the tracks of a Caterpillar tractor is quite good. Individual plates in the track find themselves being lowered to the ground at the front and raised again at the back.

The constant interchange of charge in the dielectric results in dielectric loss of signal energy. Dielectric loss increases with frequency and so a long transmission line acts as a filter. Thus the term 'low-loss' cable refers primarily to the kind of dielectric used. Transmission lines which transport energy in this way have a characteristic impedance caused by the interplay of the inductance along the conductors with the parallel capacitance. One consequence of that transmission mode is that correct termination or matching is required between the line and both the driver and the receiver. When a line is correctly matched, the rolling energy rolls straight out of the line into the load and the maximum energy is available.

If the impedance presented by the load is incorrect, there will be reflections from the mismatch. An open circuit will reflect all the energy back in the same polarity as the original, whereas a short circuit will reflect all the energy back in the opposite polarity. Thus impedances above or below the correct value will have a tendency towards reflections whose magnitude depends upon the degree of mismatch and whose polarity depends upon whether the load is too high or too low. In practice it is the need to avoid reflections which is the most important reason to terminate correctly.

A perfectly square pulse contains an indefinite series of harmonics, but the higher ones suffer progressively more loss. A square pulse at the driver becomes less and less square with distance as Figure 10.3 shows. The harmonics are progressively lost until in the extreme case all that is left is the fundamental. A transmitted square wave is received as a sine wave. Fortunately data can still be recovered from the fundamental signal component. Once all the harmonics have been lost, further losses cause the amplitude of the fundamental to fall. The effect worsens with distance and it is necessary to ensure that data recovery is still possible from a signal of unpredictable level.

Fibre-optics have some advantages for digital interfacing. The advantages of this technology are numerous: the bandwidth of an optical fibre is staggering, as it is determined primarily by the response speed of the light source and sensor. The optical transmission is immune to electromagnetic interference from other sources, nor does it contribute any. This is advantageous for connections between cameras and control units, where a long cable run may be required in outside broadcast applications. The cable can be made completely from insulating materials, so that ground loops cannot occur, although many practical fibre-optic cables include electrical conductors for power and steel strands for mechanical strength.

Drawbacks of fibre-optics are few. They do not like too many connectors in a given channel, as the losses at a connection are much greater than with an electrical plug and socket. It is preferable for the only breaks in the fibre to be at the transmitting and receiving points. For similar reasons, fibre-optics are less suitable for distribution, where one source feeds many destinations. The familiar loop-through connection of

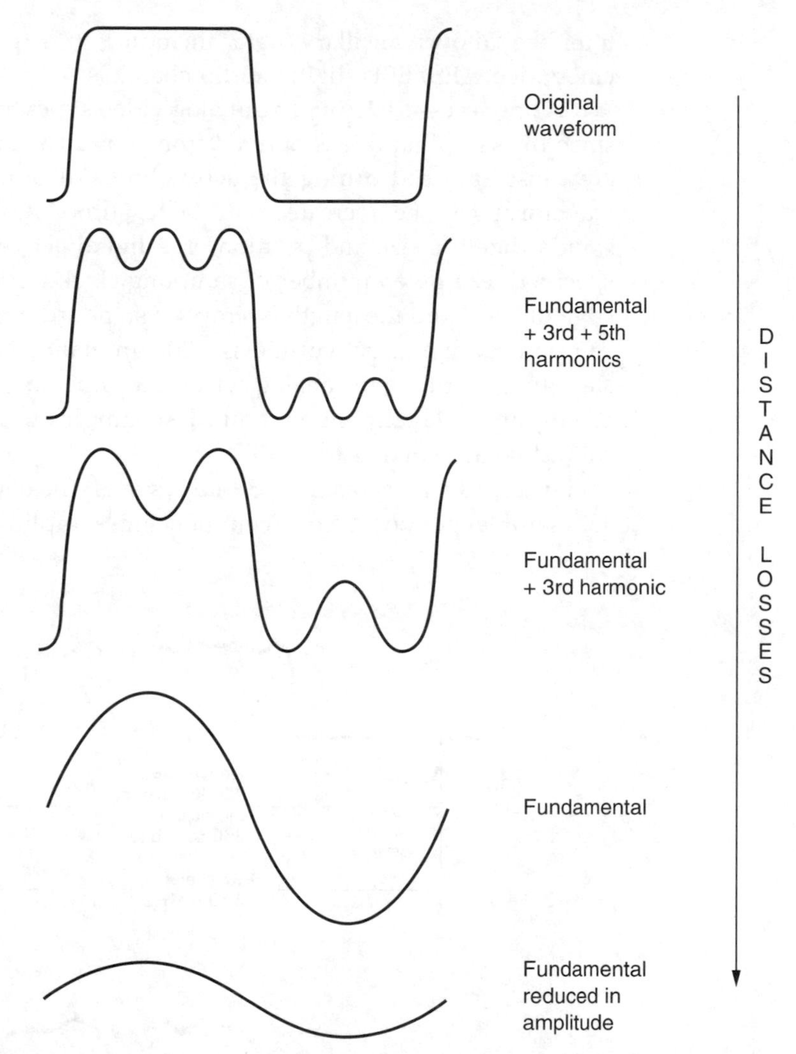

Figure 10.3 A signal may be square at the transmitter, but losses increase with frequency, and as the signal propagates, more of the harmonics are lost until only the fundamental remains. The amplitude of the fundamental then falls with further distance.

analog video is just not possible. The bidirectional open-collector or tri-state buses of electronic systems cannot be implemented with fibre-optics, nor is it easy to build a cross-point matrix.

10.3 Serial digital interface (SDI)

SDI was developed to allow up to ten-bit samples of component or composite digital video to be communicated serially.[1] 16:9 format component signals with 18 MHz sampling rate can also be handled. The interface allows ancillary data including transparent conveyance of embedded AES/EBU digital audio channels.

It is not necessary to digitize analog video syncs in component systems, since the sampling rate is derived from sync. The only useful video data are those sampled during the active line. All other parts of the video waveform can be recreated at a later time. It is only necessary to standardize the size and position of a digital active line. The position is specified as a given number of sampling clock periods from the leading edge of sync, and the length is simply a standard number of samples. The component digital active line is 720 luminance samples long. This is slightly longer than the analog active line and allows for some drift in the analog input. Ideally the first and last samples of the digital active line should be at blanking level.

Figure 10.4 shows that in 625-line systems[2] the control system waits for 132 sample periods before commencing sampling the line. Then 720

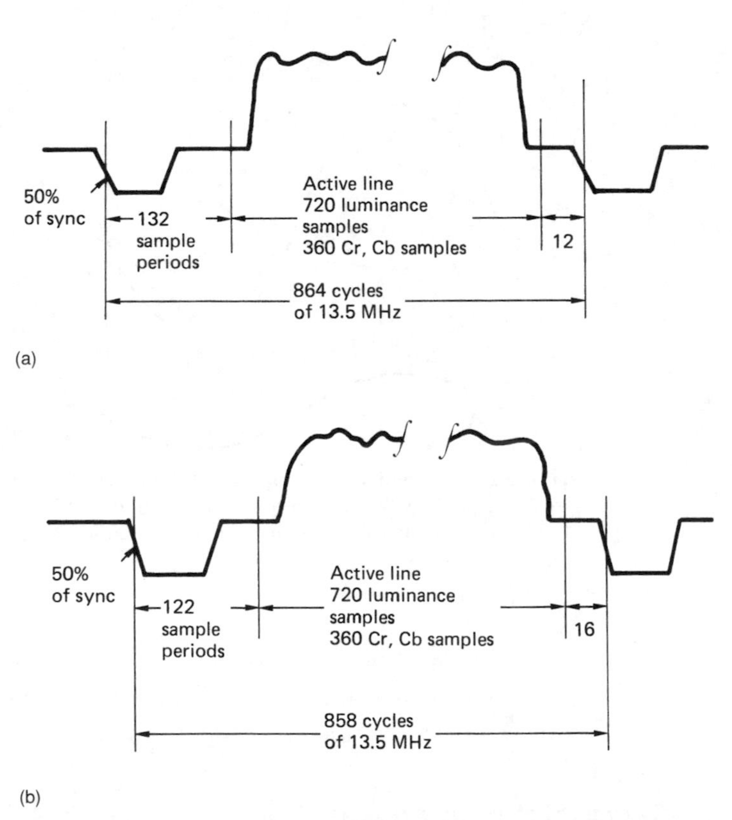

Figure 10.4 (a) In 625-line systems to CCIR-601, with 4:2:2 sampling, the sampling rate is exactly 864 times line rate, but only the active line is sampled, 132 sample periods after sync. (b) In 525 line systems to CCIR-601, with 4:2:2 sampling, the sampling rate is exactly 858 times line rate, but only the active line is sampled, 122 sample periods after sync. Note active line contains exactly the same quantity of data as for 50 Hz systems.

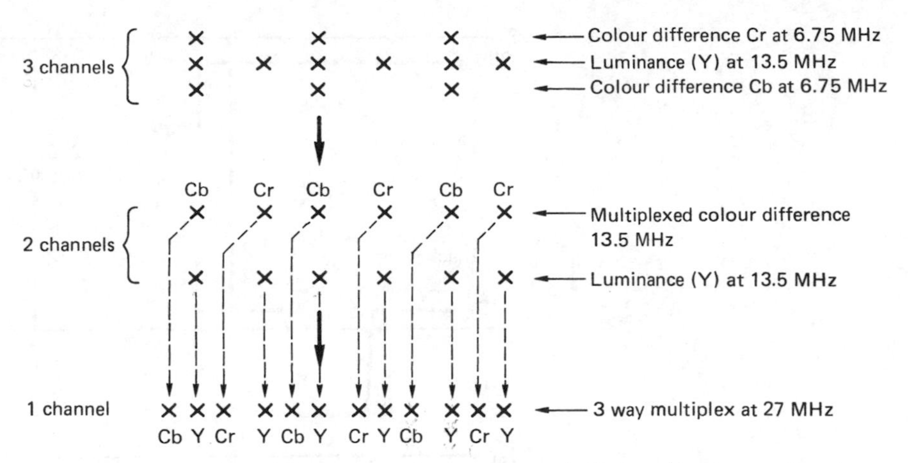

Figure 10.5 The colour difference sampling rate is one-half that of luminance, but there are *two* colour difference signals, C_r and C_b hence the colour difference data rate is equal to the luminance data rate, and a 27 MHz interleaved format is possible in a single channel.

luminance samples and 360 of each type of colour difference sample are taken; 1440 samples in all. A further 12 sample periods will elapse before the next sync edge, making 132 + 720 + 12 = 864 sample periods. In 525-line systems[3] the analog active line is in a slightly different place and so the controller waits 122 sample periods before taking the same digital active line samples as before. There will then be 16 sample periods before the next sync edge, making 122 + 720 + 16 = 858 sample periods.

Figure 10.5 shows the luminance signal sampled at 13.5 MHz and two colour difference signals sampled at 6.75 MHz. Three separate signals with different clock rates are inconvenient and so multiplexing can be used. If the colour difference signals are multiplexed into one channel, then two 13.5 MHz channels will be required. If these channels are multiplexed into one, a 27 MHz clock will be required. The word order will be:

C_b, Y, C_r, Y, etc.

In order unambiguously to deserialize the samples, the first sample in the line is always C_b.

In addition to specifying the location of the samples, it is also necessary to standardize the relationship between the absolute analog voltage of the waveform and the digital code value used to express it so that all machines will interpret the numerical data in the same way. These relationships are in the voltage domain and are independent of the line standard used.

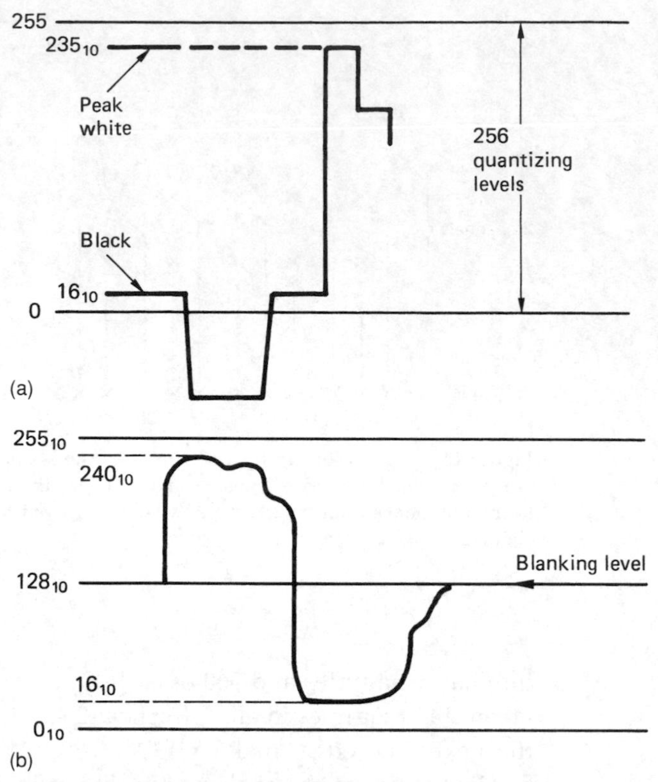

Figure 10.6 (a) The luminance signal fits into the quantizing range with a little allowance for excessive gain. Here black is at 16_{10} and peak white is at 235_{10}. The sync pulse goes outside the quantizing range but this is of no consequence as it is not transmitted. (b) The colour difference signals use offset binary where blanking level is at 128_{10}, and the peaks occur at 16_{10} and 240_{10} respectively.

Figure 10.6 shows how the luminance signal fits into the quantizing range of an eight-bit system. Black is at a level of 16_{10} and peak white is at 235_{10} so that there is some tolerance of imperfect analog signals. The sync pulse will clearly go outside the quantizing range, but this is of no consequence as conventional syncs are not transmitted. The visible voltage range fills the quantizing range and this gives the best possible resolution.

The colour difference signals use offset binary, where 128_{10} is the equivalent of blanking voltage. The peak analog limits are reached at 16_{10} and 240_{10} respectively allowing once more some latitude for maladjusted analog inputs.

Note that the code values corresponding to all ones and all zeros, i.e. the two extreme ends of the quantizing range are not allowed to occur in the active line as they are reserved for synchronizing. Convertors must be followed by circuitry which catches these values and forces the LSB to a different value if out of range analog inputs are applied.

The peak-to-peak amplitude of Y is 220 quantizing intervals, whereas for the colour difference signals it is 225 intervals. There is thus a small gain difference between the signals. This will be cancelled out by the opposing gain difference at any future DAC, but must be borne in mind when digitally converting to other standards.

As conventional syncs are not sent, horizontal and vertical synchronizing is achieved by special bit patterns sent with each line. Immediately before the digital active line location is the *SAV* (start of active video) pattern, and immediately after is the *EAV* (end of active video) pattern. These unique patterns occur on every line and continue throughout the vertical interval.

Each sync pattern consists of four symbols. The first is all ones and the next two are all zeros. As these cannot occur in active video, their detection reliably indicates a sync pattern. The fourth symbol is a data byte which contains three data bits, *H*, *F* and *V*. These bits are protected by four redundancy bits which form a seven-bit Hamming codeword for the purpose of detecting and correcting errors. Figure 7.10(a) shows the structure of the sync pattern. The sync bits have the following meanings:

H is used to distinguish between *SAV*, where it is set to 0 and *EAV* where it is set to 1.
F defines the state of interlace and is 0 during the first field and 1 during the second field. *F* is only allowed to change at *EAV*. In interlaced systems, one field begins at the centre of a line, but there is no sync pattern at that location so the field bit changes at the end of the line in which the change took place.
V is 1 during vertical blanking and 0 during the active part of the field. It can only change at *EAV*.

Figure 10.7(b) (top) shows the relationship between the sync pattern bits and 625-line analog timing, whilst below is the relationship for 525 lines.

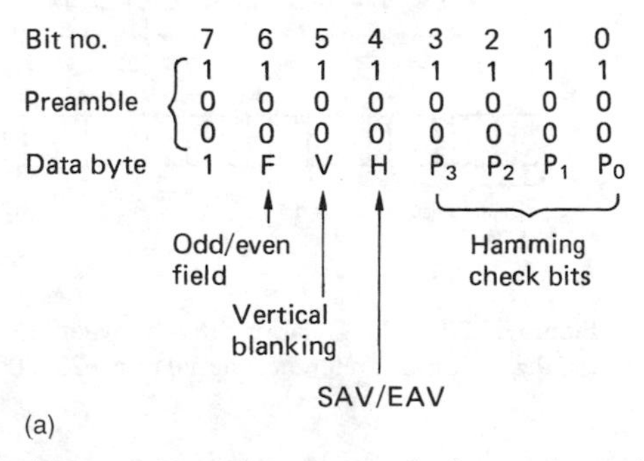

Figure 10.7(a) The 4-byte synchronizing pattern which precedes and follows every active line sample block has this structure.

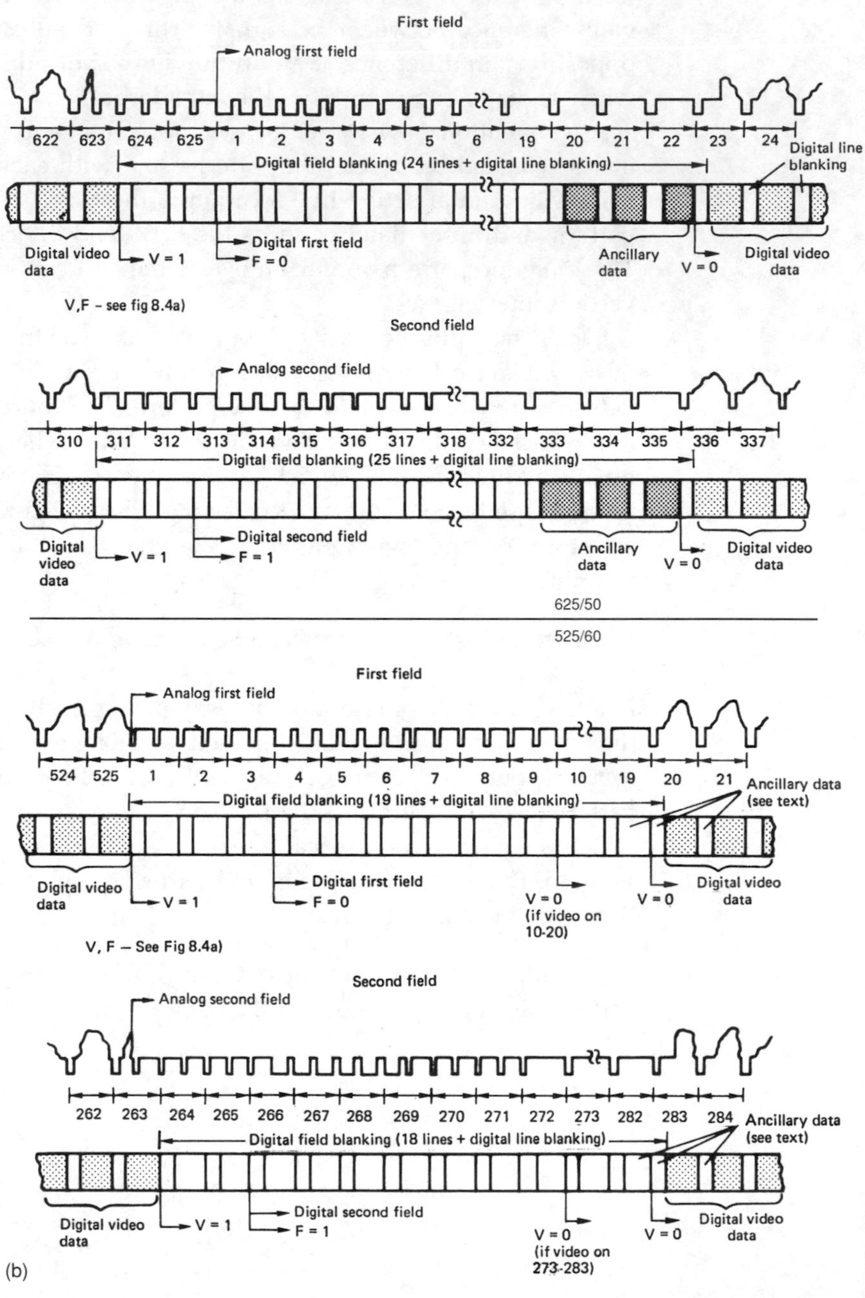

Figure 10.7(b) The relationships between analog video timing and the information in the digital timing reference signals for 625/50 (above) and 525/60 (below).

Only the active line is transmitted and this leaves a good deal of spare capacity. The two line standards differ on how this capacity is used. In 625 lines, only the active line period may be used on lines 20 to 22 and 333 to 335.[2] Lines 20 and 333 are reserved for equipment self-testing. In 525 lines there is considerably more freedom and ancillary data may be inserted anywhere there is no active video, either during horizontal blanking, vertical blanking, or both.[3]

The all zeros and all ones codes are reserved for synchronizing, and cannot be allowed to appear in ancillary data. In practice only seven bits of the eight-bit word can be used as data; the eighth bit is redundant and gives the byte odd parity. As all ones and all zeros are even parity, the sync pattern cannot then be generated accidentally.

Chapter 8 introduced the concepts of DC components and uncontrolled clock content in serial data for recording and the same issues are important in interfacing, leading to a coding requirement. SDI uses convolutional randomizing, in which the signal sent down the channel is the serial data waveform which has been convolved with the impulse response of a digital filter. On reception the signal is deconvolved to restore the original data. Figure 10.8(a) shows that the filter is an infinite impulse response (IIR) filter which has recursive paths from the output back to the input. As it is a one-bit filter its output cannot decay, and once excited, it runs indefinitely. The filter is followed by a transition generator which consists of a one-bit delay and an exclusive-OR gate. An input 1 results in an output transition on the next clock edge. An input 0 results in no transition.

A result of the infinite impulse response of the filter is that frequent transitions are generated in the channel which leads to sufficient clock content for the phase-locked loop in the receiver. Transitions are converted back to 1s by a differentiator in the receiver. This consists of a one-bit delay with an exclusive-OR gate comparing the input and the output. When a transition passes through the delay, the input and the output will be different and the gate outputs a 1 which enters the de-convolution circuit.

Figure 10.8(b) shows that in the de-convolution circuit a data bit is simply the exclusive-OR of a number of channel bits at a fixed spacing. The deconvolution is implemented with a shift register having the exclusive-OR gates connected in a reverse pattern to that in the encoder. The same effect as block randomizing is obtained, in that long runs are broken up and the DC content is reduced, but it has the advantage over block randomizing in that no synchronizing is required to remove the randomizing, although it will still be necessary for deserialization. Clearly the system will take a few clock periods to produce valid data after commencement of transmission, but this is no problem on a permanent wired connection where the transmission is continuous.

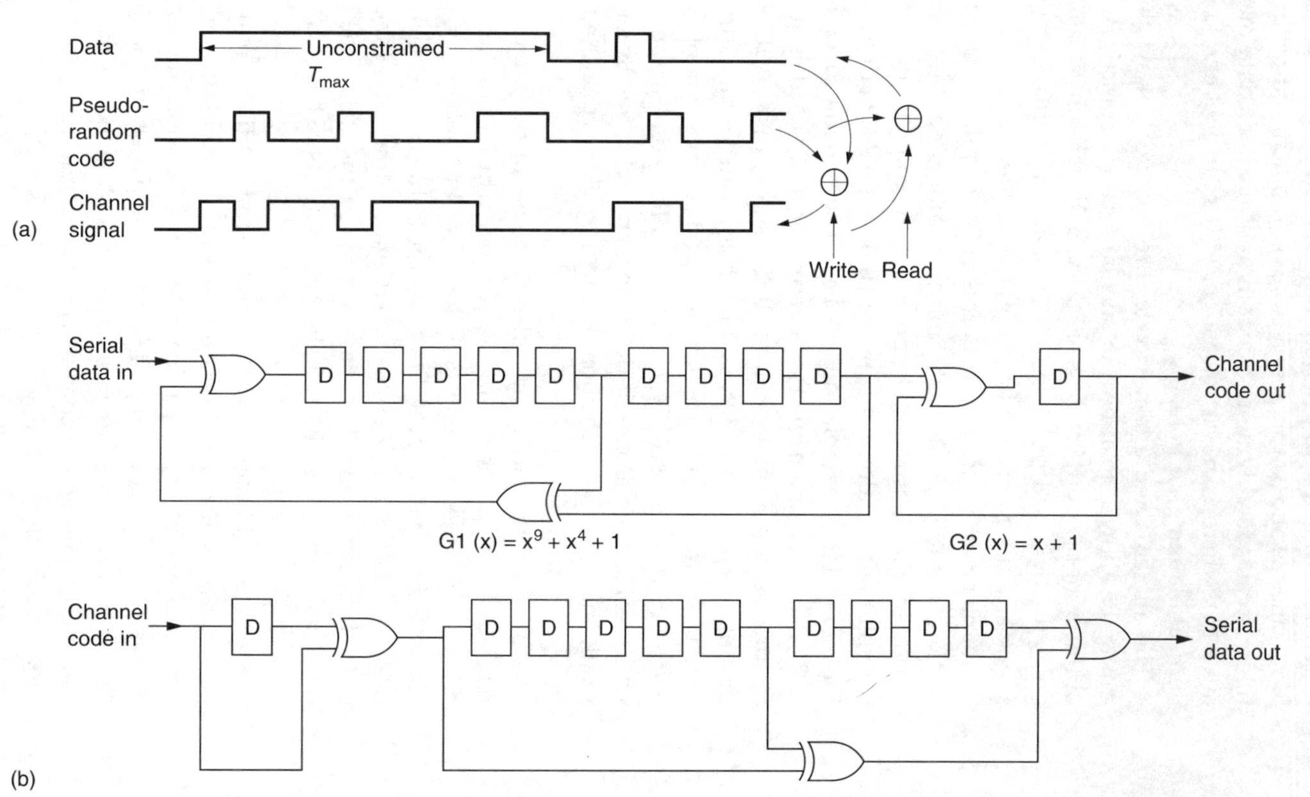

Figure 10.8 (a) Modulo-2 addition with a pseudo-random code removes unconstrained runs in real data. Identical process must be provided on replay. (b) Convolutional randomizing encoder, at top, transmits exclusive OR of three bits at a fixed spacing in the data. One bit delay, far right, produces channel transitions from data ones. Decoder, below, has opposing one bit delay to return from transitions to data levels, followed by an opposing shift register which exactly reverses the coding process.

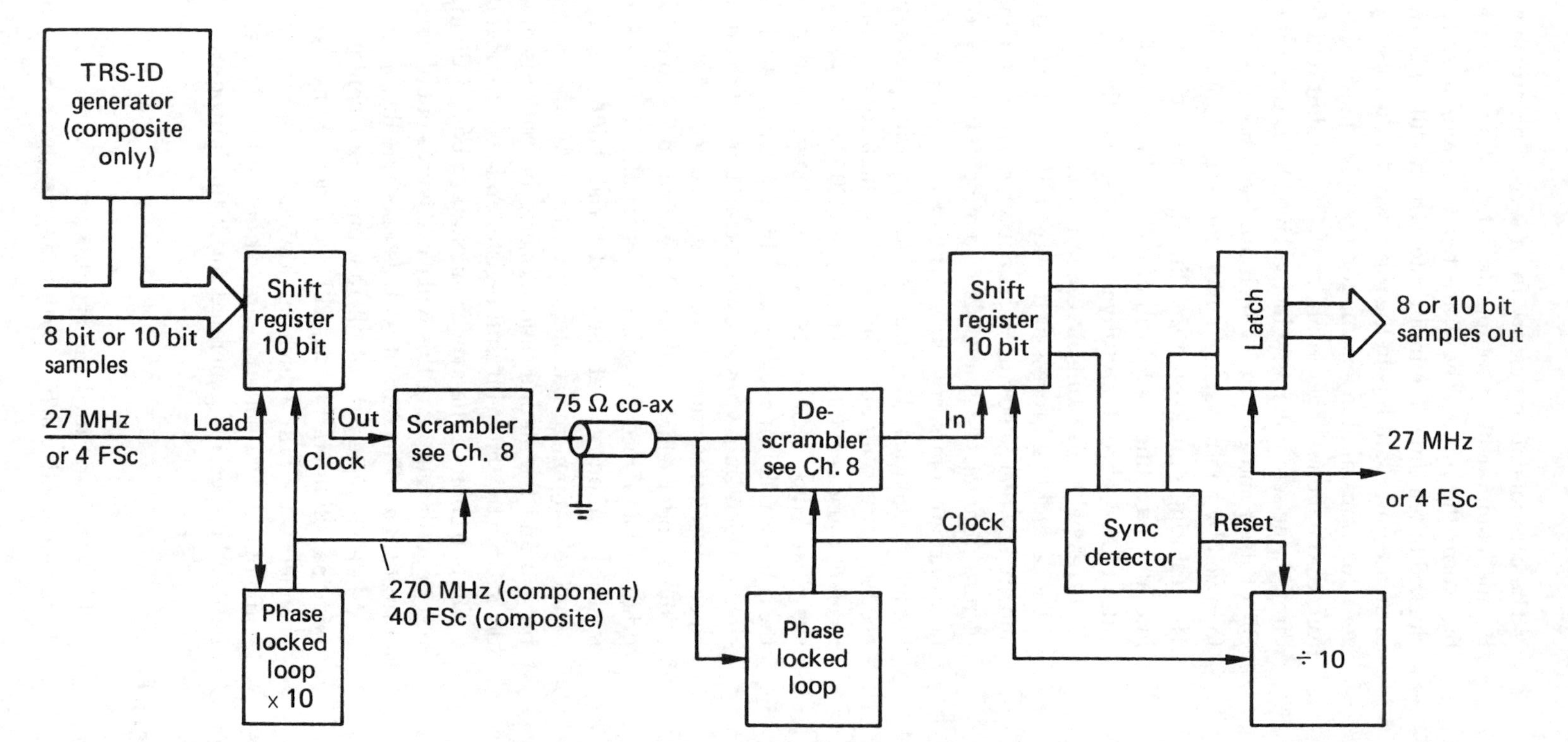

Figure 10.9 Major components of a serial scrambled link. Input samples are converted to serial form in a shift register clocked at ten times the sample rate. The serial data are then scrambled for transmission. On reception, a phase-locked loop recreates the bit rate clock and drives the de-scrambler and serial-to-parallel conversion. On detection of the sync pattern, the divide-by-ten counter is rephased to load parallel samples correctly into the latch. For composite working the bit rate will be 40 times subcarrier, and a sync pattern generator (top left) is needed to inject TRS-ID into the composite data stream.

The components necessary for a serial link are shown in Figure 10.9. Parallel component or composite data having a wordlength of up to ten bits form the input. These are fed to a ten-bit shift register which is clocked at ten times the input rate, which will be 270 MHz or $40 \times F_{sc}$. If there are only eight bits in the input words, the missing bits are forced to zero for transmission except for the all-ones condition which will be forced to ten ones. The serial data from the shift register are then passed through the scrambler, in which a given bit is converted to the exclusive-OR of itself and two bits which are five and nine clocks ahead. This is followed by another stage, which converts channel ones into transitions. The resulting signal is fed to a line driver which converts the logic level into an alternating waveform of 800 milliVolts peak to peak. The driver output impedance is carefully matched so that the signal can be fed down 75 Ω co-axial cable using BNC connectors.

The scrambling process at the transmitter spreads the signal spectrum and makes that spectrum reasonably constant and independent of the picture content. It is possible to assess the degree of equalization necessary by comparing the energy in a low-frequency band with that in higher frequencies. The greater the disparity, the more equalization is needed. Thus fully automatic cable equalization is easily achieved. The receiver must generate a bit clock at 270 MHz or $40 \times F_{sc}$ from the input signal, and this clock drives the input sampler and slicer which converts the cable waveform back to serial binary. The local bit clock also drives a circuit which simply reverses the scrambling at the transmitter. The first stage returns transitions to ones, and the second stage is a mirror image of the encoder which reverses the exclusive-OR calculation to output the original data. Since transmission is serial, it is necessary to obtain word synchronization, so that correct deserialization can take place.

In the component parallel input, the *SAV* and *EAV* sync patterns are present and the all-ones and all-zeros bit patterns these contain can be detected in the thirty-bit shift register and used to reset the deserializer.

On detection of the synchronizing symbols, a divide-by-ten circuit is reset, and the output of this will clock words out of the shift register at the correct times. This output will also become the output word clock.

It is a characteristic of all randomizing techniques that certain data patterns will interact badly with the randomizing algorithm to produce a channel waveform which is low in clock content. These so-called pathological data patterns[4] are extremely rare in real program material, but can be specially generated for testing purposes.

10.4 SDTI

SDI is closely specified and is only suitable for transmitting 2:1 interlaced 4:2:2 digital video in 525/60 or 625/50 systems. Since the development of

SDI, it has become possible economically to compress digital video and the SDI standard cannot handle this. SDTI (serial data transport interface) is designed to overcome that problem by converting SDI into an interface which can carry a variety of data types whilst retaining compatibility with existing SDI router infrastructures.

SDTI[5] sources produce a signal which is electrically identical to an SDI signal and which has the same timing structure. However, the digital active line of SDI becomes a data packet or *item* in SDTI. Figure 10.10 shows how SDTI fits into the existing SDI timing. Between EAV and SAV (horizontal blanking in SDI) an ancillary data block is incorporated. The structure of this meets the SDI standard, and the data within describe the contents of the following digital active line.

The data capacity of SDTI is about 200 Mbits/s because some of the 270 Mbits/s is lost due to the retention of the SDI timing structure. Each digital active line finishes with a CRCC (cyclic redundancy check character) to check for correct transmission.

SDTI raises a number of opportunities, including the transmission of compressed data at faster than real time. If a video signal is compressed at 4:1, then one quarter as much data would result. If sent in real time the bandwidth required would be one quarter of that needed by uncompressed video. However, if the same bandwidth is available, the compressed data could be sent in 1/4 of the usual time. This is particularly advantageous for data transfer between compressed

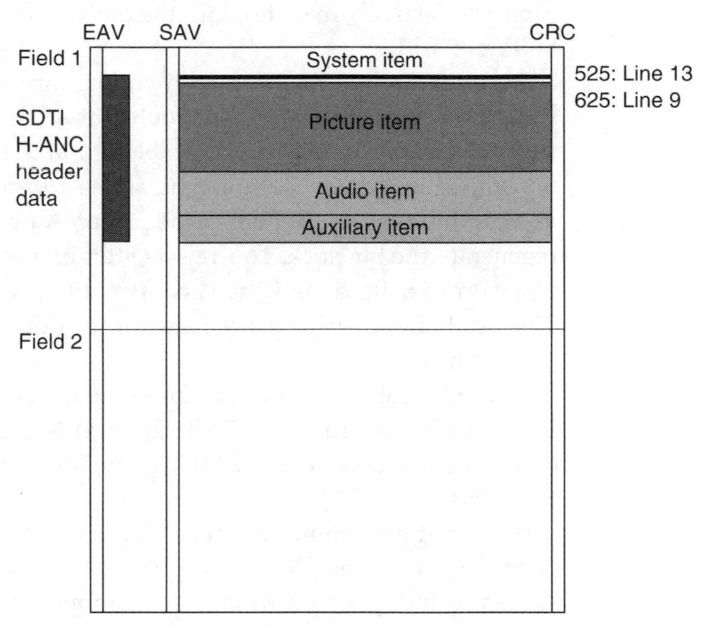

Figure 10.10 SDTI is a variation of SDI which allows transmission of generic data. This can include compressed video and non-real-time transfer.

camcorders and non-linear editing workstations. Alternatively, four different 50 megabit/s signals could be conveyed simultaneously.

Thus an SDTI transmitter takes the form of a multiplexer which assembles packets for transmission from input buffers. The transmitted data can be encoded according to MPEG, MotionJPEG, Digital Betacam or DVC formats and all that is necessary is that compatible devices exist at each end of the interface. In this case the data are transferred with bit accuracy and so there is no generation loss associated with the transfer. If the source and destination are different, i.e. having different formats or, in MPEG, different group structures, then a conversion process with attendant generation loss would be needed.

10.5 Serial digital routing

Digital routers have the advantage that they need cause no loss of signal quality as they simply pass on a series of numbers. Analog routers inevitably suffer from crosstalk and noise however well made, and this reduces signal quality on every pass. A serial digital router is potentially very inexpensive as it is a single-pole device. It can be easier to build than an analog router because the digital signal is more resistant to crosstalk.

A digital router can be made using analog switches, so that the input waveform is passed from input to output. This is not the best approach, as the total length of cable which can be used is restricted as the input cable is effectively in series with the output cable and the analog losses in both will add.

A better approach is for each router input to reclock and slice the waveform back to binary. The router itself then routes logic levels and relaunches a clean signal. The cables to and from the router can be of maximum length as the router is also a repeater. It is not necessary to unscramble the serial signal at the router. A phase-locked loop is used to regenerate the bit clock. This rejects jitter on the incoming waveform. The waveform is sliced, and the slicer output is sampled by the local clock. The result is a clean binary waveform, identical to the original driver waveform.

The only parameter of any consequence is the bit rate. 4:2:2 runs at 270 Mbits/s, PAL runs at 177 Mbits/s and NTSC runs at 143 Mbits/s. 16:9 component video with 18 MHz luminance sampling rate results in 360 Mbits/s.

Some routers require a link or DIP switch to be set in each input in order to centre the VCO at the correct frequency. Otherwise they are standards independent which allows more flexibility and economy. With a mixed standard router, it is only necessary to constrain the control software so that inputs of a given standard can only be routed to outputs connected to devices of the same standard, and one router can then handle component, SDTI and composite signals simultaneously.

10.6 Timing in digital installations

The issue of signal timing has always been critical in analog video, but the adoption of digital routing relaxes the requirements considerably. Analog vision mixers need to be fed by equal-length cables from the router to prevent propagation delay variation. In the digital domain this is no longer an issue as delay is easily obtained and each input of a digital vision mixer can have its own local timebase corrector. Provided signals are received having timing within the window of the inputs, all inputs are retimed to the same phase within the mixer.

Figure 10.11 shows how a mixing suite can be timed to a large SDI router. Signals to the router are phased so that the router output is aligned to station reference within a microsecond or so. The delay in the router may vary with its configuration but only by a few microseconds. The mixer reference is set with respect to station reference so that local signals

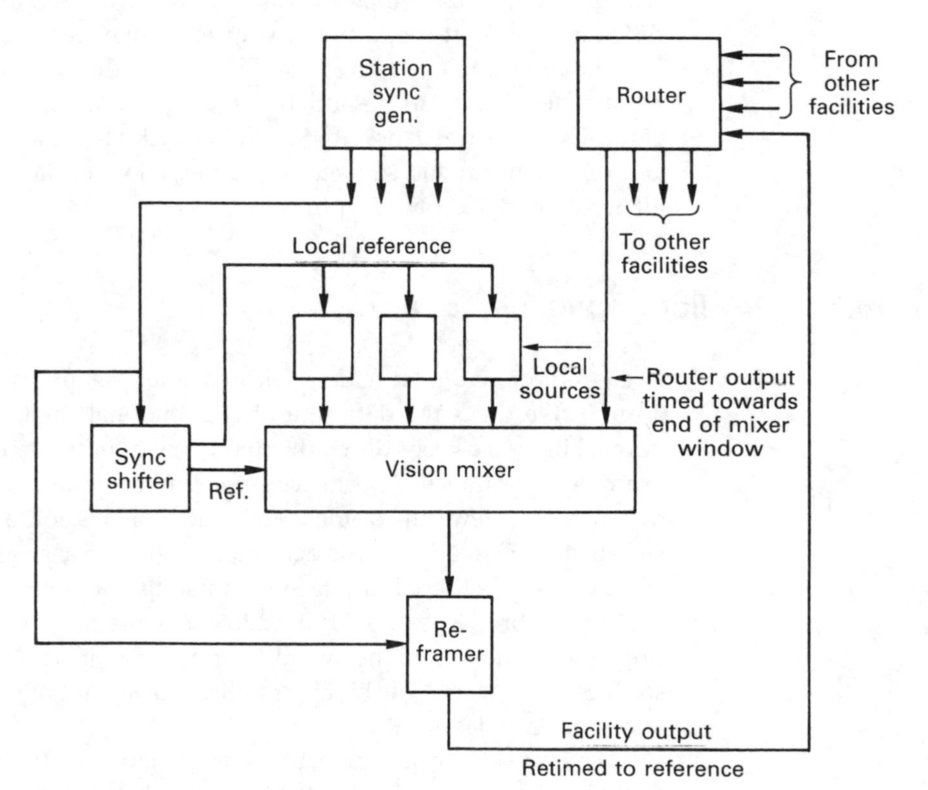

Figure 10.11 In large systems some care is necessary with signal timing as is shown here. See text for details.

arrive towards the beginning of the input windows and signals from the router (which, having come further, will be the latest) arrive towards the end of the windows. Thus all sources can be retimed within the mixer and any signal can be mixed with any other. Clearly the mixer introduces delay, and the signal fed back to the router experiences further delay. In order to send the mix back to the router a frame synchronizer is needed on the output of the suite. This introduces somewhat less than a frame of delay so that by the time the signal has re-emerged from the router it is aligned to station reference once more, but a frame late.

10.7 HD component parallel interface

The 1250/50 HDTV standard uses a CCIR-601-related sampling rate of 72 MHz for luminance and 36 MHz for colour difference.[6] The HD picture is 16:9 aspect ratio and has twice the resolution of 4:3. Thus the conventional sampling rate of 13.5 MHz is multiplied by $4 \times 4/3$ to give 72 MHz. There are 1920 luminance pixels in the active line. The two colour difference sample streams are multiplexed to give a second data stream at 72 MHz. Unlike the 4:3 standards, the luminance and colour difference are kept separate to avoid excessive frequencies in the cable. With ten-bit samples and two parallel signals, the data are in a twenty-bit parallel format. Transmission is by differential drive with a separate clock. The clock frequency at 36 MHz is exactly half the bit rate. The lower clock frequency reduces losses and allows a longer cable.[7] The signals are all accommodated on a 50-pin D-connector.

10.8 HD fibre-optic interface

High-definition digital video without the use of compression needs around five times the data rate of conventional digital video. Whilst a parallel interface keeps down the frequency of individual conductors, the example in section 10.7 needs a 50-pin connector and is only suitable for distances of a few tens of metres. Serial transmission on electrical cable suffers from skin effect at the extremely high bit rates needed. A practical solution to a HD serial interface is to use fibre-optics.

Optical fibre has extremely low loss per metre and a nearly constant propagation speed and the only difficulty is the provision of suitable light sources and sensors. The bit rate required is beyond the ability of an LED, and requires a laser.

Lasers are basically unstable (see Chapter 12) and need to be modulated with care. Figure 10.12 shows that the transfer function of a laser has a threshold which is strongly temperature dependent. If the

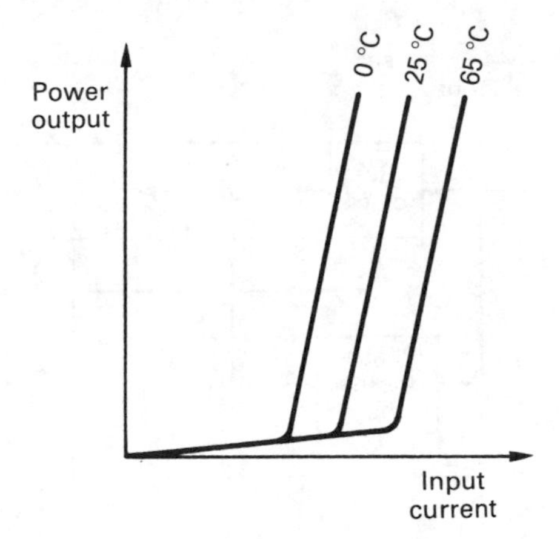

Figure 10.12 The transfer function of a laser shows strong temperature dependency and a threshold.

drive current is allowed to fall below the threshold, transient oscillation may occur when the current is reapplied. Careful biasing is required, using feedback from a local photo-detector.

In a fibre-optic serial interface developed for HDTV,[8] the same scrambling technique as SDI has been used. Figure 10.13 shows the modulator. The input is 2 × ten-bit parallel HDTV at 72 MHz. In order to reduce power consumption, the randomizer is implemented as a five-bit parallel system, prior to serializing for transmission at 1.44 GBits/s. The receiving PIN diode is AC coupled to eliminate drift. As the scrambled signal is not DC-free, adaptive slicing is needed to eliminate baseline wander. The demodulation system is shown in Figure 10.14.

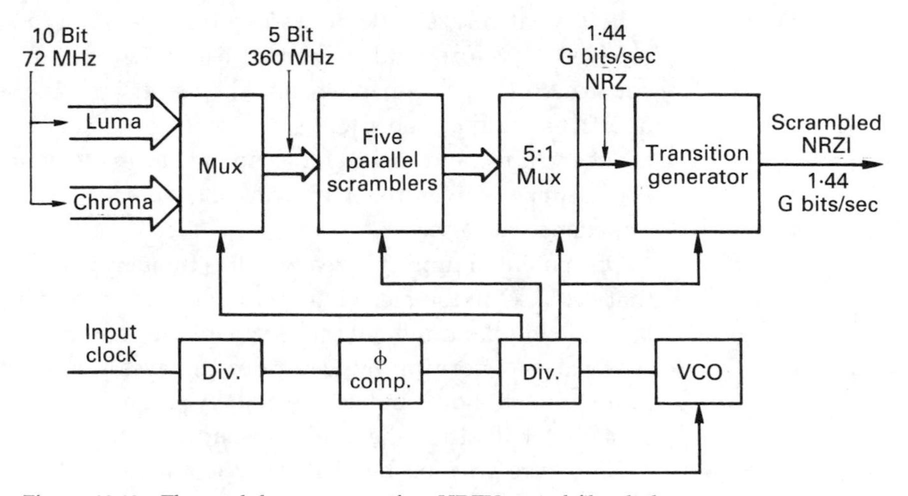

Figure 10.13 The modulator system of an HDTV optical fibre link.

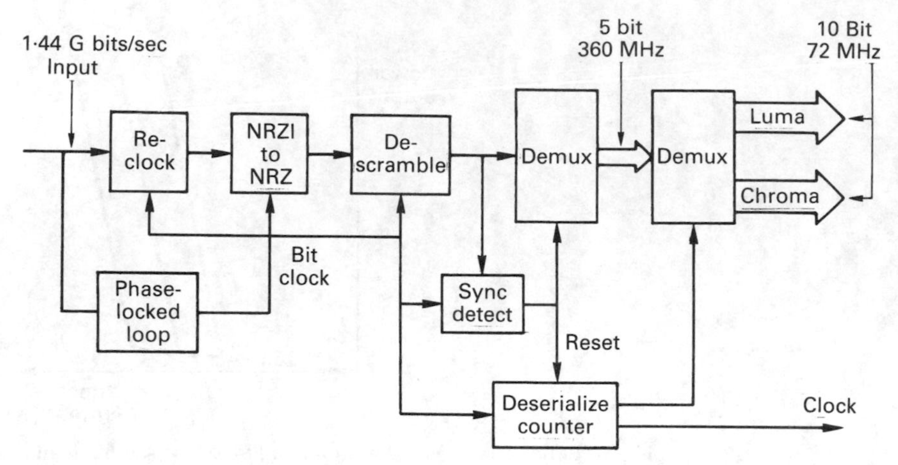

Figure 10.14 The demodulation system of the link of Figure 10.13.

10.9 Testing digital video interfaces

Upon installing a new interface it is necessary to test it for data integrity. It is not adequate to examine the picture quality. The very nature of digital transmission gives beneficial resistance to minor noises and cable loss with no effect on the data and thus no visible picture artifacts. The down side of this characteristic is that a digital system can be on the verge of failure and there are no visible symptoms. If one sees a perfect picture on a monitor, it is not possible to say whether the data which produced it was received with a good noise margin which will resist future degradations or whether it was only just received and is likely to fail in the near future. Some more objective testing means is necessary if any confidence is to be had in the long-term reliability of an installation.

It is relatively simple to assess the eye pattern generated by the received waveform and many SDI test units offer such a display. After equalization the eye opening should be clearly visible, and the size of the opening should be consistent with the length and type of cable used. If it is not, the noise and/or jitter margin may be inadequate. Inspection of the eye pattern waveform may reveal evidence of reflections due to impedance mismatches.

One potential problem area which is frequently overlooked is to ensure that the VCO in the receiving phase-locked loop is correctly centred. If it is not, it will be running with a static phase error and will not sample the received waveform at the centre of the eyes. The sampled bits will be more prone to noise and jitter errors. A pathological sequence from a test generator will stress the VCO by sending minimal clock content, or the VCO can be stressed by artificially lengthening the cable using a cable clone.[9]

Inspecting the eye pattern is a good technique for establishing that the basic installation is sound and has proper signal levels, termination impedance and equalization, but is not very good at detecting infrequent impulsive noise. Contact noise from electrical power installations such as air conditioners is unlikely to be visible with an eye pattern display.

The original SDI standard had no provisions for data integrity checking. EDH is an option for serial digital which goes a long way to rectifying the problem.[10,11] Figure 10.15 shows an EDH-equipped SDI (serial digital interface) transmission system. At the first transmitter, the data from one field are transmitted and simultaneously fed to a Cyclic Redundancy Check (CRC) generator. The CRC calculation is a mathematical division by a polynomial and the result is the remainder. The remainder is transmitted in a special ancillary data packet which is sent early during the vertical interval, before any switching takes place in a router.[12] The first receiver has an identical CRC generator which performs a calculation on the received field. The ancillary data extractor identifies the EDH packet and demultiplexes it from the main data stream. The remainder from the ancillary packet is then compared with the locally calculated remainder.

If the transmission is error-free, the two values will be identical. In this case no further action results. However, if as little as one bit is in error in the data, the remainders will not match. The remainder is a sixteen-bit word and guarantees to detect up to sixteen bits in error anywhere in the field. Greater numbers of errors are not guaranteed to be detected, but this is of little consequence as enough fields in error will be detected to indicate that there is a problem.

Should a CRC mismatch indicate an error in this way, two things happen. First, an optically isolated output connector on the receiving equipment will present a low impedance for a period of 1 to 2 ms. This will result in a pulse in an externally powered circuit to indicate that a field contained an error. An external error monitoring system wired to this connector can note the occurrence in a log or sound an alarm or whatever it is programmed to do. As the data are incorrectly received, the fact must also be conveyed to subsequent equipment.

It is not permissible to pass on a mismatched remainder. The centre unit in Figure 10.15 must pass on the data as received, complete with errors, but it must calculate a new CRC which matches the erroneous data. When received by the third unit in Figure 10.15, there will then only be a CRC mismatch if the transmission between the second and third devices is in error. This is correct as the job of the CRC is only to locate faulty hardware and clearly if the second link is not faulty the CRC comparison should not fail. However, the third device still needs to know that there is a problem with the data, and this is the job of the error flags which also reside in the EDH packet. One of these flags is called edh

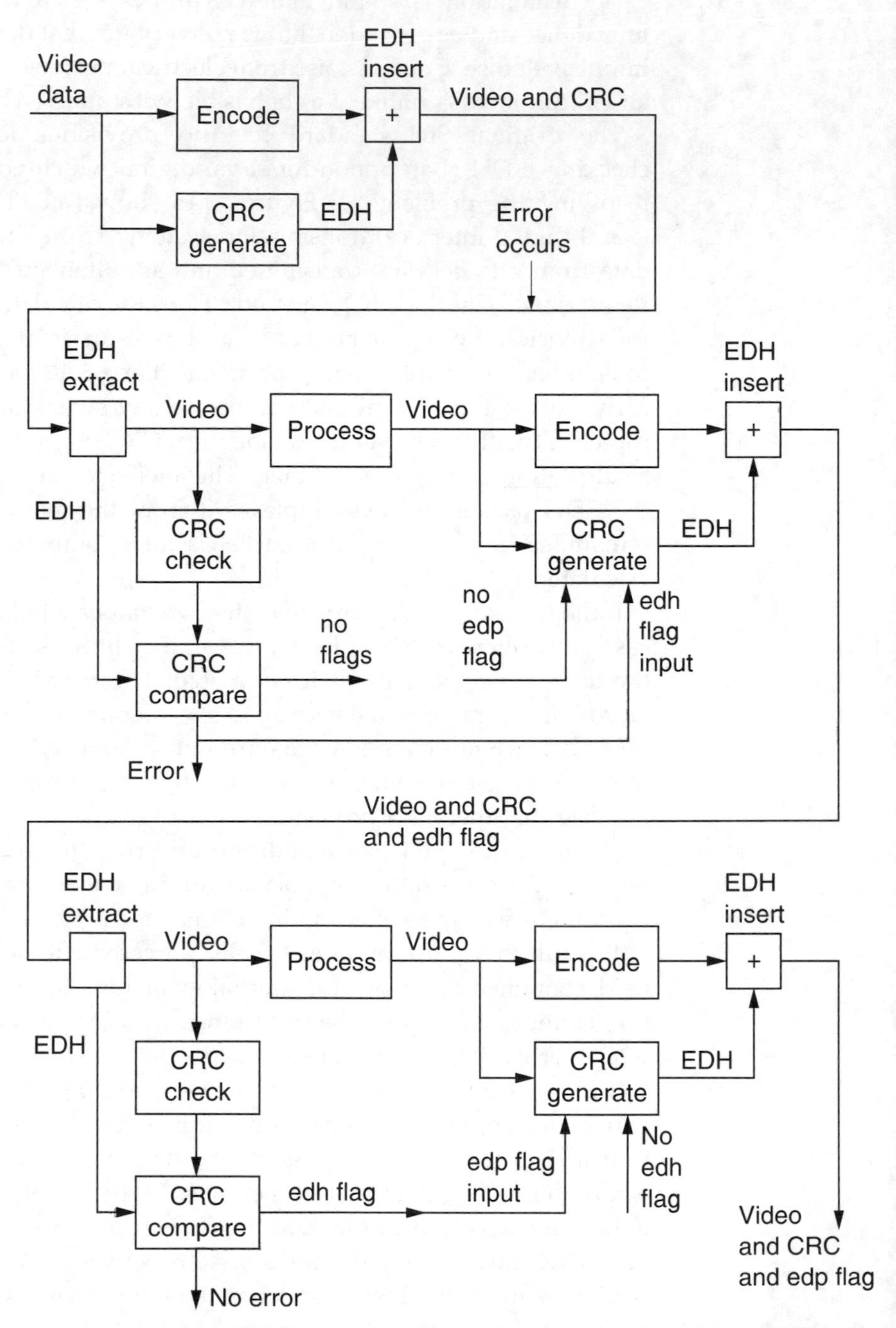

Figure 10.15 A typical EDH system illustrating the way errors are detected and flagged. See text for details.

(error detected here) and this will be asserted by the centre device in Figure 10.15. The last device in Figure 10.15 will receive edh and transmit eda (error detected already). There are also flags to handle hardware failures (e.g. over temperature or diagnostic failure). The idh (internal error detected here) and ida (internal error detected already) handle this function. Locally detected hardware errors drive the error output socket to a low-impedance state constantly to distinguish from the pulsing of a CRC mismatch.

A slight extra complexity is that error checking can be performed in two separate ways. One CRC is calculated for the active picture only, and another is calculated for the full field. Both are included in the EDH packet which is shown in Figure 10.16. The advantage of this arrangement is that whilst regular program material is being passed in active

Data item	b9 msb	b8	b7	b6	b5	b4	b3	b2	b1	b0 lsb
Ancillary data header, word 1 – component	0	0	0	0	0	0	0	0	0	0
Ancillary data header, word 2 – component	1	1	1	1	1	1	1	1	1	1
Ancillary data header, word 3 – component	1	1	1	1	1	1	1	1	1	1
Auxiliary data flag – composite	1	1	1	1	1	1	1	1	0	0
Data ID (1F4)	0	1	1	1	1	1	0	1	0	0
Block number	1	0	0	0	0	0	0	0	0	0
Data count	0	1	0	0	0	1	0	0	0	0
Active picture data word 0 crc<5:0>	$\overline{P}$	P	c_5	c_4	c_3	c_2	c_1	c_0	0	0
Active picture data word 1 crc<11:6>	$\overline{P}$	P	c_{11}	c_{10}	c_9	c_8	c_7	c_6	0	0
Active picture data word 2 crc<15:12>	$\overline{P}$	P	V	0	c_{15}	c_{14}	c_{13}	c_{12}	0	0
Full-field data word 0 crc<5:0>	$\overline{P}$	P	c_5	c_4	c_3	c_2	c_1	c_0	0	0
Full-field data word 1 crc<11:6>	$\overline{P}$	P	c_{11}	c_{10}	c_9	c_8	c_7	c_6	0	0
Full-field data word 2 crc<15:12>	$\overline{P}$	P	V	0	c_{15}	c_{14}	c_{13}	c_{12}	0	0
Auxiliary data error flags	$\overline{P}$	P	0	ues	ida	idh	eda	edh	0	0
Active picture error flags	$\overline{P}$	P	0	ues	ida	idh	eda	edh	0	0
Full-field error flags	$\overline{P}$	P	0	ues	ida	idh	eda	edh	0	0
Reserved words (7 total)	1	0	0	0	0	0	0	0	0	0
Checksum	$\overline{S8}$	S8	S7	S6	S5	S4	S3	S2	S1	S0

Error flags
All error flags indicate only the status of the previous field; that is, each flag is set or cleared on a field-by-field basis. A logical 1 is the set state and a logical 0 is the unset state. The flags are defined as follows:

edh – error detected here: Signifies that a serial transmission data error was detected. In the case of ancillary data, this means that one or more ANC data blocks did not match its checksum.

eda – error detected already: Signifies that a serial transmission data error has been detected somewhere upstream. If device B receives a signal from device A and device A has set the edh flag, when device B retransmits the data to device C, the eda flag will be set and the edg flag will be unset if there is no further error in the data.

idh – internal error detected here: Signifies that a hardware error unrelated to serial transmission has been detected within a device. This is provided specifically for devices which have internal data error checking facilities, as an error reporting mechanism.

ida – internal error detected already: Signifies that an idh flag was received and there was a hardware device failure somewhere upstream.

ues – unknown error status: Signifies that a serial signal was received from equipment not supporting this error-detection mechanism.

Checkword values
Each checkword value consists of 16 bits of data calculated using the CRC-CCITT polynomial generation method. The equation and a conceptual logic diagram are shown below:

Checkword (16-bit) = $x^{16} + x^{12} + x^5 + 1$

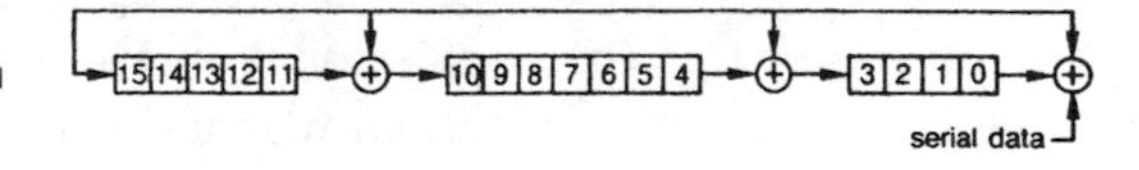

Figure 10.16 The contents of the EDH packet which is inserted in ancillary data space after the associated field.

picture, test patterns can be sent in vertical blanking which can be monitored separately. Thus if active picture is received without error but full field gives an error, the error must be outside the picture. It is then possible to send, for example, pathological test patterns during the vertical interval which stress the transmission system more than regular data to check the performance margin of the system. This can be done alongside the picture information without causing any problems.

In a large system, if every SDI link is equipped with EDH, it is possible for automatic error location to be performed. Each EDH-equipped receiver is connected to a monitoring system which can graphically display on a map of the system the location of any transmission errors. If a suitable logging system is used, it is not necessary for the display to be in the same place as the equipment. In the event of an error condition, the logging system can communicate with the display by dial-up modem or dedicated line over any distance. Logging allows infrequent errors to be counted. Any increase in error rate indicates a potential failure which can be rectified before it becomes serious.

10.10 Introduction to the AES/EBU interface

The AES/EBU digital audio interface, originally published in 1985,[13] was proposed to embrace all the functions of existing formats in one standard. The goal was to ensure interconnection of professional digital audio equipment irrespective of origin. The EBU ratified the AES proposal with the proviso that the optional transformer coupling was made mandatory and led to the term AES/EBU interface, also called EBU/AES by some Europeans. The contribution of the BBC to the development of the interface must be mentioned here. Alongside the professional format, Sony and Philips developed a similar format now known as SPDIF (Sony Philips Digital Interface) intended for consumer use. This offers different facilities to suit the application, yet retains sufficient compatibility with the professional interface so that, for many purposes, consumer and professional machines can be connected together.[14,15]

The AES concerns itself with professional audio and accordingly has had little to do with the consumer interface. Thus the recommendations to standards bodies such as the IEC (International Electrotechnical Commission) regarding the professional interface came primarily through the AES whereas the consumer interface input was primarily from industry, although based on AES professional proposals. The IEC and various national standards bodies naturally tended to combine the two into one standard such as IEC 958[16] which refers to the professional interface and the consumer interface. This process has been charted by Finger.[17]

10.11 The electrical interface

It was desired to use the existing analog audio cabling in such installations, which would be 600 Ω balanced line screened, with one cable per audio channel, or in some cases one twisted pair per channel with a common screen. The interface has to be self-clocking and self-synchronizing, i.e. the single signal must carry enough information to allow the boundaries between individual bits, words and blocks to be detected reliably. To fulfil these requirements, the AES/EBU and SPDIF interfaces use FM channel code (see Chapter 8) which is DC-free, strongly self-clocking and capable of working with a changing sampling rate. Synchronization of deserialization is achieved by violating the usual encoding rules.

The use of FM means that the channel frequency is the same as the bit rate when sending data ones. Tests showed that in typical analog audio-cabling installations, sufficient bandwidth was available to convey two digital audio channels in one twisted pair. The standard driver and receiver chips for RS-422A[18] data communication (or the equivalent CCITT-V.11) are employed for professional use, but work by the BBC[19] suggested that equalization and transformer coupling were desirable for longer cable runs, particularly if several twisted pairs occupy a common shield. Successful transmission up to 350 m has been achieved with these techniques.[20] Figure 10.17 shows the standard configuration. The output impedance of the drivers will be about 110 Ω, and the impedance of the cable and receiver should be similar at the frequencies of interest. The driver was specified in AES-3–1985 to produce between 3 and 10 V pk–pk into such an impedance but this was changed to between 2 and 7 V in AES-3–1992 better to reflect the characteristics of actual RS-422 driver chips.

In Figure 10.18, the specification of the receiver is shown in terms of the minimum eye pattern (see Chapter 8) which can be detected without

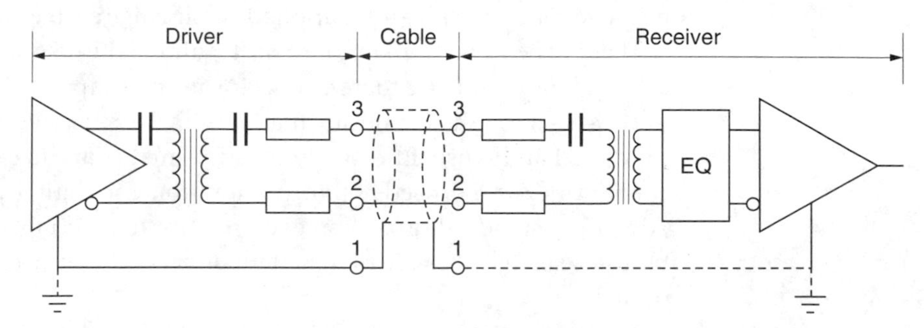

Figure 10.17 Recommended electrical circuit for use with the standard two-channel interface.

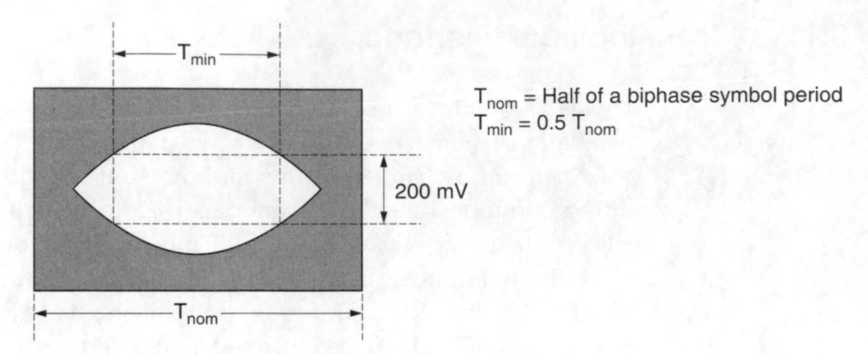

Figure 10.18 The minimum eye pattern acceptable for correct decoding of standard two-channel data.

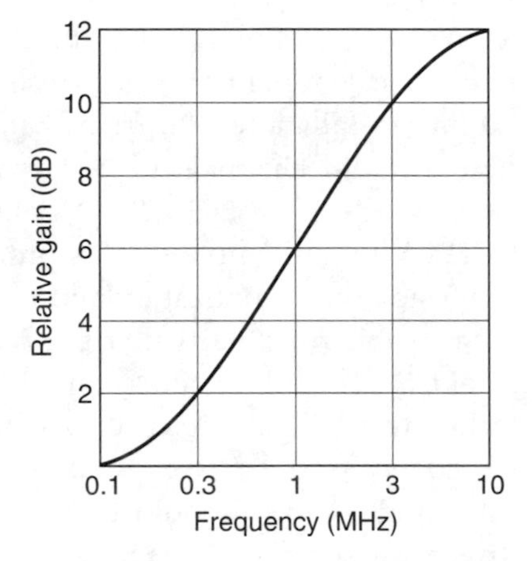

Figure 10.19 EQ characteristic recommended by the AES to improve reception in the case of long lines.

error. It will be noted that the voltage of 200 mV specifies the height of the eye opening at a width of half a channel bit period. The actual signal amplitude will need to be larger than this, and even larger if the signal contains noise. Figure 10.19 shows the recommended equalization characteristic which can be applied to signals received over long lines.

The purpose of the standard is to allow the use of existing analog cabling, and as an adequate connector in the shape of the XLR is already in wide service, the connector made to IEC 268 Part 12 has been adopted for digital audio use. Effectively, existing analog audio cables having XLR connectors can be used without alteration for digital connections. The AES/EBU standard does, however, require that suitable labelling should be used so that it is clear that the connections on a particular unit are digital.

There is a separate standard[21] for a professional interface using coaxial cable for distances of around 1000 m. This is simply the AES/EBU protocol but with a 75 Ω coaxial cable carrying a one volt signal so that it can be handled by analog video distribution amplifiers. Impedance converting transformers allow balanced 110 Ω to unbalanced 75 Ω matching.

10.12 Frame structure

In Figure 10.20 the basic structure of the professional and consumer formats can be seen. One subframe consists of thirty-two bit-cells, of which four will be used by a synchronizing pattern. Subframes from the two audio channels, A and B, alternate on a time-division basis, with the least significant bit sent first. Up to twenty-four-bit sample wordlength can be used, which should cater for all conceivable future developments, but normally twenty-bit maximum-length samples will be available with four auxiliary data bits, which can be used for a voice-grade channel in a professional application.

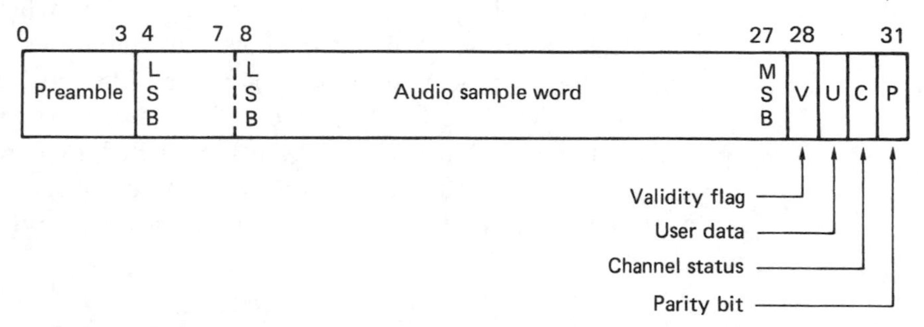

Figure 10.20 The basic subframe structure of the AES/EBU format. Sample can be twenty bits with four auxiliary bits, or twenty-four bits. LSB is transmitted first.

The format specifies that audio data must be in two's complement coding. If different wordlengths are used, the MSBs must always be in the same bit position otherwise the polarity will be misinterpreted. Thus the MSB has to be in bit 27 irrespective of wordlength. Shorter words are leading-zero filled up to the twenty-bit capacity. The channel status data included from AES-3–1992 signalling of the actual audio wordlength used so that receiving devices could adjust the digital dithering level needed to shorten a received word which is too long or pack samples onto a disk more efficiently.

Four status bits accompany each subframe. The validity flag will be reset if the associated sample is unreliable. Whilst there have been many aspirations regarding what the V-bit could be used for, in practice a single bit cannot specify much, and if combined with other V-bits to make a word, the time resolution is lost. AES-3–1992 described the V-bit as indicating that the information in the associated subframe is 'suitable for conversion to an analog signal'. Thus it might be reset if the interface was being used for non-audio data as is done, for example, in CD-I players.

The parity bit produces even parity over the subframe, such that the total number of ones in the subframe is even. This allows for simple detection of an odd number of bits in error, but its main purpose is that it causes successive sync patterns to have the same polarity, which can be used to improve the probability of detection of sync. The user and channel-status bits are discussed later.

Two of the subframes described above make one frame, which repeats at the sampling rate in use. The first subframe will contain the sample from channel A, or from the left channel in stereo working. The second subframe will contain the sample from channel B, or the right channel in stereo. At 48 kHz, the bit rate will be 3.072 MHz, but as the sampling rate can vary, the clock rate will vary in proportion.

In order to separate the audio channels on receipt the synchronizing patterns for the two subframes are different as Figure 10.21 shows. These sync patterns begin with a run length of 1.5 bits which violates the FM channel coding rules and so cannot occur due to any data combination. The type of sync pattern is denoted by the position of the second transition which can be 0.5, 1.0 or 1.5 bits away from the first. The third transition is designed to make the sync patterns DC-free.

The channel status and user bits in each subframe form serial data streams with one bit of each per audio channel per frame. The channel

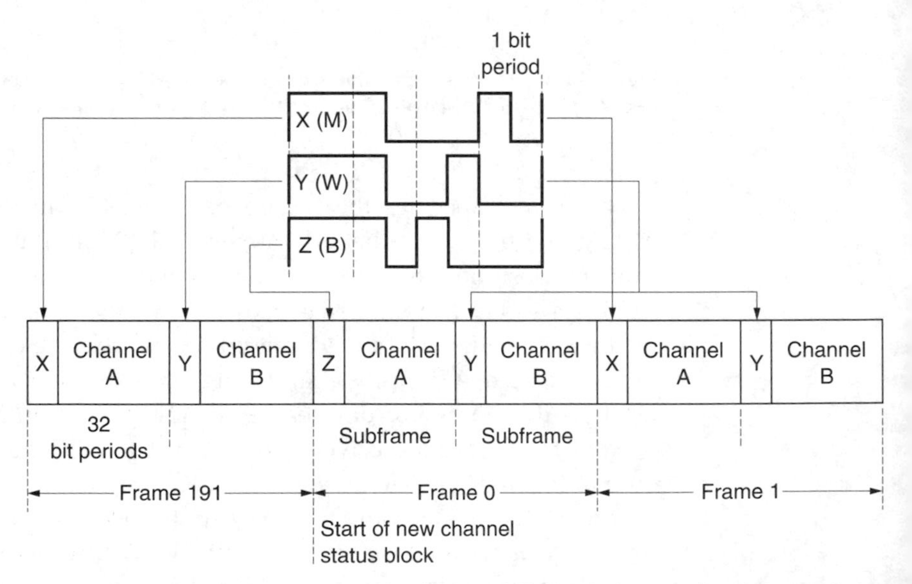

Figure 10.21 Three different preambles (X, Y and Z) are used to synchronize a receiver at the start of subframes.

status bits are given a block structure and synchronized every 192 frames, which at 48 kHz gives a block rate of 250 Hz, corresponding to a period of 4 ms. In order to synchronize the channel-status blocks, the channel A sync pattern is replaced for one frame only by a third sync pattern which is also shown in Figure 10.21. The AES standard refers to these as X,Y and Z whereas IEC 958 calls them M,W and B. As stated, there is a parity bit in each subframe, which means that the binary level at the end of a subframe will always be the same as at the beginning. Since the sync patterns have the same characteristic, the effect is that sync patterns always have the same polarity and the receiver can use that information to reject noise. The polarity of transmission is not specified, and indeed an accidental inversion in a twisted pair is of no consequence, since it is only the transition that is of importance, not the direction.

In both the professional and consumer formats, the sequence of channel-status bits over 192 subframes builds up a 24-byte channel-status

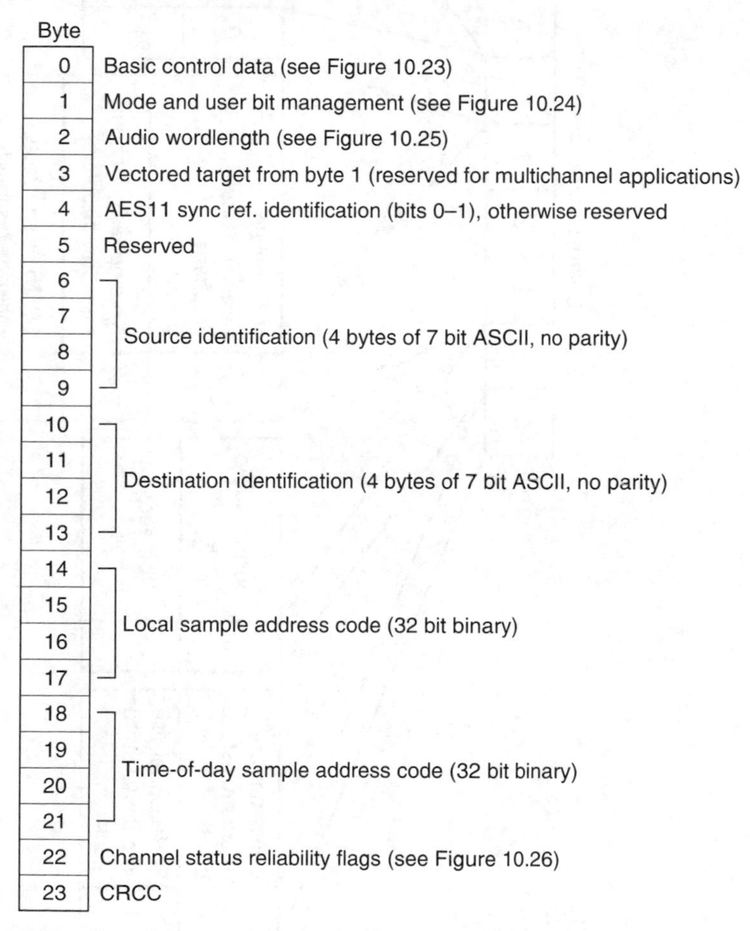

Figure 10.22 Overall format of the professional channel-status block.

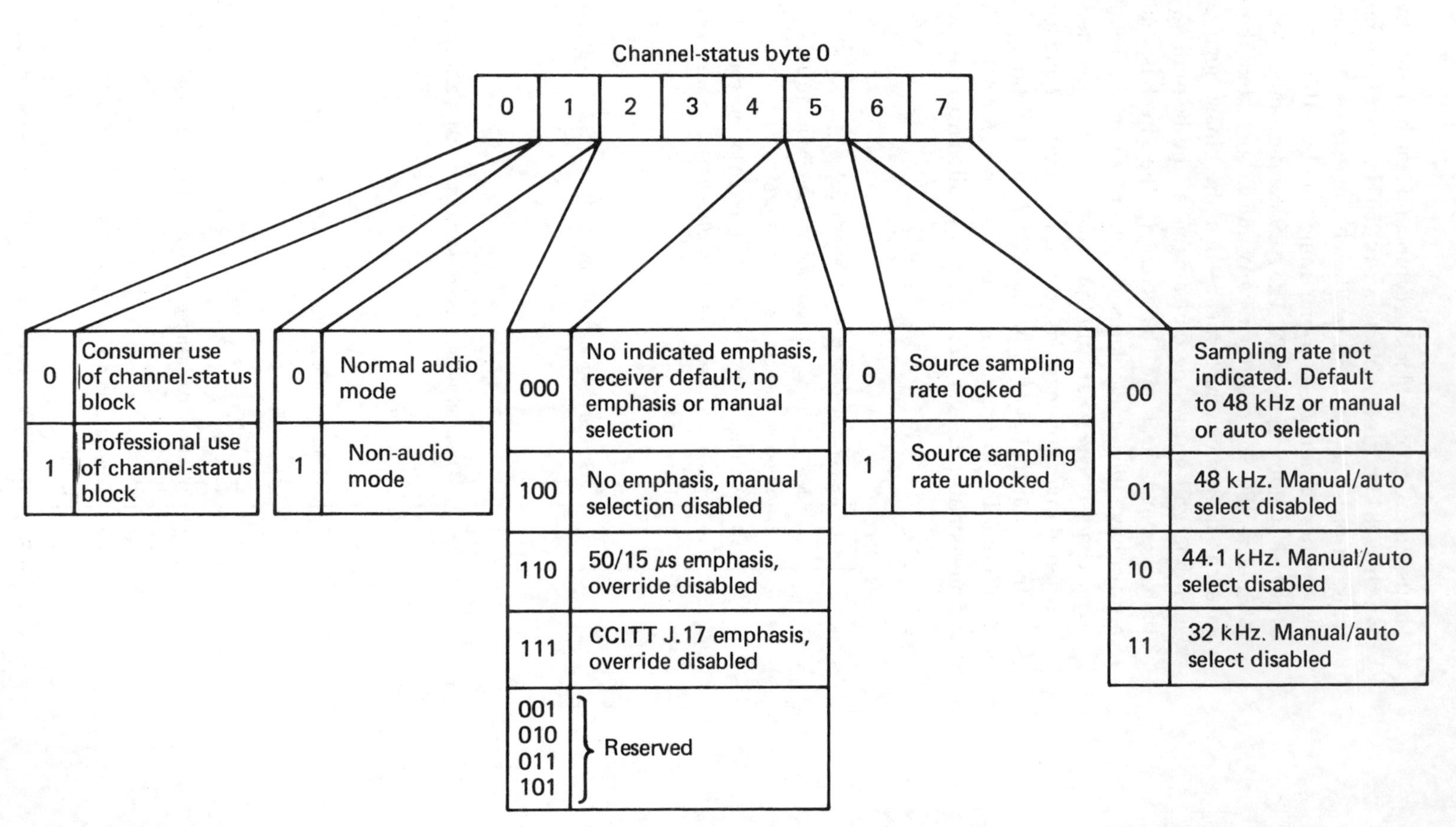

Figure 10.23 The first byte of the channel-status information in the AES/EBU standard deals primarily with emphasis and sampling-rate control.

block. However, channel status data are completely different between the two applications. The professional channel status structure is shown in Figure 10.22. Byte 0 determines the use of emphasis and the sampling rate, with details in Figure 10.23. Byte 1 determines the channel usage mode, i.e. whether the data transmitted are a stereo pair, two unrelated mono signals or a single mono signal, and details the user bit handling. Figure 10.24 gives details. Byte 2 determines wordlength as in Figure 10.25. This was made more comprehensive in AES-3–1992. Byte 3 is applicable only to multichannel applications. Byte 4 indicates the suitability of the signal as a sampling rate reference and will be discussed in more detail later in this chapter.

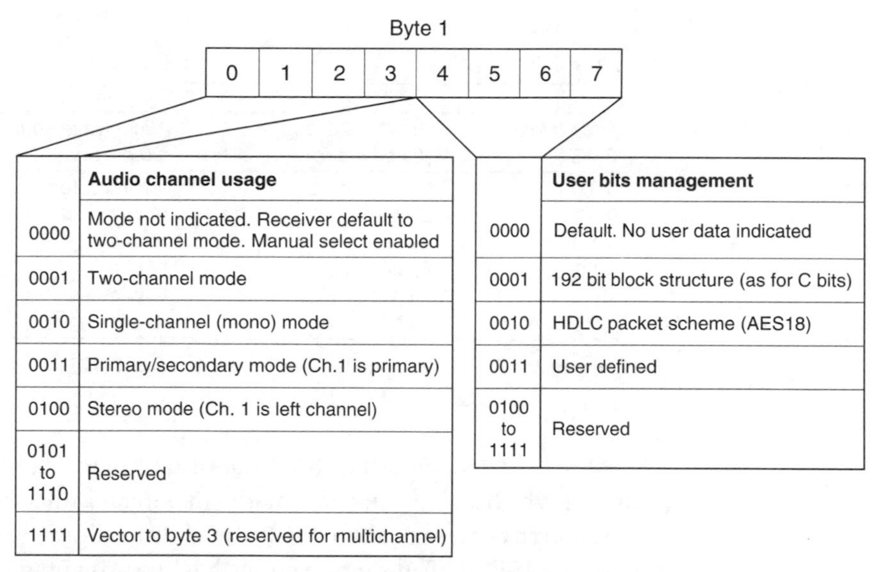

Figure 10.24 Format of byte 1 of professional channel status.

There are two slots of four bytes each which are used for alphanumeric source and destination codes. These can be used for routing. The bytes contain seven-bit ASCII characters (printable characters only) sent LSB first with the eighth bit set to zero acording to AES-3–1992. The destination code can be used to operate an automatic router, and the source code will allow the origin of the audio and other remarks to be displayed at the destination.

Bytes 14–17 convey a thirty-two-bit sample address which increments every channel status frame. It effectively numbers the samples in a relative manner from an arbitrary starting point. Bytes 18–21 convey a similar number, but this is a time-of-day count, which starts from zero at midnight. As many digital audio devices do not have real-time clocks built in, this cannot be relied upon.

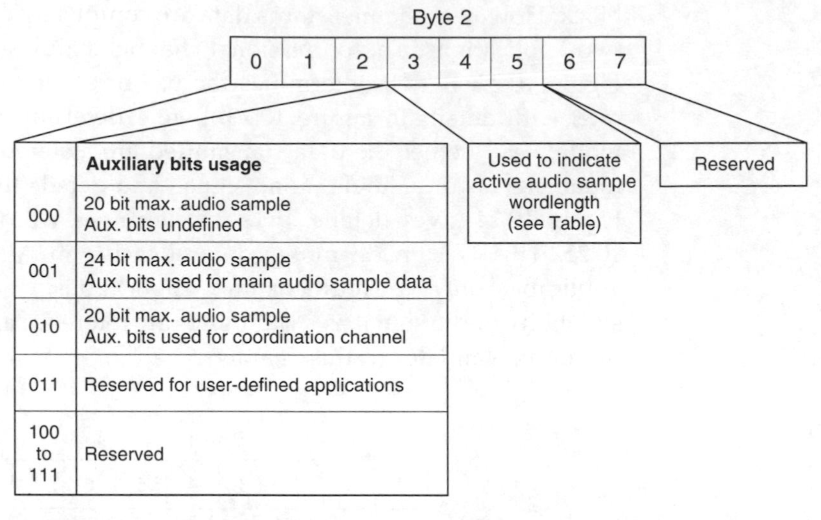

Bits states 3 4 5	*Audio wavelength (24 bit mode)*	*Audio wavelength (20 bit mode)*
0 0 0	Not indicated	Not indicated
0 0 1	23 bits	19 bits
0 1 0	22 bits	18 bits
0 1 1	21 bits	17 bits
1 0 0	20 bits	16 bits
1 0 1	24 bits	20 bits

Figure 10.25 Format of byte 2 of professional channel status.

AES-3–92 specified that the time-of-day bytes should convey the real time at which a recording was made, making it rather like timecode. There are enough combinations in thirty-two bits to allow a sample count over 24 hours at 48 kHz. The sample count has the advantage that it is universal and independent of local supply frequency. In theory if the sampling rate is known, conventional hours, minutes, seconds, frames timecode can be calculated from the sample count, but in practice it is a lengthy computation and users have proposed alternative formats in which the data from EBU or SMPTE timecode is transmitted directly in these bytes. Some of these proposals are in service as *de facto* standards.

The penultimate byte contains four flags which indicate that certain sections of the channel-status information are unreliable (see Figure 10.26). This allows the transmission of an incomplete channel-status block where the entire structure is not needed or where the information is not available. For example, setting bit 5 to a logical one would mean that no origin or destination data would be interpreted by the receiver, and so it need not be sent.

The final byte in the message is a CRCC which converts the entire channel-status block into a codeword (see Chapter 9). The channel status

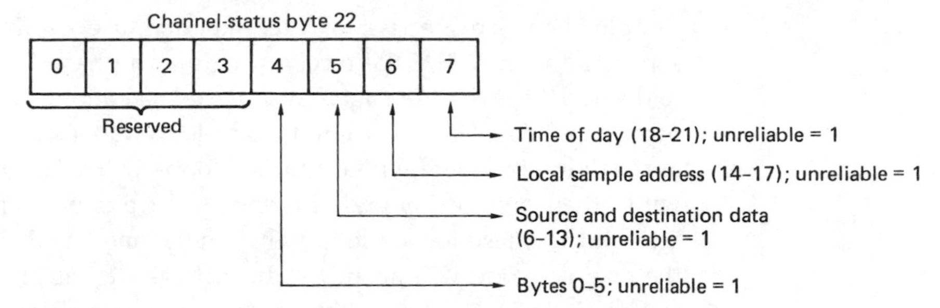

Figure 10.26 Byte 22 of channel status indicates if some of the information in the block is unreliable.

message takes 4 ms at 48 kHz and in this time a router could have switched to another signal source. This would damage the transmission, but will also result in a CRCC failure so the corrupt block is not used. Error correction is not necessary, as the channel status data are either stationary, i.e. they stay the same, or change at a predictable rate, e.g. timecode. Stationary data will only change at the receiver if a good CRCC is obtained.

10.13 Timing tolerance of serial interfaces

There are three parameters of interest when conveying audio down a serial interface, and these have quite different importance depending on the application. The parameters are:

1 The jitter tolerance of the serial FM data separator.
2 The jitter tolerance of the audio samples at the point of conversion back to analog.
3 The timing accuracy of the serial signal with respect to other signals.

The serial interface is a digital interface, in that it is designed to convey discrete numerical values from one place to another. If those samples are correctly received with no numerical change, the interface is perfect. The serial interface carries clocking information, in the form of the transitions of the FM channel code and the sync patterns and this information is designed to enable the data separator to determine the correct data values in the presence of jitter. It was shown in Chapter 8 that the jitter window of the FM code is half a data bit period in the absence of noise. This becomes a quarter of a data bit when the eye opening has reached the minimum allowable in the professional specification as can be seen from Figure 10.18. If jitter is within this limit, which corresponds to about 80 nanoseconds pk–pk the serial digital interface perfectly reproduces the sample data, irrespective of the intended use of the data. The data separator of an AES/EBU receiver requires a phase-locked loop in order to decode the serial message. This phase-locked loop will have jitter of its own, particularly if it is a digital phase-locked loop where the phase steps are of finite size. Digital phase-locked loops are easier to implement along with other logic in integrated circuits. There is no point in making the jitter of the phase-locked loop vanishingly small as the jitter tolerance of the channel code will absorb it. In fact the digital phase-locked loop is simpler to implement and locks up quicker if it has larger phase steps and therefore more jitter.

This has no effect on the ability of the interface to convey discrete values, and if the data transfer is simply an input to a digital recorder no other parameter is of consequence as the data values will be faithfully recorded. However, it is a further requirement in some applications that a sampling clock for a convertor is derived from a serial interface signal.

The jitter tolerance of convertor clocks is measured in hundreds of picoseconds. Thus a phase-locked loop in the FM data separator of a serial receiver chip is quite unable to drive a convertor directly as the jitter it contains will be as much as a thousand times too great.

Figure 10.27 shows how an outboard convertor should be configured. The serial data separator has its own phase-locked loop which is less jittery than the serial waveform and so recovers the audio data. The serial data are presented to a shift register which is read in parallel to a latch when an entire sample is present by a clock edge from the data separator. The data separator has done its job of correctly returning a sample value

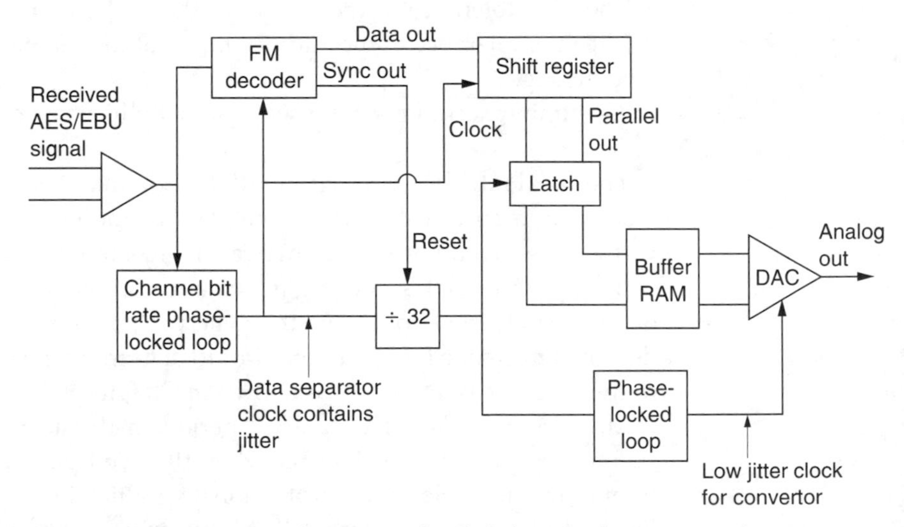

Figure 10.27 In an outboard convertor, the clock from the data separator is not sufficiently free of jitter and additional clock regeneration is necessary to drive the DAC.

to parallel format. A quite separate phase-locked loop with extremely high damping and low jitter is used to regenerate the sampling clock. This may use a crystal oscillator or it may be a number of loops in series to increase the order of the jitter filtering. In the professional channel status, bit 5 of byte 0 indicates whether the source is locked or unlocked. This bit can be used to change the damping factor of the phase-locked loop or to switch from a crystal to a varicap oscillator. When the source is unlocked, perhaps because a recorder is in varispeed, the capture range of the phase-locked loop can be widened and the increased jitter is accepted. When the source is locked, the capture range is reduced and the jitter is rejected.

The third timing criterion is only relevant when more than one signal is involved as it affects the ability of, for example, a mixer to combine two inputs.

In order to decide which criterion is most important, the following may be helpful. A single signal which is involved in a data transfer to a recording medium is concerned only with eye-pattern jitter as this affects the data reliability. A signal which is to be converted to analog is concerned primarily with the jitter at the convertor clock. Signals which are to be mixed are concerned with the eye-pattern jitter and the relative timing. If the mix is to be monitored, all three parameters become important.

10.14 Embedded audio in SDI

In component SDI, there is provision for ancillary data packets to be sent during blanking.[22,23] The high clock rate of component means that there is capacity for up to sixteen audio channels sent in four Groups. Composite SDI has to convey the digitized analog sync edges and bursts and only sync tip is available for ancillary data. As a result of this and the lower clock rate composite has much less capacity for ancillary data than component although it is still possible to transmit one audio data packet carrying four audio channels in one Group.

As was shown above, the data content of the AES/EBU digital audio subframe consists of Valid, User and Channel status bits, a twenty-bit sample and four auxiliary bits which optionally may be appended to the main sample to produce a twenty-four-bit sample. The AES recommends sampling rates of 48, 44.1 and 32 kHz, but the interface permits variable sampling rates. SDI has various levels of support for the wide range of audio possibilities and these levels are defined in Figure 10.28. The default or minimum level is Level A which operates only with a video-synchronous 48 kHz sampling rate and transmits V,U,C and the main twenty-bit sample only. As Level A is a default it need not be signalled to

	Level	Description
	A (Default)	Synchronous 48 kHz, 20 bit audio, 48 sample buffer
Needs audio control packet	B	Synchronous 48 kHz for composite video. 64 sample buffer to receive 20 bits from 24 bit data
Needs audio control packet	C	Synchronous 48 kHz 24 bit with extended packets
Needs audio control packet	D	Asynchronous audio
Needs audio control packet	E	44.1 kHz audio
Needs audio control packet	F	32 kHz audio
Needs audio control packet	G	32–48 kHz variable sampling rate
Needs audio control packet	H	Audio frame sequence
Needs audio control packet	I	Time delay tracking
Needs audio control packet	J	Non-coincident channel status Z bits in a pair

Figure 10.28 The different levels of implementation of embedded audio. Level A is default.

a receiver as the presence of IDs in the ancillary data is enough to ensure correct decoding. However, all other levels require an Audio Control Packet to be transmitted to teach the receiver how to handle the embedded audio data. The Audio Control Packet is transmitted once per field in the second horizontal Ancillary space after the video switching point before any associated audio sample data. One Audio Control Packet is required per Group of audio channels.

If it is required to send twenty-four-bit samples, the additional four bits of each sample are placed in Extended Data Packets which must directly follow the associated Group of audio samples in the same ancillary data space.

There are thus three kinds of packet used in embedded audio: the Audio Data Packet which carries up to four channels of digital audio, the Extended Data Packet and the Audio Control Packet.

In component systems, ancillary data begin with a reversed TRS or sync pattern. Normal video receivers will not detect this pattern and so ancillary data cannot be mistaken for video samples. The ancillary data TRS consists of all zeros followed by all ones twice. There is no separate TRS for ancillary data in composite. Immediately following the usual TRS, there will be an Ancillary Data Flag whose value must be $3FC_{16}$. Following the ancillary TRS or Data Flag is a Data ID word which contains one of a number of standardized codes which tell the receiver how to interpret the ancillary packet. Figure 10.29 shows a list of ID codes for various types of packets. Next come the Data Block Number and the Data Block Count parameters. The Data Block Number increments by 1

	Group 1	Group 2	Group 3	Group 4
Audio data	2FF	1FD	1FB	2F9
Audio CTL	1EF	2EE	2ED	1EC
Ext. data	1FE	2FC	2FA	1F8

Figure 10.29 The different packet types have different ID codes as shown here.

on each instance of a block with a given ID number. On reaching 255 it overflows and recommences counting. Next, a Data Count parameter specifies how many symbols of data are being sent in this block. Typical values for the Data Count are 36_{10} for a small packet and 48_{10} for a large packet. These parameters help an audio extractor to assemble contiguous data relating to a given set of audio channels.

Figure 10.30 shows the structure of the audio data packing. In order to prevent accidental generation of reserved synchronizing patterns, bit 9 is the inverse of bit 8 so the effective system wordlength is nine bits. Three nine-bit symbols are used to convey all the AES/EBU subframe data except for the four auxiliary bits. Since four audio channels can be conveyed, there are two 'Ch' or channel number bits which specify the audio channel number to which the subframe belongs. A further bit, Z, specifies the beginning of the 192-sample channel status message. V, U and C have the same significance as in the normal AES/EBU standard, but the P bit reflects parity on the three nine-bit symbols rather than the AES/EBU definition. The three-word sets representing audio an audio sample will then be repeated for the remaining three channels in the packet but with different combinations of the CH bits.

Address / Bit	x3	x3 + 1	x3 + 2
B9	$\overline{B8}$	$\overline{B8}$	$\overline{B8}$
B8	A (2^5)	A (2^{14})	P
B7	A (2^4)	A (2^{13})	C
B6	A (2^3)	A (2^{12})	U
B5	A (2^2)	A (2^{11})	V
B4	A (2^1)	A (2^{10})	A MSB (2^{19})
B3	A LSB (2^0)	A (2^9)	A (2^{18})
B2	CH (MSB)	A (2^8)	A (2^{17})
B1	CH (LSB)	A (2^7)	A (2^{16})
B0	Z	A (2^6)	A (2^{15})

Figure 10.30 AES/EBU data for one audio sample is sent as three nine-bit symbols. A = audio sample. Bit Z = AES/EBU channel status block start bit.

One audio sample in each of the four channels of a Group requires twelve video sample periods and so packets will contain multiples of twelve samples. At the end of the packet a checksum is calculated on the entire packet contents.

If twenty-four-bit samples are required, Extended Data Packets must be employed in which the additional four bits of each audio sample in an AES/EBU frame are assembled in pairs according to Figure 10.31. Thus for every twelve symbols conveying the four twenty-bit audio samples of one group in an Audio Data Packet two extra symbols will be required in an Extended Data Packet.

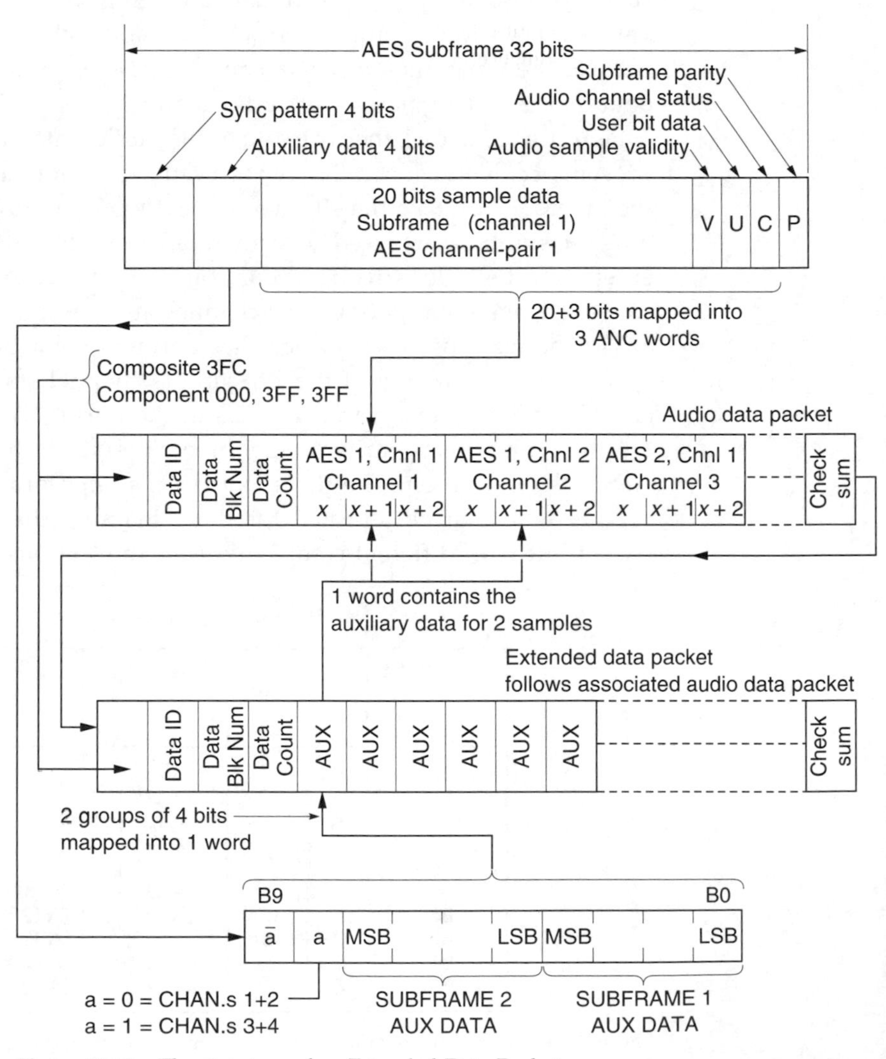

Figure 10.31 The structure of an Extended Data Packet.

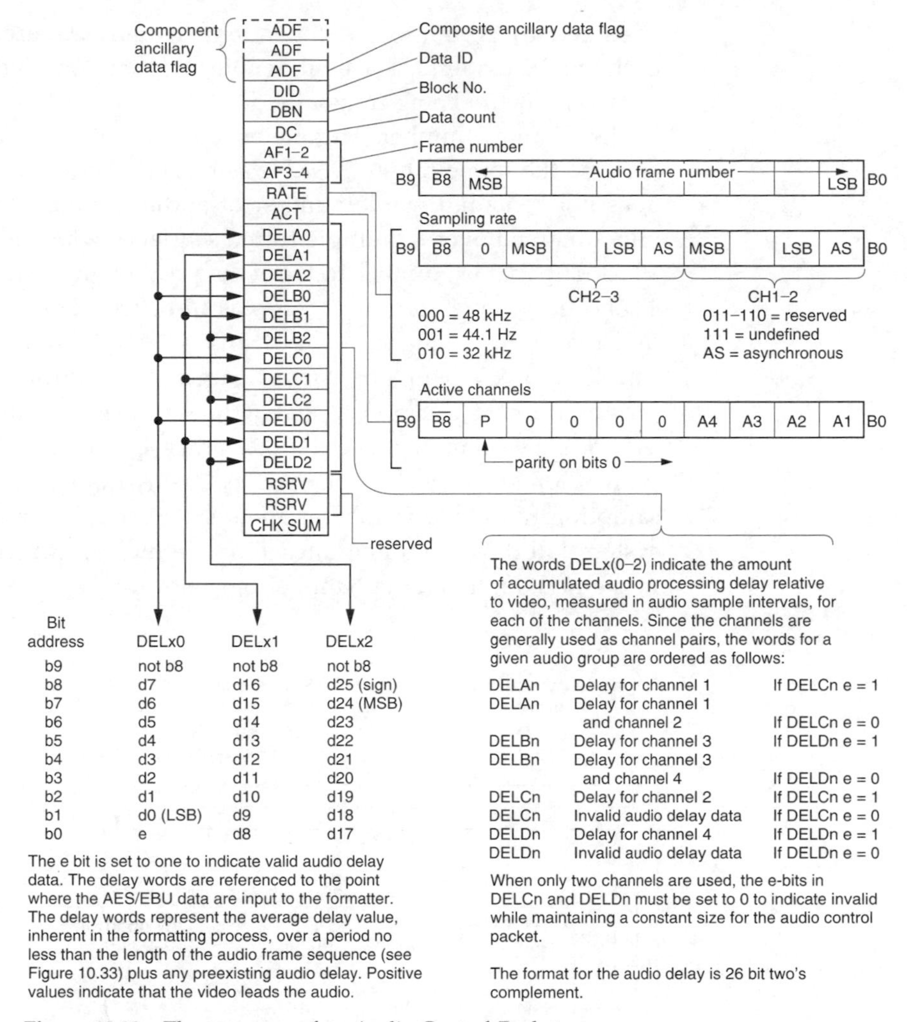

Figure 10.32 The structure of an Audio Control Packet.

The Audio Control Packet structure is shown in Figure 10.32. Following the usual header are symbols representing the Audio Frame Number, the Sampling Rate, the Active Channels, the Processing Delay and some reserved symbols. The Sampling Rate parameter allows the two AES/EBU channel pairs in a Group to have different sampling rates if required. The Active Channel parameter simply describes which channels in a Group carry meaningful audio data. The Processing Delay parameter denoted the delay the audio has experienced measured in audio sample periods. The parameter is a twenty-six-bit two's complement number requiring three symbols for each channel. Since the four audio channels in a Group are generally channel pairs, only two delay parameters are needed. However, if four independent channels are used, one parameter each will be required. The e-bit denotes whether four individual channels or two pairs are being transmitted.

The Frame Number Parameter comes about in 525-line systems because the frame rate is 29.97 Hz not 30 Hz. The resultant frame period does not contain a whole number of audio samples. An integer ratio is only obtained over a multiple frame sequence which is shown in Figure 10.33. The Frame Number conveys the position in the frame sequence. At 48 kHz odd frames hold 1602 samples and even frames hold 1601 samples in a five-frame sequence. At 44.1 and 32 kHz the relationship is not so simple and to obtain the correct number of samples in the sequence certain frames (exceptions) have the number of samples altered. At 44.1 kHz the frame sequence is 100 frames long whereas at 32 kHz it is 15 frames long. As the two channel pairs in a Group can have different sampling rates, two frame parameters are required per Group. In 50 Hz systems all three sampling rates allow an integer number of samples per frame and so the Frame Number is irrelevant.

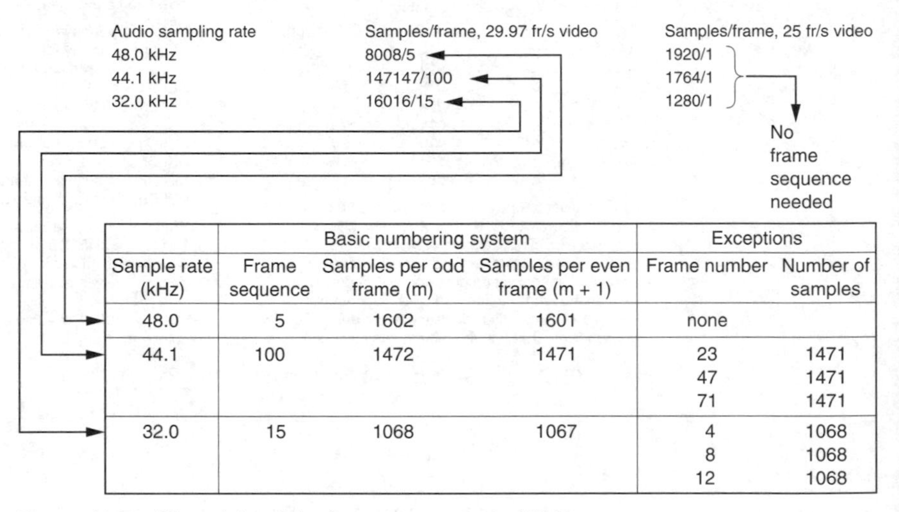

	Basic numbering system			Exceptions	
Sample rate (kHz)	Frame sequence	Samples per odd frame (m)	Samples per even frame (m + 1)	Frame number	Number of samples
48.0	5	1602	1601	none	
44.1	100	1472	1471	23 47 71	1471 1471 1471
32.0	15	1068	1067	4 8 12	1068 1068 1068

Figure 10.33 The origin of the frame sequences in 525-line systems.

As the ancillary data transfer is in bursts, it is necessary to provide a little RAM buffering at both ends of the link to allow real-time audio samples to be time compressed up to the video bit rate at the input and expanded back again at the receiver. Figure 10.34 shows a typical audio insertion unit in which the FIFO buffers can be seen. In such a system all that matters is that the average audio data rate is correct. Instantaneously there can be timing errors within the range of the buffers. Audio data cannot be embedded at the video switch point or in the areas reserved for

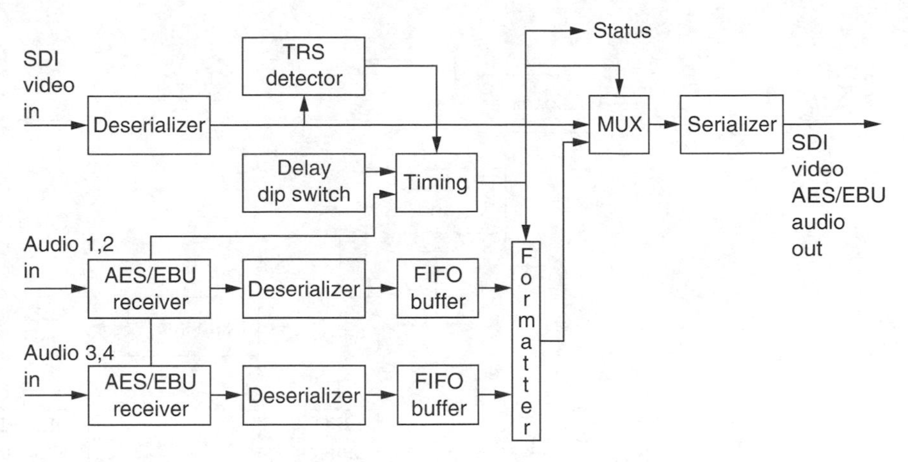

Figure 10.34 A typical audio insertion unit. See text for details.

EDH packets, but provided that data are evenly spread throughout the frame twenty-bit audio can be embedded and retrieved with about 48 audio samples of buffering. If the additional four bits per sample are sent this requirement rises to 64 audio samples. The buffering stages cause the audio to be delayed with respect to the video by a few milliseconds at each insertion. Whilst this is not serious, Level I allows a delay tracking mode which allows the embedding logic to transmit the encoding delay so a subsequent receiver can compute the overall delay. If the range of the buffering is exceeded for any reason, such as a non-synchronous audio sampling rate fed to a Level A encoder, audio samples are periodically skipped or repeated in order to bring the delay under control.

It is permitted for receivers which can only handle twenty-bit audio to discard the four-bit sample extension data. However, the presence of the extension data requires more buffering in the receiver. A device having a buffer of only 48 samples for Level A working could experience an overflow due to the presence of the extension data.

In 48 kHz working, the average number of audio samples per channel is just over three per video line. In order to maintain the correct average audio sampling rate, the number of samples sent per line is variable and not specified in the standard. In practice a transmitter generally switches between packets containing three samples and packets containing four samples per channel per line as required to keep the buffers from overflowing. At lower sampling rates either smaller packets can be sent or packets can be omitted from certain lines.

As a result of the switching, ancillary data packets in component video occur mostly in two sizes. The larger packet is 55 words in length of which 48 words are data. The smaller packet contains 43 words of which

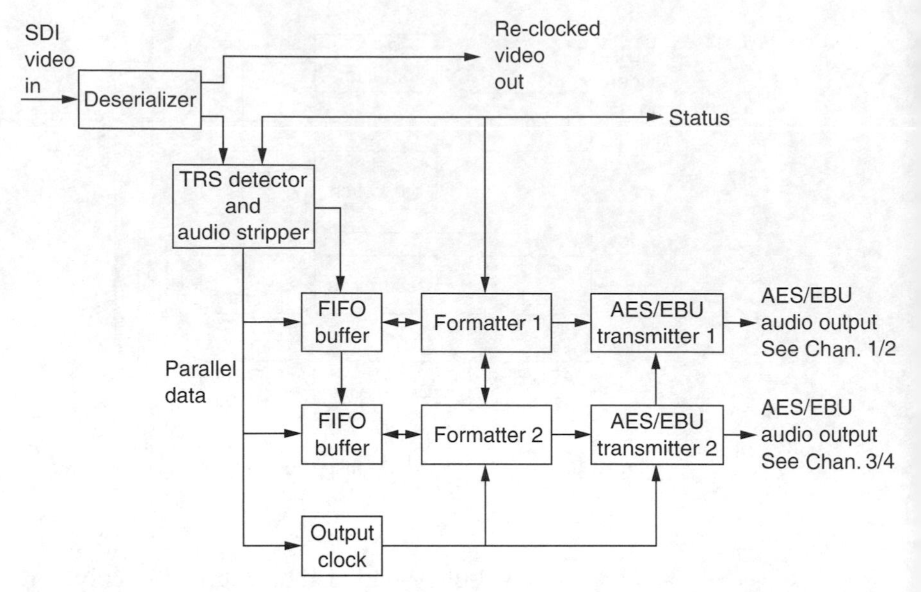

Figure 10.35 A typical audio extractor. Note the FIFOs for timebase expansion of the audio samples.

36 are data. There is space for two large packets or three small packets in the horizontal blanking between EAV and SAV.

A typical embedded audio extractor is shown in Figure 10.35. The extractor recognizes the ancillary data TRS or flag and then decodes the ID to determine the content of the packet. The Group and channel addresses are then used to direct extracted symbols to the appropriate audio channel. A FIFO memory is used to timebase expand the symbols to the correct audio sampling rate.

10.15 FireWire

FireWire[24]is actually an Apple Computers Inc. trade name for the interface which is formally known as IEEE 1394–1995. It was originally intended as a digital audio network, but grew out of recognition. FireWire is more than just an interface as it can be used to form networks and if used with a computer effectively extends the computer's data bus. Figure 10.36 shows that devices are simply connected together as any combination of daisy-chain or star network.

Any pair of devices can communicate in either direction, and arbitration ensures that only one device transmits at once. Intermediate devices simply pass on transmissions. This can continue even if the intermediate device is powered down as the FireWire carries power to keep repeated functions active.

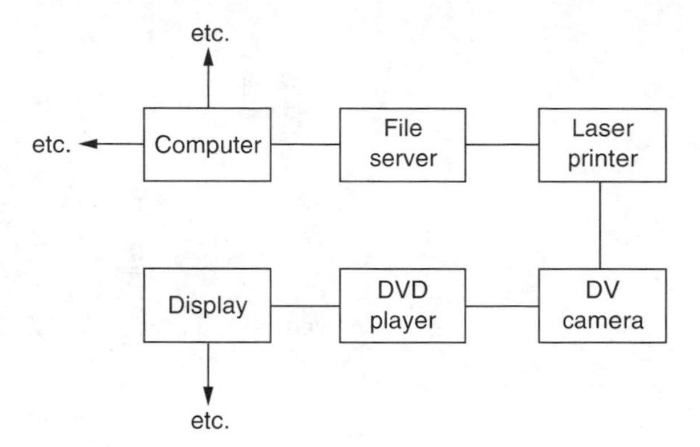

Figure 10.36 FireWire supports radial (star) or daisy-chain connection. Two-port devices pass on signals destined for a more distant device – they can do this even when powered down.

Communications are divided into *cycles* which have a period of 125 μs. During a cycle, there are 64 time slots. During each time slot, any one node can communicate with any other, but in the next slot, a different pair of nodes may communicate. Thus FireWire is best described as a time-division multiplexed (TDM) system. There will be a new arbitration between the nodes for each cycle.

FireWire is eminently suitable for video/computer convergent applications because it can simultaneously support asynchronous transfers of non-real-time computer data and isochronous transfers of real-time audio/video data. It can do this because the arbitration process allocates a fixed proportion of slots for isochronous data (about 80 per cent) and these have a higher priority in the arbitration than the asynchronous data. The higher the data rate a given node needs, the more time slots it will be allocated. Thus a given bit rate can be guaranteed throughout a transaction; a prerequisite of real-time A/V data transfer.

It is the sophistication of the arbitration system which makes FireWire remarkable. Some of the arbitration is in hardware at each node, but some is in software which only needs to be at one node. The full functionality requires a computer somewhere in the system which runs the isochronous bus management arbitration. Without this only asynchronous transfers are possible. It is possible to add or remove devices whilst the system is working. When a device is added the system will recognize it through a periodic learning process. Essentially every node on the system transmits in turn so that the structure becomes clear.

The electrical interface of FireWire is shown in Figure 10.37. It consists of two twisted pairs for signalling and a pair of power conductors. The twisted pairs carry differential signals of about 220 mV swinging around a common mode voltage of about 1.9 V with an impedance of 112 Ω. Figure

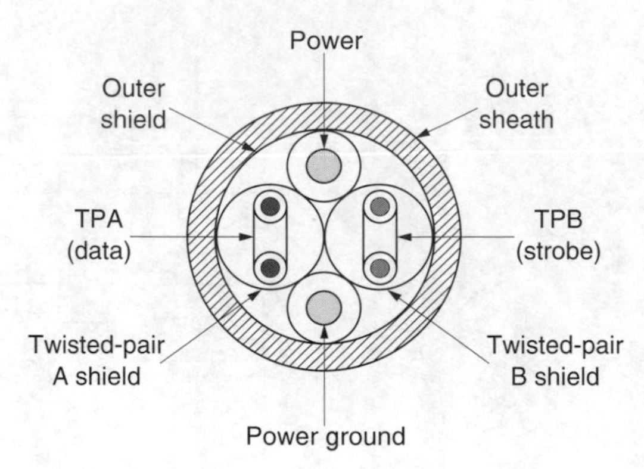

Figure 10.37 FireWire uses twin twisted pairs and a power pair.

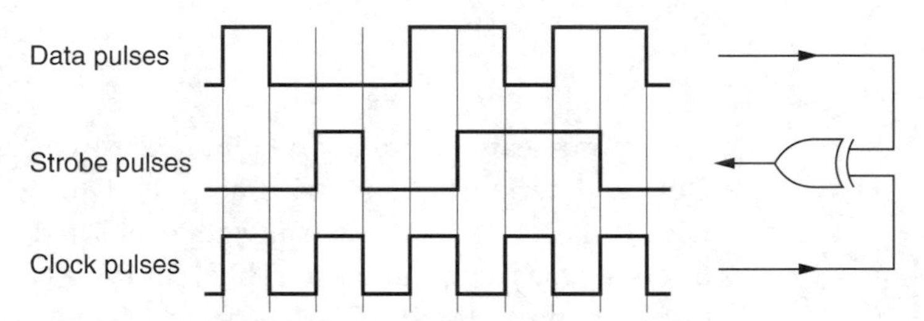

Figure 10.38 The strobe signal is the X-OR of the data and the bit clock. The data and strobe signals together form a self-clocking system.

10.38 shows how the data are transmitted. The host data are simply serialized and used to modulate twisted pair A. The other twisted pair (B) carries a signal called a *strobe*, which is the exclusive-OR of the data and the clock. Thus whenever a run of identical bits results in no transitions in the data, the strobe signal will carry transitions. At the receiver another exclusive-OR gate adds data and strobe to recreate the clock.

This signalling technique is subject to skew between the two twisted pairs and this limits cable lengths to about 10 m between nodes. Thus FireWire is not a long-distance interface technique, instead it is very useful for interconnecting a large number of devices in close proximity. Using a copper interconnect, FireWire can run at 100, 200 or 400 Mbits/s, depending on the specific hardware. It is proposed to create an optical-fibre version which would run at gigabit speeds.

References

1. SMPTE 259M – 10-bit 4:2:2 Component and 4FSc NTSC Composite Digital Signals – Serial Digital Interface
2. EBU Doc. Tech. 3246
3. SMPTE 125M, Television – Bit Parallel Digital Interface – Component Video Signal 4:2:2

4. Eguchi, T., Pathological check codes for serial digital interface systems. Presented at SMPTE Conference, Los Angeles, October 1991
5. SMPTE 305M – Serial Data Transport Interface
6. Specifications of Eureka 95 digital coding and interfaces, Draft version, November 1992
7. Schiffler, W., The implementation of the European HDTV (HDI) standard in studio equipment. *ITS Montreux Symposium Record*, 354–364 (1993)
8. Fischer, J. and Jost, J., Sophi-chip set for serial digital HDTV – Basic research. *ITS Montreux Symposium Record*, 365–375 (1993)
9. Rumsey, F. and Watkinson, J. *The Digital Interface Handbook*, Oxford: Focal Press (1994)
10. Elkind, R. and Fibush, D., Proposal for error detection and handling in studio equipment. Presented at 25th SMPTE Television Conference, Detroit, 1991
11. SMPTE RP 165 – Error detection checkwords and status flags for use in bit-serial digital interfaces for television.
12. SMPTE RP 168 – Definition of vertical interval switching point for synchronous video switching
13. Audio Engineering Society, AES recommended practice for digital audio engineering – serial transmission format for linearly represented digital audio data. *J. Audio Eng. Soc.*, **33**, 975–984 (1985)
14. EIAJ CP-340 *A Digital Audio Interface*, Tokyo: EIAJ (1987)
15. EIAJ CP-1201 *Digital Audio Interface* (revised), Tokyo: EIAJ (1992)
16. IEC 958 *Digital Audio Interface*, first edition, Geneva: IEC (1989)
17. Finger, R., AES3–1992: the revised two channel digital audio interface. *J. Audio.Eng. Soc.*, **40**, 107–116 (1992)
18. EIA RS-422A. Electronic Industries Association, 2001 Eye Street NW, Washington, DC 20006, USA
19. Smart, D.L., Transmission performance of digital audio serial interface on audio tie lines. *BBC Designs Dept Technical Memorandum*, 3.296/84
20. European Broadcasting Union, Specification of the digital audio interface. *EBU Doc. Tech.*, 3250
21. Rorden, B. and Graham, M., A proposal for integrating digital audio distribution into TV production. *J. SMPTE* 606–608 (September, 1992)
22. SMPTE 259M – 10-bit 4:2:2 Component and 4FSc NTSC Composite Digital Signals – Serial Digital Interface
23. Wilkinson, J.H., Digital audio in the digital video studio – a disappearing act? Presented at 9th International AES Conference, Detroit, 1991
24. Wicklegren, I.J., The facts about FireWire. *IEEE Spectrum*, 19–25 (1997)

11

Digital video tape

11.1 Introduction

This chapter considers the techniques necessary to build and control a rotary head digital video tape recorder (DVTR). In addition to LSI signal processing, the DVTR depends heavily on mechanical engineering of the highest quality, and several extraordinarily accurate servo systems.

Tape-based recording suffers from the disadvantage that tape is linear and random access is relatively slow. For editing purposes hard disk-based systems have essentially taken over from tape in the television industry except in a few special cases. However, tape technology has a number of advantages which will ensure its survival in some form. The cost per bit of disk recording remains a couple of orders of magnitude more than that of tape. Data on tape is also more compact than on disk and so for archiving tape has no equal.

In linear applications the slow access of tape is no drawback, hence its popularity for ENG. This popularity has been boosted by the development of DVTRs which can play back at typically four times normal speed so that the data can rapidly be transferred to a disk.

Playout of programs to air is also a linear process at which tape excels. Consequently it is increasingly common to find hybrid systems in which disks are used for editing and for playout of trailers or commercials, with tape used for playout of programs. However, with the editing done on disk, the editing requirement on tape no longer exists and so a generic data tape format can be used rather than a DVTR.

11.2 Compression in DVTRs

Since the development of the first DVTR formats, economical compression hardware has become available and this can be applied to DVTRs. As compression reduces the amount of data to be recorded, there is an immediate economic benefit in the reduction in tape consumption this allows. This reduction can be used in two ways: either the playing time can be extended or the cassette can be made smaller, allowing a miniaturized format. The DVC (digital video cassette) format is a good example of the latter. Compression also improves the access time as a given video clip uses a shorter length of tape so the transport appears to wind faster.

In many cases the video recording will need to be edited and this places constraints on the type of compression which can be used. As was seen in Chapter 6, temporal coding makes pictures interdependent and this causes difficulties in editing. For this reason most compressed DVTRs, including Digital Betacam, D-9 and DVC, use only spatial coding so that the pictures are then independent and full edit freedom exists.

In a compressed DVTR an encoder is needed in the record process and a decoder in the playback process. Thus the number of encoders and decoders required is equal. There is then no advantage in using an asymmetrical coding scheme such as MPEG. DVTRs work better with symmetrical codes which is why MPEG is little used in tape applications.

Video compression is lossy and the reproduced pixel data are not identical to the original, re-introducing generation loss to the digital domain. The losses may not be visible on a first generation, but after several generations they may become visible. Thus a production format might be expected to run at a higher bit rate on tape than an ENG format which in turn might use a higher bit rate than a consumer format.

In production formats such as Digital Betacam and D-9 the compression factor is relatively mild such that with most program material generation loss is not an issue. In formats such as DVC the compression factor is higher and so fewer generations are possible, but for ENG purposes and for straightforward production excellent results are possible. The DVCPRO format supports the same 25 Mbits/s rate of DVC, but later introduced a 50 Mbits/s version, giving better results.

Where the finest quality is required, an uncompressed recorder can be used as it has no generation loss at all. The D-1 format was uncompressed, but uneconomic, whereas the uncompressed D-5 format offered lower running costs. D-5 can record the full bit rate of standard definition component video, and so with compression it can record digital HDTV.

11.3 Why rotary heads?

The attractions of the helical recorder are that the head-to-tape speed and hence bandwidth are high, whereas the linear tape speed is not. The space between tracks is determined by the linear tape speed, not by multitrack head technology. Chapter 8 showed how rotary head machines make better use of the tape area, particularly if azimuth recording is used. The high head speed raises the frequency of offtape signals, and since output is proportional to frequency, playback signals are raised above head noise even with very narrow tracks.

With stationary heads, the offtape frequency is proportional to the linear tape speed and this makes recovery of data at other than the correct speed extremely difficult as data separators tend to work well only over a narrow speed range. In contrast, with rotary heads the scanning speed dominates head-to-tape speed and variations in the linear speed of the tape have a smaller effect. Thus picture-in-shuttle and slow motion are readily accommodated in a rotary head machine.

11.4 Helical geometry

Figure 11.1 shows the general arrangement of the two major categories of rotary-head recorder. In transverse-scan recorders, relatively short tracks are recorded almost at right angles to the direction of tape motion by a rotating headwheel containing four or six heads. In helical-scan recorders, the tape is wrapped around the drum in such a way that it enters and leaves in two different positions along the drum axis. This causes the rotating heads to record long slanting tracks. In both approaches, the pitch of the tracks is determined by the linear tape speed. The space between tracks can easily be made much smaller than in stationary-head recorders, in fact it can be zero or even negative.

The use of rotary heads was instrumental in the development of the first analog video recorders. As video signals consist of discrete lines and frames, it was possible to conceal the interruptions in the tracks of a rotary-head machine by making them coincident with the time when the CRT was blanked during flyback. The first video recorders developed by Ampex used the transverse-scan approach, with four heads on the rotor: hence the name quadruplex which was given to this system. The tracks were a little shorter than the 2-inch width of the tape, and several sweeps were necessary to build up a video frame. The change-overs between the heads were made during the horizontal synchronizing pulses.

Variable-speed operation in rotary head machines is obtained by deflecting the playback heads along the drum axis to follow the tape tracks. Periodically the heads will need to jump in order to omit or repeat

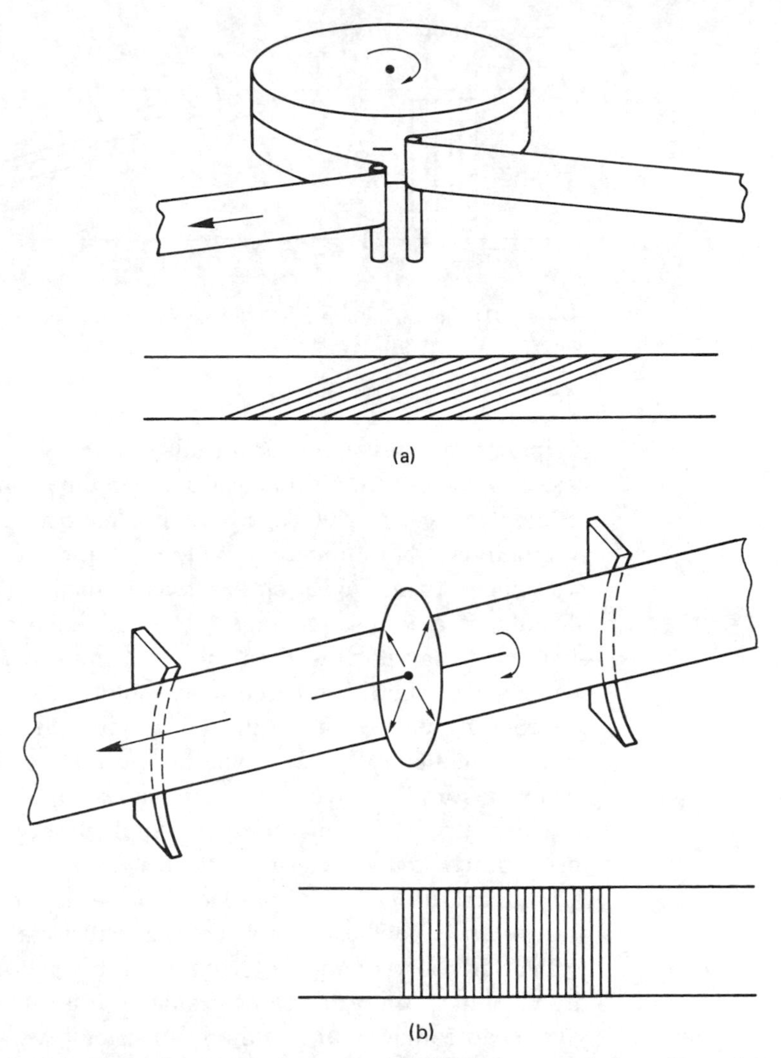

Figure 11.1 Types of rotary-head recorder. (a) Helical scan records long diagonal tracks. (b) Transverse scan records short tracks across the tape.

a field. In transverse scan the number of tracks needed to accommodate one field is high, and the head deflection required to jump fields was virtually impossible to achieve.

In helical scanning, the tracks become longer, so fewer of them are needed to accommodate a field. Figure 11.2 shows that the head displacement needed is then reduced. In digital recorders fields are inevitably segmented into a number of tracks. Uncompressed formats will require more tracks than compressed formats.

Figure 11.3 shows the two fundamental approaches to drum design. In the rotating upper drum system, one or more heads are mounted on the

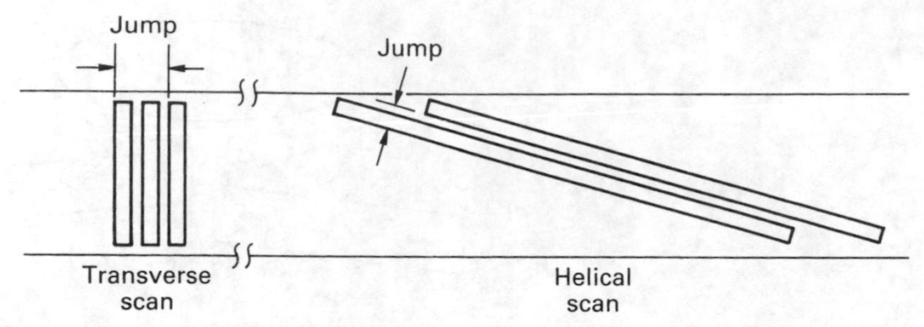

Figure 11.2 Helical scan results in longer tracks so a shorter distance needs to be jumped to miss out or repeat a field.

periphery of a revolving drum (also known as the scanner). The fixed base of the drum carries a helical ramp in the form of a step (see inset) or, for greater wear resistance, a hardened band which is suitably attached. Alternatively, both top and bottom of the drum are fixed, and the headwheel turns in a slot between them. Both approaches have advantages and disadvantages. The rotating upper drum approach is simpler to manufacture than the fixed upper drum, because the latter requires to be rigidly and accurately cantilevered out over the headwheel. The rotating part of the drum will produce an air film which raises its effective diameter. The rotor will thus need to be made a slightly different diameter from the lower drum, so that the tape sees a constant diameter as it rises up the drum. There will be plenty of space inside the rotor to install individually replaceable heads.

The fixed upper drum approach requires less power, since there is less air resistance. The headwheel acts as a centrifugal pump, and supplies air to the periphery of the slot where it lubricates both upper and lower drums, which are of the same diameter. It is claimed that this approach gives head contact which is more consistent over the length of the track, but it makes the provision of replaceable heads more difficult, and it is generally necessary to replace the entire headwheel as an assembly.

The presence of the air film means that the tape surface is not normally in contact with the cylindrical surface of the drum, but is merely guided by one edge following the helical step. The thickness of the air film is that where the pumping effect of the rotating drum reaches equilibrium with the tape tension. Figure 11.4 shows that the heads project out of the drum by a distance which must exceed the air film thickness in order to deform the tape slightly. In the absence of such a deformation there would be no contact pressure. The creation and collapse of the deformities in the tape results in appreciable acoustic output, in the form of an irritating buzz.

Two sets of fixed guides, known as the entry and exit guides, lead the tape to and from the ramp. In DVTRs the tape passes between half-way and three-quarters of the way around the drum. The total angle of drum

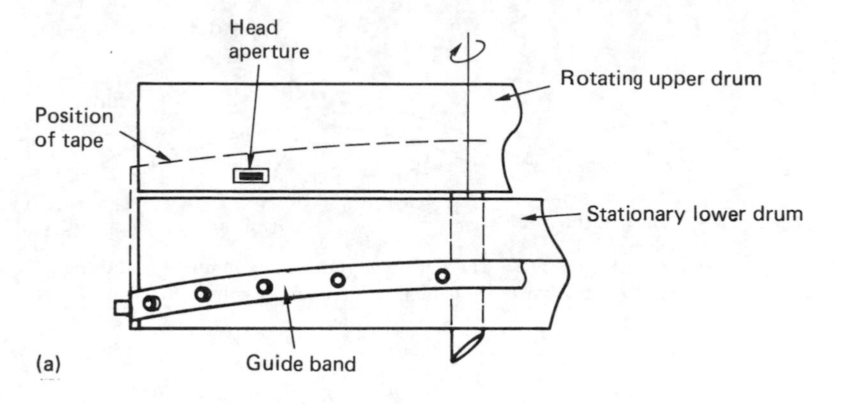

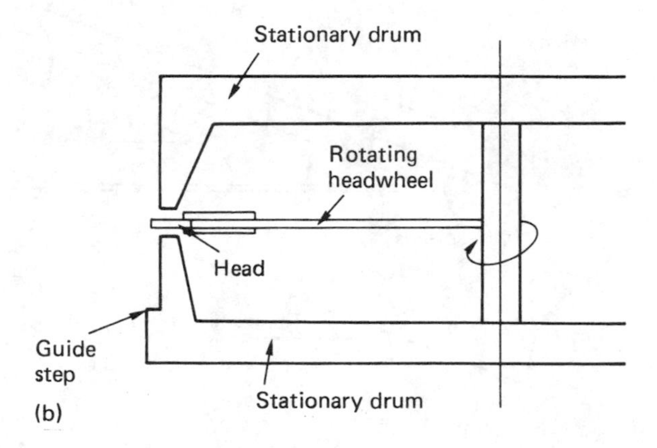

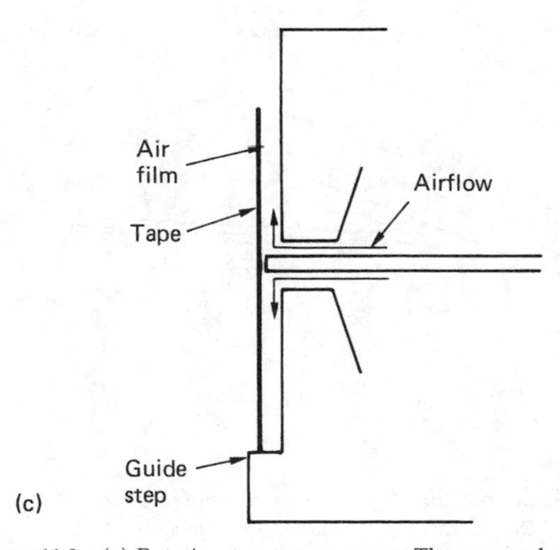

Figure 11.3 (a) Rotating-top-type scanner. The upper drum is slightly smaller than the lower drum due to the air film it develops. Helical band guides lower edge of tape. (b) Stationary-top scanner, where headwheel rotates in a slot between upper and lower drums. (c) Airflow in stationary-top scanner. Headwheel acts as a centrifugal pump, producing air film between tape and drums.

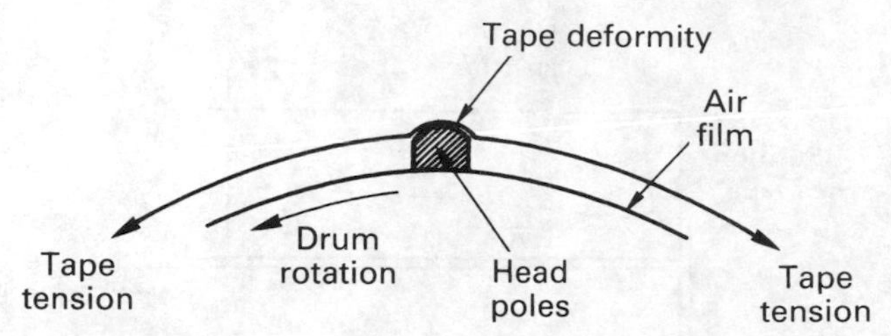

Figure 11.4 The tape is supported on an air film and the heads must project by a greater distance to achieve contact pressure.

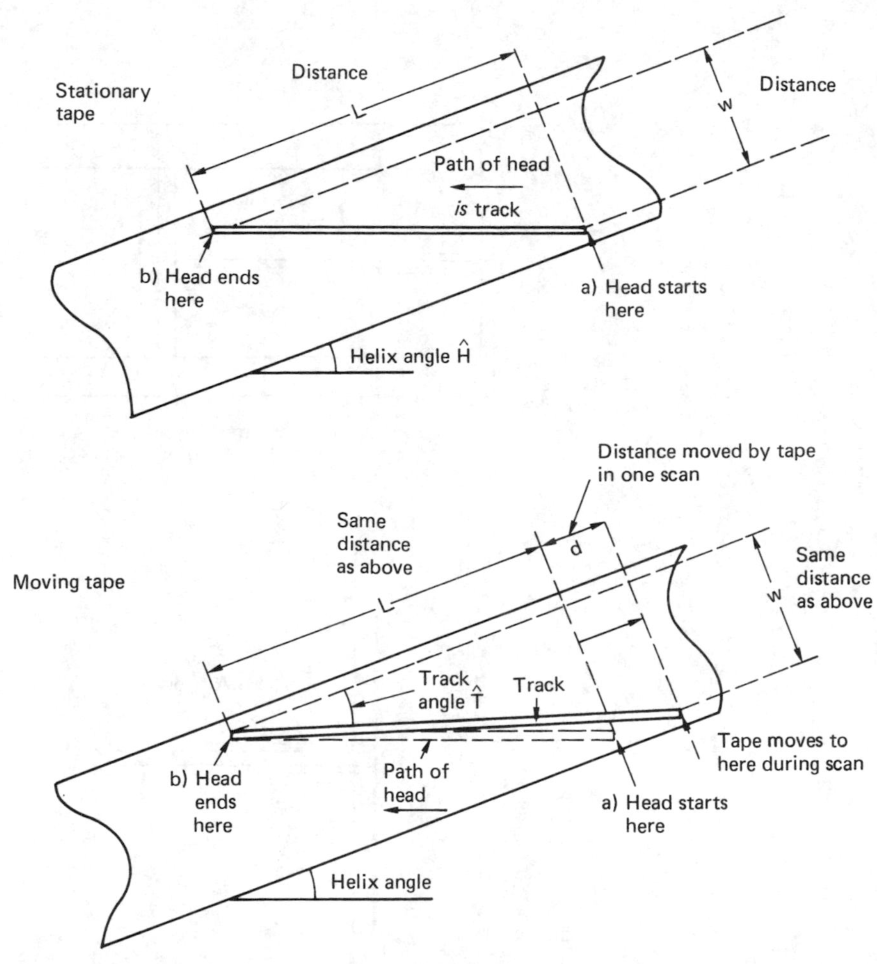

Figure 11.5(a) and (b) When the tape is moving the track laid down will be at an angle which is different from the helix angle, because the tape is in a different place at the end of the scan than at the beginning. Track length can be considered as two components L and w for stationary case, and $L + d$ and w for moving case. Hence:

$$\frac{w}{L} = \tan \hat{H} \quad \text{and} \quad \frac{w}{L + d} = \hat{T}$$

rotation for which the heads touch the tape is called the *mechanical wrap angle* or the *head contact angle*. The angle over which a useful recorded track is laid down is always somewhat shorter than the mechanical wrap angle.

It will be evident that as the tape is caused to travel along the drum axis by the ramp, it actually takes on the path of a helix. The head rotates in a circular path, and so will record diagonal tracks across the width of the tape. So it is not really the scanning which is helical, it is the tape path.

In practical machines the tape may travel up the ramp (D-2) or down the ramp (D-3). In addition, the drum can rotate either with (D-3) or against (D-2) the direction of linear tape motion.

If the tape is stationary, as in Figure 11.5(a) the head will constantly retrace the same track, and the angle of the track can be calculated by measuring the rise of the tape along the drum axis, and the circumferential distance over which this rise takes place. The latter can be obtained from the diameter of the drum and the wrap angle. The tangent of the *helix angle* is given by the rise over the distance. It can also be obtained by measuring the distance along the ramp corresponding to the wrap angle, and this dimension will give the sine of the helix angle.

However, when the tape moves, the angle between the tracks and the edge of the tape will not be the helix angle. Figure 11.5(b) shows an

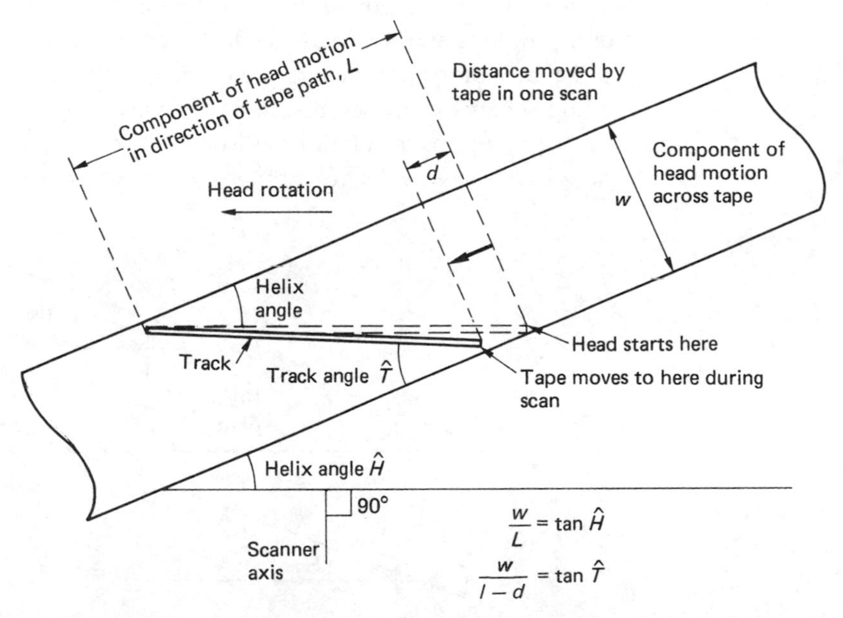

Figure 11.5(c) When the tape moves, the track laid down will not be at the helix angle because the tape is in different places at the beginning and end of the scan. In D-3 the scanner rotates *with* the tape motion, so that the track angle is greater than the helix angle. The diagram is drawn with the observer on the scanner axis looking outwards at the magnetic surface of the tape.

example in which tape climbs up the drum which rotates against the direction of tape travel (D-2). When the head contact commences, the tape will be at a given location, but when the head contact ceases, the tape will have moved a certain linear distance and the resultant track will be longer.

Figure 11.5(c) shows what happens when the tape moves down the drum in the same direction as head rotation. In this case the track is shorter.

In order to obtain the *track angle* it is necessary to take into account the tape motion. The length of the track resulting from scanning a stationary tape, which will be at the helix angle, can be resolved in to two distances at right angles as shown in Figure 11.6. One of these is across the tape width, the other is along the length of the tape. The tangent of the helix angle will be the ratio of these lengths. If the length along the tape is corrected by adding or subtracting the distance the tape moves during one scan, depending upon whether tape motion aids or opposes drum rotation, the new ratio of the lengths will be the tangent of the track angle. In practice this process will often be reversed, because it is the track angle which is standardized, and the transport designer has to find a helix angle which will produce it. It can also be seen from Figure 11.5 that when a normally recorded tape is stopped, the head must be deflected by a triangular waveform or ramp in order to follow the track.

The relative direction of drum and tape motion is unimportant, as both configurations work equally well. Often the direction is left undefined until many other parameters have been settled. If the segmentation and coding scheme proposed results in the wavelengths on tape being on the short side, opposite rotation will be chosen, as this lengthens the tracks and with it the wavelength. Sometimes the direction of rotation is a given;

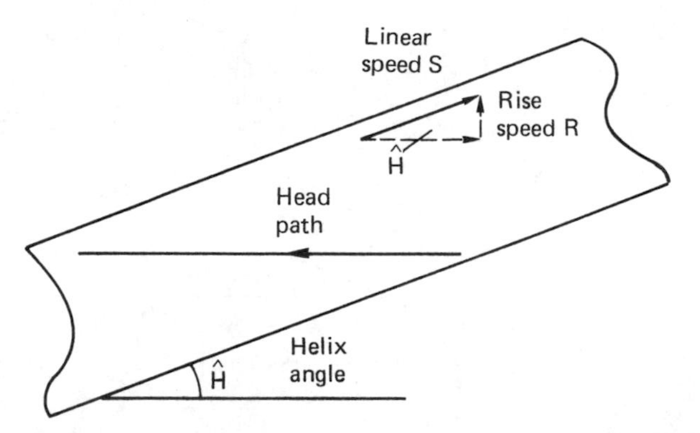

Figure 11.6 The linear speed of the tape S can be resolved into the speed of the tape along the scanner axis which is the rise speed R.$R = S \sin \hat{H}$. Dividing the rise speed by the head passing frequency gives the track pitch. Head passing frequency is simply the number of heads multiplied by the scanner rotational frequency. Thus track pitch is proportional to tape speed.

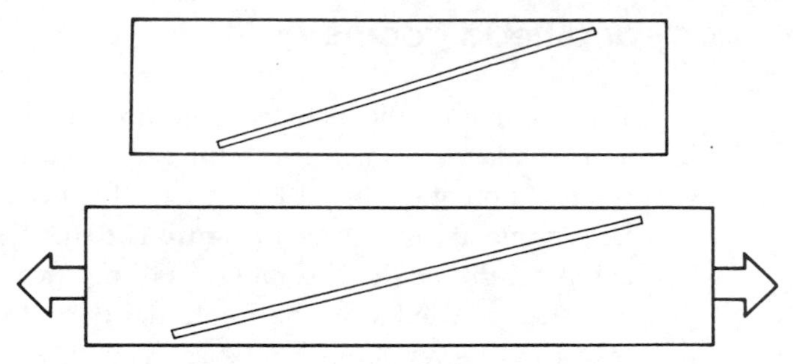

Figure 11.7 Tape is flexible, so change of tension will result in change of length. This causes a change in track angle known as skew shown exaggerated here.

in Digital Betacam, playback of analog tapes was a requirement, so the rotation direction had to be the same as in analog Betacam.

Provided that the tape linear speed and drum speed remain the same, the theoretical track angle will remain the same. In practice this will only be the case if the tape tension is constant. Figure 11.7 shows that if the back tension changes, the effective length of the tape will also change, and with it the track angle, a phenomenon known as *skew*. It will be evident that skew causes a tracking error which increases towards the ends of the track. Tension errors also affect the air film thickness and head contact pressure. All digital rotary head recorders need some form of tape tension servo to maintain the tape tension around the drum constant irrespective of the size of the pack on the supply hub.

In practice tension servos can only control the tension of the tape entering the drum. Since there will be friction between the tape and the drum, the tape tension will gradually increase between the entrance and exit guides, and so the tape will be extended more towards the exit guide. When the tape subsequently relaxes, it will be found that the track is actually curved. The ramp can be made to deviate from a theoretical helix to counteract this effect, or all drums can be built to the same design, so that effectively the track curvature becomes part of the format. Another possibility is to use some form of embedded track-following system, or a system like azimuth recording which tolerates residual tracking errors.

When the tape direction reverses, the sense of the friction in the drum will also reverse, so the back tension has to increase to keep the average tension the same as when going forward.

When verifying the track angle produced by a new design, it is usual to develop a tape which it has recorded using magnetic fluid, and take measurements under a travelling microscope. With the very small track widths of digital recorders, it is usually necessary to compensate for skew by applying standard tension to the tape when it is being measured, or by computing a correction factor for the track angle from the modulus of elasticity of the tape.

11.5 Track and head geometry

The *track pitch* is the distance, measured at right angles to the tracks, from a given place on one track to the same place on the next. The track pitch is a function of the linear tape speed, the head-passing frequency and the helix angle. It can be seen in Figure 11.6 that the linear speed and the helix angle determine the rise rate (or fall rate) with respect to the drum axis, and the knowledge of the rotational period of the drum will allow the travel in one revolution to be calculated. If the travel is divided by the number of active heads on the drum, the result will be the track pitch. If everything else remains equal, the track pitch is proportional to the linear tape speed. Note that the track angle will also change with tape speed.

Figure 11.8(a) shows that in guard band recording used on early machines, the track pitch is equal to the width of the track plus the width of the guard band. The track width is determined by the width of the head poles, and the linear tape speed will be high enough so that the desired guard band is obtained. In guard band recording, the erase head is wider than the record head, which in turn is wider than the replay head, as shown in Figure 11.9(a). This ensures that despite inevitable misalignments, the entire area to be recorded is erased and the playback head is entirely over a recorded track.

In azimuth recording the situation is different, depending upon whether it is proposed to use flying erase heads. In Figure 11.8(b) the head width is greater than the track pitch, so that the tape does not rise far enough for one track to clear the previous one. Part of the previous track will be overwritten, so that the track width and the track pitch become identical. As Figure 11.9(b) shows, this approach guarantees that, when re-recording, previous tracks are fully overwritten as the overlapping heads cover the entire tape area at least once, in places twice. For some purposes, flying erase heads are not then necessary.

The alternative shown in Figure 11.8(c) is for the track width to be exactly the same as the track pitch so there is no overlapping. In this case misalignment during re-recording can allow a thin strip of a previous track to survive unless a flying erase head is used. This is shown in Figure 11.9(c).

If it is necessary to play back at variable speed, each of the heads will need to be moveable. To reduce the number of actuators needed, the heads can be mounted in pairs (or sets of four), provided that the heads are mounted in different planes. The separation between the head planes depends on the track pitch and the angular separation betwen the heads. If this geometry is correct, the same format will result, but the time at which the various tracks are laid down will be different, as the figure shows. The tracks which are effectively written together are known as *segments*.

(a)

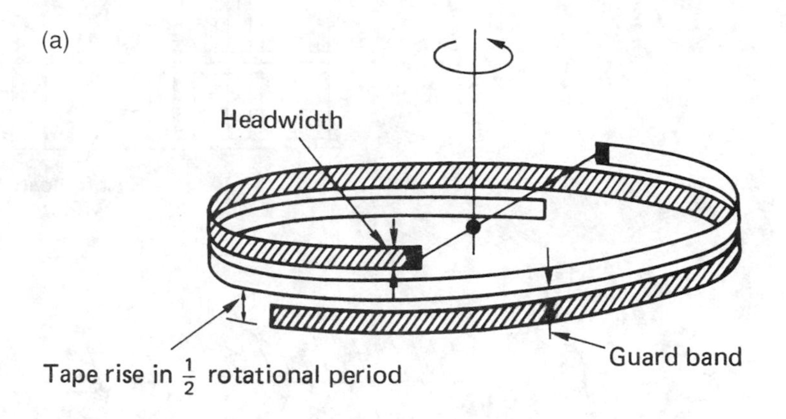

(b)

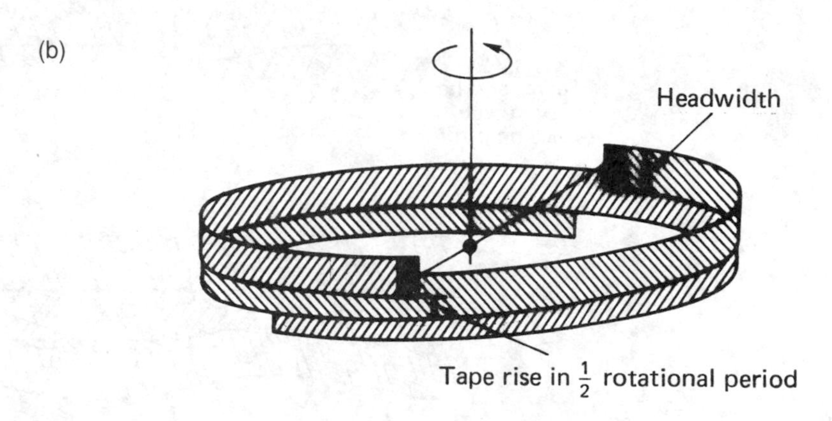

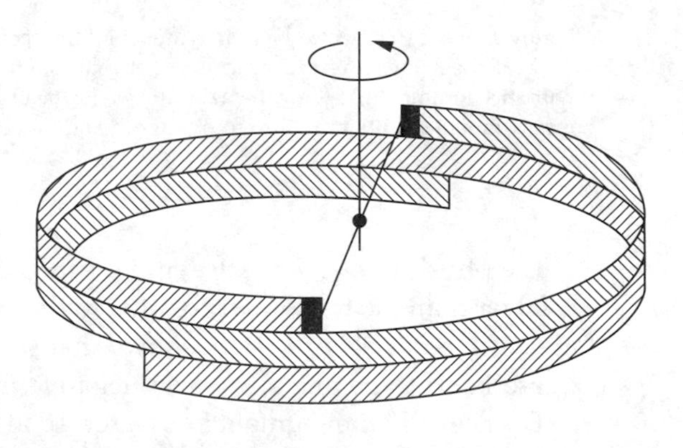

(c)

Figure 11.8 In the absence of azimuth recording, guard bands (a) must be left between the tracks. With azimuth recording, the record head may be wider than (b) or of the same width as (c) the track.

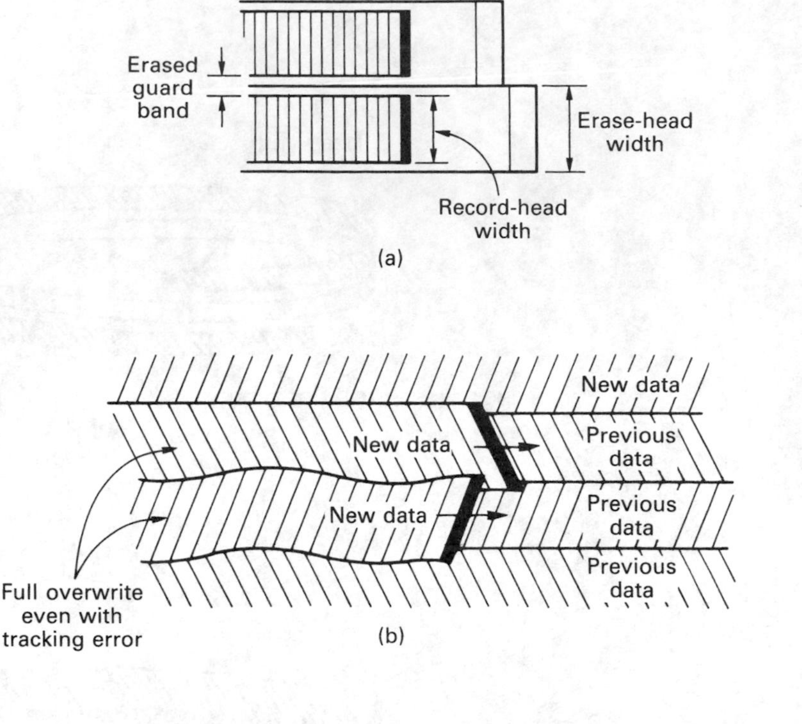

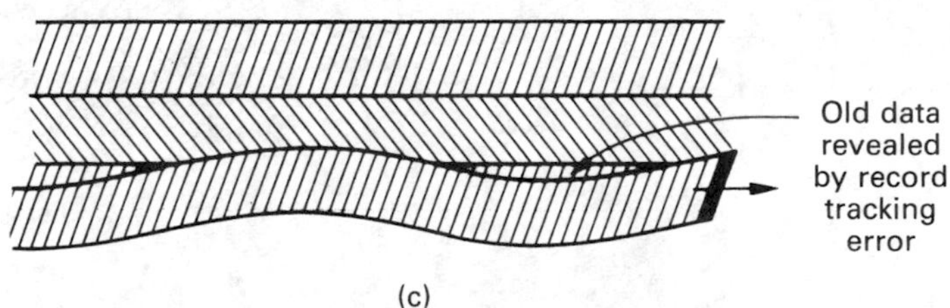

Figure 11.9 (a) The erase head is wider than the record head in guard-band recording. At (b) if the azimuth record head is wider than the track, full overwrite is obtained even with misalignment. At (c) if the azimuth record head is the same width as the track, misalignment results in failure to overwrite and an erase head becomes necessary.

The physical stagger between the heads is necessary because the head is larger than the track it writes, due to the poles being milled away in the area of the gap. Figure 11.10 shows that staggering the heads allows the guard band to be any size, even negative in azimuth applications.

The heads in the segment are staggered and if the recording waveforms in the two heads are synchronous, diagonal tracks will begin in slightly different places along the tape as is evident in the D-2 format. Alternatively it is possible to delay the record signal to one of the heads so that both tracks begin and end in the same place. This is done in D-3/D-5 and as a result the edit gaps between sectors line up in adjacent

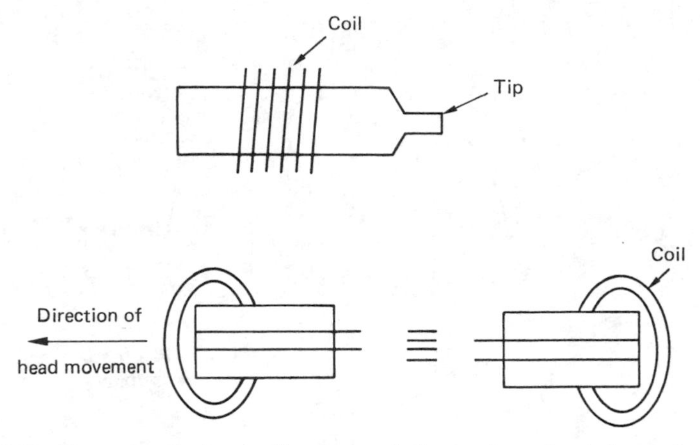

Figure 11.10 Staggering the heads allows any relative track spacing to be used.

tracks allowing the use of a single flying erase head shared between the tracks.

DVTRs vary considerably in the number of heads installed in the drum. In read–modify–write or pre-read, the tape tracks are read by heads prior to the record heads for editing purposes. In some cases, the pre-read function is obtained by deflecting the track-following playback heads so that they precede the record heads. Clearly there is then no confidence replay in this editing mode.

In formats which use compression, the coding and decoding processes introduce significant additional delay into the read–modify–write loop and the pre-read head must then be physically further advanced. The additional distance required will be beyond the deflection range of the track-following actuator and extra heads will be necessary.

Figure 11.11(a) shows an early D-2 drum in which a two pairs of record heads are mounted opposite one another so that they function alternately with a 180° wrap. The replay heads pairs are mounted at 90° on track-following actuators.

Figure 11.11(b) shows a D-3 drum which has a pair of flying erase heads in addition to the record and play heads. In D-3 the two tracks in the segment are aligned by delaying the record signals, so one flying erase head functions for two tracks.

The D-5 format is backward compatible with D-3 tapes, but can also record the higher data rate of component digital video by doubling the data throughput of the drum. This is done by having four parallel tracks per segment, requiring four heads on each base. To play a D-3 tape, only two of the heads are used. In order to work in components, the drum speed is unchanged, but the tape linear speed is doubled. Figure 11.11(c) shows a D-5 drum in which the general arrangement is the same as for D-3, but each base is fitted with four heads.

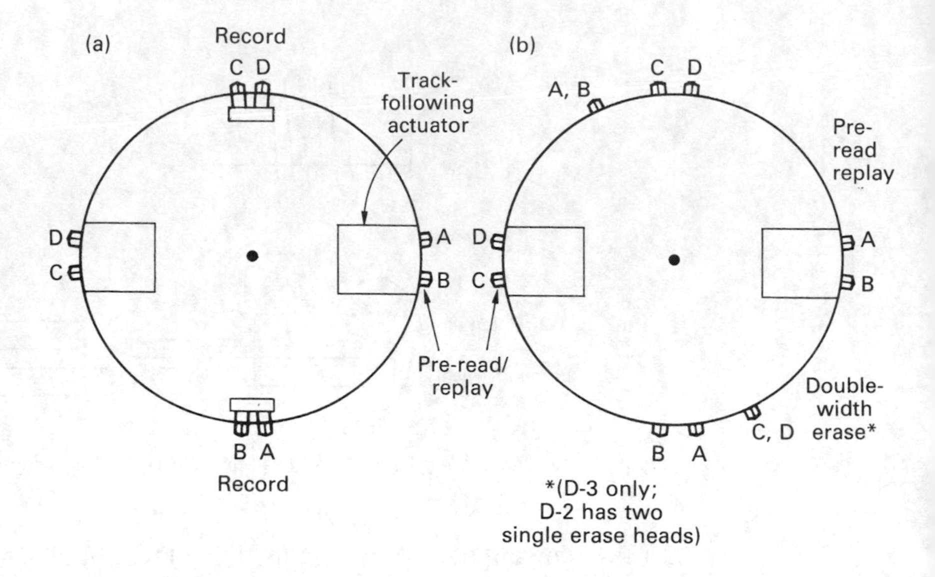

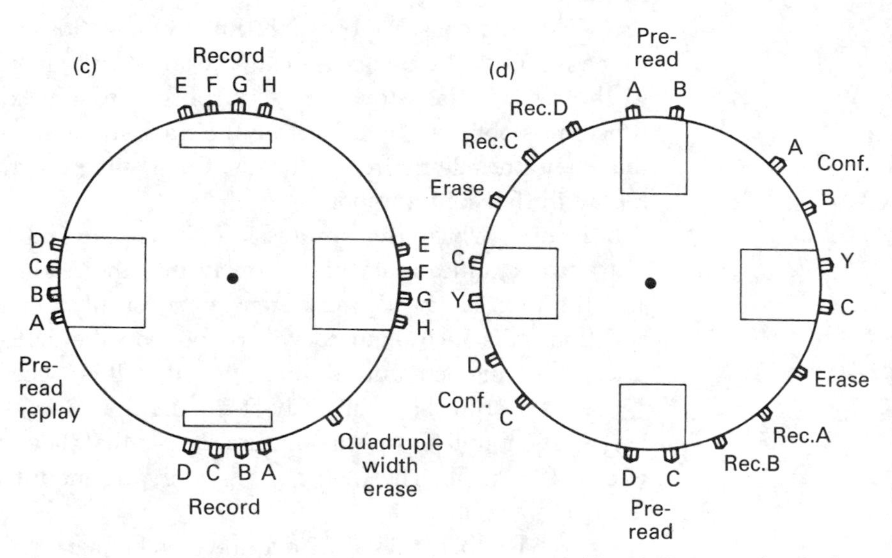

Figure 11.11 Various drum configurations contrasted. At (a) early D-2 recorders had no flying erase. At (b) later D-2 and D-3 drums have flying erase. At (c) D-5 uses four tracks per segment. Two heads of D-5 head base will register with D-3 segments. At (d) analog-compatible Digital Betacam has separate confidence replay heads as well as track-following analog heads.

Figure 11.11(d) shows the drum of an analog compatible Digital Betacam. The two tracks in each segment are aligned by delays so that double-width flying erase heads can be used. The erase, record A and record B heads corresponding to one segment can be seen to the bottom-right of the diagram, with the same configuration diametrically opposite.

Fixed confidence replay heads are arranged nearly at right angles to the record heads. These heads trace the tracks which have just been recorded. At the top and bottom of the diagram can be seen the advanced playback heads which are also used for track following. Finally there are two more track-following actuators which carry pairs of heads that are the width of tracks on analog Betacam tapes.

11.6 Track-following systems

Making a helical scan recording is simply a matter of spinning the drum and moving the tape at the right speed. On replay, it is necessary to follow the slanting tracks accurately. This is achieved in two ways. In early formats replay tracking is achieved using a linear control track. This is recorded with a stationary head along the edge of the tape. The control track contains one pulse for every head sweep across the tape. The drum also generates a pulse once per head sweep. Figure 11.12(a) shows that the capstan is driven by a phase comparator which compares pulses from the drum and the control track head. When these coincide, the rotary head must be over the slant track.

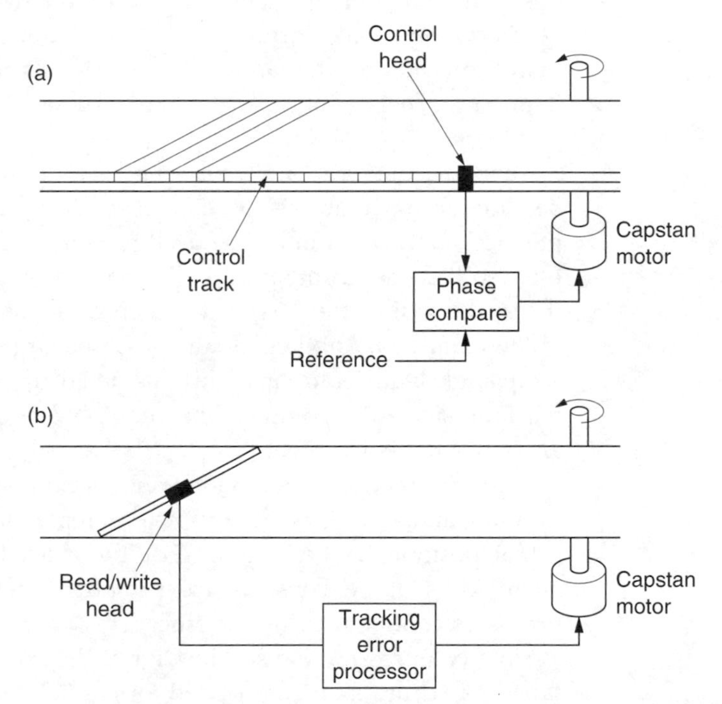

Figure 11.12 Video recorder track following. In older formats (a) a control (CTL) track carries one pulse per slant track and phase comparison of these pulses drives the capstan. In recent formats (b) the CTL track is dispensed with. Instead the video head measures its own tracking error and uses this to drive the capstan.

The control track approach was inherited from analog video recorders in which the tracks are wider than digital tracks. In high-density digital recording the control track approach is not sufficiently accurate. Figure 11.12(b) shows that if the rotary head can sense its position relative to the track this tracking error can be used to drive the capstan. This is the approach taken in the DVC format. There is then no need for a linear track.

For playback at normal speed the tracking error need only be established once or twice per track, but for variable-speed operation the tracking error needs to be measured continuously.

11.7 Time compression and segmentation

The length of the track laid on the tape is a function of the drum diameter and the wrap angle. Figure 11.13 shows a number of ways in which the same length of track can be put on the tape. A small drum with a large wrap angle works just as well as a large drum with a small wrap angle. If the tape speed, hence the track rate, is constant, and for the purposes of this comparison the number of heads on the drum remains constant, the heads on the smaller drum will take a longer time to traverse the track than the heads on the larger drum. This has the effect of lowering the frequencies seen by the heads, and is responsible for the 270° wrap of D-1 recorders.

Changing the wrap angle also means that the helix angle will need to be different so that the same track angle is achieved in all cases. In practice, a larger drum can be used to mount more heads which work in turn, so that the drum speed will then be less than before. Alternatively the space inside the larger drum may be used to incorporate track-following heads. Another advantage of the large drum is that the reduced wrap angle is easier to thread up. A small drum is attractive for a portable machine since it is compact and requires less power. The greater wrap angle has to be accepted.

In the digital domain the video field is expressed by a given number of samples, and as long as these appear on replay in the right sequence, their actual position in the recording or the exact time at which they were recorded is of no consequence provided that the record and replay processes complement one another.

Time compression is used in Digital Betacam in order to play analog tapes. The drum diameter needed for Digital Betacam was larger than for the analog format in order to accommodate the high bit rate with the segmentation chosen. However, the helix angle of a DB transport is such that when analog Betacam tapes are played at the correct linear speed, the

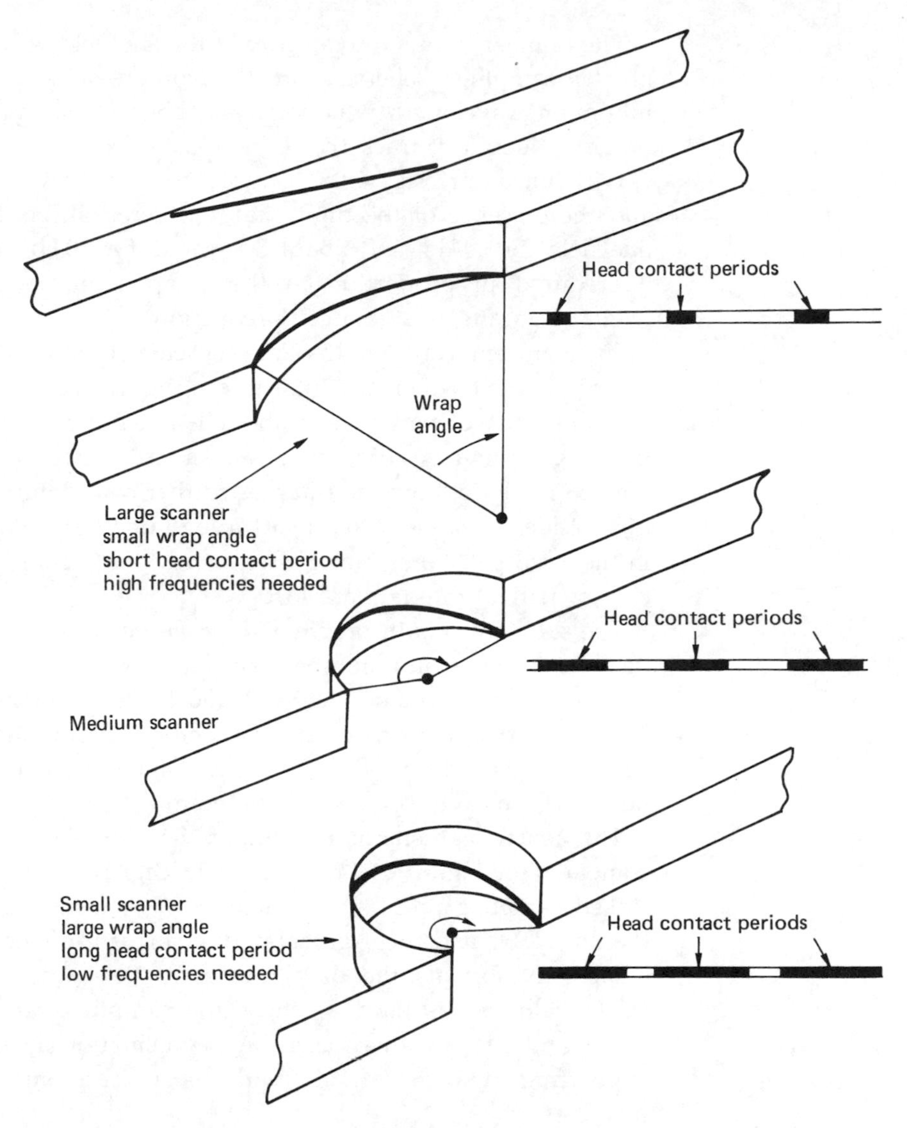

Figure 11.13 A standard track can be put on tape with a variety of scanners. A large wrap allows the lowest frequencies at the head but increases complexity of threading mechanism and increases tape path friction. The helix angle will be different in each case in order to match the tape rise speed to the head peripheral speed – giving the same track angle.

track angle matches the head path. Thus the analog heads follow the tracks. As the drum is larger than the original, the head-to-tape speed is too high and the analog Betacam signal is time compressed as it replays. It is expanded in the TBC. The Digital Betacam format has to take the analog-compatible helix angle as a given.

The number of bits to be recorded for one field will be a direct result of the sampling scheme and the compression factor chosen. The minimum wavelength which can reliably be recorded will then determine the length of track that is necessary.

With uncompressed 4:2:2 sampling, and allowing for redundancy and the presence of the digital audio, there will be about 4.5 Mbits in one field for 50 Hz, and about 3.75 Mbits for 60 Hz. These figures can be reduced in proportion to the compression factor. For example, Digital Betacam uses about 2:1 reduction.

Since current wavelengths are restricted to about 0.7–0.9 μm (which records two bits) then the total track lengths needed in the absence of compression will be between about 1 and 2 metres. Clearly no-one in their right mind is going to design a drum large enough to put the whole of one field on one track. A further consideration is that it is not a good idea to pass all the data through one head. Eddy current losses in the head get worse at high frequencies, and should the single head clog or fail, all the data are lost.

The solution to both problems is *segmentation* where the data for one field are recorded in a number of head sweeps. With segmentation it is easy to share the data between more than one head; a technique known as distribution. In the D-1 and D-5 formats four heads are necessary, whereas in D-2, D-3 and Digital Betacam, two heads are necessary, and the formats reflect these restrictions.

The actual segmentation techiques used in the various formats are compared in Figure 11.14. Composite digital formats use subcarrier-locked sampling rates, so the data rate is substantially higher in PAL than in NTSC and the segmentation changes to reflect the data rate. In component formats the data rate stays the same whether 50 Hz or 60 Hz field rate is used as the same sampling rate is used in both line standards. In this case the segmentation is used to simplify supporting two field rates with as much common hardware as possible.

For these reasons D-2 and D-3, which are both composite digital, use the same segmentation approach. In PAL, data for each field is divided into four segments, each of which requires two tracks, one of each azimuth type. There are thus eight tape tracks per field. In NTSC the data rate is less and three segments suffice so the field is accommodated in six tracks.

The Digital Betacam format has a similar data rate to digital NTSC after the compression process and also requires three segments or six tracks per field in the 60 Hz version. However, the 50 Hz version uses four segments or eight tracks per field, so a common segment rate cannot be used because there is no common factor at these lower rates.

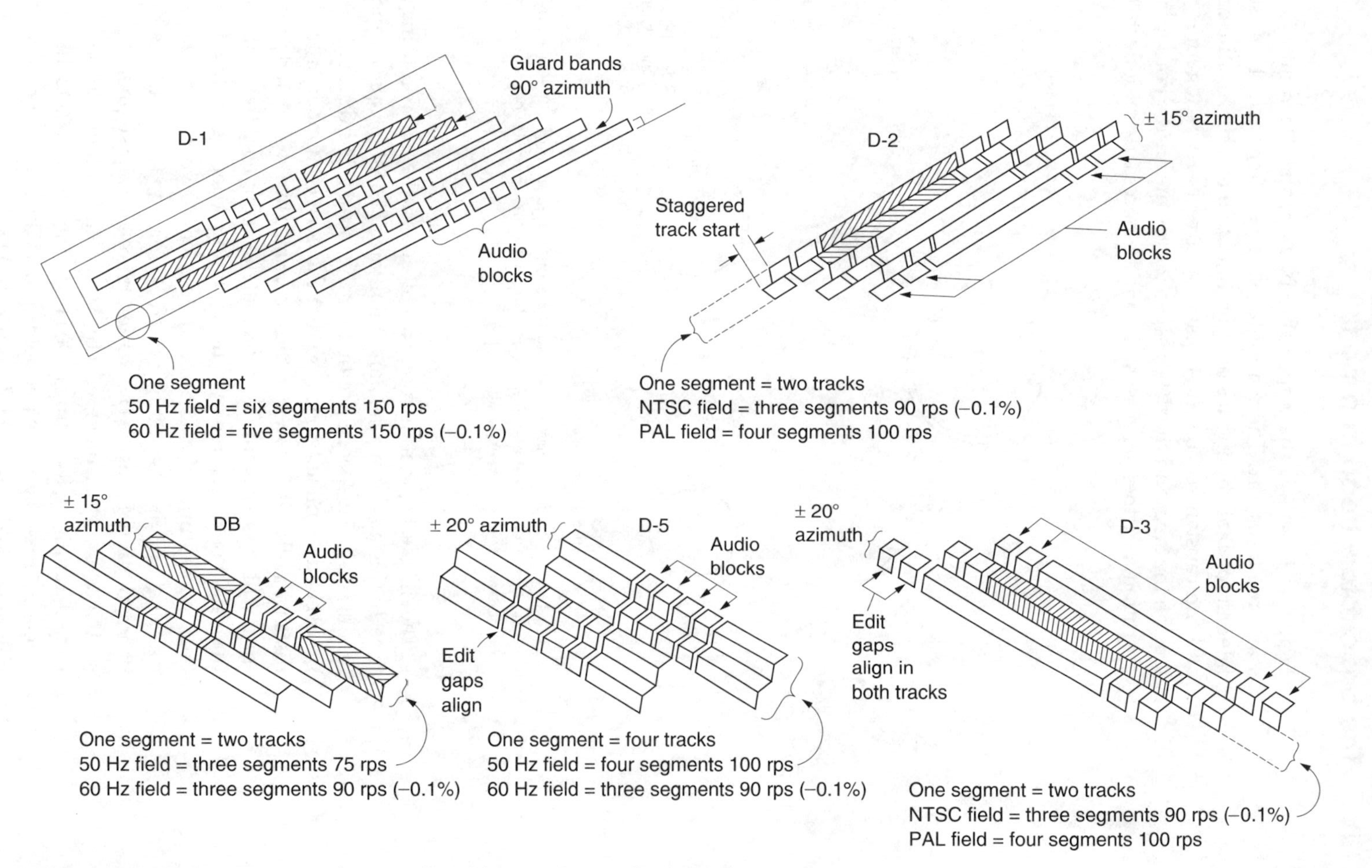

Figure 11.14 The segmentation of various digital formats. See text for details.

11.8 The basic rotary head transport

Figure 11.15 shows the important components of a rotary head helical scan tape transport. There are four servo systems which must correctly interact to obtain all modes of operation: two reel servos, the drum servo and the capstan servo. The capstan and reel servos together move and tension the tape, and the drum servo moves the heads. For variable-speed operation a further servo system will be necessary to deflect the heads.

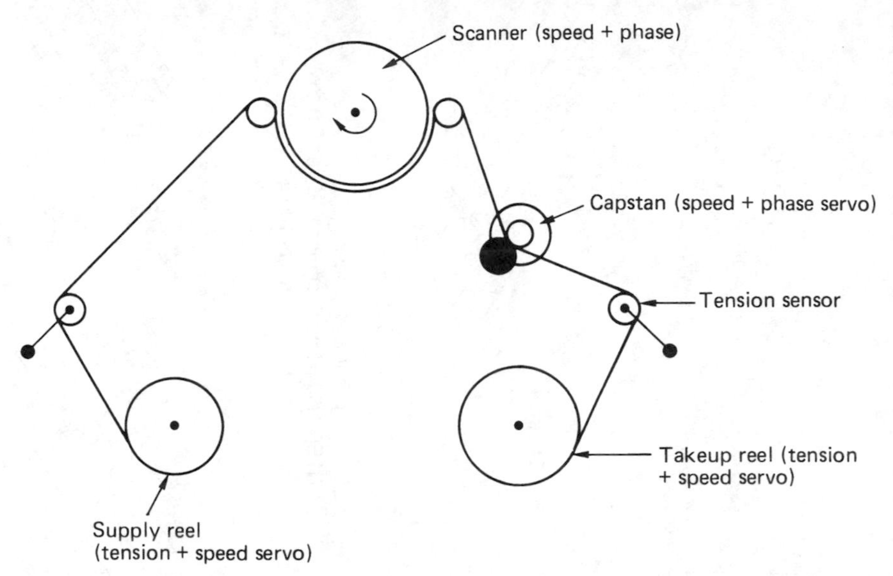

Figure 11.15 The four servos essential for proper operation of a helical-scan DVTR. Cassette-based units will also require loading and threading servos, and for variable speed a track-following servo will be necessary.

The simplest operating mode to consider is the first recording on a blank tape. In this mode, the capstan will rotate at constant speed, and drive the tape at the linear speed specified for the format. The drum must rotate at a precisely determined speed, so that the correct number of tracks per unit distance will be laid down on the tape. Since in a segmented recording each track will be a constant fraction of a television field, the drum speed must ultimately be determined by the incoming video signal to be recorded. To take the example of a D-5 recorder having two record head assemblies carrying four heads each, eight tracks or two segments result from one drum rotation. In the 50 Hz version there are four segments per field, and so the drum must make exactly two complete revolutions in one field period, requiring it to run at 100 Hz. In the case of 60 Hz, there are twelve tracks or three segments per field, and so the drum must turn at one and a half times field rate, or a little under 90 Hz.

11.9 Servos

In various modes of operation, the capstan and/or the drum will need to have accurate control of their rotational speed. During crash record (a mode in which no attempt is made to lock to a previous recording on the tape) the capstan must run at an exact and constant speed. When the drum is first started, it must be brought to the correct speed before phase lock can be attempted. The principle of speed control commonly used will be examined here. It should be pointed out that all the hardware used in these diagrams may not actually exist, but may instead be replaced by suitable instructions in a microprocessor, interfaced by ports to motors and sensors.

Figure 11.16(a) shows that the motor whose speed is to be controlled is fitted with a toothed wheel or slotted disk. For convenience, the number of slots will usually be some power of two. A sensor, magnetic or optical, will produce one pulse per slot, and these will be counted by a binary divider. A similar counter is driven by a reference frequency. This may often be derived by multiplying the input video field rate in a phase-locked loop. The operation of a phase-locked loop was described in Chapter 3.

The outputs of the two counters are taken to a full adder, whose output drives a DAC which in turn drives the motor. The bias of the motor amplifier is arranged so that a DAC code of one half of the quantizing range results in zero drive to the motor, and smaller or larger codes will result in forward or reverse drive.

If the count in the tacho divider lags the count in the reference divider, the motor will receive increased power, whereas if the count in the tacho divider leads the count in the reference divider, the motor will experience reverse drive, which slows it down. The result is that the speed of the motor is exactly proportional to the reference frequency. In principle the system is a phase-locked loop, where the voltage-controlled oscillator has been replaced by a motor and a frequency-generating (FG) wheel.

In a DVTR the rotational phase of the drum and capstan must be accurately controlled. In the case of the drum, the phase must be controlled so that the heads reach the beginning of a track at a time which is appropriate to the station reference video fed to the machine.

A slightly more complex version of the speed control system of Figure 11.16(a) is required, as will be seen in (b). In addition to a toothed wheel or slotted disk, the motor carries a reference slot which produces a rotational phase reference commonly called *once-round tach*. This reference pre-sets the tach divider, so that the tach count becomes an accurate binary representation of the actual angle of rotation.

A similar divider is fed by a reference clock as before, and pre-set at appropriate intervals. The reference clock has the same frequency as the

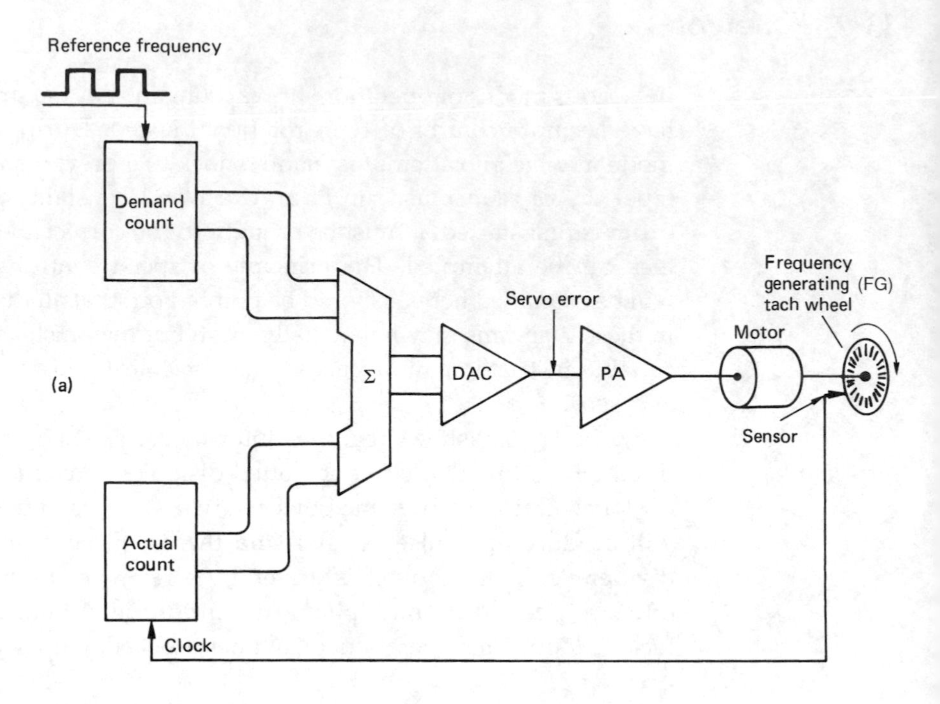

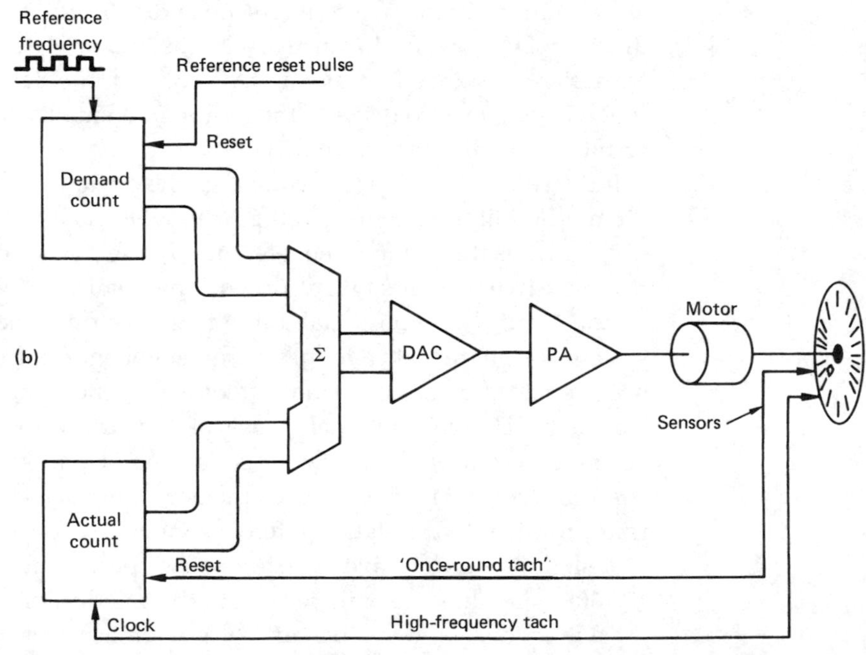

Figure 11.16 (a) Motor speed control using frequency generator on motor shaft. Pulses from FG are compared with pulses from reference to derive servo error. (b) An additional sensor resets the actual count once per revolution, so it counts motor phase angle. Demand count is reset by reference timing. Thus motor phase is locked to reference.

tooth-passing frequency of the tacho at normal speed. In a PAL D-2 machine which needs two drum rotations per field, the counter will need to be pre-set twice per field, or, more elegantly, the counter will have an additional high-order bit which does not go to the adder, and then it can be pre-set once per field, since disconnection of the upper bit will cause two repeated counts of half a field duration each, with an overflow between.

The adder output will increase or decrease the motor drive until the once-round tach occurs exactly opposite the reference pre-set pulse, because in this condition the sum of the two inputs to the adder is always zero. The binary count of the tach counter can be used to address a rotation phase PROM. This will be programmed to generate signals which enable the different sectors of the recorded format to be put in the correct place on the track. For example, if it is desired to edit one audio channel without changing any other part of a recording, the record head must be enabled for a short period at precisely the correct drum angle. The drum phase PROM will provide the timing information needed.

When the tape is playing, the phase of the control track must be locked to reference segment phase in order to achieve accurate tracking. Figure 11.17 shows that a similar configuration to the drum servo is used, but

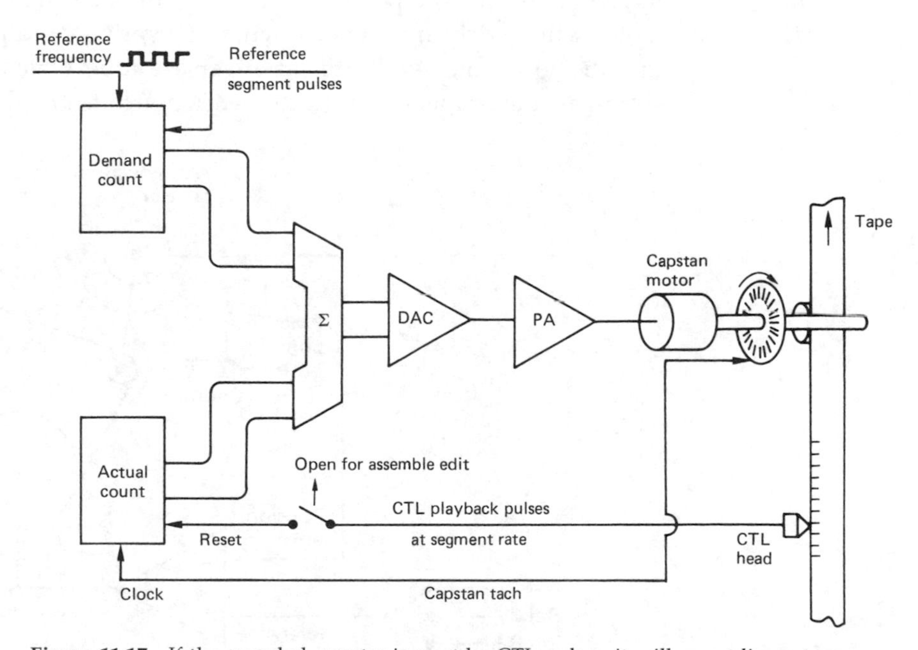

Figure 11.17 If the encoded counter is reset by CTL pulses, it will count linear tape phase, i.e. distance between CTL pulses. Controlling demand counter from reference segment pulses phase locks tape CTL track to segment rate resulting in correct tracking. At an assemble edit the reset is disabled and the capstan servo makes a smooth transition to the velocity control mode of Figure 11.6(a).

there is no once-round tach on the capstan wheel. Instead, the tach counter is reset by the segment pulses obtained by replaying the control track. Since the speed of the control track is proportional to the capstan speed, resetting the tach count in this way results in a count of control track phase. The reference counter is reset by segment rate pulses, which can be obtained from the drum, and so the capstan motor will be driven in such a way that the phase error between control track pulses and reference pulses is minimized. In this way the rotary heads will accurately track the diagonal tracks.

During an assemble edit, the capstan will phase lock to control track during the pre-roll, but must revert to constant speed mode at the in-point, since the control track will be recorded from that point. This transition can be obtained simply by disabling the capstan tach counter reset, which causes the system to revert to the speed control servo of Figure 11.16.

Figure 11.18 shows a typical back-tension control system. A swinging guide called a tension arm is mounted on a position sensor which can be optical or magnetic. A spring applies a force to the tension arm in a direction which would make the tape path longer. The position sensor output is connected to the supply reel motor amplifier which is biased in such a way that when the sensor is in the centre of its travel the motor will receive no drive. The polarity of the system is arranged so that the reel motor will be driven in reverse when the spring contracts. The result is that the reel motor will apply a reverse torque which will extend the spring as it attempts to return the tension arm to the neutral position.

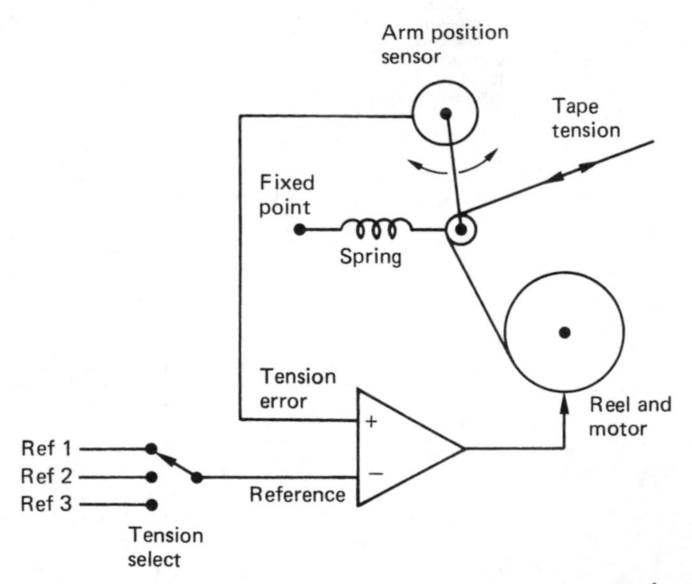

Figure 11.18 Tension servo controls reel motor current so that motor extends sprung arm. Sensor determines extension of spring and thus tape tension.

If the system is given adequate loop gain, a minute deflection of the arm will result in full motor torque, so the arm will be kept in an essentially constant position. This means that the spring will have a constant length, and so the tape tension will be constant. Different tensions can be obtained by switching offsets into the feedback loop, which result in the system null being at a different extension of the spring.

The error in the servo will only be small if the reel can accelerate fast enough to follow tape motion. In a machine with permanent capstan engagement, the capstan can accelerate much faster than the reels can. If not controlled, this would lead to the tension arm moving to one or other end of its travel, resulting in tape stretch, or a slack loop which will subsequently snatch tight with the same results. A fixed acceleration limit on the capstan servo would prevent the problem, but it would have to be set to allow a full reel on a large cassette to keep up. The motion of a small cassette with vastly reduced inertia would be made artificially ponderous. The solution is to feed back the displacement of the tension sensors to the capstan servo. If the magnitude of the tension error from either tension arm is excessive, the capstan acceleration is cut back. In this way the reels are accelerated as fast as possible without the tension becoming incorrect.

In a cassette-based recorder, the tension arms usually need to be motorized so that they can fit into the mouth of the cassette for loading, and pull tape out into the threaded position. It is possible to replace the spring which provides tape tension with a steady current through the arm motor. By controlling this current, the tape tension can be programmed.

It is most important in a cassette to prevent the tape running out at speed. The tape is spliced to a heavy leader at each end, and this leader is firmly attached to the reel hub. In an audio cassette, there is sufficient length of this leader to reach to the other reel, and so the impact of a run-off can be withstood. This approach cannot be used with a video cassette, because the heavy leader cannot be allowed to enter the drum, as it would damage the rotating heads.

Instead the transport must compute the tape remaining. This may be done by measuring the linear speed of the tape and comparing it with the rotational speed of the reel. The linear speed will be the capstan speed in a permanent capstan drive transport, a timer roller will be necessary in a pinch roller type transport. Alternatively, tape remaining may be computed by comparing the speed of the two reels. The rotational speed of the reels will be obtained from a frequency generator on the reel motors.

To prevent runoffs, the reel pack radius of the reel which is unwinding profiles the allowable shuttle speed, so that as the pack radius falls, the tape speed falls with it. Thus when the tape finally runs out, it will be travelling at very low speed, and the photo-electric sensor which detects the leader will be able to halt the transport without damage.

11.10 The cassette

The main advantages of a cassette are that the medium is better protected from contamination whilst out of the transport, and that an unskilled operator or a robotic elevator can load the tape. Cassettes do not offer any advantage for storage except for the ease of handling. In fact a cassette takes up more space than a tape reel, because it must contain two such reels, only one of which will be full at any one time. In some cases it is possible to reduce the space needed by using flangeless hubs and guiding liner sheets as is done in the Compact Cassette and DAT, or pairs of hubs with flanges on opposite sides, as in U-matic. Whilst such approaches are acceptable for consumer and industrial products, they are inappropriate for professional units which are expected to wind at high speeds in automation and editing systems. Accordingly the digital video cassette contains two fully flanged reels side by side. The centre of each hub is fitted with a thrust pad and when the cassette is not in the drive a spring acts on this pad and presses the lower flange of each reel firmly against the body of the cassette to exclude dust. When the cassette is in the machine the relative heights of the reel turntables and the cassette supports are such that the reels seat on the turntables before the cassette comes to rest. This opens a clearance space between the reel flanges and the cassette body by compressing the springs. This should be borne in mind if a machine is being tested without the cassette elevator as a suitable weight must be placed on the cassette in order to compress the springs.

The use of a cassette means that it is not as easy to provide a range of sizes as it is with open reels. Simply putting smaller reels in a cassette with the same hub spacing does not produce a significantly smaller cassette. The only solution is to specify different hub spacings for different sizes of cassette. This gives the best volumetric efficiency for storage, but it does mean that the transport must be able to reposition the reel drive motors if it is to play more than one size of cassette.

Digital Betacam offers two cassette sizes, whereas many other formats offer three. If the small, medium and large digital video cassettes are placed in a stack with their throats and tape guides in a vertical line, the centres of the hubs will be seen to fall on a pair of diagonal lines going outwards and backwards. This arrangement was chosen to allow the reel motors to travel along a linear track in machines which accept more than one size.

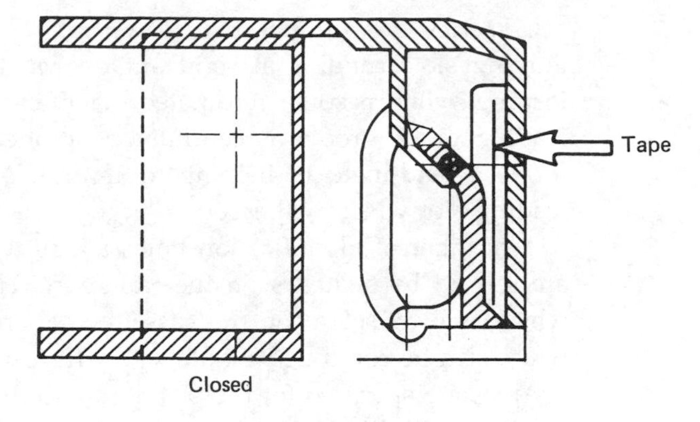

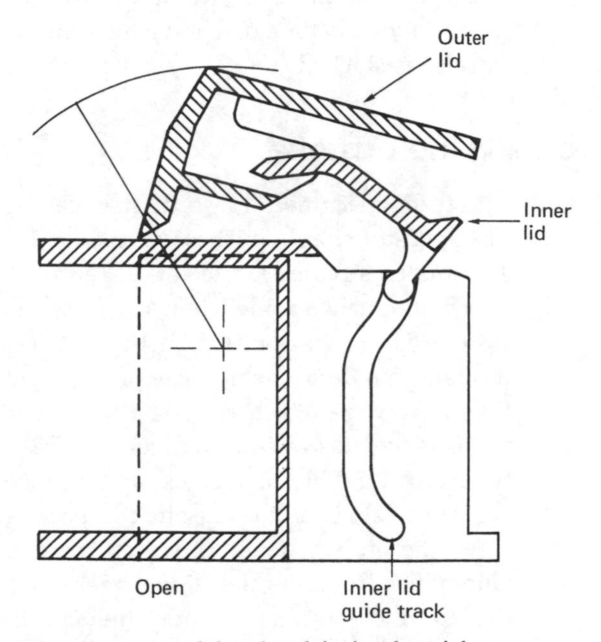

Figure 11.19 When the cassette lid is closed, both sides of the tape are protected by covers. As the outer lid swings up on a pivot, the inner lid is guided around the back of the tape by a curved track.

Often the cassette has a double-door arrangement shown in Figure 11.19. When the cassette is removed, both sides of the tape in the throat are covered. The inner door is guided by a curved track. Generally the door has a lock which prevents accidental opening, and this is released by a pin when the cassette enters the transport.

Cassettes also need hub locks which prevent unwanted rotation in storage or transit. The lock may be released by a lever operated by the act of opening the door. Alternatively the brake may be released by a central post in the transport.

Cassettes may be designed for front- or side-loading, and so have guiding slots running at right angles. Most studio recorders are front loading, whereas some automated machines use side-loading. The front-loading guide groove is centrally positioned and the threading throat forms a lead-in to it, helping to centralize the smaller cassettes in a loading slot which will accept a large cassette.

A number of identification holes are provided in the cassettes which are sensed by switches on the transport. These may be break-off tabs which will be set when the cassette is made, or resettable user holes, which can be controlled with a screwdriver.

Areas are specified for the positioning of labels, and these are recessed to prevent fouling of elevators or guides. Automated machines will use the end location for a barcode which can be read by the elevator while the cassette is stacked.

11.11 Loading the cassette

In a studio machine, all cassette sizes can be used without any adjustment. The operator simply pushes the cassette into the aperture in the machine. The smaller sizes are located at the centre of the aperture by sprung guides which are pushed aside when a large cassette is used. The presence of the cassette is sensed optically, or by switches, and by the same means, the machine can decide which size of cassette has been inserted. The cassette fits into a cage which is driven by a toothed rack so that it can move inwards and down. Alternatively the cassette may be driven by rubber belts. In portable machines and camcorders not all sizes will be accommodated and the cassette compartment may be closed by hand.

As the hub spacing differs between cassette sizes, the transport automatically moves the sliding reel motors to the correct position. The final part of horizontal travel unlocks the cassette door. The cassette elevator is then driven downwards and the door is opened. The opened cassette is then lowered onto the reel motors, and locating dowels on the transport register the cassette body. The hub brakes are released either by a post entering the cassette as it is lowered or by the act of opening the door. The several guides and tension arms and the pinch roller, where applicable, are positioned so that the throat of the cassette drops over them with the front run of tape between them and the head drum.

It is inherent in helical scan recorders that the tape enters and leaves the head drum at different heights. In open-reel recorders, the reels are simply mounted at different heights, but in a cassette this is not practicable. The tape must be geometrically manipulated in some way.

Various methods exist to achieve the helical displacement of the tape. Some VTR transports have used conical posts, but these have the disadvantages that there is a considerable lateral force against the edge of the tape, and that they cannot be allowed to revolve or the tape would climb off them. Angled pins do not suffer side-force, but again cannot be allowed to revolve. The friction caused by non-rotating pins can be reduced by air lubrication from a compressor, as used in Ampex transports, or by vibrating them ultrasonically with a piezo-electric actuator, as is done in certain Sony transports.

In the Panasonic D-3/D-5 transports, as with any VCR, the guides start inside the cassette, and move to various positions as threading proceeds. The sequence can be followed in Figure 11.20. The entry-side threading is

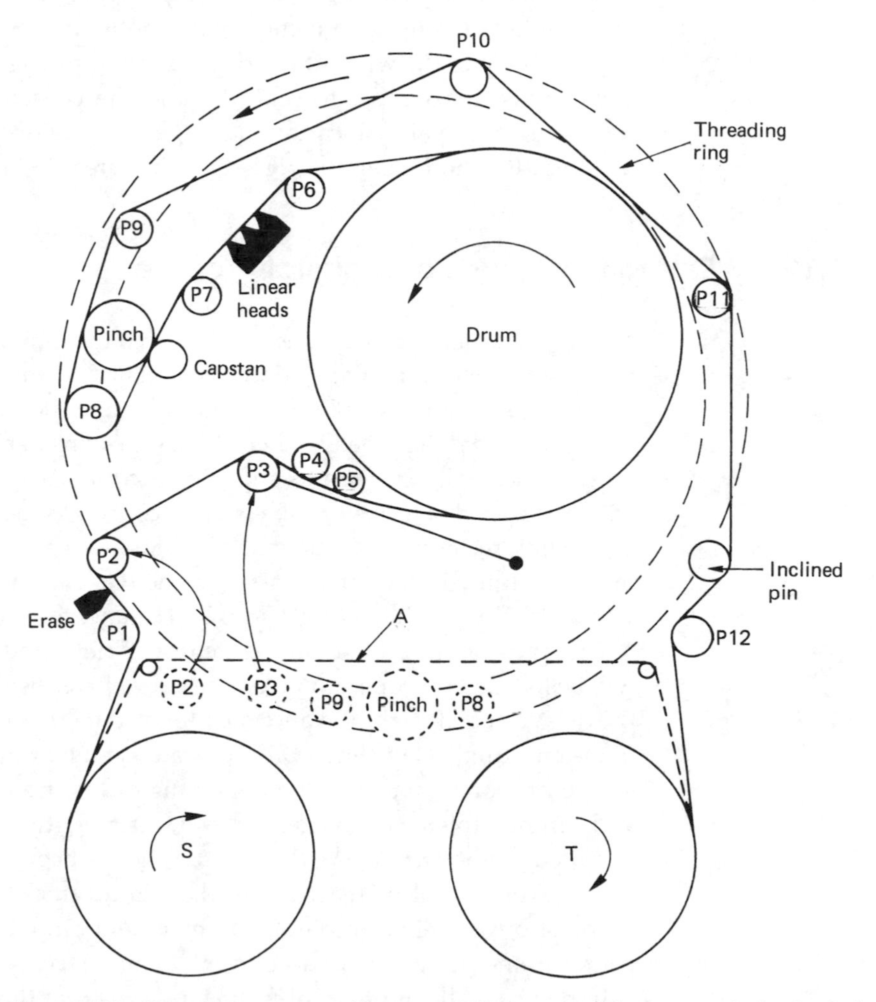

Figure 11.20 Threading sequence of Panasonic D-3 transport. When cassette is first loaded, tape runs straight between the cassette pins (dotted line shown at A) in front of the moving guide pins P2, P3, P9, P8 and the pinch roller. Guides P2 and P3 wrap the entry side by swinging on arms. P3 is in fact the tension arm. The pinch roller and its guides travel anticlockwise on the large threading ring. As the ring rotates P10 and P11 guide the return loop. The inclined pin hinges over the edge of the tape and does not need to be in the cassette mouth at threading start.

performed by guide P2 which swings anticlockwise on an arm to wrap the tape onto fixed guide P1 and the full-width erase head and by the tension arm guide P3 which swings clockwise to bring the tape across the drum entry guides P4 and P5. The drum wrapping and exit-side threading is performed by guides which move in a circular path on a threading ring, which rotates around the drum and capstan. When the cassette is initially lowered, guide P8, the pinch roller and guide P9 are inside the front run of tape. As the threading ring turns anticlockwise, these guides take a loop of tape from the cassette and begin to wrap it around the drum. As the threading ring proceeds further, guide P10 and then guide P11 come into contact with the return loop, and the leading guide completes the wrap of the drum and wraps fixed exit guide P6, the fixed heads, fixed capstan guide P7 and the capstan. The pinch roller completes its travel by locating in a cage which is operated by the pinch solenoid. It is no longer supported by the threading ring.

11.12 Operating modes of a digital recorder

The digital video recorder is expected to do much more than record and replay in a modern installation. The main operating modes will be examined here, followed by the more obscure modes.

In *crash record* the tape is either blank or is considered to be blank, and no reference will be made to any previous information on the tape. As with analog video recorders, if a crash recording is made in the middle of an existing recording, the playback will lose lock at the transition. In this mode the full-width erase head will be active, and the entire tape format will be laid down, including audio blocks and control track.

There are a fixed number of segments in a field, and so the drum speed will be locked to incoming video. The drum phase will be determined by the delay needed to shuffle and encode the data in a segment. As a result of the encoding delay, the recording heads on the drum will begin tracing a segment some time after the beginning of the input field. The capstan will simply rotate at constant speed, driving the tape at the speed specified in the format. As the rotating heads begin a diagonal track, a segment pulse will be recorded in the control track. Once per field, the segment pulse will be accompanied by a field pulse. There will also be a colour frame pulse recorded at appropriate intervals.

In playback, the number of frames played from the tape will be equal to the number of station reference frames. In a progressively scanned format all that is required is that the segmentation on tape is properly undone, so that all the segments in one picture are assembled together. In an interlaced format there is an additional requirement that the interlace sequence is synchronous with reference, so that an odd field comes from tape when there is an odd field in the station reference. In composite formats, the off-tape field must also be the same type of field in the four- or eight-field sequence as exists in the reference.

The error-correction and de-interleaving processes on replay require a finite time in which to operate, as will the decoder in the case of a compressed format. The transport must play back segments ahead of real time, so that after the necessary processing delays the timing aligns with that of the reference. The drum speed will be synchronous with reference, but the drum phase will be offset by the decoding delay.

The capstan and tape form part of a phase-locked loop where the phase of the control track pulses is compared with reference pulses. In order to achieve framing, the phase comparison must be between reference frame pulses and offtape frame pulses. In this way the linear motion of the tape is phased such that reference and offtape framing are the same and de-segmentation will be correct. The capstan speed must vary during this part of the lock-up sequence so that segments can be skipped. The picture may be blanked unless the machine has provision for operating at non-normal speed. In composite machines the process will be extended to align with the colour-framing sequence.

Once framing is achieved, the capstan will switch to a different mode in order to obtain accurate tracking. The phase comparison will now be between offtape segment pulses and pulses generated by the rotation of the drum. If the phase error between these is used to modify the capstan drive, the error can be eliminated, since the capstan drives the tape which produces the control track segment pulses. Eliminating this timing error results in the rotating heads following the tape tracks properly. Artificially delaying or advancing the reference pulses from the drum will result in a tracking adjustment.

11.13 Editing

In an assemble edit, new material is appended to an existing recording in such a way that the video timing structure continues unbroken over the edit point. In a composite recording, the subcarrier phase will also need to be continuous.

The recording will begin at a vertical interval, and it will be necessary to have a pre-roll in order to synchronize tape and drum motion to the signal to be recorded. The procedure during the pre-roll is similar to that of framed playback, except that the input video timing is used. The goal of the pre-roll process is to ensure that the record head is at the beginning of the first segment in a field exactly one encode delay after the beginning of the desired field in the input video signal. At this point the record head will be turned on. After the assemble point, there will not necessarily be a control track, and the capstan must smoothly enter constant speed mode without a disturbance to tape motion. The control track will begin to record in a continuation of the phase of the existing control track. This process will then continue indefinitely until halted.

An insert differs from an assemble in that a short part of a recording is replaced by new material somewhere in the centre. An insert can only be made on a tape which has a continuous control track, because the in- and out-point edits must both be synchronous. The pre-roll will be exactly as for an assemble edit, and the video record process will be exactly the same, but the capstan control will be achieved by playing back the control track at all times, so that the new video tracks are laid down in exactly the same place as the previous recording. At the out-point, the record heads will be turned off at the end of the last segment in a field.

As the tracks used in DVTRs are so narrow, some thought has to be given to the results of mechanical tolerances when edits are performed. In transports with no flying erase, the record head width is typically 10 per cent larger than the track width, to ensure full overwrite. As a result a correctly aligned insert edit leaves a track which is slightly narrower than normal at the out point. This is not a serious problem as the result of a narrow track is that the signal-to-noise ratio is a little less so the error-correction system works harder. There is, however, a potential problem if repeated edits are made in the same area of a recording. The problem is overcome in studio transports by making the record heads the same width as the track pitch and using a flying erase head which is slightly wider than the recorded segment.

It will be evident that if a DVTR were relieved of the requirement to edit, the track pitch could be reduced significantly, improving the recording density and reducing tape cost. When used in conjunction with disk-based editors, tape storage no longer needs to support editing and high-density data recorders can be used instead of DVTRs.

11.14 Variable-speed replay

It was seen in section 11.4 that the movement of the tape results in the tracks having an angle different from the helix angle. The rotary head will only be able to follow the track properly if the tape travels at the correct speed. At all other speeds, the head will move at an angle to the tracks. If the head is able to move along the drum axis as it turns, it will be possible to follow whole tracks at certain speeds by moving the head as a function of the rotational angle of the drum. A ramp or triangle waveform is necessary to deflect the head. The slope of the ramp is proportional to the speed *difference*, since at normal speed the difference is zero and no deflection is needed. Clearly the deflection cannot continue to grow forever, because the head will run out of travel, and it must then jump to miss out some tracks and reduce the deflection.

There are a number of issues to be addressed in providing a track-following system. The use of segmented formats means that head jumps necessary to omit or repeat one or more fields must jump over several segments. This requires a mechanical head-positioning system which has the necessary travel and will work reliably despite the enormous acceleration experienced at the perimeter of the drum. As there are generally two head bases, two such systems are needed, and they must be independently controlled since they are mounted in opposition on the drum and contact the tape alternately. Also required is a control system which will ensure that jumps only take place at the end of a field to prevent a picture from two different fields being displayed.

When the tape speed is close to three times normal, most of the time a two field jump will be necessary at the end of every field. Figure 11.21 shows the resulting ramp deflection waveform (for one head pair only). Since the control track phase is random, an offset of up to $\pm\frac{1}{2}$ field will be superimposed on the deflection, so that if a two-field jump must always be possible, a total deflection of three fields must be available. This, along with the segmentation employed, determines the mechanical travel of the heads.

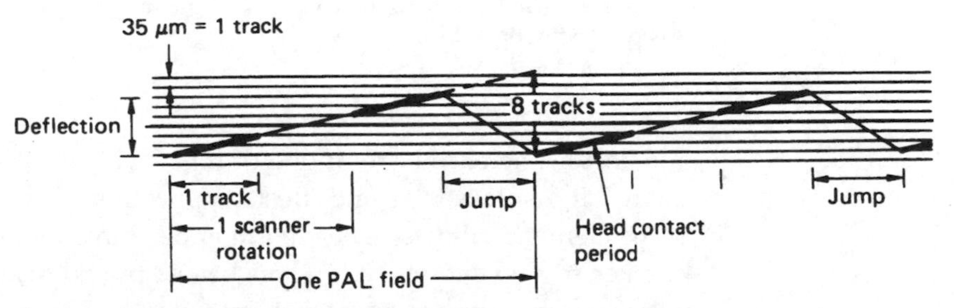

Figure 11.21 Head deflection waveform for PAL D-3 at +3 × normal speed. This requires a jump of two fields between every field played, and this corresponds to eight tracks. Since head pairs trace the tape alternately, the head has half a revolution in which to jump, so the actual jump is only six tracks. However, the vertical position of the above waveform is subject to an uncertainty of plus or minus two tracks because capstan phase is random during variable speed.

In Sony and Panasonic machines, the actuator used for head deflection is based on piezo-electric elements as used in numerous previous products. Figure 11.22 shows the construction of the Sony dynamic tracking head. A pair of parallel piezo-electric bimorphs is used, to ensure that head zenith is affected as little as possible by deflection. The basic principle of the actuator is that an applied voltage causes the barium titanate crystal to shrink. If two thin plates are bonded together to create

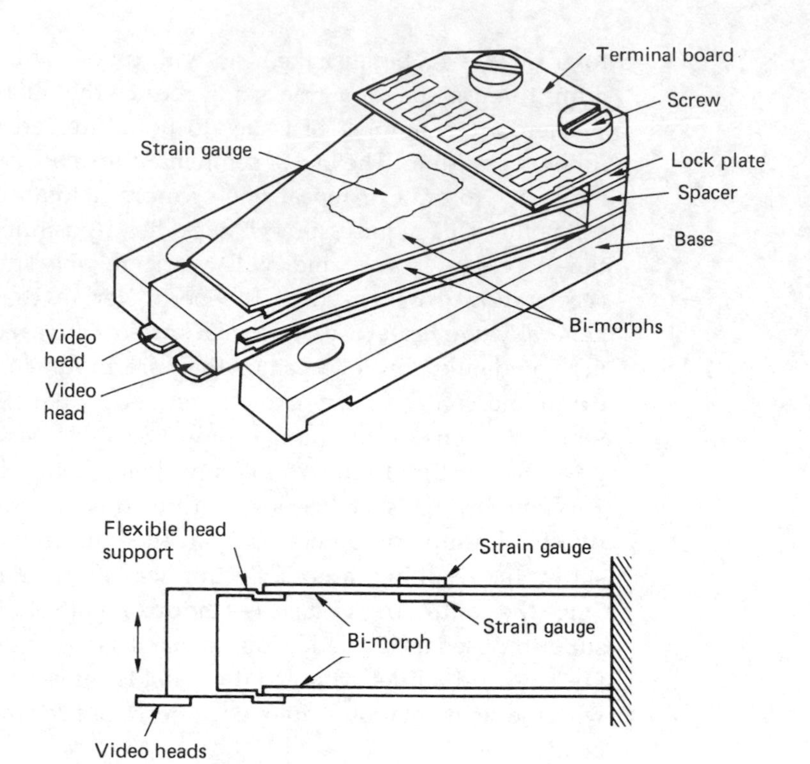

Figure 11.22 The dynamic tracking head of a Sony D-2 transport uses a pair of parallel bimorphs to maintain head zenith angle. Note use of strain gauges for feedback. (Courtesy Sony Broadcast).

a bimorph, the shrinkage of one of them will result in bending. A stale sandwich displays the same effect. Application of voltage to one or other of the elements allows deflection in either direction, although care must be taken to prevent reverse voltage being applied to an element since this will destroy the inherent electric field.

In view of the high-g environment, which attempts to restore the actuator to the neutral position, high deflection voltages are necessary. The deflection amplifiers are usually static, and feed the drum via slip rings which are often fitted on the top of the drum. When worn, these can spark and increase the error rate. Piezo-electric actuators display hysteresis, and some form of position feedback is necessary to allow a linear system. This is obtained by strain gauges which are attached to one of the bimorphs, and can be seen in the figure. When power is first applied to the transport, the actuator is supplied with a gradually decaying sinusoidal drive signal, which removes any unwanted set from the bimorph.

In Ampex D-2 and DCT machines, the piezo-electric actuators was not considered adequate for the larger travel demanded in a segmented

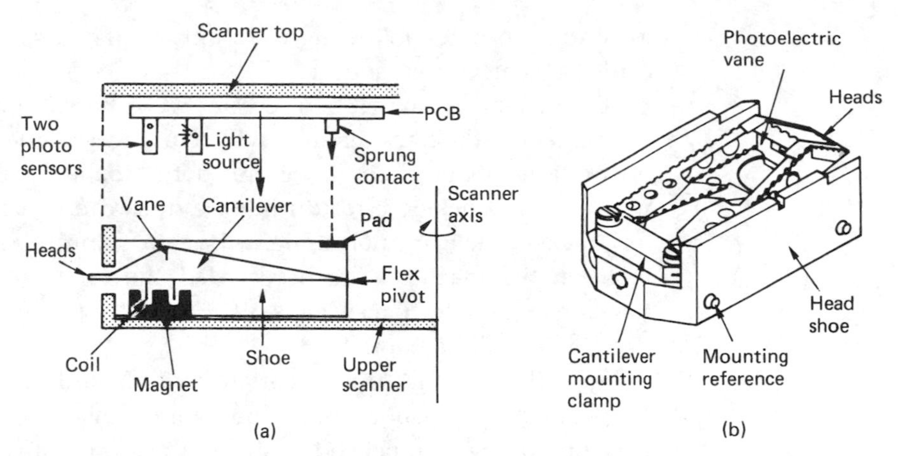

Figure 11.23 (a) Showing the concept of moving-coil head deflection. The cantilever is flexurally pivoted at the opposite end to the heads, and position feedback is obtained from a vane on the cantilever which differentially varies the light falling on two photosensors. (b) The appearance of an actual replaceable AST head assembly. (Courtesy Ampex.)

format,[1] and a moving-coil actuator has been developed. Now that rare earth magnets are available, which offer high field strength with low mass, the moving-coil actuator becomes attractive because it allows a low-mass cantilever which has higher resonant frequencies and requires less force to deflect. The moving coil is inherently a low-impedance device requiring a current drive which is easier to provide than the high voltages needed by piezo-electric devices.

Figure 11.23(a) shows the concept of the moving coil head and (b) shows the appearance of the actual unit used in D-2. The cantilever is folded from thin sheet metal which is perforated to assist the folding process. The resulting structure is basically a torsion box supported on a wide flexural pivot. This means that it can bend up and down, but it cannot twist, since twisting would introduce unwanted azimuth errors. The cantilever carries the actuator coil, and is supported in a metal shoe which carries the magnet.

Positional feedback of the cantilever deflection is obtained by a photo-electric system. This has one light source and two sensors between which moves a blade that is integral with the cantilever. When the cantilever deflects, a differential signal results from the sensors. The photo-electric sensor is mounted inside the top of the drum, and automatically aligns with the moving blade when the top is fitted.

Track following is a means of actively controlling the relationship between the replay head and the track so that the track is traced more accurately than it would be by purely mechanical means. This can be applied to systems operating at normal speed, in order to allow interchange with high recording density. In earlier formats working at lower-density track following is an option, and satisfactory interchange could be achieved without it.

The tracking error will be used on two different levels. First, the DC component of the tracking error will be used to set the average elevation of the head about which the ramp deflection will take place. Second, variations of tracking error during the track can be used to compensate for tracks which are not straight, due to some relative misalignment between the machine which recorded the tape and the player. The tracking error-detection systems differ between manufacturers and these differences will be noted.

Figure 11.24 shows three relationships of the head to the track, and corresponding signal output. The waveforms correspond to the RF envelope of the channel-coded signal. Case (a) and case (c) display the same output, although the tracking error has the opposite sense. Simple processing of the RF level only gives the magnitude of the error, not the sense.

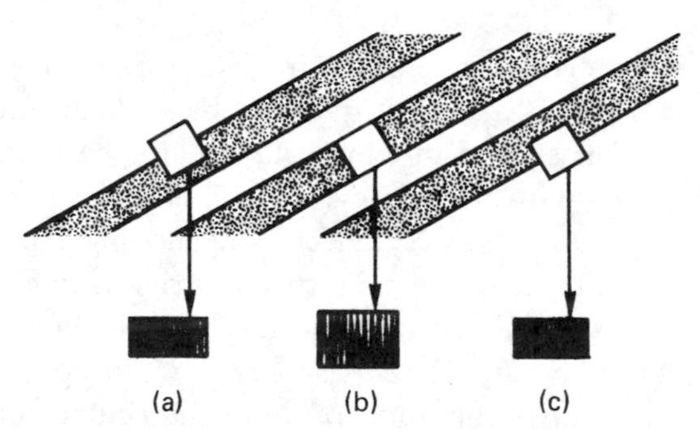

Figure 11.24 Effect of tracking error on playback signal. Signal amplitude in (a) and (c) is identical, despite sense of tracking error. Maximum signal occurs with correct alignment as in (b).

In order to extract the sense of the tracking error, it is necessary to move the tracking head, to see if the error becomes greater or smaller. The process of manually tuning an AM radio is very similar. In the Sony machines, the head elevation is changed by small steps between segments, and if the steps cause the average RF level to fall, the direction of the steps is reversed. This process is known colloquially as 'bump and look', and was originally developed for the analog Betacam products.

In the Ampex machines, the head is subject to a sinusoidal oscillation or dither, as was done in C-format machines. One field scan contains many cycles of dither. The effect on the RF envelope, as shown in Figure 11.25, is an amplitude modulation of the carrier, which has little effect on the video, due to the insensitivity of the digital recording system to amplitude effects. Figure 11.25(a) shows that the effect of dither on a

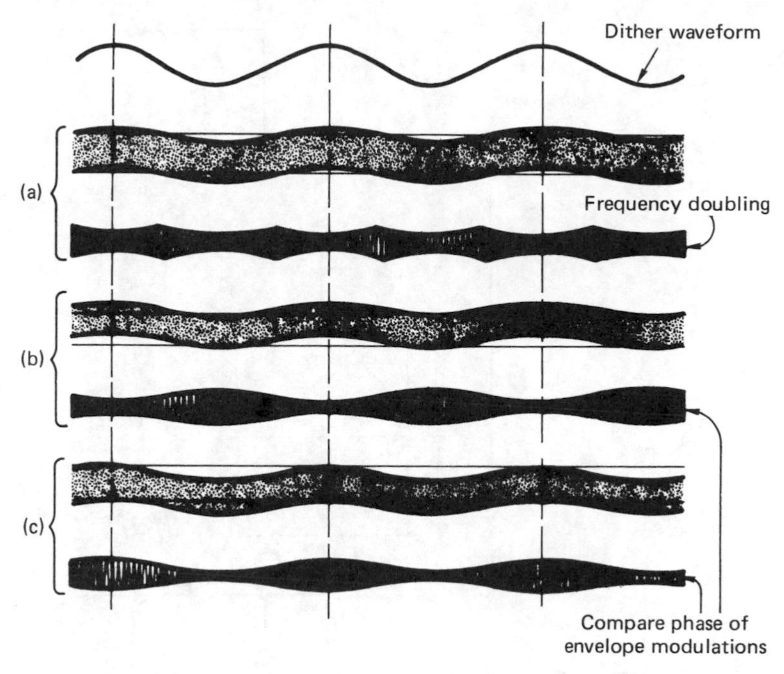

Figure 11.25 Top, dither waveform which causes head to oscillate across track. At (a) is optimum alignment, showing frequency doubling in RF envelope. With head above track centre, as in (b), RF amplitude increases as head reaches lowest point, whereas reverse applies in case (c).

correctly aligned head is a frequency doubling in the RF envelope, (b) and (c) show the effect of the head off-track. Both cases appear similar, but the phase of the envelope modulation is different, and can be used to extract the sense of the tracking error.

In Figure 11.26 the RF is detected to obtain a level, which is fed to a phase-sensitive rectifier, whose reference is the dither drive signal. The output of the phase-sensitive rectifier is a tracking-error signal which contains both magnitude and sense and rather a lot of harmonics of dither. Careful choice of the dither frequency allows easier cancelling of the dither harmonics. The tracking error is averaged over a pair of segments to cancel the harmonics and to produce an elevation error. The tracking error will also be sampled at several points down the track to see whether there is a consistent curvature in the tracking. This can be reduced by adding a correction curve to the deflection waveform which will adjust itself until the best envelope is obtained over the whole track length.

A different approach to measuring track curvature is used in Sony and Panasonic transports. Deflecting a head which uses azimuth recording causes timing changes. If a pair of heads of opposite azimuth are tracing the appropriate tracks, tracking error will result in differential timing

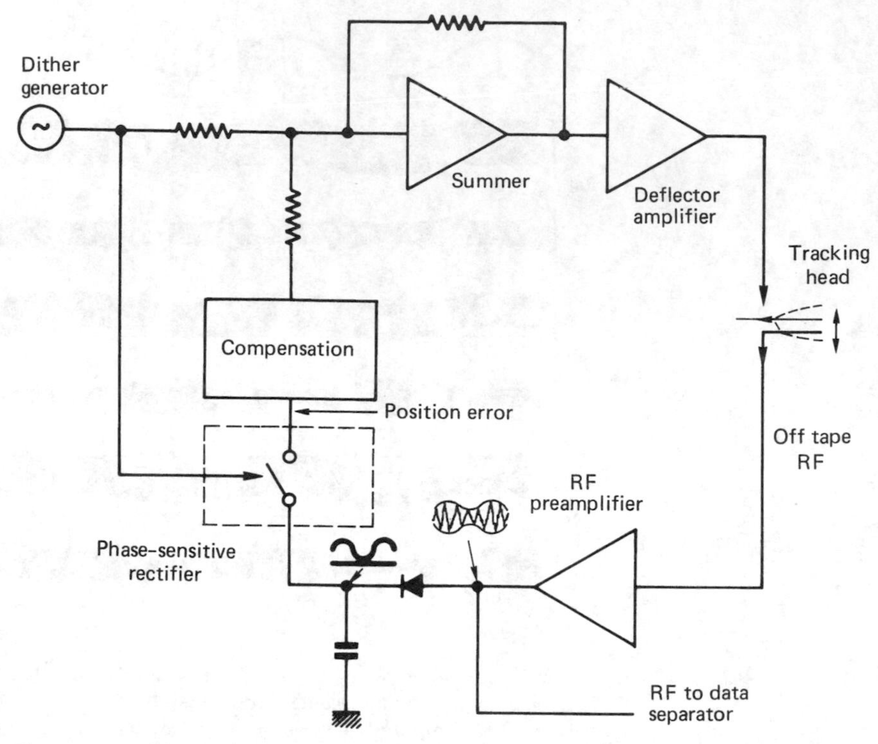

Figure 11.26 The tracking error is extracted from the RF envelope by a phase-sensitive rectifier.

changes. This can be measured by comparing the time at which sync patterns are detected in the two channels. Figure 11.27 illustrates the principle. This system will detect tracking errors due to track distortion. Unfortunately, manufacturing tolerances in the physical stagger between the heads on both the machine which made the recording and the player will combine to put a permanent offset in the tracking error derived using

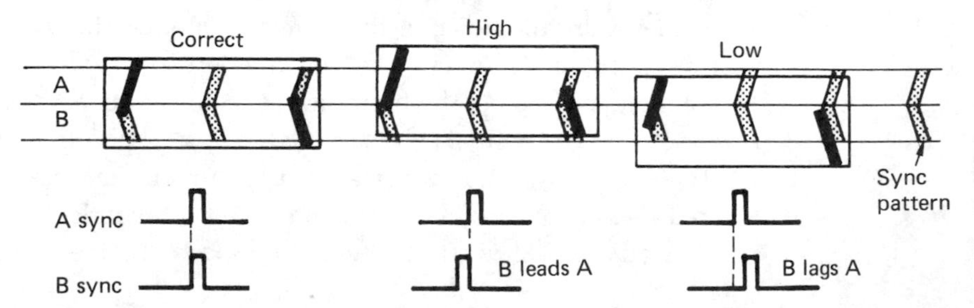

Figure 11.27 In the Sony D-2 DT system the tracking error is detected by comparing the times at which sync patterns are detected by the two heads on each arm. Owing to the use of azimuth recording a component of transverse movement results in a movement of the effective position of the head along the track.

this principle. It is necessary to calibrate the system each time it is used. The bump and look system maximizes the RF level at the beginning of the track, and the control system then inserts a changing offset into the sync-timing comparator until it too gives zero tracking error at the beginning of the track. The offset is then held at that value. This process removes the static deflection and allows the tracking error to be used to deflect the heads within the track.

The positional feedback generated by the observation of the dithered RF envelope or the sync pattern phase is not sufficiently accurate to follow the tracks unassisted except at normal speed where it can be used as an interchange aid. In variable speed the deflection of the head is predicted to produce a feedforward signal which adds to the head deflection. The feedback then only has to correct for the difference between the feedforward and the actuality. When the tape travels at the wrong speed, the track angle changes, and so the head needs to deflect by a distance proportional to the angle it has rotated in order to follow the track. The deflection signal will be in the form of a ramp, which becomes steeper as the tape speed deviates more from normal. The actual speed of the capstan can be used to generate the slope of the ramp. It is also possible to predict the average or static deflection of the heads from the control track phase.

The same deflection ramp slope will be fed to both head pairs so that they will read alternate segments for an entire field. At some point it will be necessary to jump the heads to reduce the deflection, and this requires some care. The heads are 180° opposed, and contact the tape alternately. The jump takes place while the head is out of contact with the tape. The two head pairs will have to jump at different times, half a drum revolution apart. Clearly if one head jumps, the second head must also jump, otherwise the resulting picture will have come from two fields. As the jumps are half a revolution apart, it follows that the decision to jump must be made half a drum revolution *before* the end of the current field. The decision is made by extrapolating the deflection ramp forward to see what the deflection *will be* when the end of the field is reached. If the deflection will exceed half a field, it can be reduced by jumping one field. If the deflection will exceed one field, it can be reduced by jumping two fields.

Tape tension changes can cause the track width to vary minutely. This is normally of no consequence, but when taken over all the tracks in a segment the error may be significant. It is possible to compare the tracking error before and after a jump, and if the error becomes greater, the jump distance was inappropriate for the tape being played. It is possible to modify the distance jumped simply by changing the number of pulses fed to the ramp counter during the jump command. In this way the jump distance can optimize itself for the tape being played.

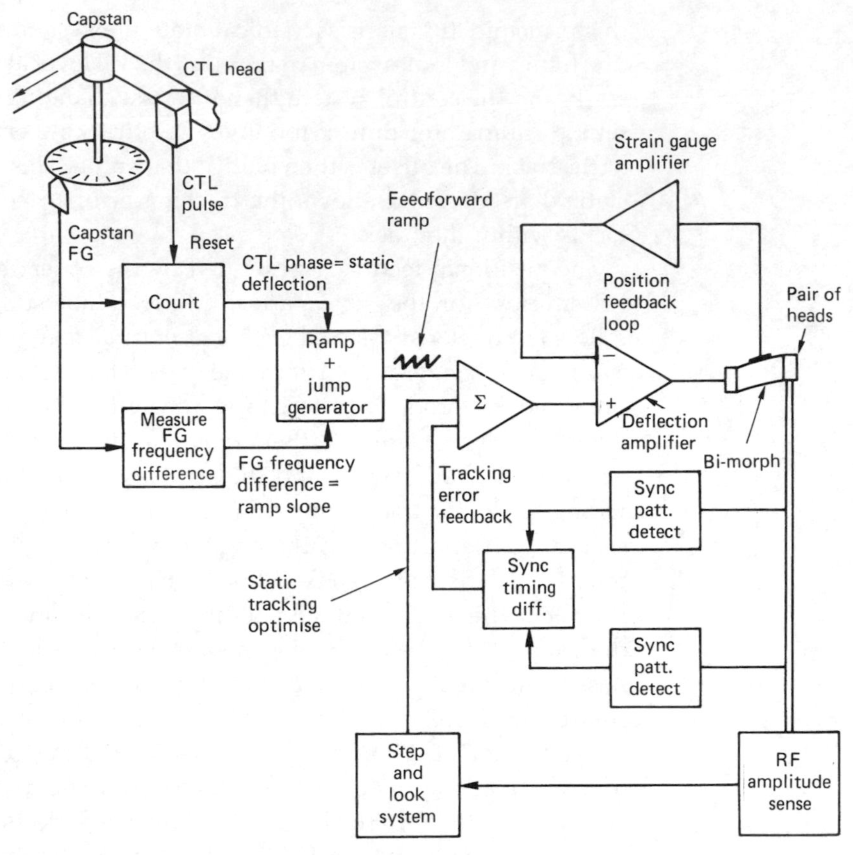

Figure 11.28 The DT system used by Sony D-2 machines combines ramp feedforward from the capstan speed and control track with tracking feedback from the differential sync detection of Figure 11.27. In addition the bump and look system slightly offsets the heads by a different amount each scan and assesses the RF amplitude to find an optimum static deflection. This is really a slow dither process. Note that this is a conceptual diagram. In practice many of these stages are carried out by microprocessor.

Figure 11.28 shows a typical track-following system. The Capstan FG and control track are used together for elevation prediction, and capstan FG is used alone for ramp slope prediction. The bump and look system is controlled by the RF detector, which is also used to calibrate the differential sync pattern timing detector. In most DVTRs there are two moving heads and so two of these systems are necessary. They are largely independent except for a common jump control system which ensures that when one head jumps during the last segment of a field the other head will follow once it has played that segment.

When a head jumps, it will move to a track which began at a different place along the tape, and so the timing in that track will not be the same as in the track which the head left. In a segmented format, the effect is magnified by the number of tracks in a field. To support a speed range from –1× to +3×, it has been seen that an overall head travel of three fields is necessary. From the start of a given segment to the start of the same segment three fields away may be a distance of several millimetres. The wrap of the scanner must be extended to allow tape contact over a greater angle, and the replay circuitry must be able to accommodate the timing uncertainty which amounts to about 5 per cent of the segment period. Clearly the timing error can be eliminated by the timebase correction processes within the playback system.

The mechanical aspects of the rotary head DVTR were considered in Chapter 8. The transport simply guides the heads repeatably along tape tracks so that waveforms can be recorded and replayed. The signal system is responsible for recording and reproducing those waveforms as digital video and audio.

11.15 DVTR signal systems

The signal system of a DVTR lies between the signal input/output connectors and the heads, and in production machines is divided into separate record and replay sections so that both processes can take place at the same time. The data to be recorded are subject to some variation as Figure 11.29 shows. The input may be composite or component, and

Standard	Lines/field	Pixels/line	Pixels/field	Bit rate
4:2:2 50 Hz	300	720 Y 360 C_r 360 C_b	432 000	173 M bits/s (8 bit) 216 M bits/s (10 bit)
4:2:2 60 Hz	250	720 Y 360 C_r 360 C_b	360 000	173 M bits/s (8 bit) 216 M bits/s (10 bit)
16:9 50 Hz	300	960 Y 480 C_r 480 C_b	576 000	230 M bits/s (8 bit) 288 M bits/s (10 bit)
16:9 60 Hz	250	960 Y 480 C_r 480 C_b	480 000	230 M bits/s (8 bit) 288 M bits/s (10 bit)
$4F_{sc}$ PAL	304	948	288 192	115 M bits/s (8 bit) 144 M bits/s (10 bit)
$4F_{sc}$ NTSC	255	768	195 840	94 M bits/s (8 bit) 117 M bits/s (10 bit)

Figure 11.29 The data to be recorded in DVTR formats are subject to a great deal of variation as shown here.

compression may or may not be employed. The various DVTR formats largely use the same processing stages, but there are considerable differences in the order in which these are applied.

The input data rate of a component digital machine can be directly deduced from the CCIR-601 standard. In every active line, 720 luminance samples and 360 of each type of colour difference sample are produced, making a total of 1440 samples per line (see Chapter 10). In 625/50, lines 11–310 and lines 324–623 are recorded, making 300 lines per field, or 15 000 lines per second. The D-1 format records these data directly. In 525/60, lines 14–263 and lines 276–525 are recorded, making 250 lines per field, which results in the same data rate of 15 000 lines per second.

In both cases the video data rate will be 1500 × 1440 samples/s, or 21.6 million samples/s. If 18 MHz sampling is used, the data rate will be correspondingly higher. The actual bit rate recorded will be in excess of

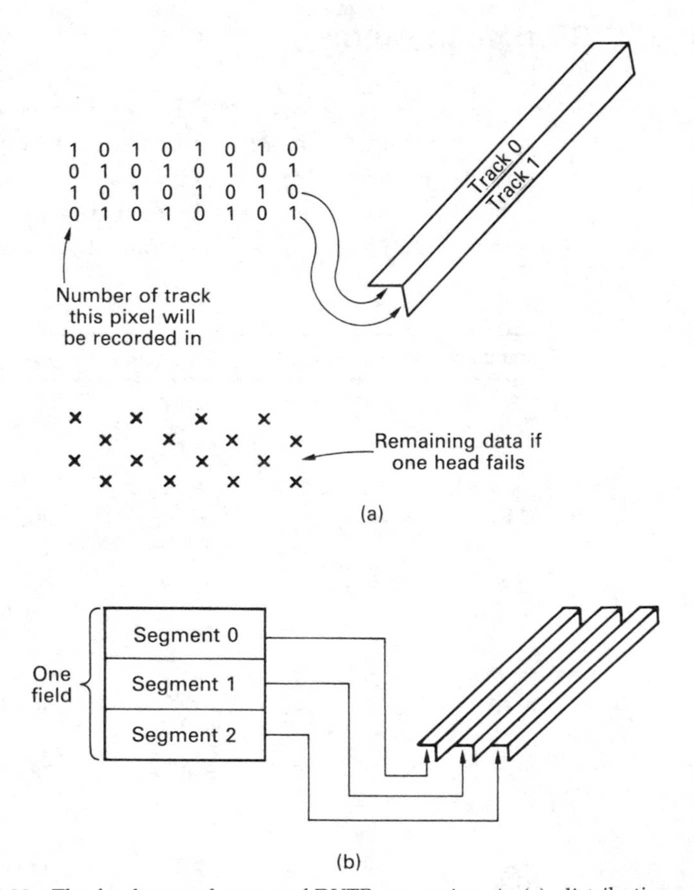

Figure 11.30 The fundamental stages of DVTR processing. At (a), distribution spreads data over more than one track to make concealment easier and to reduce the data rate per head. At (b) segmentation breaks video fields into manageable track lengths.

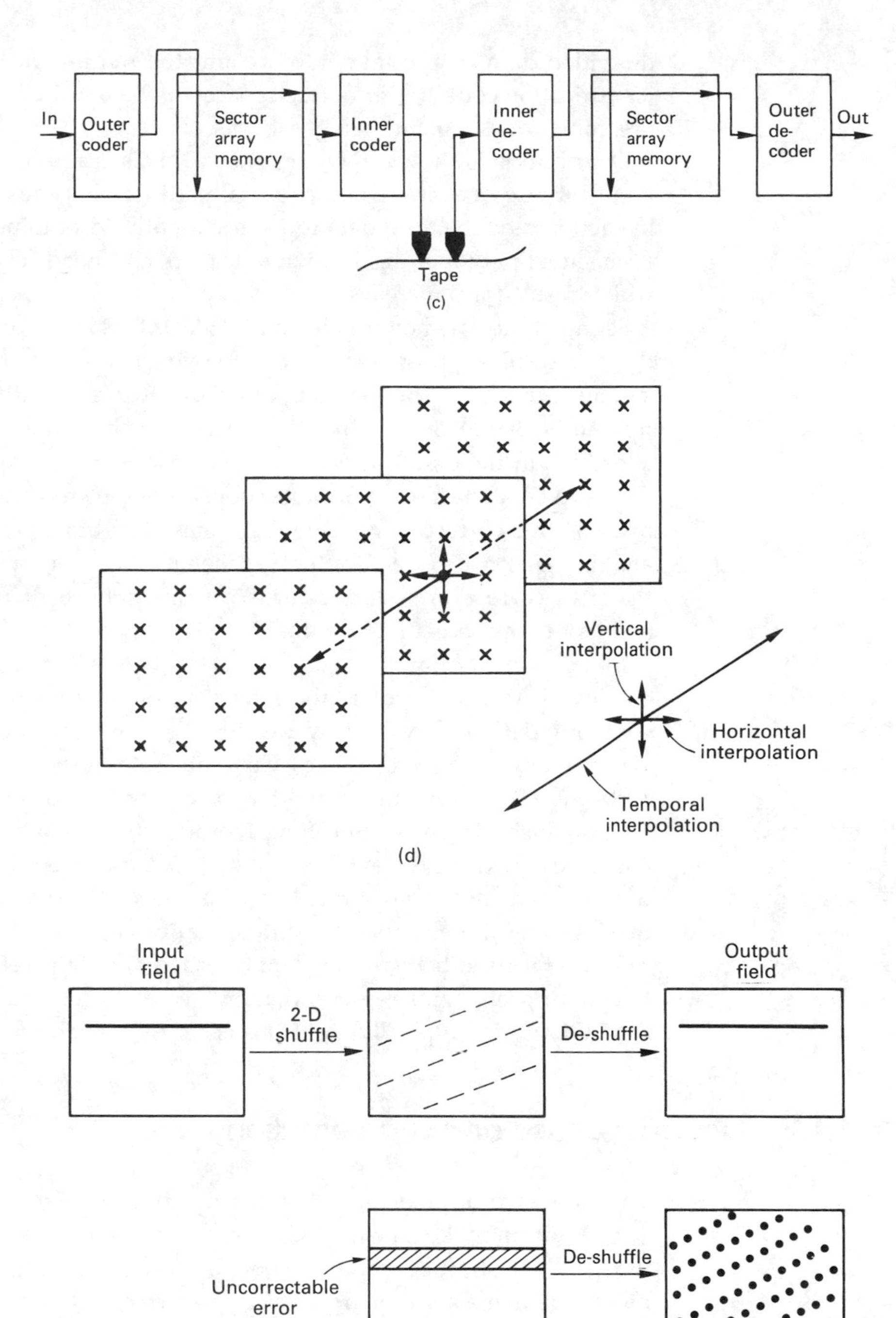

Figure 11.30 continued Product codes (c) correct mixture of random and burst errors. Correction failure requires concealment which may be in three dimensions as shown in (d). Irregular shuffle (e) makes concealments less visible.

the video data rate to take into account the redundancy, the audio and identification codes. Further track space will be required to accommodate preambles and synchronizing patterns.

Distribution is shown in Figure 11.30(a). This is a process of sharing the input bit rate over two or more signal paths so that the bit rate recorded in each is reduced. The data are subsequently recombined on playback. Each signal path requires its own tape track and head. The parallel tracks which result form a *segment*.

Segmentation is shown in Figure 11.30(b). This is the process of sharing the data resulting from one video field over several segments. The replay system must have some means to ensure that associated segments are reassembled into the original field. This may be done by incorporating addresses in the data tracks.

Figure 11.30(c) shows a product code. Data to be recorded are protected by two error-correcting codeword systems at right angles; the inner code and the outer code (see Chapter 9). When it is working within its capacity the error-correction system returns corrupt data to their original value and its operation is undetectable.

If errors are too great for the correction system, concealment will be employed. Concealment is the *estimation* of missing data values from surviving data nearby. Nearby means data on vertical, horizontal or time axes as shown in Figure 11.30(d). Concealment relies upon distribution, as all tracks of a segment are unlikely to be simultaneously lost, and upon the *shuffle* shown in (e). Shuffling reorders the pixels prior to recording and is reversed on replay. The result is that uncorrectable errors due to dropouts are not concentrated, but are spread out by the de-shuffle, making concealment easier. A different approach is required where data reduction is used because the data recorded are not pixels representing a point, but coefficients representing an area of the image. In this case it is the DCT blocks (typically eight pixels across) which must be shuffled.

11.16 Product codes and segmentation

There are two approaches to error correction in segmented recordings. The input may be segmented first, then each segment becomes an independent shuffled product code. This requires less RAM to implement, but it means that from an error-correction standpoint each tape track is self-contained and must deal alone with any errors encountered.

In some formats, such as D-3, following distribution the entire field is used to produce one large shuffled product code in each channel. The product code is then segmented for recording on tape. Although more RAM is required to assemble the large product code, the result is that outer codewords on tape spread across several tracks and redundancy in one track can compensate for errors in another. The result is that size of a single-burst error which can be fully corrected is increased. As RAM is now cheaper than when the first formats were designed, this approach is becoming more common.

11.17 Distribution

Component formats require three signal components to be recorded in parallel. These components must be distributed between the number of signal channels used. In D-1 and D-5 no compression is employed and four signal channels are required to handle the bit rate. The distribution strategy is based on the requirement for concealment of uncorrected errors. This requires samples from all three video components to be distributed over all four recording channels. If this is done the loss of a channel causes an equal loss to each video component rather than a greater loss concentrated in one component.

In composite or compressed formats, the data rate is less than in component formats and it is sufficient to distribute data between two heads. Where compression is employed, the data to be recorded consist of blocks of coefficients. It is these blocks which enter the distribution process. The contents of the DCT blocks cannot be distributed because of the requirement of picture-in-shuttle. When a single sync block is recovered in shuttle, it must include all the coefficients of the DCT blocks it contains otherwise the pixel data cannot be re-created in that screen area. Thus the distribution and shuffle processes take the DCT block as the minimum unit of data.

11.18 The shuffle strategy

Error correction in DVTRs is more complex than in other forms of recording because of the number of conflicting requirements. Chapter 9 showed that interleave is necessary to break up burst errors, and that a product code gives an efficient and reliable system. Unfortunately the product code produces a regular structure of rows and columns, and if such a system is overwhelmed, regular patterns of uncorrected errors are produced, and the effectiveness of concealment is reduced because the eye can still perceive the regular structure.

An irregular interleave, or *shuffle* can be designed to disperse contiguous errors on the tape track over two dimensions on the screen, and the concealment is then more effective. For the best concealment, the shuffle should work over the entire field and this will also help the provision of a picture in shuttle.

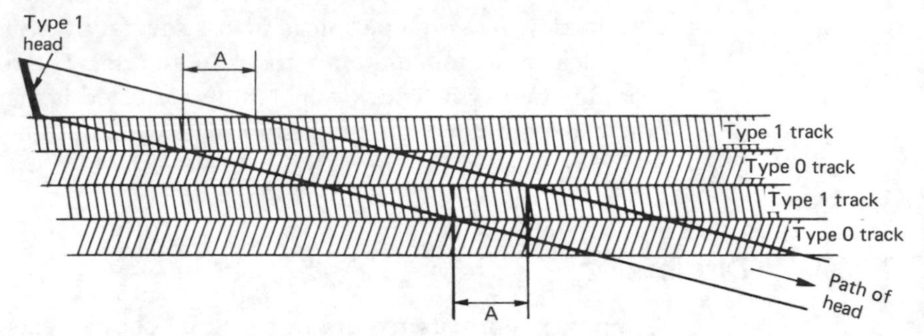

Figure 11.31 During shuttle, the heads cross tracks randomly as shown here (exaggerated). Owing to the use of azimuth recording, a head can only play a track of its own type (0 or 1) but as the head is typically 50 per cent wider than the track it is possible to recover normal signal for the periods above marked A. If sync blocks are shorter than this period, they can be picked up intact. In fact a slightly longer pickup will be possible because the replay system may tolerate a less-than-perfect signal. In any case the final decision is made by checking that the sync block recovered contains valid or correctable codewords.

In shuttle the track-following process breaks down, and the heads cross tracks randomly so that the recovery of entire segments is impossible. A picture of some sort is needed in shuttle, in order to assist in the location of wanted material, but the quality does not need to be as high as in other modes.

Figure 11.31 shows that in shuttle, the path of the head crosses tape tracks obliquely. The use of azimuth recording allows the replay head to be about 50 per cent wider than the track, and so a useful replay signal results for a reasonable proportion of the head path, at the times where the head is near the centreline of a track of the same azimuth type, interrupted by noise when the heads cross tracks of the wrong azimuth type. The sectors are broken into short elements called *sync blocks* which are smaller than the length of track that can be recovered at typical shuttle speeds. Each sync block contains a Reed–Solomon codeword, and so it is possible to tell whether the sync block was correctly recovered or not, without reference to any other part of the recording.

If a sync block is read properly, it is used to update a frame store which refreshes the picture monitor. Sync blocks which are not read correctly will not update the frame store, and so if concealment cannot be used, data from an earlier field will be used to refresh the display.

Using this mechanism, the picture in shuttle is composed of data from different segments of many off-tape fields. A problem with segmented recordings is that at certain tape speeds beating occurs between the passage of tracks and the scanner rotation so that some parts of the segments are never recovered and the picture has to be refreshed with stale data no longer representative of the recording. There are two solutions to the problem. The use of a field-based shuffle helps because one off-tape sync block then contains samples from all over the screen. A field-based shuffle is complex to implement and D-3 was the first format to use it.

The beating effect is mechanical, and results in certain parts of a track being inaccessible. If the data within successive tracks are ordered differently, beating will cause loss from all over the picture once more. Thus in a segment-based shuffle, each track in the field will use a different shuffle. In a field-based shuffle the two channels will use different shuffles. In composite formats the data in successive fields of the colour frame sequence may be rotated by differing amounts. Obtaining a shuttle picture is easier if the shuffle is random, but a random shuffle is less effective for concealment, because it can result in variable density of uncorrected errors, which are harder to conceal than the constant error density resulting from a maximum distance shuffle.

Essentially the field of data to be shuffled are considered to be a two-dimensional memory, and the shuffle is achieved by address mapping in each dimension. The address generator used should not be unnecessarily complicated and it will be simpler if the rows and columns of the memory are mapped in turn. Clearly the shuffle process on recording has to be exactly reversed on replay.

In address mapping, samples in each channel of a segment are at addresses which are initially sequential. The samples are selected non-sequentially by generating a suitable address sequence. When data reduction is used, individual pixels cannot be shuffled, but instead it is the addresses of DCT blocks which are mapped.

In principle, the address sequence could be arbitrarily generated from a lookup table, but the memory arrays are large, and this approach would be extremely expensive to implement. It is desirable that the address generation should be algorithmic to reduce cost, but an algorithm must be found which gives irregular results.

The solution adopted in DVTR formats is to obtain a pseudo-random address sequence by using address multiplication by a prime number in modulo arithmetic for one of the mapping axes. This is easier than it sounds, as the simple example of Figure 11.32(a) shows. In this example, 16 pixels are to be shuffled. The pixel addresses are multiplied by 11, which is relatively prime to 16, and the result is expressed modulo-16, which is to say that if the product exceeds an integer multiple of 16, that integer multiple is subtracted. It will be seen from Figure 11.32(b) that an effective shuffle is created. Modulo arithmetic has a similar effect to quantizing, in fact a number expressed modulo is the quantizing error due to quantizing in steps equal to the modulo base as (c) shows. When the input has a large range, the quantizing error becomes random.

Generating a randomizing effect with modulo arithmetic depends upon the terms in the expression being relatively prime so that repeats are not found in the resulting shuffle, and this constrains the dimensions which can be used for the codewords.

A calculation of the type described will shuffle samples or or DCT blocks in a pseudo-random manner in one dimension, but in the other dimension of the memory a pseudo-random shuffle is undesirable as this would result in a variable-distance shuffle. A straightforward interleave is used in the other dimension to give a maximum distance shuffle. Figure 11.33 shows a conceptual example. When samples are read in rows for recording they retain the same horizontal sequence but there is an offset in the starting address which is a simple function of the row number. This

Address A	A × 11	A × 11 mod 16
0	0	0
1	11	11
2	22 (−16)	6
3	33 (−32)	1
4	44 (−32)	12
5	55 (−48)	7
6	66 (−64)	2
7	77 (−64)	13
8	88 (−80)	8
9	99 (−96)	3
10	110 (−96)	14
11	121 (−112)	9
12	132 (−128)	4
13	143 (−128)	15
14	154 (−144)	10
15	165 (−160)	5

(a)

Figure 11.32(a) The permuted addresses for a shuffle can be obtained by multiplication of the addresses by a number relatively prime to the modulo base.

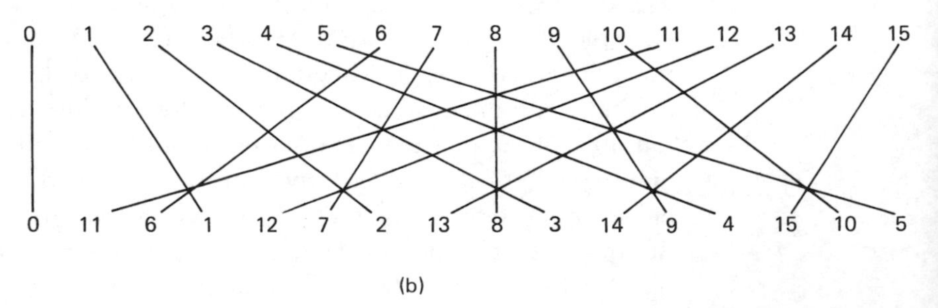

(b)

Figure 11.32(b) The shuffle which results from the calculation in Figure 11.32(a).

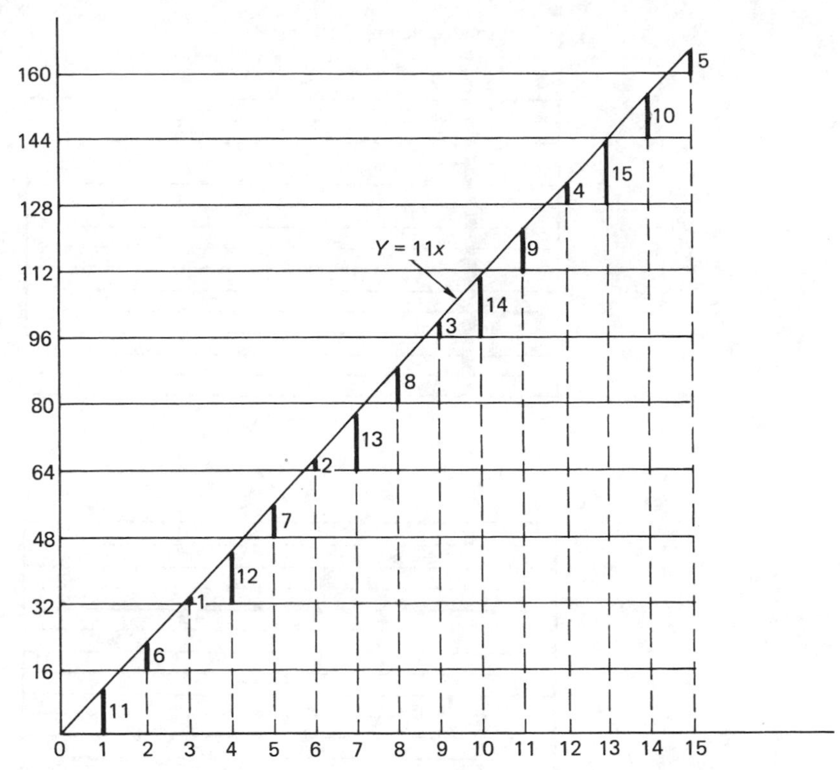

Figure 11.32(c) Modulo arithmetic is rather similar to quantizing. The results of Figure 11.32(a) can be obtained by quantizing the graph of $y = 11x$ with the quantizing intervals of 16 units.

turns a column into a series of diagonals. The combination of these two processes results in the two-dimensional shuffle. The effect of the shuffle can be seen by turning off the correction and concealment circuits on a machine and searching for a large dropout with the jog control. This will reveal the irregular diagonal structure.

11.19 The track structure

Each slant track is subdivided into sectors. In D-1, D-5, DCT and Digital Betacam there is a video sector at each end and four audio sectors in the middle. In D-2 and D-3 there are two audio sectors at each end, and one video sector in the centre. In DVC there is one audio sector and one video sector. It is necessary to be able to edit any or all of the audio channels independently of the video. For this reason short spaces are left between the sectors to allow write current to turn on and off away from wanted data. Since the write current can only turn off at the end of a sector, it follows that the smallest quantum of data which can be written in the

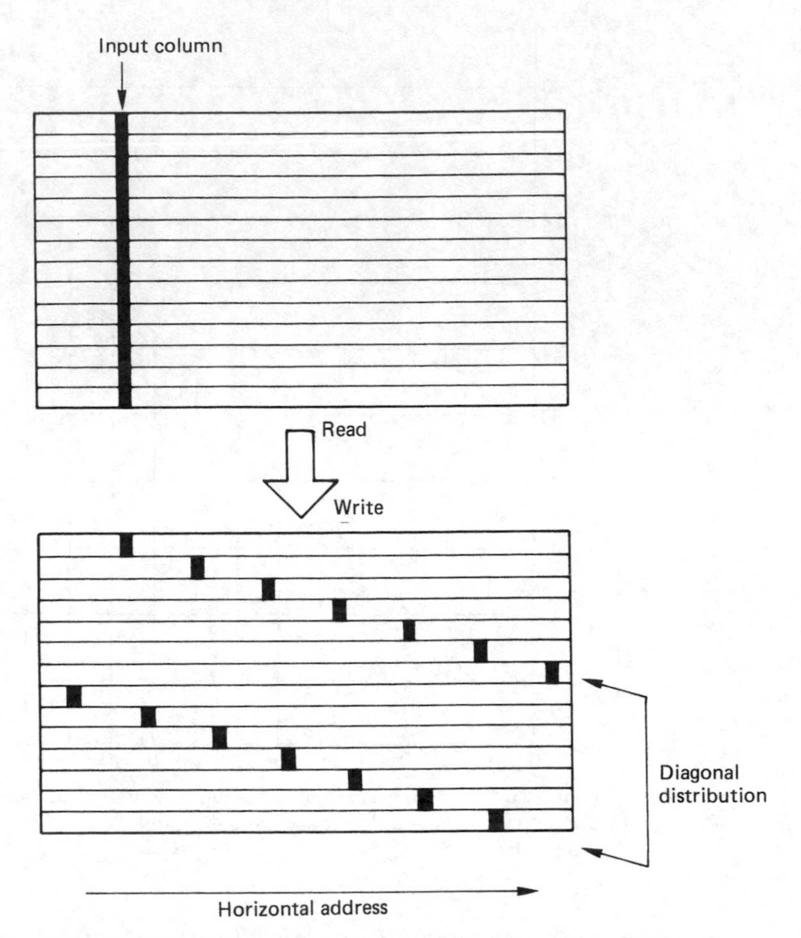

Figure 11.33 The second dimension of the shuffle is obtained by offsetting the horizontal address during transfer from one array to another. This results in a diagonal shuffle if the offset increases by a constant from one line to the next.

track at one time is a whole sector, although the smallest quantum which can be read is a sync block.

A preamble is necessary before the first sync block in a sector to allow the phase-locked loop in the data separator to lock. The preamble also contains a sync pattern and ID code so that the machine can confirm the position of the head before entering the sector proper. At the end of the sector a postamble containing an ID code is written before the write current is turned off. The pre- and postamble details of D-3 can be seen in Figure 11.34.

The space between two adjacent sectors is known as an edit gap, and it will be seen in Figure 11.34 to begin with the postamble of the previous sector and ends with the preamble of the next sector. Editing may result in a discontinuity nominally in the middle of the gap. There is some

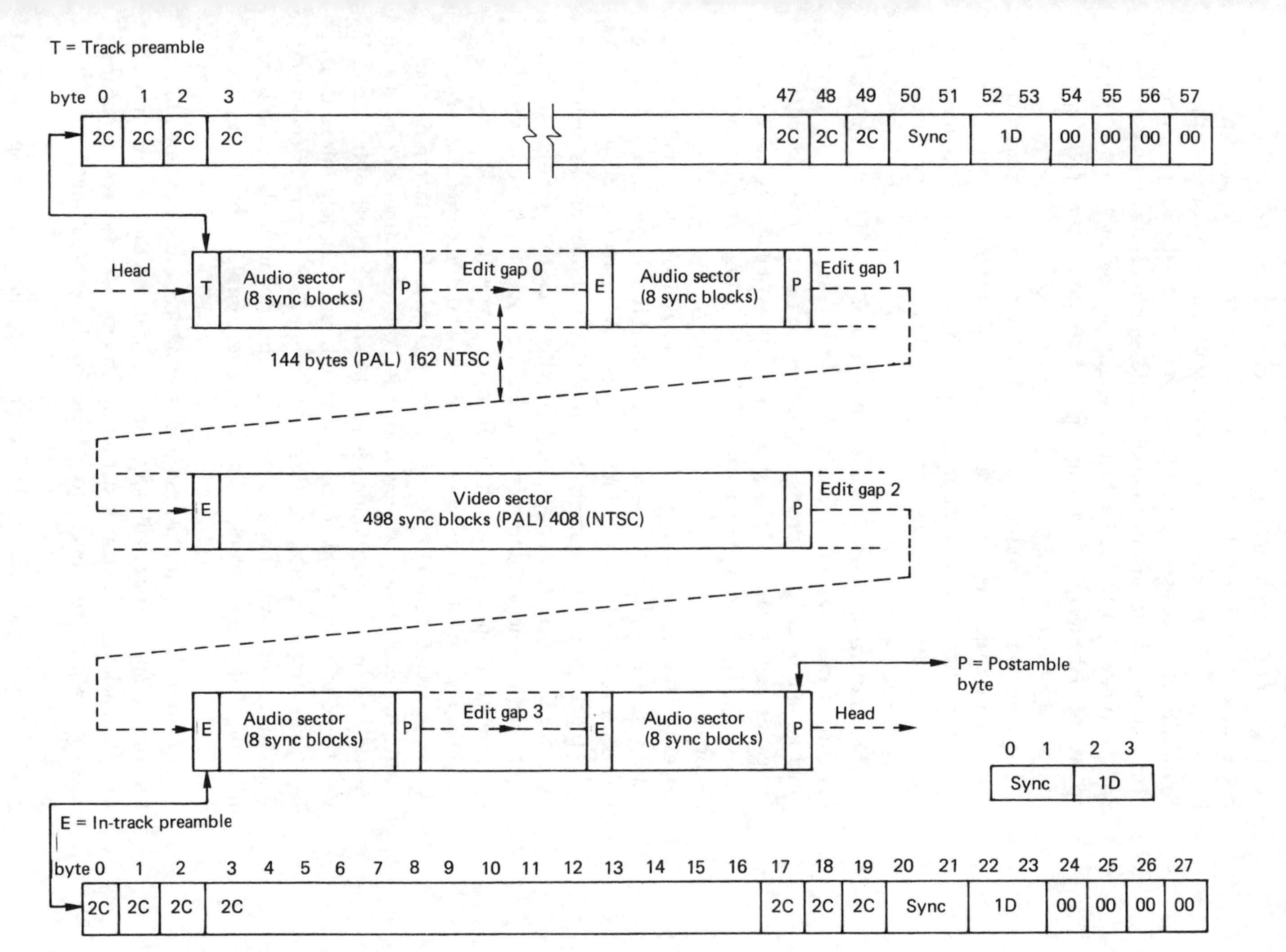

Figure 11.34 All separately recordable areas begin with a preamble E except for the first sector which has a longer track preamble T. Edit gaps are the same size as two sync blocks and begin with the postamble P of the previous sector and end with the preamble E of the next sector.

latitude, but the new recording must not begin so early that the postamble at the end of the previous sector is damaged, and at least 20 bytes of preamble must be written before the sync pattern at the beginning of the next sector, because it will not be possible to maintain continuity of bit phase at an edit, and the PLL must resynchronize.

In practice the length of the postamble, gap and preamble combined will be made exactly equal to the length of one or two sync blocks according to the format. This simplifies the design of the sequencer which controls the track recording.

The inner codes in product codes are designed to correct random errors and to detect the presence of burst errors by declaring the whole inner block bad. There is an optimum size for an inner code: too small and the redundancy factor rises, too large and burst errors are magnified, as are errors due to losing sync.

When the product code becomes large, one row will be too long to be one inner code, and so in D-3, for example, it is split into twelve or nine inner code blocks. The dimensions of these are chosen to give adequate correction performance. One sync block contains one inner codeword and an ID code. Twelve or nine sync blocks then form one complete row of the sector array.

Figure 11.35 shows a typical sync block. A two-byte synchronizing pattern serves to phase the data separator and the deserializer. Following this is a two-byte ID code which uniquely identifies the location of this sync block in the product code. Audio and video blocks will have different ID codes. The data content of the block then follows, and the eight bytes of inner redundancy are calculated over the ID code as well as the data, so a random error in the ID can be corrected.

Whilst the size of the inner codeword is related to correction power, it is not essential for each inner code to have its own ID. In fact individually identifying each inner code would raise the recording overheads. In shuttle, a geometric calculation reveals the shortest length of track which will be recovered at maximum tape speed, and this determines that in D-2, for example, it is not necessary to identify every inner codeword. Accordingly one sync block contains two inner codewords, but only one ID code. Three sync blocks then form one complete row of the sector array.

Each sync block contains one or two inner codewords, depending on the format, and is the smallest quantum of data on the track with which the playback channel can synchronize independently. Synchronization takes place on three levels: to the bit, to the symbol and to the sync block.

The recording density in DVTRs is such that it is not possible to mechanically position the heads with sufficient accuracy to locate an individual bit. The heads are rotated with reasonable accuracy, and the

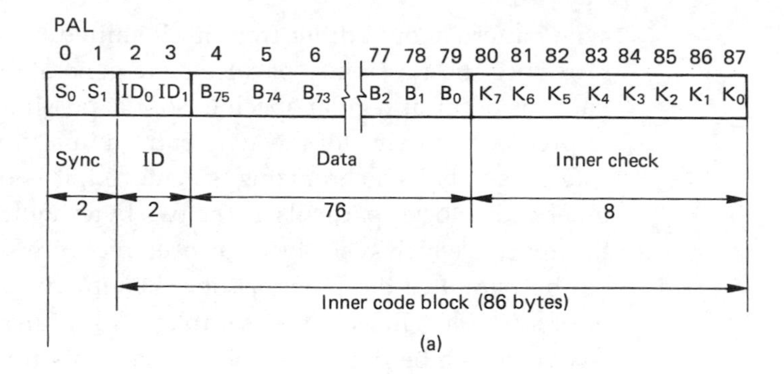

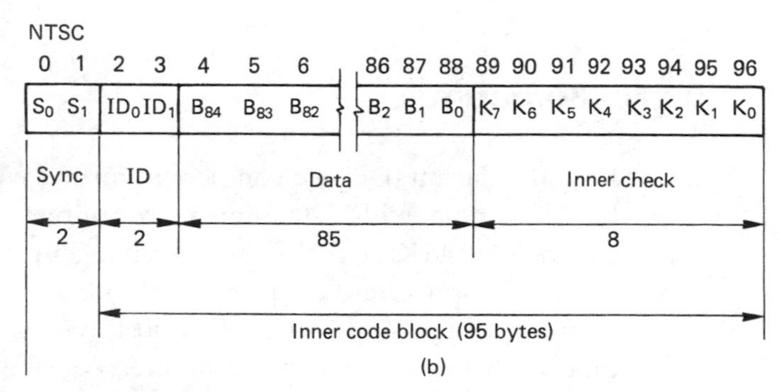

Figure 11.35 D-3 sync blocks (a) PAL and (b) NTSC. Each begins with the sync pattern 97F1 which is not randomized. The 2 byte ID, the data and the 8 bytes of inner check symbols are all randomized prior to EFM coding. The same block structure is used in both audio and video sectors.

replay signal is accepted as and when it arrives. A phase-locked loop in the data separator must lock to the bit rate and phase of the recording. At the beginning of the track a reference for the phase-locked loop known as a track preamble is recorded. This serves to set the frequency of the PLL the same as the offtape bit rate. Once the loop is locked, it can stay in synchronism by phase comparing the data transitions with its own. However, if synchronism is lost due to a dropout, it cannot be regained until the next synchronizing pattern is seen. This is one reason why the tracks are broken into short sync blocks.

Once bit synchronism has been achieved at the preamble, the serial data have to be correctly divided up into symbols, or deserialized, in order to decode the original samples. This requires word synchronization, and is achieved by the unique pattern which occurs at the beginning of every sync block. Detection of this pattern resets the counter which deserializes the data by dividing by the wordlength. If randomizing is used, as in DVC, D-1, D-3/5 and Digital Betacam, sync detection will also be used to pre-set the derandomizer to the correct starting value. The sync pattern should differ from itself shifted by as many bits as possible to reduce the possibility of false sync generation. Within a sector, sync patterns occur at regular spacing, so it is possible for the replay circuitry to predict the arrival of the sync pattern in a time window.

Once symbol synchronizing is achieved, it is then possible to read the inner code blocks, particularly the two bytes which contain the ID code as this reveals which sync block has been recovered. In normal play this will be the one after the previous one, but in shuttle, the sequence will be unpredictable. In shuttle, samples from any sync block properly recovered can be put in the correct place in a frame store by reference to the ID code.

11.20 Digital Betacam

Digital Betacam (DB) is a component format which accepts eight- or ten-bit 4:2:2 data with 720 luminance samples per active line and four channels of 48 KHz digital audio having up to twenty-bit wordlength. Spatial compression based on Discrete Cosine Transform is employed,[2] with a compression factor of almost two to one (assuming eight-bit input). The audio data are uncompressed. The cassette shell of the half-inch analog Betacam format is retained, but contains 14 micrometre metal particle tape. The digital cassette contains an identification hole which allows the transport to identify the tape type. Unlike the other digital formats, only two cassette sizes are available. The large cassette offers 124 minutes of playing time; the small cassette plays for 40 minutes.

As a result of the tradeoff between SNR and bandwidth which is a characteristic of digital recording, the tracks must be longer than in the analog Betacam format, but narrower. The drum diameter of the DB transport is 81.4 mm which is designed to produce tracks of the required length for digital recording. The helix angle of the digital drum is designed such that when an analog Betacam tape is driven past at the correct speed, the track angle is correct. Certain DB machines are fitted with analog heads which can trace the tracks of an analog tape. As the drum size is different, the analog playback signal is time compressed by about 9 per cent, but this is easily dealt with in the timebase correction process.[3] The fixed heads are compatible with the analog Betacam positioning. The reverse compatibility is for playback only; the digital machine cannot record on analog cassettes.

Figure 11.36 shows the track patterns for 525/60 and 625/50 Digital Betacam. The four digital audio channels are recorded in separate sectors of the slant tracks, and so one of the linear audio channels of the analog format is dispensed with, leaving one linear audio track for cueing.

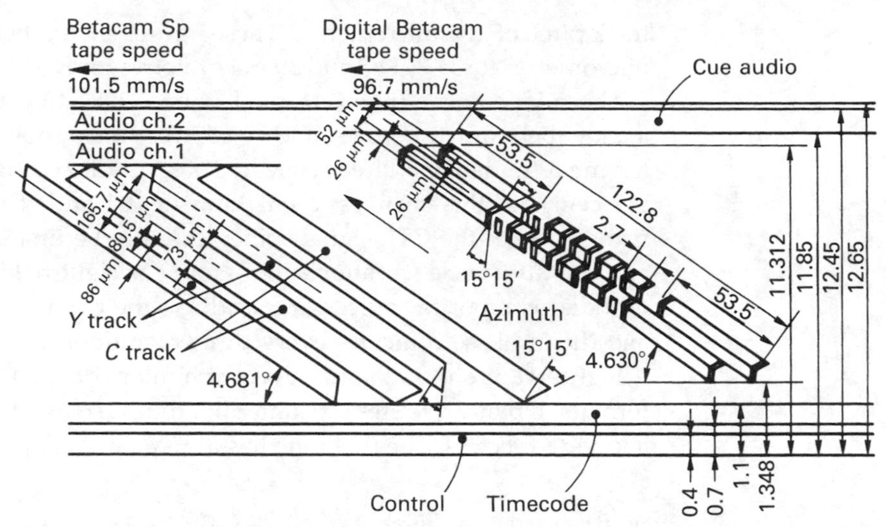

Figure 11.36 The track pattern of Digital Betacam. Control and timecode tracks are identical in location to the analog format, as is the single analog audio cue track. Note the use of a small guard band between segments.

Azimuth recording is used, with two tracks being recorded simultaneously by adjacent heads. Electronic delays are used to align the position of the edit gaps in the two tracks of a segment, allowing double-width flying erase heads to be used. Three segments are needed to record one field, requiring one and a half drum revolutions. Thus the drum speed is three times that of the analog format. However, the track pitch is less than one third that of the analog format so the linear speed of the digital tape is actually slower. In 625/50, track width is 24 micrometres, with a 4-micrometre guard band between segments making the effective track pitch 26 micrometres, whereas in 525/60 the track width is 20 micrometres with a 3.4-micrometre guard band between segments making an effective

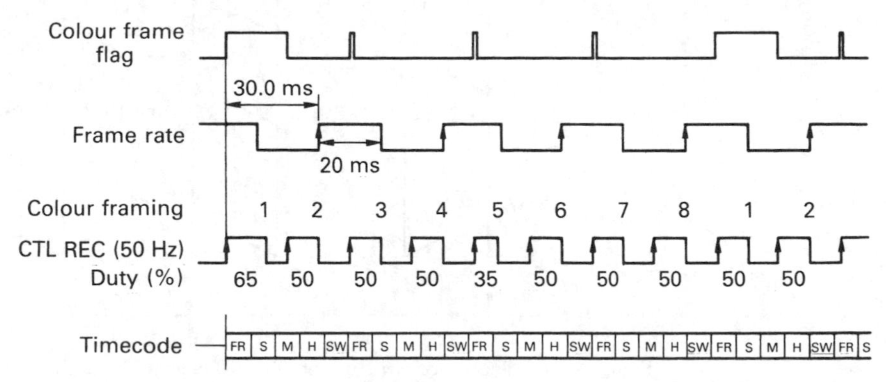

Figure 11.37 The control track of Digital Betacam uses duty cycle modulation for colour framing purposes.

track pitch of 21.7 micrometres. These figures should be compared with 18 micrometres for D-3/D-5 and 39 micrometres for D-2/DCT.

There is a linear timecode track whose structure is identical to the analog Betacam timecode, and a control track shown in Figure 11.37 having a fundamental frequency of 50 Hz. Ordinarily the duty cycle is 50 per cent, but this changes to 65/35 in field 1 and 35/65 in field 5. The rising edge of the CTL signal coincides with the first segment of a field, and the duty cycle variations allow four- or eight-field colour framing if decoded composite sources are used. As the drum speed is 75 Hz, CTL and drum phase coincides every three revolutions.

Figure 11.38 shows the track layout in more detail. Unlike other digital formats, DB incorporates tracking pilot tones recorded between the audio and video sectors. The first tone has a frequency of approximately 4 MHz

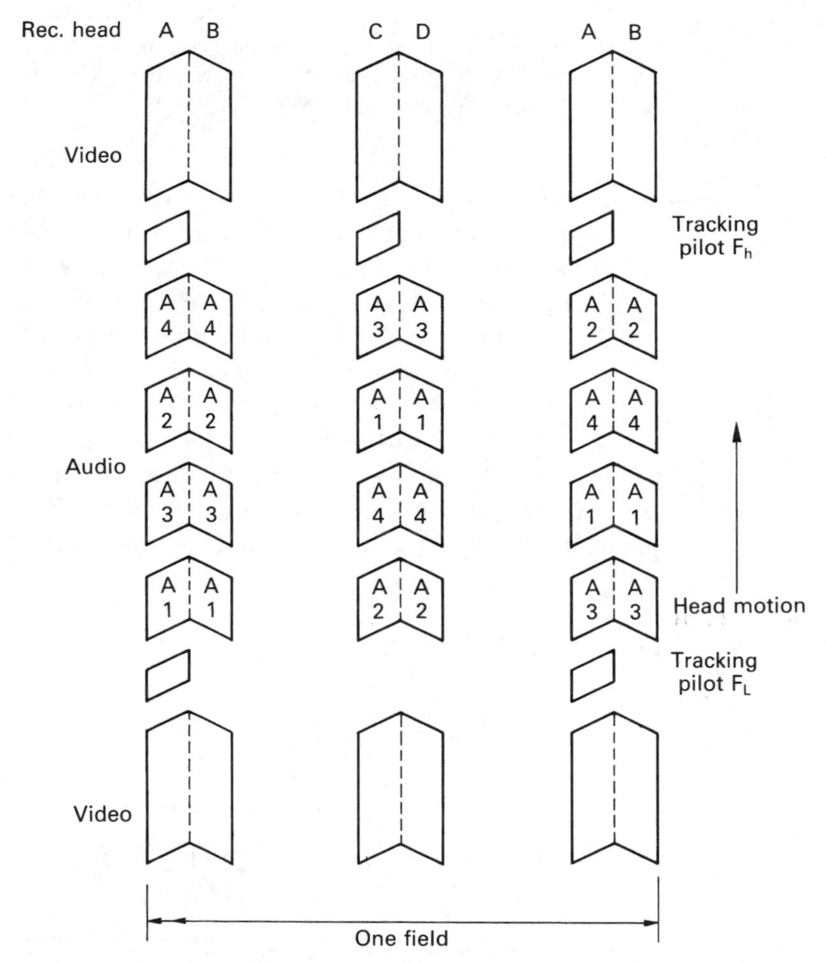

Figure 11.38 The sector structure of Digital Betacam. Note the tracking tones between the audio and video sectors which are played back for alignment purposes during insert edits.

and appears once per drum revolution. The second is recorded at approximately 400 KHz and appears twice per drum revolution. The pilot tones are recorded when a recording is first made on a blank tape, and will be re-recorded following an assemble edit, but during an insert edit the tracking pilots are not re-recorded, but used as a guide to the insertion of the new tracks.

The amplitude of the pilot signal is a function of the head tracking. The replay heads are somewhat wider than the tracks, and so a considerable tracking error will have to be present before a loss of amplitude is noted. This is partly offset by the use of a very long wavelength pilot tone in which fringing fields increase the effective track width. The low-frequency tones are used for automatic playback tracking.

With the tape moving at normal speed, the capstan phase is changed by steps in one direction then the other as the pilot tone amplitude is monitored. The phase which results in the largest amplitude will be retained. During the edit pre-roll the record heads play back the high-frequency pilot tone and capstan phase is set for largest amplitude. The record heads are the same width as the tracks and a short-wavelength pilot tone is used such that any mistracking will cause immediate amplitude loss. This is an edit-optimize process and results in new tracks being inserted in the same location as the originals. As the pilot tones are played back in insert editing, there will be no tolerance build-up in the case of multiple inserts.

Figure 11.39 shows a block diagram of the record section of DB. Analog component inputs are sampled at 13.5 and 6.75 MHz. Alternatively the input may be SDI at 270 Mbits/s which is deserialized and demultiplexed to separate components. The raster scan input is first converted to blocks which are 8 pixels wide by 4 pixels high in the luminance channel and 4 pixels by 4 in the two colour difference channels. When two fields are combined on the screen, the result is effectively an interlaced 8 × 8 luminance block with colour difference pixels having twice the horizontal luminance pixel spacing. The pixel blocks are then subject to a field shuffle. A shuffle based on individual pixels is impossible because it would raise the high-frequency content of the image and impair the effectiveness of the compression.

Instead the block shuffle helps the compression by making the average entropy of the image more constant. This happens because the shuffle exchanges blocks from flat areas of the image with blocks from highly detailed areas. The shuffle algorithm also has to consider the requirements of picture in shuttle. The blocking and shuffle take place when the read addresses of the input memory are permuted with respect to the write addresses.

Following the input shuffle the blocks are associated into sets of ten in each component and are then subject to the discrete cosine transform. The

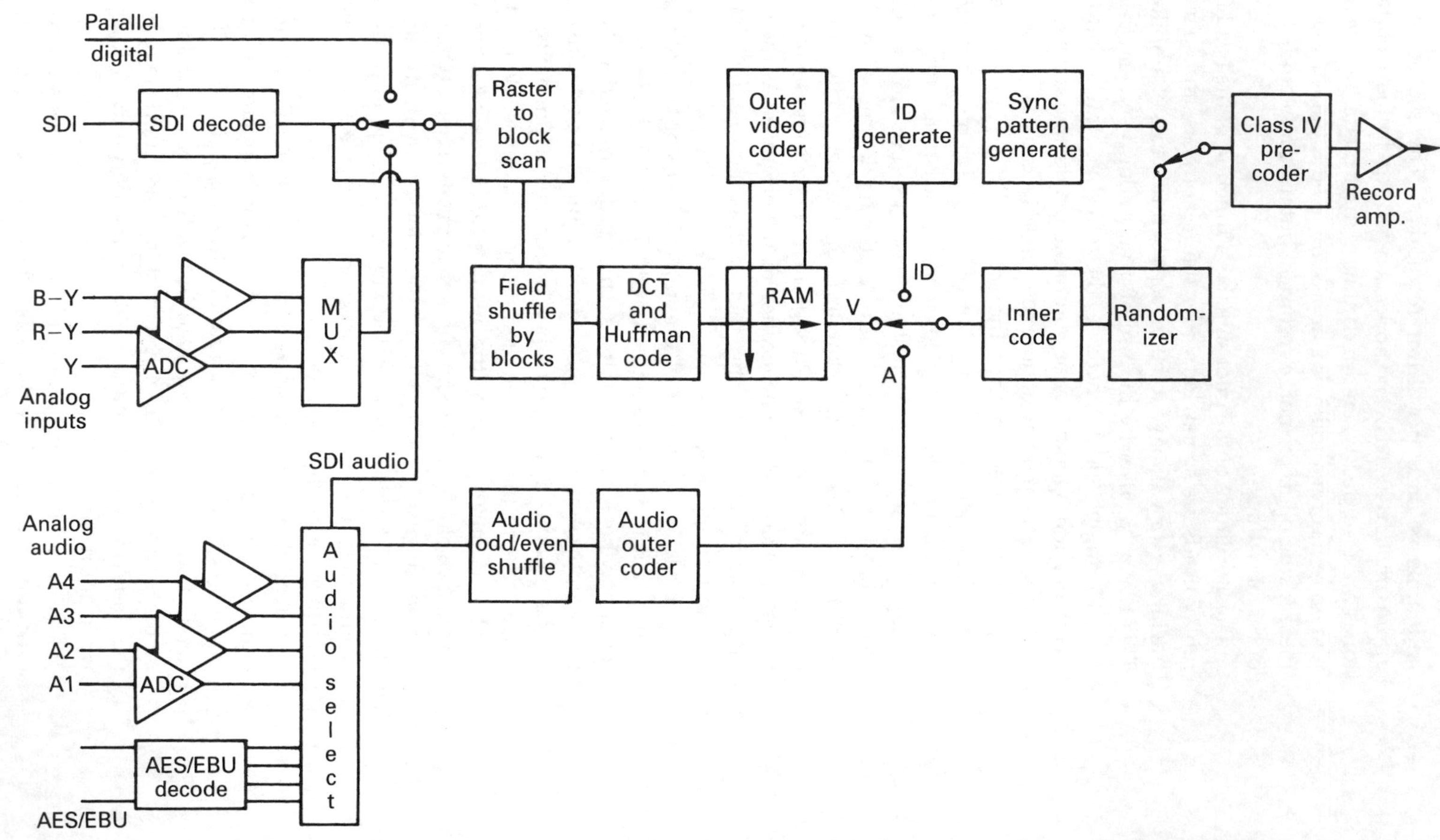

Figure 11.39 Block diagram of Digital Betacam record channel. Note that the use of compression makes this rather different to the block layout of full-bit formats.

resulting coefficients are then subject to an iterative requantizing process followed by variable-length coding. The iteration adjusts the size of the quantizing step until the overall length of the ten coefficient sets is equal to the constant capacity of an entropy block which is 364 bytes. Within that entropy block the amount of data representing each individual DCT blocks may vary considerably, but the overall block size stays the same.

The DCT process results in coefficients whose wordlength exceeds the input wordlength. As a result it does not matter if the input wordlength is eight bits or ten bits; the requantizer simply adapts to make the output data rate constant. Thus the compression factor is greater with ten-bit input, corresponding to about 2.4 to 1.

The next step is the generation of product codes as shown in Figure 11.40(a). Each entropy block is divided into two halves of 162 bytes each and loaded into the rows of the outer code RAM which holds 114 such rows, corresponding to one twelfth of a field. When the RAM is full, it is read in columns by address mapping and 12 bytes of outer Reed–Solomon redundancy are added to every column, increasing the number of rows to 126.

The outer code RAM is read out in rows once more, but this time all 126 rows are read in turn. To the contents of each row is added a two-byte ID code and then the data plus ID bytes are turned into an inner code by the addition of 14 bytes of Reed–Solomon redundancy.

Inner codewords pass through the randomizer and are then converted to serial form for class-IV partial response pre-coding. With the addition of a sync pattern of two bytes, each inner code word becomes a sync block as shown in Figure 11.40(c). Each video block contains 126 sync blocks, preceded by a preamble and followed by a postamble. One field of video data requires twelve such blocks. Pairs of blocks are recorded simultaneously by the parallel heads of a segment. Two video blocks are recorded at the beginning of the track, and two more are recorded after the audio and tracking tones.

The audio data for each channel are separated into odd and even samples for concealment purposes and assembled in RAM into two blocks corresponding to one field period. Two twenty-bit samples are stored in five bytes. Figure 11.40(b) shows that each block consists of 1458 bytes including auxiliary data from the AES/EBU interface, arranged as a block of 162 × 9 bytes. One hundred per cent outer code redundancy is obtained by adding 9 bytes of Reed–Solomon check bytes to each column of the blocks.

The inner codes for the audio blocks are produced by the same circuitry as the video inner codes on a time-shared basis. The resulting sync blocks are identical in size and differ only in the provision of different ID codes. The randomizer and pre-coder are also shared. It will be seen from Figure

(a) Video . . . 12 ECC blocks/field (2 ECC blocks/track)

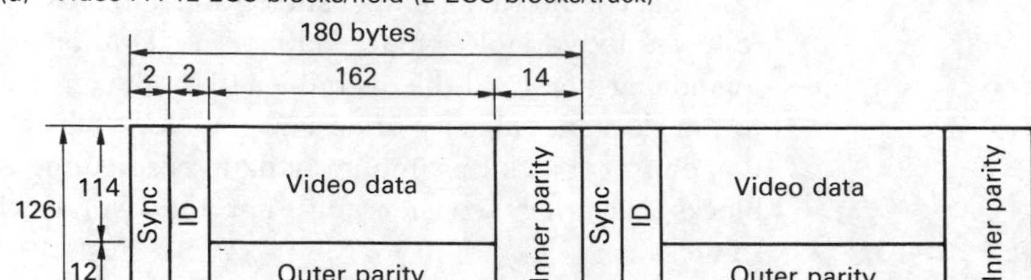

(b) Audio . . . 2 ECC blocks/ (CH. × field)

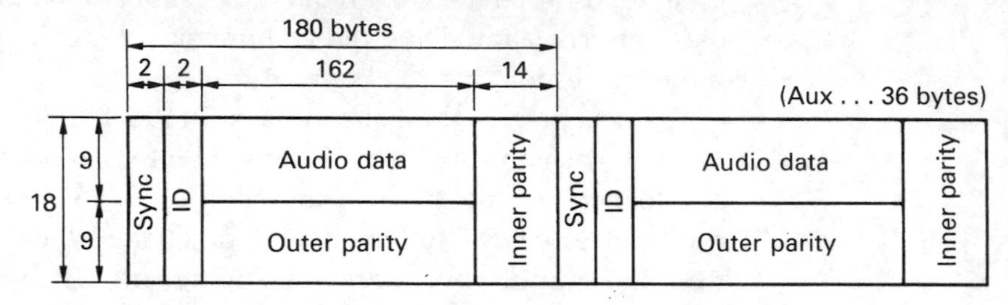

(c) Sync block

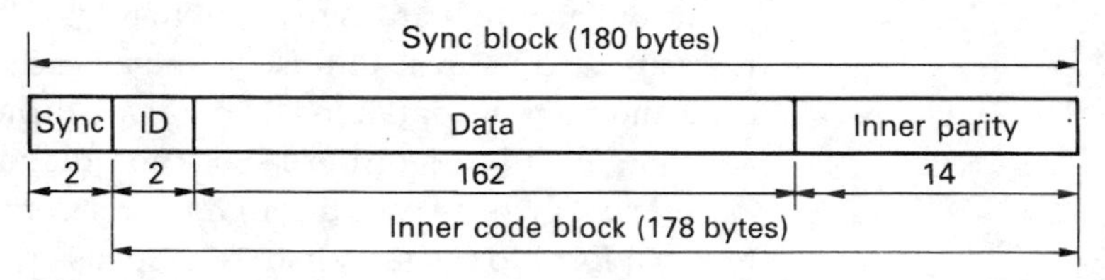

Figure 11.40 Video product codes of Digital Betacam are shown at (a); 12 of these are needed to record one field. Audio product codes are shown at (b); two of these record samples corresponding to one field period. Sync blocks are common to audio and video as shown at (c). The ID code discriminates between video and audio channels.

11.38 that there are three segments in a field and that the position of an audio sector corresponding to a particular audio channel is different in each segment. This means that damage due to a linear tape scratch is distributed over three audio channels instead of being concentrated in one.

Each audio product block results in 18 sync blocks. These are accommodated in audio sectors of six sync blocks each in three segments. The audio sectors are preceded by preambles and followed by postambles. Between these are edit gaps which allow each audio channel to be independently edited. By spreading the outer codes over three different audio sectors the correction power is much improved because data from two sectors can be used to correct errors in the third.

Figure 11.41 shows the replay channel of DB. The RF signal picked up by the replay head passes first to the Class IV partial response playback circuit in which it becomes a three-level signal as was shown in Chapter 8. The three-level signal is passed to an ADC which converts it into a digitally represented form so that the Viterbi detection can be carried out in logic circuitry. The sync detector identifies the synchronizing pattern at the beginning of each sync block and resets the block bit count. This allows the entire inner codeword of the sync block to be deserialized into bytes and passed to the inner error checker. Random errors will be corrected here, whereas burst errors will result in the block being flagged as in error.

Sync blocks are written into the de-interleave RAM with error flags where appropriate. At the end of each video sector the product code RAM will be filled, and outer code correction can be performed by reading the RAM at right angles and using the error flags to initiate correction by erasure. Following outer code correction the RAM will contain corrected data or uncorrectable error flags which will later be used to initiate concealment.

The sync blocks can now be read from memory and assembled in pairs into entropy blocks. The entropy block is of fixed size, but contains coefficient blocks of variable length. The next step is to identify the individual coefficients and separate the luminance and colour difference coefficients by decoding the Huffman-coded sequence. Following the assembly of coefficient sets the inverse DCT will result in pixel blocks in three components once more.

The pixel blocks are de-shuffled by mapping the write address of a field memory. When all the tracks of a field have been decoded, the memory will contain a de-shuffled field containing either correct sample data or correction flags. By reading the memory without address mapping the de-shuffled data are then passed through the concealment circuit where flagged data are concealed by data from nearby in the same field or from a previous field. The memory readout process is buffered from the offtape timing by the RAM and as a result the timebase correction stage is inherent in the replay process. Following concealment the data can be output as conventional raster scan video either formatted to parallel or serial digital standards or converted to analog components.

11.21 DVC and DVCPRO

This component format uses quarter-inch-wide metal evaporated (ME) tape which is only 7 micrometres thick in conjunction with compression to allow realistic playing times in miniaturized equipment. The format has jointly been developed by all the leading VCR manufacturers. Whilst

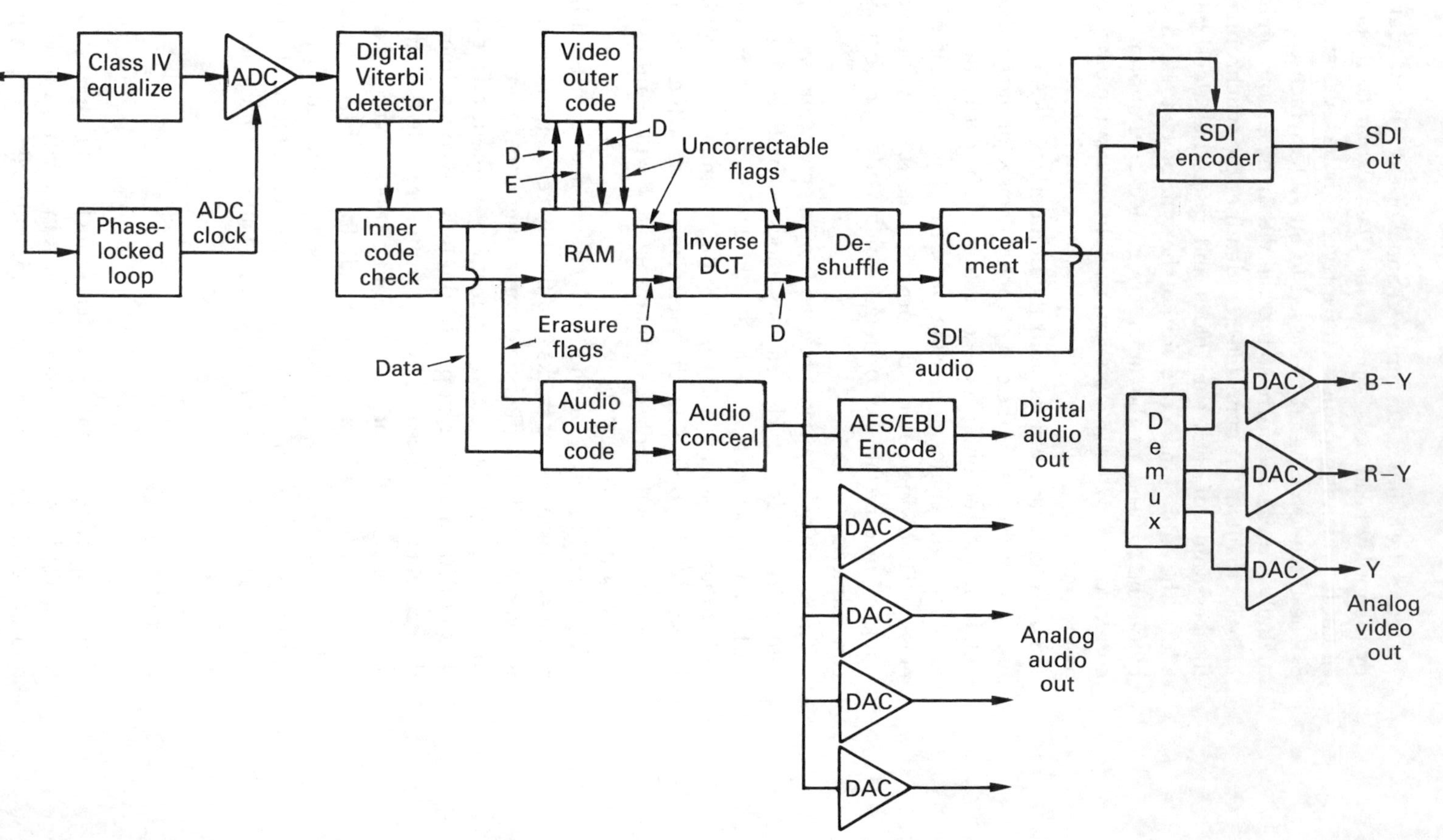

Figure 11.41 The replay channel of Digital Betacam. This differs from a full-bit system primarily in the requirement to deserialize variable-length coefficient blocks prior to the inverse DCT.

intended as a consumer format it was clear that such a format is ideal for professional applications such as news gathering and simple production because of the low cost and small size. This led to the development of the DVCPRO format. In addition to component video there are also two channels of sixteen-bit uniformly quantized digital audio at 32, 44.1 or 48 kHz, with an option of four audio channels using twelve-bit non-uniform quantizing at 32 kHz.

Figure 11.42 shows that two cassette sizes are supported. The standard size cassette offers $4\frac{1}{2}$ hours of recording time and yet is only a little larger than an audio Compact Cassette. The small cassette is even smaller than a DAT cassette yet plays for one hour. Machines designed to play both tape sizes will be equipped with moving-reel motors. Both cassettes are equipped with fixed identification tabs and a moveable write-protect tab. These tabs are sensed by switches in the transport.

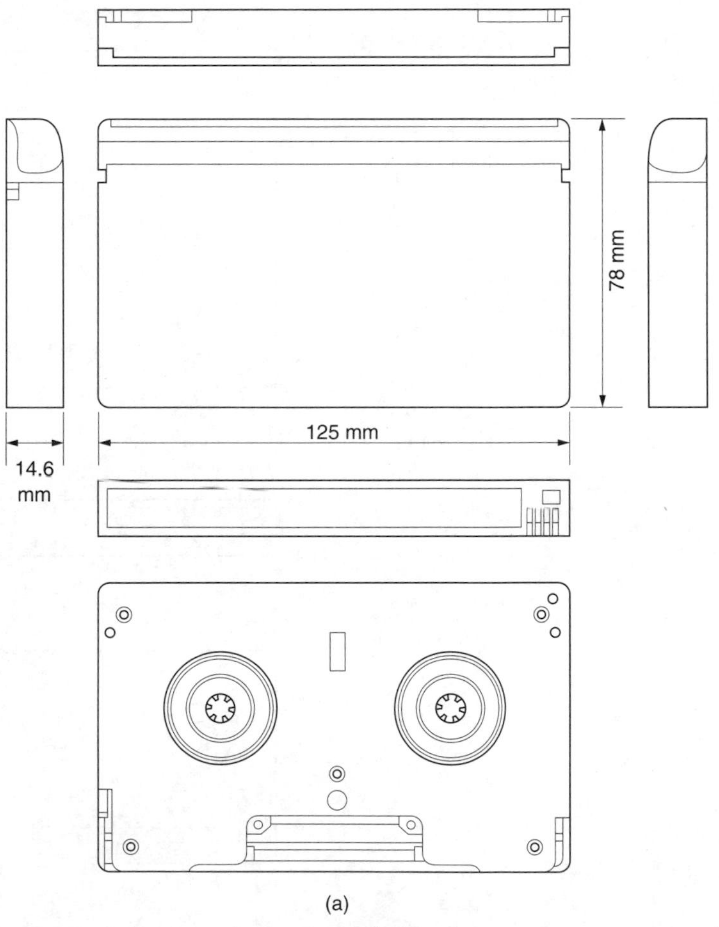

Figure 11.42 The cassettes developed for the $\frac{1}{4}$ inch DVC format. At (a) the standard cassette which holds 4.5 hours of program material.

DVC (Digital Video Cassette) has adopted many of the features first seen in small formats such as the DAT digital audio recorder and the 8 mm analog video tape format. Of these the most significant is the elimination of the control track permitted by recording tracking signals in the slant tracks themselves. The adoption of metal evaporated tape and embedded tracking allows extremely high recording density. Tracks recorded with slant azimuth are only 10 mm wide and the minimum wavelength is only 0.49 mm resulting in a superficial density of over 0.4 Megabits per square millimetre.

Segmentation is used in DVC in such a way that as much commonality as possible exists between 50 and 60 Hz versions. The transport runs at 300 tape tracks per second. Figure 11.43 shows that 50 Hz frames contain 12 tracks and 60 Hz frames contain 10 tracks.

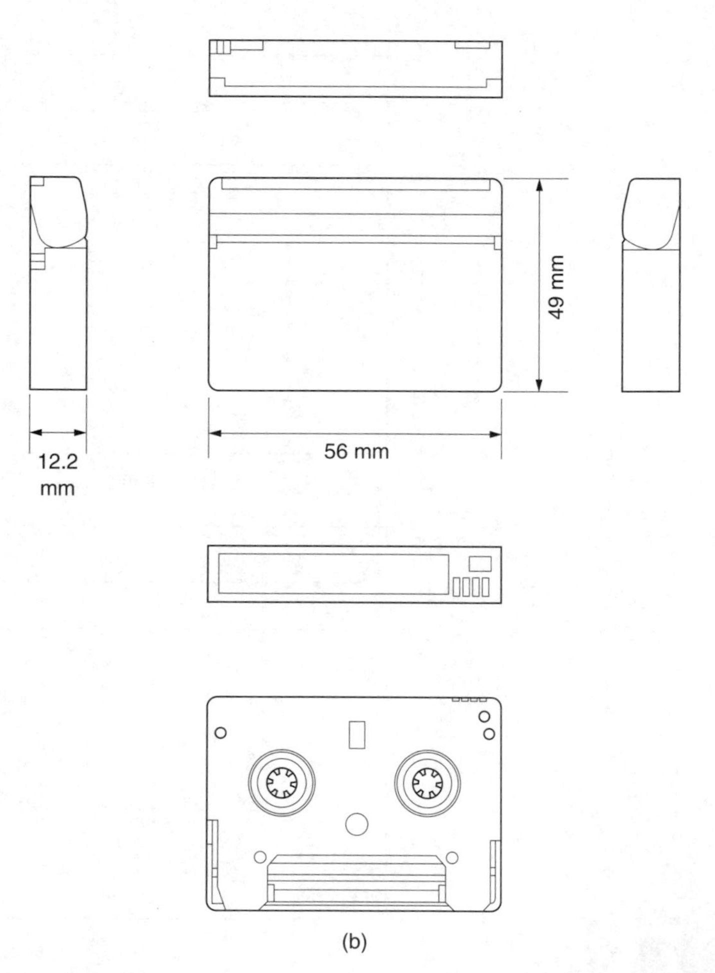

Figure 11.42 continued The small cassette, shown at (b) is intended for miniature equipment and plays for 1 hour.

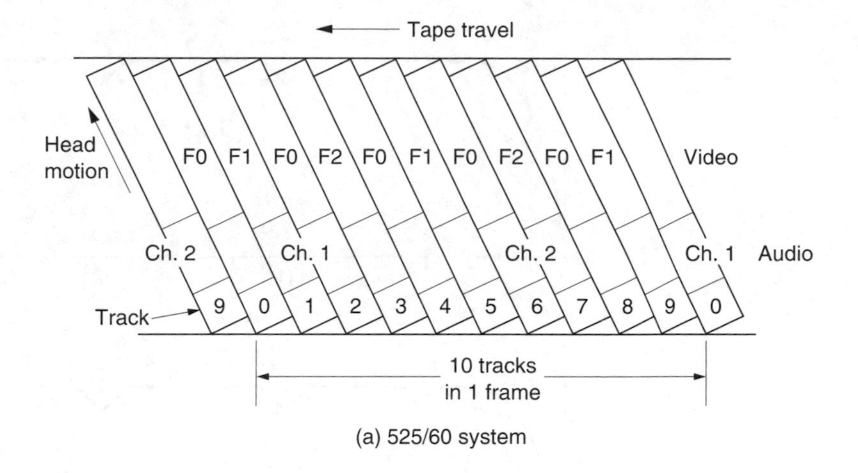

(a) 525/60 system

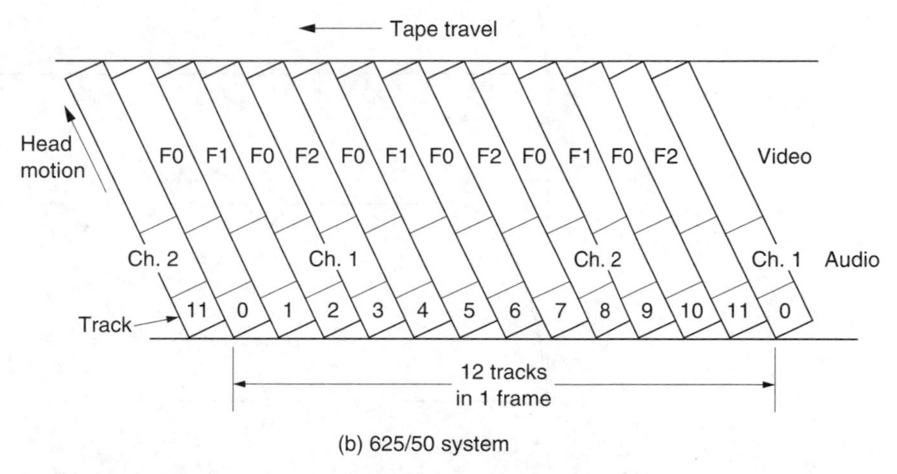

(b) 625/50 system

Figure 11.43 In order to use a common transport for 50 and 60 Hz standards the segmentation shown here is used. The segment rate is constant but 10 or 12 segments can be used in a frame.

The tracking mechanism relies upon buried tones in the slant tracks. From a tracking standpoint there are three types of track shown in Figure 11.44; F_0, F_1 and F_2. F_1 contains a low-frequency pilot and F_2 contains a high-frequency pilot. F_0 contains no pilot tone, but the recorded data spectrum contains notches at the frequencies of the two tones. Figure 11.44 also shows that every other track will contain F_0 following a four-track sequence.

The embedded tracking tones are recorded throughout the track by inserting a low frequency into the channel-coded data. Every twenty-four data bits an extra bit is added whose value has no data meaning but whose polarity affects the average voltage of the waveform. By

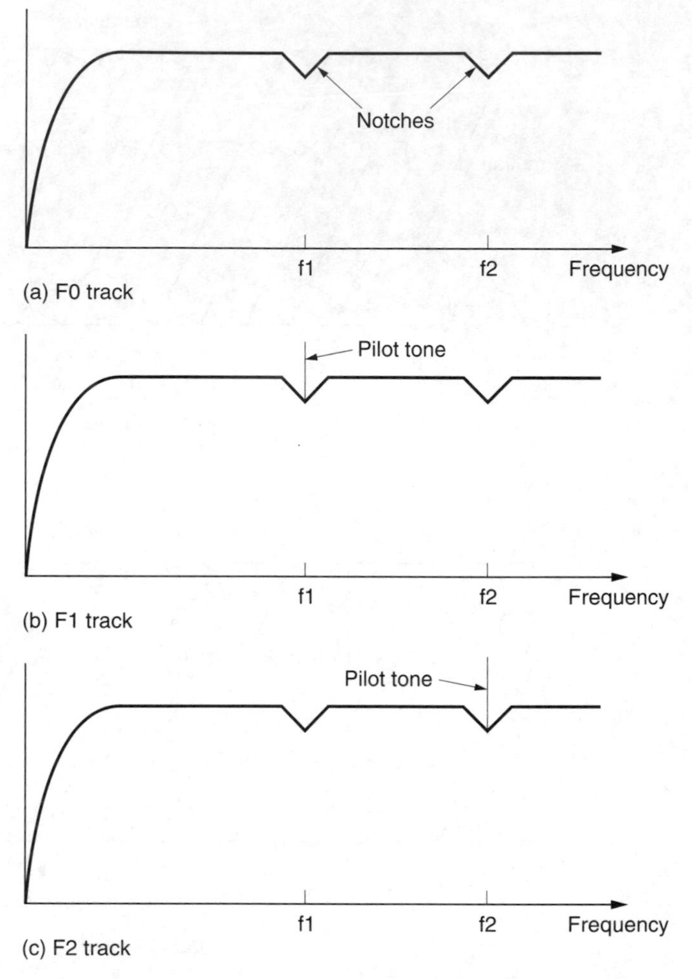

Figure 11.44 The tracks are of three types shown here. The F_0 track (a) contains spectral notches at two selected frequencies. The other two track types (b), (c) place a pilot tone in one or other of the notches.

controlling the average voltage with this bit, low frequencies can be introduced into the channel-coded spectrum to act as tracking tones. The tracking tones have sufficiently long wavelength that they are not affected by head azimuth and can be picked up by the 'wrong' head. When a head is following an F_0 type track, one edge of the head will detect F_1 and the other edge will detect F_2. If the head is centralized on the track, the amplitudes of the two tones will be identical. Any tracking error will result in the relative amplitudes of the F_1 F_2 tones changing. This can be used to modify the capstan phase in order to correct the tracking error. As azimuth recording is used requiring a minimum of two heads, one head of the pair will always be able to play a type F_0 track.

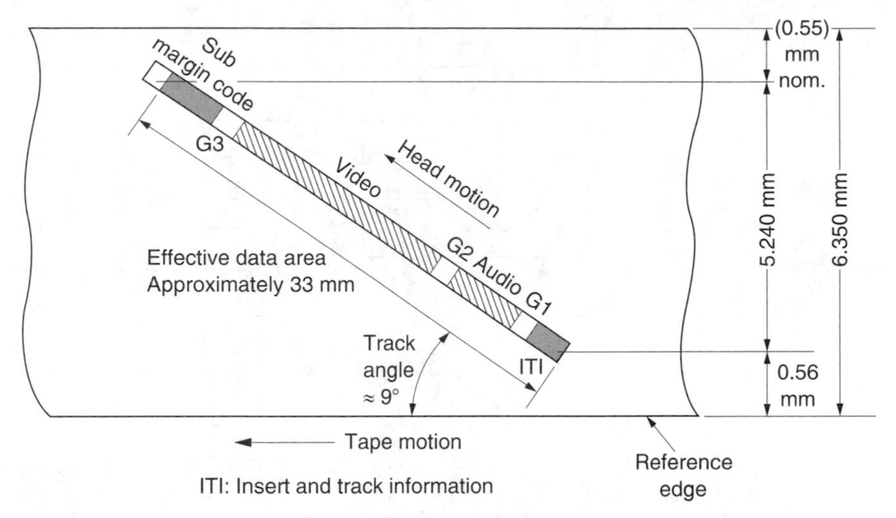

Figure 11.45 The dimensions of the DVC track. Audio, video and subcode can independently be edited. Insert and Track Information block aligns heads during insert.

In simple machines only one set of heads will be fitted and these will record or play as required. In more advanced machines, separate record and replay heads will be fitted. In this case the replay head will read the tracking tones during normal replay, but in editing modes, the record head would read the tracking tones during the pre-roll in order to align itself with the existing track structure.

Figure 11.45 shows the track dimensions. The tracks are approximately 33 mm long and lie at approximately 9° to the tape edge. A transport with a 180° wrap would need a drum of only 21 mm diameter. For camcorder applications with the small cassette this would allow a transport no larger than an audio 'Walkman'. With the larger cassette it would be advantageous to use time compression as shown in section 11.7 to allow a larger drum with partial wrap to be used. This would simplify threading and make room for additional heads in the drum for editing functions.

The audio, video and subcode data are recorded in three separate sectors with edit gaps between so that they can be independently edited in insert mode. In the case where all three data areas are being recorded in insert mode, there must be some mechanism to keep the new tracks synchronous with those that are being overwritten. In a conventional VTR this would be the job of the control track.

In DVC there is no control track and the job of tracking during insert is undertaken by part of each slant track. Figure 11.45 shows that the track begins with the insert and track information (ITI) block. During an insert edit the ITI block in each track is always read by the record head. This

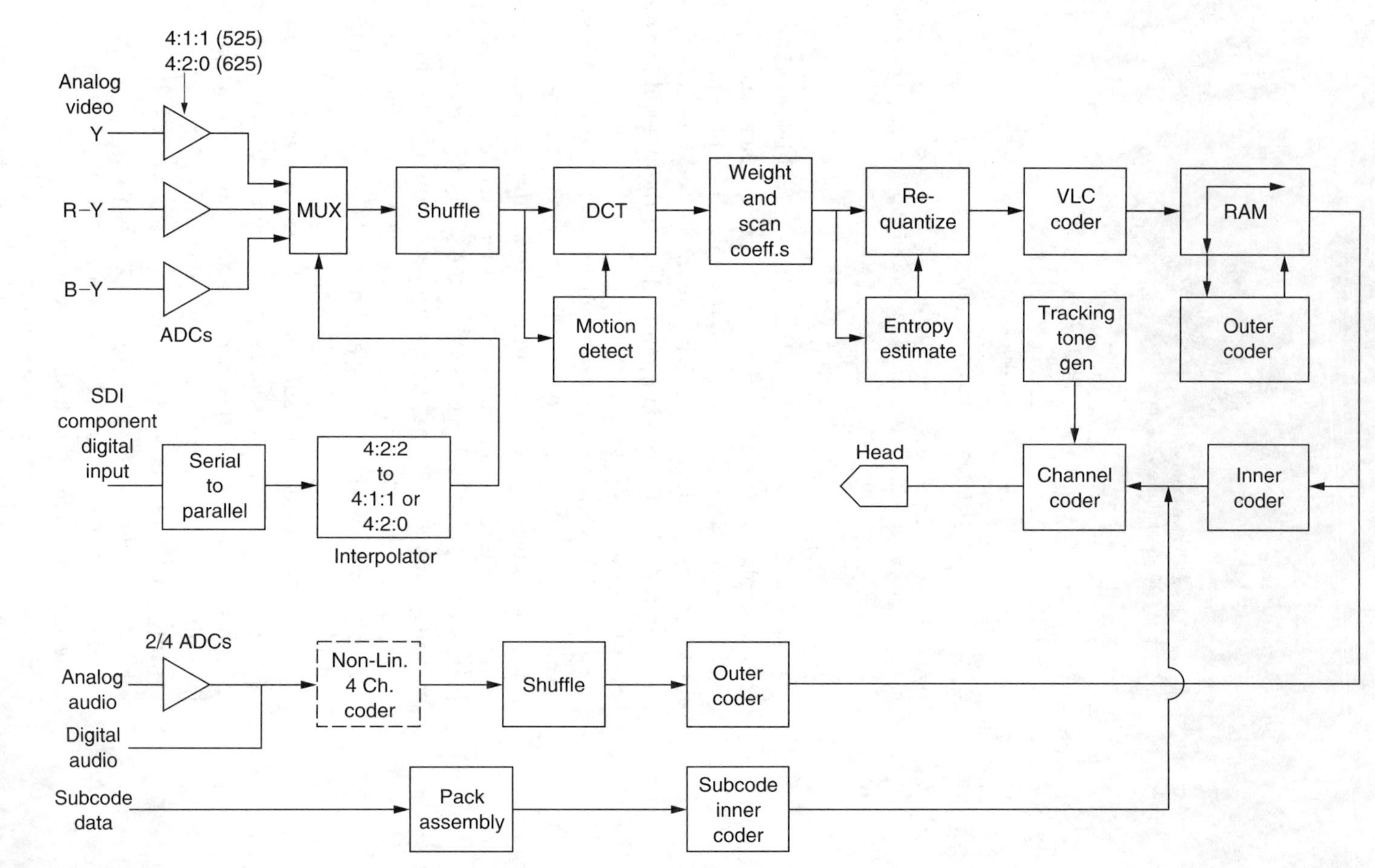

Figure 11.46 Block diagram of DVC signal system. This is similar to larger formats except that a high compression factor allows use of a single channel with no distribution.

identifies the position of the track in the segmentation sequence and in the tracking tone sequence and allows the head to identify its physical position both along and across the track prior to an insert edit. The remainder of the track can then be recorded as required.

As there are no linear tracks, the subcode is designed to be read in shuttle for access control purposes. It will contain timecodes and flags.

Figure 11.46 shows a block diagram of the DVC signal system. The input video is eight-bit component digital according to CCIR-601, but compression of about 5:1 is used. The colour difference signals are subsampled prior to compression. In 60 Hz machines, 4:1:1 sampling is used, allowing a colour difference bandwidth in excess of that possible with NTSC. In 50 Hz machines, 4:2:0 sampling is used. The colour difference sampling rate is still 6.75 MHz, but the two colour difference signals are sent on sequential lines instead of simultaneously. The result is that the vertical colour difference resolution matches the horizontal resolution. A similar approach is used in SECAM and MAC video. Studio standard 4:2:2 parallel or serial inputs and outputs can be handled using simple interpolators in the colour difference signal paths.

A 16:9 aspect ratio can be supported in standard definition by increasing the horizontal pixel spacing as is done in Digital Betacam. High-definition signals can be supported using a higher compression factor.

As in other DVTRs, the error-correction strategy relies upon a combination of shuffle and product codes. Frames are assembled in RAM, and partitioned into blocks of 8 x 8 pixels. In the luminance channel, four of these blocks cover the same screen area as one block in each colour difference signal as Figure 11.47 shows. The four luminance blocks and

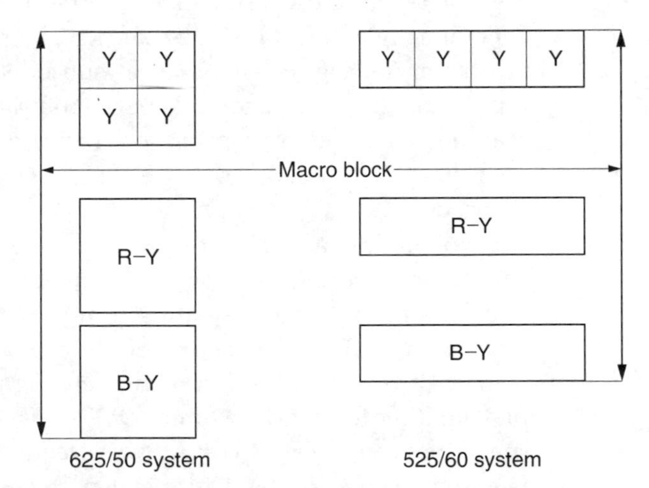

Figure 11.47 In DVC a macro block contains information from a fixed screen area. As the colour resolution is reduced, there are twice as many luminance pixels.

the two colour difference blocks are together known as a macroblock. The shuffle is based upon reordering of macroblocks. Following the shuffle, compression takes place. The compression system is DCT-based and uses techniques described in Chapter 6. Compression acts within frame boundaries to permit frame-accurate editing. This contrasts with the intra-field compression used in DB.

Intra-frame compression uses 8 × 8 pixel DCT blocks and allows a higher compression factor because advantage can be taken of redundancy between the two fields when there is no motion. If motion is detected, then moving areas of the two fields will be independently coded in 8 × 4 pixel blocks to prevent motion blur. Following the motion compensation the DCT coefficients are weighted, zig-zag scanned and requantized prior to variable-length coding. As in other compressed VTR formats, the requantizing is adaptive so that the same amount of data is output irrespective of the input picture content. The entropy block occupies one sync block and contains data compressed from five macroblocks.

The DVC product codes are shown in Figure 11.48. The video product block is shown in (a). This block is common to both 525 and 625 line formats. Ten such blocks record one 525-line frame whereas 12 blocks are required for a 625-line frame.

The audio channels are shuffled over a frame period and assembled into the product codes shown in (b). Video and audio sync blocks are identical except for the ID numbering. The subcode structure is different. and (c) shows the structure of the subcode block. The subcode is not a product block because these can only be used for error correction when the entire block is recovered. The subcode is intended to be read in shuttle where only parts of the track are recovered. Accordingly only inner codes are used and these are much shorter than the video/audio codes, containing only five data bytes, known as a pack. The structure of a pack is shown in Figure 11.49. The subcode block in each track can accommodate twelve packs. Packs are repeated throughout the frame so that they have a high probability of recovery in shuttle. The pack header identifies the type of pack, leaving four bytes for pack data, e.g. timecode.

Following the assembly of product codes, the data are then channel coded for recording on tape. A scrambled NRZI channel code is used which is similar to the system used in D-1 except that the tracking tones are also inserted by the modulation process.

In the DVCPRO format the extremely high recording density and long playing time of the consumer DVC was not a requirement. Instead ruggedness and reliable editing were required. In developing DVCPRO, Panasonic chose to revert to metal particle tape as used in most other DVTRs. This requires wider tracks, and this was achieved by increasing the tape linear speed. The wider tracks also reduce the mechanical

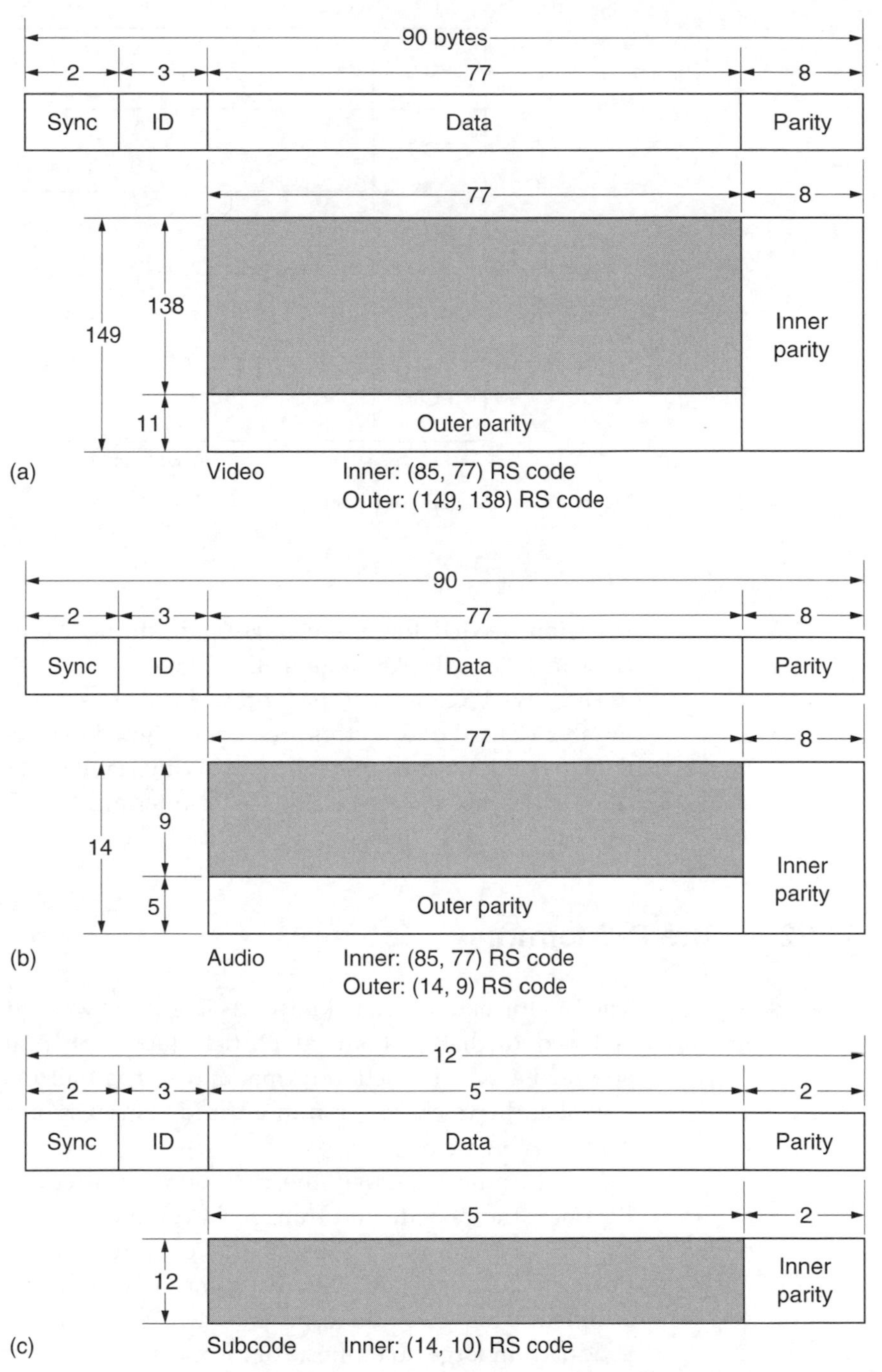

Figure 11.48 The product codes used in DVC. Video and audio codes at (a) and (b) differ only in size and use the same inner code structure. Subcode at (c) is designed to be read in shuttle and uses short sync blocks to improve chances of recovery.

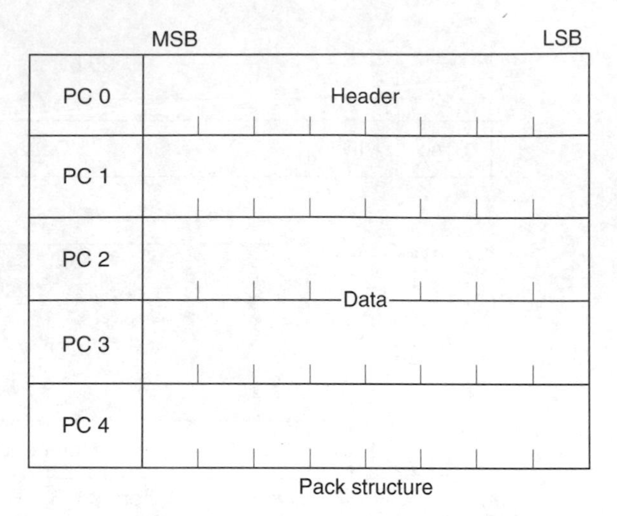

Figure 11.49 Structure of a Pack.

precision needed for interchange and editing. However, the DVCPRO transport can still play regular DVC tapes.

The DVCPRO format has proved to be extremely popular and a number of hard disk editors are now designed to import native DVCPRO data to cut down on generation loss. With a suitable tape drive this can be done at 4 × normal speed. The SDTI interface (see Chapter 10) can also carry native DVC data.

11.22 The D-9 format

The D-9 format (formerly known as Digital-S) was developed by JVC and is based on half-inch metal particle tape. D-9 is intended as a fully specified 4:2:2 production format and so has four uncompressed 48 kHz digital audio tracks and a frame-based compression scheme with a mild compression factor.

In the same way that the Sony format used technology from the Betamax, the cassette and transport design of D-9 are refinements of the analog VHS format. Whereas the compression algorithm of Digital Betacam is unique, in D-9 the same algorithm as in DVC has been adopted. This gives an economic advantage as the DVC chip set is produced in large quantities. An increasing number of manufacturers are producing equipment which can accept native DVC coded data to avoid generation loss and D-9 fits well with that concept.

In D-9 the bit rate on tape is 50 Mbits/s: twice that of the DVC format. As a result, the DCT coefficients will suffer less requantizing allowing a lower noise floor and genuine multi-generation performance. Tests performed by the EBU/SMPTE task force found that D-9 and Digital Betacam had almost identical performance.

References

1. Oldershaw, R., Design of an automatic scan tracking system for a D-2 recorder. *IEE Conf. Pub. No. 293*, 395–398 (1988)
2. Creed, D. and Kaminaga, K. Digital compression strategies for video tape recorders in studio applications. *Record of 18th ITS*, 291–301 (Montreux, 1993)
3. Huckfield, D., Sato, N. and Sato, I. Digital Betacam – The application of state of the art technology to the development of an affordable component DVTR. *Record of 18th ITS*, 180–199 (Montreux, 1993)

12

Disks

12.1 Types of disk

Once the operating speed of computers began to take strides forward, it became evident that a single processor could be made to jump between several different programs so fast that they all appeared to be executing simultaneously, a process known as multiprogramming. Computer memory remains more expensive than other types of mass storage, and so it has never been practicable to store every program or data file necessary within the computer memory. In practice some kind of storage medium is necessary where only programs which are running or are about to run are in the memory, and the remainder are stored on the medium.

The disk drive was developed specifically to offer rapid random access to stored data. Figure 12.1 shows that, in a disk drive, the data are recorded on a circular track. In floppy disks, the magnetic medium is flexible, and the head touches it. This restricts the rotational speed. In hard-disk drives, the disk rotates at several thousand rev/min so that the head-to-disk speed is of the order of 100 miles per hour. At this speed no contact can be tolerated, and the head flies on a boundary layer of air

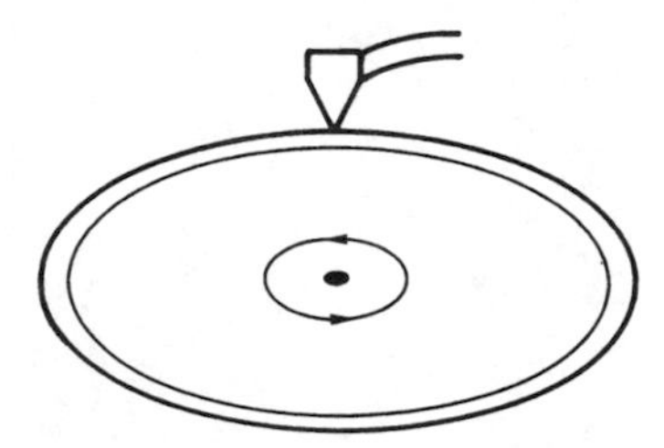

Figure 12.1 The rotating store concept. Data on the rotating circular track are repeatedly presented to the head.

turning with the disk at a height measured in microinches. The longest time it is necessary to wait to access a given data block is a few milliseconds. To increase the storage capacity of the drive without a proportional increase in cost, many concentric tracks are recorded on the disk surface, and the head is mounted on a positioner which can rapidly bring the head to any desired track.

Such a machine is termed a moving-head disk drive. An increase in capacity could be obtained by assembling many disks on a common spindle to make a disk pack. The small size of magnetic heads allows the disks to be placed close together. The positioner is usually designed so that it can remove the heads away from the disk completely, so that it can be exchanged. The exchangeable-pack moving-head disk drive became the standard for mainframe and minicomputers for a long time, and usually at least two were provided so that important data could be 'backed up' or copied to a second disk for safe keeping.

Later came the so-called Winchester technology disks, where the disk and positioner formed a compact sealed unit which allowed increased storage capacity but precluded exchange of the disk pack alone. This led to the development of high-speed tape drives which could be used as security back-up storage.

Disk drive development has been phenomenally rapid. The first flying head disks were about 3 feet across. Subsequently disk sizes of 14, 8, $5\frac{1}{4}$, $3\frac{1}{2}$ and 1.7/8 inches were developed. Despite the reduction in size, the storage capacity is not compromised because the recording density has increased and continues to increase. In fact there is an advantage in making a drive smaller because the moving parts are then lighter and travel a shorter distance, improving access time.

There are numerous types of optical disk, which have different characteristics.[1] There are, however, three broad groups which can be usefuly compared:

1 DVD and Compact Disc are read-only laser disks, which are designed for mass duplication by stamping. Stamped disks cannot be recorded. The main difference between DVD and CD is that the former uses a shorter-wavelength laser and a larger lens aperture so that the resolution of the optics is improved. This allows narrower tracks and shorter recorded wavelengths so that nearly 5 gigabytes of data can be stored on a single surface.
2 Some optical disks can be recorded, but once a recording has been made, it cannot be changed or erased. These are usually referred to as write-once-read-many (WORM) disks. The general principle is that the disk contains a thin layer of a compound which is sensitive to intense light. On recording, a relatively powerful laser changes the sensitive layer, making it darker so that it can subsequently be read at lower

power. Clearly once a recording has been made in this way it is permanent.

3 Erasable optical disks have essentially the same characteristic as magnetic disks, in that new and different recordings can be made in the same track indefinitely, but there may be a separate erase cycle needed before a new recording can be made since overwrite is not always possible.

Figure 12.2 introduces the essential subsystems of a disk drive which will be discussed here. Magnetic drives and optical drives are similar in that both have a spindle drive mechanism to revolve the disk, and a positioner to give radial access across the disk surface. In the optical drive, the positioner has to carry a collection of lasers, lenses, prisms, gratings and so on, and will be rather larger than a magnetic head. The heavier pickup cannot be accelerated as fast as a magnetic-drive positioner, and access time is slower. A large number of pickups on one positioner makes matters worse. For this reason and because of the larger spacing needed between the disks, multi-platter optical disks are uncommon.

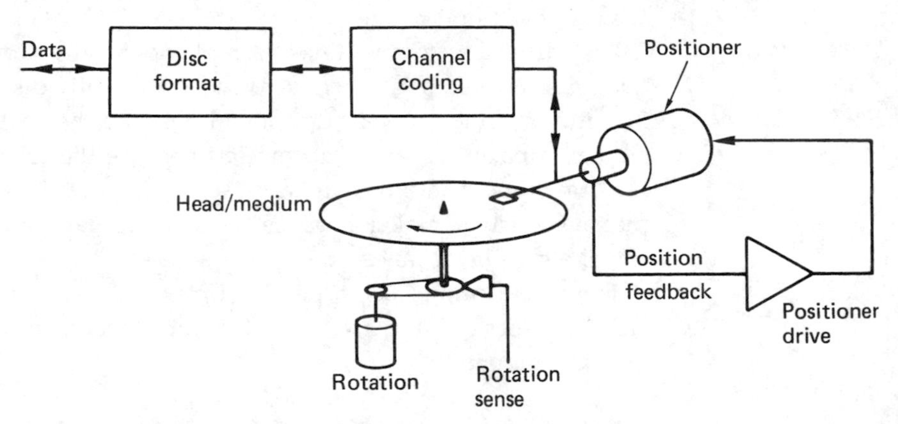

Figure 12.2 The main subsystems of a typical disk drive.

Instead 'jukebox' mechanisms have been developed to allow a large library of optical disks to be mechanically accessed by one or more drives.[2] Access time is sometimes reduced by having more than one positioner per disk; a technique adopted rarely in magnetic drives. A penalty of the very small track pitch possible in laser disks, which gives the enormous storage capacity, is that very accurate track following is needed, and it takes some time to lock onto a track. For this reason tracks on laser disks are usually made as a continuous spiral, rather than the concentric rings of magnetic disks. In this way, a continuous data transfer involves no more than track following once the beginning of the file is located.

12.2 Structure of disk

The floppy disk is actually made using tape technology, and will be discussed later. Rigid disks are made from aluminium alloy. Magnetic-oxide types use an aluminium oxide substrate, or undercoat, giving a flat surface to which the oxide binder can adhere. Later metallic disks having higher coercivity are electroplated with the magnetic medium. In both cases the surface finish must be extremely good due to the very small flying height of the head. As the head-to-disk speed and recording density are functions of track radius, the data are confined to the outer areas of the disks to minimize the change in these parameters. As a result, the centre of the pack is often an empty well. In fixed (i.e. non-interchangeable) disks the drive motor is often installed in the centre well.

The information layer of optical disks may be made of a variety of substances, depending on the working principle. This layer is invariably protected beneath a thick transparent layer of glass or polycarbonate.

Exchangeable optical and magnetic disks are usually fitted in protective cartridges. These have various shutters which retract on insertion in the drive to allow access by the drive spindle and heads. Removable packs usually seat on a taper to ensure concentricity and are held to the spindle by a permanent magnet. A lever mechanism may be incorporated into the cartridge to assist their removal.

12.3 Magnetic disk terminology

In all technologies there are specialist terms, and those relating to magnetic disks will be explained here. Figure 12.3 shows a typical multiplatter magnetic disk pack in conceptual form. Given a particular set of coordinates (cylinder, head, sector), known as a disk physical address, one unique data block is defined. A common block capacity is 512 bytes. The subdivision into sectors is sometimes omitted for special applications. A disk drive can be randomly accessed, because any block address can follow any other, but unlike a RAM, at each address a large block of data is stored, rather than a single word.

12.4 Principle of flying head

Magnetic disk drives permanently sacrifice storage density in order to offer rapid access. The use of a flying head with a deliberate air gap between it and the medium is necessary because of the high medium speed, but this causes a separation loss which restricts the linear density

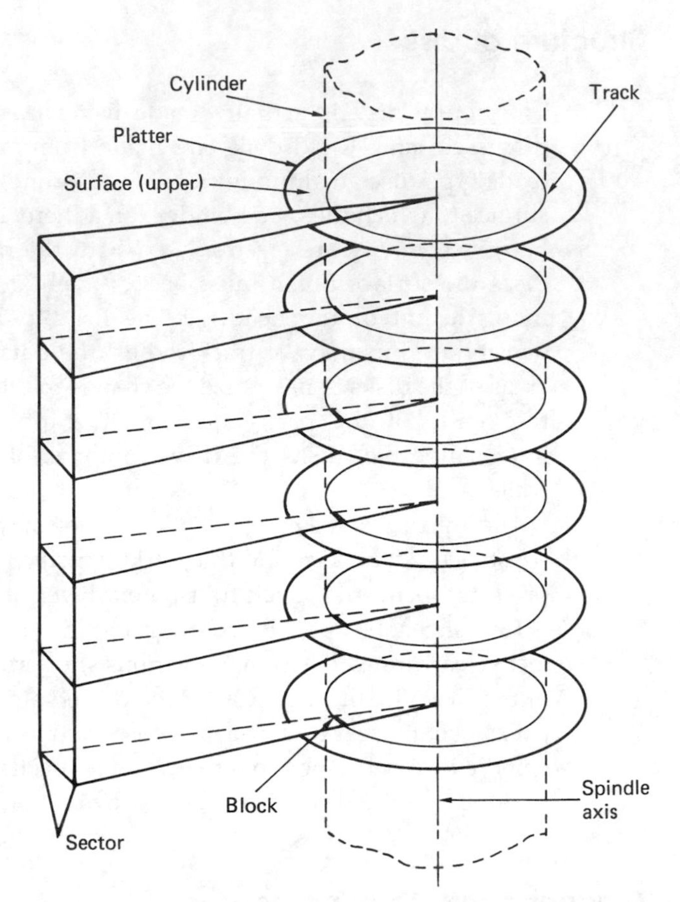

Figure 12.3 Disk terminology. Surface: one side of a platter. Track: path described on a surface by a fixed head. Cylinder: imaginary shape intersecting all surfaces at tracks of the same radius. Sector: angular subdivision of pack. Block: that part of a track within one sector. Each block has a unique cylinder, head and sector address.

available. The air gap must be accurately maintained, and consequently the head is of low mass and is mounted flexibly.

The aerodynamic part of the head is known as the slipper; it is designed to provide lift from the boundary layer which changes rapidly with changes in flying height. It is not initially obvious that the difficulty with disk heads is not making them fly, but making them fly close enough to the disk surface. The boundary layer travelling at the disk surface has the same speed as the disk, but as height increases, it slows down due to drag from the surrounding air. As the lift is a function of relative air speed, the closer the slipper comes to the disk, the greater the lift will be. The slipper is therefore mounted at the end of a rigid cantilever sprung towards the medium. The force with which the head is pressed towards the disk by the spring is equal to the lift at the designed flying height. Because of the spring, the head may rise and fall over small warps in the disk.

It would be virtually impossible to manufacture disks flat enough to dispense with this feature. As the slipper negotiates a warp it will pitch and roll in addition to rising and falling, but it must be prevented from yawing, as this would cause an azimuth error. Downthrust is applied to the aerodynamic centre by a spherical thrust button, and the required degrees of freedom are supplied by a thin flexible gimbal. The slipper has to bleed away surplus air in order to approach close enough to the disk, and holes or grooves are usually provided for this purpose in the same way that pinch rollers on some tape decks have grooves to prevent tape slip.

In exchangeable-pack drives, there will be a ramp on the side of the cantilever which engages a fixed block when the heads are retracted in order to lift them away from the disk surface.

12.5 Magnetic reading and writing

Figure 12.4 shows how disk heads are made. The magnetic circuit of disk heads was originally assembled from discrete magnetic elements. As the gap and flying height became smaller to increase linear recording density, the slipper was made from ferrite, and became part of the magnetic circuit. This was completed by a small C-shaped ferrite piece which carried the coil. Ferrite heads were restricted in the coercivity of the disk they could write without saturating. In thin-film heads, the magnetic circuit and coil are both formed by deposition on a substrate which becomes the rear of the slipper.

In a moving-head device it is not practicable to position separate erase, record and playback heads accurately. Erase is by overwriting, and reading and writing are carried out by the same head. The presence of the air film causes severe separation loss, and peak shift distortion is a major problem. The flying height of the head varies with the radius of the disk track, and it is difficult to provide accurate equalization of the replay channel because of this. The write current is often controlled as a function of track radius so that the changing reluctance of the air gap does not change the resulting record flux. Automatic gain control (AGC) is used on replay to compensate for changes in signal amplitude from the head.

Equalization may be used on recording in the form of precompensation, which moves recorded transitions in such a way as to oppose the effects of peak shift in addition to any replay equalization used. This was discussed in Chapter 5, which also introduced digital channel coding.

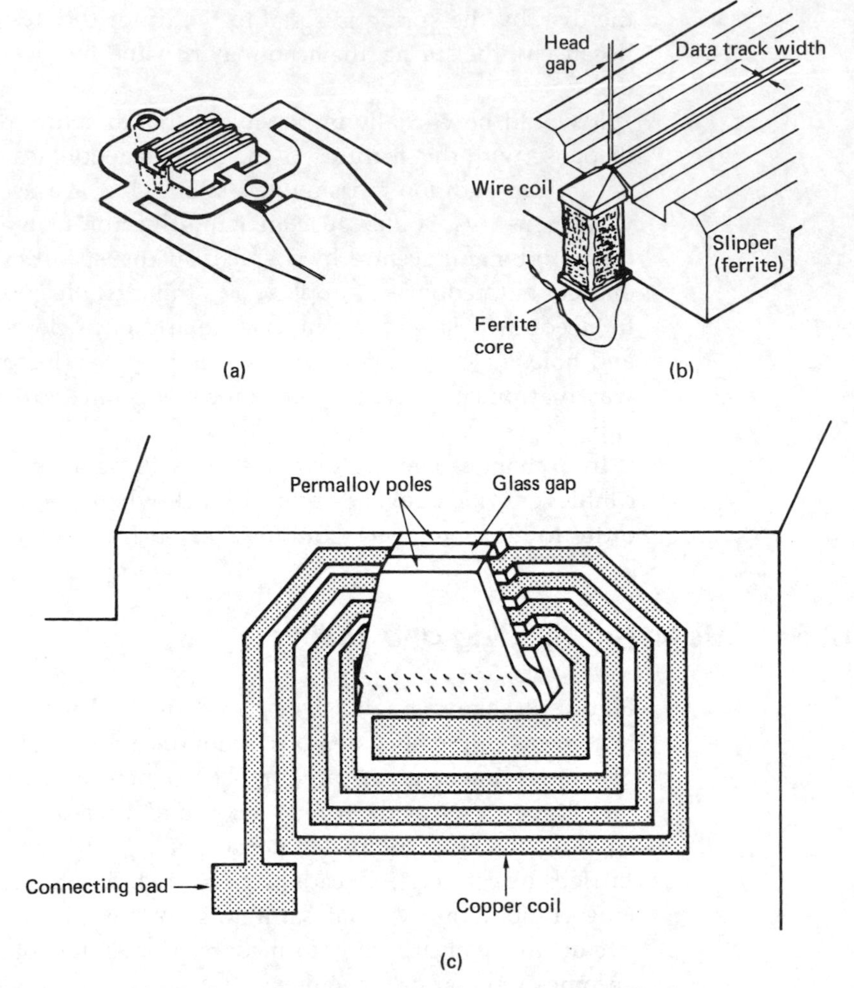

Figure 12.4 (a) Winchester head construction showing large air bleed grooves. (b) Close-up of slipper showing magnetic circuit on trailing edge. (c) Thin film head is fabricated on the end of the slipper using microcircuit technology.

Early disks used FM coding, which was easy to decode, but had a poor density ratio. The invention of MFM revolutionized hard disks, and was at one time universal. Further progress led to run-length-limited codes such as 2/3 and 2/7 which had a high density ratio without sacrificing the large jitter window necessary to reject peak shift distortion. Partial response is also becoming increasingly common in disks.

Typical drives have several heads, but with the exception of special-purpose parallel-transfer machines, only one head will be active at any one time, which means that the read and write circuitry can be shared between the heads. Figure 12.5 shows that in one approach the centre-tapped heads

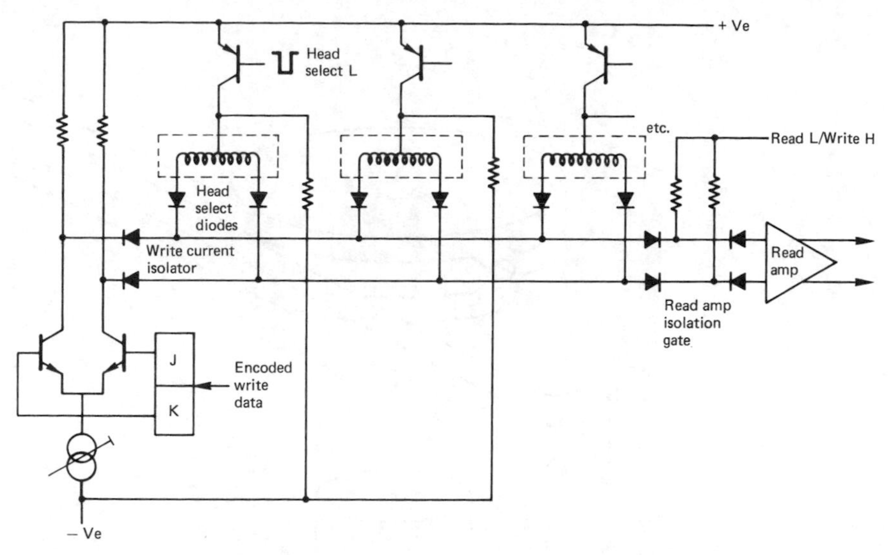

Figure 12.5 Representative head matrix.

are isolated by connecting the centre tap to a negative voltage, which reverse-biases the matrix diodes. The centre tap of the selected head is made positive. When reading, a small current flows through both halves of the head winding, as the diodes are forward-biased. Opposing currents in the head cancel, but read signals due to transitions on the medium can pass through the forward-biased diodes to become differential signals on the matrix bus. During writing, the current from the write generator passes alternately through the two halves of the head coil. Further isolation is necessary to prevent the write-current-induced voltages from destroying the read preamplifier input. Alternatively, FET analog switches may be used for head selection.

The read channel usually incorporates AGC, which will be overridden by the control logic between data blocks in order to search for address marks, which are short unmodulated areas of track. As a block preamble is entered, the AGC will be enabled to allow a rapid gain adjustment.

The high bit rates of disk drives, due to the speed of the medium, mean that peak detection in the replay channel is usually by differentiation. The detected peaks are then fed to the data separator.

12.6 Moving the heads

The servo system required to move the heads rapidly between tracks, and yet hold them in place accurately for data transfer, is a fascinating and complex piece of engineering. In exchangeable pack drives, the disk

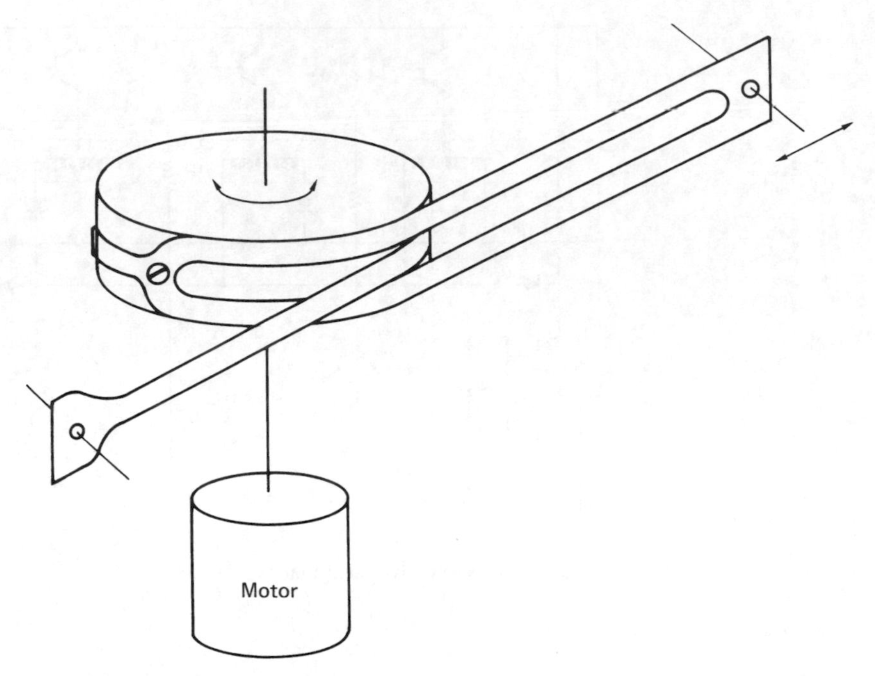

Figure 12.6 A low-cost linear positioner can be obtained using a drum and split-band drive, shown here, or with flexible wire. The ends of the band are fixed to the carriage.

positioner moves on a straight axis which passes through the spindle. The head carriage will usually have preloaded ball races which run on rails mounted on the bed of the machine, although some drives use plain sintered bushes sliding on polished rods.

Motive power on early disk drives was hydraulic, but this soon gave way to moving-coil drive, because of the small moving mass which this technique permits. Lower-cost units use a conventional electric motor as shown in Figure 12.6 which drives the carriage through steel wires wound around it, or via a split metal band which is shaped to allow both ends to be fixed to the carriage despite the centre making a full turn around the motor shaft. The final possibility is a coarse-threaded shaft or leadscrew which engages with a nut on the carriage. In very low-cost drives, the motor will be a stepping motor, and the positions of the tracks will be determined by the natural detents of the stepping motor. This has an advantage for portable drives, because a stepping motor will remain detented without power. Moving-coil actuators require power to stay on-track.

When a drive is track-following, it is said to be detented, in fine mode or in linear mode depending on the manufacturer. When a drive is seeking from one track to another, it can be described as being in coarse mode or velocity mode. These are the two major operating modes of the servo.

With the exception of stepping-motor-driven carriages, the servo system needs positional feedback of some kind. The purpose of the feedback will be one or more of the following:

1 To count the number of cylinders crossed during a seek
2 To generate a signal proportional to carriage velocity
3 To generate a position error proportional to the distance from the centre of the desired track

Magnetic and optical drives obtain these feedback signals in different ways, which will be discussed individually. Many drives incorporate a tacho which may be a magnetic moving-coil type or its complementary equivalent the moving-magnet type. Both generate a voltage proportional to velocity, and can give no positional information.

A seek is a process where the positioner moves from one cylinder to another. The speed with which a seek can be completed is a major factor in determining the access time of the drive. The main parameter controlling the carriage during a seek is the cylinder difference, which is obtained by subtracting the current cylinder address from the desired cylinder address. The cylinder difference will be a signed binary number representing the number of cylinders to be crossed to reach the target, direction being indicated by the sign. The cylinder difference is loaded into a counter which is decremented each time a cylinder is crossed. The counter drives a DAC which generates an analog voltage proportional to the cylinder difference. As Figure 12.7 shows, this voltage, known as the

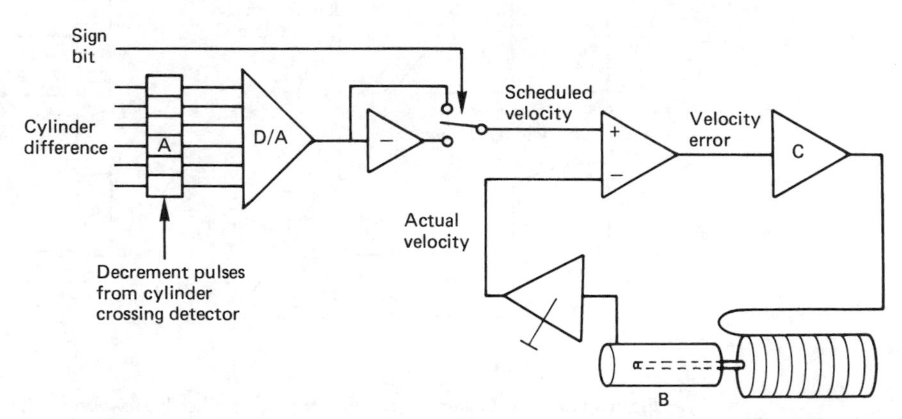

Figure 12.7 Control of carriage velocity by cylinder difference. The cylinder difference is loaded into the difference counter A. A digital-to-analog convertor generates an analog voltage from the cylinder difference, known as the scheduled velocity. This is compared with the actual velocity from the transducer B in order to generate the velocity error which drives the servo amplifier C.

scheduled velocity, is compared with the output of the carriage-velocity tacho. Any difference between the two results in a velocity error which drives the carriage to cancel the error. As the carriage approaches the target cylinder, the cylinder difference becomes smaller, with the result that the run-in to the target is critically damped to eliminate overshoot.

Figure 12.8(a) shows graphs of scheduled velocity, actual velocity and motor current with respect to cylinder difference during a seek. In the first half of the seek, the actual velocity is less than the scheduled velocity, causing a large velocity error which saturates the amplifier and provides maximum carriage acceleration. In the second half of the graphs, the scheduled velocity is falling below the actual velocity, generating a negative velocity error which drives a reverse current through the motor to slow down the carriage. The scheduled deceleration slope can clearly not be steeper than the saturated acceleration slope. Areas A and B on the graph will be about equal, as the kinetic energy put into the carriage has to be taken out.

The current through the motor is continuous, and would result in a heating problem, so to counter this, the DAC is made non-linear so that above a certain cylinder difference no increase in scheduled velocity will occur. This results in the graph of Figure 12.8(b). The actual velocity graph is called a velocity profile. It consists of three regions: acceleration,

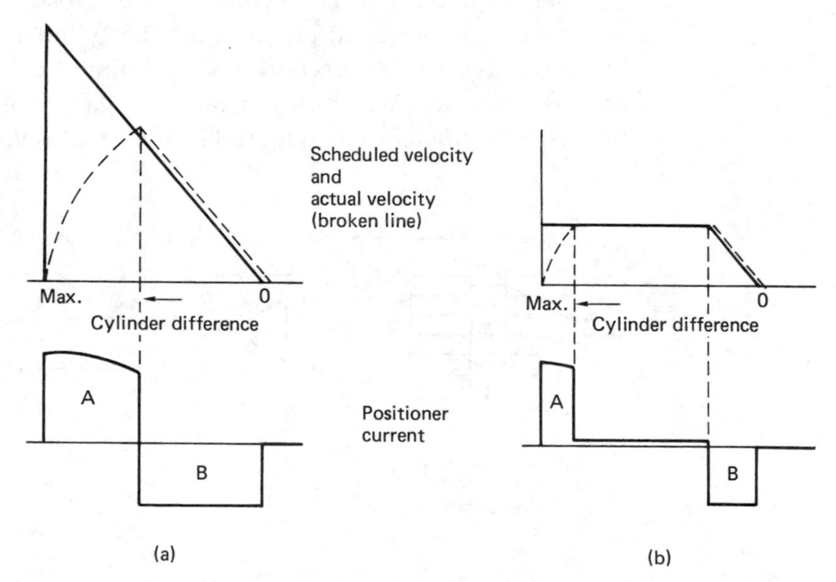

Figure 12.8 In the simple arrangement at (a) the dissipation in the positioner is continuous, causing a heating problem. The effect of limiting the scheduled velocity above a certain cylinder difference is apparent in (b) where heavy positioner current only flows during acceleration and deceleration. During the plateau of the velocity profile, only enough current to overcome friction is necessary. The curvature of the acceleration slope is due to the back EMF of the positioner motor.

where the system is saturated; a constant velocity plateau, where the only power needed is to overcome friction; and the scheduled run-in to the desired cylinder. Dissipation is only significant in the first and last regions.

A consequence of the critically damped run-in to the target cylinder is that short seeks are slow. Sometimes further non-linearity is introduced into the velocity scheduler to speed up short seeks. The velocity profile becomes a piecewise linear approximation to a curve by using non-linear feedback. Figure 12.9 shows the principle of an analog shaper or profile generator. Later machines will compute the curve in microprocessor sofware or use a PROM lookup table.

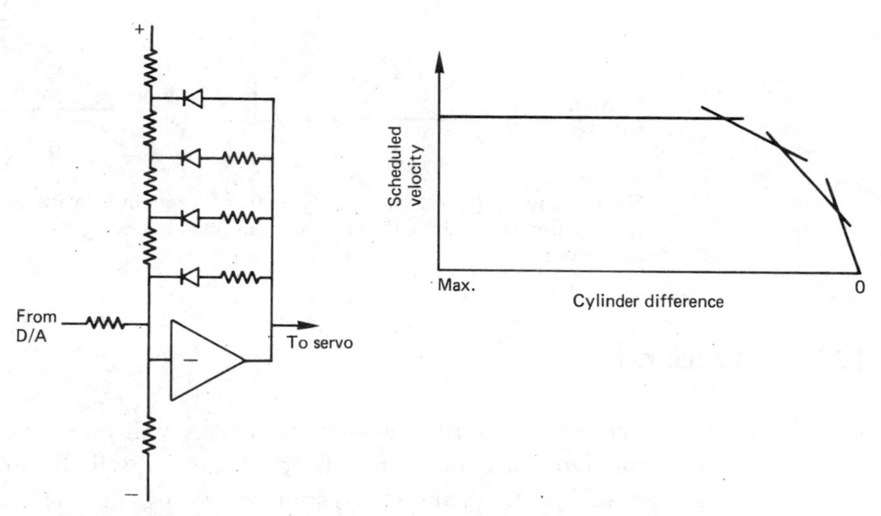

Figure 12.9 The use of voltage-dependent feedback around an operational amplifier permits a piecewise linear approximation to a curved velocity profile. This has the effect of speeding up short seeks without causing a dissipation problem on long seeks. The circuit is referred to as a shaper.

In small disk drives the amplifier may be linear in all modes of operation, resembling an audio power amplifier. Larger units may employ pulse-width-modulated drive to reduce dissipation, or even switched-mode amplifiers with inductive flywheel circuits. These switching systems can generate appreciable electromagnetic radiation, but this is of no consequence as they are only active during a seek. In track-following mode, the amplifier reverts to linear mode; hence the use of the term linear to mean track-following mode.

The input of the servo amplifier normally has a number of analog switches which select the appropriate drive and feedback signals according to the mode of the servo. A typical system is shown in Figure 12.10.

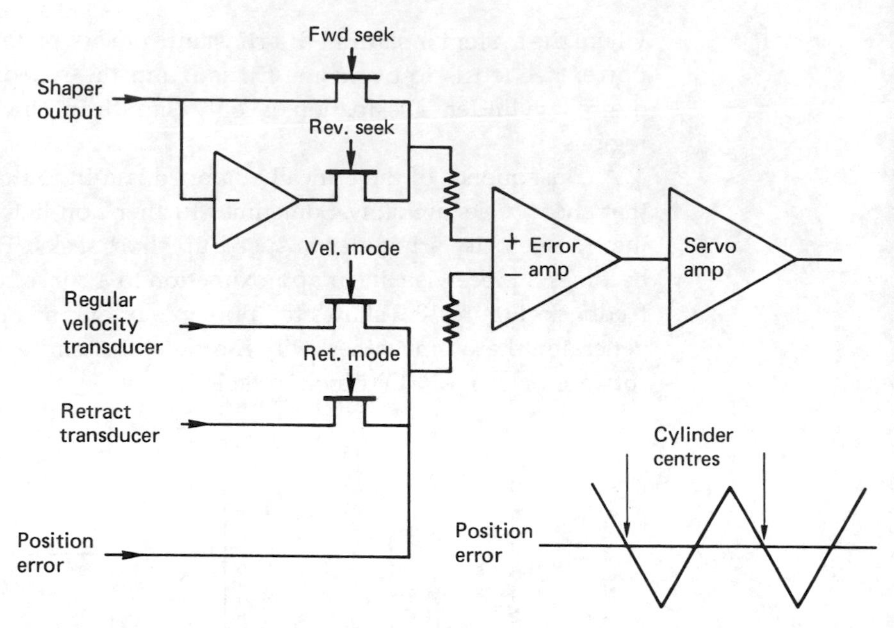

Figure 12.10 A typical servo-amplifier input stage. In velocity mode the shaper and the velocity transducer drive the error amp. In track-following mode the position error is the only input.

12.7 Rotation

The rotation subsystems of disk drives will now be covered. The track-following accuracy of a drive positioner will be impaired if there is bearing runout, and so the spindle bearings are made to a high degree of precision. Most modern drives incorporate brushless DC motors with integral speed control. In exchangeable-pack drives, some form of braking is usually provided to slow down the pack for convenient removal.

In order to control reading and writing, the drive control circuitry needs to know which cylinder the heads are on, and which sector is currently under the head. Sector information used to be obtained from a sensor which detects holes or slots cut in the hub of the disk. Modern drives will obtain this information from the disk surface, as will be seen. The result is that a sector counter in the control logic remains in step with the physical rotation of the disk. The desired sector address is loaded into a register, which is compared with the sector counter. When the two match, the desired sector has been found. This process is referred to as a search, and usually takes place after a seek. Having found the correct physical place on the disk, the next step is to read the header associated with the data block to confirm that the disk address contained there is the same as the desired address.

Rotation of a disk pack at speed results in heat build-up through air resistance. This heat must be carried away. A further important factor with exchangeable pack magnetic drives is to keep the disk area free from contaminants which might lodge between the head and the disk and cause the destructive phenomenon known as a head crash, where debris builds up on the head until it ploughs the disk surface. Optical drives have a larger separation betwen the pickup and the disk, but contamination can block the light beam and cause errors.

12.8 Servo-surface disks

One of the major problems to be overcome in the development of high-density disk drives was that of keeping the heads on track despite changes of temperature. The very narrow tracks used in digital recording have similar dimensions to the amount a disk will expand as it warms up. The cantilevers and the drive base all expand and contract, conspiring with thermal drift in the cylinder transducer to limit track pitch. The breakthrough in disk density came with the introduction of the servo-surface drive. The position error in a servo-surface drive is derived from a head reading the disk itself. This virtually eliminates thermal effects on head positioning and allows great increases in storage density.

In a multiplatter drive, one surface of the pack holds servo information which is read by the servo head. In a ten-platter pack this means that 5 per cent of the medium area is lost, but this is unimportant since the increase in density allowed is enormous. Using one side of a single-platter cartridge for servo information would be unacceptable as it represents 50 per cent of the medium area, so in this case the servo information can be interleaved with sectors on the data surfaces. This is known as an embedded-servo technique. These two approaches are contrasted in Figure 12.11.

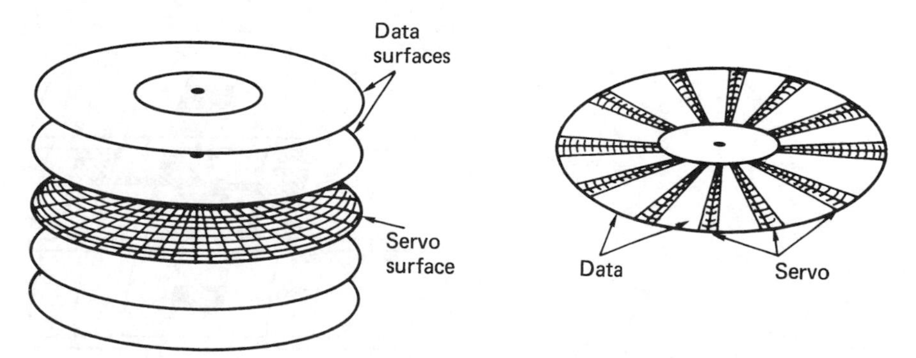

Figure 12.11 In a multiplatter disk pack, one surface is dedicated to servo information. In a single platter, the servo information is embedded in the data on the same surfaces.

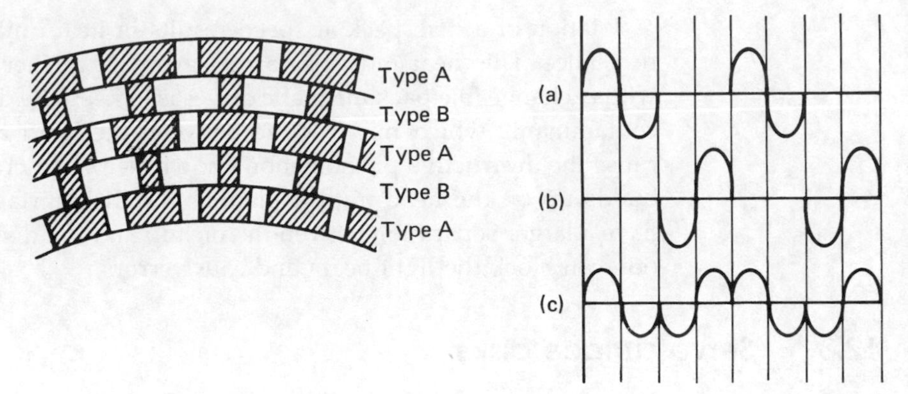

Figure 12.12 The servo surface is divided into two types of track, A and B, which are out of phase by 180° and are recorded with reverse polarity with respect to one another. Waveform (a) results when the servo head is entirely above a type A track, and waveform (b) results from reading solely a type B track. When the servo head is correctly positioned with one-half of its magnetic circuit over each track, the waveform of (c) results.

The servo surface is written at the time of disk pack manufacture, and the disk drive can only read it. Writing the servo surface has nothing to do with disk formatting, which affects the data storage areas only.

The key to the operation of the servo surface is the special magnetic pattern recorded on it. In a typical servo surface, recorded pairs of transitions, known as dibits, are separated by a space. Figure 12.12 shows that there are two kinds of track. On an A track, the first transition of the pair will cause a positive pulse on reading, whereas on a B track, the first pulse will be negative. In addition, the A-track dibits are shifted by one half-cycle with respect to the B-track dibits. The width of the magnetic circuit in the servo head is equal to the width of a servo track. During track following, the correct position for the servo head is with half of each type of track beneath it. The read/write heads will then be centred on their respective data tracks. Figure 12.13 illustrates this relationship.

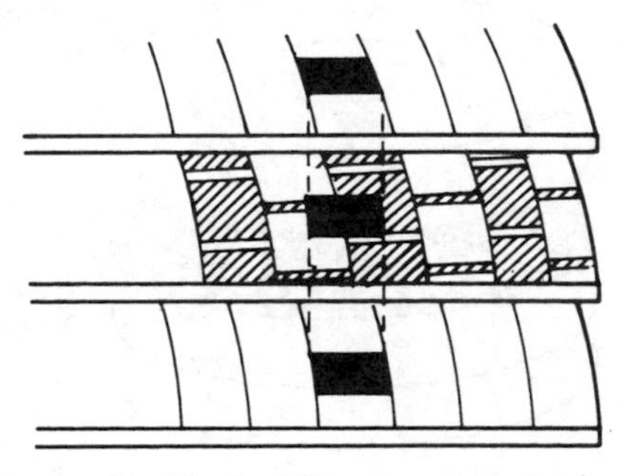

Figure 12.13 When the servo head is straddling two servo tracks, the data heads are correctly aligned with their respective tracks.

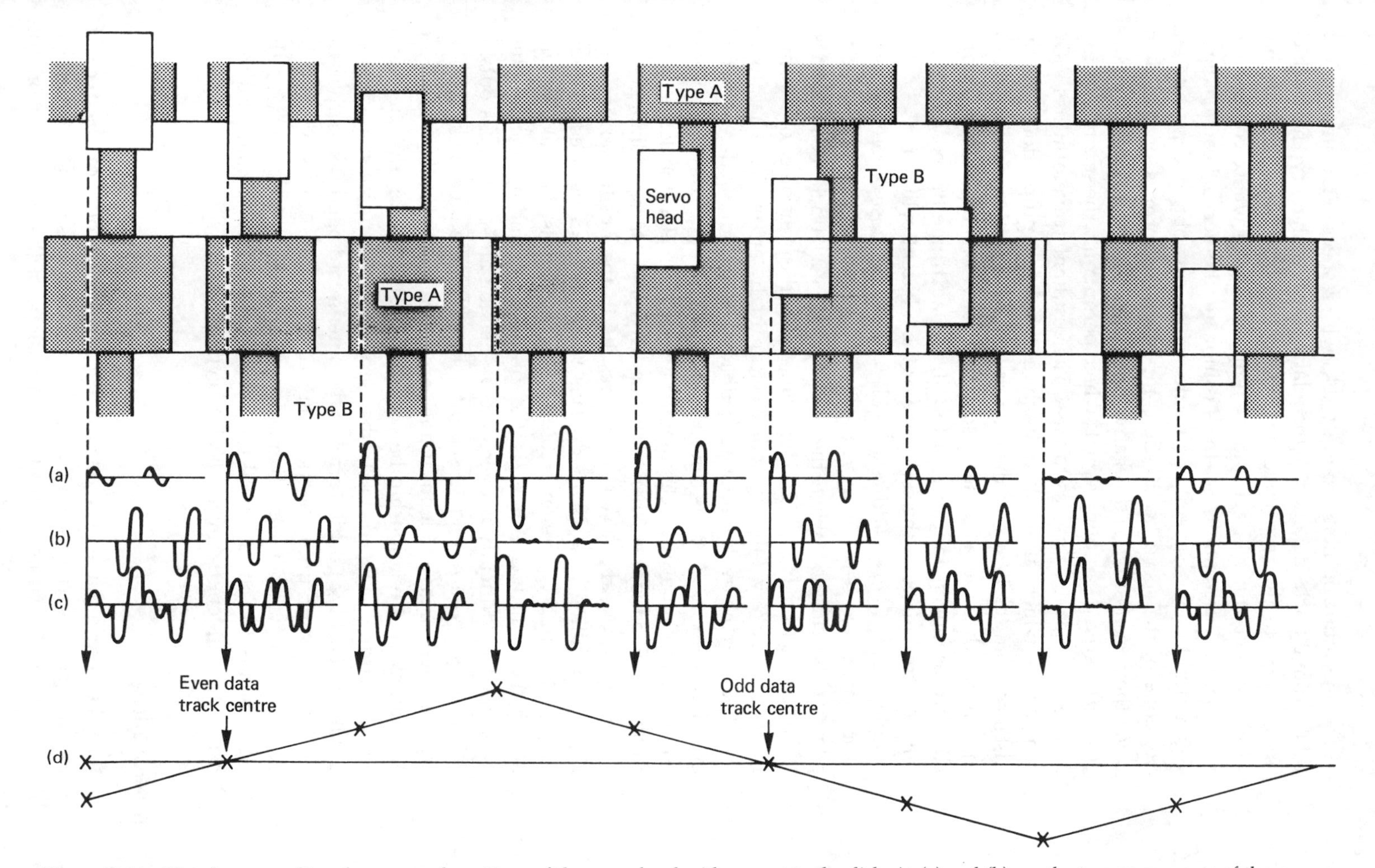

Figure 12.14 Waveforms resulting from several positions of the servo head with respect to the disk. At (a) and (b) are the two components of the waveforms, whose relative amplitudes are controlled by the relative areas of the servo head exposed to the two types of servo track. Because the servo head has only one magnetic circuit, these waveforms are not observed in practice, but are summed together, resulting in the composite waveforms shown at (c). By comparing the magnitudes of the second positive and second negative peaks in the composite waveforms, a position error signal is generated, as shown at (d).

The amplitude of dibits from A tracks with respect to the amplitude of dibits from B tracks depends on the relative areas of the servo head which are exposed to the respective tracks. As the servo head has only one magnetic circuit, it will generate a composite signal whose components will change differentially as the position of the servo head changes. Figure 12.14 shows several composite waveforms obtained at different positions of the servo head. The composite waveform is processed by using the first positive and negative pulses to generate a clock. From this clock are derived sampling signals which permit only the second positive and second negative pulses to pass. The resultant waveform has a DC component which after filtering gives a voltage proportional to the distance from the centre of the data tracks. The position error reaches a maximum when the servo head is entirely above one type of servo track, and further movement causes it to fall. The next time the position error falls to zero will be at the centreline of the adjacent cylinder.

Cylinders with even addresses (LSB = 0) will be those where the servo head is detented between an A track and a B track. Cylinders with odd addresses will be those where the head is between a B track and an A track. It can be seen from Figure 12.14 that the sense of the position error becomes reversed on every other cylinder. Accordingly, an inverter has to be switched into the track-following feedback loop in order to detent on odd cylinders. This inversion is controlled by the LSB of the desired cylinder address supplied at the beginning of a seek, such that the sense of the feedback will be correct when the heads arrive at the target cylinder.

Seeking across the servo surface results in the position-error signal rising and falling in a sawtooth. This waveform can be used to count down the cylinder difference counter which controls the seek. As with any cyclic transducer there is the problem of finding the absolute position. This difficulty is overcome by making all servo tracks outside cylinder zero type A, and all servo tracks inside the innermost cylinder type B. These areas of identical track are called guard bands, and Figure 12.15 shows the relationship between the position error and the guard bands. During a head load, the servo head generates a constant maximum positive position error in the outer guard band. This drives the carriage forward until the position error first falls to zero. This, by definition, is cylinder zero. Some drives, however, load by driving the heads across the surface until the inner guard band is found, and then perform a full-length reverse seek to cylinder zero.

12.9 Soft sectoring

It has been seen that a position error and a cylinder count can be derived from the servo surface, eliminating the cylinder transducer. The carriage velocity could also be derived from the slope of the position error, but

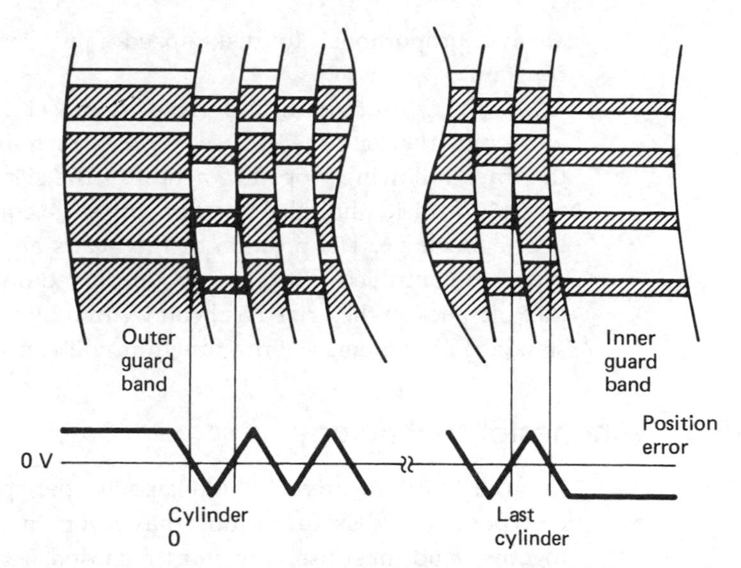

Figure 12.15 The working area of the servo surface is defined by the inner and outer guard bands, in which the position error reaches its maximum value.

there would then be no velocity feedback in the guard bands or during retraction, and so some form of velocity transducer is still necessary.

As there are exactly the same number of dibits or tribits on every track, it is possible to describe the rotational position of the disk simply by counting them. All that is needed is an unique pattern of missing dibits once per revolution to act as an index point, and the sector transducer can also be eliminated.

Unlike the read-data circuits, the servo-head circuits are active during a seek as well as when track-following, and have to be protected against interference from switched-mode positioner drivers. The main problem is detecting index, where noise could cause a 'missing' dibit to be masked. There are two solutions available: a preamplifier can be built into the servo-head cantilever, or driver switching can be inhibited when index is expected.

The advantage of deriving the sector count from the servo surface is that the number of sectors on the disk can be varied. Any number of sectors can be accommodated by feeding the dibit-rate signal through a programmable divider, so the same disk and drive can be used in numerous different applications.

In a non-servo-surface disk, the write clock is usually derived from a crystal oscillator. As the disk speed can vary due to supply fluctuations, a tolerance gap has to be left at the end of each block to cater for the highest anticipated speed, to prevent overrun into the next block on a write. In a servo-surface drive, the write clock is obtained by multiplying the dibit-rate signal with a phase-locked loop. The write clock is then always proportional to disk speed, and recording density will be constant.

Most servo-surface drives have an offset facility, where a register written by the controller drives a DAC which injects a small voltage into the track-following loop. The action of the servo is such that the heads move off-track until the position error is equal and opposite to the injected voltage. The position of the heads above the track can thus be program-controlled. Offset is only employed on reading if it is suspected that the pack in the drive has been written by a different drive with non-standard alignment. A write function will cancel the offset.

12.10 Winchester technology

In order to offer extremely high capacity per spindle, which reduces the cost per bit, a disk drive must have very narrow tracks placed close together, and must use very short recorded wavelengths, which implies that the flying height of the heads must be small. The so-called Winchester technology is one approach to high storage density. The technology was developed by IBM, and the name came about because the model number of the development drive was the same as that of the famous rifle.

Reduction in flying height magnifies the problem of providing a contaminant-free environment. A conventional disk is well protected whilst inside the drive, but outside the drive the effects of contamination become intolerable. In exchangeable-pack drives, there is a real limit to the track pitch that can be achieved because of the difficulty or cost of engineering head-alignment mechanisms to make the necessary minute adjustments to give interchange compatibility.

The essence of Winchester technology is that each disk pack has its own set of read/write and servo heads, with an integral positioner. The whole is protected by a dust-free enclosure, and the unit is referred to as a head disk assembly, or HDA.

As the HDA contains its own heads, compatibility problems do not exist, and no head alignment is necessary or provided for. It is thus possible to reduce track pitch considerably compared with exchangeable pack drives. The sealed environment ensures complete cleanliness which permits a reduction in flying height without loss of reliability, and hence leads to an increased linear density. If the rotational speed is maintained, this can also result in an increase in data transfer rate.

The HDA is completely sealed, but some have a small filtered port to equalize pressure. Into this sealed volume of air, the drive motor delivers the majority of its power output. The resulting heat is dissipated by fins on the HDA casing. Some HDAs are filled with helium which significantly reduces drag and heat build-up.

An exchangeable-pack drive must retract the heads to facilitate pack removal. With Winchester technology this is not necessary. An area of the disk surface is reserved as a landing strip for the heads. The disk surface is lubricated, and the heads are designed to withstand landing and take-off without damage. Winchester heads have very large air-bleed grooves to allow low flying height with a much smaller downthrust from the cantilever, and so they exert less force on the disk surface during contact. When the term *parking* is used in the context of Winchester technology, it refers to the positioning of the heads over the landing area.

Disk rotation must be started and stopped quickly to minimize the length of time the heads slide over the medium. A powerful motor will accelerate the pack quickly. Eddy-current braking cannot be used, since a power failure would allow the unbraked disk to stop only after a prolonged head-contact period. A failsafe mechanical brake is used, which is applied by a spring and released with a solenoid.

A major advantage of contact start/stop is that more than one head can be used on each surface if retraction is not needed. This leads to two gains: first, the travel of the positioner is reduced in proportion to the number of heads per surface, reducing access time; and, second, more data can be transferred at a given detented carriage position before a seek to the next cylinder becomes necessary. This increases the speed of long transfers. Figure 12.16 illustrates the relationships of the heads in such a system.

With contact start/stop, the servo head is always on the servo surface, and it can be used for all of the transducer functions needed

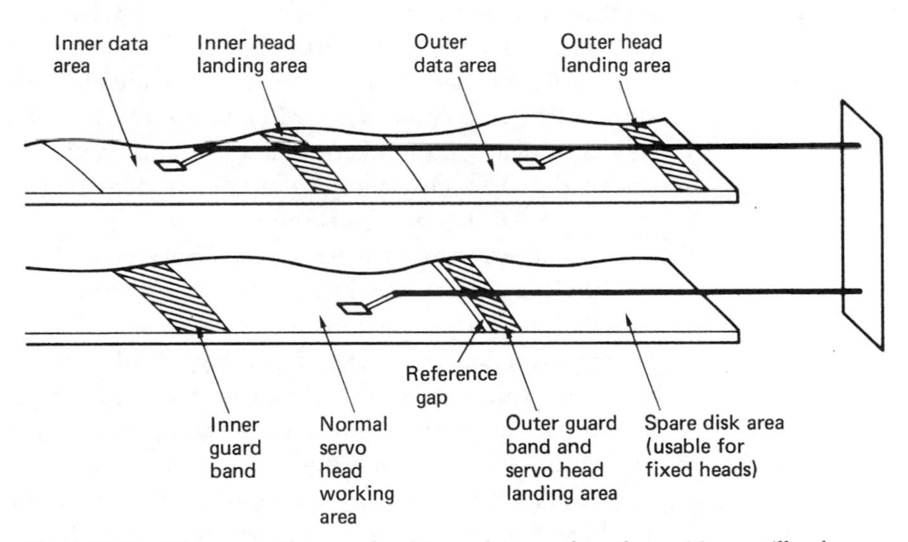

Figure 12.16 When more than one head is used per surface, the positioner still only requires one servo head. This is often arranged to be equidistant from the read/write heads for thermal stability.

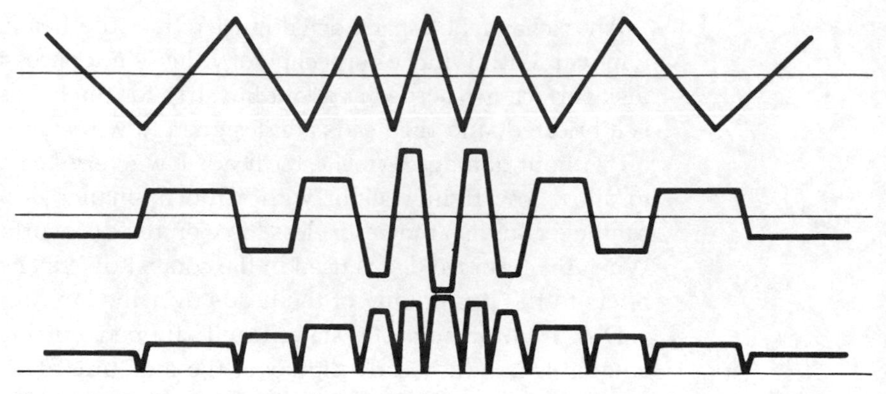

Figure 12.17 To generate a velocity signal, the position error from the servo head is differentiated and rectified.

by the drive. Figure 12.17 shows the position-error signal during a seek. The signal rises and falls as servo tracks are crossed, and the slope of the signal is proportional to positioner velocity. The position-error signal is differentiated and rectified to give a velocity feedback signal. Due to the cyclic nature of the position-error signal, the velocity signal derived from it has troughs where the derivative becomes zero at the peaks. These cannot be filtered out, as the signal is in a servo loop, and the filter would introduce an additional lag. The troughs would, however, be interpreted by the servo driver as massive momentary velocity errors which might overload the amplifier. The solution which can be adopted is to use a signal obtained by integrating the positioner-motor current which is selected when there is a trough in the differentiated position-error signal.

In order to make velocity feedback available over the entire servo surface, the conventional guard-band approach cannot be used since it results in steady position errors in the guard bands. In contact start/stop drives, the servo head must be capable of detenting in a guard band for the purpose of landing on shutdown.

A modification to the usual servo surface is used in Winchester drives, one implementation of which is shown in Figure 12.18, where it will be seen that there are extra transitions, identical in both types of track, along with the familiar dibits. The repeating set of transitions is known as a frame, in which the first dibit is used for synchronization, and a phase-locked oscillator is made to run at a multiple of the sync signal rate. The PLO is used as a reference for the write clock, as well as to generate sampling pulses to extract a position error from the composite waveform and to provide a window for the second dibit in the frame, which may or may not be present. Each frame thus contains one data bit, and succesive frames are read to build up a pattern in a shift register.

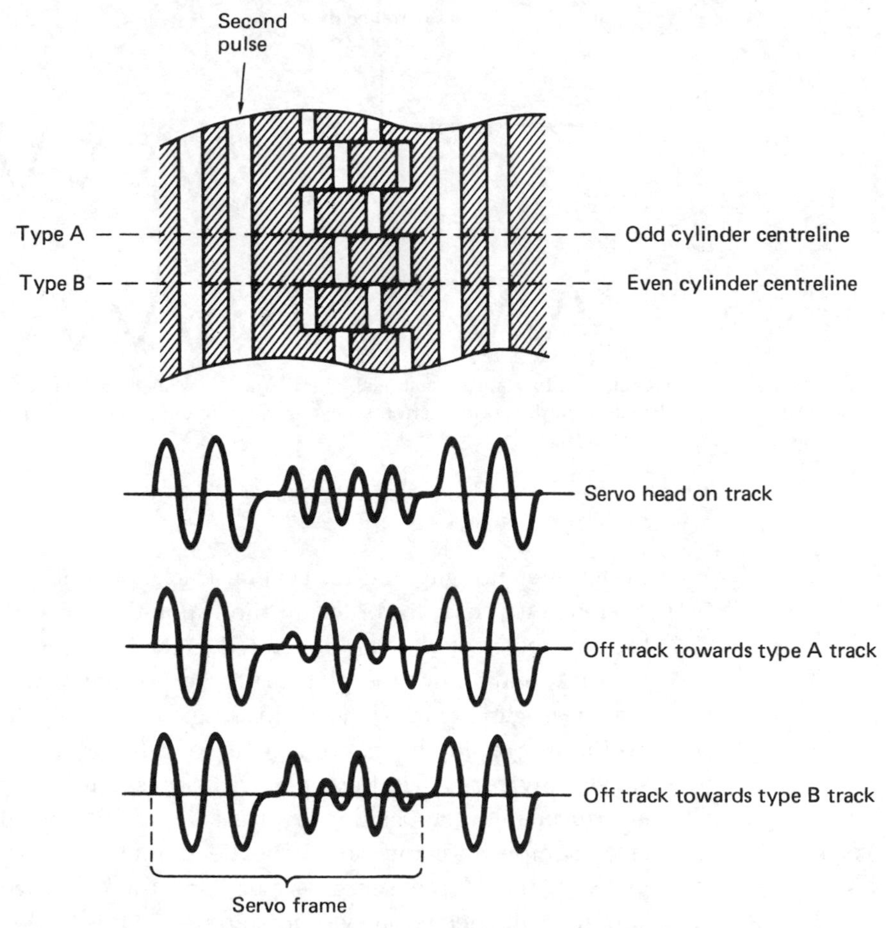

Figure 12.18 This type of servo-surface pattern has a second pulse which may be omitted to act as a data bit. This is used to detect the guard bands and index.

The parallel output of the shift register is examined by a decoder which recognizes a number of unique patterns. In the guard bands, the decoder will repeatedly recognize the guard-band code as the disk revolves. An index is generated in the same way, by recognizing a different pattern. In a contact start/stop drive, the frequency of index detection is used to monitor pack speed in order to dispense with a separate transducer. This does mean, however, that it must be possible to detect index everywhere, and for this reason, index is still recorded in the guard bands by replacing the guard-band code with index code once per revolution.

A consequence of deriving velocity information from the servo surface is that the location of cylinder zero is made more difficult, as there is no longer a continuous maximum position error in the guard band. A common solution is to adopt a much smaller area of continuous position error known as a reference gap; this is typically three servo tracks wide.

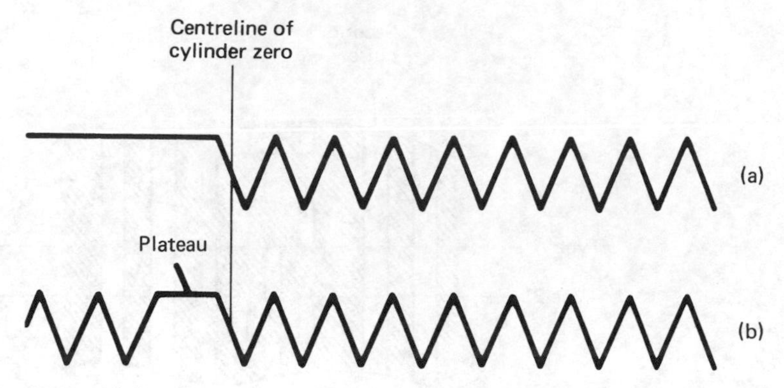

Figure 12.19 (a) Conventional guard band. (b) Winchester guard band, showing the plateau in the position error, known as the reference gap, which is used to locate cylinder zero.

In the reference gap and for several tracks outside it, there is an unique reference gap code recorded in the frame-data bits. Figure 12.19 shows the position error which is generated as the positioner crosses this area of the disk, and shows the plateau in the position-error signal due to the reference gap. During head loading, which in this context means positioning from the parking area to cylinder zero, the heads move slowly inwards. When the reference code is detected, positioner velocity is reduced, and the position error is sampled. When successive position-error samples are the same, the head must be on the position-error plateau, and if the servo is put into track-following mode, it will automatically detent on cylinder zero, since this is the first place that the position error falls to zero.

12.11 Rotary positioners

Figure 12.20 shows that rotary positioners are feasible in Winchester drives; they cannot be used in exchangeable-pack drives because of interchange problems. There are some advantages to a rotary positioner. It can be placed in the corner of a compact HDA allowing smaller overall size. The manufacturing cost will be less than a linear positioner because fewer bearings and precision bars are needed. Significantly, a rotary positioner can be made faster since its inertia is smaller. With a linear positioner all parts move at the same speed. In a rotary positioner, only the heads move at full speed, as the parts closer to the shaft must move more slowly. Figure 12.21 shows a typical HDA with a rotary positioner. The principle of many rotary positioners is exactly that of a moving-coil ammeter, where current is converted directly into torque. Alternatively

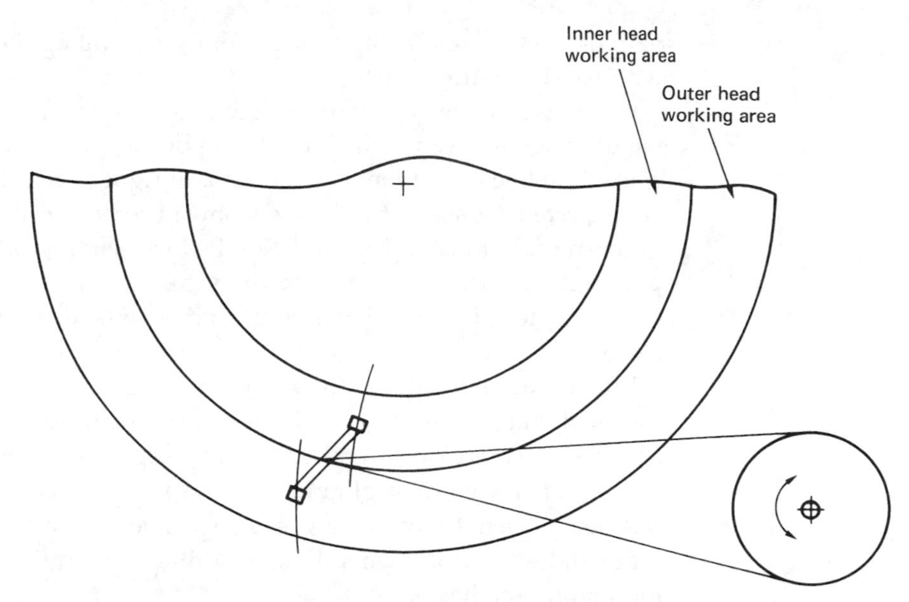

Figure 12.20 A rotary positioner with two heads per surface. The tolerances involved in the spacing between the heads and the axis of rotation mean that each arm records data in an unique position. Those data can only be read back by the same heads, which rules out the use of a rotary positioner in exchangeable-pack drives. In a head disk assembly the problem of compatibility does not arise.

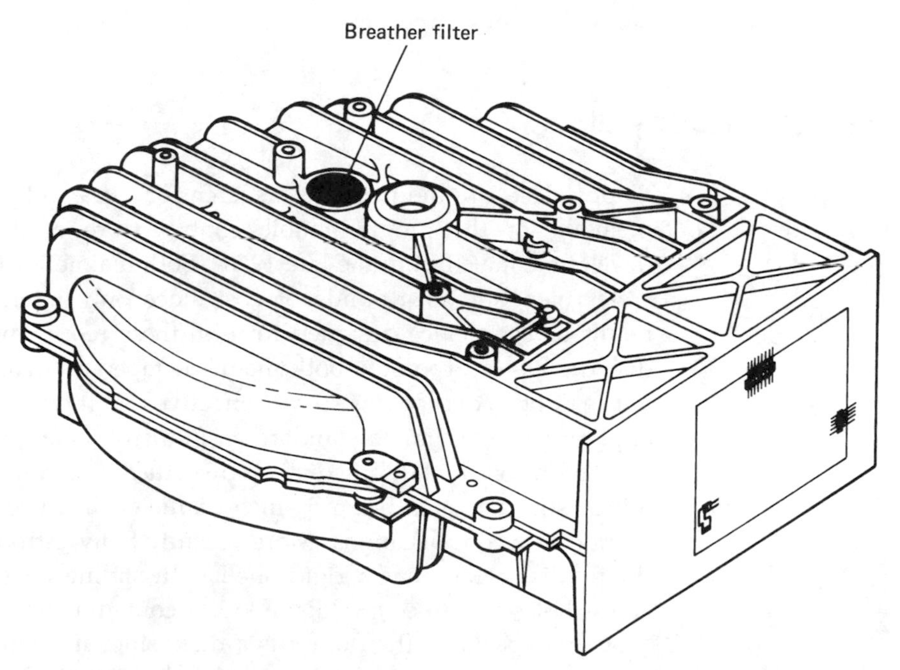

Figure 12.21 Head disk assembly with a rotary positioner. The adoption of this technique allows a very compact structure.

various configurations of electric motor or stepping motor can be used with band or wire drive.

One disadvantage of rotary positioners is that there is a component of windage on the heads which tends to pull the positioner in towards the spindle. In linear positioners windage is at right angles to motion and can be neglected. Windage can be overcome in rotary positioners by feeding the current cylinder address to a ROM which sends a code to a DAC. This produces an offset voltage which is fed to the positioner driver to generate a torque that balances the windage whatever the position of the heads.

When extremely small track spacing is contemplated, it cannot be assumed that all the heads will track the servo head due to temperature gradients. In this case the embedded-servo approach must be used, where each head has its own alignment patterns. The servo surface is often retained in such drives to allow coarse positioning, velocity feedback and index and write-clock generation, in addition to locating the guard bands for landing the heads.

Winchester drives have been made with massive capacity, but the problem of back-up is then magnified, and the general trend has been for the physical size of the drive to come down as the storage density increases in order to improve access time. Very small Winchester disk drives are now available which plug into standard integrated circuit sockets. These are competing with RAM for memory applications where non-volatility is important.

12.12 Floppy disks

Floppy disks are the result of a search for a fast yet cheap non-volatile memory for the programmable control store of a processor under development at IBM in the late 1960s. Both magnetic tape and hard disk were ruled out on grounds of cost since only intermittent duty was required. The device designed to fulfil these requirements – the floppy disk drive – incorporated both magnetic-tape and disk technologies.

The floppy concept was so cost-effective that it transcended its original application to become a standard in industry as an online data-storage device. The original floppy disk, or diskette as it is sometimes called, was 8 inches in diameter, but a $5\frac{1}{4}$-inch diameter disk was launched to suit more compact applications. More recently Sony introduced the $3\frac{1}{2}$-inch floppy disk which has a rigid shell with sliding covers over the head access holes to reduce the likelihood of contamination.

Strictly speaking, the floppy is a disk, since it rotates and repeatedly presents the data on any track to the heads, and it has a positioner to give fast two-dimensional access, but it also resembles a tape drive in that the

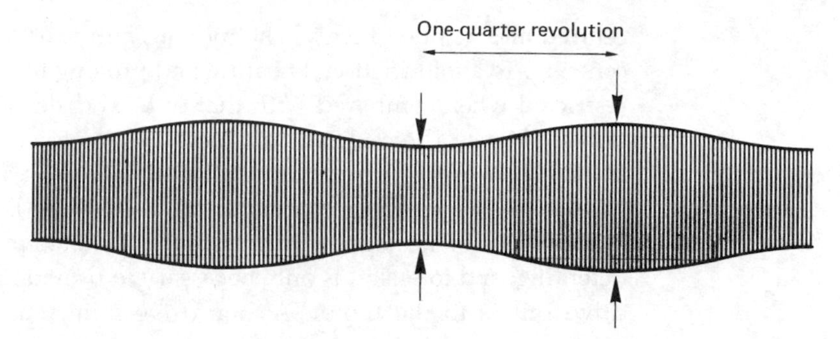

Figure 12.22 Sinusoidal amplitude modulation of floppy-disk output due to anisotropy of medium.

magnetic medium is carried on a flexible substrate which deforms when the read/write head is pressed against it.

Floppy disks are stamped from wide, thick tape, and are anisotropic, because the oxide becomes oriented during manufacture. On many disks this can be seen by the eye as parallel striations on the disk surface. A more serious symptom is the presence of sinusoidal amplitude modulation of the replay signal at twice the rotational frequency of the disk, as illustrated in Figure 12.22.

Floppy disks have radial apertures in their protective envelopes to allow access by the head. A further aperture allows a photoelectric index sensor to detect a small hole in the disk once per revolution.

Figure 12.23 shows that the disk is inserted into the drive edge-first, and slides between an upper and a lower hub assembly. One of these has a fixed bearing which transmits the drive; the other is spring-loaded and mates with the drive hub when the door is closed, causing the disk to be

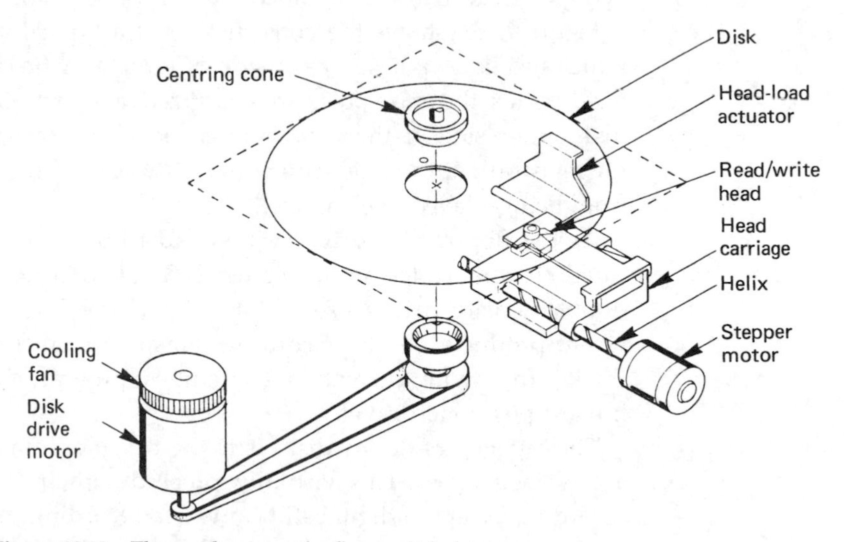

Figure 12.23 The mechanism of a floppy-disk drive.

centred and gripped firmly. The moving hub is usually tapered to assist centring. To avoid frictional heating and prolong life, the spindle speed is restricted when compared with that of hard disks. Recent drives almost univerally use direct-drive brushless DC motors. Since the rotational latency is so great, there is little point in providing a fast positioner, and the use of leadscrews driven by a stepping motor is universal.

The permanent magnets in the stepping motor provide the necessary detenting, and to seek it is only necessary to provide a suitable number of drive pulses to the motor. As the drive is incremental, some form of reference is needed to determine the position of cylinder zero. At the rearward limit of carriage travel, a light beam is interrupted which resets the cylinder count. Upon power-up, the drive has to reverse-seek until this limit is found in order to calibrate the positioner. The grinding noise this makes is a characteristic of most PCs on power-up.

One of the less endearing features of plastics materials is a lack of dimensional stability. Temperature and humidity changes affect plastics much more than metals. The effect on the anisotropic disk substrate is to distort the circular tracks into a shape resembling a dog bone. For this reason, the track width and pitch have to be generous.

The read/write head of a single-sided floppy disk operates on the lower surface only, and is rigidly fixed to the carriage. Contact with the medium is achieved with the help of a spring-loaded pressure pad applied to the top surface of the disk opposite the head. Early drives retracted the pressure pad with a solenoid when not actually transferring data; later drives simply stop the disk. In double-sided drives, the pressure pad is replaced by a second sprung head. Because of the indifferent stability of the medium, side trim or tunnel erasing is used, because it can withstand considerable misregistration.

Figure 12.24 shows the construction of a typical side-trimming head, which has erase poles at each side of the magnetic circuit. When such a head writes, the erase poles are energized, and erase a narrow strip of the disk either side of the new data track. If the recording is made with misregistration, the side-trim prevents traces of the previous recording from being played back as well.

As the floppy-disk drive is intended to be a low-cost item, sophisticated channel codes are never used. Single-density drives use FM and double-density drives use MFM, with a trend towards RLL codes in recent products. As the recording density becomes higher at the inner tracks, the write current is sometimes programmed to reduce with inward positioner travel.

The capacity of floppy disks is in the range of hundreds of kilobytes to a few megabytes. This virtually precludes their use for digital video recording except in digital still cameras. Recording time can be increased by the use of compression. Floppy disks find almost universal application

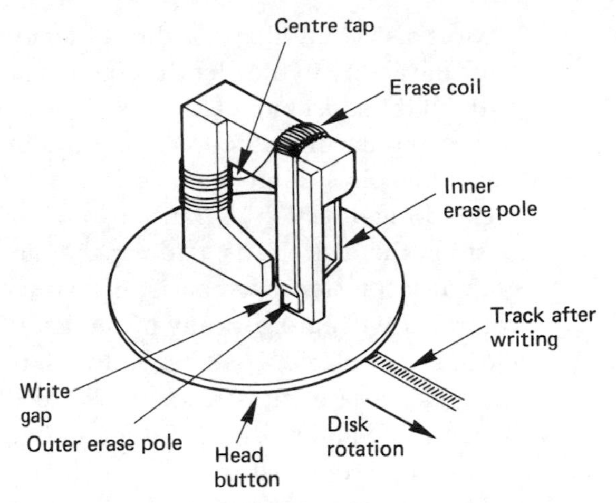

Figure 12.24 The poor dimensional stability of the plastic diskette means that tunnel erase or side trim has to be used. The extra erase poles can be seen here.

in edit-list storage, console set-up storage, and as a software-loading medium for computer-based equipment.

12.13 The disk controller

A disk controller is a unit which is interposed between the drives and the rest of the system. It consists of two main parts; that which issues control signals to and obtains status from the drives, and that which handles the data to be stored and retrieved. Both parts are synchronized by the control sequencer. The essentials of a disk controller are determined by the characteristics of drives and the functions needed, and so they do not vary greatly. It is desirable for economic reasons to use a commercially available disk controller intended for computers. Such controllers are adequate for still store applications, but cannot support the data rate required for real-time moving video unless data reduction is employed. Disk drives are generally built to interface to a standard controller interface, such as the SCSI bus. The disk controller will then be a unit which interfaces the drive bus to the host computer system.

The execution of a function by a disk subsystem requires a complex series of steps, and decisions must be made between the steps to decide what the next will be. There is a parallel with computation, where the function is the equivalent of an instruction, and the sequencer steps needed are the equivalent of the microinstructions needed to execute the instruction. The major failing in this analogy is that the sequence in a disk drive must be accurately synchronized to the rotation of the disk.

Most disk controllers use direct memory access, which means that they have the ability to transfer disk data in and out of the associated memory without the assistance of the processor. In order to cause a file transfer, the disk controller must be told the physical disk address (cylinder, sector, track), the physical memory address where the file begins, the size of the file and the direction of transfer (read or write). The controller will then position the disk heads, address the memory, and transfer the samples. One disk transfer may consist of many contiguous disk blocks, and the controller will automatically increment the disk-address registers as each block is completed. As the disk turns, the sector address increases until the end of the track is reached. The track or head address will then be incremented and the sector address reset so that transfer continues at the beginning of the next track.

This process continues until all the heads have been used in turn. In this case both the head address and sector address will be reset, and the cylinder address will be incremented, which causes a seek. A seek which takes place because of a data transfer is called an implied seek, because it is not necessary formally to instruct the system to perform it. As disk drives are block-structured devices, and the error correction is codeword-based, the controller will always complete a block even if the size of the file is less than a whole number of blocks. This is done by packing the last block with zeros.

The status system allows the controller to find out about the operation of the drive, both as a feedback mechanism for the control process and to handle any errors. Upon completion of a function, it is the status system which interrupts the control processor to tell it that another function can be undertaken.

In a system where there are several drives connected to the controller via a common bus, it is possible for non-data-transfer functions such as seeks to take place in some drives simultaneously with a data transfer in another.

Before a data transfer can take place, the selected drive must physically access the desired block, and confirm this by reading the block header. Following a seek to the required cylinder, the positioner will confirm that the heads are on-track and settled. The desired head will be selected, and then a search for the correct sector begins. This is done by comparing the desired sector with the current sector register, which is typically incremented by dividing down servo-surface pulses. When the two counts are equal, the head is about to enter the desired block.

Figure 12.25 shows the structure of a typical magnetic disk track. In between blocks are placed address marks, which are areas without transitions which the read circuits can detect. Following detection of the address mark, the sequencer is roughly synchronized to begin handling the block. As the block is entered, the data separator locks to the

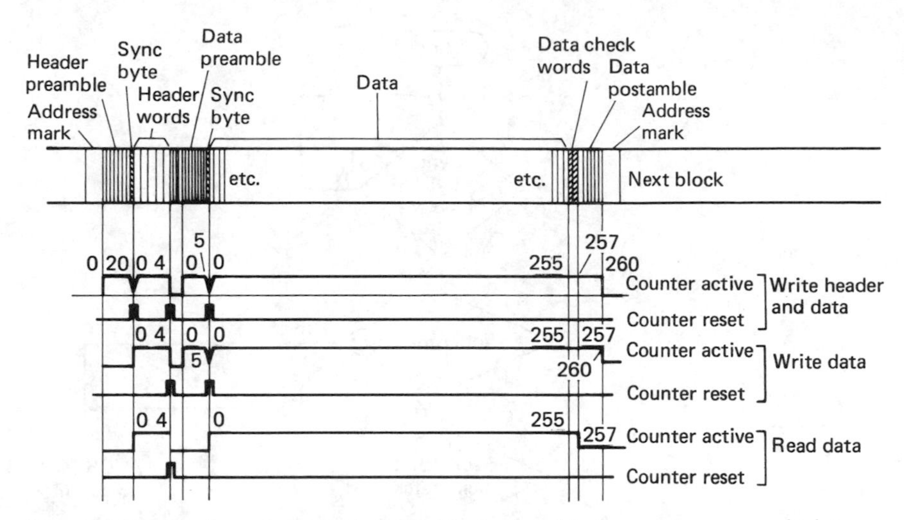

Figure 12.25 The format of a typical disk block related to the count process which is used to establish where in the block the head is at any time. During a read the count is derived from the actual data read, but during a write, the count is derived from the write clock.

preamble, and in due course the sync pattern will be found. This sets to zero a counter which divides the data-bit rate by eight, allowing the serial recording correctly to be assembled into bytes, and also allowing the sequencer to count the position of the head through the block in order to perform all the necessary steps at the right time.

The first header word is usually the cylinder address, and this is compared with the contents of the desired cylinder register. The second header word will contain the sector and track address of the block, and these will also be compared with the desired addresses. There may also be bad-block flags and/or defect-skipping information. At the end of the header is a CRCC which will be used to ensure that the header was read correctly. Figure 12.26 shows a flowchart of the position verification, after which a data transfer can proceed. The header reading is completely automatic. The only time it is necessary formally to command a header to be read is when checking that a disk has been formatted correctly.

During the read of a data block, the sequencer is employed again. The sync pattern at the beginning of the data is detected as before, following which the actual data arrive. These bits are converted to byte or sample parallel, and sent to the memory by DMA. When the sequencer has counted the last data-byte off the track, the redundancy for the error-correction system will be following.

During a write function, the header-check function will also take place as it is perhaps even more important not to write in the wrong place on a disk. Once the header has been checked and found to be correct, the

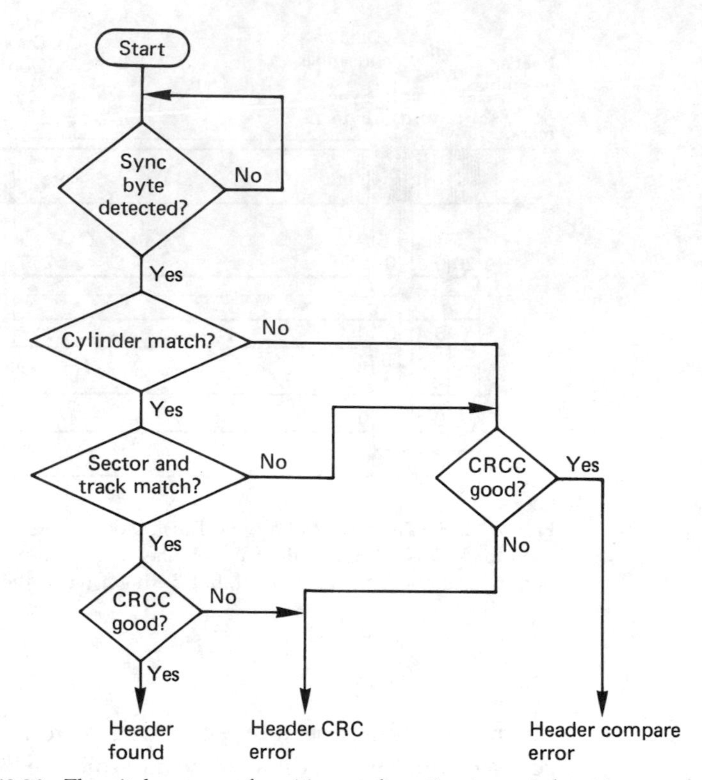

Figure 12.26 The vital process of position confirmation is carried out in accordance with the above flowchart. The appropriate words from the header are compared in turn with the contents of the disk-address registers in the subsystem. Only if the correct header has been found and read properly will the data transfer take place.

write process for the associated data block can begin. The preambles, sync pattern, data block, redundancy and postamble have all to be written contiguously. This is taken care of by the sequencer, which is obtaining timing information from the servo surface to lock the block structure to the angular position of the disk. This should be contrasted with the read function, where the timing comes directly from the data.

When video samples are fed into a disk-based system, from a digital interface or from an A/D convertor, they will be placed in a memory, from which the disk controller will read them by DMA. The continuous-input sample stream will be split up into disk blocks for disk storage.

The disk transfers must, by definition, be intermittent, because there are headers between contiguous sectors. Once all the sectors on a particular cylinder have been used, it will be necessary to seek to the next cylinder, which will cause a further interruption to the data transfer. If a bad block is encountered, the sequence will be interrupted until it has passed. The instantaneous data rate of a parallel transfer drive is made

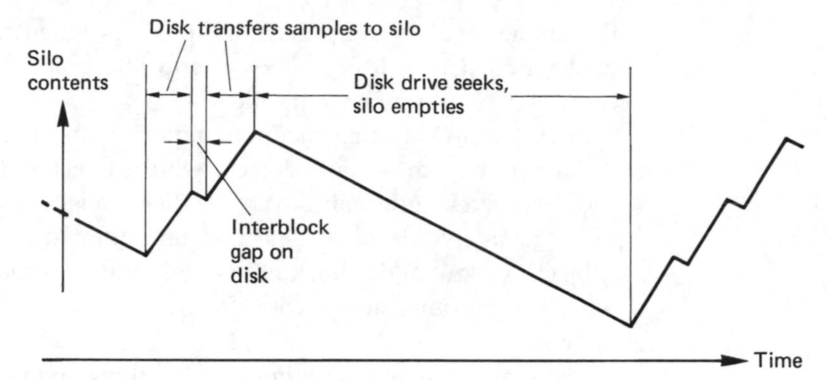

Figure 12.27 During a video replay sequence, silo is constantly emptied to provide samples, and is refilled in blocks by the drive.

higher than the continuous video data rate, so that there is time for the positioner to move whilst the video output is supplied from the FIFO memory. In replay, the drive controller attempts to keep the FIFO as full as possible by issuing a read command as soon as one block space appears in the FIFO. This allows the maximum time for a seek to take place before reading must resume. Figure 12.27 shows the action of the FIFO during reading. Whilst recording, the drive controller attempts to keep the FIFO as empty as possible by issuing write commands as soon as a block of data is present, as in Figure 12.28. In this way the amount of time available to seek is maximized in the presence of a continuous video sample input.

12.14 Defect handling

The protection of data recorded on disks differs considerably from the approach used on other media in digital video. This has much to do with

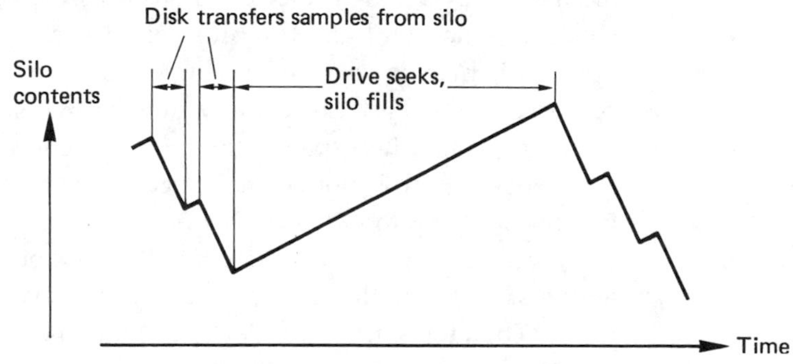

Figure 12.28 During video recording, the input samples constantly fill the silo, and the drive attempts to keep it empty by reading from it.

the intolerance of data processors to errors when compared with video data. In particular, it is not possible to interpolate to conceal errors in a computer program or a data file.

In the same way that magnetic tape is subject to dropouts, magnetic disks suffer from surface defects whose effect is to corrupt data. The shorter wavelengths employed as disk densities increase are affected more by a given size of defect. Attempting to make a perfect disk is subject to a law of diminishing returns, and eventually a state is reached where it becomes more cost-effective to invest in a defect-handling system.

There are four main methods of handling media defects in magnetic media, and further techniques needed in WORM laser disk, whose common goal is to make their presence transparent to the data. These methods vary in complexity and cost of implementation, and can often be combined in a particular system.

In the construction of bad-block files, a brand new disk is tested by the operating system. Known patterns are written everywhere on the disk, and these are read back and verified. Following this the system gives the disk a volume name, and creates on it a directory structure which keeps records of the position and size of every file subsequently written. The physical disk address of every block which fails to verify is allocated to a file which has an entry in the disk directory. In this way, when genuine data files come to be written, the bad blocks appear to the system to be in use storing a fictitious file, and no attempt will be made to write there. Some disks have dedicated tracks where defect information can be written during manufacture or by subsequent verification programs, and these permit a speedy construction of the system bad-block file.

In association with the bad-block file, many drives allocate bits in each header to indicate that the associated block is bad. If a data transfer is attempted at such a block, the presence of these bits causes the function to be aborted. The bad-block file system gives very reliable protection against defects, but can result in a lot of disk space being wasted. Systems often use several disk blocks to store convenient units of data called clusters, which will all be written or read together. Figure 12.29 shows how a bit map is searched to find free space, and illustrates how the presence of one bad block can write off a whole cluster.

In sector skipping, space is made at the end of every track for a spare data block, which is not normally accessible to the system. Where a track is found to contain a defect, the affected block becomes a skip sector. In this block, the regular defect flags will be set, but in addition, a bit known as the skip-sector flag is set in this and every subsequent block in the track. When the skip-sector flag is encountered, the effect is to add one to the desired sector address for the rest of the track, as in Figure 12.30. In this way the bad block is unused, and the the track format following the

	1	1	1	1	1	1	1	1	1	1	1	0	0	0	0	0	A
A	0	0	0	0	0	0	1	1	1	1	1	1	1	1	1	1	
	1	1	1	1	1	1	1	1	1	1	1	1	1	1	1	1	
	1	1	1	1	1	1	1	0	0	0	0	1	0	0	0	0	B
	0	0	1	1	1	1	1	1	1	1	1	1	1	0	0	0	
	0	0	0	0	0	0	0	0	e	tc.							

Figure 12.29 A disk-block-usage bit map in sixteen-bit memory for a cluster size of 11 blocks. Before writing on the disk, the system searches the bit map for contiguous free space equal to or larger than the cluster size. The first available space is the second cluster shown at A above, but the next space is unusable because the presence of a bad block B destroys the contiguity of the cluster. Thus one bad block causes the loss of a cluster.

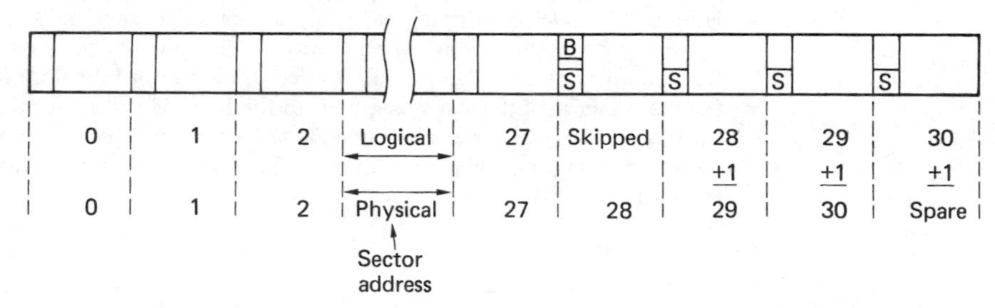

Figure 12.30 Skip sectoring. The bad block in this example has a physical sector address of 28. By setting the skip-sector flags in the header, this and subsequent logical blocks have one added to their sector addresses, and the spare block is brought into use.

bad block is effectively slid along by one block to bring into use the spare block at the end of the track. Using this approach, the presence of single bad blocks does not cause the loss of clusters, but requires slightly greater control complexity. If two bad blocks exist in a track, the second will be added to the bad-block file as usual.

The two techniques described so far have treated the block as the smallest element. In practice, the effect of a typical defect is to corrupt only a few bytes. The principle of defect skipping is that media defects can be skipped over within the block so that a block containing a defect is made usable. The header of each block contains the location of the first defect in bytes away from the end of the header, and the number of bytes from the first defect to the second defect, and so on up to the maximum of four shown in the example of Figure 12.31. Each defect is overwritten with a fixed number of bytes of preamble code and a sync pattern.

The skip is positioned so that there is sufficient undamaged preamble after the defect for the data separator to regain lock. Each defect lengthens the block, causing the format of the track to slip round. A space is left at

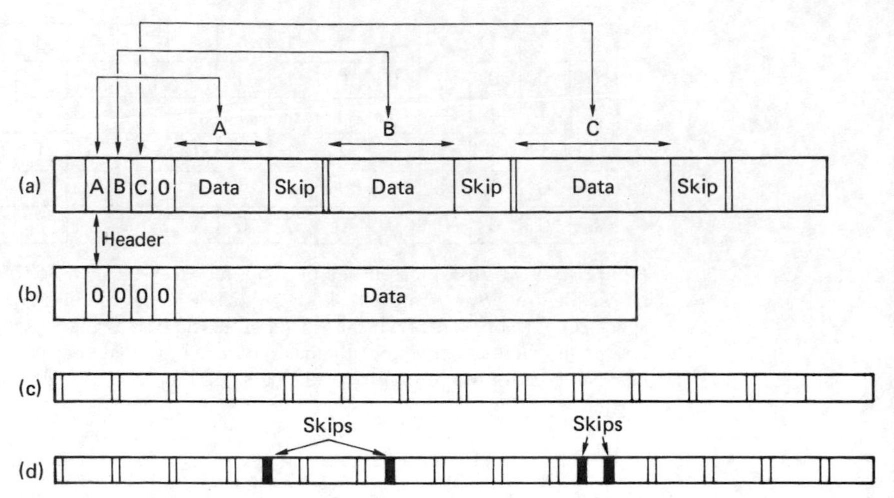

Figure 12.31 Defect skipping. (a) A block containing three defects. The header contains up to four parameters which specify how much data is to be written before each skip. In this example only three entries are needed. (b) An error-free block for comparison with (a); the presence of the skips lengthens in the block. To allow for this lengthening, the track contains spare space at the end, as shown in (c), which is an error-free track. (d) A track containing the maximum of four skips, which have caused the spare space to be used up.

the end of each track to allow a reasonable number of skips to be accommodated. Often a track descriptor is written at the beginning of each track which contains the physical position of defects relative to index. The disk format needed for a particular system can then be rapidly arrived at by reading the descriptor, and translating the physical defect locations into locations relative to the chosen sector format. Figure 12.32 shows how a soft-sectoring drive can have two different formats around the same defects using this principle.

In the case where there are too many defects in a track for the skipping to handle, the system bad-block file will be used. This is rarely necessary in practice, and the disk appears to be contiguous error-free logical and physical space. Defect skipping requires fast processing to deal with events in real time as the disk rotates. Bit-slice microsequencers are one approach, as a typical microprocessor would be too slow.

A refinement of sector skipping which permits the handling of more than one bad block per track without the loss of a cluster is revectoring. A bad block caused by a surface defect may only have a few defective bytes, so it is possible to record highly redundant information in the bad block. On a revectored disk, a bad block will contain in the data area repeated records pointing to the address where data displaced by the defect can be found. The spare block at the end of the track will be the first such place, and can be read within the same disk revolution, but out

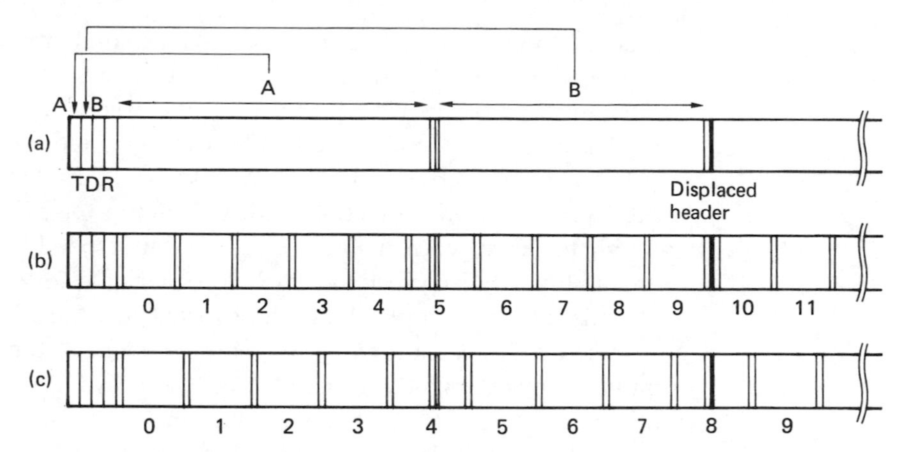

Figure 12.32 The purpose of the track descriptor record (TDR) is to keep a record of defects independent of disk format. The positions of the defects stored in the TDR (a) are used by the formatter to establish the positions relative to the format used. With the format (b), the first defect appears in sector 5, but the same defect would be in sector 4 for format (c). The second defect falls where a header would be written in (b) so the header is displaced for sector 10. The same defect falls in the data area of sector 8 in (c).

of sequence, which puts extra demands on the controller. In the less frequent case of more than one defect in a track, the second and subsequent bad blocks revector to spare blocks available in an area dedicated to that purpose. The principle is illustrated in Figure 12.33. In this case a seek will be necessary to locate the replacement block. The low probability of this means that access time is not significantly affected.

The steps outlined above are the first line of defence against errors in disk drives, and serve to ensure that, by and large, the errors due to obvious surface defects are eliminated. There are other error mechanisms

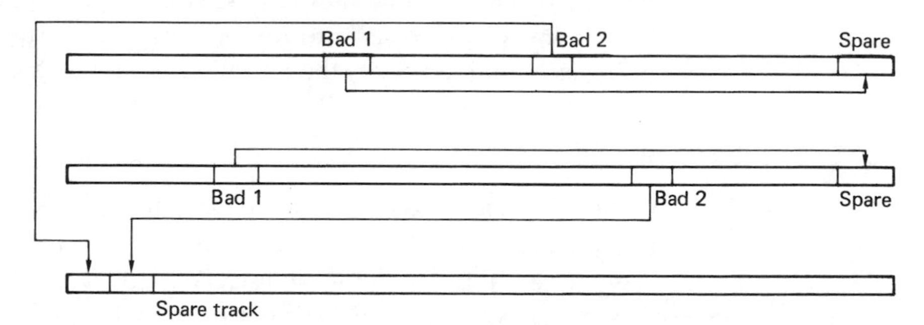

Figure 12.33 Revectoring. The first bad block in each track is revectored to the spare block at the end of the track. Unlike skip sectoring, subsequent good blocks are unaffected, and the replacement of block is read out of sequence. The second bad block on any one track is revectored to one of a number of spare tracks kept on the disk for this purpose.

in action, such as noise and jitter, which can result in random errors, and it is necessary to protect disk data against these also. The error-correction mechanisms described in Chapter 6 will be employed. In general, each data block is made into a codeword by the addition of redundancy at the end.

The error-correcting code used in disks was, for a long time, Fire code, because it allowed correction with the minimum circuit complexity. It could, however, correct only one error burst per block, and it had a probability of miscorrection which was marginal for some applications. The advances in complex logic chips meant that the adoption of a Reed–Solomon code was a logical step, since these have the ability to correct multiple error bursts. As the larger burst errors in disk drives are taken care of by verifying the medium, interleaving in the error correction sense is not generally needed. When interleaving is used in the context of disks, it usually means that the sectors along a track are interleaved so that reading them in numerical order requires two revolutions. This will slow down the data transfer rate where the drive is too fast for the associated circuitry.

In some systems, the occurrence of errors is monitored to see if they are truly random, or if an error persistently occurs in the same physical block. If this is the case, and the error is small, and well within the correction power of the code, the block will continue in use. If, however, the error is larger than some threshold, the data will be read, corrected and rewritten elsewhere, and the block will then be added to the bad-block file so that it will not be used again.

In erasable optical disks, formatting is possible to map out defects in the same way as for magnetic disks, but in WORM disks, it is not possible to verify the medium because it can only be written once. The presence of a defect cannot be detected until an attempt has been made to write on the disk. The data written can then be read back and checked for error. If there is an error in the verification, the block concerned will be rewritten, usually in the next block along the track. This verification process slows down the recording operation, but some drives have a complex optical system which allows a low-powered laser to read the track in between pulses of the writing laser, and can verify the recording as it is being made.

12.15 RAID arrays

Whilst the MTBF of a disk drive is very high, it is a simple matter of statistics that when a large number of drives is assembled in a system the time between failures becomes shorter. Disk drives are sealed units and the disks cannot be removed if there is an electronic failure. Even if this were possible the system cannot usually afford down time whilst such a data recovery takes place.

Consequently any system in which the data are valuable must take steps to ensure data integrity. This is commonly done using RAID (redundant

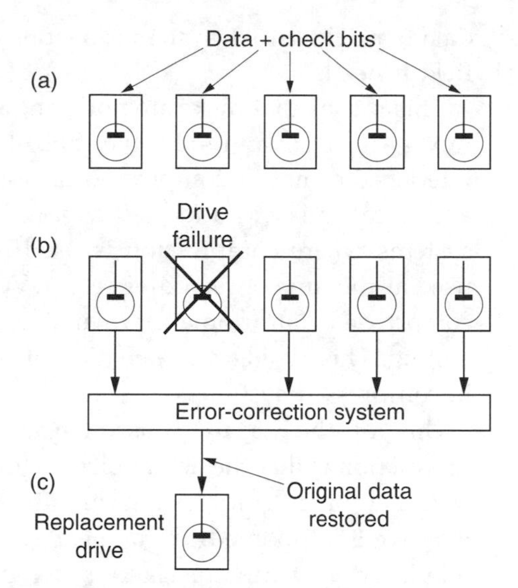

Figure 12.34 In RAID technology, data and redundancy are spread over a number of drives (a). In the case of a drive failure (b) the error-correction system can correct for the loss and continue operation. When the drive is replaced (c) the data can be rewritten so that the system can then survive a further failure.

array of inexpensive disks) technology. Figure 12.34 shows that in a RAID array data blocks are spread across a number of drives.

An error-correcting check symbol (typically Reed–Solomon) is stored on a redundant drive. The error correction is powerful enough fully to correct any error in the block due to a single failed drive. In RAID arrays the drives are designed to be hot-plugged (replaced without removing power) so if a drive fails it is simply physically replaced with a new one. The error correction system will rewrite the drive with the data which was lost with the failed unit.

When a large number of disk drives are arrayed together, it is necessary and desirable to spread files across all the drives in a RAID array. Whilst this ensures data integrity, it also means that the data transfer rate is multiplied by the number of drives sharing the data. This means that the data transfer rate can be extremely high and new approaches are necessary to move the data in and out of the disk system.

12.16 Disk servers

The disk controller will automatically divide files up into blocks of the appropriate size for recording. If any partial blocks are left over, these will be zero stuffed. Consequently disk stores are not constrained to files of a particular size. Unlike a DVTR which always stores the same amount of data per field, a disk system can store a different amount of data for each field if needs be.

This means that disks are not standards dependent. A disk system can mix 4:4:4, 4:2:2 and 4:2:0 files and it doesn't care whether the video is interlaced or not or compressed or not. It can mix 525- and 625-line files and it can mix 4:3 and 16:9 aspect ratios. This an advantage in news systems where compression is used. If a given compression scheme is used at the time of recording, e.g. DVCPRO, the video can remain in the compressed data domain when it is loaded onto the disk system for editing. This avoids concatenation of codecs which is generally bad news in compressed systems.

One of the happy consequences of the move to disk drives in production is that the actual picture format used need no longer be fixed. With computer graphics and broadcast video visibly merging, interlace may well be doomed. In the near future it will be possible to use non-interlaced HD cameras, and down-convert to a non-interlaced intermediate resolution production format.

As production units such as mixers, character generators, paint systems and DVEs become increasingly software driven, such a format is much easier to adopt than in the days of analog where the functionality was frozen into the circuitry. Following production the intermediate format can be converted to any present or future emission standard.

12.17 Disks and compression

The cost of disk storage continues to be orders of magnitude more than that of tape, but both get cheaper as time goes by. Many disk-based systems have resorted to compression in order to reduce cost or extend playing time. This is fine for offline editing, where full bit rate recordings will subsequently be conformed to the EDL, but not so good for online work, where the viewers see the result of compression. Some manufacturers, notably Quantel, have argued that for online working the full bit rate should be stored and have found ways to do it. With the falling cost of storage they may ultimately be proved right in quality applications.

The first disk-based systems used JPEG compression, designed for still pictures, as this was the first coding scheme for which chips were available. Each field was individually compressed and so there was total freedom for editing. Now that MPEG chip sets are becoming available this will start to be used. Even if the MPEG coding is restricted to *I* pictures for editing it is still slightly more efficient than motion JPEG.

Sony's SX system uses an *IB* picture sequence because it gives a significant improvement in compression factor without causing too much difficulty in editing. Wavelet compression is also of importance in disk-based systems because it is primarily a spatial coding scheme and is naturally scaleable. In other words a fuzzy but complete picture can be displayed from only the first part of the data block, and the picture gets sharper as more data are added. This means that viewable pictures can be provided at much greater speeds than normal, the disk equivalent of picture in shuttle.

12.18 Optical disk principles

In order to record MO disks or replay any optical disk, a source of monochromatic light is required. The light source must have low noise otherwise the variations in intensity due to the noise of the source will mask the variations due to reading the disk. The requirement for a low-noise monochromatic light source is economically met using a semi-conductor laser.

The semiconductor laser is a relative of the light-emitting diode (LED). Both operate by raising the energy of electrons to move them from one valence band to another conduction band. Electrons which fall back to the valence band emit a quantum of energy as a photon whose frequency is proportional to the energy difference between the bands. The process is described by Planck's Law:

$$\text{Energy difference } E = H \times f$$

where H = Planck's Constant

$$= 6.6262 \times 10^{-34} \text{ Joules/Hertz}$$

For gallium arsenide, the energy difference is about 1.6 eV, where 1 eV is 1.6×10^{-19} Joules.

Using Planck's Law, the frequency of emission will be:

$$f = \frac{1.6 \times 1.6 \times 10^{-19}}{6.6262 \times 10^{-34}} \text{Hz}$$

The wavelength will be c/f where

$$c = \text{the velocity of light} = 3 \times 10^{8} \text{ m/s.}$$

$$\text{Wavelength} = \frac{3 \times 10^{8} \times 6.6262 \times 10^{-34}}{2.56 \times 10^{-19}} \text{ m}$$

$$= 780 \text{ nanometres}$$

In the LED, electrons fall back to the valence band randomly, and the light produced is incoherent. In the laser, the ends of the semiconductor are optically flat mirrors, which produce an optically resonant cavity. One photon can bounce to and fro, exciting others in synchronism, to produce coherent light. This is known as Light Amplification by Stimulated Emission of Radiation, mercifully abbreviated to LASER, and can result in a runaway condition, where all available energy is used up in one flash. In injection lasers, an equilibrium is reached between energy input and light output, allowing continuous operation. The equilibrium is delicate, and such devices are usually fed from a current source. To avoid runaway when temperature change disturbs the equilibrium, a photosensor is often fed back to the current source. Such lasers have a finite life, and become steadily less efficient. The feedback will maintain output, and it is possible to anticipate the failure of the laser by monitoring the drive voltage needed to give the correct output.

Figure 12.35(a) shows that there are two main types of WORM disks. In the first, the disk contains a thin layer of metal; on recording, a powerful laser melts spots on the layer. Surface tension causes a hole to form in the metal, with a thickened rim around the hole. Subsequently a low-power laser can read the disk because the metal reflects light, but the hole passes it through. Computer WORM disks work on this principle. In the second, the layer of metal is extremely thin, and the heat from the laser heats the material below it to the point of decomposition. This causes gassing which raises a blister or bubble in the metal layer. Clearly once such a pattern of holes or blisters has been made, it is permanent.

Re-recordable or erasable optical disks rely on magneto-optics,[3] also known more fully as thermomagneto-optics. Writing in such a device makes use of a thermomagnetic property posessed by all magnetic materials, which is that above a certain temperature, known as the Curie temperature, their coercive force becomes zero. This means that they become magnetically very soft, and take on the flux direction of any externally applied field. On cooling, this field orientation will be frozen in the material, and the coercivity will oppose attempts to change it. Although many materials possess this property, there are relatively few which have a suitably low Curie temperature. Compounds of terbium and gadolinium have been used, and one of the major problems to be overcome is that almost all suitable materials from a magnetic viewpoint corrode very quickly in air.

There are two ways in which magneto-optic (MO) disks can be written. Figure 12.35(b) shows the first system, in which the intensity of laser is modulated with the waveform to be recorded. If the disk is considered initially to be magnetized along its axis of rotation with the

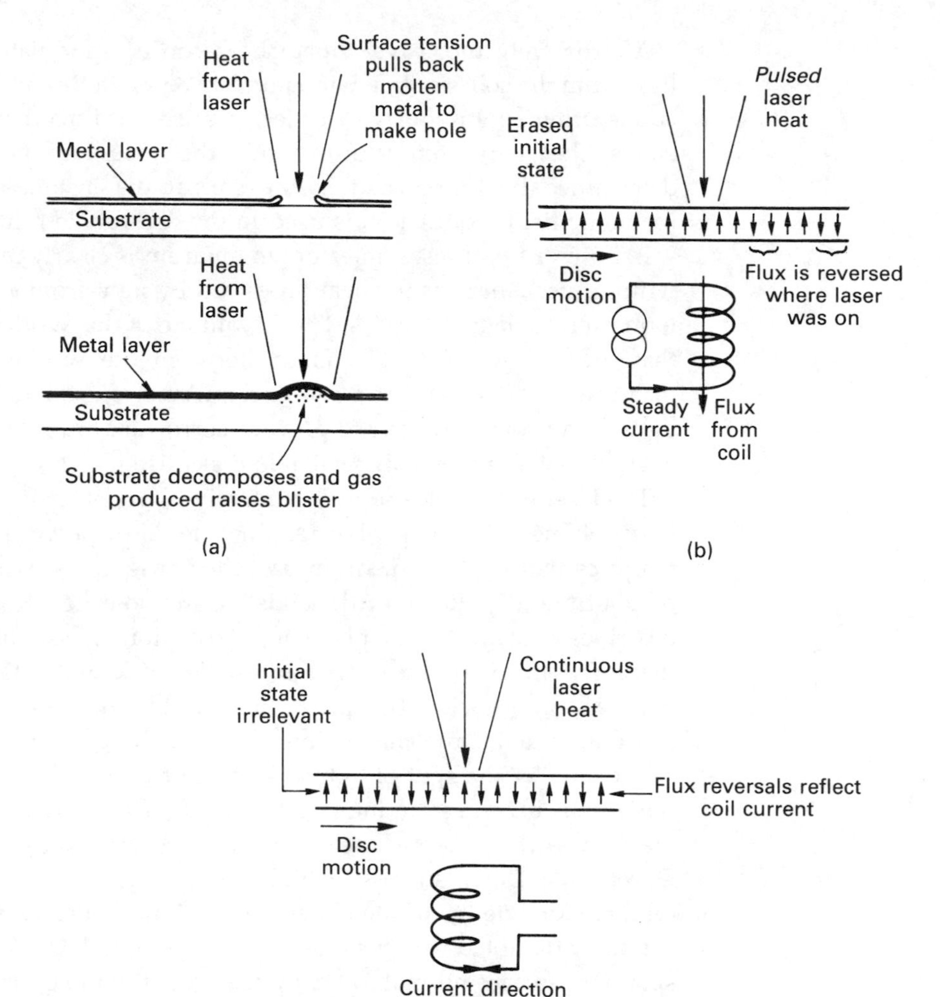

Figure 12.35 At (a) the two main types of WORM (Write-Once-Read-Many) disk. MO disks can be written by modulating the laser (b) or the magnetic field (c). See text for differences.

north pole upwards, it is rotated in a field of the opposite sense, produced by a steady current flowing in a coil which is weaker than the room-temperature coercivity of the medium. The field will therefore have no effect. A laser beam is focused on the medium as it turns, and a pulse from the laser will momentarily heat a very small area of the medium past its Curie temperature, whereby it will take on a reversed flux due to the presence of the field coils. This reversed-flux direction will be retained indefinitely as the medium cools.

Alternatively the waveform to be recorded modulates the magnetic field from the coils as shown in Figure 12.35(c). In this approach, the laser is operating continuously in order to raise the track beneath the beam above the Curie temperature, but the magnetic field recorded is determined by the current in the coil at the instant the track cools. Magnetic field modulation is used in the recordable MiniDisc.

In both of these cases, the storage medium is clearly magnetic, but the writing mechanism is the heat produced by light from a laser; hence the term thermomagneto-optics. The advantage of this writing mechanism is that there is no physical contact between the writing head and the medium. The distance can be several millimetres, some of which is taken up with a protective layer to prevent corrosion. In prototypes, this layer is glass, but commercially available disks use plastics.

The laser beam will supply a relatively high power for writing, since it is supplying heat energy. For reading, the laser power is reduced, such that it cannot heat the medium past the Curie temperature, and it is left on continuously. Readout depends on the so-called Kerr effect, which describes a rotation of the plane of polarization of light due to a magnetic field. The magnetic areas written on the disk will rotate the plane of polarization of incident polarized light to two different planes, and it is possible to detect the change in rotation with a suitable pickup.

The smaller the spot of light which can be created, the smaller can be the details carrying the information, and so more information per unit area (known in the art as the superficial recording density) can be stored. Development of a successful high-density optical recorder requires an intimate knowledge of the behaviour of light focused into small spots.

If it is attempted to focus a uniform beam of light to an infinitely small spot on a surface normal to the optical axis, it will be found that it is not possible. This is probably just as well as an infinitely small spot would have infinite intensity and any matter it fell on would not survive. Instead the result of such an attempt is a distribution of light in the area of the focal point which has no sharply defined boundary. This is called the Airy distribution[4] (sometimes pattern or disk) after Lord Airy (1835) the then Astronomer Royal. If a line is considered to pass across the focal plane, through the theoretical focal point, and the intensity of the light is plotted on a graph as a function of the distance along that line, the result is the intensity function shown in Figure 12.36. It will be seen that this contains a central sloping peak surrounded by alternating dark rings and light rings of diminishing intensity. These rings will in theory reach to infinity before their intensity becomes zero. The intensity distribution or function described by Airy is due to diffraction effects across the finite aperture of the objective. For a given wavelength, as the aperture of the objective is increased, so the diameter of the features of the Airy pattern reduces. The Airy pattern vanishes to a singularity of infinite intensity with a lens of

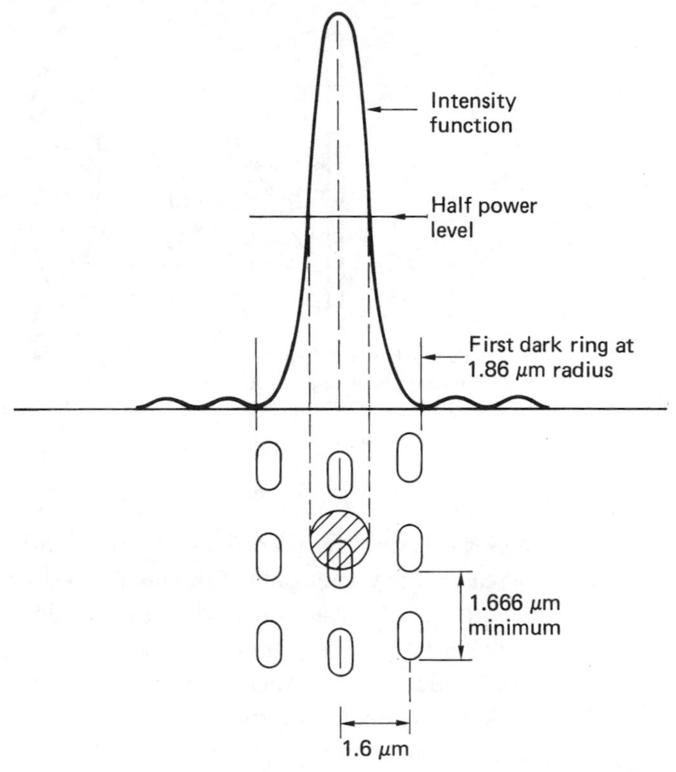

Figure 12.36 The structure of a maximum frequency recording is shown here, related to the intensity function of an objective of 0.45NA with 780 μm light. Note that track spacing puts adjacent tracks in the dark rings, reducing crosstalk. Note also that as the spot has as an intensity function it is meaningless to specify the spot diameter without some reference such as an intensity level.

infinite aperture which, of course, cannot be made. The approximation of geometric optics is quite unable to predict the occurence of the Airy pattern.

An intensity function does not have a diameter, but for practical purposes an effective diameter typically quoted is that at which the intensity has fallen to some convenient fraction of that at the peak. Thus one could state, for example, the half-power diameter.

Since light paths in optical instruments are generally reversible, it is possible to see an interesting corollary which gives a useful insight into the readout principle of CD. Considering light radiating from a phase structure, as in Figure 12.37, the more closely spaced the features of the phase structure, i.e. the higher the spatial frequency, the more oblique the direction of the wavefronts in the diffraction pattern which results and the larger the numerical aperture of the lens needed to collect the light if the resolution is not to be lost. The corollary of this is that the smaller the Airy distribution it is wished to create, the larger must be the aperture of

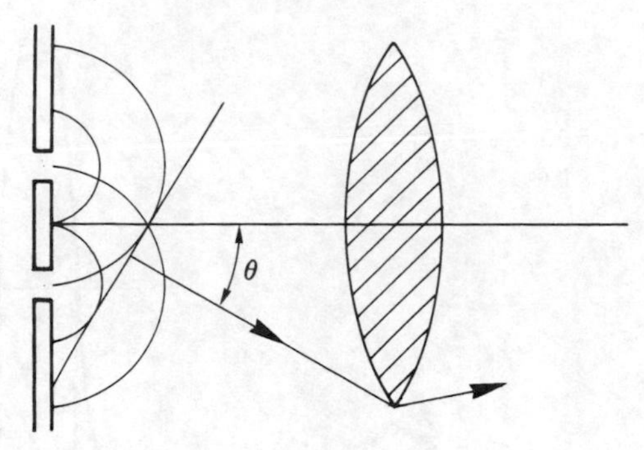

Figure 12.37 Fine detail in an object can only be resolved if the diffracted wavefront due to the highest spatial frequency is collected by the lens. Numerical aperture (NA) = sin θ, and as θ is the diffraction angle it follows that, for a given wavelength, NA determines resolution.

the lens. Spatial frequency is measured in lines per millimetre and as it increases, the wavefronts of the resultant diffraction pattern become more oblique. In the case of a CD, the smaller the bumps and the spaces between them along the track, the higher the spatial frequency, and the more oblique the diffraction pattern becomes in a plane tangential to the track. With a fixed-objective aperture, as the tangential diffraction pattern becomes more oblique, less light passes the aperture and the depth of modulation transmitted by the lens falls. At some spatial frequency, all the diffracted light falls outside the aperture and the modulation depth transmitted by the lens falls to zero. This is known as the spatial cut-off frequency. Thus a graph of depth of modulation versus spatial frequency can be drawn and which is known as the modulation transfer function (MTF). This is a straight line commencing at unity at zero spatial frequency (no detail) and falling to zero at the cut-off spatial frequency (finest detail). Thus one could describe a lens of finite aperture as a form of spatial low-pass filter. The Airy function is no more than the spatial impulse response of the lens, and the concentric rings of the Airy function are the spatial analog of the symmetrical ringing in a phase linear electrical filter. The Airy function and the triangular frequency response form a transform pair[5] as shown in Chapter 3.

When an objective lens is used in a conventional microscope, the MTF will allow the resolution to be predicted in lines per millimetre. However, in a scanning microscope the spatial frequency of the detail in the object is multiplied by the scanning velocity to give a temporal frequency measured in Hertz. Thus lines per millimetre multiplied by millimetres per second gives lines per second. Instead of a straight-line MTF falling to the spatial cut-off frequency, a scanning microscope has a temporal frequency response falling to zero at the optical cut-off frequency. Whilst this concept requires a number of idiomatic terms to be assimilated at once, the point can be made clear by a simple analogy. Imagine the evenly spaced iron railings outside a schoolyard. These are permanently fixed, and can have no temporal frequency, yet they have a spatial frequency which is the number of railings per unit distance. A small boy with a stick takes great delight in running along the railings so that his stick hits each one in turn and makes a great noise. The rate at which his stick hits the railings is the temporal frequency which results from their being scanned. This rate would increase if the boy ran faster, but it would also increase if the rails were closer together. As a consequence it can be seen that the temporal frequency is proportional to the spatial frequency multiplied by the scanning speed. Put more technically, the frequency response of an optical recorder is the Fourier transform of the Airy distribution of the readout spot multiplied by the track velocity. Figure 12.38 shows that the frequency response falls progressively from DC to the optical cutoff frequency which is given by:

$$F_c = \frac{2NA}{\text{Wavelength}} \times \text{velocity}$$

Measurements reveal that the optical response of practical devices is only a little worse than the theory predicts. This characteristic has a large bearing on the type of modulation schemes which can be successfully employed. Clearly, to obtain any noise immunity, the maximum operating frequency must be rather less than the cutoff frequency.

Lasers do not produce a beam of uniform intensity. It is more intense in the centre than it is at the edges, and this has the effect of slightly increasing the half-power diameter of the intensity function. The effect is

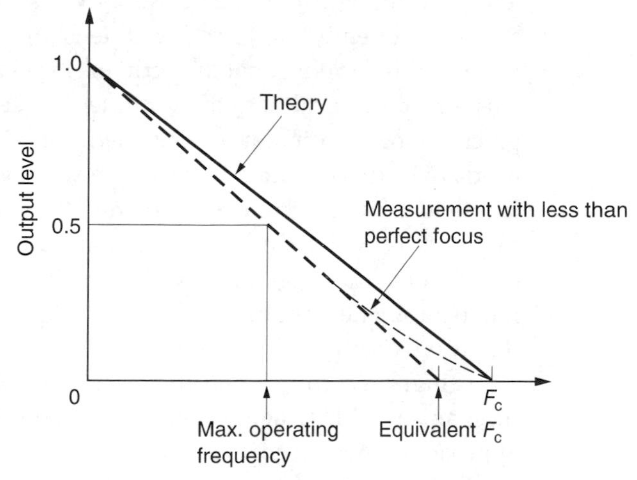

Figure 12.38 Frequency response of laser pickup. Maximum operating frequency is about half of cut-off frequency F_c.

analogous the the effect of window functions in FIR filters (see Chapter 2). The intensity function can also be enlarged if the lens used suffers from optical aberrations. This was studied by Maréchal[6] who established criteria for the accuracy to which the optical surfaces of the lens should be made to allow the ideal Airy distribution to be obtained. CD player lenses must meet the Maréchal criterion. With such a lens, the diameter of the distribution function is determined solely by the combination of Numerical Aperture (NA) and the wavelength. When the size of the spot is as small as the NA and wavelength allow, the optical system is said to be diffraction limited. Figure 12.37 showed how Numerical Aperture is defined, and illustrates that the smaller the spot needed, the larger must be the NA. Unfortunately the larger the NA, the more obliquely to the normal the light arrives at the focal plane and the smaller the depth of focus will be. This was investigated by Hopkins[7] who established the depth of focus available for a given NA. Pickups in CD use an NA of 0.45[1] whereas the later DVD uses an NA of 0.6 which requires a more accurate focus mechanism as the depth of focus will be smaller.

The intensity function will also be distorted and grossly enlarged if the optical axis is not normal to the medium. The initial effect is that the energy in the first bright ring increases strongly in one place and results in a secondary peak adjacent to the central peak. This is known as coma and its effect is extremely serious as the enlargement of the spot restricts the recording density. The larger the NA, the smaller becomes the allowable tilt of the optical axis with respect to the medium before coma becomes a problem. With a typical NA this angle is less than a degree.[1]

Numerical Aperture is defined as the cosine of the angle between the optical axis and rays converging from the perimeter of the lens. It will be apparent that there are many combinations of lens diameter and focal length which will have the same NA. As the difficulty of manufacture, and consequently the cost, of a lens meeting the Maréchal criterion increases disproportionately with size, it is advantageous to use a small lens of short focal length, mounted close to the medium and held precisely perpendicular to the medium to prevent coma. As the lens needs to be driven along its axis by a servo to maintain focus, the smaller lens will facilitate the design of the servo by reducing the mass to be driven. It is extremely difficult to make a lens which meets the Maréchal criterion over a range of wavelengths because of dispersion. The use of monochromatic light eases the lens design as it has only to be correct for one wavelength.

At high recording densities, there is literally only one scanning mechanism with which all the optical criteria can be met and this is the approach known from the scanning microscope. The optical pickup is mounted in a carriage which can move it parallel to the medium in such a way that the optical axis remains at all times parallel to the axis of rotation of the medium. The latter rotates as the pickup is driven away from the axis of rotation in such a way that a spiral track on the disk is followed. The pickup contains a short focal length lens of small diameter which must therefore be close to the disk surface to allow a large NA. All high-density optical recorders operate on this principle in which the readout of the carrier is optical but the scanning is actually mechanical.

12.19 Optical pickups

Whatever the type of disk being read, it must be illuminated by the laser beam. Some of the light reflected back from the disk re-enters the aperture of the objective lens. The pickup must be capable of separating the reflected light from the incident light. Figure 12.39 shows two systems. In

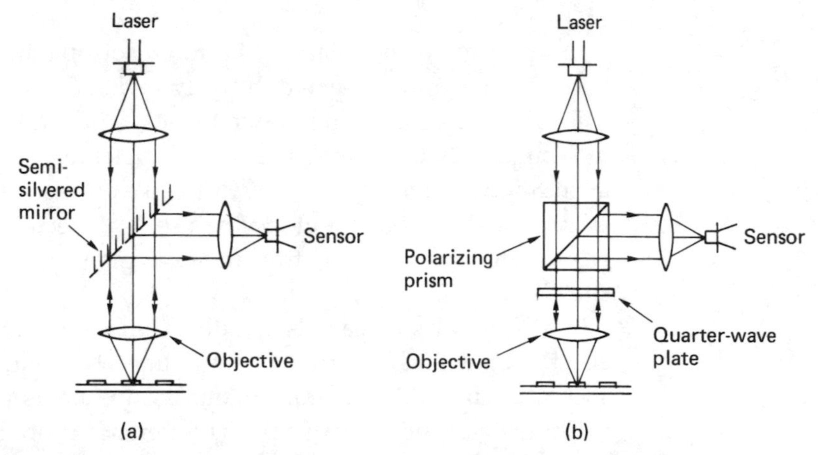

Figure 12.39 (a) Reflected light from the disk is directed to the sensor by a semisilvered mirror. (b) A combination of polarizing prism and quarter-wave plate separate incident and reflected light.

(a) an intensity beamsplitter consisting of a semisilvered mirror is inserted in the optical path and reflects some of the returning light into the photosensor. This is not very efficient, as half of the replay signal is lost by transmission straight on. In the example in (b) separation is by polarization.

In natural light, the electric-field component will be in many planes. Light is said to be polarized when the electric field direction is constrained. The wave can be considered as made up from two orthogonal components. When these are in phase, the polarization is said to be linear. When there is a phase shift between the components, the polarization is said to be elliptical, with a special case at 90° called circular polarization. These types of polarization are contrasted in Figure 12.40.

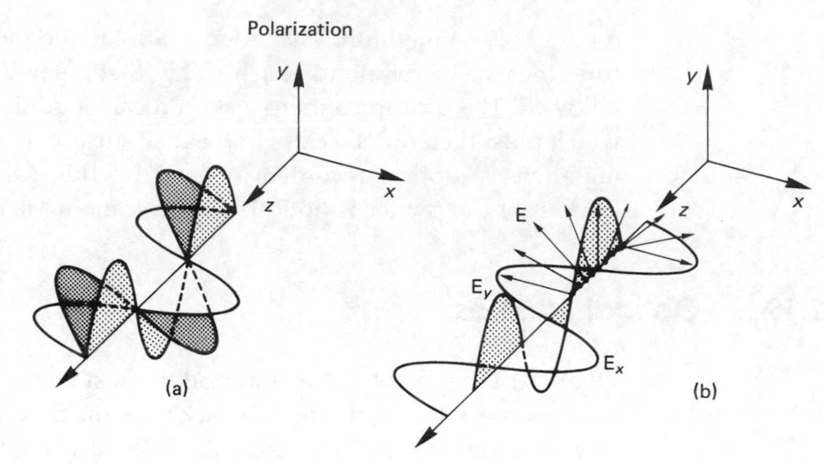

Figure 12.40 (a) Linear polarization: orthogonal components are in phase. (b) Circular polarization: orthogonal components are in phase quadrature.

In order to create polarized light, anisotropic materials are necessary. Polaroid material, invented by Edwin Land, is vinyl which is made anisotropic by stretching it while hot. This causes the long polymer molecules to line up along the axis of stretching. If the material is soaked in iodine, the molecules are rendered conductive, and short out any electric-field component along themselves. Electric fields at right angles are unaffected; thus the transmission plane is at right angles to the stretching axis.

Stretching plastics can also result in anisotropy of refractive index; this effect is known as birefringence. If a linearly polarized wavefront enters such a medium, the two orthogonal components propagate at different velocities, causing a relative phase difference proportional to the distance travelled. The plane of polarization of the light is rotated. Where the thickness of the material is such that a 90° phase change is caused, the device is known as a quarter-wave plate. The action of such a device is shown in Figure 12.41. If the plane of polarization of the incident light is at 45° to the planes of greatest and least refractive index, the two orthogonal components of the light will be of equal magnitude, and this results in circular polarization. Similarly, circular-polarized light can be returned to the linear-polarized state by a further quarter-wave plate. Rotation of the plane of polarization is a useful method of separating incident and reflected light in a laser pickup. Using a quarter-wave plate, the plane of polarization of light leaving the pickup will have been turned 45° and on return it will be rotated a further 45°, so that it is now at right angles to the plane of polarization of light from the source. The two can easily be separated by a polarizing prism, which acts as a transparent block to light in one plane, but as a prism to light in the other plane, such that reflected light is directed towards the sensor.

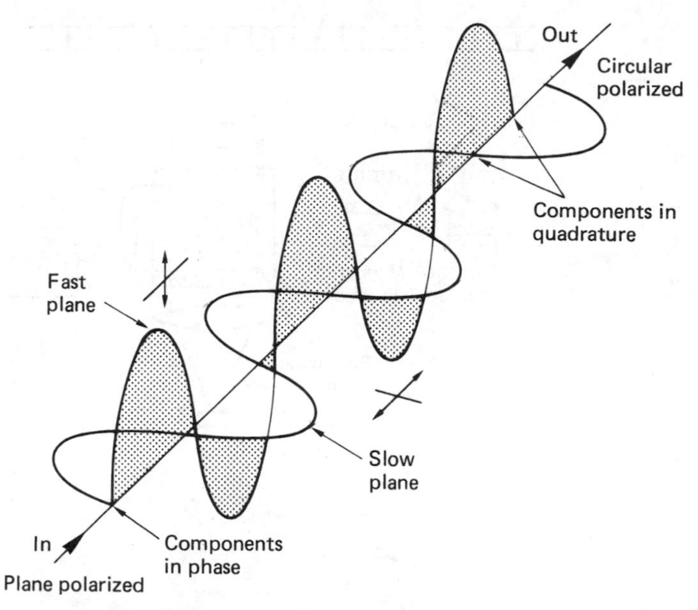

Figure 12.41 Different speed of light in different planes rotates the plane of polarization in a quarter-wave plate to give a circularly polarized output.

In LaserVision or WORM disk drives, the sensor is concerned only with the intensity of the light falling on it. When playing MO disks, the intensity does not change, but the magnetic recording on the disk rotates the plane of polarization one way or the other depending on the direction of the vertical magnetization. MO disks cannot be read with circular polarized light. Light incident on the medium must be plane polarized and so the quarter-wave plate of the CD pickup cannot be used. Figure 12.42(a) shows that a polarizing prism is still required to linearly polarize the light from the laser on its way to the disk. Light returning from the disk has had its plane of polarization rotated by approximately ±1 degree. This is an extremely small rotation. Figure 12.42(b) shows that the returning rotated light can be considered to be composed of two orthogonal components. R_x is the component which is in the same plane as the illumination and is called the *ordinary* component and R_y in the component due to the Kerr effect rotation and is known as the *magneto-optic* component. A polarizing beam splitter mounted squarely would reflect the magneto-optic component R_y very well because it is at right angles to the transmission plane of the prism, but the ordinary component would pass straight on in the direction of the laser. By rotating the prism slightly a small amount of the ordinary component is also reflected. Figure 12.42(c) shows that when combined with the magneto-optic component, the angle of rotation has increased.[8] Detecting this rotation requires a further polarizing prism or analyser as shown in

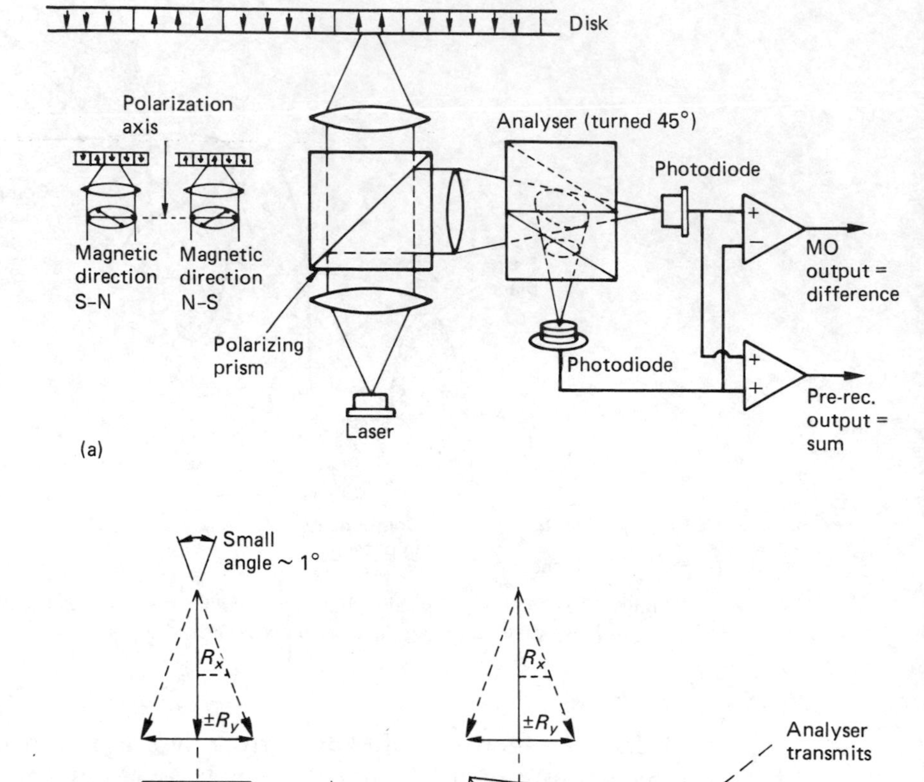

Figure 12.42 A pickup suitable for the replay of magneto-optic disks must respond to very small rotations of the plane of polarization.

Figure 12.42. The prism is twisted such that the transmission plane is at 45° to the planes of R_x and R_y. Thus with an unmagnetized disk, half of the light is transmitted by the prism and half is reflected. If the magnetic field of the disk turns the plane of polarization towards the transmission plane of the prism, more light is transmitted and less is reflected. Conversely if the plane of polarization is rotated away from the transmission plane, less light is transmitted and more is reflected. If two sensors are used, one for transmitted light and one for reflected light, the difference between the two sensor outputs will be a waveform representing the angle of polarization and thus the recording on the disk. This differential analyser eliminates common mode noise in the reflected beam.[9]

High-density recording implies short wavelengths. Using a laser focused on the disk from a distance allows short wavelength recordings to be played back without physical contact, whereas conventional magnetic recording requires intimate contact and implies a wear mechanism, the need for periodic cleaning, and susceptibility to contamination.

The information layer is read through the thickness of the disk. Figure 12.43 shows that this approach causes the readout beam to enter and leave the disk surface through the largest possible area. Typical dimensions involved are shown in the figure. Despite the minute spot size of about 1.2 mm diameter, light enters and leaves through a 0.7 mm-diameter circle. As a result, surface debris has to be three orders of magnitude larger than the readout spot before the beam is obscured. This approach has the further advantage in MO drives that the magnetic head, on the opposite side to the laser pickup, is then closer to the magnetic layer in the disk.

The bending of light at the disk surface is due to refraction of the wavefronts arriving from the objective lens. Wave theory of light suggests that a wavefront advances because an infinite number of point sources can be considered to emit spherical waves which will only add when they are all in the same phase. This can only occur in the plane of the wavefront. Figure 12.44 shows that at all other angles, interference between spherical waves is destructive.

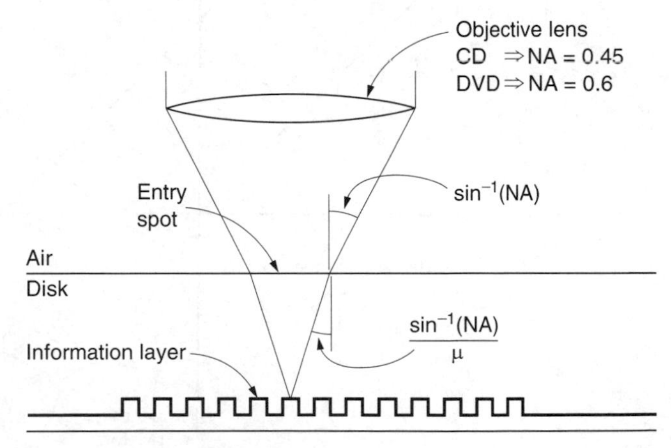

Figure 12.43 In optical disks the entry spot size is about 0.5 mm in diameter. The diameter is affected by the numerical aperture of the lens (NA) the refractive index of the disk (μ) and the disk thickness.

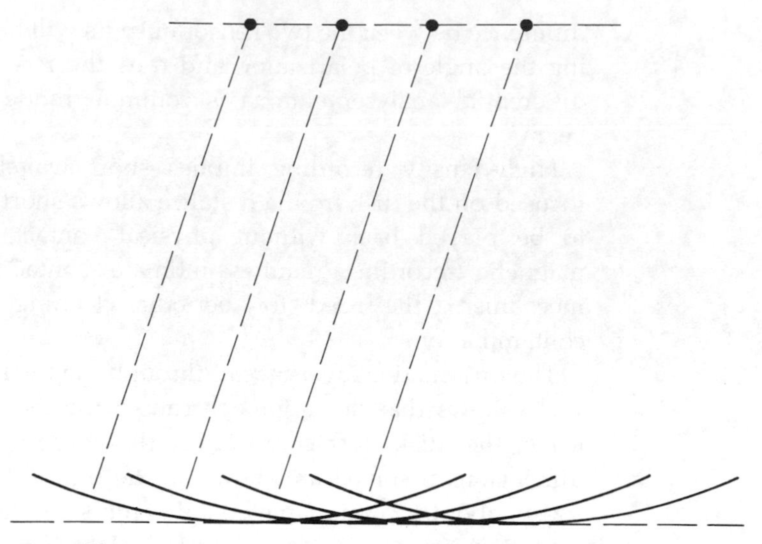

Figure 12.44 Plane-wave propagation considered as infinite numbers of spherical waves.

When such a wavefront arrives at an interface with a denser medium, such as the surface of an optical disk, the velocity of propagation is reduced; therefore the wavelength in the medium becomes shorter, causing the wavefront to leave the interface at a different angle (Figure 12.45). This is known as refraction. The ratio of velocity *in vacuo* to

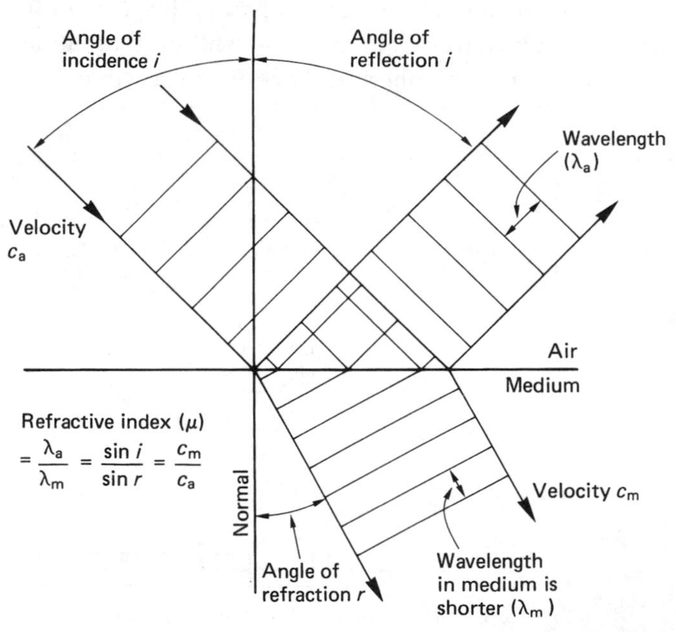

Figure 12.45 Reflection and refraction, showing the effect of the velocity of light in a medium.

velocity in the medium is known as the refractive index of that medium; it determines the relationship between the angles of the incident and refracted wavefronts. Reflected light, however, leaves at the same angle to the normal as the incident light. If the speed of light in the medium varies with wavelength, dispersion takes place, where incident white light will be split into a rainbow-like spectrum leaving the interface at different angles. Glass used for chandeliers and cut glass is chosen to be highly dispersive, whereas glass for optical instruments will be chosen to have a refractive index which is as constant as possible with changing wavelength. The use of monochromatic light in optical disks allows low-cost optics to be used as they only need to be corrected for a single wavelength.

The size of the entry circle in Figure 12.43 is a function of the refractive index of the disk material, the numerical aperture of the objective lens and the thickness of the disk.

12.20 Focus systems

The frequency response of the laser pickup and the amount of crosstalk are both a function of the spot size and care must be taken to keep the beam focused on the information layer. If the spot on the disk becomes too large, it will be unable to discern the smaller features of the track, and can also be affected by the adjacent track. Disk warp and thickness irregularities will cause focal-plane movement beyond the depth of focus of the optical system, and a focus servo system will be needed. The depth of field is related to the numerical aperture, which is defined, and the accuracy of the servo must be sufficient to keep the focal plane within that depth, which is typically ± 1 mm.

The focus servo moves a lens along the optical axis in order to keep the spot in focus. Since dynamic focus-changes are largely due to warps, the focus system must have a frequency response in excess of the rotational speed. A moving-coil actuator is often used owing to the small moving mass which this permits. Figure 12.46 shows that a cylindrical magnet assembly almost identical to that of a loudspeaker can be used, coaxial with the light beam. Alternatively a moving magnet design can be used. A rare earth magnet allows a sufficiently strong magnetic field without excessive weight.

A focus-error system is necessary to drive the lens. There are a number of ways in which this can be derived, the most common of which will be described here.

In Figure 12.47 a cylindrical lens is installed between the beam splitter and the photosensor. The effect of this lens is that the beam has no focal point on the sensor. In one plane, the cylindrical lens appears parallel-

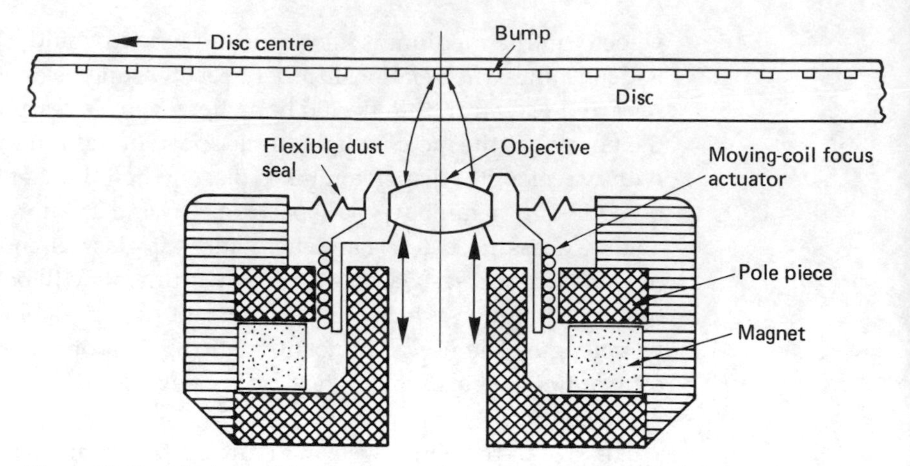

Figure 12.46 Moving-coil-focus servo can be coaxial with the light beam as shown.

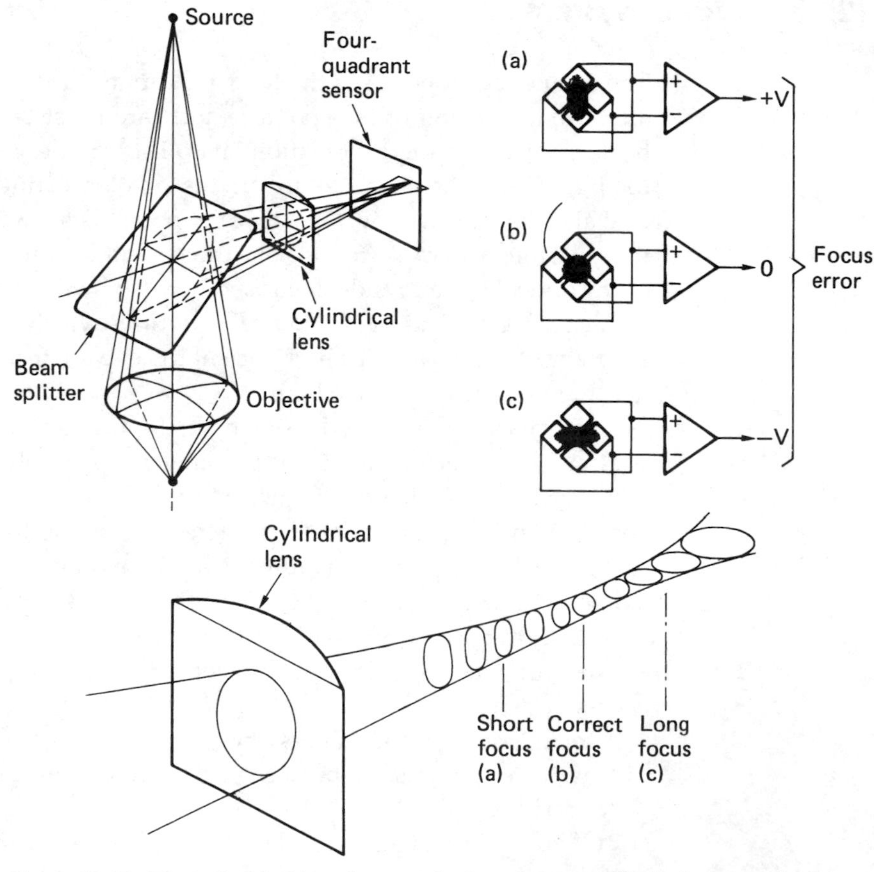

Figure 12.47 The cylindrical lens focus method produces an elliptical spot on the sensor whose aspect ratio is detected by a four-quadrant sensor to produce a focus error.

sided, and has negligible effect on the focal length of the main system, whereas in the other plane, the lens shortens the focal length. The image will be an ellipse whose aspect ratio changes as a function of the state of focus. Between the two foci, the image will be circular. The aspect ratio of the ellipse, and hence the focus error, can be found by dividing the sensor into quadrants. When these are connected as shown, the focus-error signal is generated. The data readout signal is the sum of the quadrant outputs.

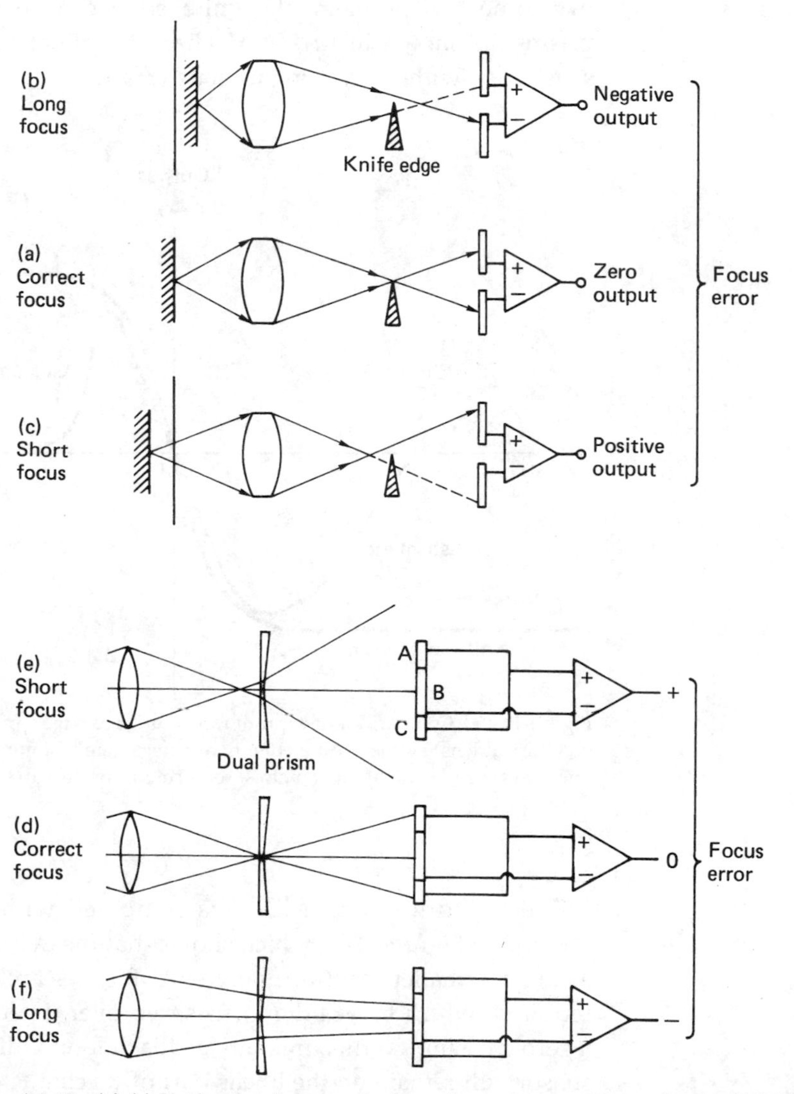

Figure 12.48 (a)-(c) Knife-edge focus-method requires only two sensors, but is critically dependent on knife-edge position. (d)-(f) Twin-prism method requires three sensors (A, B, C), where focus error is (A + C) – B. Prism alignment reduces sensitivity without causing focus error.

Figure 12.48 shows the knife-edge method of determining focus. A split sensor is also required. In (a) the focal point is coincident with the knife edge, so it has little effect on the beam. In (b) the focal point is to the right of the knife edge, and rising rays are interrupted, reducing the output of the upper sensor. In (c) the focal point is to the left of the knife edge, and descending rays are interrupted, reducing the output of the lower sensor. The focus error is derived by comparing the outputs of the two halves of the sensor. A drawback of the knife-edge system is that the lateral position of the knife edge is critical, and adjustment is necessary. To overcome this problem, the knife edge can be replaced by a pair of prisms, as shown in (d)–f). Mechanical tolerances then only affect the sensitivity, without causing a focus offset.

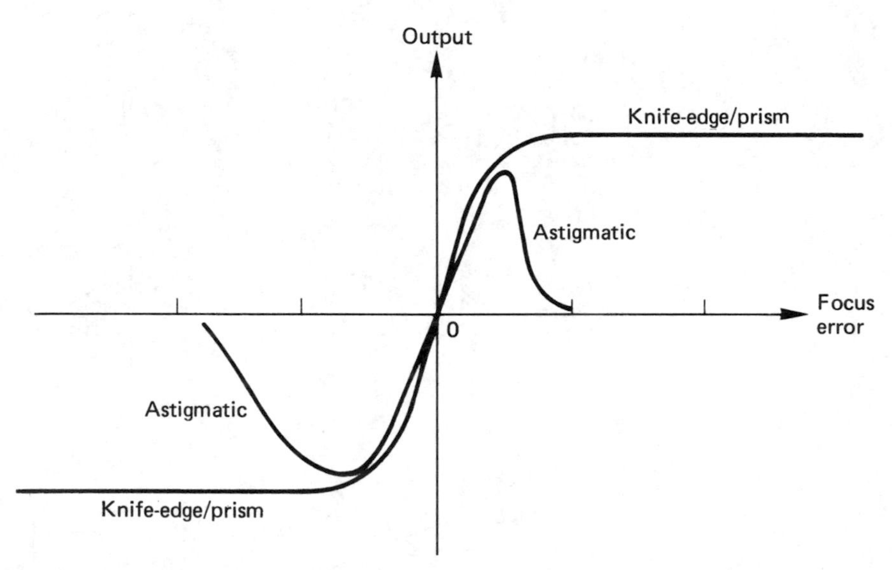

Figure 12.49 Comparison of capture range of knife-edge/prism method and astigmatic (cylindrical lens) system. Knife edge may have range of 1 mm, whereas astigmatic may only have a range of 40 micrometres, requiring a focus-search mechanism.

The cylindrical lens method is compared with the knife-edge/prism method in Figure 12.49, which shows that the cylindrical lens method has a much smaller capture range. A focus-search mechanism will be required, which moves the focus servo over its entire travel, looking for a zero crossing. At this time the feedback loop will be completed, and the sensor will remain on the linear part of its characteristic. The spiral track of CD and MiniDisc starts at the inside and works outwards. This was deliberately arranged because there is less vertical runout near the hub, and initial focusing will be easier.

12.21 Tracking systems

The track pitch of a typical optical disk is of the order of a micrometre, and this is much smaller than the accuracy to which the player chuck or the disk centre hole can be made; on a typical player, runout will swing several tracks past a fixed pickup. The non-contact readout means that there is no inherent mechanical guidance of the pickup. In addition, a warped disk will not present its surface at 90° to the beam, but will constantly change the angle of incidence during two whole cycles per revolution. Due to the change of refractive index at the disk surface, the tilt will change the apparent position of the track to the pickup, and Figure 12.50 shows that this makes it appear wavy. Warp also results in coma of the readout spot. The disk format specifies a maximum warp amplitude to keep these effects under control. Finally, vibrations induced in the player from outside, particularly in portable and automotive players, will tend to disturb tracking. A track-following servo is necessary to keep the spot centralized on the track in the presence of these difficulties. There are several ways in which a tracking error can be derived.

In the three-spot method, two additional light beams are focused on the disk track, one offset to each side of the track centreline. Figure 12.51

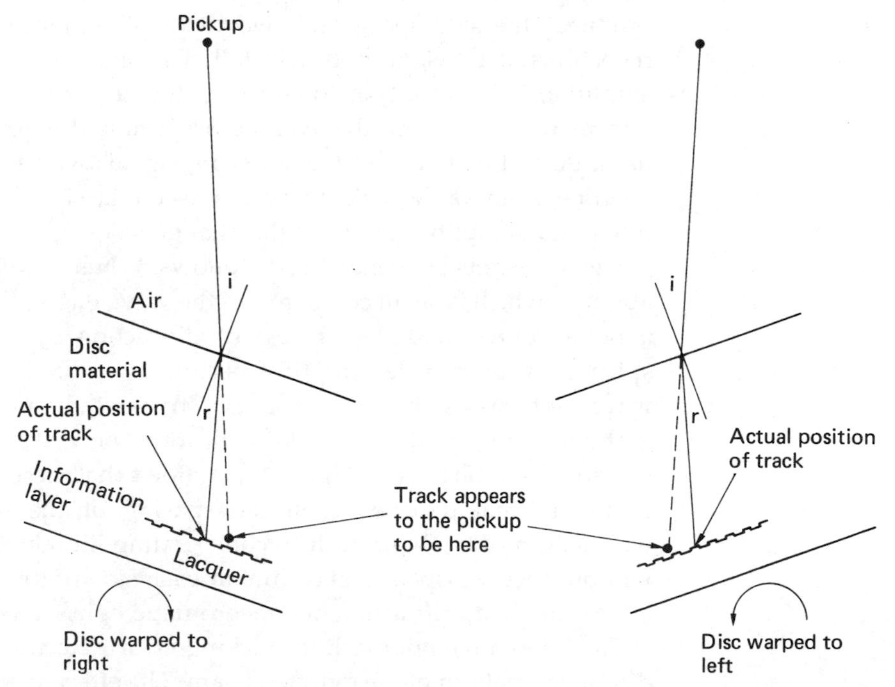

Figure 12.50 Owing to refraction, the angle of incidence (i) is greater than the angle of refraction (r). Disk warp causes the apparent position of the track (dotted line) to move, requiring the tracking servo to correct.

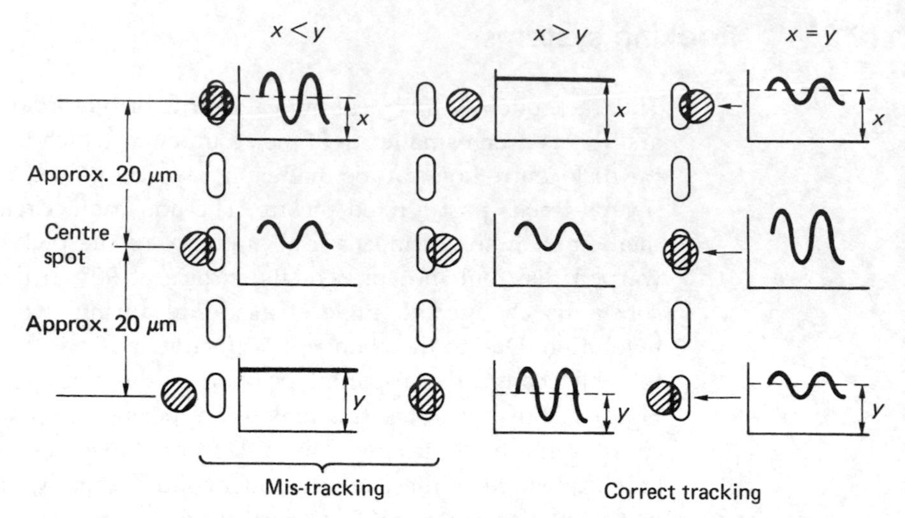

Figure 12.51 Three-spot method of producing tracking error compares average level of side-spot signals. Side spots are produced by a diffraction grating and require their own sensors.

shows that, as one side spot moves away from the track into the mirror area, there is less destructive interference and more reflection. This causes the average amplitude of the side spots to change differentially with tracking error. The laser head contains a diffraction grating which produces the side spots, and two extra photosensors onto which the reflections of the side spots will fall. The side spots feed a differential amplifier, which has a low-pass filter to reject the channel-code information and retain the average brightness difference. Some players use a delay line in one of the side-spot signals whose period is equal to the time taken for the disk to travel between the side spots. This helps the differential amplifier to cancel the channel code.

The side spots are generated as follows. When a wavefront reaches an aperture which is small compared to the wavelength, the aperture acts as a point source, and the process of diffraction can be observed as a spherical wavefront leaving the aperture as in Figure 12.52. Where the wavefront passes through a regular structure, known as a diffraction grating, light on the far side will form new wavefronts wherever radiation is in phase, and Figure 12.53 shows that these will be at an angle to the normal depending on the spacing of the structure and the wavelength of the light. A diffraction grating illuminated by white light will produce a dispersed spectrum at each side of the normal. To obtain a fixed angle of diffraction, monochromatic light is necessary.

The alternative approach to tracking-error detection is to analyse the diffraction pattern of the reflected beam. The effect of an off-centre spot is to rotate the radial diffraction pattern about an axis along the track. Figure 12.54 shows that, if a split sensor is used, one half will see greater

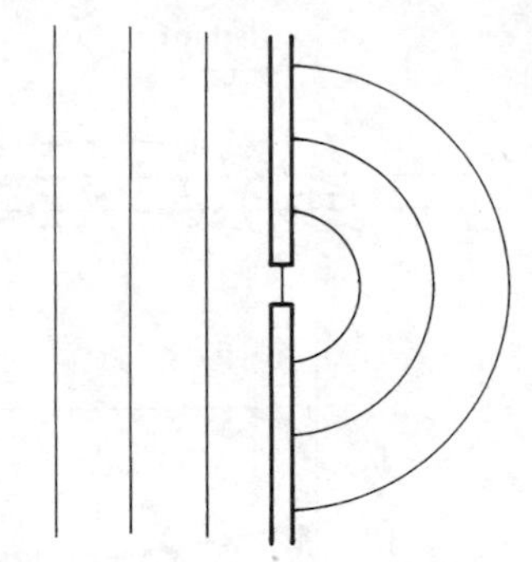

Figure 12.52 Diffraction as a plane wave reaches a small aperture.

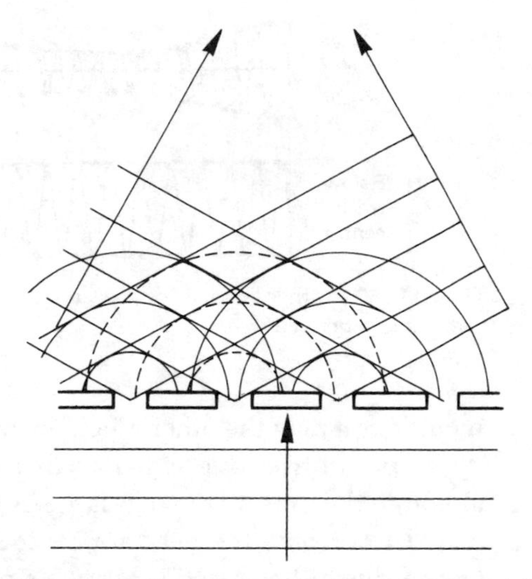

Figure 12.53 In a diffraction grating, constructive interference can take place at more than one angle for a single wavelength.

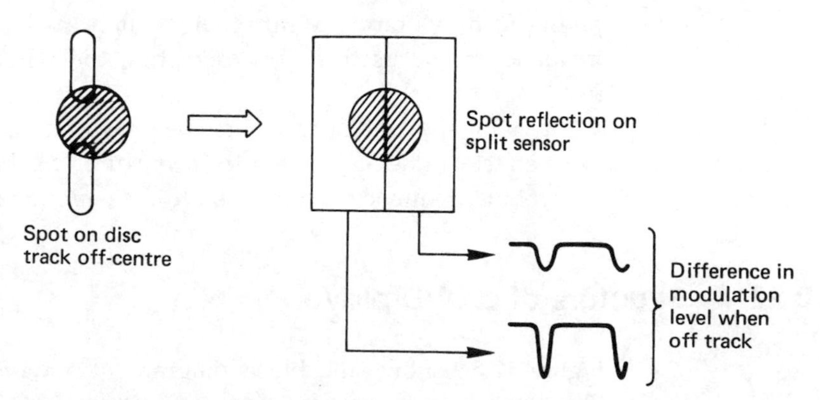

Figure 12.54 Split-sensor method of producing tracking error focuses image of spot onto sensor. One side of spot will have more modulation when off track.

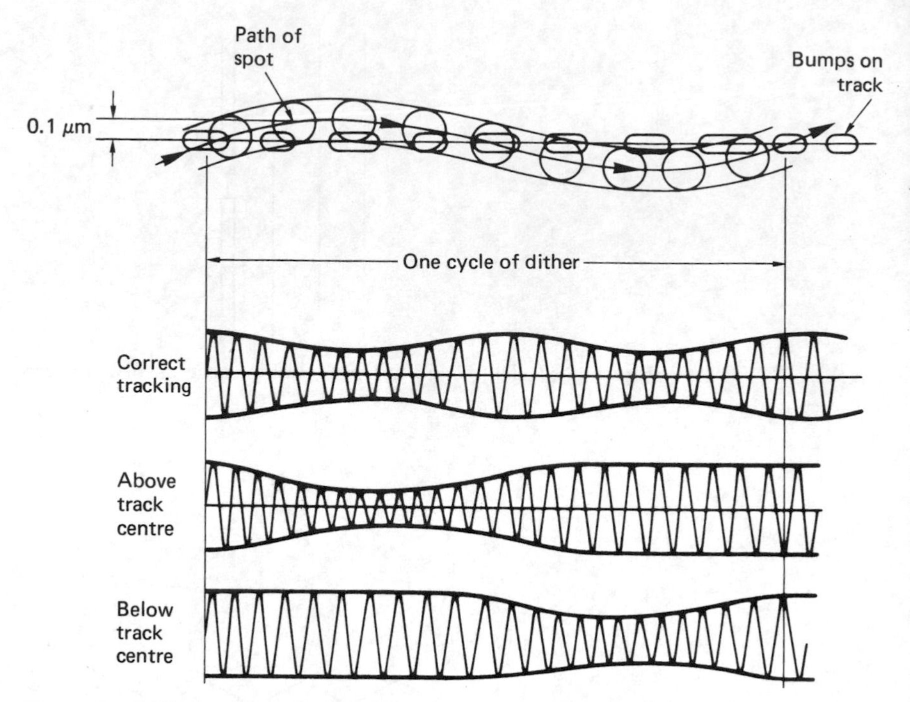

Figure 12.55 Dither applied to readout spot modulates the readout envelope. A tracking error can be derived.

modulation than the other when off-track. Such a system may be prone to develop an offset due either to drift or to contamination of the optics, although the capture range is large. A further tracking mechanism is often added to obviate the need for periodic adjustment. Figure 12.55 shows that in this dither-based system, a sinusoidal drive is fed to the tracking servo, causing a radial oscillation of spot position of about ±50 nm. This results in modulation of the envelope of the readout signal, which can be synchronously detected to obtain the sense of the error. The dither can be produced by vibrating a mirror in the light path, which enables a high frequency to be used, or by oscillating the whole pickup at a lower frequency.

Alternatively, the disk track may be produced with a wobble so that the same effect can be had with a fixed pickup. This has the advantage that the wobble frequency can be used to measure the speed of the track.

12.22 Structure of a DVD player

Figure 12.56 shows the block diagram of a typical DVD player, and illustrates the essential components. The most natural division within the block diagram is into the control/servo system and the data path. The

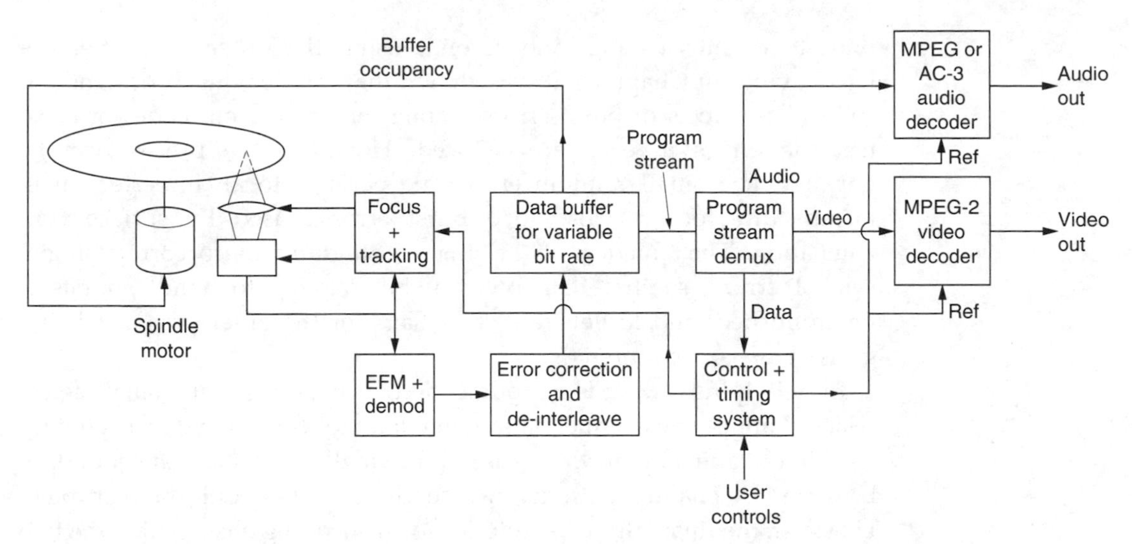

Figure 12.56 A DVD player's essential parts. See text for details.

control system provides the interface between the user and the servo mechanisms, and performs the logical interlocking required for safety and the correct sequence of operation.

The servo systems include any power-operated loading drawer and chucking mechanism, the spindle-drive servo, and the focus and tracking servos already described. Power loading is usually implemented on players where the disk is placed in a drawer. Once the drawer has been pulled into the machine, the disk is lowered onto the drive spindle, and clamped at the centre, a process known as chucking. In the simpler top-loading machines, the disk is placed on the spindle by hand, and the clamp is attached to the lid so that it operates as the lid is closed.

The lid or drawer mechanisms have a safety switch which prevents the laser operating if the machine is open. This is to ensure that there can be no conceivable hazard to the user. In actuality there is very little hazard in a DVD pickup. This is because the beam is focused a few millimetres away from the objective lens, and beyond the focal point the beam diverges and the intensity falls rapidly. It is almost impossible to position the eye at the focal point when the pickup is mounted in the player, but it would be foolhardy to attempt to disprove this.

The data path consists of the data separator, the de-interleaving and error-correction process followed by a RAM buffer which supplies the MPEG decoder. The data separator converts the EFMplus readout waveform into data. Following data separation the error-correction and de-interleave processes take place. Because of the interleave system, there are two opportunities for correction, first, using the inner code prior to de-interleaving, and second, using the outer code after de-interleaving. In Chapter 9 it was shown that interleaving is designed to spread the effects of burst errors among many different codewords, so that the errors in each are reduced. However, the process can be impaired if a small random error, due perhaps to an imperfection in manufacture, occurs close to a burst error caused by surface contamination. The function of the inner redundancy is to correct single-symbol errors, so that the power of interleaving to handle bursts is undiminished, and to generate error flags for the outer system when a gross error is encountered.

The EFMplus coding is a group code which means that a small defect which changes one channel pattern into another could have corrupted up to eight data bits. In the worst case, if the small defect is on the boundary between two channel patterns, two successive bytes could be corrupted. However, the final odd/even interleave on encoding ensures that the two bytes damaged will be in different inner codewords; thus a random error can never corrupt two bytes in one inner codeword, and random errors are therefore always correctable.

The de-interleave process is achieved by writing sequentially into a memory and reading out using a sequencer. The outer decoder will then correct any burst errors in the data. As MPEG data are very sensitive to error the error-correction performance has to be extremely good.

Following the de-interleave and outer error-correction process an MPEG program stream (see Chapter 6) emerges. Some of the program stream data will be video, some will be audio and this will be routed to the appropriate decoder. It is a fundamental concept of DVD that the bit rate of this program stream is not fixed, but can vary with the difficulty of the program material in order to maintain consistent image quality. The bit rate is changed by changing the speed of the disk. However, there is a complication because the disk uses constant linear velocity rather than constant angular velocity. It is not possible to obtain a particular bit rate with a fixed spindle speed.

The solution is to use a RAM buffer between the transport and the MPEG decoders. The RAM is addressed by counters which are arranged to overflow, giving the memory a ring structure as described in Chapter 1. Writing into the memory is done using clocks derived from the disk whose frequency rises and falls with runout, whereas reading is done by the decoder which, for each picture, will take as much data as is required from the buffer.

The buffer will only function properly if the two addresses are kept apart. This implies that the amount of data read from the disk over the long term must equal the amount of data used by the MPEG decoders. This is done by analysing the address relationship of the buffer. If the disk is turning too fast, the write address will move towards the read address;

if the disk is turning too slowly, the write address moves away from the read address. Subtraction of the two addresses produces an error signal which can be fed to the spindle motor.

The speed of the motor is unimportant. The important factor is that the data rate needed by the decoder is correct, and the system will drive the spindle at whatever speed is necessary so that the buffer neither underflows nor overflows.

The MPEG decoder will convert the compressed Elementary Streams into PCM video and audio and place the pictures and audio blocks into RAM. These will be read out of RAM whenever the time stamps recorded with each picture or audio block match the state of a time stamp counter. If bidirectional coding is used, the RAM readout sequence will convert the recorded picture sequence back to the real time sequence. The time stamp counter is derived from a crystal oscillator in the player which is divided down to provide the 90 kHz time stamp clock.

As a result the frame rate at which the disk was mastered will be replicated as the pictures are read from RAM. Once a picture buffer is read out, this will trigger the decoder to decode another picture. It will read data from the buffer until this has been completed and thus indirectly influence the disk speed.

Due to the use of constant linear velocity, the disk speed will be wrong if the pickup is suddenly made to jump to a different radius using manual search controls. This may force the data separator out of lock, or cause a buffer overflow and the decoder may freeze briefly until this has been remedied.

The control system of a DVD player is inevitably microprocessor-based, and as such does not differ greatly in hardware terms from any other microprocessor-controlled device. Operator controls will simply interface to processor input ports and the various servo systems will be enabled or overridden by output ports. Software, or more correctly firmware, connects the two. The necessary controls are Play and Eject, with the addition in most players of at least Pause and some buttons which allow rapid skipping through the program material.

Although machines vary in detail, the flowchart of Figure 12.57 shows the logic flow of a simple player, from start being pressed to pictures and sound emerging. At the beginning, the emphasis is on bringing the various servos into operation. Towards the end, the disk subcode is read in order to locate the beginning of the first section of the program material.

When track-following, the tracking-error feedback loop is closed, but for track crossing, in order to locate a piece of material, the loop is opened, and a microprocessor signal forces the laser head to move. The tracking error becomes an approximate sinusoid as tracks are crossed. The cycles of tracking error can be counted as feedback to determine

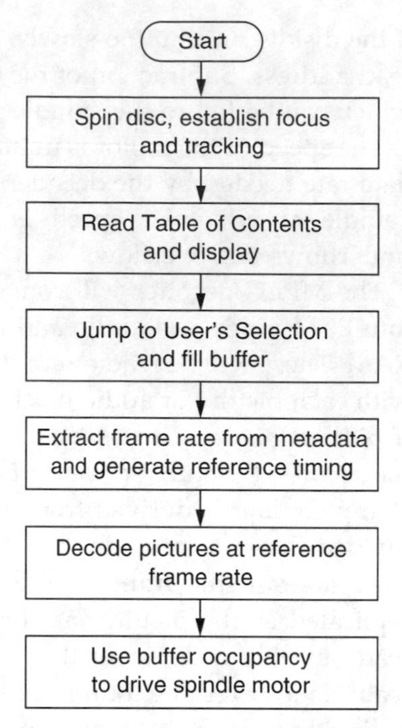

Figure 12.57 Simple processes required for a DVD player to operate.

when the correct number of tracks have been crossed. The 'mirror' signal obtained when the readout spot is half a track away from target is used to brake pickup motion and re-enable the track following feedback.

References

1. Bouwhuis, G. *et al.*, *Principles of Optical Disc Systems*, Bristol: Adam Hilger (1985)
2. Itoi, S. *et al.*, Development of the PAL digital MO disc recorder. *Montreux ITS Record*, 545 (1993)
3. Mee, C.D. and Daniel, E.D. (eds) *Magnetic Recording Vol.III*, Chapter 6, New York: McGraw-Hill (1988)
4. Airy, G.B., *Trans. Camb. Phil. Soc.*, **5**, 283 (1835)
5. Ray, S.F., *Applied Photographic Optics*, Chapter 17, Oxford: Focal Press (1988)
6. Maréchal, A., *Rev. d'Optique*, **26**, 257 (1947)
7. Hopkins, H.H., Diffraction theory of laser read-out systems for optical video discs. *J. Opt. Soc. Am.*, **69**, 4 (1979)
8. Connell, G.A.N., Measurement of the magneto-optical constants of reactive metals. *Appl. Opt.*, **22**, 3155 (1983)
9. Goldberg, N., A high density magneto-optic memory. *IEEE Trans. Magn.*, **MAG-3**, 605 (1967)

13

Digital video editing

13.1 Introduction

The term editing covers a multitude of possibilities in video production. Simple video editors work in two basic ways, by assembling or by inserting sections of material or *clips* comprising a whole number of frames to build the finished work. Assembly begins with a blank master recording. The beginning of the work is copied from the source, and new material is successively appended to the end of the previous material. Figure 13.1 shows how a master recording is made up by assembly from source recordings. Insert editing begins with an existing recording in which a section is replaced by the edit process.

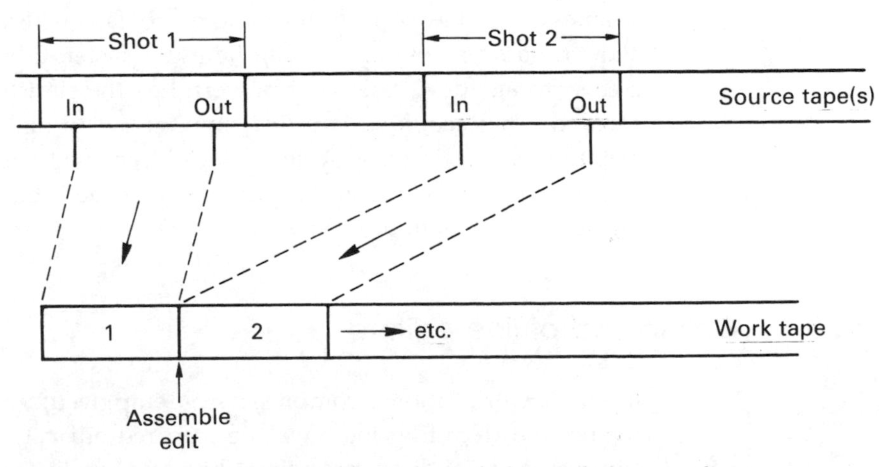

Figure 13.1 Assembly is the most basic form of editing where source clips are sequentially added to the end of a recording.

At its most general, editing is subdivided into horizontal editing, which refers to any changes with respect to the time axis, and vertical editing,[1] which is the generic term for processes which take place on an imaginary z-axis running back into the screen. These include keying, dissolves, wipes, layering and so on.[2] DVEs may also be used for editing, where a page turn or rotate effect reveals a new scene.

In all types of editing the goal is the appropriate sequence of material at the appropriate time. The first type of picture editing was physically by cutting and splicing film, in order mechanically to assemble the finished work. This approach was copied on early quadruplex video recorders; a difficult and laborious process. This gave way to electronic editing on VTRs where lengths of source tape were copied to the master. Once the speed and capacity of disk drives became sufficient, it was obvious that they would take over as editing media.

13.2 Linear and non-linear editing

When video tape was the only way of editing, it did not need a qualifying name. Now that video is stored as data, alternative storage media have become available which allow editors to reach the same goal but using different techniques. Whilst digital VTR formats copy their analog predecessors and support field-accurate editing on the tape itself, in all other digital editing, pixels from various sources are brought from the storage media to various pages of RAM. The edit is previewed by selectively processing two (or more) sample streams retrieved from RAM. Once the edit is satisfactory it may subsequently be written on an output medium. Thus the nature of the storage medium does not affect the form of the edit in any way except the amount of time needed to execute it.

Tapes only allow serial or linear access to data, whereas disks and RAM allow random access and so can be much faster. Editing using random access storage devices is very powerful as the shuttling of tape reels is avoided. The technique is called non-linear editing. This is not a very helpful name, as in these systems the editing itself is performed in RAM in the same way as before. In fact it is only the time axis of the storage medium which is non-linear.

13.3 Online and offline editing

In many workstations, compression is employed, and the appropriate coding and decoding logic will be required adjacent to the inputs and outputs. With mild compresssion, the video quality of the machine may be used directly for some purposes. This is known as *online* editing and this may be employed for the creation of news programs. Alternatively a high compression factor may be used, and the editor is then used only to create an edit decision list (EDL). This is known as *offline* editing. The EDL is subsequently used to control automatic editing of the full bandwidth source material, possibly on tape. The full-bandwidth material is *conformed* to the edit decisions taken on the compressed material.

13.4 Timecode

Timecode is essential to editing, as many different processes occur during an edit, and each one is programmed beforehand to take place at a given timecode value. Provision of a timecode reference effectively synchronizes the processes.

SMPTE standard timecode for 525/60 use is shown in Figure 13.2. EBU timecode is basically similar to SMPTE except that the frame count will reach a lower value in each second. These store hours, minutes, seconds and frames as binary-coded decimal (BCD) numbers, which are serially encoded along with user bits into an FM channel code (see Chapter 8) which is recorded on one of the linear audio tracks of the tape. The user bits are not specified in the standard, but a common use is to record the take or session number. Disks also use timecode for synchronization, but the timecode forms part of the access mechanism so that samples are retrieved by specifying the required timecode. This mechanism was detailed in Chapter 12.

13.5 Editing on recording media

Digital recording media vary in their principle of operation, but all have in common the use of error correction. This requires an interleave, or reordering, of samples to reduce the impact of large errors, and the assembling of many samples into an error-correcting codeword. Codewords are recorded in constant sized blocks on the medium. In DVTRs this will be a segment or even a field. Vertical editing requires the modification of source material within the field to pixel accuracy. This contradicts the large interleaved block-based codes of real media.

Editing to pixel accuracy simply cannot be performed directly on real media. Even if an individual pixel could be located in a block, replacing it would destroy the codeword structure and render the block uncorrectable.

The only solution is to ensure that the medium itself is only edited at block boundaries so that entire error-correction codewords are written down. In order to obtain greater editing accuracy, blocks must be read

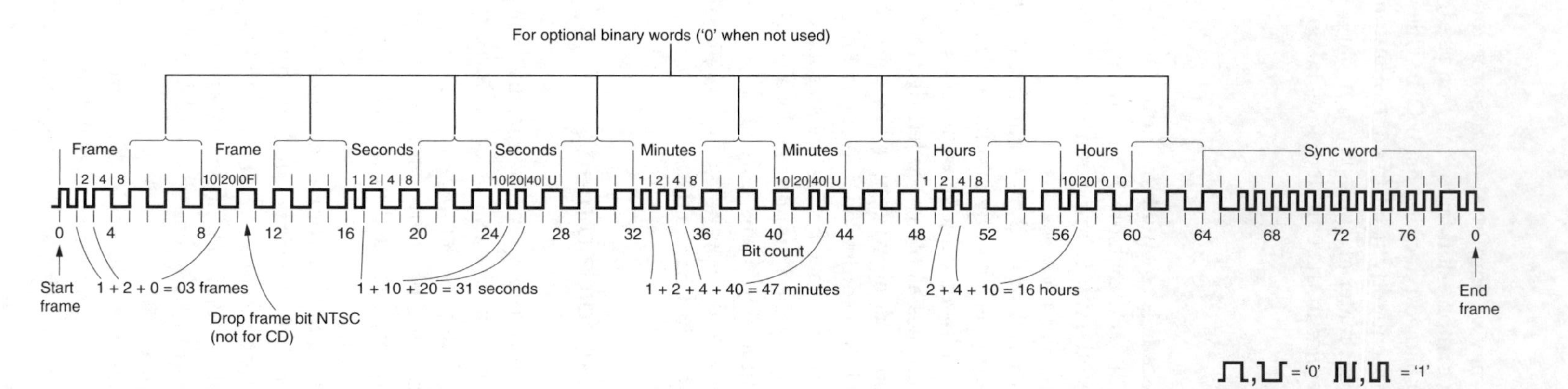

Figure 13.2 In SMPTE standard timecode, the frame number and time are stored as eight BCD symbols. There is also space for thirty-two user-defined bits. The code repeats every frame. Note the asymmetrical sync word which allows the direction of tape movement to be determined.

from the medium and de-interleaved into RAM, modified there and re-interleaved for writing back on the medium, the so-called *read–modify–write* process.

In disks, blocks are often associated into clusters which consist of a fixed number of blocks in order to increase data throughput. When clustering is used, editing on the disk can only take place by rewriting entire clusters.

13.6 A digital tape editor

The digital tape editor consists of three main areas. First, the various contributory recordings must enter the processing stage at the right time with respect to the master recording. This will be achieved using a combination of timecode, transport synchronization and RAM timebase correction. The synchronizer will take control of the various transports during an edit so that one section reaches its out-point just as another reaches its in-point.

Second, the video signal path of the editor must take the appropriate action, under timecode control, such as a dissolve or wipe, at the edit point. This requires some digital processing circuitry except in the case of the simplest editor which merely switches sources during the vertical interval.

Third, the editing operation must be supervised by a control system which coordinates the operation of the transports and the signal processing to achieve the desired result.

Figure 13.3 shows a block diagram of an editor. Each source transport must produce timecode locked to the data. The synchronizer section of the control system uses the timecode to determine the relative timing of sources and sends remote control signals to the transport(s) to make the timing correct. The master recorder is also fed with timecode in such a way that it can make a contiguous timecode track when performing assembly edits. The control system also generates a master video reference to which contributing devices must genlock in order to feed data into the edit process.

If vertical edits such as dissolves or wipes are contemplated, the master recorder must be a DVTR which has a pre-read function. Data from the pre-read heads are routed to the vision mixer and the mixer output is fed to the record input of the master recorder. The signal processor takes contributing sources and mixes or vertically edits them as instructed by the control system. The mix is then routed to the recorder.

Figure 13.4 shows the sequence of events during an edit where there is a wipe transition from the end of the previous clip on the master to the beginning of the next. Both the master recorder and the source machine

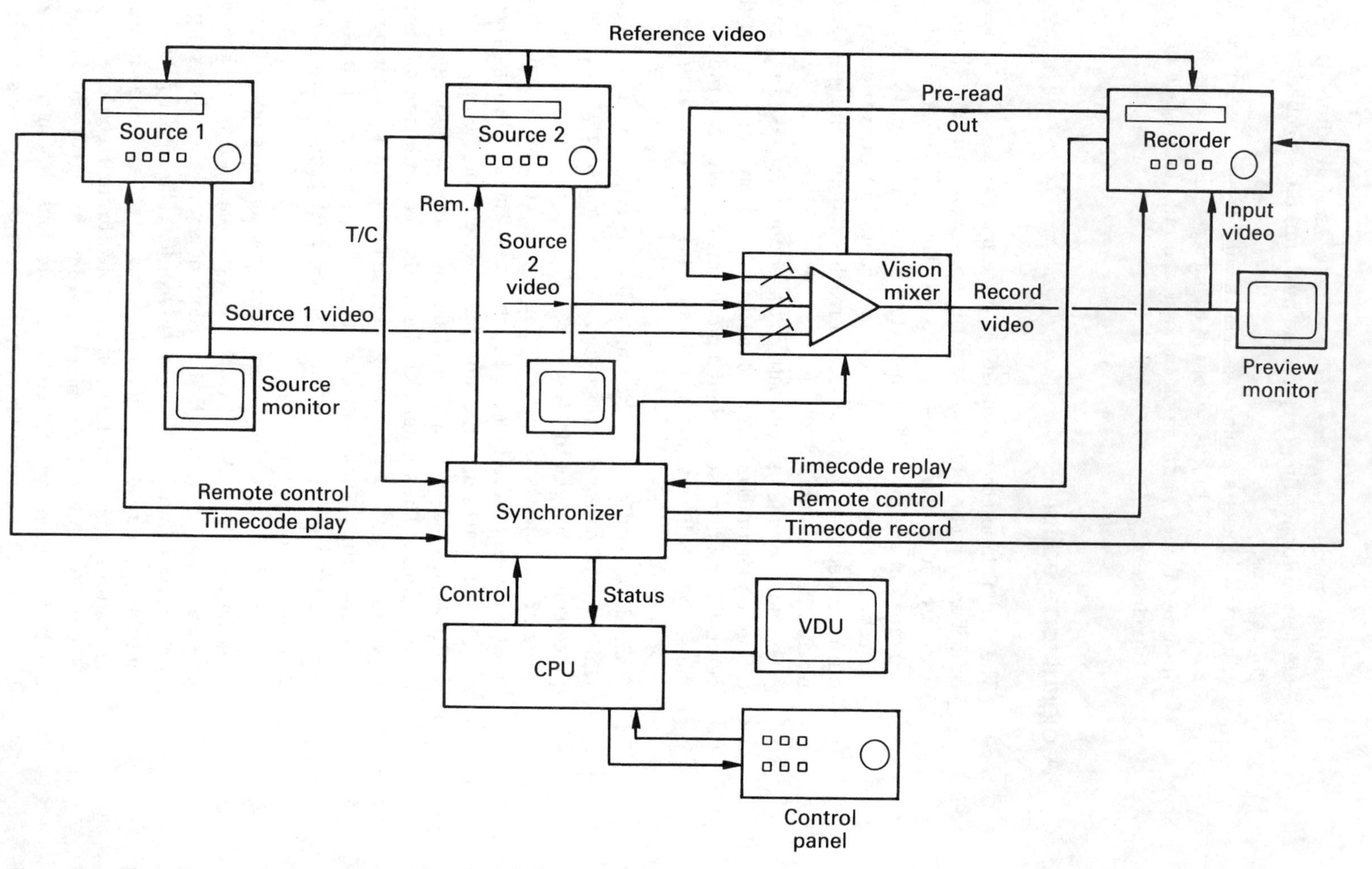

Figure 13.3 Block diagram of a tape edit controller. See text for details.

will reverse away from their respective edit points by the same number of frames, typically a few seconds. Both transports then roll and lock to reference video. If the correct timecode relationship is not obtained, the source machine will be frame slipped faster or slower until both will reach their edit points simultaneously. Initially the master recorder is simply playing back with the pre-read heads, and sending data to the vision mixer. The vision mixer is switched to the master recorder input and so the end of the first clip is seen on the monitor. When the appropriate timecode is reached, the master recorder begins recording at a field boundary. As the record input is simply the pre-read output, the tape is re-recorded with the existing data. However, once recording has commenced, the vision mixer transitions from the master pre-read signal to the source machine signal. This transition may take as long as required.

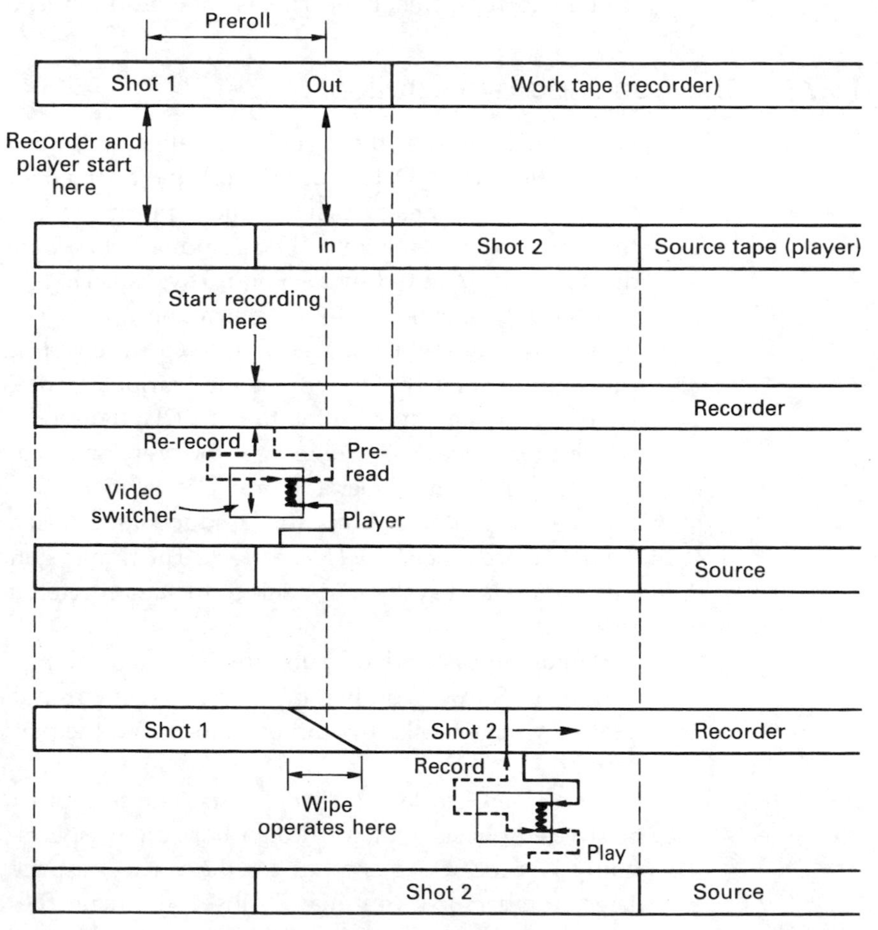

Figure 13.4 The sequence of events during an edit having a wipe between the clips.

In the middle of the transition the output fields will contain data from both tapes. At the end of the transition the vision mixer is outputting data from the source machine only, and the two machines are now dubbing. This will continue for as long as required.

A preview of the edit can be obtained by performing all the above steps with the exception of the master recorder entering record mode. The preview can be repeated as often as required and the edit points and the transition period can be changed until the desired effect is obtained without committing anything to tape.

In component DVTRs, any frame relationship between the two tapes may be used, but in the case of composite working, the edit must be constrained by the requirement for colour framing which was described in Chapter 9.

If a DVTR with three sets of heads and two playback signal paths is used, the edit can be monitored off-tape on the master recorder confidence replay channel. This is shown dotted in Figure 13.4.

13.7 The non-linear workstation

Figure 13.5 shows the general arrangement of a hard disk-based workstation. The VDU in such devices has a screen which is a montage of many different signals, each of which appear in windows. In addition to the video windows there will be a number of alphanumeric and graphic display areas required by the control system. There will also be a cursor which can be positioned by a trackball or mouse. The screen is refreshed by a frame store which is read at the screen refresh rate. The frame store can simultaneously be written by various processes to produce a windowed image. In addition to the VDU, there may be a second screen which reproduces full-size images for preview purposes.

A master timing generator provides reference signals to synchronize the internal processes. This also produces an external reference to which source devices such as VTRs can lock. The timing generator may free-run in a stand-alone system, or genlock to station reference to allow playout to air.

Digital inputs and outputs are provided, along with optional convertors to allow working in an analog environment. A compression process will generally be employed to extend the playing time of the disk storage.

Disk-based workstations fall into two categories depending on the relative emphasis of the vertical or horizontal aspects of the process. High end post-production emphasizes the vertical aspect of the editing as a large number of layers may be used to create the output image. The length of such productions is generally quite short and so disk capacity is not an issue and compression may not be employed. In contrast, a

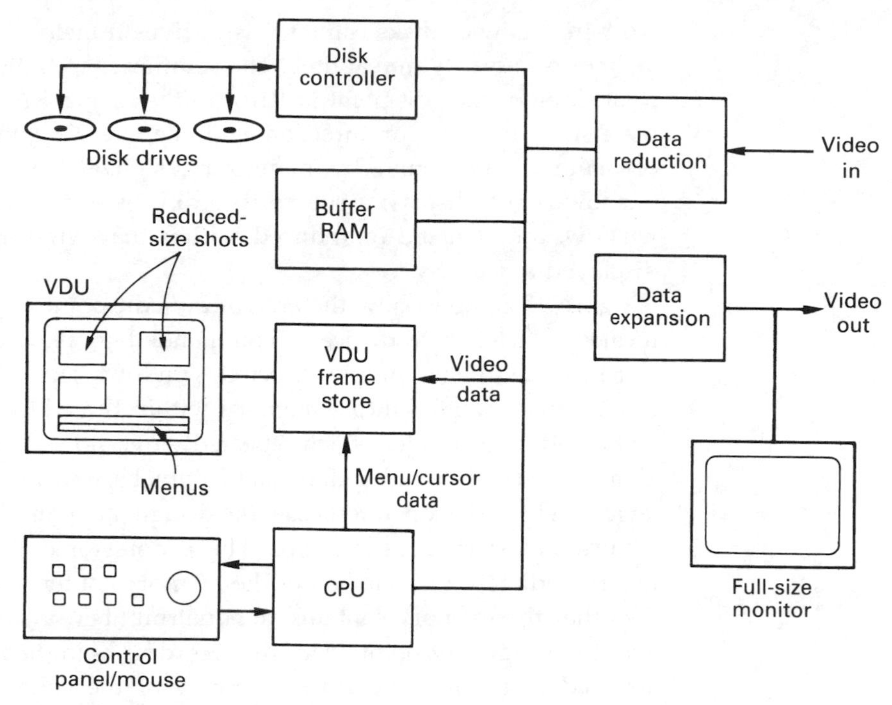

Figure 13.5 A hard-disk-based workstation. Note the screen which can display numerous clips at the same time.

general-purpose editor used for program production will emphasize the horizontal aspect of the task. Extended recording ability will be needed, and the use of compression is more likely.

The machine will be based around a high data rate bus, connecting the I/O, RAM, disk subsystem and the processor. If magnetic disks are used, these will be Winchester types, because they offer the largest capacity. Exchangeable magneto-optic disks may also be supported.

Before any editing can be performed, it is necessary to have source material online. If the source material exists on MO disks with the appropriate file structure, these may be used directly. Otherwise it will be necessary to input the material in real time and record it on magnetic disks via the data-reduction system. In addition to recording the data reduced source video, reduced-size versions of each field may also be recorded which are suitable for the screen windows.

13.8 Locating the edit point

Digital editors must simulate the 'rock and roll' process of edit-point location in VTRs where the tape is moved to and fro by the action of a jogwheel or joystick. Whilst DVTRs with track-following systems can work in this way, disks cannot. Disk drives transfer data intermittently and not necessarily in real time. The solution is to transfer the recording in the area of the edit point to RAM in the editor. RAM access can take place at any speed or direction and the precise edit point can then conveniently be found by monitoring signals from the RAM. In a window-based display, a source recording is attributed to a particular window, and will be reproduced within that window, with timecode displayed adjacently.

Figure 13.6 shows how the area of the edit point is transferred to the memory. The source device is commanded to play, and the operator watches the replay in the selected window. The same samples are continuously written into a memory within the editor. This memory is addressed by a counter which repeatedly overflows to give the memory a ring-like structure rather like that of a timebase corrector, but somewhat larger. When the operators see the rough area in which the edit is required, they will press a button. This action stops the memory writing, not immediately, but one half of the memory contents later. The effect is then that the memory contains an equal number of samples before and after the rough edit point. Once the recording is in the memory, it can be accessed at leisure, and the constraints of the source device play no further part in the edit-point location.

There are a number of ways in which the the memory can be read. If the field address in memory is supplied by a counter which is clocked at the appropriate rate, the edit area can be replayed at normal speed, or at some fraction of normal speed repeatedly. In order to simulate the analog

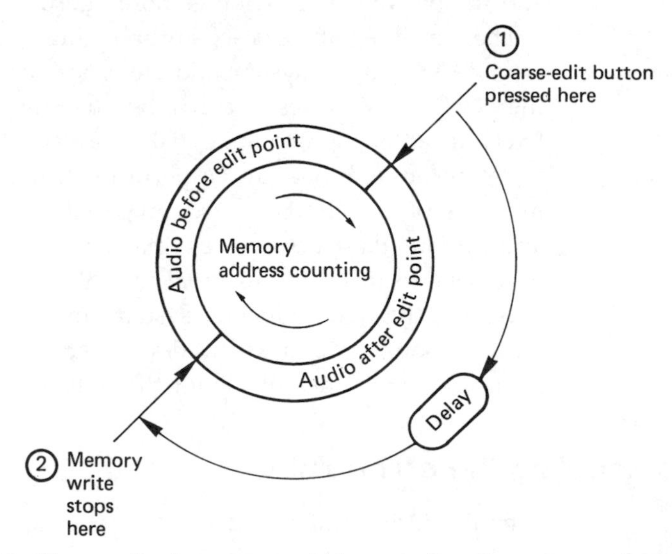

Figure 13.6 The use of a ring memory which overwrites allows storage of samples before and after the coarse edit point.

method of finding an edit point, the operator is provided with a *scrub wheel* or rotor, and the memory field address will change at a rate proportional to the speed with which the rotor is turned, and in the same direction. Thus the recording can be seen forward or backward at any speed, and the effect is exactly that of manually jogging an analog tape. The operation of a jogwheel encoder was shown in section 3.17.

If the position of the jog address pointer through the memory is compared with the addresses of the ends of the memory, it will be possible to anticipate that the pointer is about to reach the end of the memory. A disk transfer can be performed to fetch new data further up the time axis, so that it is possible to jog an indefinite distance along the source recording. Samples which will be used to make the master recording need never pass through these processes; they are solely to assist in the location of the edit points.

The act of pressing the coarse edit-point button stores the timecode of the source at that point, which is frame-accurate. As the rotor is turned, the memory address is monitored, and used to update the timecode.

Before the edit can be performed, two edit points must be determined, the out-point at the end of the previously recorded signal, and the in-point at the beginning of the new signal. The second edit point can be determined by moving the cursor to a different screen window in which video from a different source is displayed. The jogwheel will now roll this material to locate the second edit point while the first source video remains frozen in the deselected window. The editor's microprocessor stores these in an edit decision list (EDL) in order to control the automatic assemble process.

It is also possible to locate a rough edit point by typing in a previously noted timecode, and the image in the window will automatically jump to that time. In some systems, in addition to recording video and audio, there may also be text files locked to timecode which contain the dialog. Using these systems one can allocate a textual dialog display to a further window and scroll down the dialog or search for a key phrase as in a word processor. Unlike a word processor, the timecode pointer from the text access is used to jog the video window. As a result an edit point can be located in the video if the actor's lines at the desired point are known.

13.9 Editing with disk drives

Using one or other of the above methods, an edit list can be made which contains an in-point, an out-point and a filename for each of the segments of video which need to be assembled to make the final work, along with a timecode-referenced transition command and period for the vision mixer. This edit list will also be stored on the disk. When a preview of the edited work is required, the edit list is used to determine what files will be necessary and when, and this information drives the disk controller.

Figure 13.7 shows the events during an edit between two files. The edit list causes the relevant blocks from the first file to be transferred from disk to memory, and these will be read by the signal processor to produce the preview output. As the edit point approaches, the disk controller will also place blocks from the incoming file into the memory. In different areas of the memory there will simultaneously be the end of the outgoing recording and the beginning of the incoming recording. Before the edit point, only pixels from the outgoing recording are accessed, but as the transition begins, pixels from the incoming recording are also accessed, and for a time both data streams will be input to the vision mixer according to the transition period required.

The output of the signal processor becomes the edited preview material, which can be checked for the required subjective effect. If necessary the in- or out-points can be trimmed, or the crossfade period changed, simply by modifying the edit-list file. The preview can be

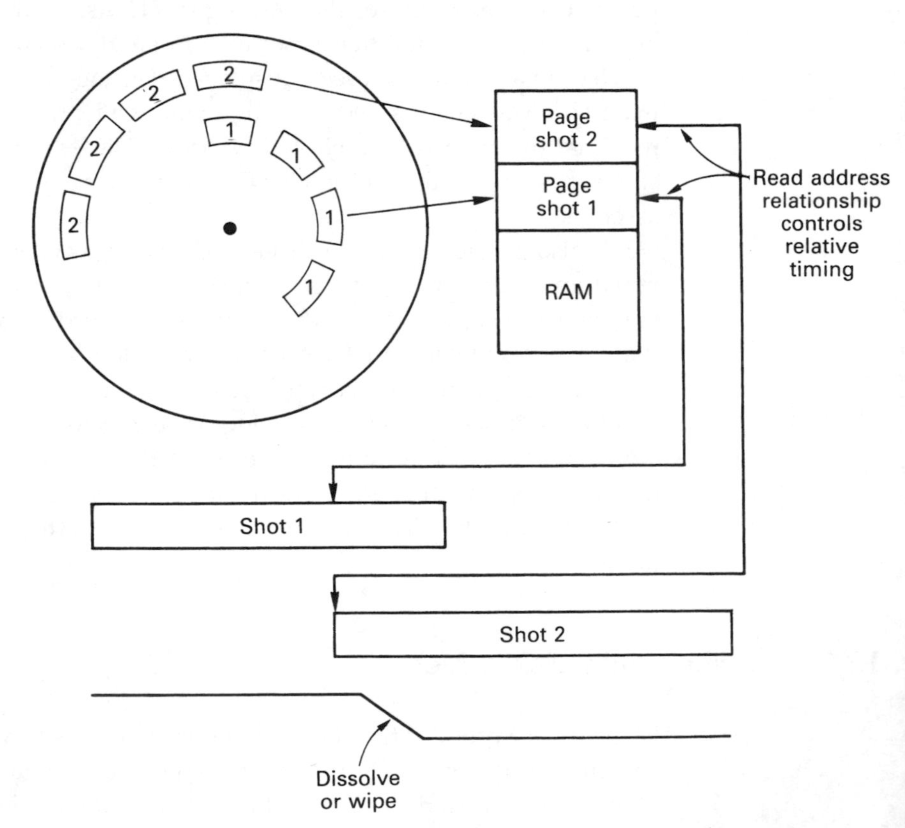

Figure 13.7 Sequence of events for a hard-disk edit. See text for details.

repeated as often as needed, until the desired effect is obtained. At this stage the edited work does not exist as a file, but is re-created each time by a further execution of the EDL. Thus a lengthy editing session need not fill up the disks.

It is important to realize that at no time during the edit process were the original files modified in any way. The editing was done solely by reading the files. The power of this approach is that if an edit list is created wrongly, the original recording is not damaged, and the problem can be put right simply by correcting the edit list. The advantage of a disk-based system for such work is that location of edit points, previews and reviews are all performed almost instantaneously, because of the random access of the disk. This can reduce the time taken to edit a program to a fraction of that needed with a tape machine.

During an edit, the disk controller has to provide data from two different files simultaneously, and so it has to work much harder than for a simple playback. If there are many close-spaced edits, the controller and drives may be hard-pressed to keep ahead of real time, especially if there are long transitions, because during a transition a vertical edit is taking place between two video signals and the source data rate is twice as great as during replay. A large buffer memory helps this situation because the drive can fill the memory with files before the edit actually begins, and thus the instantaneous sample rate can be met by allowing the memory to empty during disk-intensive periods.

Disk formats which handle defects dynamically, such as defect skipping, will also be superior to bad-block files when throughput is important. Some drives rotate the sector addressing from one cylinder to the next so that the drive does not lose a revolution when it moves to the next cylinder. Disk-editor performance is usually specified in terms of peak editing activity which can be achieved, but with a recovery period between edits. If an unusually severe editing task is necessary where the drive just cannot access files fast enough, it will be necessary to rearrange the files on the disk surface so that files which will be needed at the same time are on nearby cylinders.[2] An alternative is to spread the material between two or more drives so that overlapped seeks are possible.

Once the editing is finished, it will generally be necessary to transfer the edited material to form a contiguous recording so that the source files can make way for new work. If the source files already exist on tape the disk files can simply be erased. If the disks hold original recordings they will need to be backed up to tape if they will be required again. In large broadcast systems, the edited work can be broadcast directly from the disk file server. In smaller systems it will be necessary to output to some removable medium, since the Winchester drives in the editor have fixed media.

13.10 Digitally assisted film making

The power of non-linear editing can be applied to film making as well as to video production. There are various levels at which this can operate. Figure 13.8 shows the simplest level. Here the filming takes place as usual, and after development the uncut film is transferred to video using a telecine or datacine machine. The data are compressed and stored on a disk-based workstation. The workstation is used to make all the edit decisions and these are stored as an EDL. This will be sent to the film laboratory to control the film cutting.

In Figure 13.9 a more sophisticated proces is employed. Here the film camera viewfinder is replaced with a video camera so that a video signal is available during filming. This can be recorded on disk so that an immediate replay is available following each take. In the event that a retake is needed, the film need not be developed, reducing costs. Edit decisions can be taken before the film has been developed.

In some cases the disk database can be extended to include the assistant's notes and the film dialog. The editor can search for an edit point by having the system search for a text string. The display would then show a mosaic of all frames in which that dialog was spoken.

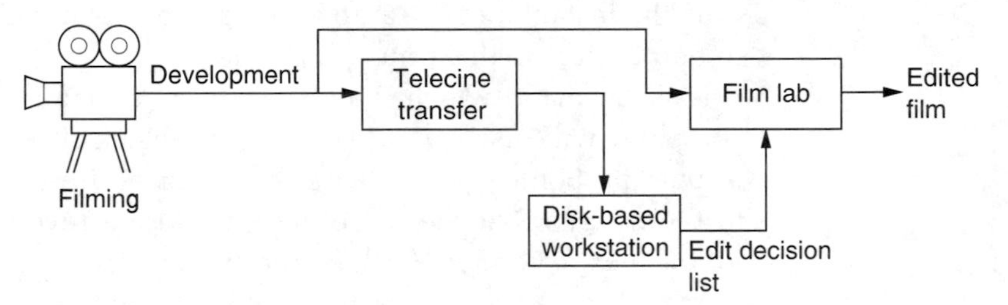

Figure 13.8 Films can be edited more quickly by transferring the uncut film to video and then to disk-based storage.

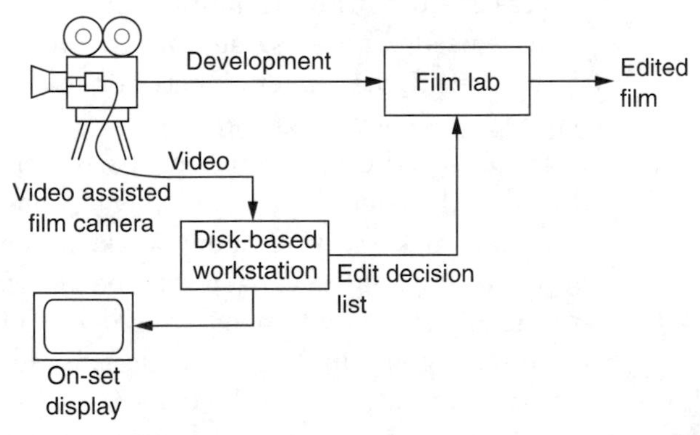

Figure 13.9 With a modified film camera which can also output video, the editing can begin before the film is developed.

Ultimately films may be made entirely electronically. When cameras of sufficient resolution and dynamic range become available, the cost of storage will be such that filming will be replaced by a direct camera-to-disk transfer. The production process will consist entirely of digital signal processing steps, resulting in a movie in the shape of a large data file.

This can be distributed to the cinema via copper or fibre-optic link, using encryption to prevent piracy and mild compression for economy. At the cinema the signal would be stored on a file server. Projection would then consist of accessing the data, decrypting, decoding the compression and delivering the data to a digital projector. This technology will change the nature of the traditional cinema out of recognition.

13.11 Automation

The economic circumstances of today's television industry dictate that increasingly equipment is replacing staff. Playout to air is a process which is suitable for extensive automation. Figure 13.10 shows the basic components of an automated playout system. A computer is required to keep track of events and this is driven by a real-time clock input forming the time reference. The time reference is constantly compared with an event list which dictates what will take place as time elapses. The event list is executed by sending commands to remote controlled signal sources and switching between them in a remote-controllable vision mixer.

The disk drive is eminently suitable for automation use as the access commands can easily be translated into disk addresses. However, disk storage is expensive and is currently limited to storage of commercials and news. Programs and movies are more economically stored on tape. Automation systems require tape to be handled mechanically under

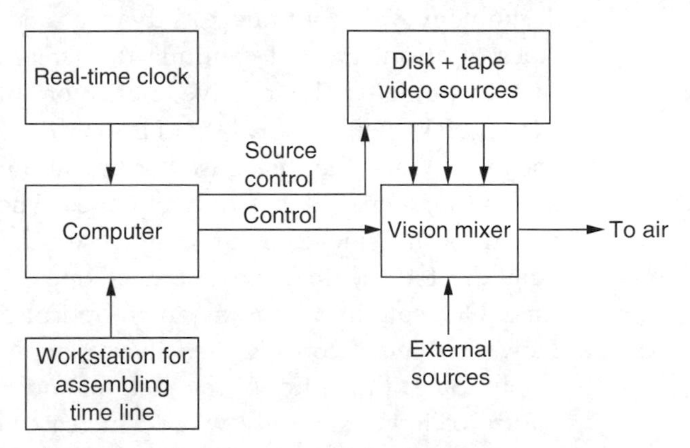

Figure 13.10 An automated system. See text for details.

computer control. Robotic tape storage systems store hundreds of cassettes in racks and move them to the tape transports on demand. The tape transfer robot consists of a picker head which can be positioned anywhere in the cassette library by an X–Y positioner driven by electric motors using positional feedback. The head contains a barcode reader which can scan the cassettes to confirm the right one has been located. It also contains a motorized gripper which can extend in the Z-axis to remove the cassette from the library and insert it in the transport.

The X–Y positioner of a large robotic cassette library moves at high speed driven by powerful motors and getting in its way will result in serious injury. The entire mechanism is enclosed behind glass doors for safety. Maintenance staff know how to disable the motors so that it is safe to open the machine.

The automation system is programmed in advance by entering events from a workstation. The automation software will indicate an error if the events do not join up, for example if it is told to broadcast a one-hour program in a 45-minute slot between two others. It is possible to preview a rehearsal of the programmed output ahead of time.

With an automation system running, the presentation staff simply check that the correct operations are taking place, and take over if there is a failure. Most robotic cassette stores have a spare tape drive so they can continue if a drive fails. Some have two X–Y robots so that if one fails the other can push it aside and continue. Most robots allow manual intervention. The positioner is disabled and the automation system displays the location of the cassettes so they can be brought to the tape drives manually.

13.12 The future

Technology will continue to advance, and as a result recording densities of all media will continue to increase, along with improvements in transfer rate. The relative merits of different media will stay the same; disks will always be faster than tape, tape will always be cheaper. Whilst the disk-based workstation has the advantage of rapid access to material, this is only obtained when the material is present on the disks. If Winchester disks are used these are irremovable and the data must come in from another source via an interface. This takes time. One solution is the magneto-optical disk which is exchangeable.[3] However, the recording density of an MO disk is limited by the resolution of the optics of the pickup, and this is governed by the wave nature of light. The only way to increase the density is to use shorter wavelengths, causing increasing difficulty in construction of the laser (and cost). Magnetic tape and disk do not suffer from such a limitation, and it may be that these media will develop further than MO media in the long term.

An alternative to exchangeable disks is to have tape and disk drives which can transfer faster than real time in order to speed up the process. The density of tape recording has a great bearing on the cost per bit. One limiting factor is the track width. In current DVTR formats, the track width is much greater than theoretically necessary because of the difficulty of mechanically tracking with the heads. This is compounded by the current insistence on editing to field accuracy on the tape itself. This is a convention inherited from analog VTRs, and it is unnecessary in digital machines. Digital VTRs employ read–modify–write, and this makes the edit precision in the data independent of the block size on tape. Future DVTRs may only be able to edit once per second, by employing edit gaps. This allows the tracks to be much narrower and the density rises. Field-accurate editing requires the block to be read intact, edited elsewhere, and written back whole. The obvious way to do this is on a disk. Thus a future approach to workstations is to integrate the DVTR and give it a format that it could not use as a stand-alone unit.

Until recently storage densities limited disk-based video to professional equipment. This restriction is no longer present. Now that disk storage suitable for digital video is available at consumer prices then non-linear technology is available to the viewer. Consumer disk-based VTRs are known as PVRs (personal video recorders). Unlike tape, which can only record or play back but not both at the same time, a

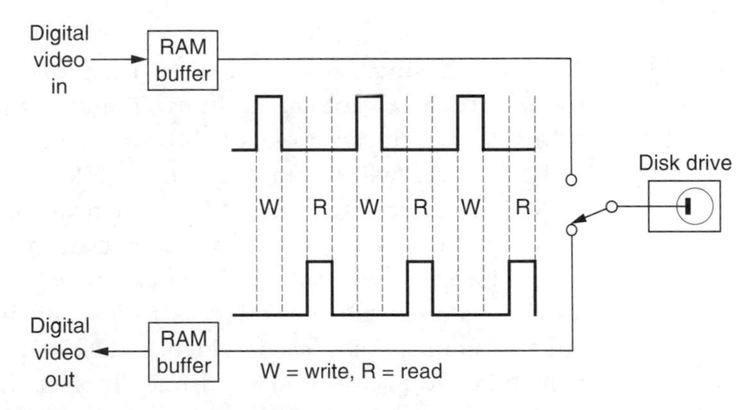

Figure 13.11 Using RAM buffers and burst transfer, a disk drive can read and write at the same time.

PVR can do both simultaneously at arbitrary points on a time line. Figure 13.11 shows how it is done. The disk drive can transfer data much faster than the required bit rate, and so it transfers data in bursts which are smoothed out by RAM buffers. The disk simply interleaves read and write functions so that it gives the impression of reading and writing simultaneously. The read and write processes can take place from anywhere on the disk.

Clearly the PVR can be used as an ordinary video recorder, but it can do some other tricks. Figure 13.12 shows the most far-reaching trick. The disk drive starts recording an off-air commercial TV station. A few minutes later the viewer starts playing back the recording. When the commercial break is transmitted, the disk drive may record it, but the viewer can skip over it using the random access of the hard

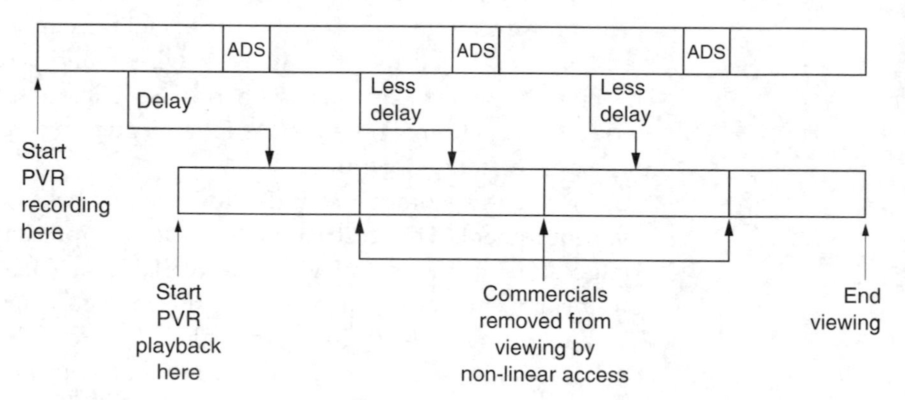

Figure 13.12 PVR allows viewing of commercial TV without any need to watch commercials.

drive. With suitable software the hard drive could skip over the commercial break automatically by simply not recording it.

Devices which can do this will be available at consumer prices soon and the prices will fall rapidly. Once PVRs become as popular as the VCR, it is difficult to see how commercial television as we know it today can survive. Industry analysts reckon on about ten years.

Paradoxically one of the technologies which has made the hard drive feasible as a consumer product is MPEG. By providing low-cost chip sets to compress and decode video, MPEG helps the PVR to work with the limited capacity of hard drives. In fact digital television broadcasting is also doing the PVR a favour. The PVR can simply record the transmitted Elementary Stream data and replay it into an MPEG decoder. In this way the PVR has no quality loss whatsoever. The picture quality will be the same as if it were off-air.

References

1. Rubin, M. The emergence of the desktop: Implications to offline editing. *Record of 18th ITS*, 384–389 (Montreux 1993)
2. Trottier, L. Digital video compositing on the desktop. *Record of 18th ITS*, 564–570 (Montreux 1993)
3. Lamaa, F. Open media framework interchange. *Record of 18th ITS*, 571–597 (Montreux 1993)

14

Digital television broadcasting

14.1 Background

Digital television broadcasting relies on the combination of a number of fundamental technologies. These are: MPEG-2 compression to reduce the bit rate, multiplexing to combine picture and sound data into a common bitstream, digital modulation schemes to reduce the RF bandwidth needed by a given bit rate and error correction to reduce the error statistics of the channel down to a value acceptable to MPEG data.

MPEG-compressed video is highly sensitive to bit errors, primarily because they confuse the recognition of variable-length codes so that the decoder loses synchronization. However, MPEG is a compression and multiplexing standard and does not specify how error correction should be performed. Consequently a transmission standard must define a system which has to correct essentially all errors such that the delivery mechanism is transparent.

Essentially a transmission standard specifies all the additional steps needed to deliver an MPEG transport stream from one place to another. This transport stream will consist of a number of Elementary Streams of video and audio, where the audio may be coded according to MPEG audio standard or AC-3. In a system working within its capabilities, the picture and sound quality will be determined only by the performance of the compression system and not by the RF transmission channel. This is the fundamental difference between analog and digital broadcasting. In analog television broadcasting, the picture quality may be limited by composite video encoding artifacts as well as transmission artifacts such as noise and ghosting. In digital television broadcasting the picture quality is determined instead by the compression artifacts and interlace artifacts if interlace has been retained.

If the received error rate increases for any reason, once the correcting power is used up, the system will degrade rapidly as uncorrected errors enter the MPEG decoder. In practice, decoders will be programmed to recognize the condition and blank or mute to avoid outputting garbage. As a result, digital receivers tend either to work well or not at all.

It is important to realize that the signal strength in a digital system does not translate directly to picture quality. A poor signal will increase the number of bit errors. Provided that this is within the capability of the error-correction system, there is no visible loss of quality. In contrast, a very powerful signal may be unusable because of similarly powerful reflections due to multipath propagation.

Whilst in one sense an MPEG transport stream is only data, it differs from generic data in that it must be presented to the viewer at a particular rate. Generic data is usually asynchronous, whereas baseband video and audio are synchronous. However, after compression and multiplexing audio and video are no longer precisely synchronous and so the term *isochronous* is used. This means a signal which was at one time synchronous and will be displayed synchronously, but which uses buffering at transmitter and receiver to accommodate moderate timing errors in the transmission.

Clearly another mechanism is needed so that the time axis of the original signal can be re-created on reception. The time stamp and program clock refrence system of MPEG does this.

14.2 Overall system block

Figure 14.1 shows that the concepts involved in digital television broadcasting exist at various levels which have an independence not found in analog technology. In a given configuration a transmitter can radiate a given payload data bit rate. This represents the useful bit rate and does not include the necessary overheads needed by error correction, multiplexing or synchronizing. It is fundamental that the transmission system does not care what this payload bit rate is used for. The entire capacity may be used up by one high-definition channel, or a large number of heavily compressed channels may be carried. The details of this data usage are the domain of the *transport stream*. The multiplexing of transport streams is defined by the MPEG standards, but these do not define any error correction or transmission technique.

At the lowest level in Figure 14.2 the source coding scheme, in this case MPEG compression, results in one or more Elementary Streams, each of which carries a video or audio channel. Elementary Streams are multiplexed into a transport stream. The viewer then selects the desired Elementary Stream from the transport stream. Metadata in the transport

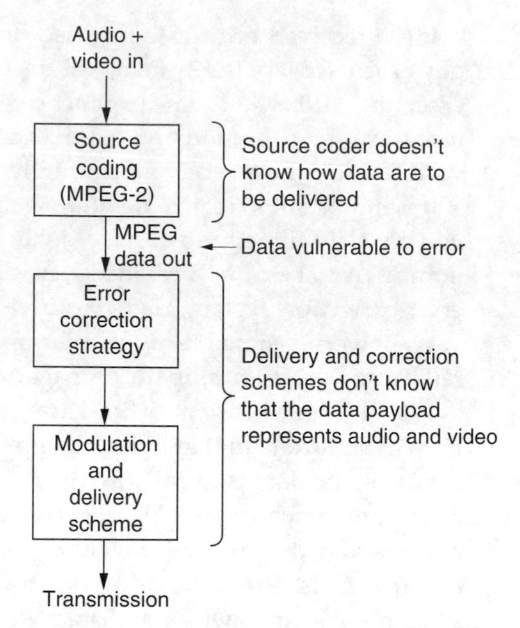

Figure 14.1 Source coder doesn't know delivery mechanism and delivery doesn't need to know what the data mean.

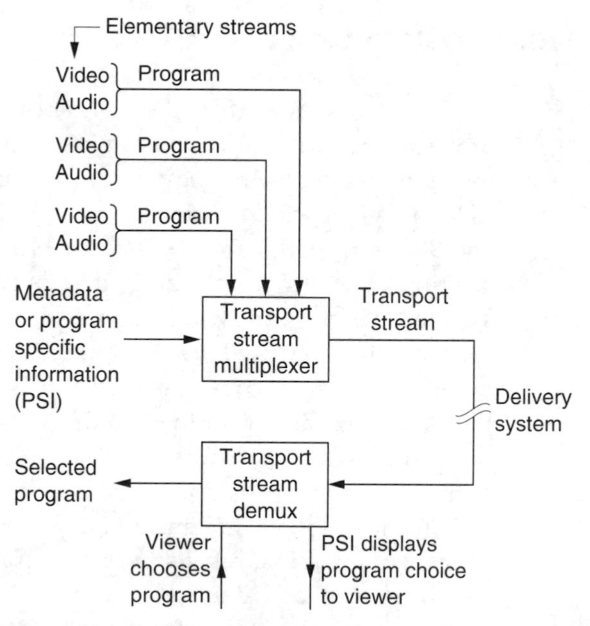

Figure 14.2 Program Specific Information helps the demultiplexer to select the required program.

stream ensures that when a video Elementary Stream is chosen, the appropriate audio Elementary Stream will automatically be selected.

14.3 Packets and time stamps

The video Elementary Stream is an endless bitstream representing pictures which take a variable length of time to transmit. Bidirection coding means that pictures are not necessarily in the correct order. Storage and transmission systems prefer discrete blocks of data and so Elementary Streams are packetized to form a PES (Packetized Elementary Stream). Audio Elementary Streams are also packetized. A packet is shown in Figure 14.3. It begins with a header containing an unique packet start code and a code which identifies the type of data stream. Optionally the packet header also may contain one or more *time stamps* which are used for synchronizing the video decoder to real time and for obtaining lip-sync.

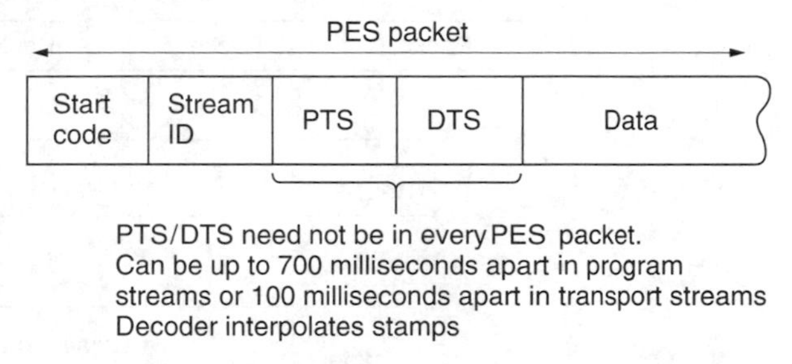

Figure 14.3 A PES packet structure is used to break up the continuous Elementary Stream.

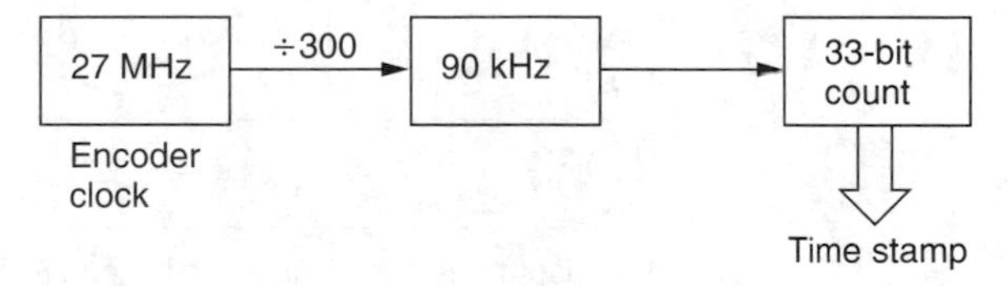

Figure 14.4 Time stamps are the result of sampling a counter driven by the encoder clock.

Figure 14.4 shows that a time stamp is a sample of the state of a counter which is driven by a 90 kHz clock. This is obtained by dividing down the master 27 MHz clock of MPEG-2. This 27 MHz clock must be locked to the video frame rate and the audio sampling rate of the program concerned. There are two types of time stamp: PTS and DTS. These are abbreviations for presentation time stamp and decode time stamp. A presentation time stamp determines when the associated picture should be displayed on the screen, whereas a decode time stamp determines when it should be decoded. In bidirectional coding these times can be quite different.

Audio packets only have presentation time stamps. Clearly if lip-sync is to be obtained, the audio sampling rate of a given program must have been locked to the same master 27 MHz clock as the video and the time stamps must have come from the same counter driven by that clock.

In practice the time between input pictures is constant and so there is a certain amount of redundancy in the time stamps. Consequently PTS/DTS need not appear in every PES packet. Time stamps can be up to 100 ms apart in transport streams. As each picture type (*I*,*P* or *B*) is flagged in the bitstream, the decoder can infer the PTS/DTS for every picture from the ones actually transmitted.

The MPEG-2 transport stream is intended to be a multiplex of many TV programs with their associated sound and data channels, although a

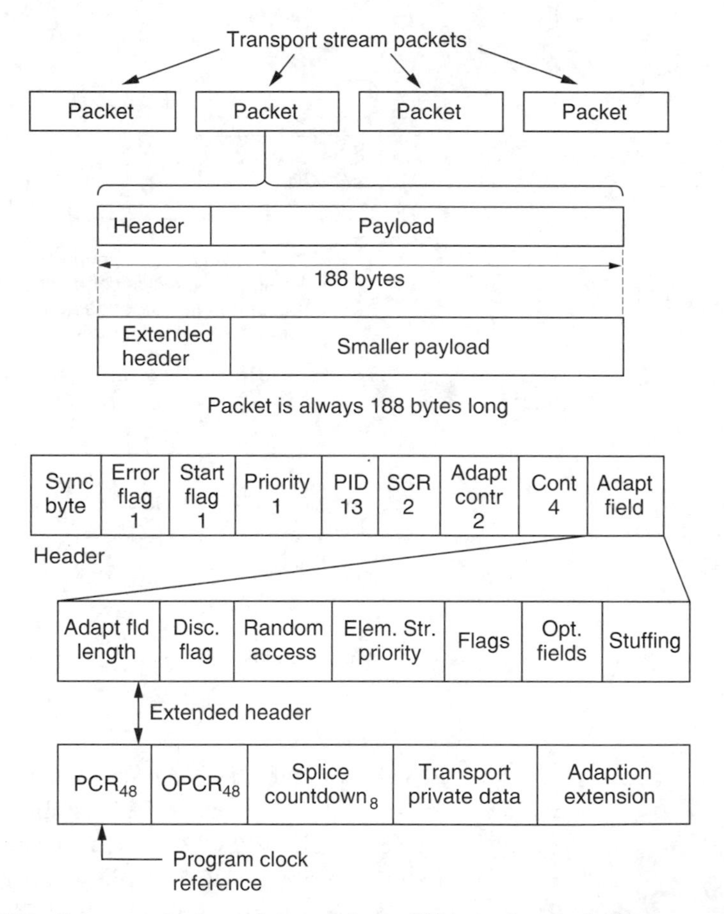

Figure 14.5 Transport stream packets are always 188 bytes long to facilitate multiplexing and error correction.

single program transport stream (SPTS) is possible. The transport stream is based upon packets of constant size so that multiplexing, adding error-correction codes and interleaving in a higher layer is eased. Figure 14.5 shows that these are always 188 bytes long.

Transport stream packets always begin with a header. The remainder of the packet carries data known as the payload. For efficiency, the normal header is relatively small, but for special purposes the header may be extended. In this case the payload gets smaller so that the overall size of the packet is unchanged. Transport stream packets should not be confused with PES packets which are larger and which vary in size. PES packets are broken up to form the payload of the transport stream packets.

The header begins with a sync byte which is an unique pattern detected by a demultiplexer. A transport stream may contain many different Elementary Streams and these are identified by giving each an unique thirteen-bit Packet Identification Code or PID which is included in the header. A multiplexer seeking a particular Elementary Stream simply checks the PID of every packet and accepts only those which match.

In a multiplex there may be many packets from other programs in between packets of a given PID. To help the demultiplexer, the packet header contains a continuity count. This is a four-bit value which increments at each new packet having a given PID.

This approach allows statistical multiplexing as it does matter how many or how few packets have a given PID; the demux will still find them. Statistical multiplexing has the problem that it is virtually impossible to make the sum of the input bit rates constant. Instead the multiplexer aims to make the average data bit rate slightly less than the maximum and the overall bit rate is kept constant by adding 'stuffing' or null packets. These packets have no meaning, but simply keep the bit rate constant. Null packets always have a PID of 8191 (all ones) and the demultiplexer discards them.

14.4 Program Clock Reference

A transport stream is a multiplex of several TV programs and these may have originated from widely different locations. It is impractical to expect all the programs in a transport stream to be genlocked and so the stream is designed from the outset to allow unlocked programs. A decoder running from a transport stream has to genlock to the encoder and the transport stream has to have a mechanism to allow this to be done independently for each program. The synchronizing mechanism is called Program Clock Reference (PCR).

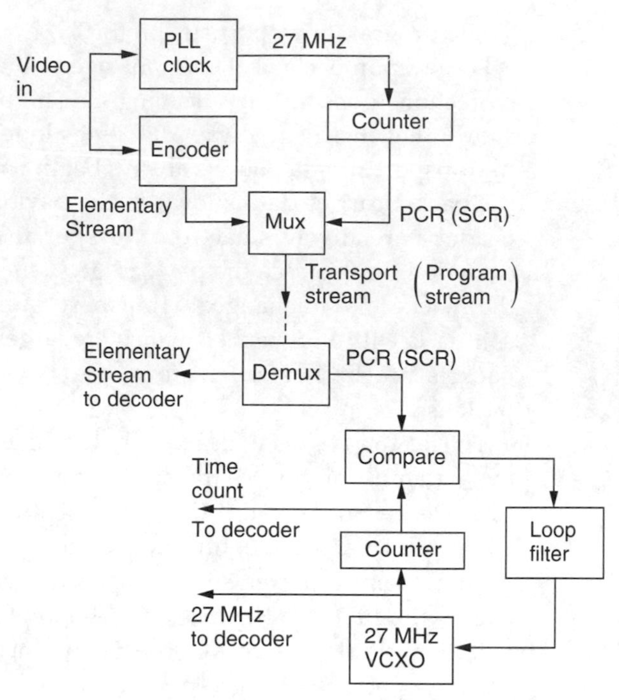

Figure 14.6 Program or System Clock Reference codes regenerate a clock at the decoder. See text for details.

Figure 14.6 shows how the PCR system works. The goal is to re-create at the decoder a 27 MHz clock which is synchronous with that at the encoder. The encoder clock drives a forty-eight-bit counter which continuously counts up to the maximum value before overflowing and beginning again.

A transport stream multiplexer will periodically sample the counter and place the state of the count in an extended packet header as a PCR (see Figure 14.5). The demultiplexer selects only the PIDs of the required program, and it will extract the PCRs from the packets in which they were inserted.

The PCR codes are used to control a numerically locked loop (NLL) described in section 3.6. The NLL contains a 27 MHz VCXO (Voltage-controlled Crystal Oscillator). This is a variable-frequency oscillator based on a crystal which has a relatively small frequency range.

The VCXO drives a forty-eight-bit counter in the same way as in the encoder. The state of the counter is compared with the contents of the PCR and the difference is used to modify the VCXO frequency. When the loop reaches lock, the decoder counter would arrive at the same value as is contained in the PCR and no change in the VCXO would then occur. In practice the transport stream packets will suffer from transmission jitter and this will create phase noise in the loop. This is removed by the loop filter so that the VCXO effectively averages a large number of phase errors.

A heavily damped loop will reject jitter well, but will take a long time to lock. Lockup time can be reduced when switching to a new program if the decoder counter is jammed to the value of the first PCR received in the new program. The loop filter may also have its time constants shortened during lockup. Once a synchronous 27 MHz clock is available at the decoder, this can be divided down to provide the 90 kHz clock which drives the time stamp mechanism.

The entire timebase stability of the decoder is no better than the stability of the clock derived from PCR. MPEG-2 sets standards for the maximum amount of jitter which can be present in PCRs in a real transport stream.

Clearly if the 27 MHz clock in the receiver is locked to one encoder it can only receive Elementary Streams encoded with that clock. If it is attempted to decode, for example, an audio stream generated from a different clock, the result will be periodic buffer overflows or underflows in the decoder. Thus MPEG defines a program in a manner which relates to timing. A program is a set of Elementary Streams which have been encoded with the same master clock.

14.5 Program Specific Information (PSI)

In a real transport stream, each Elementary Stream has a different PID, but the demultiplexer has to be told what these PIDs are and what audio belongs with what video before it can operate. This is the function of PSI which is a form of metadata. Figure 14.7 shows the structure of PSI. When a decoder powers up, it knows nothing about the incoming transport stream except that it must search for all packets with a PID of zero. PID zero is reserved for the Program Association Table (PAT). The PAT is transmitted at regular intervals and contains a list of all the programs in this transport stream. Each program is further described by its own Program Map Table (PMT) and the PIDs of of the PMTs are contained in the PAT.

Figure 14.7 also shows that the PMTs fully describe each program. The PID of the video Elementary Stream is defined, along with the PID(s) of the associated audio and data streams. Consequently when the viewer selects a particular program, the demultiplexer looks up the program number in the PAT, finds the right PMT and reads the audio, video and data PIDs. It then selects Elementary Streams having these PIDs from the transport stream and routes them to the decoders.

Program 0 of the PAT contains the PID of the Network Information Table (NIT). This contains information about what other transport

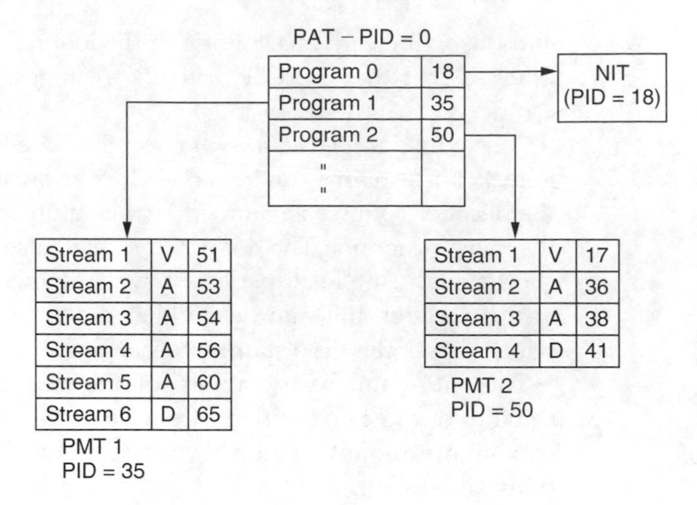

Figure 14.7 MPEG-2 Program Specific Information (PSI) is used to tell a demultiplexer what the transport stream contains.

streams are available. For example, in the case of a satellite broadcast, the NIT would detail the orbital position, the polarization, carrier frequency and modulation scheme. Using the NIT a set-top box could automatically switch between transport streams.

Apart from 0 and 8191, a PID of 1 is also reserved for the Conditional Access Table (CAT). This is part of the access control mechanism needed to support pay-per-view or subscription viewing.

14.6 Multiplexing

A transport stream multiplexer is a complex device because of the number of functions it must perform. A fixed multiplexer will be considered first. In a fixed multiplexer, the bit rate of each of the programs must be specified so that the sum does not exceed the payload bit rate of the transport stream. The payload bit rate is the overall bit rate less the packet headers and PSI rate.

In practice the programs will not be synchronous to one another, but the transport stream must produce a constant packet rate given by the bit rate divided by 188 bytes, the packet length. Figure 14.8 shows how this is handled. Each Elementary Stream entering the multiplexer passes through a buffer which is divided into payload-sized areas. Note that periodically the payload area is made smaller because of the requirement to insert PCR.

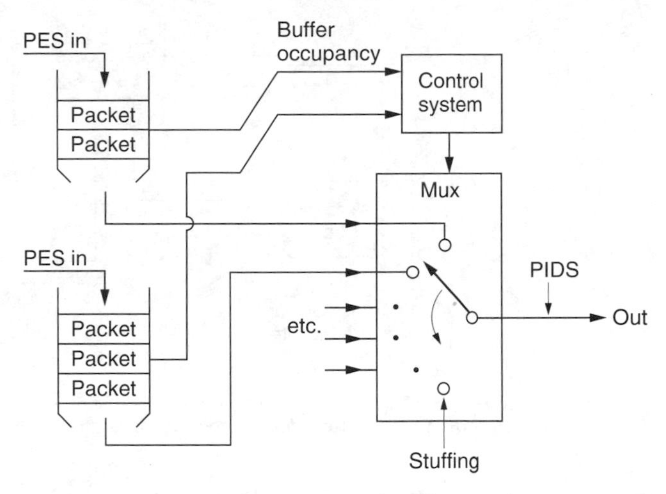

Figure 14.8 A transport stream multiplexer can handle several programs which are asynchronous to one another and to the transport stream clock. See text for details.

MPEG-2 decoders also have a quantity of buffer memory. The challenge to the multiplexer is to take packets from each program in such a way that neither its own buffers nor the buffers in any decoder either overflow or underflow. This requirement is met by sending packets from all programs as evenly as possible rather than bunching together a lot of packets from one program. When the bit rates of the programs are different, the only way this can be handled is to use the buffer contents indicators. The fuller a buffer is, the more likely it should be that a packet will be read from it. This a buffer content arbitrator can decide which program should have a packet allocated next.

If the sum of the input bit rates is correct, the buffers should all slowly empty because the overall input bit rate has to be less than the payload bit rate. This allows for the insertion of Program Specific Information. Whilst PATs and PMTs are being transmitted, the program buffers will fill up again. The multiplexer can also fill the buffers by sending more PCRs as this reduces the payload of each packet. In the event that the multiplexer has sent enough of everything but still can't fill a packet then it will send a null packet with a PID of 8191. Decoders will discard null packets and as they convey no useful data, the multiplexer buffers will all fill whilst null packets are being transmitted.

The use of null packets means that the bit rates of the Elementary Streams do not need to be synchronous with one another or with the transport stream bit rate. As each Elementary Stream can have its own PCR, it is not necessary for the different programs in a transport stream to be genlocked to one another; in fact they don't even need to have the same frame rate.

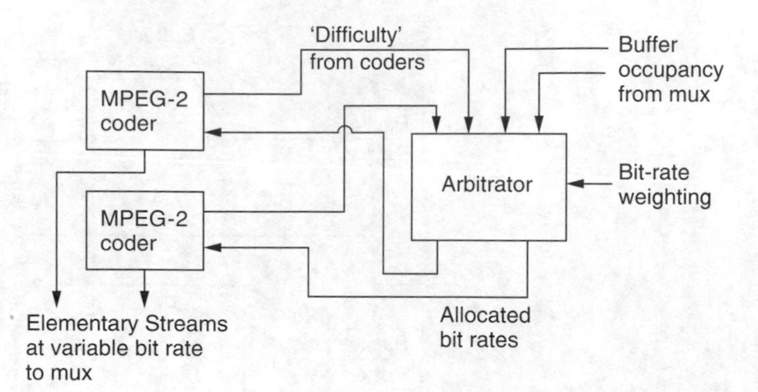

Figure 14.9 A statistical multiplexer contains an arbitrator which allocates bit rate to each program as a function of program difficulty.

This approach allows the transport stream bit rate accurately to be defined and independent of the timing of the data carried. This is important because the transport stream bit rate determines the spectrum of the transmitter and this must not vary.

In a statistical multiplexer or STATMUX, the bit rate allocated to each program can vary dynamically. Figure 14.9 shows that there must be tight connection between the STATMUX and the associated compressors. Each compressor has a buffer memory which is emptied by a demand clock from the statmux. In a normal, fixed bit rate, coder the buffer content feeds back and controls the requantizer. In statmuxing this process is less severe and only takes place if the buffer is very close to full, because the degree of coding difficulty is also fed to the statmux.

The statmux contains an arbitrator which allocates more packets to the program with the greatest coding difficulty. Thus if a particular program encounters difficult material it will produce large prediction errors and begin to fill its output buffer. As the statmux has allocated more packets to that program, more data will be read out of that buffer, preventing overflow. Of course, this is only possible if the other programs in the transport stream are handling typical video.

In the event that several programs encounter difficult material at once, clearly the buffer contents will rise and the requantizing mechanism will have to operate.

14.7 Remultiplexing

In real life a program creator may produce a transport stream which carries all of its programs simultaneously. A service provider may take in several such streams and create its own transport stream by selecting different programs from different sources. In an MPEG-2 environment

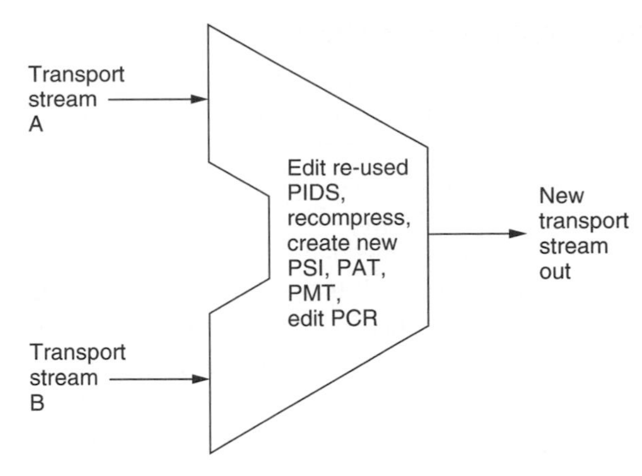

Figure 14.10 A remultiplexer creates a new transport stream from selected programs in other transport streams.

this requires a remultiplexer, also known as a transmultiplexer. Figure 14.10 shows what a remultiplexer does.

Remultiplexing is easier when all the incoming programs have the same bit rate. If a suitable combination of programs is selected it is obvious that the output transport stream will always have sufficient bit rate. Where statistical multiplexing has been used, there is a possibility that the sum of the bit rates of the selected programs will exceed the bit rate of the output transport stream. To avoid this, the remultiplexer will have to employ recompression.

Recompression requires a partial decode of the bitstream to identify the DCT coefficients. These will then be requantized to reduce the bit rate until it is low enough to fit the output transport stream.

Remultiplexers have to edit the Program Specific Information (PSI) such that the Program Association Table (PAT) and the Program Map Tables (PMT) correctly reflect the new transport stream content. It may also be necessary to change the packet-identification codes (PIDs) since the incoming transport streams could inadvertently have used the same values.

When Program Clock Reference (PCR) data are included in an extended packet header, they represent a real-time clock count and if the associated packet is moved in time the PCR value will be wrong. Remultiplexers have to re-create a new multiplex from a number of other multiplexes and it is inevitable that this process will result in packets being placed in different locations in the output transport stream than they had in the input. In this case the remultiplexer must edit the PCR values so that they reflect the value the clock counter would have had at the location at which the packet now resides.

14.8 Modulation techniques

A key difference between analog and digital transmission is that the transmitter output is switched between a number of discrete states rather than continuously varying. The process is called channel coding which is the digital equivalent of modulation. A good code minimizes the channel bandwidth needed for a given bit rate. This quality of the code is measured in bits/s/Hz and is the equivalent of the density ratio in recording. Figure 14.11 shows, not surprisingly, that the less bandwidth required, the better the signal-to-noise ratio has to be. The figure shows the theoretical limit as well as the performance of a number of codes which offer different balances of bandwidth/noise performance.

Where the SNR is poor, as in satellite broadcasting, the amplitude of the signal will be unstable, and phase modulation is used. Figure 14.12 shows that phase-shift keying (PSK) can use two or more phases. When four phases in quadrature are used, the result is Quadrature

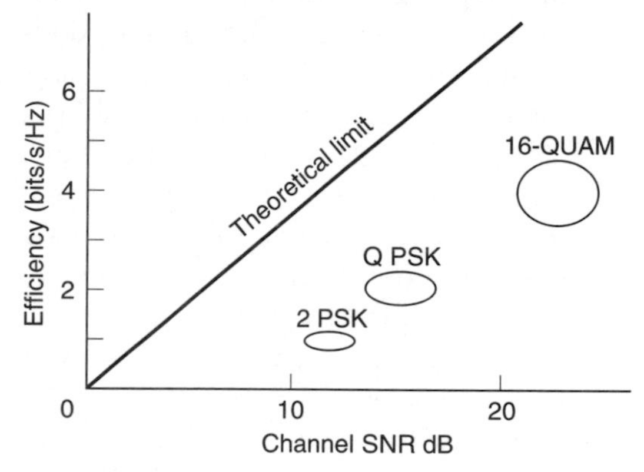

Figure 14.11 Where a better SNR exists, more data can be sent down a given bandwidth channel.

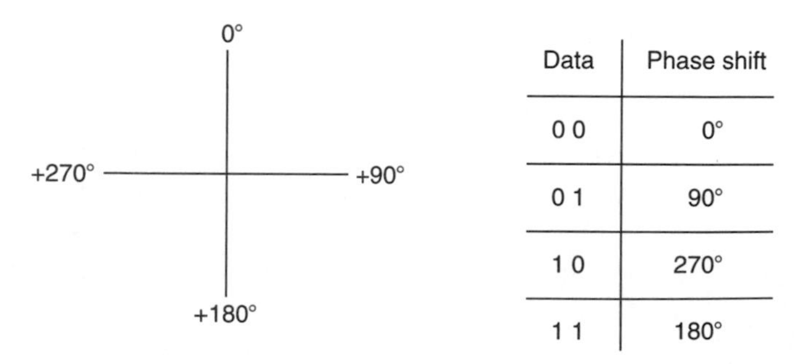

Data	Phase shift
0 0	0°
0 1	90°
1 0	270°
1 1	180°

Figure 14.12 Differential quadrature phase shift keying (DQPSK).

Phase Shift Keying or QPSK. Each period of the transmitted waveform can have one of four phases and therefore conveys the value of two data bits. 8-PSK uses eight phases and can carry three bits per symbol where the SNR is adequate. PSK is generally encoded in such a way that a knowledge of absolute phase is not needed at the receiver. Instead of encoding the signal phase directly, the data determine the magnitude of the phase shift between symbols. A QPSK coder is shown in Figure 14.13.

In terrestrial transmission more power is available than, for example, from a satellite and so a stronger signal can be delivered to the receiver. Where a better SNR exists, an increase in data rate can be had using multi-level signalling or *m*-ary coding instead of binary. Figure 14.14 shows that the ATSC system uses an eight-level signal (8-VSB) allowing three bits to be sent per symbol. Four of the levels exist with

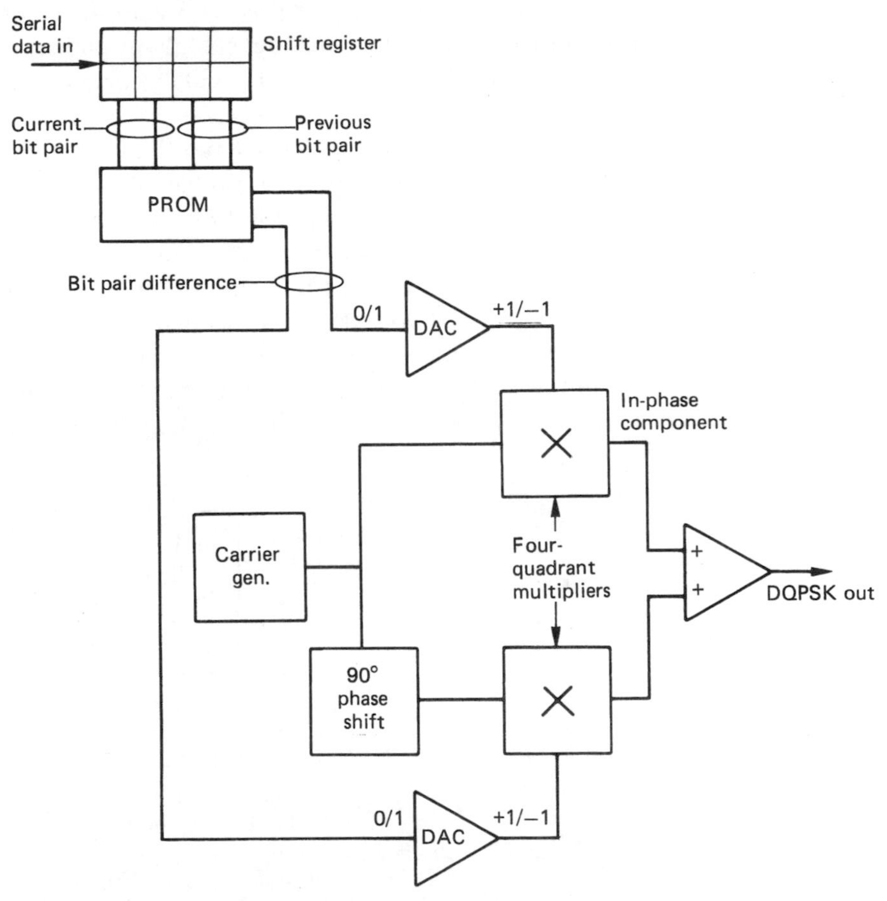

Figure 14.13 A DQPSK coder conveys two bits for each modulation period. See text for details.

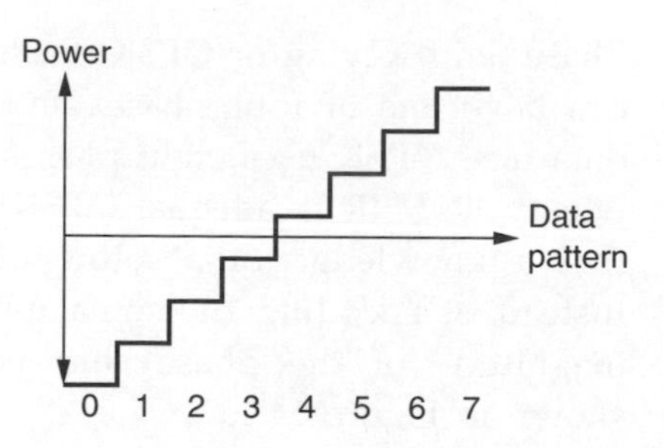

Figure 14.14 In 8VSB the transmitter operates in 8 different states enabling three bits to be sent per symbol.

normal carrier phase and four exist with inverted phase so that a phase-sensitive rectifier is needed in the receiver. Clearly the data separator must have a three-bit ADC which can resolve the eight signal levels. The gain and offset of the signal must be precisely set so that the quantizing levels register precisely with the centres of the eyes. The transmitted signal contains sync pulses which are encoded using specified code levels so that the data separator can set its gain and offset.

Multi-level signalling systems have the characteristic that the bits in the symbol have different error probability. Figure 14.15 shows that a small noise level will corrupt the low-order bit, whereas twice as much noise will be needed to corrupt the middle bit and four times as much will be needed to corrupt the high-order bit. In ATSC the solution is that the lower two bits are encoded together in an inner error-correcting scheme so that they represent only one bit with similar reliability to the top bit. As a result the 8-VSB system actually delivers two data bits per symbol even though eight-level signalling is used.

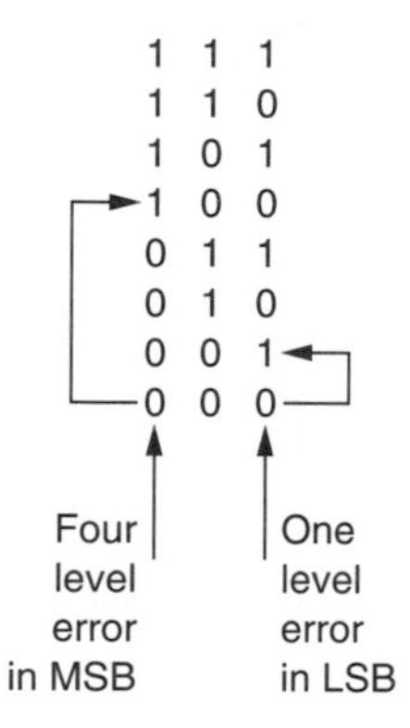

Figure 14.15 In multi-level signalling the error probability is not the same for each bit.

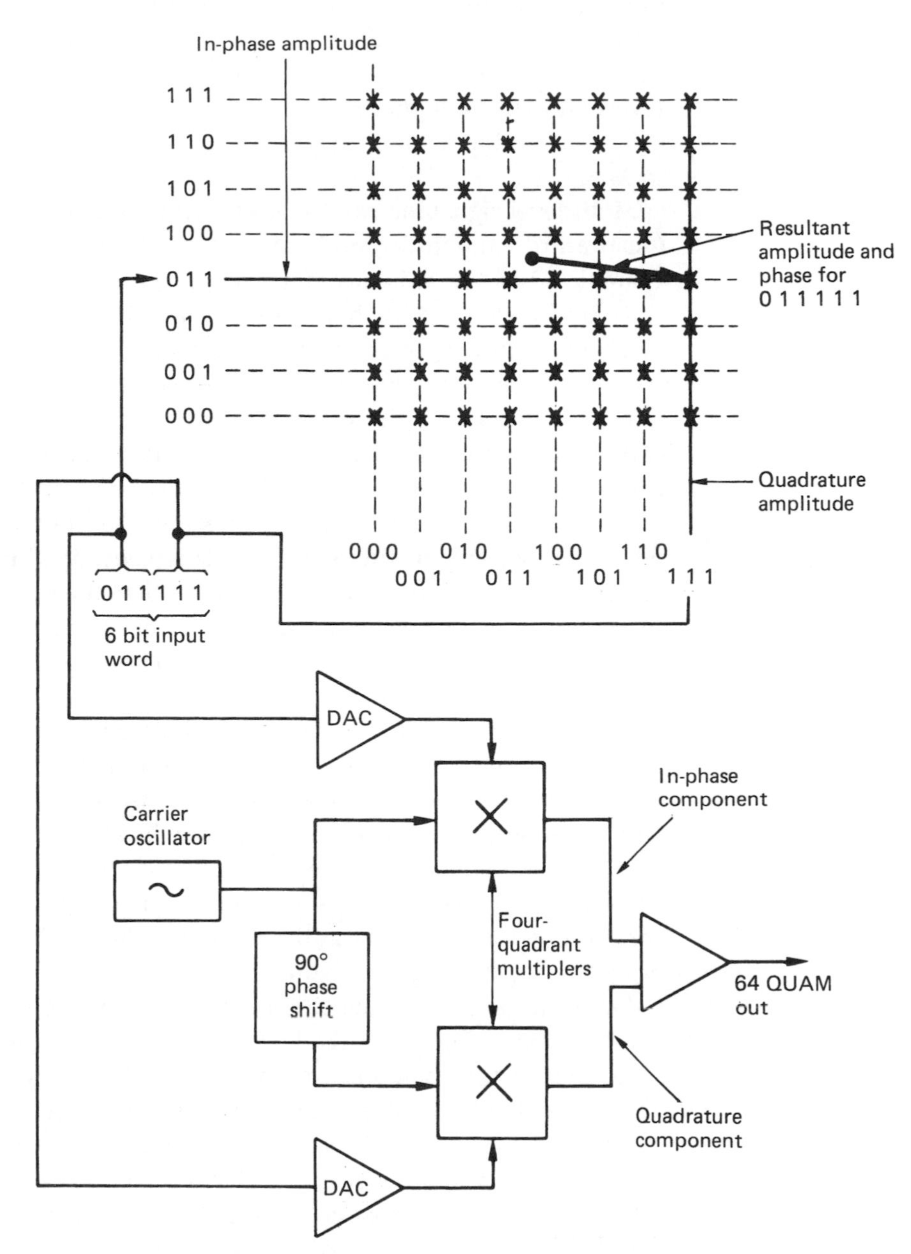

Figure 14.16 In 64-QUAM, two carriers are generated with a quadrature relationship. These are independently amplitude modulated to eight discrete levels in four quadrant multipliers. Adding the signals produces a QUAM signal having 64 unique combinations of amplitude and phase. Decoding requires the waveform to be sampled in quadrature like a colour TV subcarrier.

The modulation of the carrier results in a double-sideband spectrum, but following analog TV practice most of the lower sideband is filtered off leaving a vestigial sideband only, hence the term 8-VSB. A small DC offset is injected into the modulator signal so that the four in-phase levels are slightly higher than the four out-of-phase levels. This has the effect of creating a small pilot at the carrier frequency to help receiver locking.

Multi-level signalling can be combined with PSK to obtain multi-level Quadrature Amplitude Modulation (QUAM). Figure 14.16 shows the example of 64-QUAM. Incoming six-bit data words are split into two three-bit words and each is used to amplitude modulate a pair of sinusoidal carriers which are generated in quadrature. The modulators are four-quadrant devices such that 2^3 amplitudes are available, four of which are in phase with the carrier and four are antiphase. The two AM carriers are linearly added and the result is a signal which has 2^6 or 64 combinations of amplitude and phase. There is a great deal of similarity between QUAM and the colour subcarrier used in analog television in which the two colour difference signals are encoded into one amplitude and phase-modulated waveform. On reception, the waveform is sampled twice per cycle in phase with the two original carriers and the result is a pair of eight-level signals. 16-QUAM is also possible, delivering only four bits per symbol but requiring a lower SNR.

The data bit patterns to be transmitted can have any combinations whatsoever, and if nothing were done, the transmitted spectrum would be non-uniform. This is undesirable because peaks cause interference with other services, whereas energy troughs allow external interference in. The randomizing technique of Section 8.17 is used to overcome the problem. The process is known as energy dispersal. The signal energy is spread uniformly throughout the allowable channel bandwidth so that it has less energy at a given frequency.

A pseudo-random sequence generator is used to generate the randomizing sequence. Figure 14.17 shows the randomizer used in DVB. This sixteen-bit device has a maximum sequence length of 65 535 bits, and is preset to a standard value at the beginning of each set of eight transport stream packets. The serialized data are XORed with the LSB of the Galois field, which randomizes the output which then goes to the modulator. The spectrum of the transmission is now determined by the spectrum of the PRS.

On reception, the de-randomizer must contain the identical ring counter which must also be set to the starting condition to bit accuracy. Its output is then added to the data stream from the demodulator. The randomizing will effectively then have been added twice to the data in modulo-2, and as a result is cancelled out leaving the original serial data.

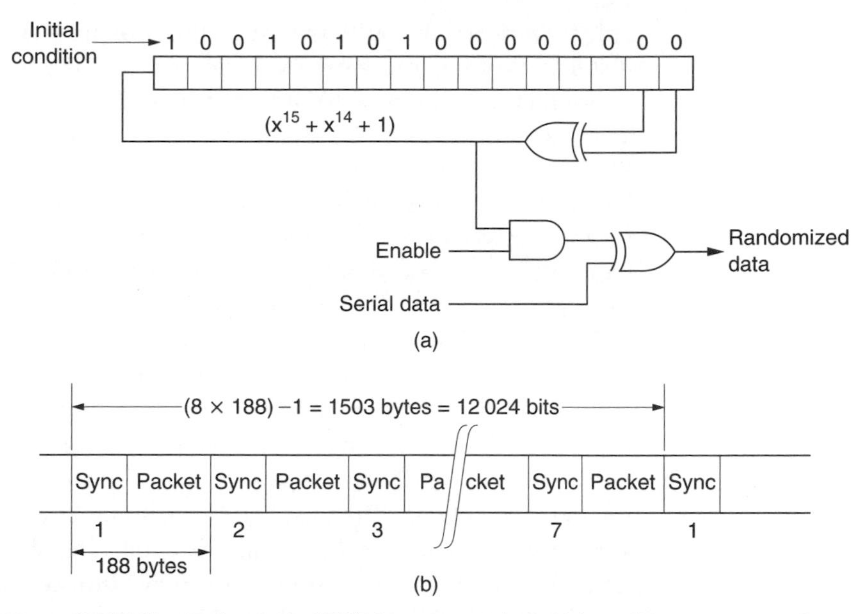

Figure 14.17 The randomizer of DVB is pre-set to the initial condition once every 8 transport stream packets. The maximum length of the sequence is 65 535 bits, but only the first 12 024 bits are used before resetting again (b).

14.9 OFDM

The way that radio signals interact with obstacles is a function of the relative magnitude of the wavelength and the size of the object. AM sound radio transmissions with a wavelength of several hundred metres can easily diffract around large objects. The shorter the wavelength of a transmission, the larger objects in the environment appear to it and these objects can then become reflectors. Reflecting objects produce a delayed signal at the receiver in addition to the direct signal. In analog television transmissions this causes the familiar ghosting. In digital transmissions, the symbol rate may be so high that the reflected signal may be one or more symbols behind the direct signal, causing intersymbol interference. As the reflection may be continuous, the result may be that almost every symbol is corrupted. No error-correction system can handle this. Raising the transmitter power is no help at all as it simply raises the power of the reflection in proportion.

The only solution is to change the characteristics of the RF channel in some way to either prevent the multipath reception or to prevent it being a problem. The RF channel includes the modulator, transmitter, antennae, receiver and demodulator.

As with analog UHF TV transmissions, a directional antenna is useful with digital transmission as it can reject reflections. However, directional antennae tend to be large and they require a skilled permanent installation. Mobile use on a vehicle or vessel is simply impractical.

Another possibility is to incorporate a ghost canceller in the receiver. The transmitter periodically sends a standardized known waveform known as a training sequence. The receiver knows what this waveform looks like and compares it with the received signal. In theory it is possible for the receiver to compute the delay and relative level of a reflection and so insert an opposing one. In practice if the reflection is strong it may prevent the receiver finding the training sequence.

The most elegant approach is to use a system in which multipath reception conditions cause only a small increase in error rate which the error correction system can manage. This approach is used in DVB. Figure 14.18(a) shows that when using one carrier with a high bit rate, reflections can easily be delayed by one or more bit periods, causing interference between the bits. Figure 14.18(b) shows that instead, OFDM sends many carriers each having a low bit rate. When a low bit rate is used, the energy in the reflection will arrive during the same bit period as the direct signal. Not only is the system immune to multipath reflections, but the energy in the reflections can actually be used. This characteristic

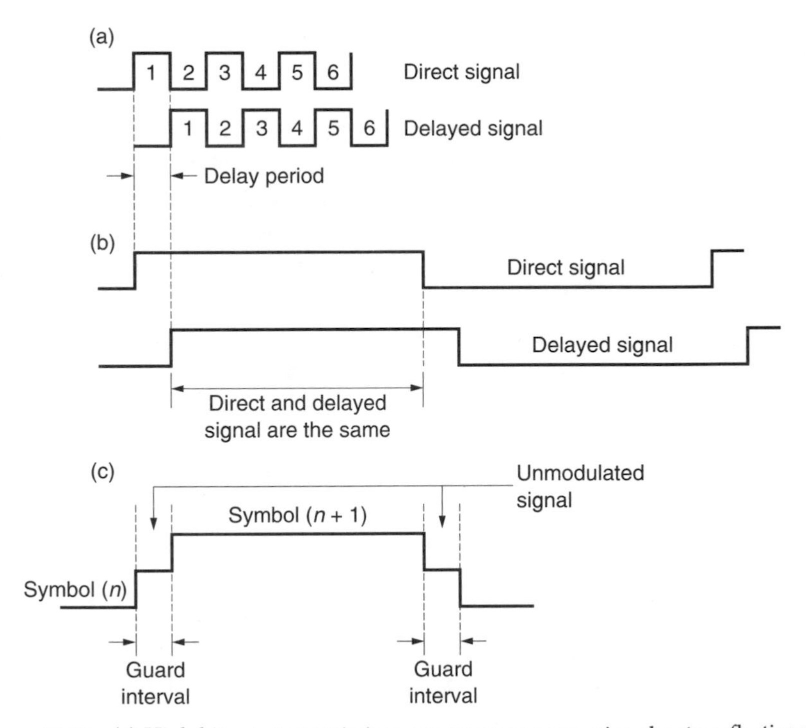

Figure 14.18 (a) High bit rate transmissions are prone to corruption due to reflections. (b) If the bit rate is reduced the effect of reflections is eliminated, in fact reflected energy can be used. (c) Guard intervals may be inserted between symbols.

can be enhanced by using guard intervals shown in (c). These reduce multipath bit overlap even more.

Note that OFDM is not a modulation scheme, and each of the carriers used in a OFDM system still needs to be modulated using any of the digital coding schemes described above. What OFDM does is to provide an efficient way of packing many carriers close together without mutual interference.

A serial data waveform basically contains a train of rectangular pulses. The transform of a rectangle is the function $\sin x/x$ and so the baseband pulse train has a $\sin x/x$ spectrum. When this waveform is used to modulate a carrier the result is a symmetrical $\sin x/x$ spectrum centred on the carrier frequency. Figure 14.19(a) shows that nulls in the spectrum appear spaced at multiples of the bit rate away from the carrier. Further carriers can be placed at spacings such that each is centred at the nulls of the others as is shown in (b). The distance between the carriers is equal to 90° or one quadrant of $\sin x$. Due to the quadrant spacing, these carriers are mutually orthogonal, hence the term orthogonal frequency division. A large number of such carriers (in practice, several thousand) will be interleaved to produce an overall spectrum which is almost rectangular and which fills the available transmission channel.

When guard intervals are used, the carrier returns to an unmodulated state between bits for a period which is greater than the period of the reflections. Then the reflections from one transmitted bit decay during the guard interval before the next bit is transmitted. The use of guard

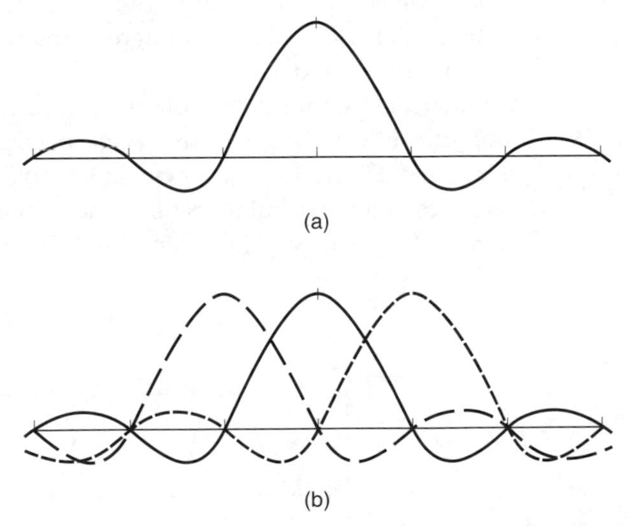

Figure 14.19 In OFDM the carrier spacing is critical, but when correct the carriers become independent and most efficient use is made of the spectrum. (a) Spectrum of bitstream has regular nulls. (b) Peak of one carrier occurs at null of another.

intervals reduces the bit rate of the carrier because for some of the time it is radiating carrier not data. A typical reduction is to around 80 per cent of the capacity without guard intervals.

This capacity reduction does, however, improve the error statistics dramatically, such that much less redundancy is required in the error-correction system. Thus the effective transmission rate is improved. The use of guard intervals also moves more energy from the sidebands back to the carrier. The frequency spectrum of a set of carriers is no longer perfectly flat but contains a small peak at the centre of each carrier.

The ability to work in the presence of multipath cancellation is one of the great strengths of OFDM. In DVB, more than 2000 carriers are used in single transmitter systems. Provided there is exact synchronism, several transmitters can radiate exactly the same signal so that a single-frequency network can be created throughout a whole country. SFNs require a variation on OFDM which uses over 8000 carriers.

With OFDM, directional antennae are not needed and, given sufficient field strength, mobile reception is perfectly feasible. Of course, directional antennae may still be used to boost the received signal outside of normal service areas or to enable the use of low-powered transmitters.

An OFDM receiver must perform Fast Fourier Transforms (FFTs) on the whole band at the symbol rate of one of the carriers. The amplitude and/or phase of the carrier at a given frequency effectively reflects the state of the transmitted symbol at that time slot and so the FFT partially demodulates as well.

In order to assist with tuning in, the OFDM spectrum contains pilot signals. These are individual carriers which are transmitted with slightly more power than the remainder. The pilot carriers are spaced apart through the whole channel at agreed frequencies which form part of the transmission standard.

Practical reception conditions, including multipath reception, will cause a significant variation in the received spectrum and some equalization will be needed. Figure 14.20 shows what the possible spectrum looks like in the presence of a powerful reflection. The signal has almost been cancelled at certain frequencies. However, the FFT performed in the receiver is

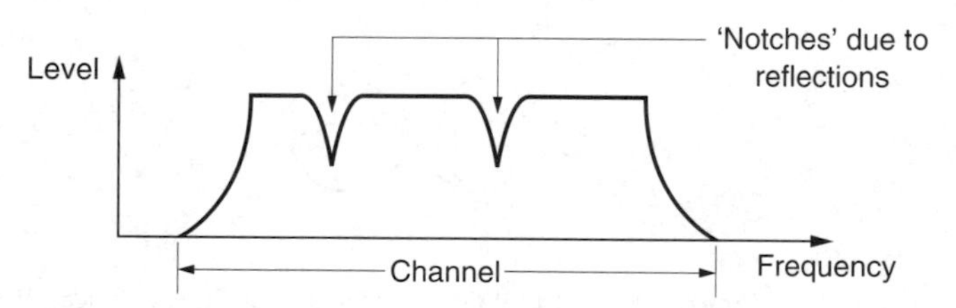

Figure 14.20 Multipath reception can place notches in the channel spectrum. This will require equalization at the receiver.

effectively a spectral analysis of the signal and so the receiver computes for free the received spectrum. As in a flat spectrum the peak magnitude of all the coefficients would be the same (apart from the pilots), equalization is easily performed by multiplying the coefficients by suitable constants until this characteristic is obtained.

Although the use of transform-based receivers appears complex, when it is considered that such an approach simultaneously allows effective equalization the complexity is not significantly higher than that of a conventional receiver which needs a separate spectral analysis system just for equalization purposes.

The only drawback of OFDM is that the transmitter must be highly linear to prevent intermodulation between the carriers. This is readily achieved in terrestrial transmitters by derating the transmitter so that it runs at a lower power than it would in analog service. This is not practicable in satellite transmitters which are optimized for efficiency so OFDM is not really suitable for satellite use.

14.10 Error correction

As in recording, broadcast data suffer from both random and burst errors and the error-correction strategies of digital television broadcasting have to reflect that. Figure 14.21 shows a typical system in which inner and outer codes are employed. The Reed–Solomon codes are universally used for burst-correcting outer codes, along with an interleave which will be convolutional rather than the block-based interleave used in recording media. The inner codes will not be R–S, as more suitable codes exist for

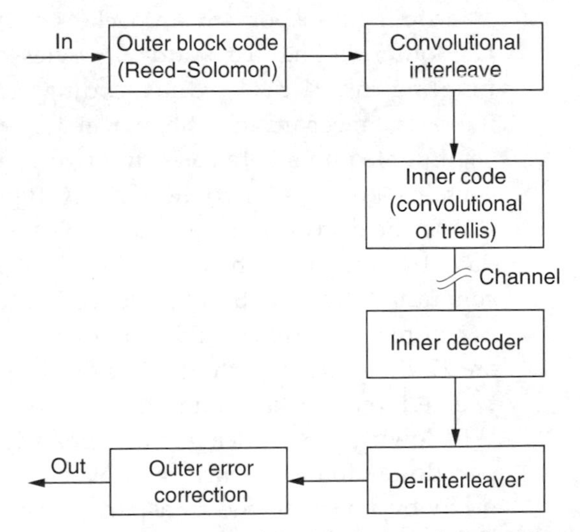

Figure 14.21 Error-correcting strategy of digital television broadcasting systems.

the statistical conditions prevalent in broadcasting. DVB uses a parity-based variable-rate system in which the amount of redundancy can be adjusted according to reception conditions. ATSC uses a fixed-rate parity-based system along with trellis coding to overcome co-channel interference from analog NTSC transmitters.

14.11 DVB

The DVB system is subdivided into systems optimized for satellite, cable and terrestrial delivery. This section concentrates on the terrestrial delivery system. Figure 14.22 shows a block diagram of a DVB-T transmitter.

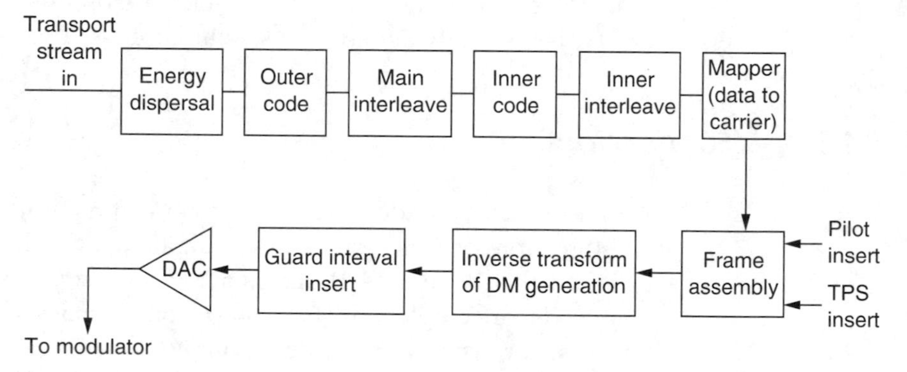

Figure 14.22 DVB-T transmitter block diagram. See text for details.

Incoming transport stream packets of 188 bytes each are first subject to R–S outer coding. This adds 16 bytes of redundancy to each packet, resulting in 204 bytes. Outer coding is followed by interleaving. The interleave mechanism is shown in Figure 14.23. Outer code blocks are commutated on a byte basis into twelve parallel channels. Each channel contains a different amount of delay, typically achieved by a ring-buffer RAM. The delays are integer multiples of 17 bytes, designed to skew the data by one outer block ($12 \times 17 = 204$). Following the delays, a commutator reassembles interleaved outer blocks. These have 204 bytes as before, but the effect of the interleave is that adjacent bytes in the input are 17 bytes apart in the output. Each output block contains data from twelve input blocks making the data resistant to burst errors.

Following the interleave, the energy-dispersal process takes place. The pseudo-random sequence runs over eight outer blocks and is synchronized by inverting the transport stream packet sync symbol in every eighth block. The packet sync symbols are not randomized.

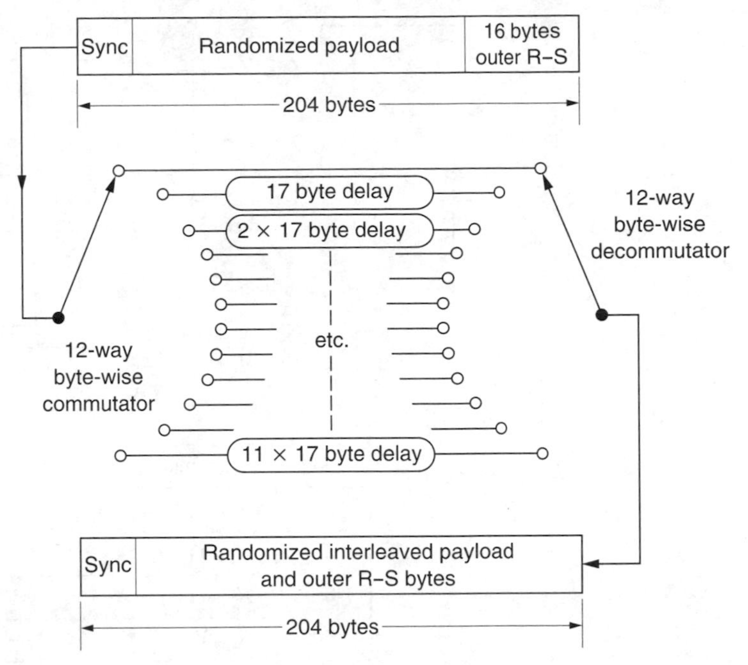

Figure 14.23 The interleaver of DVB uses 12 incrementing delay channels to reorder the data. The sync byte passes through the undelayed channel and so is still at the head of the packet after interleave. However, the packet now contains non-adjacent bytes from 12 different packets.

The inner coding process of DVB is shown in Figure 14.24. Input data are serialized and pass down a shift register. Exclusive-OR gates produce convolutional parity symbols *X* and *Y*, such that the output bit rate is twice the input bit rate. Under the worst reception conditions, this 100 per cent redundancy offers the most powerful correction with the penalty that a low data rate is delivered. However, Figure 14.24 also shows that a variety of inner redundancy factors can be used from 1/2 down to 1/8 of the transmitted bit rate. The *X,Y* data from the inner coder are subsampled, such that the coding is punctured.

The DVB standard allows the use of QPSK, 16-QUAM or 64-QUAM coding in an OFDM system. There are five possible inner code rates, and four different guard intervals which can be used with each modulation scheme, Thus for each modulation scheme there are twenty possible transport stream bit rates in the standard DVB channel, each of which requires a different receiver SNR. The broadcaster can select any suitable balance between transport stream bit rate and coverage area. For a given transmitter location and power, reception over a larger area may require a channel code with a smaller number of bits/s/Hz and this reduces the bit rate which can be delivered in a standard channel. Alternatively a higher amount of inner redundancy means that the proportion of the

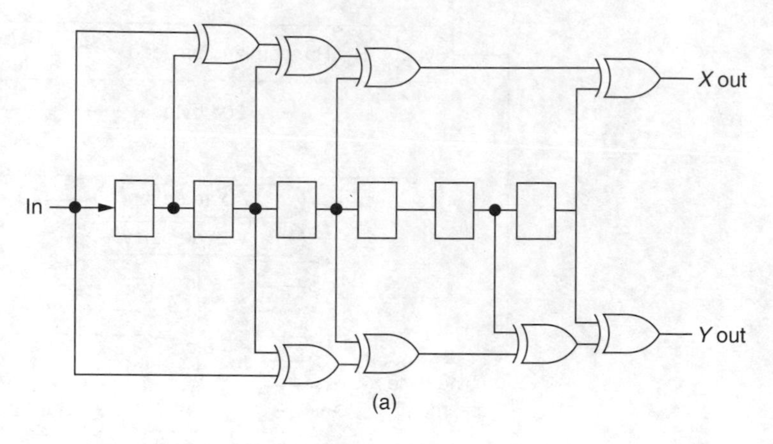

Rate	Transmitted sequence
1/2	X_1 Y_1
2/3	X_1 Y_1 Y_2
3/4	X_1 Y_1 Y_2 X_3
5/6	X_1 Y_1 Y_2 X_3 Y_4 X_6
7/8	X_1 Y_1 Y_2 Y_3 Y_4 X_5 X_6 X_7

(b)

Figure 14.24 (a) The mother inner coder of DVB produces 100 per cent redundancy, but this can be punctured by subsampling the *X* and *Y* data to give five different code rates as (b) shows.

transmitted bit rate which is data goes down. Thus for wider coverage the broadcaster will have to send fewer programs in the multiplex or use higher compression factors.

14.12 The DVB receiver

Figure 14.25 shows a block diagram of a DVB receiver. The off-air RF signal is fed to a mixer driven by the local oscillator. The IF output of the mixer is bandpass filtered and supplied to the ADC which outputs a digital IF signal for the FFT stage. The FFT is initially analysed to find the higher-level pilot signals. If these are not in the correct channels the local oscillator frequency is incorrect and it will be changed until the pilots emerge from the FFT in the right channels. The data in the pilots will be decoded in order to tell the receiver how many carriers, what inner redundancy rate, guard-band rate and modulation scheme are in use in the remaining carriers. The FFT magnitude information is also a measure of the equalization required.

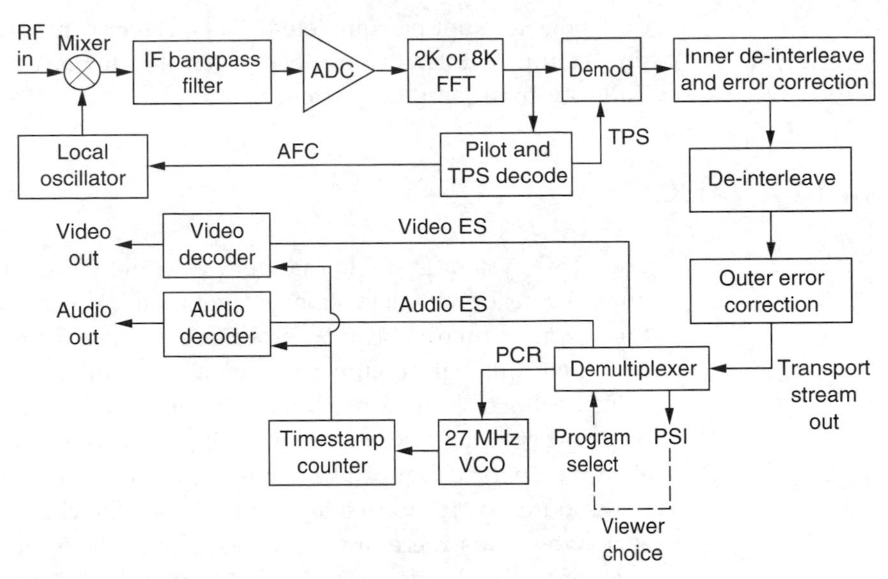

Figure 14.25 DVB receiver block diagram. See text for details.

The FFT outputs are demodulated into 2K or 8K bitstreams and these are multiplexed to produce a serial signal. This is subject to inner error correction which corrects random errors. The data are then de-interleaved to break up burst errors and then the outer R–S error-correction operates. The output of the R–S correction will then be derandomized to become an MPEG transport stream once more. The derandomizing is synchronized by the transmission of inverted sync patterns.

The receiver must select a PID of 0 and wait until a program association table (PAT) is transmitted. This will describe the available programs by listing the PIDs of the program map tables (PMT). By looking for these packets the receiver can determine what PIDs to select to receive any video and audio Elementary Streams.

When an Elementary Stream is selected, some of the packets will have extended headers containing program clock reference (PCR). These codes are used to synchronize the 27 MHz clock in the receiver to the one in the MPEG encoder of the desired program. The 27 MHx clock is divided down to drive the time stamp counter so that audio and video emerge from the decoder at the correct rate and with lip sync.

It should be appreciated that time stamps are relative, not absolute. The time stamp count advances by a fixed amount each picture, but the exact count is meaningless. Thus the decoder can only establish the frame rate of the video from time stamps, but not the precise timing. In practice the receiver has finite buffering memory between the demultiplexer and the MPEG decoder. If the displayed video timing is too late, the buffer will tend to overflow whereas if the displayed video timing is too early the decoding may not be completed. The receiver can advance or retard the time stamp counter during lock-up so that it places the output timing mid-way between these extremes.

14.13 ATSC

The ATSC system is an alternative way of delivering a transport stream, but it is considerably less sophisticated than DVB, and supports only one transport stream bit rate of 19.28 Mbits/s. If any change in the service area is needed, this will require a change in transmitter power.

Figure 14.26 shows a block diagram of an ATSC transmitter. Incoming transport stream packets are randomized, except for the sync pattern, for energy dispersal. Figure 14.27 shows the randomizer.

The outer correction code includes the whole packet except for the sync byte. Thus there are 187 bytes of data in each codeword and 20 bytes of R–S redundancy are added to make a 207-byte codeword. After outer coding, a convolutional interleaver shown in Figure 14.28 is used. This reorders data over a time span of about 4 ms. Interleave simply exchanges content between packets, but without changing the packet structure.

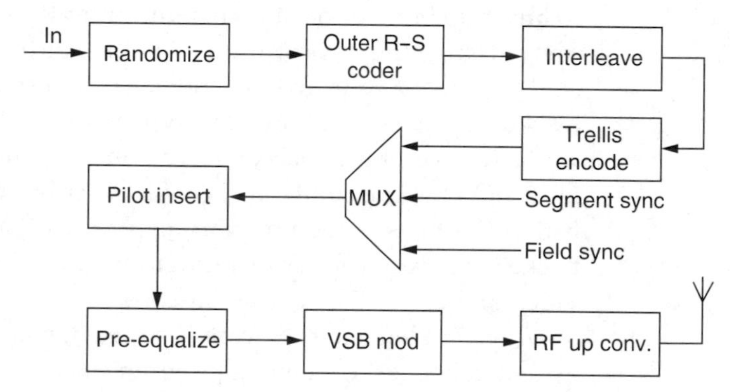

Figure 14.26 Block diagram of ATSC transmitter. See text for details.

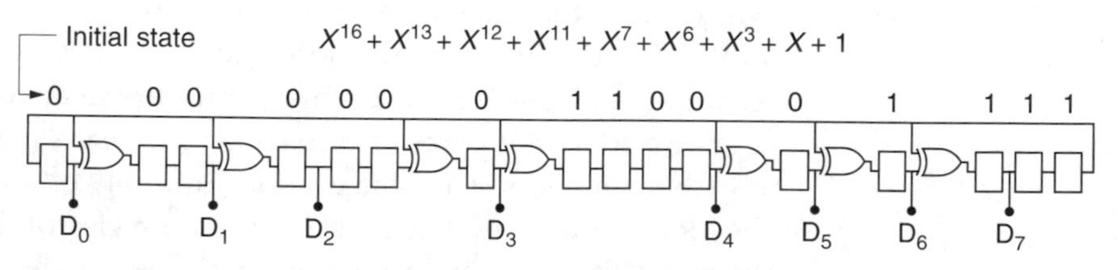

Figure 14.27 The randomizer of ATSC. The twisted ring counter is pre-set to the initial state shown each data field. It is then clocked once per byte and the eight outputs D_0–D_7 are X-ORed with the data byte.

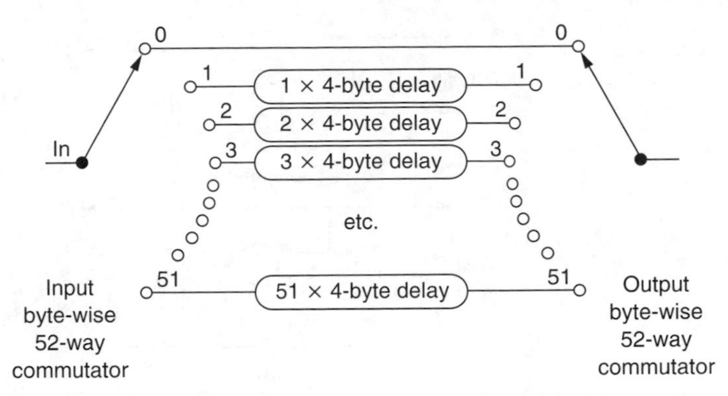

Figure 14.28 The ATSC convolutional interleaver spreads adjacent bytes over a period of about 4 ms.

Figure 14.29 shows that the result of outer coding and interleave is a data frame which is divided into two fields of 313 segments each. The frame is tranmitted by scanning it horizontallly a segment at a time. There is some similarity with a traditional analog video signal here, because there is a sync pulse at the beginning of each segment and a field sync which occupies two segments of the frame. *Data segment sync* repeats every 77.3 ms, a segment rate of 12 933 Hz, whereas a frame has a period of 48.4 ms. The field sync segments contain a training sequence to drive the adaptive equalizer in the receiver.

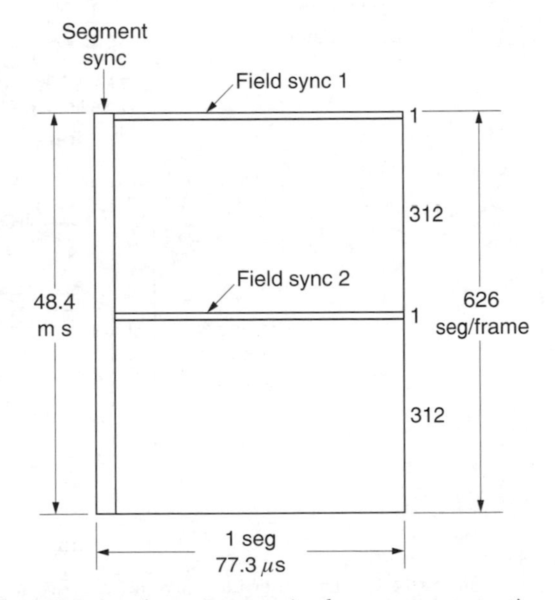

Figure 14.29 The ATSC data frame is transmitted one segment at a time. Segment sync denotes the beginning of each segment and the segments are counted from the field sync signals.

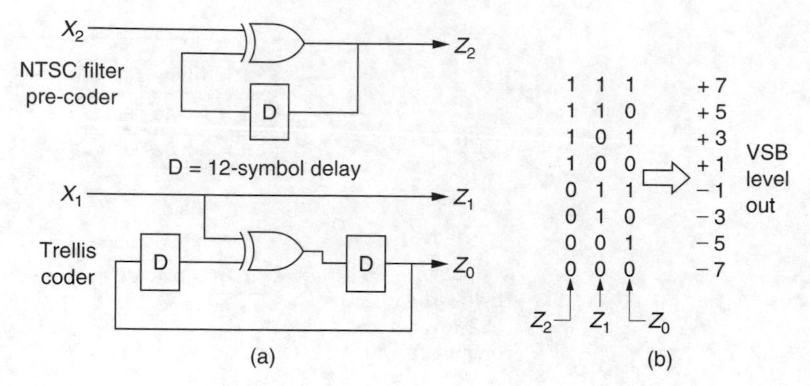

Figure 14.30 (a) The precoder and trellis coder of ATSC converts two data bits X_1, X_2 to three output bits Z_0, Z_1, Z_2. (b) The Z_0, Z_1, Z_2 output bits map to the eight-level code as shown.

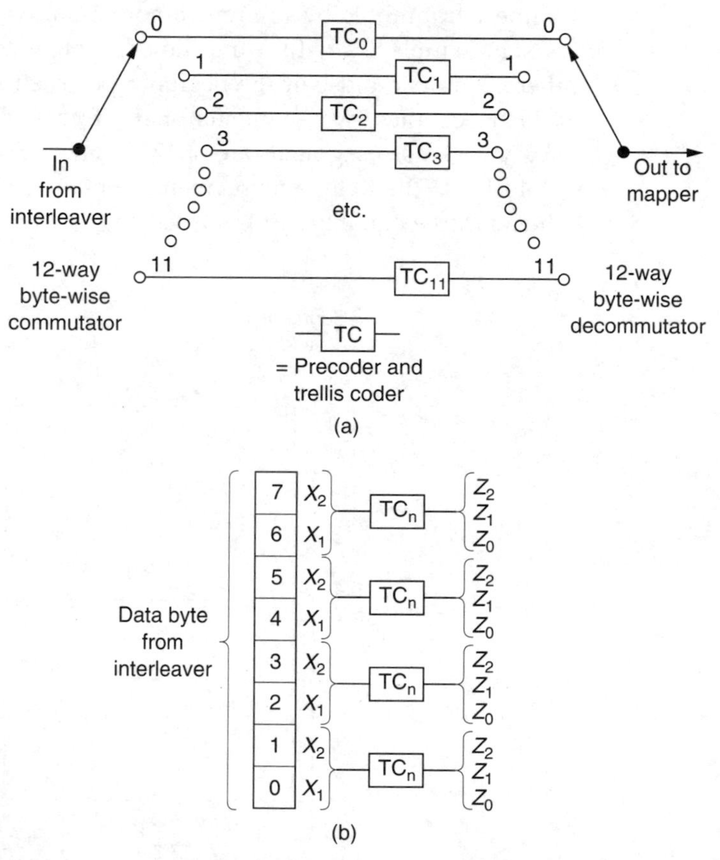

Figure 14.31 The inner interleave (a) of ATSC makes the trellis coding operate as twelve parallel channels working on every twelfth byte to improve error resistance. The interleave is byte-wise, and, as (b) shows, each byte is divided into four di-bits for coding into the tri-bits Z_0, Z_1, Z_2.

The data content of the frame is subject to trellis coding which converts each pair of data bits into three channel bits inside an inner interleave. The trellis coder is shown in Figure 14.30 and the interleave in Figure 14.31. Figure 14.30 also shows how the three channel bits map to the eight signal levels in the 8-VSB modulator.

Figure 14.32 shows the data segment after eight-level coding. The sync pattern of the transport stream packet, which was not included in the error-correction code, has been replaced by a segment sync waveform. This acts as a timing reference to allow deserializing of the segment, but as the two levels of the sync pulse are standardized, it also acts as an amplitude reference for the eight-level slicer in the receiver.

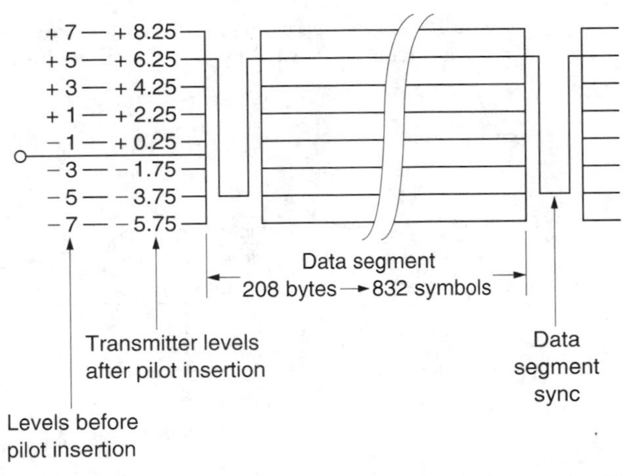

Figure 14.32 ATSC data segment. Note the sync pattern which acts as a timing and amplitude reference. The 8 levels are shifted up by 1.25 to create a DC component resulting in a pilot at the carrier frequency.

The eight-level signal is subject to a DC offset so that some transmitter energy appears at the carrier frequency to act as a pilot. Each eight-level symbol carries two data bits and so there are 832 symbols in each segment. As the segment rate is 12 933 Hz, the symbol rate is 10.76 MHz and so this will require 5.38 MHz of bandwidth in a single sideband.

Figure 14.33 shows the transmitter spectrum. The lower sideband is vestigial and an overall channel width of 6 MHz results.

Figure 14.34 shows an ATSC receiver. The first stages of the receiver are designed to lock to the pilot in the transmitted signal. This then allows the eight-level signal to be sampled at the right times. This process will allow location of the segment sync and then the field sync signals. Once the receiver is synchronized, the symbols in each segment can be decoded. The inner or trellis coder corrects for random errors, then following

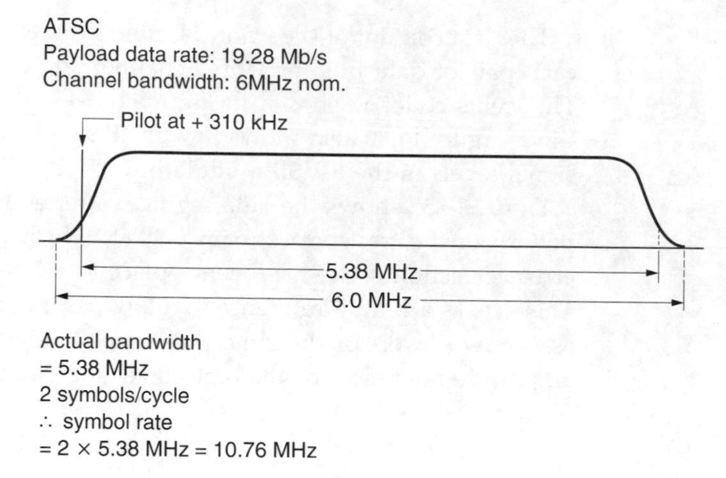

Figure 14.33 The spectrum of ATSC and its associated bit and symbol rates. Note pilot at carrier frequency created by DC offset in multi-level coder.

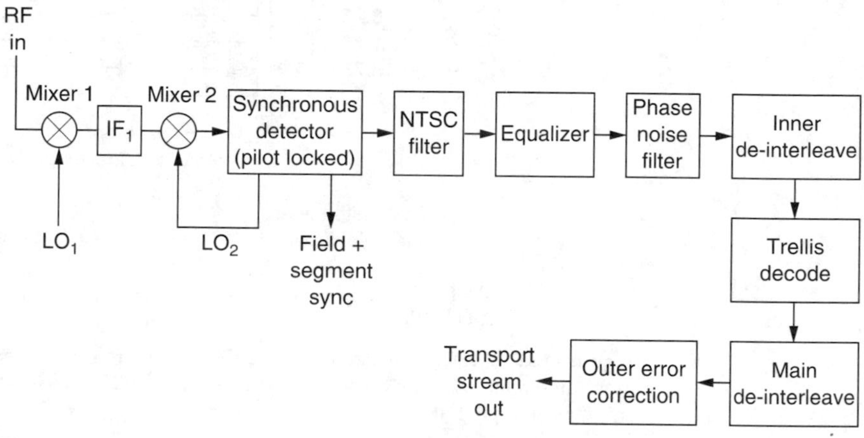

Figure 14.34 An ATSC receiver. Double conversion can be used so that the second conversion stage can be arranged to lock to the transmitted pilot.

de-interleave the R–S coder corrects burst errors, After derandomizing, standard transport stream sync patterns are added to the output data.

In practice ATSC transmissions will experience co-channel interference from NTSC transmitters and the ATSC scheme allows the use of an NTSC rejection filter. Figure 14.35 shows that most of the energy of NTSC is at the carrier, subcarrier and sound carrier frequencies. A comb filter with a suitable delay can produce nulls or notches at these frequencies. However, the delay-and-add process in the comb filter also causes another effect. When two eight-level signals are added together, the result is a sixteen-level signal. This will be corrupted by noise of half the level that would corrupt an eight-level signal. However, the sixteen-level

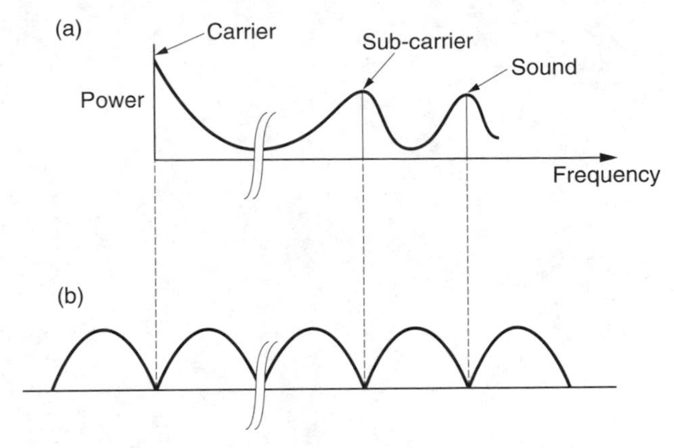

Figure 14.35 Spectrum of typical analog transmitter showing (a) maximum power at carrier, subcarrier and audio carrier. A comb filter (b) with a suitable delay can notch out NTSC interference. The pre-coding of ATSC is designed to work with the necessary receiver delay.

signal contains redundancy because it corresponds to the combinations of four bits whereas only two bits are being transmitted. This allows a form of error correction to be used.

The ATSC inner precoder results in a known relationship existing between symbols independent of the data. The time delays in the inner interleave are designed to be compatible with the delay in the NTSC rejection comb filter. This limits the number of paths the received waveform can take through a time/voltage graph called a trellis. Where a signal is in error it takes a path sufficiently near to the correct one that the correct one can be implied.

ATSC uses a training sequence sent once every data field, but is otherwise helpless against multipath reception as tests have shown. In urban areas, ATSC must have a correctly oriented directional antenna to reject reflections. Unfortunately the American viewer has been brought up to believe that television reception is possible with a pair of 'rabbit's ears' on top of the TV set and ATSC will not work like this. Mobile reception is not practicable. As a result the majority of the world's broadcasters appear to be favouring an OFDM-based system.

Index

Absolute accuracy, convertor, 203
Absolute digital sum value, 453, 455, 458
AC-3 system *see* Dolby AC-3 system
Accumulator *see* Discrete time integrator
Accuracy, conversion, 203
Adders, binary, 95–8
Address, pixel, 235–6
Address generation/reading, 237–43
  accumulators, 236
  equations, 251, 252
  fast/slow, 244
  horizontal, 254
  mapping, 234–5, 615
  for perspective operation, 250–56
  slope, 256
  vertical, 253
AES/EBU interface, 471, 546
  AES-3–1985 standard, 547
  AES-3–1992 standard, 547, 550, 553, 554
  electrical requirements, 547–9
  FM channel code, 443–5, 547
  frame structure, 549–555
  protocol, 548–9
  short-cable, 524
  signal levels, 400
  sync patterns, 471
  use of cyclic code, 497, 498
Airy distribution pattern, 684–5
Aliasing, 167–9
Ampex:
  track error sensing, 604
  transverse scan approach, 570
Amplitude bars, colour difference signals, 75
Analog systems, 2–6
  composite video compression schemes, 306
  monochrome video spectrum, 175
Analog-to-digital conversion (ADC), 4, 14, 199–201
  audio, 380–85
Anti-aliasing filter, 169, 171, 175, 207
Anti-image filter, 167
Aperture corrector, tube camera, 59
Aperture effect, television, 58–60
Apple Computers Inc., FireWire interface, 564–6
Aqueous humour, human eye, 34
Arithmetic logic unit (ALU), 101
Artifacts *see* Coding artifacts
ASCII symbols, use in character generator, 263, 264
ASPEC (Adaptive Spectral Perceptual Entropy Coding) system, 402, 412
Aspect ratio, television, 179
Astigmatism, 34
ATLANTIC research project, concatenation loss, 358–9
ATSC system, 752–7
Audio, digital, 32–3, 398, 400–401
Audio control packet, 558, 561

Audio data packet, 558–9
Audio extractor, 564
Auditory masking, 368
Automatic gain control (AGC), disk, 420–21, 647
Automation, in television industry, 721–2
Azimuth recording, 424–5, 578, 579, 580

B-adjacent code (burst correction), 501
Background:
  DVE, 256–9
  handling, 288, 290
  strobing effect, 77
Bad-block file system, 674, 676
Bandwidth, 55–6
  critical, 368, 369, 412
Bartlett window, 129
Baseline wander, 432
Basilar membrane, function, 364–6, 367–8
Bi-phase mark code *see* FM code
Bidirectional communication, 12
Bidirectional decoding process, 349–50
Bidirectional encoding process, 324–6, 336–9
Bimorph, 601–602
Binary system, 6–8
  circuitry for adding, 95–8
  code for digital video use, 80–83
  offset, 82–3
Binary-coded decimal (BCD) numbers, for timecode, 709
Birefringence, 690
Bit error rate (BER), effect on sensitivity, 474
Bit rate reduction *see* Compression techniques
Bits (BInary digiTS), 6
  errors, 20
Black crushing, 51
Black-level:
  clamping, 46, 47, 218
  set-up voltage, 44
Blackman window, 129
Blanking process, 44–6, 83, 224–5
Block coding, 404–406
  interleaving, 512–13
  matching technique, 272–4
Bootstrap program, 99
Brightness control, 51–2
Brillouin zone, 174, 176
Buffer, decoder, 347–8
Bump and look tracking process, Betacam, 604, 607
Burst correction, 497, 500–501
  codes *see* Reed–Solomon (R-S) codes
Burst errors, 20, 22
Butterfly (FFT element), 148, 149, 152
Bytes (eight-bit word), 7

Cable, low-loss, 525–6
Camcorder, 28, 29
Camera shake, compensation, 294–6
Capacitance, 524
Carriage velocity, disk drive, 651–2
Cassettes, 594–6
  loading, 596–8
Cathode ray tube (CRT):
  emission persistence, 53
  monochrome, 49–53
CCIR-601 standard *see* ITU-601 standard
Central processing unit (CPU), 100–101
Channel coding, 26, 437–46, 738
Channel equalization, 429–30
Channel pattern selection, 456–7, 460
Channel symbols, 454
Channel-status block format, 551, 552, 553–5
Character generator, 263–8
Charge-coupled device (CCD) camera, 53–5
Chien search (error position location), 504
Chroma keying, 227–8
Chroma solarizing, 228
Chroma subcarrier, 10
Chromaticity diagram, CIE, 64, 66, 68, 212
Cinemascope films, 238

Class numbers, code, 454
Clock edge, 87
Clock jitter, 179–82
Clocking system, channel coding, 436, 438
Coarse quantizing, 352
Code digital sum (CDS), 453, 454
Code rate, 446
Codebook (conversion table), 446, 448
Codecs, 304
  handling noise, 307
Coder, 304, 307–308
Codewords, 22
  CRC, 502
  error-correction, 477, 480, 482, 491
Coding:
  artifacts, 307, 308–309, 352–4
  bidirectional, 324–6
  classes, 454
  gain, 304, 315, 404–406
  lossless, 305
  lossy, 305
  perceptive, 305
  sub-band, 406–409
  variable length, 309
Colorimetry, 61–8
  basics, 9–10
Colour:
  in digital domain, 210–16
  in televsion, 212
Colour bars, test signal, 74–5
Colour bleed, 352
Colour difference signals, 9, 71–5, 82, 213–15
  processing, 218, 219
Colour displays, cathode ray tube, 69–70
Colour matching processes, 61
Colour mixture curves, 62–3
Colour vision, 60
Combiners *see* Concentrators
Commission Internationale d'Eclairage (CIE), chromaticity diagram, 64, 66, 68, 212
Compact Disc (CD), 643
  EFM code, 459–61
  enhanced, 391
  sampling rate, 372
Compander, 304
Companding, digital, 403–404
Composite video *see* Monochrome video
Compression factor, 304, 410
Compression techniques, 26–7
  advantages, 303
  asymmetrical, 304
  basics, 304–311
  interlacing, 305
  subjectively lossless, 308
  tape recorders, 569
  time *see* Time compression
  use in broadcasting, 313–14
  video, 26–7
Computer systems, 98–100
Concatenation loss, 354, 355, 357–8
Concealment, tape errors, 612
Concentrators, 221–4
Conditional access table (CAT), 734
Contact start/stop, Winchester technology, 661, 662
Contouring effect, 187, 232
Contrast control, 52–3
Contrast enhancement, 228, 230, 232
Contrast image (CI), 56–7
Contrast sensitivity, eye, 38
Control track, tape recorder, 583–4, 591–2
Convergence controls, electron beam, 70
Conversion:
  analog-to-digital, 199–201, 380–85
  digital-to-analog, 197–9, 373–80
  system quality, 201–203
Convertors:
  alternative structures, 385–7
  differential, 387
  flash, 199–201, 381, 388
  integrator-type, 376, 379–80
  noise-shaping, 387
  one-bit, 396–8
  outboard, 556
  sigma delta, 387, 397–8
  weighted-current, 378
Convolution:
  in filtering process, 119, 120, 121
  in frequency domain, 122

Convolutional code, 483, 484
Convolutional interleaving, 513–14, 515
  block completed, 514, 516
Convolutional randomizing, 533
Correction, by erasure, 510–12
Correction PROM, 224–5
Correlation surface, 277, 279
Corti, organ of *see* Organ of Corti
Co-siting, luminance samples, 177–8
Critical bandwidth, 368, 369, 412
Critical flicker frequency (CFF), human eye, 37
Cross-fading, 218, 225, 517, 519, 520
Cross-interleaving, 514
Crosstalk (interference), 179, 181
Crossword code, 482
Cutter *see* Crossfading
Cyclic codes, 491–3
  analysis, 494–6
  applications, 497
Cyclic redundancy check (CRC), 493, 496, 543, 545
  codewords, 502
Cylindrical lens method, focus determination, 698

D-3/D-5 formats, 177, 440–41, 449, 580, 581
D-9 format, 177, 569, 640–41
D-type latch element, 91, 93
Data reduction *see* Compression techniques
Data separation, 26, 430
DC coefficient, 332
DC-free coded waveform, 433–4
De-convolution circuit, 533
De-interlacing technique, 297–8
  by motion compensation, 299
Decimation process, 136, 205
Decode time stamp (DTS), 729–30
Decoder, 304
  bidirectional, 349–50
  buffer, 347
  composite, 353, 354
  MPEG-2, 349–52
Defect handling, disks, 673–8
Defect skipping, 675, 676
Definition, picture, 55
Delay control, 104–105
Delta modulation, 387
Demultiplexing, principles, 111
Density ratio (DR), channel coding, 441, 442
Detent, digital coding, 435, 440
Dibits, 656
Dielectric loss, 525
Dielectrics, 523
Difference picture, 318
Differential non-linearity deviation, convertor, 202
Differential pulse code modulation (DPCM), 385, 386, 387
  sigma, 387–8, 392–3
Differential quadrature phase shift keying (DQPSK), 739
Differential vector coding, 340–41
Differentiation, digital, 98
Diffraction process, 700, 701
  grating, 700, 701
Digital audio broadcasting (DAB), MUSICAM system, 401
Digital Betacam (DB), 622–9, 630
  4:2:2 format, 177
  cassettes, 594
  compression factor, 569
  drum configuration, 582
  frequency response, 466, 468
  time compression, 584–5, 586
Digital sum value (DSV), 453, 456, 460
  absolute, 453, 455
  end, 457
Digital systems:
  advantages, 10–11
  introduction, 2–6
Digital VHS (DVHS), 30
Digital video:
  advantages, 10–11
  applications, 11–12
  processes, 14–15
Digital video broadcasting (DVB), 748, 749
  DVB-T transmitter, 748
  receiver, 750–52
  sampling rate, 372

Digital video cassette (DVC), 629, 631–40
Digital video disk (DVD), 30, 312, 643
  coding, 403
  EFM code, 459–61
  player structure, 702–706
Digital video tape recorder (DVTR), 568
  address mapping, 615
  amorphous heads, 426
  audio recording, 400–401
  audio sampling rate, 373
  block diagram, 30, 31
  composite heads, 426
  compression techniques, 27, 569, 584–7
  data distribution, 613
  future developments, 723
  head design, 425–6
  laminated heads, 426
  operating modes, 598–9
  oversampling, 205, 206
  product codes, 612
  segmentation, 612–13
  shuffle strategy, 613–17
  signal systems, 609–12
  speed control, 589–93
  track and head geometry, 578–83
  track structure, 617–22
Digital-S format *see* D-9 format
Digital-to-analog conversion (DAC), 14, 197–9
  audio, 373–80
  R-2R structure, 198
Digitizing tablet, 269
Discrete cosine transform (DCT), 153–4
  blocks, 613
  coefficient block, 351
  MPEG, 315
  MPEG-2 spatial compression, 327–31
Discrete Fourier transform (DFT), 123, 143, 146
  for phase correlation, 275
Discrete frequency transform (DFT), 407
Discrete time integrator, in filtering, 98
Disk controller, 669–73
Disk servers, 679–80
Disks:
  compression systems, 680–81
  defect handling, 673–8
  drive basics, 642–4
  floppy *see* Floppy disks
  magneto-optical *see* Magneto-optical disks
  optical *see* Optical disks
  physical address, 645
  recording, 28–9
  rigid, 645
  structure, 645, 646
  types, 642–4
Display flicker, reduction, 209
Distortion, signal, 187
Distribution technique, tape recording, 586
Dither, 187–8, 188–91, 398
  digital, 191–3
  Gaussian pdf, 196–7
  rectangular pdf, 193–5
  tracking system use, 702
  triangular pdf, 195–6
Dolby:
  AC-3 system, 33, 403, 413–14
  Surround Sound, 32
Drift (tolerance build-up), 359
Drives, exchangeable-pack, 660
Drooping signals, 382, 383
Dual-current-source DAC, for ADC conversion, 382, 384–5
DVCPRO format, 569, 638, 640
DVE, 250, 295, 296–7
  backgrounds, 256–9
DVHS (digital VHS), 30
Dynamic convergence, electron beams, 70
Dynamic element matching, 375, 376–8
Dynamic range, improvement, 394
Dynamic resolution, human eye, 37, 76–7

EBU/AES interface *see* AES/EBU interface
Edit decision list (EDL), 313, 709, 717

Edit point location, 715–17
Editing process, 707–708
  assembly, 707
  with disk drives, 717–19
  edit gap, 618
  error correction, 709
  film making, 720–21
  horizontal, 708
  linear/non-linear, 708
  online/offline, 708–709
  on recording media, 709, 711
  timecode, 709
  vertical, 708
Effects machines *see* Special effects
EFMplus coding, DVD, 703–704
Electrical interfaces, 547–9
Electromagnetic energy effects, 523
Elementary stream, 27, 347, 348–9
Encoder, rotary incremental, 117
Encryption basis, 463
End of block (EOB) symbol, 332
Energy dispersal process, 742
Entropy, signal, 307–309
Equalization characteristic signal, 548
Equalization technique, 429–30, 465–6, 647
Equivalent rectangular bandwidth (ERB), 370
Erasable optical disks, formatting, 678
Erasure flags *see* Error flags
Error concealment, 21, 27
  by interpolation, 479–80
Error correction, 20–22, 469–70
  basics, 477–8
  broadcasting data, 747–8
  editing process, 709
  error detection by parity checking, 480–82
  error handling, 478–9
  message sensitivity, 474–5
Error detected here (EDH) packet, 543–6
Error flags, 516, 543–5
Error mechanisms, 476–7
Exchangeable disks, future use, 722–3
Exchangeable-pack drives, 647, 649, 655, 660
Expander, 304
Extended data packet, 558, 560
Eye *see* Human eye

Faders:
  absolute linear, 115–16
  analog, 114–15
  digital, 114
False codes, 115
Faraday effect, 429
Fast Fourier transform (FFT), 143, 147
  butterfly, 148
Feedback, servo system, 651
Fibre-optics:
  for digital interfacing, 526–7
  HDTV interface, 540–41
Field pictures, 342–5
FIFO chip *see* Silo (FIFO chip)
Figure of merit (FoM), 442
File transfer *see* Picture transport
Filler (fast digital mixer), 268
Film effects *see* Telecine system
Film making, digitally-assisted, 720–21
Film weave, 354
Filters, 117–19
  anti-aliasing, 169, 171, 175
  anti-image, 167
  decimator, 205
  design, 171–2
  digital, 119
  finite-impulse response (FIR), 126–32
  folding, 132, 133
  high-order loop, 394
  infinite-impulse response (IIR), 126
  interpolator, 205
  low-pass, 167, 171
  reconstruction, 167, 169
  recursive, 126
  spatial, 167
  switched capacitor, 119
Finite time resolution, human ear, 369
Finite-impulse response (FIR) filters, 126–32
Fire code (burst correction), 501
  use on disks, 678

FireWire interface, 564–6
arbitration system, 565
Firmware (ROM programs), 99
Flash convertor, 199–201, 381, 388
Flash guns, MPEG-2 hazard, 354
Floating point notation, 251, 404–406
Floppy disks, 645, 666–9
Flux-measuring heads, 423
Flyback, scanning, 41
Flying head, 645–7
FM channel code, 443–5, 547
Focus servo systems, laser, 695–8
Folded filters, 132
Forward spatial coder, 337
Fourier Transform, 123, 143–52
Frame, Winchester technology, 662
Frame number, 562
Frame pictures, 342–5
Frame store, 16
Frequency domain, 122–3
Frequency resolution, 368
Frequency response, 4, 465–6
limit, 5
optical recorder, 687
Full adder circuit, 95
Future technology, 13–14, 722–4

Gain control, by multiplication, 113–14, 117
Gain error, convertor, 202
Galois field (GF), 158–9, 505, 506, 508, 512
Gamma function/correction, 38–40
in cathode ray tube, 50–51
in digital domain, 210
Gate functions, 90, 91
Gaussian distribution, 162, 163
Gaussian pdf dither, 196–7
Generation loss, 359
Genlocking, 93
Gibb's phenomenon, 128, 129
Gradient matching technique, 274–5
Graphic art/paint systems, 269–70
Graphics system, 262
Gratings, fader, 116
Gray code, 115, 116
Great circle routes (mapping), 234
Group code recording (GCR), 446
Group codes, 446–8
2/3 code, 448–9
error detection, 461
Group patterns, synchronizing, 471–2
Group of pictures (GOP), 317, 323, 325, 349
MPEG structures, 327
MPEG-2, 306
Sequence, 349
Group-delay error, 117–18
Guardband, 658, 663
recording, 578, 579, 580

Half adder circuit, 95
Hamming codes, 485–7
Hamming distance, 487–91, 500–501
Hamming window, 129
Head crash, disk drive, 655
Head disk assembly (HDA), 660, 664
Head geometry, tape recorders, 578–83
Headroom:
analog, 398
artificial digital, 400
Helical scanning geometry, 570–77
High-definition television (HDTV), 313
component parallel interface, 40
fibre-optic interface, 540–41
High-order loop filters, 394
Hue, colour, 66
Huffman codes, 309, 310, 332, 333, 349
Human ear, 362
frequency response, 368–9
structure, 363–6, 368
time discrimination, 369
Human eye:
colour perception, 9, 60–61, 73, 306
functions and responses, 34–8
optic flow axis, 270, 736
sensitivity to noise, 330
temporal aperture effect, 60
tracking, 76–7
Human visual system (HVS):
quality perception, 306
see also Human eye
Hypermetropia, 34

Identification code (ID), packet, 112
IEEE, 1394–1995 interface, 564
Illuminants, chromaticity diagram, 67, 68
Image correlation process, 281–3
Image rotation, MPEG-2, 354
Image sampling spectra, 173
Image-stabilizing cameras, 294–6
Infinite impulse response (IIR) filter, 240, 533
Information bus, encoding, 359
Information content, sample, 307
Information rate, 306–307
Infra-red wavelengths, 60
Input integrator, 387
Integral linearity deviation, convertor, 202
Inter-coding, MPEG, 314, 315, 316
Inter-symbol interference, 424, 430
Interfaces:
  digital, 522
  electrical *see* Electrical interfaces
  principles, 523–7
  testing, 542–5
Interlace twitter, 77
Interlaced scanning, 77–9
Interlacing, 42, 297, 305, 306, 341–6
  prediction mechanism, 352
Interleaving, 22–5
  with burst-correcting codes, 512–14
  cross-, 514
  editing process, 517, 519, 520
  for error correction, 487
International Electrotechnical Commission (IEC), 546
  coding schemes, 401
  IEC, 958 standard, 546, 551
International Standards Organization (ISO):
  ISO/IEC compression standards, 401
  see also MPEG, JPEG
Internet video, 32
Interpolation *see* Sampling-rate conversion
Interpolator filter, 205
Interrupts, computer, 101–104
Intersymbol interference, 465
Intra (*I*) pictures, 317, 318, 323
Intra-coding coding, MPEG, 314, 315, 316, 323
Inverse transform:
  one-dimensional, 329, 330
  phase correlation, 277, 278
IRE unit (synchronizing), 42
ITU-601 standard, sampling rates, 177, 178, 218, 522

Jitter, 4
  causes, 430
  clock, 179–82
  margin, 441, 442
  overcoming, 15
  rejection, 435–7, 441
  tolerance, 555–7
Jogwheel encoder, 114, 117, 717
JPEG (Joint Photographic Experts Group), 314
  compression for disks, 680
Judder, 354
  elimination, 286, 290

Kaiser window, 129
Kell factor, 55
Kerr effect, 429, 684
Key processor, 222, 223
Keyframes, 261
Keying process, 225–6
  chroma, 227–8
  soft, 226
Knife-edge method, focus determination, 698
Knots *see* Keyframes

LASER (Light Amplification by Stimulated Emission of Radiation), 682
  beam intensity, 687–8
  in fibre-optic interface, 540
Latch element, 91
Leaky predictor, 316
Least significant bit (LSB), 7, 184
Lempel-Ziv-Welch (LZW) codes, 310

Lens, human eye, 34
Level meters, 398–400
Light-emitting diode (LED), 681, 682
Linear fader, 114, 115
Linear quantizing *see* Uniform quantizing
Linear-phase systems, 276
Linearity, transfer function, 184
Litz wire conductors, 523
Logging, error-rate, 546
Logic elements, 87, 88–91
  gate functions, 90, 91
Look-up tables, 39, 101
Lossless coding, 305, 309, 310
Lossy coding, 305
Loudness, 366–7
  contours, 366
Low-loss cable, 525–6
Low-pass filter, 167, 171
Luma, 40
  digital, 210
  transcoding signals, 72
Luminance, 40
  digital spectrum, 176
  gamma corrected, 40
  processing, 219
  reversal, 228
Luminance signal, 529, 530

Macroblocks (screen areas), 320–23
Magnetic disks, 28, 645
  multiplatter pack, 645
Magnetic recording, 419–24, 647–9
  early techniques, 442
Magneto-optic (MO) disk, 29, 428–9, 681, 682, 691, 693
  future use, 722–3
Magneto-resistive (M-R) head, 423
Manchester code *see* FM code
Mapping, distortion, 234–5
Maréchal criterion, 688
Masking, human ear discrimination, 369
Mechanical interleaving, 513, 514
Media forms, digital, 418
Memory elements, 91–3
Mercator map projection, 234
Metal evaporated (ME) tape, 629
Metal oxide semiconductor (MOS), 93
Microinstructions, look-up table, 101
Miller code, 445
  see also Modified frequency modulation (MFM)
Minimum eye pattern, receiver specification, 547–8
Minimum transition parameter, 447, 449
Modified discrete cosine transform (MDCT), 413
Modified frequency modulation (MFM), 445–6
Modulation techniques, 738–42
Modulation transfer function (MTF), 56–8, 686
Modulo arithmetic, 157–8, 615–16, 617
Mole (information signal), 359
Monochrome television camera, 53–5
Monochrome video, 10, 42, 354
Monotonicity, convertor error, 202–203
Morse code, 309
Mosaicing effect, 232, 233
Most significant bit (MSB), 7, 84
Motion compensation, 270–72, 317
  conversion standards, 286–91
  MPEG-2, 319, 320
  telecine system, 292–4
Motion estimation, 270
  block matching, 272–4
  gradient matching, 274–5
  in interlaced system, 344–5
  phase correlation, 275–9, 280–86
  sensing, 297
  vectors, 287–8, 289
Motion JPEG, 314
Motion portrayal, 76–7
Moving Pictures Experts Group *see* MPEG
Moving-coil actuators/drives, 603, 650
Moving-head device, 649–53
  disk drive, 643
MPEG, 311
  audio coding, 402, 403
  audio compression, 401–13
  coder, 346–8
  compression, 10, 27, 680, 726

MPEG – *continued*
  decoder subcarrier waveform problem, 353–4
  future developments, 724
  inter-coding, 314, 315, 316
  intra-coding, 314, 315, 316, 323
  Layer I, 402, 410–11
  Layer II, 402, 411–13
  Layer III, 402
  spatial coding, 314–15
  standards, 27, 33, 311, 312
  statistical multiplexing, 113
  temporal coding, 314
MPEG-1, standards, 402
MPEG-2, 311, 312
  4:2:0 sampling, 178
  allowable structures, 355
  artifact levels, 353
  averaging process, 324
  bitstream decoding, 28, 359
  buffer size, 347
  compressor techniques, 357
  decoder, 349–52, 735
  elementary stream, 348–9
  encoding, 332, 348–9
  field pictures, 342–5
  frame pictures, 342–5
  group of pictures, 306
  handling redundancy, 318
  image rotation, 354
  interlacing, 305, 342, 346
  judder, 354
  macroblocks (screen areas), 320–23
  motion compensation, 319, 320–22, 345–6
  noise sensitivity, 331
  output bitstream protocol, 336
  prediction error code, 335
  redundancy, 340
  remultiplexer, 737
  RLC/VLC coding, 332–3, 336
  Simple Profile, 323
  spatial compression, 327–31
  standards, 402
  switcher, 361
  top/bottom field handling, 341–2
MPEG/Audio group, 401, 402
Multiplatter drive, 655
Multiplexing:
  packet, 111–12
  principles, 111
  statistical, 112–13, 309
  transport stream, 734–6
Multiplication:
  in binary circuits, 114
  for gain control, 113
Multiprogramming, 642
MUSICAM system, 401, 402, 407, 411
Myopia, 34

Naive concatenation, 355
Near-instantaneous companding *see* Floating point block coding
Negative (low true) logic, 89–90
Network information table (NIT), 733–4
NICAM, 728 system, 32
  sampling rate, 372
  sync pattern, 471
Nodes *see* Keyframes
Noise, 161–3
  eye sensitivity, 330
  Gaussian distribution, 477
  immunity in digital circuitry, 417
  quantizing *see* Quantizing noise
  reduction, 300–301
Noise pumping, 352, 353
Noise shaping, 387, 388–91, 394–5
Non-interlacing, 680
Non-planar effects, 259–60
Non-return to zero (NRZ) recording, 442–3
Non-return to zero invert (NRZI) recording, 443
Non-subtractive dither, 188
NTSC:
  interference with ATSC transmissions, 756
  primary colours, 10, 65, 68
  raster lines, 79
  segmentation technique, 586
  use of interlace, 42
Numerical aperture (NA), lens, 685, 686, 688

Numerically locked loop (NLL), 94, 95
Nyquist rate, 203, 205
Nyquist's theorem, 168

OFDM, 744–7
Offline editing, 313
Offset binary, 82–3
Offset error, convertor, 201
One-bit convertors, 396–8
Online editing, 313
Open-collector system, 89–90
Operating levels, digital audio, 398, 400
Operating system, computer, 99
Optic flow axis, 76, 318
Optical cut off frequency, 687
Optical disks, 29, 426–8, 643
  erasable, 644
  principles, 681–9
  tracking systems, 699–702
Optical pickups, 688, 689–95
Organ of Corti, 365
Oversampling, 203–10
  audio, 388, 393–4
Overwriting, tape recording, 425

Packet identification code (PID), 731, 751
Packet multiplexing, 111–12
Packetized elementary stream (PES), 729
Packing (null packets), 112
PAL:
  primary colours, 10, 65
  raster lines, 79
  segmentation technique, 586
  use of interlace, 42
Pan, effect on window vectors, 284, 285
Panasonic:
  D-3/D-5 transports, 597
  DVCPRO, 638
  head deflection actuator, 601
  track curvature measurement, 605–606
Parallel transmission, binary signals, 8
Parity checking, 480–81, 482
Parking, Winchester technology, 661
Pattern sensitivity, 463
  channel selection, 457
Pay-per-view system, 463
Peak-shift distortion, 424, 430
Perceptive coding, 305, 306, 310
  sound quality, 362–3
Perceptual filtering, 391, 392
Persistence, cathode ray tube (CRT), 53
Persistence of vision, human eye, 37, 60
Personal video recorders (PVRs), 723–4
Perspective effect, 221–4, 246–7
  manipulation, 246–8
  rotation, 246–50
Phase correlation technique, 275–9
  motion-estimation system, 280–86
  one-dimensional, 277
Phase definition, 143–4
Phase linearity, 276
Phase locking, basilar membrane, 366
Phase margin, 441
Phase-locked loop, 93–4
Phase-shift keying (PSK), 738
Phon (loudness unit), 366
Photoptic vision, 36
Physical address, 235
Pickups, optical *see* Optical pickups
Picture cells (pixels), 4, 235
  square, 215, 216
Picture in shuffle, 613, 614
Picture transport, 357
Picture-type flags, 326
Piecewise linear graph, 251, 255
Piezo-electric actuators, 601–602
Pixels *see* Picture cells (pixels)
Place theory, resonance prediction, 365
Planar video effects, 232–4
Planck's Law, 681
Plasma displays, cathode ray tube, 70–71
Plastics, stretching, 690
Playout to air process, automation, 721

PLUGE (Picture Line-Up GEnerator) signal, 52–3
Polarization, light, 689–90
Polaroid material, 690
Positive (high true) logic, 89
Posterizing effect, 232
Pre-roll process, tape recording, 599–600
Preamble, 618–20
  timing reference pattern, 437
Precision-in-line (PIL), CRT, 69
Precompensation, 647
Predicted (*P*) pictures, 317, 318, 319, 323
Prediction error, 318, 320–21
Predictive coding, 335
Presbyopia, 34
Presentation time stamp (PTS), 729–30
Primary colours, 61–2
Primitive element, 505
Probability processes, 161–3
Processing, digital, 87–8
Processor *see* Central processing unit (CPU)
Product codes, 22, 482, 514–17, 518, 612
Program association table (PAT), 733, 737, 751
Program clock reference (PCR), 731–3, 737, 751
Program (computer instructions), 98
Program counter (PC), 100, 102
Program map table (PMT), 733, 737, 751
Program specific information (PSI), 733–4, 737
Program stream, 27
Programmable logic array (PLA) decoder, 461
Progressive scanning, 78
PROM, correction, 224
Propagation delay, 87
Pseudo-random sequences (PRS), 124, 126, 159, 161, 188
  DVB generator, 161, 742, 743
  technique, 462–3
Pseudo-rotation, non-perspective, 246
Pseudo-video system, 372
Psychoacoustics, 362, 363
Pulse code modulation (PCM), 3, 8, 165, 205
  alternative convertors, 385, 386
  bit rate, 306
  see also Differential pulse code modulation (DPCM)
Pulse cross-monitor, 53
Pulse pair, 124
Punctured codes, 496, 497
Purity adjustment, scan coils, 70

QMF band splitting, 412
Quadrature amplitude modulation (QUAM), 741, 742
Quadrature phase shift keying (QPSK), 738–9
Quantel, 680
Quantizing error, 184–8
  noise, 185–6
Quantizing process, 182–4
  coarse, 352
  distortion for special effects, 228, 231
  mid-tread, 184
  serial, 381–2
Quincunx sampling, 174

Ramp integrator, 382, 384
Random access memory (RAM), 16
  addressing, 105
  in computer, 99
  data bit storage, 93
Randomizing, 462–5, 742
Raster lines, 79
Ratcheting, 225
Rate conversion *see* Sampling-rate conversion
RC network, 119
Read-only memory (ROM), 16
  disks, 424–5
Reclocking, 87–8
Recompression, 359
Reconstruction filter, 167, 169

Recording:
  digital channels, 417–18
  disk-based, 28–9
  magnetic, 419–24
Rectangular pdf dither, 193–5
Recursion process, 275
Recursive device, 301
Redundancy, 20, 307
  bidirectional coding process, 340
  bit, 480
  MPEG-2, 340
Redundant array of inexpensive disks (RAID), 678–9
Reed–Solomon (R-S) codes, 159, 484, 501, 502–504, 747
  calculations, 504–508
  generator polynomial calculations, 521
  use on disks, 678
Reference gap, Winchester technology, 663–4
Reflection, 694, 695
Refraction, 694–5
Reissner's membrane, 364
Remez exchange algorithm, 129
Remultiplexing, 736–7
Rendering process, computer, 98
Requantizing, 191, 220, 331
  with noise-shaping, 389, 390, 391
Resolution:
  response limit, 5
  in spatial domain, 4
Retina, 34–6
Return to zero (RZ) recording, 442
Revectoring, 676–7
Reverse gamma function, 39
RGB signals, 9, 71–5, 211, 212
  colour space, 63–4
Rigid disk *see* Disk, rigid
Ripple logic, 107
Rods and cones, human eye, 34–6
Rolling-up transform, 259, 260
Rotary positioners, Winchester technology, 664–6
Rotary-head recorders, 30, 570–71, 571–7
  tape transport, 588
Rotation subsystems, disk drive, 654–5
Routers, serial digital, 538
RS-422A data communication standard, 547
Run-length coding (RLC), 332–3, 334
Run-length-limited (RLL) codes, 447–8, 451
Run/size parameters, 333

Saccadic motion, human eye, 36
Sampling, 3–6, 165–7
  4:2:0 format, 178
  4:2:1 format, 178
  4:2:2 format, 177, 178
  quincunx matrix, 174
  theory, 168
  two-dimensional spectra, 172–6
Sampling clock, 176, 436, 438
  jitter, 179–82
Sampling-rate conversion, 132–43, 238, 240
  applications, 132, 134
  for audio, 371–3
  by an integer factor, 135–7
  categories, 134–5
  choices, 176–9
  for error concealment, 479–80
  fractional-ratio, 140
  impulse response, 139–40
  rate increase, 137–40
  variable-ratio, 135
Saturation, colour, 66
Sawtooth waveform, 41
Scanning, 1, 40–42
  interlaced, 77–9
  progressive, 78
  zig-zag, 332, 333
Scotopic vision, 36
Scrambling process, 465, 535, 536
Search process, disk drive, 654
SECAM system, colour spectrum, 10
SECDED characteristic, MOS memories, 487
Sector skipping, 674
Seek process, disk drive, 651
Segments, tape track, 578, 584–7, 612
Semiconductor laser, 681
Separability principle, 236

Sequence, GOP, 349
Serial data transport interface (SDTI), 536–8
Serial digital interface (SDI), 20, 358, 527–36
  component, 557
  composite, 557
  decoding, 353
  embedded audio, 557–64
  routing, 358
  timing tolerance, 555–7
  use of cyclic codes, 497
Serial scrambled link, 465, 535, 536
Servo systems, 589–93, 649–51
Servo-amplifier, input, 653, 654
Servo-surface disks, 655–8
  soft sectoring, 658–60
Shannon sampling theory, 168, 391
Shift register, 93
Shuffle process, 25
  strategy, 613–17
Side-trim, floppy disks, 668
Sidebands, 167
Sigma-delta convertors, 387, 397–8
Signal quality, digital, 727
Signal timing, digital, 539–40
Signal-to-noise ratio (SNR), 186–8, 195, 196, 197, 477
Signalling systems, multi-level, 740
Silo (FIFO chip), 104, 106–109
Skew, 244, 245, 577
  rotation, 244–6
Skin effect (current flow), 523
Skip sector flag, 674, 675
Slew rate limited, convertors, 8, 387
Slicing process, 8, 341, 430, 431–4
Slipper (flying head), 646–7
Smooth blanking, 259
SMPTE, C phosphors, 65
SMPTE/EBU:
  standard colour difference signal, 75
  timecode, 443, 709, 710
Soft keying, 226
Soft sectoring, disks, 658–60
Software, 98
  verification process, 99
Solarizing (contrast enhancement), 228, 230, 232
Sony:
  CX-20017 convertor diagram, 378
  DT system, 608–609
  dynamic tracking head, 601–602
  NT sampling rate, 372
  SX system, 680–81
  track curvature measurement, 605–606
  track-following system, 608
Sony Philips Digital Interface (SPDIF), 546
  FM channel code, 547
Sound, definition, 362
Sound pressure level (SPL), 366
Source address, 250
Spatial aliasing, 174–5
Spatial coding:
  complete system, 335–6
  MPEG, 314–15
Spatial compression, MPEG-2, 327–31
Spatial filter, 167
Spatial frequency, 686, 687
Spatial luminance gradient, 274
Spatial sampling, 3_4
  eye function, 36
Special effects, by quantizing distortion, 228–32
Spectrum shaping, by channel coding, 26
Speed control *see* Servo systems
Spline algorithms, 261
Split sensor tracking method, 700–701
Square pixel, 215, 216
Square wave pulse, signal losses, 526, 527
Stack pointer, 102
Stacker programs, 305
Standards convertor, motion-compensated, 286–91
Static convergence, electron beams, 70
Statistical multiplexer (STATMUX), 736
Stepped edge effect, character, 264
Stepping motor, 650
STFT, 151
Storage elements, 87, 91–3
Strobe signal, 566
Stuffing (null) packets, 112

Sub-band coding, 406–409
Subroutine, computer, 102
Subtractive dither, 188
Symbol interleaving, 512–14
Synchronization (sync), 16–17, 42–6, 470
  blocks, 614, 620–22, 628–9
  patterns, 471, 472, 531
  separation, 47–8

Tape editing, 599–600
Tape editor, digital, 711–14
Target frequency search, 145–7
Telecine system, motion-compensated, 292–4
Television broadcasting:
  analog, 726
  colour displays, 69–70
  colour primaries, 211–13
  digital, 30, 32, 726–8
  elementary streams, 727–9
  future developments, 13–14
  monochrome camera, 53–5
  simple colour, 61–3
Temporal aperture effect, 59–60
Temporal coding, 336
  MPEG, 314
Temporal filtering, human eye, 37
Temporal frequency, 686, 687
Temporal sampling, 3–4
Ternary signal, 466, 469
Testing:
  data integrity, 542
  DV interfaces, 542–5
  SDI units, 542, 543
  VCO centring, 542
Thermomagneto-optics, 682, 684
Three-spot tracking method, 699–700
Time compression, 17–20
  Digital Betacam system, 584–5, 586
  tape recorders, 584–7
Time domain aliasing cancellation (TDAC), 413
Time instability *see* Jitter
Time stamps, 326, 729–30, 751
Time uncertainty *see* Jitter
Time-division multiplexed (TDM) system, 565
Timebase correction (TBC), 4, 104
Timecode, editing process, 709
Timers, programmable, 104
Timing accuracy, serial signal, 555, 557
Timing reference signal (TRS), 46, 558
Timing tolerance, serial interfaces, 555–7
Totem-pole output configuration, 90
Track descriptor record (TDR), 677
Track geometry:
  angle, 576
  curvature measurement, 605–606
  pitch, 578
  tape recorders, 578–83
Track-following systems, 462, 583–4, 601–604
  Sony DT system, 608–609
Track/hold circuit, 381, 382, 384
Tracking error measurement, 604–607
Tracking systems:
  dither-based method, 702
  optical disk, 699–702
Transfer functions, 201, 228, 229
Transform coding, 406–407
Transform pair, 124
Transforms, 122–3
  controlling, 260–62
Transistor transistor logic (TTL), 89–90
Transitions, digital coding, 435
Transmultiplexer *see* Remultiplexer
Transparency processing, 221–4
Transport stream, 27, 30
  multiplexer, 734–6
  packets, 730–31
Transposition process, 236–7
Transverse scan, 570–71
Trellis (time/voltage graph), 757
Tri-state bus systems, 90
Triangular pdf dither, 195–6
Twisted-ring counter, 159, 161, 162
Two's complement coding system, 84–7, 470

Ultra-violet wavelengths, 60
Uniform quantizing, 184

V (data-valid) bit, 106–107
Validity flag, 549
Variable speed replay, tape recorders, 600–609
Variable time window, 406
Variable-length coding (VLC), 332–3, 334
Velocity profile, disk drive, 652–3
VHS format:
 digital, 30
 hi-fi sound, 32
Video buffer verifier (VBV), 347, 348
Video compression, 26–7
Video signals:
 definition, 1
 switching waveform, 225
 waveforms, 43, 44, 45
Video switcher, 225
Video-On-Demand technology, 313
Virtual address, 235
Viruses, computer, 99
Vision mixer, simple digital design, 218–20
Visual accommodation, 34
Visual display unit (VDU), 262
Viterbi decoding, 467–70
Vitreous humour, human eye, 34
Voltage-controlled crystal oscillator (VCXO), 732
Voltage-controlled oscillator (VCO), 93–4
 centring, 437

Wave table, 328
Waveforms:
 audio, 406
 Fourier analysis, 143–4
 numerical description, 6
 sawtooth, 41
 square, 58, 526, 527
 time domain, 406
 video signals, 1, 43, 44, 45, 225
Wavelet compression, for disks, 681
Wavelet transform, 154–7
 MPEG, 315
Winchester technology, 648, 660–66
 disks, 643
Windage, disk drive, 666
Window, RAM, 17
Window functions, 128–30
Window margin, 441
Wire frame effect, character, 268
Wordlength, 220
 truncation, 191
Workstation, non-linear, 714–15
WORM (Write Once Read Many) disks, 427, 643, 682, 683
 verification, 678
Wrap angle, tape, 584
Write clock, 659

Yeltsin walk, 344

Zero-order hold system, 59
Zero-run-length, 332–3
Zig-zag scanning, 332, 333
Zoom, effect on window vectors, 284, 285

http://www.focalpress.com

Visit our web site for:

- ❑ The latest information on new and forthcoming Focal Press titles
- ❑ Technical articles from industry experts
- ❑ Special offers
- ❑ Our email news service

Join our Focal Press Bookbuyers' Club

As a member, you will enjoy the following benefits:

- ❑ Special discounts on new and best-selling titles
- ❑ Advance information on forthcoming Focal Press books
- ❑ A quarterly newsletter highlighting special offers
- ❑ A 30-day guarantee on purchased titles

Membership is FREE. To join, supply your name, company, address, phone/fax numbers and email address to:

USA
Christine Degon, Product Manager
Email: christine.degon@bhusa.com
Fax: +1 781 904 2620
Address: Focal Press,
225 Wildwood Ave, Woburn,
MA 01801, USA

Europe and rest of World
Elaine Hill, Promotions Controller
Email: elaine.hill@repp.co.uk
Fax: +44 (0)1865 314572
Address: Focal Press, Linacre House,
Jordan Hill, Oxford,
UK, OX2 8DP

Catalogue

For information on all Focal Press titles, we will be happy to send you a free copy of the Focal Press catalogue:

USA
Email: christine.degon@bhusa.com

Europe and rest of World
Email: carol.burgess@repp.co.uk
Tel: +1(0)1865 314693

Potential authors

If you have an idea for a book, please get in touch:

USA
Terri Jadick, Associate Editor
Email: terri.jadick@bhusa.com
Tel: +1 781 904 2646
Fax: +1 781 904 2640

Europe and rest of World
Beth Howard, Editorial Assistant
Email: beth.howard@repp.co.uk
Tel: +44 (0)1865 314365
Fax: +44 (0)1865 314572